molecular weight in amu = molar weight in gr

representative group

valence electrons

Noble Gases

					18 VIIIA
13 IIIA	14 IVA	15 VA	16 VIA	17 VIIA Halogens	2 **He** helium 4.00260
5 **B** boron 10.81	6 **C** carbon 12.011	7 **N** nitrogen 14.0067	8 **O** oxygen 15.9994	9 **F** fluorine 18.998403	10 **Ne** neon 20.179
13 **Al** aluminum 26.98154	14 **Si** silicon 28.0855	15 **P** phosphorus 30.97376	16 **S** sulfur 32.06	17 **Cl** chlorine 35.453	18 **Ar** argon 39.948

10 VIIIB	11 IB	12 IIB						
28 **Ni** nickel 58.70	29 **Cu** copper 63.546	30 **Zn** zinc 65.38	31 **Ga** gallium 69.72	32 **Ge** germanium 72.59	33 **As** arsenic 74.9216	34 **Se** selenium 78.96	35 **Br** bromine 79.904	36 **Kr** krypton 83.80
46 **Pd** palladium 106.4	47 **Ag** silver 107.868	48 **Cd** cadmium 112.41	49 **In** indium 114.82	50 **Sn** tin 118.69	51 **Sb** antimony 121.75	52 **Te** tellurium 127.60	53 **I** iodine 126.9045	54 **Xe** xenon 131.30
78 **Pt** platinum 195.09	79 **Au** gold 196.9665	80 **Hg** mercury 200.59	81 **Tl** thallium 204.37	82 **Pb** lead 207.2	83 **Bi** bismuth 208.9804	84 **Po** polonium (209)	85 **At** astatine (210)	86 **Rn** radon (222)

64 **Gd** gadolinium 157.25	65 **Tb** terbium 158.9254	66 **Dy** dysprosium 162.50	67 **Ho** holmium 164.9304	68 **Er** erbium 167.26	69 **Tm** thulium 168.9342	70 **Yb** ytterbium 173.04	71 **Lu** lutetium 174.97
96 **Cm** curium (247)	97 **Bk** berkelium (247)	98 **Cf** californium (251)	99 **Es** einsteinium (254)	100 **Fm** fermium (257)	101 **Md** mendelevium (258)	102 **No** nobelium (255)	103 **Lr** lawrencium (260)

■ metals   ■ metalloids   ■ nonmetals

General, Organic, and Biological Chemistry

General, Organic, and Biological Chemistry

David G. Lygre
Central Washington University

Brooks/Cole Publishing Company
ITP™ *An International Thomson Publishing Company*

Pacific Grove · Albany · Bonn · Boston · Cincinnati · Detroit · London
Madrid · Melbourne · Mexico City · New York · Paris · San Francisco
Singapore · Tokyo · Toronto · Washington

Sponsoring Editor: *Lisa J. Moller*
Marketing Team: *Connie Jirovsky and Kathleen Bowers*
Editorial Associate: *Beth Wilbur*
Permissions Editor: *Linda Rill*
Production Editor: *Penelope Sky*
Manuscript Editor: *Peter Fong*
Interior and Cover Design: *Roy R. Neuhaus*

Interior Illustration: *Tech-Graphics*
Cover Photo: *Brian Bailey/Tony Stone Images*
Art Coordinator: *Lisa Torri*
Photo Editor: *Larry Molmud*
Photo Researcher: *Chris Pullo*
Typesetting: *Progressive Information Technologies*
Printing and Binding: *Quebecor Printing/Hawkins*

COPYRIGHT © 1995 by Brooks/Cole Publishing Company
A Division of International Thomson Publishing Inc.
I(T)P The ITP logo is a trademark under license.

For more information, contact:

BROOKS/COLE PUBLISHING COMPANY
511 Forest Lodge Road
Pacific Grove, CA 93950
USA

International Thomson Publishing Europe
Berkshire House 168-173
High Holborn
London WC1V 7AA
England

Thomas Nelson Australia
102 Dodds Street
South Melbourne, 3205
Victoria, Australia

Nelson Canada
1120 Birchmount Road
Scarborough, Ontario
Canada M1K 5G4

International Thomson Editores
Campos Eliseos 385, Piso 7
Col. Polanco
11560 México D. F. México

International Thomson Publishing GmbH
Königswinterer Strasse 418
53227 Bonn
Germany

International Thomson Publishing Asia
221 Henderson Road
#05-10 Henderson Building
Singapore 0315

International Thomson Publishing Japan
Hirakawacho Kyowa Building, 3F
2-2-1 Hirakawacho
Chiyoda-ku, Tokyo 102
Japan

All rights reserved. No part of this work may be reproduced, stored in a retrieval system, or transcribed, in any form or by any means—electronic, mechanical, photocopying, recording, or otherwise—without the prior written permission of the publisher, Brooks/Cole Publishing Company, Pacific Grove, California 93950

Printed in the United States of America

10 9 8 7 6 5 4 3

Library of Congress Cataloging-in-Publication Data

Lygre, David G.
    General, organic, and biological chemistry/David G. Lygre.
         p.   cm.
    Includes index.
    ISBN 0-534-24252-9
    1. Chemistry. I. Title.
QD33.L96 1994
540—dc20   94-8083
                                                    CIP

To my parents and stepfather:
Esther, Gerald, and Carlyle

Preface

Chemistry teachers see their subject as a fascinating way of making sense of how nature works. The more chemistry we know, the better we understand how the human body works and how our environment operates.

This book is designed to help students see chemistry the way teachers do. I've used a clear, conversational style so students will enjoy reading. My goal is to have them finish the course feeling satisfied by how much they have learned. I want them to look at the world with new eyes that see its underlying chemistry.

Organization and Content

This text is suitable for a one-year course in general, organic, and biological chemistry, primarily for health science majors. Although most students have some chemistry background, none is assumed. Appendixes A–C review the mathematics used in chemistry calculations.

The three-part organization is traditional, but I have included four short chapters (24–27) on metabolism, and two unusual chapters. Chapter 18, on macromolecules and polymers, bridges organic and biological chemistry. In Chapter 29, on medical drugs, I show many practical connections between health and chemistry, associations that in fact permeate the text. Biochemistry is discussed in every chapter, to show how chemistry works in our bodies and in our daily lives. For example, principles of nutrition are referred to repeatedly, instead of being limited to a single chapter. Chemistry Spotlights throughout the book highlight interesting applications of chemistry and biochemistry.

Learning Aids

Students will benefit from the following features.

- *Chapter Openers:* Each chapter begins with general questions that provide an overview of the following material, and with a story that illustrates either a practical application of chemistry or a significant event in the history of the field.
- *Chemistry Spotlights:* Boxes in each chapter contain interesting applications of chemistry, often in relation to health, or descriptions of chemical discoveries; they also show the human side of scientists.
- *Margin Notes:* Brief comments in the margins summarize a main point, give extra historical information, or explain the origin of a term.
- *Key Terms:* When a new term is defined it is shown in boldfaced type. These terms are listed at the end of the chapter, where they are cross-referenced back to the text.

- *Examples:* In every chapter are practice examples with answers that help students test their understanding of the material they've just studied.
- *Chapter Summaries:* Summaries reinforce the main points of each chapter.
- *End-of-Chapter Exercises:* Numerous exercises conclude each chapter; answers to the even-numbered exercises are in the back of the book. *Discussion Exercises* require more complex responses or personal opinions.
- *Graphics:* Many diagrams and photographs illustrate chemical ideas and show chemistry in the real world. In many structural formulas and equations, color highlights the most important part of the molecule.
- *Index and Glossary:* I have provided both an extensive index and a glossary of key terms that is cross-referenced to the text.
- *Appendixes:* Appendixes A–C review the mathematics needed for chemistry calculations. Appendix D, unique in a text of this type, gives information about each element: its name and symbol; who discovered it, and where, and when; the origin of the name; and its appearance at room temperature. Appendix E describes cell structure.

Ways to Use This Book

Because different teachers emphasize different aspects of the course, this text can be used with considerable flexibility. Although I have provided ample material for a one-year course, it can be condensed as necessary in several ways. Most courses will include all the chapters in Part I, General Chemistry. However, Chapter 10, on nuclear chemistry, may be abbreviated if time is short. Part II, Organic Chemistry, can be streamlined if much of Chapter 18, on macromolecules and polymers, is omitted. Part III, Biochemistry, can withstand reduced emphasis on metabolism and optional coverage of the last two chapters, on body fluids and medical drugs.

Supplementary Materials

- *Solutions Manual:* This useful supplement contains answers to all the end-of-chapter exercises.
- *Transparencies:* Master sheets for making overhead transparencies of many key diagrams are available from the publisher.
- *Instructor's Manual:* A test bank of questions is available either in hard copy or on disk for IBM-compatible or Macintosh systems.
- *College Chemistry in the Laboratory, Fifth Edition,* by Morris Hein, Leo R. Best, Robert L. Miner, and James M. Ritchey (Brooks/Cole) contains 42 experiments and is recommended for a laboratory that may accompany the lecture course.
- *Alchemist: The Chemical Equation Balancer,* by Stephen J. Townsend and Joyce C. Brockwell. This Macintosh software will help students develop their equation balancing skills.

Acknowledgments

I appreciate the superb work of people at Brooks/Cole Publishing Company. Chemistry editor Lisa Moller and editorial associate Beth Wilbur coordinated reviews and juggled many other tasks with competence and good humor. Peter Fong skillfully edited the manuscript and Chris Pullo was tenacious in obtaining excellent photographs. Roy Neuhaus used his artistic talents to design the text, and Penelope Sky moved the book through production with persistence and skill.

All the work that has gone into this project reminds me of Samuel Johnson's remark, "What is written without effort is in general read without pleasure." I hope the obverse is just as true. I would appreciate knowing what you like and dislike about this book, in order to improve it in future editions. Please write, telephone, or fax me with your responses.

David Lygre
Department of Chemistry
Central Washington University
Ellensburg, WA 98926
509/963-2817 telephone
509/963-1050 fax

Reviewers

I'm greatly indebted to the following reviewers. Their excellent criticism let me improve the text significantly. Any remaining errors and shortcomings are my own.

Marvin Albinak, Essex Community College
Norman Bhacca, Louisiana State University
David Dever, Macon College
Thomas Donnelley, Loyola University
Jerry Driscoll, University of Utah
Wes Fritz, College of Dupage
Donald Glover, Bradley University
Mark Greenberg, Colorado State University
Leland Harris, University of Arizona
Alton Hassell, Baylor University
Melvin Henley, Murray State University
Merrill Hugo, Shasta College
Neil Kestler, Louisiana State University
Ross Klinck, Adirondack Community College
Melvin Merken, Worcester State College
Donald Pon, Foothill College
Joan B. Seif, Mohawk Valley Community College
Rick Steiner, University of Utah
Donald Williams, Hope College
Linda Wilson, Middle Tennessee State University

Brief Contents

I GENERAL CHEMISTRY   1

1. Introduction to Chemistry   2
2. Measurements   12
3. Elements and Atoms   32
4. Compounds: Ions and Molecules   60
5. Chemical Reactions   88
6. Solids, Liquids, and Gases   114
7. Solutions   138
8. Reaction Rates and Equilibrium   162
9. Acids and Bases   184
10. Nuclear Chemistry   212

II ORGANIC CHEMISTRY   237

11. Introduction to Organic Chemistry   238
12. Alkanes and Alkyl Halides   256
13. Unsaturated and Aromatic Hydrocarbons   280
14. Alcohols, Phenols, Thiols, and Ethers   306
15. Aldehydes and Ketones   326
16. Carboxylic Acids and Esters   348
17. Nitrogen-Containing Organic Compounds   370
18. Polymers   388

III BIOCHEMISTRY   404

19. Carbohydrates   406
20. Lipids   428
21. Proteins   448
22. Enzymes   472
23. Nucleic Acids   492
24. Metabolism and Energy   520
25. Carbohydrate Metabolism   538
26. Lipid Metabolism   556
27. Metabolism of Nitrogen Compounds   570
28. Body Fluids   584
29. Medical Drugs   600

APPENDIXES

A. Significant Figures   618
B. Exponents and Scientific Notation   621
C. Converting from One Unit to Another   625
D. The Elements   630
E. Cell Structure   636

Chemistry Spotlights

The Hall Process: It Can Pay to Listen to Your Chemistry Teacher   3

What Is Normal Body Temperature?   13

Density and Body Fat   20

Dmitri Mendeleev   33

The Ultimate Particles of Matter?   40

Cadmium: A Toxic Metal   47

Table Salt   61

Lithium Ion and Manic Depression   70

Mercury: A Dangerous Metal   71

N_2O: A Chemical of Many Uses   77

Carbon Monoxide: A Deadly, Polar Molecule   82

The Haber Process   89

Exercise and Lactic Acid   97

Metabolism and Respiratory Quotient   104

Calories and Food   109

Earth's Atmosphere and Life: Then and Now   115

Keeping Cool   126

Breathing and the Gas Laws   129

Liquid Nitrogen and Cadavers   130

Nitrogen Supersaturation and the Bends   139

Detergents   148

Edema   157

Treating Patients with Oxygen   163

Temperature and Aging   171

Getting Oxygen to Your Cells   176

Equilibrium and Steady State   177

Antacids   185

Corrosive Acids and Bases   191

Acid Deposition   198

Acidosis and Alkalosis   207

Marie Curie   213

The Discovery of Radioactivity   215

Synthesizing Isotopes   225

J. Robert Oppenheimer and the Manhattan Project   231

The Vital Force   239

The Right Shape   248

Hydrocarbons and Skin Care   257

Chlorofluorocarbons and the Ozone Layer   276

The Birth of Birth Control Pills   281

Vitamin A and Scientific Knowledge   287

The Last Word on Reaction Mechanisms   292

August Kekulé and the Whirling Snakes   295

Thiol Groups and Toxic Metals   307

Treating Alcohol Poisoning   318

Sex Secrets of the Gypsy Moth   322

Chemistry Spotlights

Understanding Undertaking 327

Diabetes and Glucose Testing 338

Kidney Stones and Gout 349

Esters That Kill 361

The Invention of Dynamite 371

Cocaine 377

Synthesis of Nylon 389

The Accidental Discovery of Teflon 395

Sweet Taste 407

Optical Isomerism and Life 417

John Dillinger and the Potato Pistol 424

Anabolic Steroids 429

Fats and Isomers 433

Lipid Storage Diseases 437

Sickle-Cell Anemia 449

Cancer and Enzyme Inhibition 473

Albert Szent-Györgyi and Vitamin C 480

Genetic Engineering 493

Discovery of an Anticancer Drug 502

Reverse Transcriptase 506

A New Way to Tell Whodunit 516

Cyanide Poisoning and How to Treat It 521

Snake Venom and Oxidation Phosphorylation 532

Glycogen Loading 539

The Glucose Tolerance Test 552

Some Unusual Polyunsaturated Fatty Acids 557

Parkinson's Disease 571

Blood Typing 585

A New Treatment for Hemoglobin Disorders? 589

Kidney Dialysis Methods 596

The Discovery of Penicillin 601

Alzheimer's Disease 610

Contents

I GENERAL CHEMISTRY 1

1 Introduction to Chemistry 2

1.1 **The Nature of Chemistry and Science** 4

1.2 **How Science Works** 5

1.3 **Applications of the Scientific Method** 7

1.4 **Limitations of Science** 8

Summary 9
Key Terms 9
Exercises 10
Discussion Exercises 10

2 Measurements 12

2.1 **Uncertainty in Measurements** 14

2.2 **The Metric System and International System of Units (SI)** 15

2.3 **Units of Measurement: Mass, Length, Volume, and Density** 15

2.4 **Units of Measurement: Energy, Temperature, and Time** 21

2.5 **Converting between Units** 25

Summary 28
Key Terms 28
Exercises 28
Discussion Exercises 30

3 Elements and Atoms 32

3.1 **Classifying Matter** 34

3.2 **Elements and Atoms: The Connection** 37

- **3.3** *Atomic Particles and Atomic Mass* 38
- **3.4** *The Periodic Table of the Elements* 43
- **3.5** *Electron Arrangements in Atoms* 47
- **3.6** *Electron Configuration and the Periodic Table* 54
    - *Summary* 56
    - *Key Terms* 56
    - *Exercises* 56
    - *Discussion Exercises* 59

4 Compounds: Ions and Molecules 60

- **4.1** *Stable Electron Structures: 8 and 2 as Special Numbers* 62
- **4.2** *Ionic Compounds: Forming Ions* 63
- **4.3** *Metallic Bonding* 70
- **4.4** *Covalent Compounds: Forming Molecules* 72
- **4.5** *Shapes of Molecules* 77
- **4.6** *Polar and Nonpolar Molecules* 78
    - *Summary* 83
    - *Key Terms* 84
    - *Exercises* 84
    - *Discussion Exercises* 86

5 Chemical Reactions 88

- **5.1** *Writing Chemical Reactions* 90
- **5.2** *Oxidation-Reduction Reactions* 93
- **5.3** *Some Practical Oxidation-Reduction Reactions* 95
- **5.4** *Moles* 98
- **5.5** *Stoichiometry* 102
- **5.6** *Chemical Reactions and Energy Changes* 108
    - *Summary* 110
    - *Key Terms* 110
    - *Exercises* 111
    - *Discussion Exercises* 113

6 Solids, Liquids, and Gases 114

- **6.1** *Physical States of Matter* 116

- **6.2** *Changes in Physical State* 119
- **6.3** *Physical Properties of Gases* 126
- **6.4** *The Universal Gas Law* 131
  - *Summary* 134
  - *Key Terms* 134
  - *Exercises* 134
  - *Discussion Exercises* 137

7 Solutions 138

- **7.1** *Water: A Remarkable Substance* 140
- **7.2** *Forming Solutions* 141
- **7.3** *Colloids and Suspensions* 146
- **7.4** *Concentrations of Solutions* 147
- **7.5** *Colligative Properties and Water Balance in the Body* 154
  - *Summary* 157
  - *Key Terms* 158
  - *Exercises* 158
  - *Discussion Exercises* 161

8 Reaction Rates and Equilibrium 162

- **8.1** *First and Second Laws of Thermodynamics* 164
- **8.2** *Reaction Rates* 166
- **8.3** *Changing Reaction Rates* 169
- **8.4** *Dynamic Chemical Equilibrium* 172
- **8.5** *Le Châtelier's Principle* 175
  - *Summary* 180
  - *Key Terms* 180
  - *Exercises* 180
  - *Discussion Exercises* 182

9 Acids and Bases 184

- **9.1** *Acids and Bases: Definitions, Properties, and Names* 186
- **9.2** *Strength of Acids and Bases* 189
- **9.3** *pH* 194

- 9.4 Reactions of Acids 197
- 9.5 Acid-Base Titrations 200
- 9.6 Buffers 204
- 9.7 pH in the Body 205
  - Summary 208
  - Key Terms 209
  - Exercises 209
  - Discussion Exercises 211

10 Nuclear Chemistry 212

- 10.1 Types of Radioactivity 214
- 10.2 Writing Nuclear Equations 216
- 10.3 Half-Life 218
- 10.4 Effects of Radioactivity 221
- 10.5 Medical Uses of Radioactivity 223
- 10.6 Nuclear Fission and Fusion 227
  - Summary 232
  - Key Terms 233
  - Exercises 233
  - Discussion Exercises 235

II ORGANIC CHEMISTRY 237

11 Introduction to Organic Chemistry 238

- 11.1 Carbon: A Unique Element 240
- 11.2 Shapes of Organic Molecules 241
- 11.3 Representing Carbon Chains and Rings 245
- 11.4 Forms of the Element Carbon 248
- 11.5 Types of Organic Compounds: Functional Groups 250
  - Summary 252
  - Key Terms 253

Exercises 253
Discussion Exercises 255

12 Alkanes and Alkyl Halides 256

12.1 **Alkanes** 258

12.2 **Structural Isomerism** 260

12.3 **Condensed Structural Formulas** 263

12.4 **Naming Alkanes** 264

12.5 **Physical Properties and Chemical Reactions of Alkanes** 268

12.6 **Fossil Fuels and Petrochemicals** 271

12.7 **Alkyl Halides** 274

Summary 276
Key Terms 277
Exercises 277
Discussion Exercises 279

13 Unsaturated and Aromatic Hydrocarbons 280

13.1 **Alkenes: Names and Structures** 282

13.2 **Alkynes: Names and Structures** 285

13.3 **Geometric Isomers** 285

13.4 **Physical Properties and Chemical Reactions of Alkenes and Alkynes** 287

13.5 **Aromatic Hydrocarbons** 295

13.6 **Physical Properties and Chemical Reactions of Aromatic Compounds** 299

Summary 301
Key Terms 302
Exercises 302
Discussion Exercises 305

14 Alcohols, Phenols, Thiols, and Ethers 306

14.1 **Alcohols: Names and Structures** 308

14.2 **Phenols: Names and Structures** 310

14.3 **Physical Properties and Uses of Alcohols and Phenols** 311

14.4 **Chemical Reactions of Alcohols and Phenols** 314

14.5 **Thiols** *319*

14.6 **Ethers** *320*

    Summary 323

    Key Terms 323

    Exercises 323

    Discussion Exercises 325

15 Aldehydes and Ketones 326

15.1 **Aldehydes: Names and Structures** *328*

15.2 **Ketones: Names and Structures** *331*

15.3 **Physical Properties and Uses of Aldehydes and Ketones** *333*

15.4 **Chemical Reactions of Aldehydes and Ketones** *335*

    Summary 344

    Key Terms 345

    Exercises 345

    Discussion Exercises 347

16 Carboxylic Acids and Esters 348

16.1 **Carboxylic Acids: Names and Structures** *350*

16.2 **Acid Properties of Carboxylic Acids** *352*

16.3 **Physical Properties and Uses of Carboxylic Acids and Their Salts** *354*

16.4 **Chemical Reactions of Carboxylic Acids** *356*

16.5 **Esters: Names and Structures** *359*

16.6 **Physical Properties and Uses of Esters** *359*

16.7 **Chemical Reactions of Esters** *362*

16.8 **Phosphate Esters and Anhydrides** *364*

    Summary 365

    Key Terms 366

    Exercises 366

    Discussion Exercises 368

17 Nitrogen-Containing Organic Compounds 370

17.1 **Amines: Names and Structures** *372*

17.2 **Physical Properties and Uses of Amines** *374*

17.3 *Chemical Reactions of Amines* 376

17.4 *Amides: Names and Structures* 379

17.5 *Physical Properties, Uses, and Chemical Reactions of Amides* 380

17.6 *Other Organic Nitrogen Compounds* 384

Summary 385
Key Terms 385
Exercises 385
Discussion Exercises 387

18 Polymers 388

18.1 *Polymers: Basic Terms* 390

18.2 *Addition Polymers* 392

18.3 *Condensation Polymers* 396

18.4 *Other Polymers* 400

Summary 401
Key Terms 401
Exercises 401
Discussion Exercises 403

III BIOCHEMISTRY 404

19 Carbohydrates 406

19.1 *Introduction to Carbohydrates* 408

19.2 *Monosaccharides* 409

19.3 *Optical Isomers* 412

19.4 *Chemical Reactions of Carbohydrates* 418

19.5 *Disaccharides and Polysaccharides* 421

Summary 424
Key Terms 425
Exercises 425
Discussion Exercises 427

20 Lipids 428

- **20.1** *Fats and Oils* 430
- **20.2** *Chemical Reactions of Triglycerides and Fatty Acids* 433
- **20.3** *Phospholipids and Sphingolipids* 435
- **20.4** *Steroids* 437
- **20.5** *Lipoproteins* 439
- **20.6** *Other Important Lipids* 440
- **20.7** *Cell Membranes* 442
  - *Summary* 445
  - *Key Terms* 445
  - *Exercises* 445
  - *Discussion Exercises* 447

21 Proteins 448

- **21.1** *Amino Acids* 450
- **21.2** *Essential Amino Acids* 454
- **21.3** *Primary Protein Structure* 456
- **21.4** *Higher Levels of Protein Structure* 459
- **21.5** *Classification of Proteins* 463
- **21.6** *A Few Important Peptides and Proteins* 465
  - *Summary* 469
  - *Key Terms* 469
  - *Exercises* 470
  - *Discussion Exercises* 471

22 Enzymes 472

- **22.1** *Enzymes: Effective and Specific Catalysts* 474
- **22.2** *Coenzymes and Vitamins* 477
- **22.3** *Naming and Classifying Enzymes* 481
- **22.4** *Rates of Enzyme-Catalyzed Reactions* 482
- **22.5** *Regulation of Enzyme Activity* 485
- **22.6** *Medical Uses of Enzymes* 487

Summary 489
Key Terms 490
Exercises 490
Discussion Exercises 491

23 Nucleic Acids 492

23.1 *Structures of DNA and RNA 494*

23.2 *Synthesis of DNA 499*

23.3 *Synthesis of RNA 503*

23.4 *Protein Synthesis 507*

23.5 *Mutations 512*

23.6 *Genetic Engineering 513*
Summary 517
Key Terms 517
Exercises 518
Discussion Exercises 519

24 Metabolism and Energy 520

24.1 *Energy Balance 522*

24.2 *Metabolism: An Overview 524*

24.3 *Citric Acid Cycle 526*

24.4 *Electron Transport and Oxidative Phosphorylation 530*

24.5 *Energy Production and Use in the Body 533*
Summary 534
Key Terms 535
Exercises 535
Discussion Exercises 537

25 Carbohydrate Metabolism 538

25.1 *Carbohydrate Digestion 540*

25.2 *Glycolysis 541*

25.3 *Gluconeogenesis 547*

25.4 *Glycogen Metabolism 548*

25.5 *Pentose Phosphate Pathway 550*

25.6 Blood Glucose Levels 550
    Summary 553
    Key Terms 553
    Exercises 554
    Discussion Exercises 555

26 Lipid Metabolism 556

26.1 Lipid Digestion 558

26.2 Catabolism of Triglycerides 559

26.3 Fatty Acid Synthesis 562

26.4 Cholesterol Synthesis 565
    Summary 567
    Key Terms 567
    Exercises 567
    Discussion Exercises 569

27 Metabolism of Nitrogen Compounds 570

27.1 Protein Digestion 572

27.2 Amino Acid Metabolism 573

27.3 Urea Cycle 577

27.4 Nucleotide and Heme Metabolism 578
    Summary 581
    Key Terms 581
    Exercises 581
    Discussion Exercises 583

28 Body Fluids 584

28.1 Composition of Blood 586

28.2 Functions of Blood 588

28.3 Immune System 590

28.4 Blood Clotting 591

28.5 Hormones 592

28.6 Kidney Functions 593

28.7 Urine 596
    Summary 597
    Key Terms 598

Exercises 598
Discussion Exercises 599

29 Medical Drugs 600

29.1 Antibiotics 602
29.2 Antimetabolic Drugs 606
29.3 Nervous System Drugs 608
29.4 Treating Hormone Disorders 611
29.5 Drugs and Athletic Performance 612

Summary 615
Key Terms 615
Exercises 615
Discussion Exercises 617

APPENDIXES

A Significant Figures 618

B Exponents and Scientific Notation 621

C Converting from One Unit to Another 625

D The Elements 630

E Cell Structure 636

Answers to Even-Numbered Exercises 638

Index 653

General, Organic, and Biological Chemistry

I

GENERAL CHEMISTRY

1

Introduction to Chemistry

1. What is science? What is chemistry? Why is chemistry classified as a science?
2. What are scientific hypotheses, theories, and laws? What is the scientific method?
3. What are some limitations of science?

The Hall Process: It Can Pay to Listen to Your Chemistry Teacher

In 1885, aluminum cost more than silver and could be produced only on a small scale. Today it costs less than 50 cents per pound, thanks to Charles Martin Hall (Figure 1.1), who studied chemistry at Oberlin College in Ohio. In his freshman year, Hall began going to his chemistry professor, Frank F. Jewett, to buy chemicals and small items of equipment for his home laboratory. Jewett took the young man under his wing and gave him work space in the college laboratory.

One day, Jewett remarked that if anyone could invent a way to produce aluminum economically on a commercial scale, that person would benefit humanity and also grow rich. Hall turned to a classmate and said, "I'm going for that metal."

Hall began working on this project during his senior year. The problem was to find a way to isolate aluminum from its ore, bauxite, which consists of aluminum oxide and small amounts of other substances. How could he separate aluminum from the oxygen in aluminum oxide and remove the other ingredients? Metals such as iron and tin are freed from their oxide or sulfide ores by heating at high temperature in the presence of coke (impure carbon) or carbon monoxide. But this didn't work with bauxite.

After graduation, Hall borrowed equipment and continued to test his ideas in a woodshed behind his father's parsonage (Figure 1.2). After developing and testing several hypotheses, he discovered that passing large amounts of electrical current through molten (melted) bauxite freed the aluminum. Adding another material (called cryolite) to the molten bauxite dissolved other substances in the ore, leaving pure aluminum. In 1886—at age 23—he walked into Jewett's office with a dozen small globules of aluminum made by the electrolytic reduction process that now bears his name.

Hall went on to cofound Pittsburgh Reduction Company, which later became Aluminum Company of America (ALCOA). He died a multimillionaire in 1914, bequeathing large gifts to his community, to Oberlin College, and to higher education in general.

Figure 1.1 Charles M. Hall (1863–1914). (Courtesy ALCOA)

Figure 1.2 Replica of the woodshed laboratory where Charles M. Hall discovered the electrolytic reduction process for making aluminum. (Courtesy ALCOA)

When you look at a cat (Figure 1.3), what do you see? Many people see a small animal with fur, white whiskers, four legs, pink tongue, round eyes, pointed ears, and a tail. If it's your cat, maybe you see a friend, a companion who likes to be with you.

An anatomist can look at a cat and see its underlying bone structure, how the skeleton is put together so that the cat moves with style and grace. A physiologist can picture the inside organs and tissues, and how they work together to keep the cat alive.

Figure 1.3 A cat. What do you see?

A chemist sees other things. When she looks at the cat's fur, she envisions protein molecules, which fold and curl in wondrous ways to keep from dissolving in water. (Otherwise, it would be hazardous for cats to go out in the rain.) When a chemist sees the cat breathe in, she can picture oxygen binding to hemoglobin, traveling in the cat's bloodstream to various organs, and then reacting with carbohydrates, fats, and proteins to produce carbon dioxide and water. When the cat exhales, she pictures carbon dioxide streaming out.

People, like all living things, are chemical and biological systems. Our bodies do not work by magic; they operate in harmony with the laws of nature. Of course, people are much more than physical beings. But if we want to understand the physical part, we need to understand the underlying biology and chemistry.

Table 1.1 Types of Science

Biological Sciences
Biology
Botany
Microbiology
Zoology

Physical Sciences
Astronomy
Chemistry
Geology
Physics

Social Sciences
Economics
History
Political science
Psychology
Sociology

1.1 THE NATURE OF CHEMISTRY AND SCIENCE

Science is the study of nature, an attempt to understand how nature works. It includes learning how people behave and interact *(social sciences)*, how living things work *(biological sciences)*, and the underlying operations of nature *(physical sciences)* (Table 1.1).

Chemistry is the study of the structure, properties, and changes of matter. **Matter** is anything that has mass (weight) and occupies space. In other words, chemistry is the study of materials and how they change.

EXAMPLE 1.1 Is chemistry a social, biological, or physical science?

SOLUTION Chemistry is a physical science.

Figure 1.4 Each item in this picture has a chemical structure that helps determine its characteristics.

Look around. Everything you can see and touch has a chemical structure (Figure 1.4). So does air, and a few other materials you can't see. Each piece of matter has properties—such as color, texture, and melting point—that depend on its chemical structure. This is true of a pencil, a rug, a cat—even your own body.

Chemical changes are all around us. Plants use sunlight to make oxygen gas, sugar, and other materials; caterpillars change into butterflies (Figure 1.5); and teenage boys sprout whiskers. People also make nature's materials into new substances such as nylon, computer chips, and artificial limbs.

Figure 1.5 A caterpillar chemically changed into this butterfly.

1.2 HOW SCIENCE WORKS

Laws, Hypotheses, and Theories. Scientists make observations and do experiments to measure what happens in nature. Their findings are called **data** or *facts*. They use this information to propose general statements about how nature works.

One type of statement is a scientific law. A **law** summarizes what is consistently observed in nature. For example, Isaac Newton (Figure 1.6) noticed that apples and other objects fall to earth. He used these and other observations to

5

Figure 1.6 Isaac Newton (1642–1727), one of the greatest physicists who ever lived, formulated the law of gravity. He also invented calculus and described the nature and properties of light.

formulate a general statement about how objects are attracted to each other and to earth: the law of gravity.

Since a law summarizes what consistently happens in nature, you can use laws to predict what will happen in a new situation. You can use the law of gravity, for example, to predict that if you jump from a tree, you will fall to the ground.

Other scientific statements explain *how* or *why* things happen. A scientific **hypothesis** is an educated guess, a possible explanation for what is observed in nature. If you watched identical amounts of water and gasoline evaporate in a room, for example, you would see the gasoline disappear first. Why does this happen? Your explanation would be a hypothesis.

If your hypothesis is useful, it should predict what will happen in new situations. What if you interchange the containers for gasoline and water? What if you watch gasoline and water evaporate while you keep them both cold? What if you watch mercury and ether evaporate under the same conditions? In each case, what does your hypothesis predict will happen?

Scientists test a hypothesis by doing experiments. If the results don't fit the prediction, the hypothesis needs to be changed. But if many experiments by different scientists do fit a hypothesis, it becomes a scientific theory. A **theory** is a hypothesis that has been widely accepted because of extensive testing.

EXAMPLE 1.2 Classify each of the following as a scientific hypothesis, theory, or law:
(a) Bats use sound instead of vision to find their way around.
(b) Gravity causes hair to turn gray.
(c) Matter cannot be created or destroyed, but it can change forms.

SOLUTION (a) theory (a widely accepted explanation); (b) hypothesis (an explanation not widely accepted); (c) law (a statement of what consistently happens)

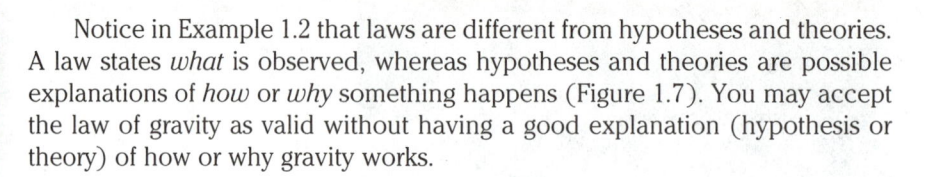

Notice in Example 1.2 that laws are different from hypotheses and theories. A law states *what* is observed, whereas hypotheses and theories are possible explanations of *how* or *why* something happens (Figure 1.7). You may accept the law of gravity as valid without having a good explanation (hypothesis or theory) of how or why gravity works.

Scientific Method. The way scientists gather data and develop scientific laws, hypotheses, and theories is called the **scientific method.** It is a logical, common-sense approach that includes these steps:

- Determine the question about nature to be answered.
- Find all the information about the question that may be relevant. Think about new information that could be obtained by doing experiments. Do the experiments.
- Think about the results of the experiments, and summarize them in the form of a scientific law. Or propose a hypothesis, a possible explanation of the data.
- Test a hypothesis further. Modify it if necessary to fit the data. Perhaps it can become a scientific theory.

Figure 1.7 You can observe that a can of Coca-Cola sinks in water and a can of Diet Coke floats. Can you devise a hypothesis to explain why? What experiments could you do to test your hypothesis?

Knowledge comes in many ways. Sometimes a law stimulates scientists to propose hypotheses to explain how the law works. Other times a scientist follows a hunch or a sudden flash of insight and does experiments to test that idea. For example, Charles Martin Hall used logical thinking, insight, and long hours of experimenting to discover how to make aluminum metal from its ore.

Occasionally an accident or totally unexpected experimental result spurs scientists to discover something new. For example, accidents seen and investigated by thoughtful scientists led to the discoveries of radioactivity (see Chapter 10) and penicillin (see Chapter 29).

In reality, scientists use many methods—not just one—to learn about nature. But sooner or later they do experiments to test their ideas. It isn't enough just to like an idea, or to believe in it. A new idea has to stand the test of many experiments by many people.

Good scientists are curious, creative, and skeptical about new ideas. They rigorously test hypotheses and theories. Devising a decisive experiment or proposing a hypothesis requires logical reasoning. It also takes imagination, intuition, insight, and creativity. These qualities are as important in science as in poetry, art, music, and other great adventures of the human mind.

1.3 APPLICATIONS OF THE SCIENTIFIC METHOD

People use the scientific method in many ways. A meteorologist, for example, uses current information and a knowledge of weather patterns to predict to

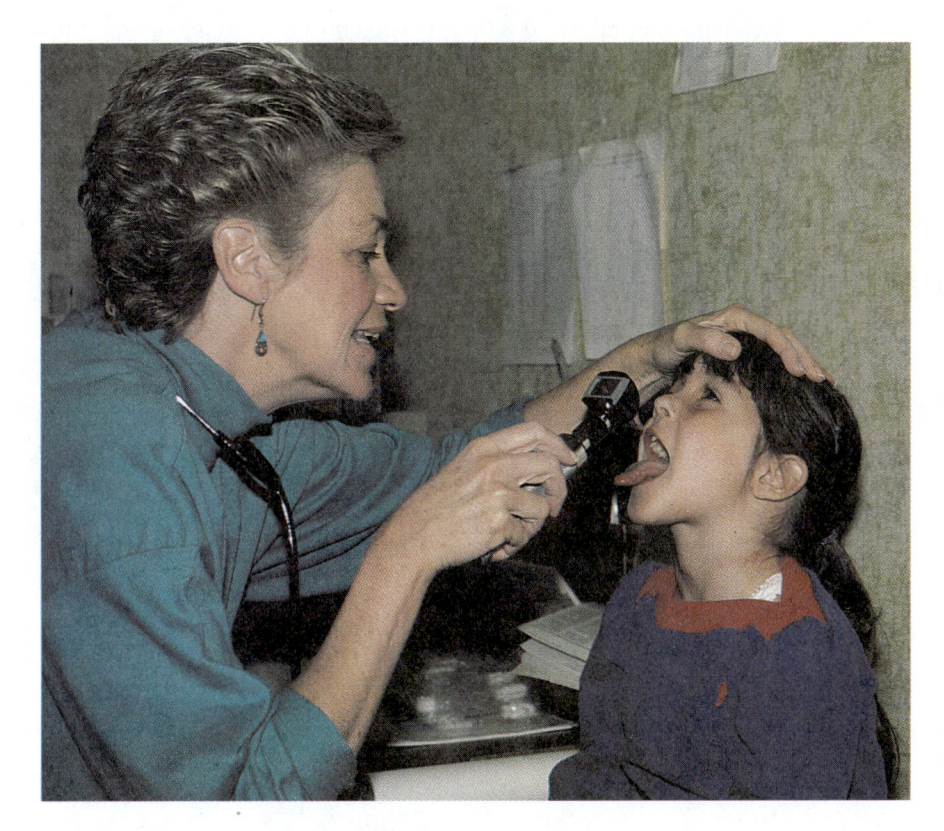

Figure 1.8 A physician uses the scientific method to diagnose her patient's illness.

morrow's weather. The forecast is a hypothesis that is tested when the next day arrives.

A meteorologist may also predict that the sun will rise tomorrow at 6:32 A.M. and set at 7:18 P.M. He bases this on scientific law—the consistent, orderly behavior of the earth and the sun in space.

A botanist uses the scientific method to find out why tomatoes don't grow well in a garden. She might test the soil for acidity and for nutrients such as nitrogen, phosphorus, and potassium. If the nitrogen level is low, she might hypothesize that a lack of nitrogen in the soil is causing the problem. The hypothesis is tested by adding nitrogen fertilizer to the soil and seeing how well the tomatoes grow.

A physician uses the scientific method to diagnose diseases (Figure 1.8). If a patient complains of a runny nose, cough, sore throat, nasal congestion, and headache, the physician must determine whether the patient has a cold, flu, bacterial infection, or something else entirely. The lack of a fever and joint and muscle aches make it more likely that the patient has a cold. The physician can test this hypothesis by taking a throat culture and by seeing how the patient responds to medication for cold symptoms.

1.4 LIMITATIONS OF SCIENCE

Scientists assume the natural world is orderly and consistent, that it operates the same way today as it did centuries ago, and as it will tomorrow (Figure 1.9). They try to discover this order, and use this knowledge to predict what will or will not happen in the natural world.

This consistency keeps favorite—but unlikely—ideas from getting out of hand. Other scientists will always try to repeat or extend important experiments. If nature is consistent, all scientists who can duplicate the experimental conditions should get the same results. If under the same conditions one person reports the formation of a red substance, while another reports a blue substance, either the conditions aren't the same or one of the reports is wrong.

Figure 1.9 Earth seen from space. If the laws of nature once were different than they are today, scientists have little hope of understanding the earth's formation.

Limits. Scientists often make careful measurements to test laws, hypotheses, and theories, but they cannot prove absolutely that these ideas are true. A theory, for example, is simply the best explanation anyone has discovered so far to account for something that happens in nature. The theory survives only as long as it is *useful* in describing and making predictions about nature. If new information doesn't fit, the theory must be changed.

Science is limited in several important ways:

- Our sensory capacities are limited. Even with sophisticated machines, we are limited in what we can see and measure. In addition, our measurements contain errors.
- We are limited in our ability to communicate our observations and ideas about nature (Figure 1.10).
- We are limited in our ability to think logically and objectively.

Figure 1.10 How completely can you communicate the contents of this photo to someone who has not seen it?

- Physical sciences (such as chemistry) deal with the physical world and the operations of nature within that world. They do not deal with supernatural concepts, and they cannot give definitive answers to questions that have a nonmaterial dimension. For example, the physical sciences are of limited use in dealing with such questions as:

Why are we on earth? What is our purpose?
At what stage in development does a human being begin to exist?
Is euthanasia appropriate in certain situations?
What is beauty, consciousness, intuition, love, God?

Bill Moyers: What is it you want to know about the universe?

Steven Weinberg (physicist): It's very simple. I just want to know one thing, which is why things are the way they are.

Source: A World of Ideas by Bill Moyers

SUMMARY

Science is the study of how nature works. Chemistry is the study of the structure, properties, and changes of matter. Chemists and other scientists learn about the physical world by gathering data and formulating scientific laws, hypotheses, and theories. This approach is called the scientific method. Scientists also use their intuition, insight, imagination, and creativity in this process. In several important ways, science is limited.

KEY TERMS

Chemistry (1.1)
Data (1.2)
Hypothesis (1.2)
Law (1.2)
Matter (1.1)
Science (1.1)
Scientific method (1.2)
Theory (1.2)

EXERCISES

Even-numbered exercises are answered at the back of this book.

Nature of Chemistry and Science

1. Classify each of the following as a social science, biological science, or physical science: **(a)** astronomy, **(b)** geology, **(c)** economics, **(d)** botany, **(e)** psychology, **(f)** medicine.
2. Which of the following are classified as matter: **(a)** apple, **(b)** air, **(c)** anxiety, **(d)** blood, **(e)** noise, **(f)** fatigue, **(g)** hair, **(h)** competition, **(i)** book.
3. Scientists try to do experiments in which they keep everything constant except one factor—the variable. In which area—social science, biological science, or physical science—do you think it is generally most difficult to do a study with a single variable? Explain why.

Laws, Hypotheses, and Theories

4. Which of the following are classified as laws, rather than hypotheses or theories:
   (a) The sun will set tonight.
   (b) Certain viruses cause AIDS.
   (c) Ancient meteorites caused dinosaurs to become extinct.
   (d) What goes up must come down.
   (e) Heat flows from warm bodies to cold bodies.
   (f) Matter is made up of atoms.
   (g) Smoking causes cancer.
   (h) The earth revolves around the sun in about 365.25 days.
   (i) Many health problems experienced by Vietnam veterans resulted from their exposure to the pesticide Agent Orange.
5. Which of the statements in exercise 4 may be classified as hypotheses or theories?
6. A physician frequently bases a diagnosis on certain symptoms in the patient. Is such a diagnosis more like a hypothesis, theory, or law?
7. Would the idea that humans evolved from apes most likely be classified as a hypothesis, theory, or law? Why?
8. In the 19th century, John Dalton, an English schoolteacher, tried to explain the law of conservation of mass by proposing that all elements are made up of tiny, indestructible particles called atoms. Was Dalton's proposal classified as a hypothesis, theory, or law? Why?
9. Are scientific laws ever broken? Explain.
10. With testing, a hypothesis may become a theory. With further testing, may a theory become a law? Why?

Scientific Method

11. Which of the following statements about the scientific method are true:
    (a) This method is widely used by chemists.

(b) This is the one method for discovering something new about nature.
   (c) It is a useful way to try to discover the ultimate particles from which all matter is composed.

12. Which of the following statements about the scientific method are true:
    (a) It is a useful way to discover whether or not God created the earth.
    (b) Social scientists use the scientific method.
    (c) This method helps scientists learn the absolute truth about what happens in nature.

13. Suppose a chemist in a laboratory measures the boiling point of water as 100°C (212°F). Then she takes the same water and thermometer to a cabin in the mountains and measures water boiling at 95°C (203°F). Is at least one of the measurements necessarily incorrect? Explain.

DISCUSSION EXERCISES

1. How useful is scientific information in determining at what stage of development a human being begins to exist?
2. Do you think fraud is more, less, or equally likely to occur in science compared with other disciplines? Explain why.

2
Measurements

1. How are measurements expressed numerically? What are accuracy and precision, and how are they different? What are significant figures? What is scientific notation?
2. What is the metric system? What is the International System of Units (SI)?
3. What are some important units that chemists use to measure matter and energy?
4. What is the factor-label method? How do you use it to convert from one unit of measurement to another?

What Is Normal Body Temperature?

When you take someone's temperature (Figure 2.1), you usually read the thermometer to a tenth of a degree Fahrenheit (°F). Most of us would say that normal body temperature is 98.6°F. But that isn't entirely correct.

In 1868, Carl Wunderlich published a paper based on more than 1 million measurements of normal body temperature. In the 25,000 adults measured, "normal" oral body temperatures varied with the time of day and method of taking the temperature. Because of this variation, Wunderlich reported the average oral temperature only to the nearest degree on the Celsius scale—37°C.

When you calculate the equivalent Fahrenheit temperature by one of the methods shown in Section 2.4, your calculator displays 98.6 degrees as the answer. Though 98.6°F is commonly given as normal body temperature, the 0.6 is not significant. The original measurement of 37°C was properly expressed to only two figures (called *significant figures*). Converting from one unit (°C) into another (°F) doesn't change the number of significant figures, because it doesn't change the fineness of the original measurement.

The lack of significance of the 0.6 in 98.6°F was shown in a study reported in 1992 by the *Journal of the American Medical Association*. Seven hundred measurements of normal body temperature gave an average value of 98.2°F.

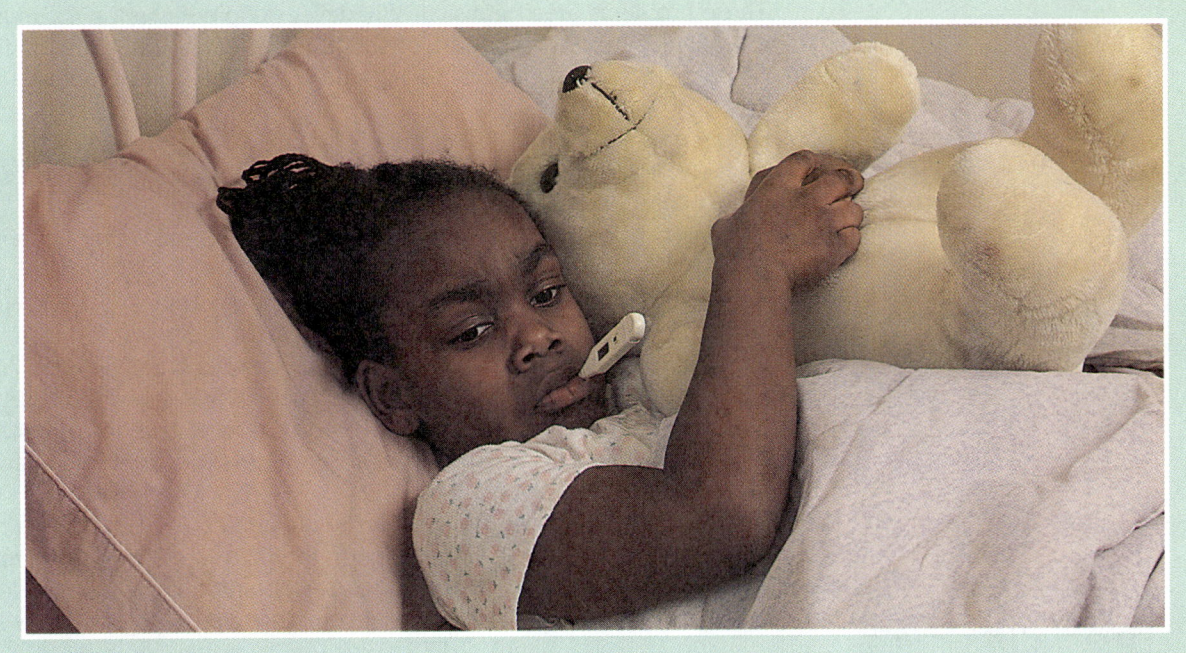

Figure 2.1

A new car has a 2.4 liter engine. Five hundred people puff their way through a 10-kilometer run. A student uses a set of metric tools to repair her motorcycle. A nurse injects 25 cubic centimeters (cc) of a drug solution to help an unconscious patient, whose blood glucose level is 50 grams per deciliter. A soft drink machine dispenses a can of diet cola that has less than 1 Calorie; a can of the same drink in Australia is a low-joule cola (Figure 2.2).

When you use a ruler, balance, watch, thermometer, or other device to find a distance, mass, time, or temperature, you are making a *measurement*. You need to express the measurement in terms of some unit(s). It doesn't do much good to say that something weighs 12. Twelve what? Pounds, grams, stones, grains, drams, kilograms, carats, tons, ounces, pennyweights, or scruples?

2.1 UNCERTAINTY IN MEASUREMENTS

How hard is it to make a perfect measurement? Think, for example, about trying to measure your height.

First, you might have trouble deciding how "tall" to stand, how to hold your head, and where exactly the top of your head is. Next, you would have to use some device, such as a ruler, to measure the distance. In how much detail could you make the measurement—to the nearest one-fourth, or one-sixteenth, or one-millionth of an inch? (This would depend partly on the markings on your ruler.) Whatever your final fraction, you could only estimate the last, tiniest part.

And what if your ruler wasn't quite correct? What if each inch marked on it was actually 1.01 inches long? You might have measured carefully, down to a tiny fraction, but it still wouldn't be quite right.

These problems aren't unusual. Whenever people make a measurement, their results depend on the nature of what they are measuring, the quality of their measuring instruments, and their ability to use those instruments.

Accuracy and Precision. **Accuracy** is how close a measurement is to the actual value. If you measure your height as 6 feet, for example, the accuracy depends on how close to 6 feet tall you actually are. In order to improve accuracy, scientists often make several measurements and calculate the average value.

Precision refers to the reproducibility of a measurement. If you keep getting exactly 6 feet as your height, your measurement has good precision.

Figure 2.3 illustrates the difference between accuracy and precision. A high degree of precision sometimes indicates accuracy. But not always. For example, if you used a faulty ruler, or if you didn't read it properly, you would consistently get the same, incorrect height (good precision, poor accuracy).

Significant Figures. Scientists express the uncertainty in measured values in terms of **significant figures.** Depending on the thermometer you use, for example, you may be able to measure your temperature to the nearest degree Celsius (°C), or 0.1°C, or even 0.01°C. But your final digit, whatever it is, will be an estimate. A measurement of 37.1°C means the value could be 37.2°C or 37.0°C, but 37.1°C is the best estimate.

A measured value of 37.1°C has three significant figures: 3, 7, and 1. If you use a thermometer that only allows measurements to the nearest 0.1°C, you have no basis to report the temperature as 37.12157°C; all digits past the first 1

Figure 2.2 A low-joule diet drink sold in Australia.

have no meaning and are not significant. If you then calculate the Fahrenheit temperature that equals 37.1°C, you cannot express that answer in more than three significant figures either (no matter what the calculator displays), for the calculation has done nothing to change the amount of detail in the original measurement.

Appendix A at the back of this book gives information about which digits in a number are significant, and how to handle significant figures in calculations.

Scientific Notation. Scientists often work with very large or small numbers. For example, the mass of the smallest unit (an atom) of iron is about 0.000000000000000000000911 g. Scientists often convert such numbers into a more convenient, standard form, called **scientific notation.**

To write a number in scientific notation, you need to write it as a number between 1 and 9.99×10 to the appropriate power. In other words, you write one nonzero number to the left of the decimal point. The mass of an iron atom, then, is written as 9.11×10^{-23} g. The exponent means that the number represented is 9.11 with the decimal point moved 23 places to the left.

Scientific notation is a convenient way to write very large and small numbers and to show clearly the number of significant figures. Appendix B at the back of this book reviews how to work with exponents and how to write numbers in scientific notation.

EXAMPLE 2.1 Write the following in scientific notation: (a) 65,095, (b) 0.00000000000031, (c) 582, (d) 0.707

SOLUTION (a) 6.5095×10^4, (b) 3.1×10^{-13}, (c) 5.82×10^2, (d) 7.07×10^{-1}

2.2 THE METRIC SYSTEM AND INTERNATIONAL SYSTEM OF UNITS (SI)

Standards of measurement are arbitrary—they work only if everyone uses the same units. Yet some units (such as gallons) are different sizes in different countries. Another drawback is that units were chosen in different ways and different places. As a result, the numerical relationships between them are inconvenient and hard to remember. Do you know, for example, how many inches there are in a mile, or cups in a gallon, or carats in an ounce?

A new, unified system of units, called the **metric system,** was adopted in France in 1875. Nearly everyone world wide (Figure 2.4) now uses these units. This system was updated and revised in 1960 and called the **International System of Units (SI).** Most of the units used in this book are the same in the metric and SI systems. The main exceptions are the liter, milliliter, and calorie, which are metric but not SI units.

2.3 UNITS OF MEASUREMENT: MASS, LENGTH, VOLUME, AND DENSITY

You use different units depending on the size of the quantity you are measuring. You would say the distance to a nearby town is a certain number of miles, not inches. But you would use inches, not miles, to measure the distance around

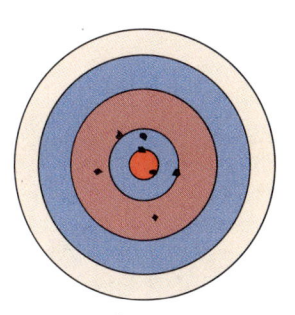

good accuracy
poor precision

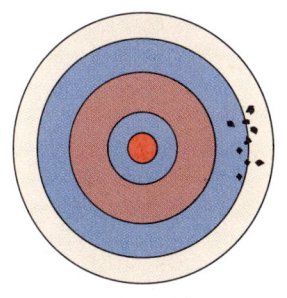

good precision
poor accuracy

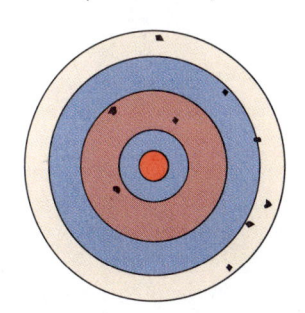

poor precision
poor accuracy

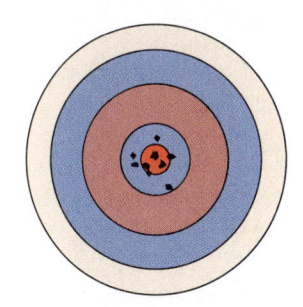

good precision
good accuracy

Figure 2.3 Bullet holes on a target illustrate the difference between accuracy (getting a bull's-eye) and precision (consistent readings).

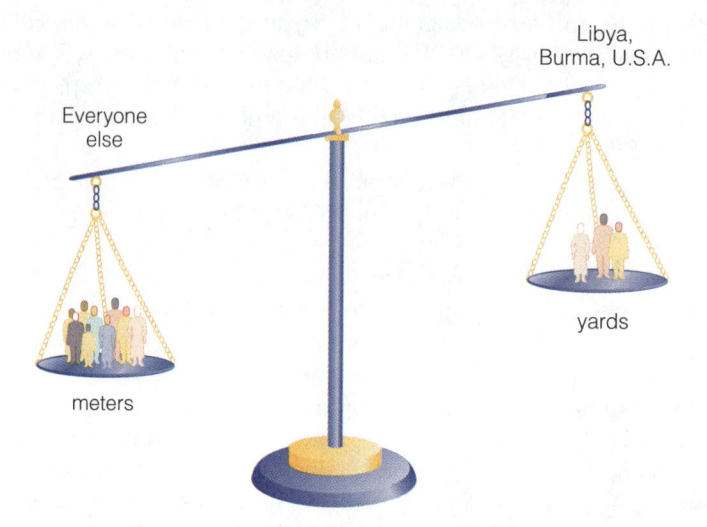

Figure 2.4

your waist. You could interconvert between inches and miles, but only if you remember how many inches are in a foot or yard, and how many feet or yards are in a mile.

One advantage of the metric system is that it uses prefixes to change the size of a unit by multiples of 10. Table 2.1 lists common prefixes, their numerical values, and their abbreviations. You can use these prefixes with any of the metric units discussed next. Consider, for example, the prefix *milli-*, which means 1/1000. A millimeter is 1/1000 of a meter; a milligram is 1/1000 of a gram; and a milliliter is 1/1000 of a liter.

Table 2.1 Common Prefixes for the Metric System

Prefix	Abbreviation	Numerical Factor	Exponential Factor*
mega-	M	1,000,000	10^6
kilo-	k	1,000	10^3
deci-	d	1/10 (0.1)	10^{-1}
centi-	c	1/100 (0.01)	10^{-2}
milli-	m	1/1000 (0.001)	10^{-3}
micro-	μ	1/1,000,000 (0.000001)	10^{-6}
nano-	n	1/1,000,000,000 (0.000000001)	10^{-9}

*To review the use of exponents, see Appendix B at the back of this book.

Because of these prefixes, you can convert from one metric unit to another simply by moving the decimal point. For example, a distance of 41.1 meters is 41,100 millimeters, or 4,110 centimeters, or 0.0411 kilometers. Many examples of this will follow.

Mass. Recall that matter (Section 1.1) is anything that occupies space and has mass; it is anything that is material. How much matter does your body have? To answer this, you would probably weigh yourself on a scale. If you did this on the moon, you would weigh only about one-sixth as much as you weigh on earth. Yet in both cases your body would have the same amount of matter.

Mass is the amount of matter in an object. **Weight** is a measure of the gravitational attraction of earth (or the moon) for an object. In the metric system, mass is measured in *grams (g)*. The official SI unit of mass is the *kilogram (kg)*, which equals 1000 g. Another common unit is the *milligram (mg)*, which equals 1/1000 or 0.001 g.

On earth, a kilogram weighs about 2.2 pounds (Table 2.2). A 121-pound jockey, for example, weighs 55 kg. A quarter-pound hamburger patty weighs about 110 g, and a 1-pound box of sugar weighs 453.6 g (Figure 2.5).

Table 2.2 English and Metric Units

Mass
1 lb = 454 g
1 kg = 2.20 lb

Length
1 in. = 2.54 cm
1 m = 39.4 in.
1 mile = 1.61 km

Volume
1 L = 1.06 qt

EXAMPLE 2.2
(a) How many milligrams (mg) are in 1 gram (g)?
(b) How many g are in 1 kilogram (kg)?

SOLUTION
(a) 1 mg = 1/1000 g (Table 2.1)
multiply both sides by 1000
1000 mg = 1 g
(b) 1 kg = 1000 g (Table 2.1)

Chemists use a device called a *balance* (Figure 2.6) to measure the mass of an object. The idea is something like balancing a pencil on your finger: if you put something on one end of the pencil, you have to put something of equal mass on the other end to bring the pencil back into balance.

In a balance, pieces of known mass are used to balance the mass of the object being measured (Figure 2.7). The total mass of these pieces equals the mass of the object. With this device, you get the same value whether the object is measured on earth or on the moon. In each place, gravity attracts the object and the pieces used to balance it with equal force.

Figure 2.5 One pound equals 453.6 g.

Length and Volume. The standard metric unit of length is the *meter (m)*. Other common units include the *kilometer (km;* Figure 2.8), equal to 1000 m; the *centimeter (cm)*, equal to 1/100 or 0.01 m; and the *millimeter (mm)*, equal to 1/1000 or 0.001 m. A meter is a little longer than a yard (Table 2.2; Figure 2.9).

EXAMPLE 2.3 How many meters (m) are in (a) 1 centimeter (cm) (b) 1 kilometer (km)?

SOLUTION (a) 1 cm = 1/100 (or 0.01) m (Table 2.1)
(b) 1 km = 1000 m (Table 2.1)

Volume is the amount of space matter occupies. It is a measure of length in three dimensions. The volume of a brick, for example, is its height × width × length. The volume of a room with dimensions of 2.7 m, 3.2 m, and 4.3 m is (2.7 m) × (3.2 m) × (4.3 m) = 37 m^3.

Since the meter is the standard metric unit of length, you might expect (correctly) that the *cubic meter* (m^3) is the SI unit of volume. But many objects are much smaller than a cubic meter. They are measured using a smaller unit, the *cubic centimeter* (cm^3, occasionally called a *cc*). One m^3 equals 1,000,000 cm^3.

Notice that you multiply the units as well as the numbers, so the final unit is m × m × m = m^3.

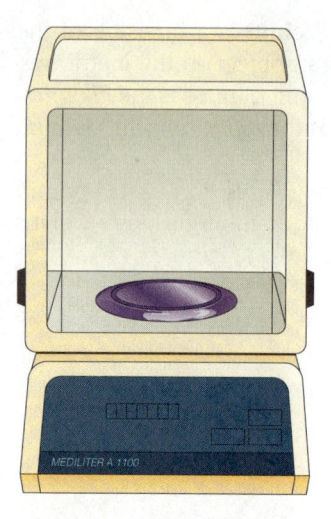

Figure 2.6 A balance is used to measure mass.

Figure 2.7 (Top) The object to be measured is unbalanced. (Bottom) The object is balanced by pieces having a total mass equal to the mass of the object.

Another common unit, particularly for measuring the volume of liquids and gases, is the *liter (L)*. A liter is slightly larger than a quart (Table 2.2; Figure 2.10). The *milliliter (mL)*, which is 1/1000 of a liter, is also commonly used for small volumes. A convenient relationship you should remember is that 1 cm³ = 1 mL.

Figure 2.8 The U.S. is slowly converting to the metric system.

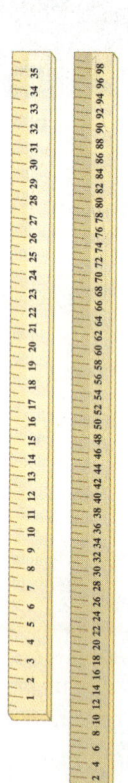

Figure 2.9 A yardstick (left) is about 10% shorter than a meterstick (right).

EXAMPLE 2.4
(a) What is the volume in cm³ of an ice cube with dimensions of 2.5 cm, 1.8 cm, and 2.0 cm?
(b) Is a liter (L) or a milliliter (mL) larger? How much larger?

SOLUTION
(a) (2.5 cm)(1.8 cm)(2.0 cm) = 9.0 cm³
(b) 1 mL = 1/1000 L (Table 2.1);
Multiply both sides by 1000: 1000 mL = 1 L.
Thus, 1 L is 1000 times larger than a mL.

Density and Specific Gravity. **Density** is a measurement of the amount of mass in a given amount of volume. Its formula is:

$$\text{density} = \frac{\text{mass}}{\text{volume}}$$

Densities of gases, which are much less than those of solids and liquids, are usually expressed as g/L.

In the metric system, the usual units of density are g/mL for a liquid or g/cm³ for a solid. Since 1 mL = 1 cm³, these two units can be used interchangeably.[1]

A golf ball and a ping–pong ball are about the same size; that is, they have about the same volume. But the golf ball has much more mass (and weight) than the ping–pong ball. Since the golf ball has more mass packed into the same volume, it has a greater density (more g/cm³) than the ping–pong ball.

To determine the density of a substance, you need to know the mass and volume of a particular sample. Then you simply calculate how many g there are per 1 mL or per 1 cm³. For example, 3.20 mL of mercury has a mass of 43.5 g. The density of mercury is:

$$\text{density of mercury} = \frac{43.5 \text{ g}}{3.20 \text{ mL}} = 13.6 \text{ g/mL}$$

Table 2.3 lists the densities of several solids and liquids.

EXAMPLE 2.5
(a) A cube of copper with a height, width, and depth of 2.0 cm has a mass of 71.3 g. What is its density?
(b) A 3.0 mL sample of octane has a mass of 2.1 g. What is the density of octane?

SOLUTION
(a) Volume of copper = (2.0 cm) × (2.0 cm) × (2.0 cm) = 8.0 cm³.

Therefore, density = $\frac{71.3 \text{ g}}{8.0 \text{ cm}^3}$ = 8.9 g/cm³

(b) density = $\frac{2.1 \text{ g}}{3.0 \text{ mL}}$ = 0.70 g/mL

Figure 2.10 This quart of oil has a volume a little less than 1 liter.

Specific gravity is the ratio of the density of a substance to the density of water:

$$\text{specific gravity} = \frac{\text{density of a substance}}{\text{density of water}}$$

Since the denominator (density of water) is 1.00 g/mL, the specific gravity of a substance is numerically the same as its density. The only difference is that specific gravity doesn't have units; the units cancel because the same units are in the numerator and denominator.

You can calculate the specific gravity of mercury, for example, as follows:

$$\text{specific gravity} = \frac{\text{density of mercury}}{\text{density of water}} = \frac{13.6 \text{ g/mL}}{1.00 \text{ g/mL}} = 13.6$$

Table 2.3 Densities

mercury	13.6 g/mL
lead	11.3 g/cm³
iron	7.86 g/cm³
iodine	4.93 g/cm³
diamond	3.52 g/cm³
table salt	2.17 g/cm³
bone	1.80 g/cm³
chloroform	1.48 g/mL
whole blood	1.06 g/mL
water	1.00 g/mL
ethyl alcohol	0.79 g/mL
ethyl ether	0.71 g/mL
octane	0.70 g/mL

---

[1] Since substances typically become more dense as they cool, the temperature technically should be listed for any density (or specific gravity) value. Density (and specific gravity) values listed in this book are the approximate values at room temperature (20°C).

CHEMISTRY SPOTLIGHT

Density and Body Fat

Muscle and bone are more dense (about 1.06 g/cm^3 and 3.0 g/cm^3, respectively) than water (1.00 g/mL) and thus sink in water. Fat floats, however, because it is less dense (about 0.90 g/cm^3) than water. In young men and women who aren't athletes, fat typically provides about 15% and 25%, respectively, of body weight. Athletes usually have a lower percentage of body fat while older people have a slightly higher percentage.

The most accurate way to estimate body fat is from body density. Since density has the units g/cm^3, the total mass and volume of the body must be measured. Mass is measured using a scale. Measuring the volume of a person is harder because of the irregular shape.

Body volume is measured by hydrostatic (underwater) weighing (Figure 2.11). The person sits in a chair suspended from a scale and beneath the water surface in a pool or tank. After exhaling as much air as possible, the person is submerged in water for 5–10 seconds and weighed. This is done several times to get more accurate results.

People weigh less underwater because they displace water, which in turn buoys the body. The amount of water displaced (and its buoyant force) depends on the *volume* of the body underwater. The decrease in weight underwater equals the person's volume (once corrections are made for the volume of air remaining in the lungs, and for the exact density of water at its temperature in the tank or pool). The general formula is:

$$\text{volume} = \frac{\text{weight} - \text{underwater weight}}{\text{density of water}} - \text{volume remaining in lungs}$$

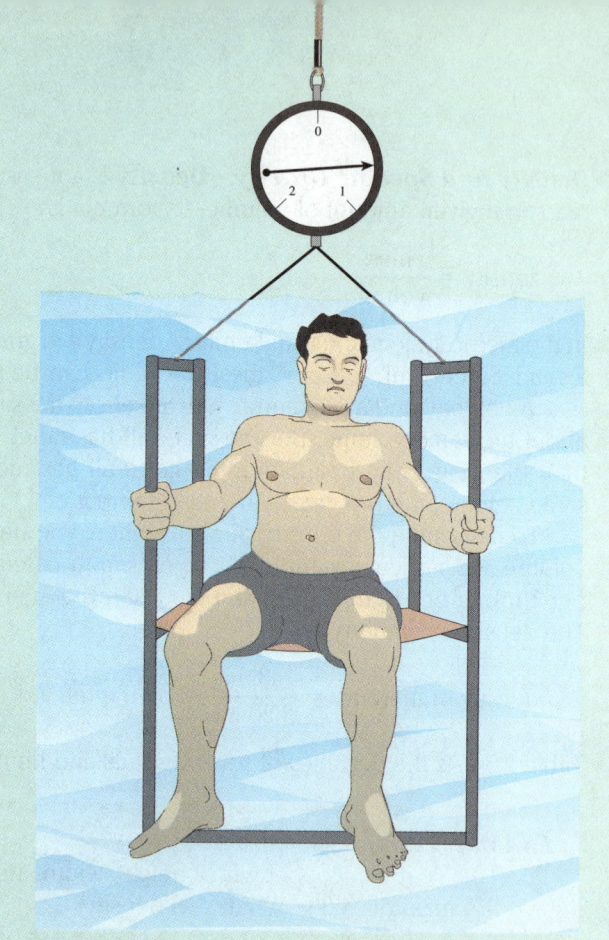

Figure 2.11 Hydrostatic (underwater) weighing to measure the volume of the body to estimate percent body fat.

You calculate body density by:

$$\text{density} = \frac{\text{weight (g)}}{\text{volume (cm}^3\text{)}}$$

Several empirical formulas have been devised to give the best mathematical fit between density and percent body fat. One of them is:

$$\text{percent body fat} = \frac{495}{\text{density}} - 450$$

From this formula, you can see mathematically that people with higher densities (due to more muscle and bone) have lower percent body fat. An average density for a male is about 1.07 g/cm^3; women average about 1.04 g/cm^3. So women on average are more buoyant in water and have a higher percent body fat.

Specific gravity is measured with a *hydrometer* (Figure 2.12). The lower the specific gravity, the deeper this instrument sinks into a liquid. You may have used a hydrometer to measure the specific gravity of antifreeze solution or battery fluid.

Laboratory technicians in hospitals use a specialized hydrometer (called a urinometer) to measure the specific gravity of urine (Figure 2.13). The normal value is in the range 1.004 to 1.030. An unusually high value (such as 1.050) indicates excess dissolved material in the urine, possibly caused by dehydration. A low value (such as 1.002) may indicate damaged kidneys that cannot excrete wastes effectively.

2.4 UNITS OF MEASUREMENT: ENERGY, TEMPERATURE, AND TIME

Energy. The light and heat of a burning lump of coal and the force you use to lift this book are examples of energy. **Energy** is the ability to do work. Energy is absorbed or given off when matter moves or changes internally, such as boiling or freezing. Energy is also the heat that flows automatically from a hot object to a cooler one. When you touch a hot stove, you painfully experience this energy flow. **Heat,** then, is energy traveling to a cooler object.

Scientists classify energy into two categories: kinetic energy and potential energy. **Kinetic energy** is energy from motion. Wind, flowing water, heat, light, electricity, and speeding bullets all have kinetic energy. **Potential energy,** in contrast, is energy stored in an object. Water ready to flow over a dam, a piece of coal ready to burn, and uranium fuel in a nuclear reactor all have potential energy (Figure 2.14). Once released, potential energy changes into kinetic energy. Coal burning at a power plant, for example, heats water to turn a turbine in a generator that makes electricity, which travels over wires to provide heat, light, and energy to our homes and appliances.

Matter has kinetic and potential energy. Its tiny particles (atoms) are in constant motion (kinetic energy), and it has the potential to release energy when it reacts with other substances. You convert the chemical (potential) energy in your food into kinetic energy so that you can dance, breathe, think, cry, and do all the other things that show you're alive.

Most chemists use the *joule (J)* to measure heat, or any form of energy. Another common unit is the *calorie[2] (cal)*, which is larger than a joule by the factor:

1 cal = 4.18 J

Larger units of energy are the *kilojoule (kJ)*, equal to 1000 J, and the *kilocalorie (kcal)*, equal to 1000 cal. In dietary tables, a Calorie (with a capital C) is actually a kcal. So a soft drink that has 160 Calories actually has 160 kcal, or 160,000 calories of energy.

Figure 2.12 The specific gravity of a liquid is the numerical value on the hydrometer at the surface of the liquid.

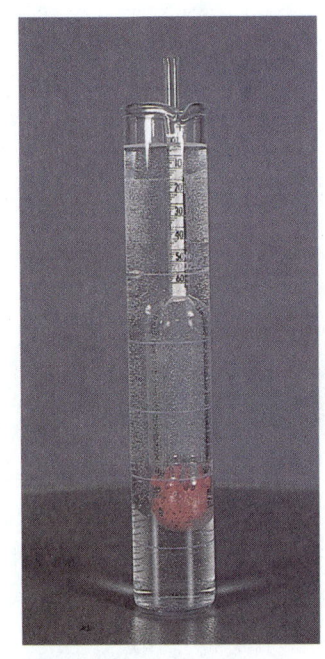

Figure 2.13 A urinometer is used to measure the specific gravity of urine.

> **EXAMPLE 2.6** What is the energy content in calories of a piece of pie that has 275 Calories?
>
> **SOLUTION** 275,000 cal

The calorie is not an SI unit, and in many countries it has been replaced by the joule (Figure 2.2). We will use both calories and joules in this text.

[2] One calorie is the amount of heat energy it takes to raise the temperature of 1 g of water by 1°C.

Figure 2.14 Potential energy changes into kinetic energy as water plunges over this dam.

Energy Requirements in the Body. Getting enough energy is the top priority for your metabolism. If the food you eat doesn't provide enough energy, your body will break down its stored carbohydrate, fat, and muscle. This will keep you alive, but at a lower energy level than normal.

Your **basal metabolic rate (BMR)** is the energy you use while awake and at rest. BMR is about 1.00 kcal (Cal) per hour per kg body weight for males, and 0.95 kcal/hr·kg for females. The energy you need each day to maintain your body weight is your BMR plus 30% if you are sedentary, 40% if you are moderately active, or 50% if you are very active.

> **EXAMPLE 2.7** Calculate the daily energy needed in kcal (Cal) for a moderately active, 154-lb (70.0-kg) male to maintain his weight.
>
> **SOLUTION**
> BMR = 1.00 kcal (Cal)/kg·hr × 70.0 kg × 24 hr = 1680 kcal (Cal)
> Total energy = 1680 + 0.40(1680) = 2350 kcal (Cal)

In order to lose weight, you must consume less energy than your body spends, while maintaining adequate amounts of vitamins and minerals. You can do this by decreasing your energy intake, increasing your activity level, or both. For example, if you consumed 1000 fewer Cal per day than your body spent, you would have a deficit of about 7000 Cal per week. One pound of body fat corresponds to 3500 Cal, so a deficit of 7000 Cal would cause a weight loss of 2 lb.

People use many other methods to lose weight. They try appetite suppressants (such as amphetamines), laxatives, diuretics, various surgical methods, and an incredible array of specialized diets. These usually don't produce permanent, satisfactory results.

Social pressures to be thin have also driven people—especially young women—to eating disorders. **Bulimia** is a disorder in which people go through cycles of eating binges followed by purging (vomiting) to expel food before it is metabolized (Figure 2.15). People with **anorexia nervosa** have an extreme fear of gaining weight. They avoid food and purge when they do eat. For some people, this condition is fatal.

Figure 2.15 A person with bulimia purges to reduce body weight.

Temperature. **Temperature** is a measurement of the average kinetic energy of a sample. It indicates the *intensity* of energy. A glass of warm milk, for example, has more kinetic energy than a glass of cold milk; its particles move with greater average energy than the same particles in cold milk.

EXAMPLE 2.8 Suppose you remove a cup of water from a lake and heat it until it boils. Compare the cup of boiling water with the water in the lake in terms of (a) temperature, (b) average kinetic energy, and (c) total kinetic energy.

SOLUTION The boiling water has (a) higher temperature, (b) greater average kinetic energy, and (c) less total kinetic energy (because the greater amount of water in the lake offsets its lower average kinetic energy).

The SI unit for temperature is the *kelvin (K)*. The Kelvin scale is related to the *Celsius* and *Fahrenheit* scales as shown in Figure 2.16. Notice that the Kelvin scale has no negative readings because it has a value of zero at the lowest possible temperature (called absolute zero). Since its values are always 273 higher than the Celsius reading, you convert a Celsius temperature into kelvins by:

°C + 273 = K

Notice that the Kelvin scale doesn't use a degree sign before K.

EXAMPLE 2.9 Normal body temperature is 37°C. What is this temperature in kelvins?

SOLUTION 37°C + 273 = 310 K

In comparing the Fahrenheit and Celsius scales (Figure 2.16), notice three things:

- The distance between the freezing and boiling points of water is 180° Fahrenheit, but only 100° Celsius; so the size of the degrees differs by a factor of 180/100 = 1.8.
- Numerical readings on the Fahrenheit scale tend to be larger.
- The scales meet at −40°, the only temperature that is numerically the same on both scales.

24   Part I   General Chemistry

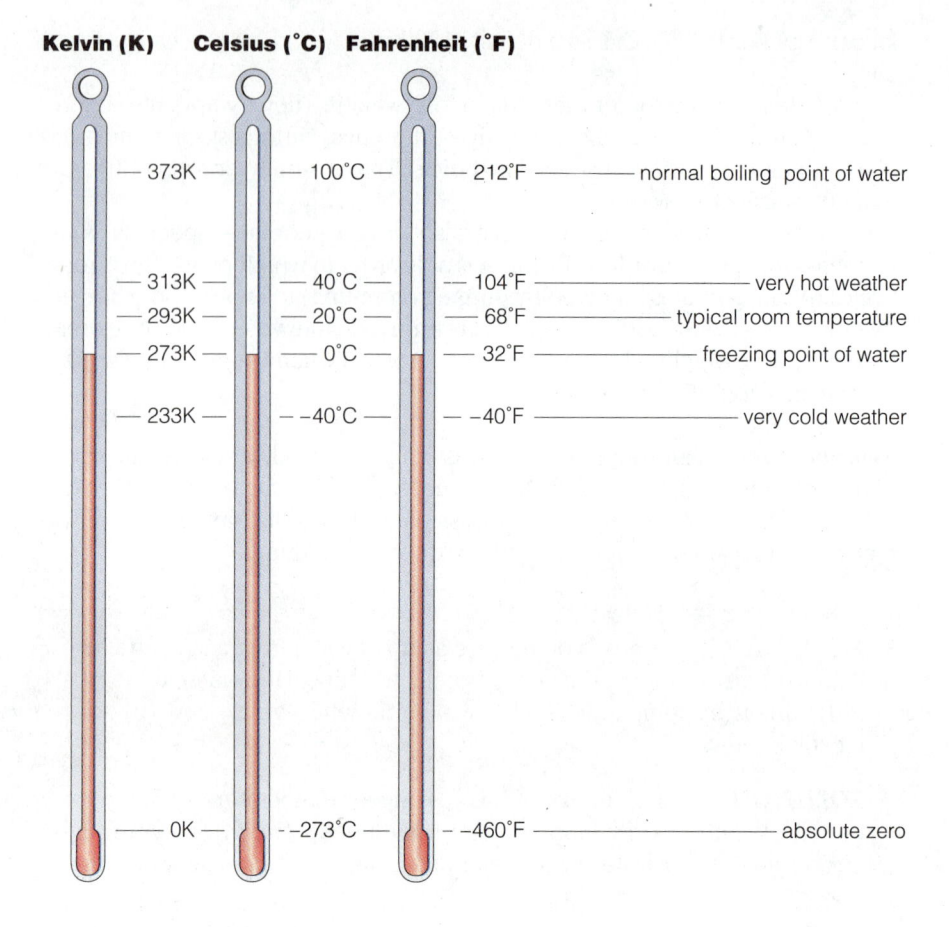

Figure 2.16 Comparison of the Kelvin, Celsius, and Fahrenheit temperature scales.

Kelvin (K): 373K, 313K, 293K, 273K, 233K, 0K
Celsius (°C): 100°C, 40°C, 20°C, 0°C, −40°C, −273°C
Fahrenheit (°F): 212°F, 104°F, 68°F, 32°F, −40°F, −460°F

- 212°F — normal boiling point of water
- 104°F — very hot weather
- 68°F — typical room temperature
- 32°F — freezing point of water
- −40°F — very cold weather
- −460°F — absolute zero

You can use these relationships to interconvert Celsius and Fahrenheit temperatures by the following method:[3]

1. add 40° to the original temperature, then
2. multiply (if going from °C to °F) or divide (if going from °F to °C) by 1.8, then
3. subtract 40° to get the final answer

On February 1, 1951, Dorothy Mae Stevens had the lowest documented body temperature—16.0°C, or 60.8°F—for a person who recovered.

EXAMPLE 2.10 Suppose a patient has a temperature of 95°F. What is this temperature in (a) °C and (b) kelvins?

SOLUTION (a) step 1: 95 + 40 = 135   (b) 35°C + 273 = 308 K
step 2: 135/1.8 = 75
step 3: 75 − 40 = 35°C

---

[3] Another, more common method uses an adjustment of 32, based on 0°C = 32°F. Formulas using this method are:
°C = (°F − 32°)/1.8
°F = (°C × 1.8) + 32°

> **EXAMPLE 2.11** The melting points for mercury and copper are −39°C and 1083°C, respectively. What are these temperatures in (a) °F and (b) kelvins?
>
> **SOLUTION** for mercury: (a) −38°F, (b) 234 K
> for copper: (a) 1981°F, (b) 1356 K

Time. The standard SI unit of time is the *second (sec)*. A few prefixes are used with it, mostly to make smaller units. For example, a millisec is 1/1000, or 0.001 sec. Can you suggest why time isn't put entirely into units that are multiples of 10?

2.5 CONVERTING BETWEEN UNITS

The Factor-Label Method. You often need to convert from one unit to another—inches to feet, pounds to ounces, cents to dollars, meters to kilometers. You know intuitively that you divide the number of inches by 12 to change to feet, and you multiply the number of feet by 12 to change to inches. But when you use less familiar, metric units, you may not be as sure how to convert.

A very helpful way to change from one unit to another is called the **factor-label method** (or *dimensional analysis*). The basic idea is to multiply the starting value by *conversion factors* that change the initial units to the desired units.

Conversion factors come from equal quantities of two separate units of measurement. For example, 1 foot = 12 inches. Rearranging the equation, we could also say $\frac{1 \text{ foot}}{12 \text{ inches}} = 1$ and $\frac{12 \text{ inches}}{1 \text{ foot}} = 1$. Since their numerators and denominators are equal (their ratios equal 1), $\frac{1 \text{ foot}}{12 \text{ inches}}$ and $\frac{12 \text{ inches}}{1 \text{ foot}}$ are both conversion factors. You can use either one to change from one unit to the other.

You can use the following steps of the factor-label method to calculate the number of inches in 5.5 feet.

Step 1: *Identify the initial units:* feet
Step 2: *Identify the desired units:* inches
Step 3: *Identify conversion factors to change from the initial units to the desired units.* You can go directly from feet to inches using $\frac{12 \text{ inches}}{1 \text{ foot}}$ or $\frac{1 \text{ foot}}{12 \text{ inches}}$
Step 3a: *Write out a map to show the sequence of units in the conversion, making sure you know a conversion factor for each step.* In this problem, the map is a simple, one-step conversion: feet → inches
Step 4: *Write the initial value and unit, then multiply it by the appropriate conversion factors to change to the desired units.* To cancel the initial unit, you need to have it both in a numerator and in a denominator. So select the conversion factor that puts the initial unit in the desired location for you to cancel it.

$5.5 \text{ feet} \times \frac{1 \text{ foot}}{12 \text{ inches}}$ doesn't work

$5.5 \text{ feet} \times \frac{12 \text{ inches}}{1 \text{ foot}}$ does work, so use it

Use the conversion factor that has the initial unit in the denominator.

Step 5: *Calculate the answer, expressing the correct number and unit:*

$$5.5 \text{ feet} \times \frac{12 \text{ inches}}{1 \text{ foot}} = \frac{5.5 \times 12 \text{ inches}}{1} = 66 \text{ inches}$$

Since you multiply the initial value by conversion factors that equal 1, you are not changing the amount of that value; only the units change. When you arrange the units to cancel, you automatically do the right thing (multiply or divide) with the numbers.

Often you need more than one conversion factor. For example, you convert 862 inches to miles by the following steps:

Step 1: Initial unit: inches
Step 2: Desired unit: miles
Step 3: Conversion factors: $\dfrac{12 \text{ inches}}{1 \text{ foot}}$ or $\dfrac{1 \text{ foot}}{12 \text{ inches}}$, $\dfrac{5280 \text{ feet}}{1 \text{ mile}}$ or $\dfrac{1 \text{ mile}}{5280 \text{ feet}}$

Map: inches → feet → miles

Step 4: $862 \text{ inches} \times \dfrac{1 \text{ foot}}{12 \text{ inches}} \times \dfrac{1 \text{ mile}}{5280 \text{ feet}}$

(Here you arrange the second conversion factor to cancel feet and leave the desired unit, miles.)

Step 5: $862 \text{ inches} \times \dfrac{1 \text{ foot}}{12 \text{ inches}} \times \dfrac{1 \text{ mile}}{5280 \text{ feet}} = 0.0136 \text{ miles}$

Conversions between metric units are easier because the units differ by multiples of 10. By knowing the numerical value of the prefixes in Table 2.1, you can immediately write conversion factors to operate within the metric system. For example, you convert 862 millimeters (mm) to meters (m) by the following steps:

Step 1: Initial unit: mm
Step 2: Desired unit: m
Step 3: Conversion factors: 1 mm = 1/1000 m, so 1000 mm = 1 m

You can use $\dfrac{1000 \text{ mm}}{1 \text{ m}}$ or $\dfrac{1 \text{ m}}{1000 \text{ mm}}$

Map: mm → m

Step 4: $862 \text{ mm} \times \dfrac{1 \text{ m}}{1000 \text{ mm}}$

Step 5: $862 \text{ mm} \times \dfrac{1 \text{ m}}{1000 \text{ mm}} = 0.862 \text{ m}$

If you want to convert 1705 centimeters (cm) to kilometers (km), you don't have a conversion factor to do this in one step. But you can use the map cm → m → km because you know conversion factors (from Table 2.1) for both of these steps. So write the initial value, arrange the conversion factors to cancel units, leaving the desired unit (km), and complete the calculation:

$$1705 \text{ cm} \times \frac{1 \text{ m}}{100 \text{ cm}} \times \frac{1 \text{ km}}{1000 \text{ m}} = \frac{1705}{100 \times 1000} \text{ km} = 0.01705 \text{ km}$$

Notice that in converting from cm to km, you essentially moved the decimal point five places to the left. Most other metric conversions also involve simply moving the decimal point to the left or right. For example, if you weigh

68 kg (about 150 lb), you can calculate your weight in grams (g) by:

$$68 \text{ kg} \times \frac{1{,}000 \text{ g}}{1 \text{ kg}} = 68{,}000 \text{ g}$$

EXAMPLE 2.12
(a) What is the mass in kg of a 226 g jar of coffee?
(b) What is the volume in mL of a 2.2 L engine? What is its volume in cubic centimeters (cm^3)?

SOLUTION

(a) $226 \text{ g} \times \dfrac{1 \text{ kg}}{1000 \text{ g}} = 0.226 \text{ kg}$

(b) $2.2 \text{ L} \times \dfrac{1000 \text{ mL}}{1 \text{ L}} = 2200 \text{ mL}$

$2.2 \text{ L} \times \dfrac{1000 \text{ mL}}{1 \text{ L}} \times \dfrac{1 \text{ cm}^3}{1 \text{ mL}} = 2200 \text{ cm}^3$

(Remember that $1 \text{ cm}^3 = 1 \text{ mL}$ [Section 2.2].)

Since density relates mass (g) to volume (mL or cm^3), it provides conversion factors among these units. The density of mercury is 13.6 g/mL (Table 2.3). You can use this conversion factor to calculate the mass in g of 4.11 mL of mercury (Hg):

$$4.11 \text{ mL Hg} \times \frac{13.6 \text{ g Hg}}{1 \text{ mL Hg}} = 55.9 \text{ g Hg}$$

Using one more conversion factor, you can calculate the mass in kg of 4.11 mL of Hg using the map, mL Hg → g Hg → kg Hg:

$$4.11 \text{ mL Hg} \times \frac{13.6 \text{ g Hg}}{1 \text{ mL Hg}} \times \frac{1 \text{ kg Hg}}{1000 \text{ g Hg}} = 0.0559 \text{ kg Hg}$$

Since the density of water is 1.00 g/mL (Table 2.3), you can easily interconvert between mass and volume for any sample of water. For example, 28.6 mL of water will have a mass of 28.6 g:

$$28.6 \text{ mL water} \times \frac{1.00 \text{ g water}}{1.00 \text{ mL water}} = 28.6 \text{ g water}$$

Similarly, 412 g of water will have a volume of 412 mL:

$$412 \text{ g water} \times \frac{1.00 \text{ mL water}}{1.00 \text{ g water}} = 412 \text{ mL water}$$

When you can measure what you are speaking about, and express it in numbers, you know something about it; but when you cannot measure it, when you cannot express it in numbers, your knowledge is of a meagre and unsatisfactory kind.
Source: William Thomson (Lord Kelvin) 1824–1907

You will need to use the factor-label method for many calculations in later chapters, so learn how to use this method now. Appendix C has more information, including questions and detailed examples, for interconverting units.

SUMMARY

The last digit in a measured value typically is an estimate and is uncertain. Measured values and calculated values need to be expressed in terms of the appropriate number of significant figures. Very large or small numbers are written in scientific notation, using exponents to locate the decimal point.

Scientists, and most other people in the world, make measurements using units of the metric system or an updated version called the International System of Units (SI). The system uses prefixes to change the size of basic units by factors of 10.

Units of time, mass, and distance are the second (sec), gram (g), and meter (m), respectively. Units for volume include the cubic meter (m^3) and the liter (L). Density is expressed as g/mL or g/cm^3. Specific gravity is the ratio of the density of a substance to the density of water; it has no units. Temperatures are measured in units of degrees Celsius (°C) or kelvins (K). Units of heat energy include the joule (J), calorie (cal) and dietary Calorie (Cal).

A convenient way to interconvert units is the factor-label method, in which you arrange conversion factors to cancel the initial unit(s) and leave the desired unit(s).

KEY TERMS

Accuracy (2.1)
Anorexia nervosa (2.4)
Basal metabolic rate (BMR) (2.4)
Bulimia (2.4)
Density (2.3)
Energy (2.4)
Factor-label method (2.5)

Heat (2.4)
International System of Units (SI) (2.2)
Kinetic energy (2.4)
Mass (2.3)
Metric system (2.2)
Potential energy (2.4)

Precision (2.1)
Scientific notation (2.1)
Significant figures (2.1)
Specific gravity (2.3)
Temperature (2.4)
Volume (2.3)
Weight (2.3)

EXERCISES

Even-numbered exercises are answered at the back of this book.

Significant Figures and Scientific Notation

1. How many significant figures are present in the following:

    (a) 581.102 m
    (b) 3100 kg
    (c) 0.00401 cm^3
    (d) 1,000,001 mL
    (e) 18.3040 mg
    (f) 0.1002 L
    (g) 710,301,000 mm
    (h) 0.123086 kcal
    (i) 0.0000021 km

2. How many significant figures are present in the following:

    (a) 160 kg
    (b) 545,100 m
    (c) 0.001010 J
    (d) 1,000,000 mL
    (e) 0.0000000004089 g
    (f) 1.000000 km

3. Express the numbers in exercise 1 in scientific notation.

4. Express the numbers in exercise 2 in scientific notation.

5. How many significant figures should the calculated answers have in each of the following:
   (a) $\dfrac{27.11 \times 0.0063}{93.114}$ (c) $\dfrac{3.1416 \times 23{,}000}{88.81020}$
   (b) 312×4.000

6. How many numbers past the decimal point should the calculated answers have in each of the following:
   (a) $\begin{aligned}743.12\\-161.905\end{aligned}$ (c) $\begin{aligned}0.1062\\+3.219\end{aligned}$
   (b) $\begin{aligned}1{,}014\\-68.217\end{aligned}$ (d) $\begin{aligned}3{,}899.0000\\-5.555\end{aligned}$

7. Do the calculations in exercise 6 and express the answers with the appropriate number of significant figures and in scientific notation.

8. Do the calculations in exercise 5 and express the answers with the appropriate number of significant figures and in scientific notation.

9. There are 1000 mg per g. There are 1000 people at a concert. Does the number *1000* have the same number of significant figures in both statements? Explain.

10. Criticize the following statements:
    (a) There are dozens of 7,000-m (22,965.87-ft) peaks in Tibet.
    (b) The high temperature today is 80°F (26.67°C).
    (c) The earthquake in Australia caused more than $2,000,000 (U.S. $2,547,368.21) in damage.
    (d) Those remains are 10,000 years old. Next year, they will be 10,001 years old.

11. The value of pi (the ratio of a circle's circumference to its diameter) is about 3.14. However, pi has been calculated to more than 2 billion digits beyond the decimal point. Suppose you used that value of pi to calculate the area of a circle with a radius of 2.4 cm. The area of a circle equals pi × (radius)². How many significant figures would your answer have?

12. Do the following calculations and express the answers in scientific notation:
    (a) $\dfrac{(6.441 \times 10^{16})(10^{-4})}{38{,}000{,}000}$
    (b) $\dfrac{(46.15 + 183.035)(7.14 \times 10^{-7})}{(8.12 \times 10^{-3})(7.77 \times 10^{5})}$
    (c) $\dfrac{(16{,}312.1 - 12.604)(9.80 \times 10^{4})}{(4.64 \times 10^{-6})(10^{8})}$

13. Do the following calculations and express the answers in scientific notation:
    (a) $\dfrac{(4.12 \times 10^{-3})^{2}}{(8.3 \times 10^{-4})(10^{7})}$
    (b) $\dfrac{(147.12 + 1.63054)(1.99 \times 10^{5})}{(16.1417 - 23.6)(4.33 \times 10^{7})}$
    (c) $\dfrac{(3.16 \times 10^{3})(8.82 \times 10^{23})}{(1.17 \times 10^{6})(6.82 \times 10^{10})}$

Additional exercises on the use of exponents and scientific notation are in Appendix A.

Metric System

14. What is the new, revised version of the metric system called?

15. What is the main numerical advantage of the metric system?

Mass, Length, and Volume

16. If you used a balance to measure the mass of a watch, would you get the same answer on the moon as on earth? Why?

17. If you used a scale to measure the mass of a lamp, would you get the same answer on the moon as on earth? Why?

18. Write the numerical value of the following prefixes as a regular number and as an exponential number (see Appendix B): (a) nano-, (b) kilo-, (c) centi-, (d) micro-, (e) milli-.

19. Is an increase in length of 1 cm a larger or smaller change than an increase of 1 mm?

20. The length of your thumb is about 5 (a) km, (b) mm, (c) m, (d) cm, (e) dm.

21. If you could buy 5 g of a drug at the rate of (a) $4.85/g or (b) $1000/kg, which rate would cost you less?

22. A typical can of Diet Coke has on the label "12 FL. OZ. (355 mL)." Is a fluid ounce a measure of mass or volume?

23. Complete the following:
    (a) 16.1 m = _____ km   (d) 495 μg = _____ g
    (b) 62 g = _____ mg   (e) 21.4 m = _____ cm
    (c) 0.075 L = _____ mL (f) 68 cal = _____ kcal

24. Complete the following:
    (a) 1 cm = _____ m   (d) 29 J = _____ kJ
    (b) 1 kg = _____ g   (e) 33.3 m = _____ cm
    (c) 1 mL = _____ L   (f) 84 g = _____ μg

25. The tallest person to play in the National Basketball Association is 7 feet, 7 inches (91 inches) tall. Using the relationship 1 inch = 2.54 cm (Table 2.2), calculate his height in cm and in m.

26. Using the relationships 1 L = 1.06 quart, and 4 quarts = 1 gallon, calculate the mL in 1.00 gallon of gasoline.

27. A book is 15 cm wide, 21 cm long, and 4.0 cm thick. Calculate its volume in **(a)** cm^3, **(b)** L, and **(c)** m^3.

28. Suppose a full can of soft drink has a mass of 361 g and the empty can has a mass of 17 g. What is the mass of the soft drink in **(a)** kg and **(b)** micrograms (μg)?

29. If your blood sugar level is 84 mg of glucose per deciliter (dL) of blood, how many mg of glucose do you have in 1.00 liter of blood?

30. Suppose you have a cold and want to take a high dose of vitamin C. If you have 250-mg tablets of vitamin C, how many tablets would you take per day to get a total dose of 2.00 g?

31. A drug to treat fungus infections has a concentration of 10.0 mg per mL of solution. How many mL of solution will provide a dose of 0.045 g?

32. The maximum dose of carbocaine, a local anesthetic used in dentistry and medicine, is 5.0 mg/kg body weight in young children. What would be the maximum dose in mg of carbocaine for a child who weighs 28 pounds? (See Table 2.2.)

33. To get a daily dose of the antibiotic tetracycline of 25 mg/kg body weight, how many mg of tetracycline should a 154-pound person take each day? (See Table 2.2.)

Density and Specific Gravity

34. A 5.00 mL sample of ethyl alcohol has a mass of 3.95 g. What is the density of ethyl alcohol?

35. What is the specific gravity of ethyl alcohol based on the information in exercise 34?

36. A solid 75.0-cm^3 block of table salt has a mass of 163 g. What is its density?

37. Lead has a density of 11.8 g/cm^3. What would be the mass of a cube of lead 2.0 cm on each side?

38. Using the information in exercise 37, calculate the mass of 156 cm^3 of lead.

39. What is the mass in kg of 83.3 mL of water?

40. A 27.0 mL sample of urine has a mass of 27.8 g. What is its specific gravity? Is this value in the normal range?

41. If you drop them both into water, a can of Diet Coke will float while a can of regular Coca-Cola will sink (see Figure 1.7). Rationalize why.

42. The percent body fat of runners was measured before a 100-mile run on a hot day. After completing the run, some runners had a *higher* percent body fat. Rationalize why.

Energy and Temperature

43. On a typical day, a student may consume 1860 Calories of food. Express this amount in calories (cal), joules (J), and kilojoules (kJ).

44. When you eat food, is its energy content mostly kinetic energy or potential energy?

45. Does food actually contain Calories? Explain.

46. How many kJ are in a 325-Calorie hamburger?

47. Calculate the basal metabolic rate (BMR) in **(a)** kJ and **(b)** Cal for a 154-pound (70.0-kg) male.

48. Calculate in **(a)** kJ and **(b)** Cal the daily energy requirement to maintain the weight of a 70.0-kg person (see exercise 47) who is sedentary.

49. What should the average daily Caloric intake be for the person in exercise 48 to lose 8.0 pounds of weight in 5 weeks?

50. Ethyl ether boils at 34.6°C. What is its boiling point in **(a)** kelvins and **(b)** °F? Could a sample boil inside your body?

51. The highest recorded body temperature from which someone recovered is 46.5°C. What is this temperature in °F?

52. A person with hypothermia may have a body temperature as low as 89°F. What is this temperature in **(a)** °C and **(b)** kelvins?

53. Complete the following:
    (a) −105°F = _____ °C **(c)** 612 K = _____ °C
    (b) 87°C = _____ K   **(d)** 612 K = _____ °F

54. Complete the following:
    (a) −24°F = _____ °C   **(c)** 245 K = _____ °C
    (b) 91°C = _____ K   **(d)** 245 K = _____ °F

55. If you took the temperature of a person three times, and got readings of 35.1°C, 35.0°C, and 35.1°C, would your measurement have a high degree of precision? Do you think it would have a high degree of accuracy? Why?

Each gram of glucose provides 4.0 Cal of energy. Use this information to answer the next two, more challenging exercises.

56. If your blood carries enough glucose to provide 20.0 Cal and your body contains 5.1 L of blood, what is the concentration of glucose in your blood in mg glucose per dL of blood?

57. If your blood glucose level is 92 mg/dL blood and your blood carries enough glucose to provide a total of 81 kJ, what is the volume of your blood in L?

DISCUSSION EXERCISES

1. Are measurements using metric units more accurate than those using English units? Are they more precise? Explain both answers.

2. What are the arguments for and against the United States converting to the metric system? Do you favor the change? Defend your answer.

3. Calculate the number of hours in a century using each of the following maps: **(a)** hours → days → weeks → months → years → centuries; **(b)** hours → days → weeks → years → centuries; **(c)** hours → days → years → centuries. Compare the answers and the significant figures in the answers. Which answer is most accurate? Why?

3

Elements and Atoms

1. How do scientists classify matter? What are mixtures, compounds, and elements, and how do they differ?
2. What are the law of conservation of mass and the law of definite proportions? How did they lead to the theory that matter is composed of atoms?
3. What are the basic particles in atoms? How are they related to the atomic number and atomic mass of an element?
4. What is the periodic table of the elements? What are some major groups of elements?
5. How are electrons arranged in atoms? How does this relate to the periodic table of the elements?

Dmitri Mendeleev

Dmitri Mendeleev (Figure 3.1), a chemistry professor at the University of St. Petersburg in Russia, was fascinated by the elements. He wrote the symbol and relative atomic mass of each one on a set of cards. Then he laid the cards on his desk, and tried to arrange them in order of increasing atomic mass, while also keeping elements with similar properties in groups. From this exercise, he discovered a pattern now called the periodic table of the elements (see inside front cover).

Mendeleev's discovery made him famous. So did some of his other activities. In 1876, he divorced his wife and married an art student. Although this was bigamy according to the Russian Orthodox Church, no one took action against him. Czar Alexander II explained: "Mendeleev has two wives, but I have only one Mendeleev."

In 1890, however, he went too far, presenting the education minister with a petition to relax the rules restricting women in their pursuit of science careers. After 175 students were arrested during this protest, Mendeleev resigned.

In the 1906 Nobel prize balloting Mendeleev missed receiving the award in chemistry by one vote. His marital and political behavior may have been a factor. He died a few months later, and didn't get another chance. He was honored posthumously, however, when element 101—synthesized in 1955—was named *mendelevium* (Md).

Figure 3.1 Dmitri Mendeleev (1834–1907).

Figure 3.2 Gold.

I magine cutting a gold bar in half with a knife (Figure 3.2). Then imagine cutting one of the two pieces in half again, and then one of the next pieces, and the next. If your knife—and your cutting technique—were perfect, could you keep subdividing the gold forever? Or would you reach some final piece that you couldn't cut any smaller without changing the gold into something else?

Now compare a gold bar with a wooden chair, a piece of silk, a grain of table salt, and the skin on your hand. What are these objects made of? Is each material unique, or do the same basic units make up all matter?

3.1 CLASSIFYING MATTER

Types of Matter. When you look at a piece of matter, you can classify it as homogeneous or heterogeneous. *Homogeneous matter* has the same appearance, composition, and properties. If you take samples from a bottle of apple juice (Figure 3.3), each one will look and taste the same.

Heterogeneous matter, by contrast, is not uniform throughout. A pizza, for example, might have pieces of pepperoni here, mushrooms there, a layer of crust, and a topping of cheese and tomato sauce (Figure 3.3). Individual components, such as the cheese, might be homogeneous, but the pizza itself is heterogeneous.

Figure 3.3 Homogeneous (left) and heterogeneous (right) matter.

Figure 3.4 shows how chemists classify matter. Heterogeneous matter can be separated into its homogeneous components. Homogeneous matter is either a pure substance or a uniform mixture of pure substances, called a *solution*. A **pure substance** has a fixed composition that doesn't vary from one part to another. Water is a pure substance made from hydrogen and oxygen. Its mass is always 88.9% oxygen and 11.1% hydrogen. If that proportion changed, the material wouldn't be water.

Elements and Atoms Chapter 3 35

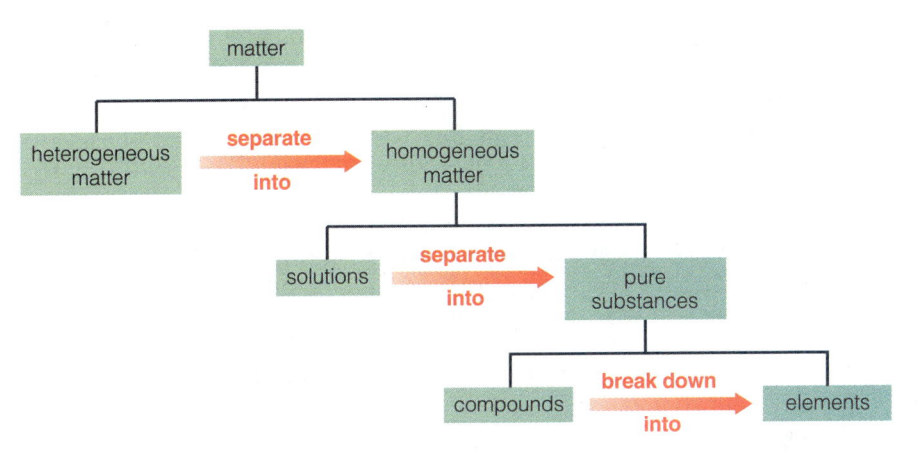

Figure 3.4 A classification of matter.

EXAMPLE 3.1 Classify each of the following as homogeneous or heterogeneous matter: (a) an unlined sheet of paper, (b) your eye, (c) a painting, (d) orange sherbert, (e) your watch, (f) copper wire, (g) sugar crystals, (h) shoes.

SOLUTION Items (a), (d), (f), and (g) are homogeneous; (b), (c), (e), and (h) are heterogeneous.

Unlike a pure substance, however, a homogeneous mixture can vary in proportion. For example, solutions of sugar water can contain different amounts of sugar (within limits) dissolved in water.

EXAMPLE 3.2 Classify each of the following as a pure substance or a mixture: (a) a cup of coffee, (b) salt, (c) mercury, (d) ether, (e) urine, (f) green paint, (g) a gold bar.

SOLUTION Items (b), (c), (d), and (g) are pure substances; (a), (e), and (f) are mixtures.

Chemists classify pure substances as elements or compounds. **Elements** cannot be broken down into simpler substances (Figure 3.5). You could cut a

Water was considered an element for many years, until a method (called electrolysis) was discovered to decompose it into hydrogen and oxygen.

Figure 3.5 Phosphorus, sulfur, bromine, and carbon (left to right, respectively) are elements.

piece of copper (an element) into tiny pieces, or melt it, or pass electricity through it, but it would still be copper.

Compounds, however, are made from two or more elements and can be broken down into those parts. Passing electricity through water produces hydrogen gas and oxygen gas, the elements that form water (Figure 3.6). Water is therefore a compound.

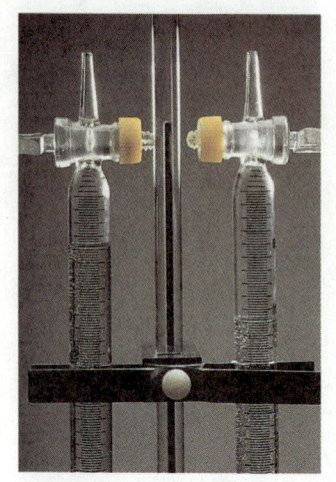

Figure 3.6 Electrolysis decomposes liquid water into its constituent elements, hydrogen gas (right), and oxygen gas (left).

EXAMPLE 3.3 Do you recognize any of the pure substances in Example 3.2 as elements?

SOLUTION Items (c) mercury and (g) gold are elements; (b) salt and (d) ether are compounds. (Later in this chapter you will learn the identities of all the elements.)

Physical and Chemical Changes. Matter changes both physically and chemically. **Chemical changes** produce different pure substances. Making grapes into wine is a chemical change. Your body converting food into muscle, autumn leaves turning red and gold (Figure 3.7), and trees burning in a forest fire are also chemical changes.

Physical changes can alter the size, shape, or physical state of matter (solid, liquid, or gas), but they don't produce different pure substances. Melting ice, cutting a piece of wood, and spreading sunscreen on your skin are all physical changes. Stirring salt into water, or separating salt water into salt and water, are also physical changes; neither produces a new substance.

EXAMPLE 3.4 Classify each of the following as a physical or a chemical change: (a) a puddle of water evaporates on a warm day; (b) a droplet of mercury falls from a broken thermometer; (c) an old car rusts; (d) a chef mixes the ingredients of a fruit salad; (e) your body converts excess carbohydrates into fat.

SOLUTION Items (a), (b), and (d) are physical changes; (c) and (e) are chemical changes.

Figure 3.7 Chemical changes produce the colors in these autumn leaves.

3.2 ELEMENTS AND ATOMS: THE CONNECTION

Let's return to the question that begins this chapter: Could you keep subdividing a sample of a single element, such as a gold bar, forever? Or would you eventually reach something that you couldn't divide without destroying the gold? To answer this question, we need to go back two centuries.

Two Important Laws. In the late 1700s, a French scientist named Antoine Lavoisier (Figure 3.8) did many experiments in which he carefully measured the mass both of reacting materials and the products they formed. He found that the mass of the reactants always equaled the mass of the products. He stated this as a scientific law (Section 1.2), called the **law of conservation of mass:** *Matter cannot be created or destroyed, but it can change forms.*[1]

A few years later Joseph Proust, a French chemist, summarized the results of many experiments in the **law of definite proportions:** *Elements in a compound occur in a fixed (definite) proportion by mass.* For example, 11.1% of the mass of water (a compound) always comes from the element hydrogen and 88.9% always comes from the element oxygen. Similarly, table sugar (sucrose; Figure 3.9) gets 42.1%, 6.4%, and 51.5% of its mass from the elements carbon, hydrogen, and oxygen, respectively.

Figure 3.8 Antoine Lavoisier (1743–1794) formulated the law of conservation of mass. To earn a living he was a tax collector for King Louis XVI—not a popular thing to do, especially during the French revolution. As a result, Lavoisier lost his head on a guillotine.

Figure 3.9 Table sugar is a compound composed of the elements carbon, hydrogen, and oxygen combined in a definite proportion.

EXAMPLE 3.5 Which scientific law—the law of conservation of mass, or the law of definite proportions—is illustrated by each of the following statements?
   (a) Table salt is composed of 39.3% sodium and 60.7% chlorine.
   (b) 39 g of potassium reacts with 18 g of water to form products that have a total mass of 57 g.

SOLUTION (a) law of definite proportions (b) law of conservation of mass

---

Atomic Theory. These two laws describe what consistently happens when elements and compounds react to form other substances. But something was missing: an explanation of *why* this happens.

John Dalton, an English schoolmaster (Figure 3.10), proposed an answer in

Figure 3.10 John Dalton (1766–1844) proposed that elements are composed of atoms. Although his Quaker beliefs kept him from accepting honors for his work, he was widely known and respected. His funeral was attended by 40,000 people.

---

[1] This law, and the law of conservation of energy (Section 8.1), were later revised when it was discovered that mass and energy are interconverted under certain conditions (see Section 10.6).

Democritus, a Greek philosopher, originated the concept of atoms about 400 B.C. He called them atomos, *which means "not cuttable."*

the early 1800s. His ideas, called the **atomic theory,** could be summarized like this:

- All elements are composed of ultimate, indivisible units called **atoms.**
- All atoms of an element are identical, and different from the atoms of other elements.
- When elements and compounds react, individual atoms of elements interact in simple, whole-number ratios.

Dalton's atomic theory explained the law of conservation of mass. Since atoms cannot be created, destroyed, or divided, each atom (and its mass) in the reacting substances remains in the products (Figure 3.11).

Dalton's theory also explained the law of definite proportions. Since each element in a compound provides a specific number of atoms with a particular weight (Figure 3.11), each element provides a fixed (definite) proportion of the mass in a compound.

Dalton's ideas about atoms explained the results of experiments so well that they eventually gained wide acceptance.

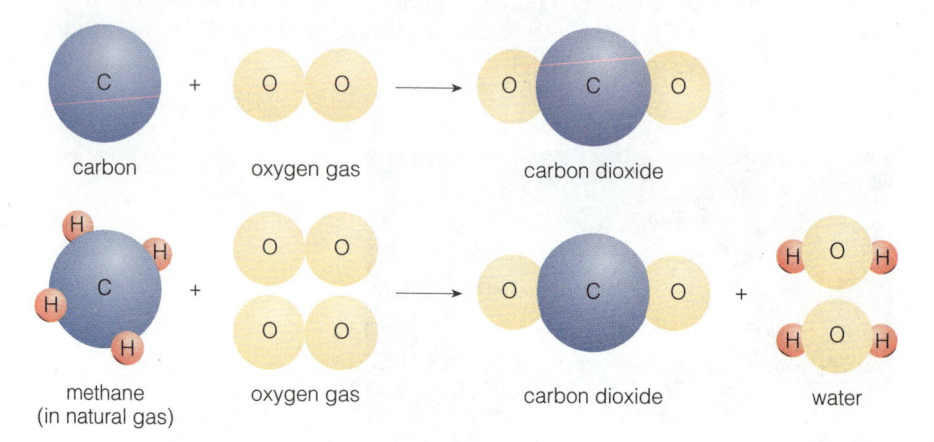

Figure 3.11 (Top) When coal (mostly carbon) burns (reacts with oxygen gas), atoms of carbon and oxygen combine to form carbon dioxide. (Bottom) When natural gas (which contains carbon and hydrogen atoms) burns, the atoms of methane and oxygen rearrange to form carbon dioxide and water. In both reactions, the reactants and products have the same number of each type of atoms, and each substance has a simple, whole-number ratio of atoms.

3.3 ATOMIC PARTICLES AND ATOMIC MASS

Atomic Structure. What does an atom look like? Atoms are so tiny that no person has seen one in detail. Even Figure 3.12, which shows one of the best photographs of atoms, reveals little about an atom's structure.

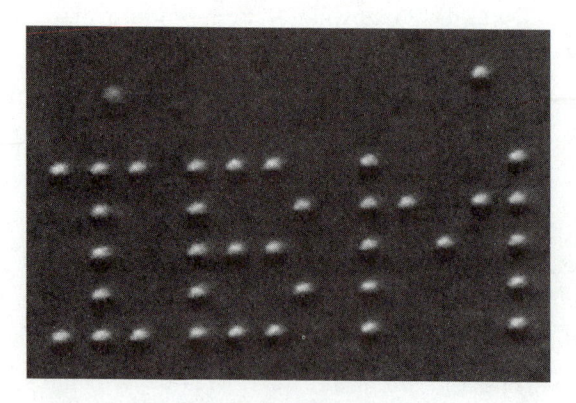

Figure 3.12 A scanning tunneling microscope photo of xenon atoms. Each sphere is an atom.

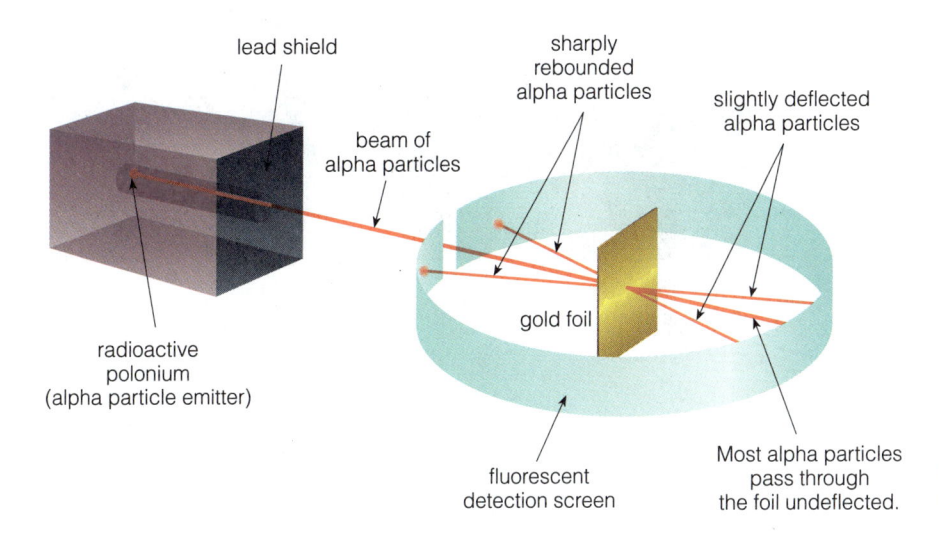

Figure 3.13 Rutherford's gold foil experiment.

Dalton thought of atoms as hard spheres, something like the picture in Figure 3.12. It turns out, however, that atoms are made of several different parts, called *subatomic particles*.

An important clue came in 1910. Ernest Rutherford, a New Zealand-born scientist working in England, devised an experiment to shoot positively charged alpha particles (a type of radioactivity; see Section 10.1) at a very thin layer of atoms in gold foil. Most of the alpha particles whisked through the foil undeflected or only slightly deflected (Figure 3.13). To Rutherford's great surprise, however, a few bounced off the foil at sharp angles. "It was the most incredible event that ever happened in my life," said Rutherford. "It was as though you had fired a 15-inch artillery shell at a piece of tissue paper and it came back and hit you."

Because most alpha particles passed through the gold atoms unaffected, Rutherford concluded that an atom is mostly empty space. He reasoned that the few positively charged particles that were sharply deflected must have collided with, or come very close to, a tiny but concentrated mass of positive electrical charge. He proposed that the positive charge—and virtually all the mass of an atom—are in a tiny space at its center called the **nucleus.**

Later discoveries led to a more detailed picture. The atom consists of three major particles: **protons,** with an electrical charge of +1; **electrons,** with an electrical charge of −1; and **neutrons,** with no electrical charge. Protons and neutrons have similar masses and are in the nucleus. Electrons have a much smaller mass (about 1/1840 the mass of a proton) and are outside the nucleus. Table 3.1 summarizes these properties.

Table 3.1 Subatomic Particles

Particle	Electrical Charge	Approximate Mass (g)	(amu*)	Location in Atom
Electron (*e*)	−1	9.110×10^{-28}	0	outside the nucleus
Neutron (*n*)	0	1.675×10^{-24}	1	in the nucleus
Proton (*p*)	+1	1.673×10^{-24}	1	in the nucleus

* Atomic mass units. Actual masses in amu are 0.00055, 1.0087, and 1.0073 for an electron, neutron, and proton, respectively.

CHEMISTRY SPOTLIGHT

The Ultimate Particles of Matter?

Figure 3.14 Tracks of particles from a high-energy collision in a Stanford linear accelerator bubble chamber.

Scientists have concluded that atoms are the ultimate unit of an element. If you break up an atom, the products no longer have properties of the element.

Yet atoms themselves are made up of particles—protons, neutrons, and electrons. Are these the ultimate particles of matter? Or are subatomic particles composed of still smaller pieces of matter?

Experiments producing high-speed collisions of nuclear matter have revealed an astonishing array of other particles—leptons, mesons, baryons, and dozens of others (Figure 3.14).

Antimatter—including positively charged electrons and negatively charged protons—also exists. Current theory holds that the ultimate particles of matter are tiny units called quarks, which come in six varieties. Protons and neutrons, for example, are each composed of three quarks. But might quarks themselves be composed of still smaller particles?

The way to find out is to build more powerful machines and smash atomic particles even more thoroughly. This is being done. But can we ever know for sure that the ultimate particle of matter has been found?

Electrons occupy most of an atom's space. Imagine that you could enlarge an atom so that its nucleus was the size of a BB. If you then put that BB on the 50-yard line of the Rose Bowl, the rest of the stadium would be the space needed for the atom's electrons.

Atomic Number. Atoms of different elements have different numbers of protons in their nuclei. But all atoms of the same element have the same number of protons. The number of protons is the **atomic number** of that element. For example, the atomic number of sodium is 11, and all sodium atoms have 11 protons. Each atom of oxygen has 8 protons, so its atomic number is 8.

This relationship provides another definition of an element: An element is a substance composed of atoms with the same atomic number.

Another relationship is important here. Although atoms have positively charged protons and negatively charged electrons, they have no net electrical charge. This is because atoms have equal numbers of protons and electrons— the amount of positive charge equals the amount of negative charge. So atomic number is both the number of protons and the number of electrons in an atom of that element.

EXAMPLE 3.6 The atomic number for uranium is 92, and the atomic number for calcium is 20. How many protons and electrons are present in an atom of (a) uranium and (b) calcium?

SOLUTION (a) 92 protons and 92 electrons; (b) 20 protons and 20 electrons.

Isotopes. Dalton's idea that all atoms of an element are identical wasn't quite right. All atoms of an element have the same number of protons and electrons, but they may differ in their number of neutrons. These different forms are called **isotopes.**

Since neutrons (like protons) have significant mass (Table 3.1), isotopes of an element differ in mass. They have the same chemical properties, however. This is because chemical properties depend on an atom's number of electrons (see Chapter 4), which is the same in all isotopes of an element.

You can identify isotopes by **mass number,** the sum of their protons and neutrons. Chlorine, for example, has atomic number 17, so its atoms all have 17 protons and 17 electrons. Some atoms of chlorine have 18 neutrons and some have 20. The first isotope has a mass number of 35 (17 protons + 18 neutrons) and is called chlorine-35. The second isotope has a mass number of 37 (17 protons + 20 neutrons) and is called chlorine-37. Another way to represent isotopes, where X represents the symbol for the element, is:

$$^{\text{mass number}}_{\text{atomic number}}X$$

For example, chlorine-35 and chlorine-37 could be written as $^{35}_{17}Cl$ and $^{37}_{17}Cl$, respectively. Table 3.2 shows isotopes of hydrogen and nitrogen.

Most natural elements consist of two or more isotopes.

Table 3.2 Isotopes of the Elements Hydrogen and Nitrogen

Isotope		Symbol and Name	Percentage Abundance in Nature	Isotope		Symbol and Name	Percentage Abundance in Nature
0n, 1p	1e	^1_1H, hydrogen-1	99.985	7n, 7p	7e	$^{14}_7N$, nitrogen-14	99.63
1n, 1p	1e	^2_1H, hydrogen-2 or deuterium (D)	0.015	8n, 7p	7e	$^{15}_7N$, nitrogen-15	0.37
2n, 1p	1e	^3_1H, hydrogen-3 or tritium (T)	Trace				

EXAMPLE 3.7 A natural sample of carbon (C) contains three isotopes having 6, 7, and 8 neutrons, respectively. Represent these three isotopes as shown previously.

SOLUTION carbon-12 or $^{12}_{6}C$; carbon-13 or $^{13}_{6}C$; carbon-14 or $^{14}_{6}C$.

From the atomic number and mass number of an isotope, you can determine how many protons, neutrons, and electrons are in each atom by the following:

atomic number = number of protons (p)
= number of electrons (e)

number of neutrons (n) = mass number ($n + p$) − atomic number (p)

EXAMPLE 3.8 How many protons, neutrons, and electrons are in an atom of each of the following: lead-204, lead-207, and lead-208. Lead has atomic number 82.

SOLUTION All have 82 protons and 82 electrons. Lead-204 has 122 neutrons (204-82); lead-207 has 125 neutrons (207-82); lead-208 has 126 neutrons (208-82).

Atomic Mass. You'd have a hard time finding a balance fine enough to weigh a single atom. One atom of carbon-12, for example, weighs about 1.99×10^{-23} g. To give you some idea of how tiny atoms are, the period at the end of this sentence is big enough to hold about 10^{18} atoms.

But it is possible to measure the *relative* masses of atoms. Two centuries ago, Dalton and others determined the relative weights of atoms of different elements from experiments in which elements combine to form compounds.

Consider, for example, the reaction in which oxygen combines with carbon to form carbon monoxide, a colorless, odorless, and dangerous gas made of one atom of carbon and one atom of oxygen. Careful measurements reveal that 3 g of carbon requires 4 g of oxygen to form 7 g of carbon monoxide. Why does this proportion occur? Because an oxygen atom is 4/3 as heavy as a carbon atom.

Such experiments revealed that hydrogen is the lightest element, with a relative atomic mass of 1. Carbon has a relative atomic mass of 12, and oxygen (4/3 as heavy as carbon) has a relative atomic mass of 16.

EXAMPLE 3.9 A 2.19 g sample of hydrogen reacts with 41.0 g of fluorine to form a compound called hydrogen fluoride. If equal numbers of hydrogen and fluorine atoms react, what is the mass of a fluorine atom if a hydrogen atom has a relative mass of 1.00?

SOLUTION Since 19.0 (41.0/2.19) times as much mass comes from fluorine as from hydrogen, a fluorine atom has a relative mass of 19.0. Mathematically, you could set up a ratio and solve for x:

$$\frac{41.0 \text{ g fluorine}}{2.19 \text{ g hydrogen}} = \frac{x \text{ (relative mass of fluorine atom)}}{1.00 \text{ (relative mass of hydrogen atom)}}$$

$$x = \frac{(41.0)(1.00)}{2.19} = 19.0$$

Eventually scientists decided to use carbon-12 as the standard for atomic mass. A carbon-12 atom is assigned a mass of exactly 12 *atomic mass units* (*amu*):

$$1 \text{ amu} = \frac{\text{mass of one carbon-12 atom}}{12}$$

One amu is about 1.67×10^{-24} g.

Using this definition, protons and neutrons each have a mass of about 1 amu, while an electron's mass is *nearly* 0 amu (Table 3.1). As a result, the masses of isotopes come out very close to their mass numbers. For example, carbon-12 has a mass of exactly 12 amu (by definition), and carbon-13 has a mass of 13.003 amu.

Because most natural elements are a mixture of isotopes, their **atomic mass** (often called *atomic weight*) is an *average*. A natural sample of carbon, for example, is about 98.892% carbon-12, 1.108% carbon-13, and a trace of carbon-14. You can calculate the average atomic mass of an element by multiplying the mass (in amu) of each isotope by the fraction it comprises in the sample (using its percentage and moving the decimal two places to the left), and then adding those values. For carbon:

The average atomic mass (weight) of carbon is 12.011 amu, but no individual atom of carbon weighs 12.011 amu.

$(.98892) \times (12 \text{ amu}) + (.01108) \times (13.003 \text{ amu}) = 12.011 \text{ amu}$

EXAMPLE 3.10 A natural sample of bromine is 50.69% bromine-79 and 49.31% bromine-81. What is the average atomic mass of bromine? (Assume for this calculation that the mass of each isotope of bromine is exactly equal to its mass number.)

SOLUTION $(.5069) \times (79 \text{ amu}) + (.4931) \times (81 \text{ amu}) = 79.99 \text{ amu}$

EXAMPLE 3.11 A natural sample of chlorine is 75.77% chlorine-35 (atomic mass 34.97) and 24.23% chlorine-37 (atomic mass 36.96). What is the average atomic mass of chlorine?

SOLUTION $(0.7577)(34.97) + (0.2423)(36.93) = 35.45 \text{ amu}$

3.4 THE PERIODIC TABLE OF THE ELEMENTS

Discovery of the Periodic Table. Scientists try to discover order in nature. In the 1800s, the list of known elements had grown so large that it was hard to keep track of each element and its properties. Could the elements be arranged into groups that had similar properties?

In 1869 Dmitri Mendeleev (see Figure 3.1), a Russian chemist, and Lothar Meyer, a German physicist, independently discovered a pattern: When they arranged elements in order of increasing atomic mass in rows of appropriate lengths, elements with similar properties were grouped in the same column. The modern version of this arrangement, called the **periodic table of the elements,** is shown inside the front cover of this book.

The original versions of the periodic table contained blanks when the next known element had properties that fit better in another space. Three of the most famous "missing" elements were gallium (Ga), scandium (Sc), and germanium (Ge). Mendeleev boldly predicted that new elements would be discovered to fill those blanks. Because he knew the properties of other elements in the same group, Mendeleev predicted the properties the unknown elements would have. His predictions came true, many within his lifetime.[2]

The Modern Periodic Table. Look at the periodic table inside the front cover of this book. Each of the 109 elements is represented by a symbol. The symbol usually is the first letter, or sometimes the first two letters, of the element's name. Notice that the first letter is always capitalized while the next—if there is one—is lowercase. Table 3.3 lists a few examples, as well as some important exceptions, often based on older names used for the element.

Table 3.3 Symbols and Names of Some Common Elements

First Letter	First Two Letters	Two Letters	Original Latin Name
H Hydrogen	He Helium	Mg Magnesium	Na Sodium (Natrium)
B Boron	Li Lithium	Cl Chlorine	K Potassium (Kalium)
C Carbon	Be Beryllium	Cr Chromium	Fe Iron (Ferrum)
N Nitrogen	Al Aluminum	Mn Manganese	Cu Copper (Cuprum)
O Oxygen	Si Silicon	Zn Zinc	Ag Silver (Argentum)
F Fluorine	Ar Argon	As Arsenic	Sn Tin (Stannum)
P Phosphorus	Ca Calcium	Sr Strontium	Au Gold (Aurum)
I Iodine	Br Bromine	Cs Cesium	Pb Lead (Plumbum)
S Sulfur	Se Selenium	Cd Cadmium	Hg Mercury (Hydrargyrum)

atomic number

26
Fe — symbol
iron

55.847

atomic mass

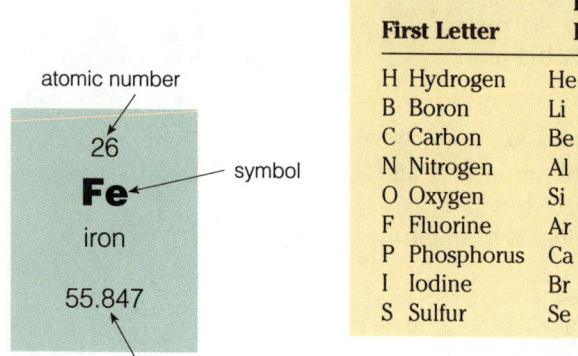

Figure 3.15 How to read the periodic table.

The smaller, whole number in the periodic table is the atomic number. The larger number, usually with a decimal point, is the average atomic mass of the element. Figure 3.15 shows how to read this information in the periodic table.

EXAMPLE 3.12 Using the periodic table, find the symbol, name, and atomic mass for the element with atomic number (a) 16 and (b) 80.

SOLUTION   (a) S, sulfur, 32.06; (b) Hg, mercury, 200.59.

If you look carefully, you can find a few places in the periodic table where an element with the next higher atomic number has a lower atomic mass.

The modern periodic table lists the elements in order of increasing atomic number instead of increasing atomic mass. Each horizontal row is called a **period.** Each vertical column is called a **group,** or *family,* of elements; these elements have similar properties.

Each group has a label. One system, consisting of a Roman numeral and the letter A or B, is the most widely used. A newer, international system numbers the groups 1–18, going from left to right. Both are listed.

Notice the staircase pattern of lines on the right side of the periodic table. Elements to the left (except hydrogen) are metals. **Metals** typically are shiny and solid (except mercury), conduct electricity, and can be shaped into sheets,

---

[2] Mendeleev predicted that gallium (Ga), for example, would have an atomic mass of about 68 (actual mass 69.9) and a specific gravity of 5.9 (actual value 5.94). He also accurately predicted the formation of several compounds of gallium, as well as the properties of those compounds.

Elements and Atoms   Chapter 3   45

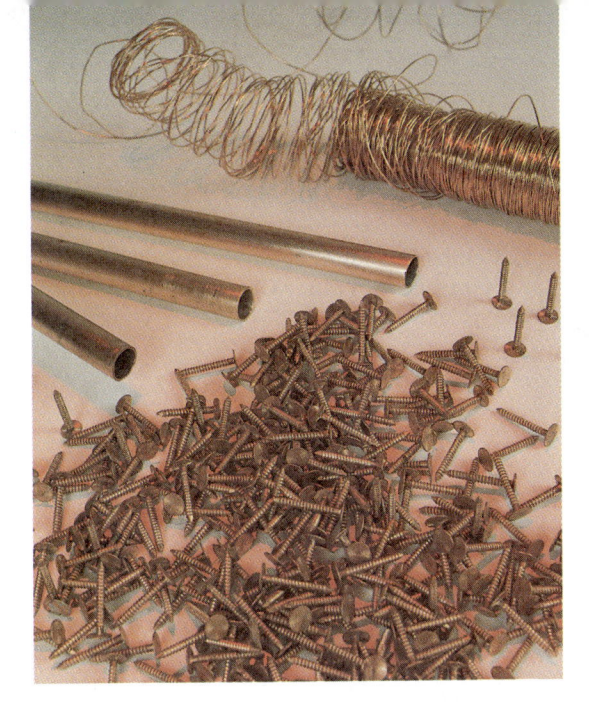

Figure 3.16 Copper, a metal, can be drawn into wire or shaped into other forms.

wires, or other forms (Figure 3.16). **Nonmetals** lack these qualities and (except for hydrogen) are to the right of the staircase. Most elements next to the staircase are **metalloids;** they have properties intermediate between metals and nonmetals.

A few groups are known by special names. They include:

- Group 1/IA (except hydrogen): **Alkali Metals**
- Group 2/IIA: **Alkaline Earth Metals**
- Groups 3-12/all B groups: **Transition Elements**
- Group 17/VIIA: **Halogens**
- Group 18/VIIIA: **Noble Gases**

The two rows across the bottom of the periodic table are subgroups of the transition elements, sometimes called *inner transition elements*. They are separated because inserting them where they should go, according to their atomic number, would make the table too wide to fit conveniently on a page.

EXAMPLE 3.13 List the element and symbol that is (a) in Group VIA and period 4; (b) an alkaline earth metal in period 5; (c) a halogen in the same period as the lightest transition element.

SOLUTION (a) selenium (Se); (b) strontium (Sr); (c) bromine (Br)

Elements in the same group have similar, but not identical, properties. Table 3.4 shows trends within groups. Properties such as melting and boiling points and chemical reactivity show trends with increasing atomic number. You will learn of other patterns in later chapters.

Minerals contain elements (such as chlorine) in the form of ions (such as chloride), which we discuss in Chapter 4.

Elements in the Body. Table 3.5 shows the abundance of elements in the human body. Notice that six elements (O, C, H, N, Ca, and P) make up 99% of the body. The remaining 1%, however, includes a wide variety of elements in trace amounts; only a few of the most abundant are included in Table 3.5.

Table 3.6 lists the Recommended Daily Allowances (RDAs) of minerals established by the Food and Nutrition Board of the National Academy of

Table 3.5 Elements in the Human Body

Element	Approx. Composition (% by Mass)
Oxygen (O)	65.0
Carbon (C)	18.0
Hydrogen (H)	10.0
Nitrogen (N)	3.0
Calcium (Ca)	2.0
Phosphorus (P)	1.1
Potassium (K)	0.35
Sulfur (S)	0.25
Sodium (Na)	0.15
Chlorine (Cl)	0.15
Magnesium (Mg)	0.05
Iron (Fe)	0.004
Iodine (I)	0.0004

Table 3.4 Comparison of Properties in Two Groups

Element	Appearance	Melting Point (°C)	Boiling Point (°C)	Chemical Reactivity
Group IA (1)—Alkali Metals				
Lithium	Silver metal	181	1342	Very high
Sodium	Silver metal	98	883	Very high
Potassium	Silver metal	63	760	Extremely high
Rubidium	Silver metal	39	686	Extremely high
Cesium	Silver metal	28	669	Extremely high
Group VIIA (17)—Halogens				
Fluorine	Green-yellow gas	−220	−188	Extremely high
Chlorine	Green-yellow gas	−101	−35	Very high
Bromine	Dark red liquid	−7	59	High
Iodine	Violet solid	114	184	High

Sciences. Sodium (Na), potassium (K), and chloride (Cl) are abundant in body fluids. Potassium levels are highest inside cells, whereas Na and Cl are the major minerals in blood and other extracellular fluids. Sodium and potassium help transmit nerve impulses and control water levels in the body.

Calcium (Ca) and phosphorus (P) are major components of bone and teeth. Ca is also important in muscle contraction, nerve action, and blood clotting. Magnesium (Mg) affects nerve and muscle action and is essential for many metabolic reactions in the body.

Iron (Fe) is an essential component of hemoglobin in red blood cells. An iron deficiency causes *anemia,* which is a shortage of red blood cells. Dietary iron in animal products is mostly in a form that is absorbed from the intestine.

Table 3.6 Minerals Required in Human Nutrition

Mineral	Recommended Daily Allowance (RDA)*	Sources	Possible Deficiency Effects
potassium	2000–5800 mg	fruits, milk	muscle weakness
chloride	1700–5100 mg	salted foods	muscle cramps, apathy
sodium	1100–3300 mg	salted foods	muscle cramps, apathy
phosphorus	800–1500 mg	milk, cheese, grains	muscle and bone weakness
calcium	800 mg	milk, cheese, legumes	osteoporosis, muscle cramps
magnesium	400 mg	grains, leafy vegetables	diarrhea, hypocalcemia
zinc	15 mg	grains, meat, cheese	poor growth
iron	12 mg	raisins, eggs, meat, legumes	anemia
manganese	2.5–5.0 mg	nuts, fruits	poor hair and nail growth
fluoride	1.5–4.0 mg	fluoridated water and toothpaste	dental caries
copper	2–3 mg	meat, water	anemia
molybdenum	0.15–0.5 mg	organ meats, legumes	poor growth
chromium	0.05–0.2 mg	meat, beer	hyperglycemia
selenium	0.05–0.2 mg	meat, seafood	muscle weakness
cobalt	0.05–1.8 mg	meat, milk	pernicious anemia
iodide	150 ug	seafood, dairy products	goiter

* For an adult male in good health

CHEMISTRY SPOTLIGHT

Cadmium: A Toxic Metal

Figure 3.17 Cadmium (Cd).

Cadmium (Cd; Figure 3.17) is used in plastics, paints, nickle-cadmium batteries, and in electroplating metals to prevent corrosion. It pollutes the air when coal is burned, and is a contaminant in phosphate fertilizers and wastes from mining zinc. About 4–8% of the cadmium you eat or drink stays in your body, mostly in the liver and kidneys, and takes a year or more to expel. Your body retains 40–50% of the cadmium you inhale. Tobacco smoke is the main source of cadmium for most people.

Cadmium is toxic. It damages the heart, arteries, kidneys, and lungs, and causes bones to become brittle. Much of this toxicity comes from cadmium's chemical similarity to zinc (Zn), which is in the same group (12/IIB) in the periodic table. Your body needs zinc in trace amounts, especially to bind to certain enzymes (see Chapter 22) and help them carry out chemical reactions. Cadmium is enough like zinc to replace it in those enzymes. Tissues then deteriorate because their cadmium-containing enzymes can't carry out needed chemical reactions.

Iron from plant sources is not absorbed as readily unless other substances, such as ascorbic acid (vitamin C), are present in the intestine.

Fluoride prevents tooth decay and may be a useful component of bone. Iodide prevents goiter and is incorporated into the thyroid hormone, thyroxin (Figure 3.18). Other substances needed in trace amounts include copper (Cu), cobalt (Co), chromium (Cr), molybdenum (Mo), manganese (Mn), zinc (Zn), and selenium (Se).

3.5 ELECTRON ARRANGEMENTS IN ATOMS

The periodic table was a remarkable accomplishment. It put a large amount of information about individual elements into a pattern of 18 groups, making it much easier to remember.

But something is still missing. *Why* do the elements in the same group have similar properties? Their atomic numbers and atomic masses don't give a clue. To find the answer, we need to know a bit more about the structure of atoms. We return now to that story.

Electrons are very important in determining an element's properties. After all, when atoms react, it is their electrons that first come together.

What more is there to know about electrons? Are the 16 electrons in a sulfur (S) atom, for example, identical in every way? Are they all outside the nucleus whizzing around randomly at some particular speed? It turns out that the answer is "No." Electrons differ in energy.

Figure 3.18 This person with goiter has an enlarged thyroid gland due to a lack of iodide in the diet.

Figure 3.19 Niels Bohr (1885–1962) proposed a solar system model of the atom. He received the Nobel prize in physics in 1922.

The Bohr Model. Danish physicist Niels Bohr (Figure 3.19) proposed in 1913 that the electrons in an atom are restricted to certain, specific amounts of energy, called *energy levels*. The amounts of energy are unevenly spaced, like the shelves of a bookcase whose lower shelves are farther apart than the upper shelves (Figure 3.20).

Electrons can move from one energy level to another, but only if they gain or lose exactly the right amount of energy. An electron in its lowest energy level is in its *ground state*. When an electron absorbs just the right amount of energy to move to a higher energy level, it is said to be in an *excited state*.

Scientists can measure these energy changes. When elements are heated, for example, their electrons absorb energy and go into excited states. When those electrons return to their ground state they give off energy, often in the form of light. You can see this directly by throwing certain crystals into a fireplace and watching the colored flames. Copper compounds, for example, emit blue flames while strontium compounds give off an intense red. Fireworks also show colors from electrons returning to their ground state (Figure 3.21).

If light from an element goes through a prism, its visible components separate and appear as separate colors. Each element emits a different pattern of light, like a distinctive "fingerprint." Figure 3.22 shows the pattern for hydrogen.

Each line of light from an element represents a specific amount of energy. In the visible spectrum, energy increases going in the direction from red to violet. Bohr calculated the energy changes in hydrogen as its electron returns from excited states to lower energy levels. Those amounts of energy matched the lines of color given off by hydrogen (Figure 3.23).

Bohr envisioned electrons orbiting the nucleus like planets orbiting the sun. We can picture each orbit as corresponding to an energy level (Figure 3.24). The nearest orbit holds electrons with the lowest energy. Electrons in orbits farther from the nucleus have more energy.

Bohr's model was an important step forward, but it had a fatal flaw: It

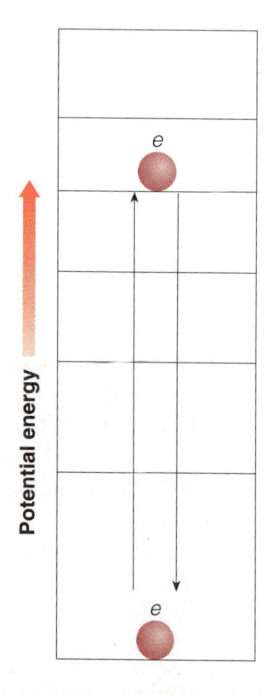

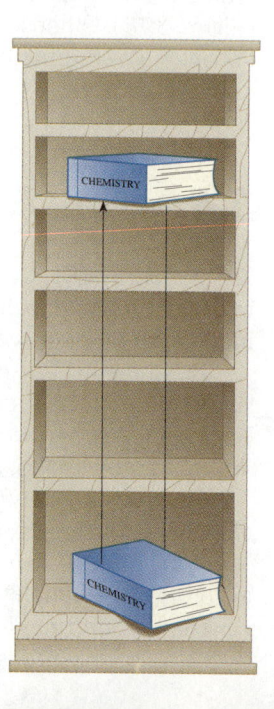

Figure 3.20 Energy levels for an electron (left) are like a bookcase (right) with only specific, unevenly spaced levels allowed. Moving from one level to another requires a gain or loss of a specific amount of energy.

sixth energy level: excited state

fifth energy level: excited state

fourth energy level: excited state

third energy level: excited state

second energy level: excited state

first energy level: ground state

Figure 3.21 Colors are emitted when electrons return from excited states to their ground state.

worked only for one element. The model correctly accounted for the energies emitted when the electron in a hydrogen atom returns from excited states to its ground state. But it didn't account so well for energies emitted by other elements.

The Quantum Mechanical Model. Although scientists can measure energy changes when electrons go from one energy level to another, they cannot measure the actual energy and location of an electron simultaneously. It turns out that measuring an electron's energy changes its location, and measuring its location changes its energy.

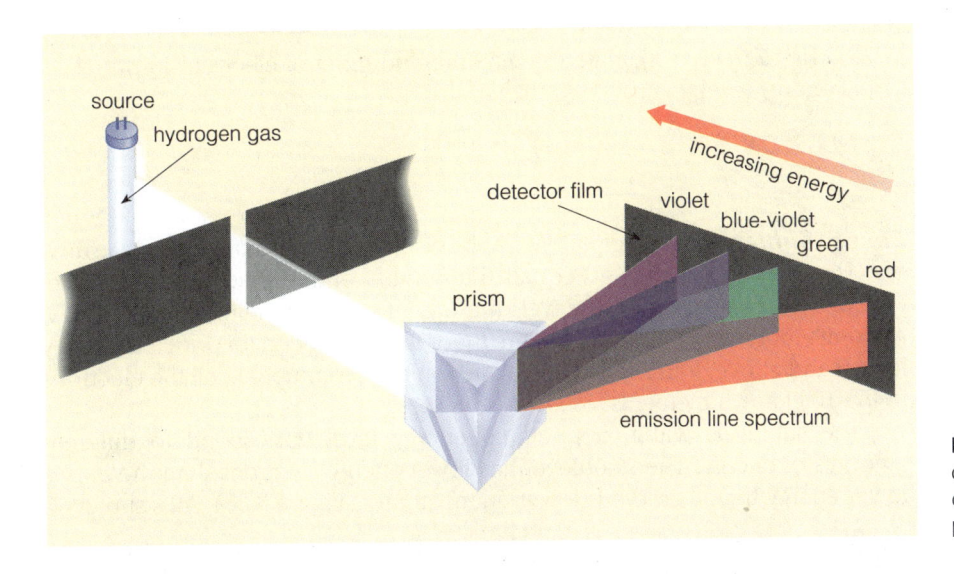

Figure 3.22 Specific lines of color appear when light from hydrogen gas goes through a prism.

Figure 3.23 Colors from hydrogen when its electron returns to the second energy level from a higher energy level. An electron returning to the first energy level (not shown) emits higher-energy, ultraviolet radiation.

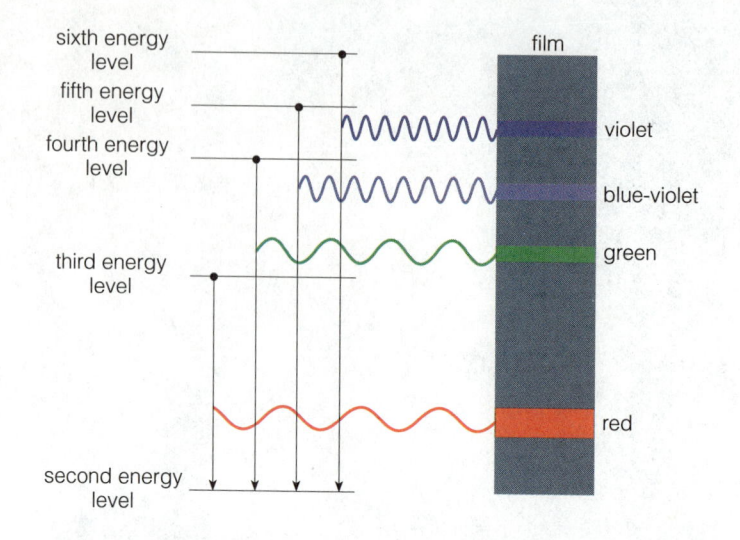

Here a quantum *is a tiny, specific unit of energy, such as that emitted when an electron goes from a higher to a lower energy state.*

This discovery, known as the *uncertainty principle*, forced scientists to abandon the idea that electrons travel in specific orbits. Since the location of an electron with a specific energy cannot be known with certainty, the location can be described only in terms of probability.

In 1926 a revised model of atomic structure, called the **quantum mechanical model,** was developed primarily by Austrian physicist Erwin Schrödinger (Figure 3.25).

Principal Energy Levels. Table 3.7 summarizes the arrangement of electrons according to the quantum mechanical model. The main energy levels, called **principal energy levels,** are designated by numbers (1, 2, 3, . . .), with higher numbers representing higher energy levels.

The maximum number of electrons in a principal energy level equals $2n^2$, where n is the number of the principal energy level. To calculate the maximum number of electrons in principal energy level 3, for example, simply substitute 3 for n in the equation:

$$2(3^2) = 2(3 \times 3) = 18$$

EXAMPLE 3.14 Calculate the maximum number of electrons in principal energy level 4.

SOLUTION $2(4^2) = 2(4 \times 4) = 32$

Figure 3.24 A fluorine-19 atom has 9 protons and 10 neutrons in its nucleus, surrounded by 9 electrons in two orbits. Each orbit corresponds to an energy level.

Energy Sublevels. Principal energy levels have subdivisions called **sublevels.** The number of sublevels equals the number of the principal energy level (Table 3.7). Four types of sublevels are identified by the letters *s, p, d,* and *f.* Although more than four sublevels are theoretically needed for principal energy level 5 and above, four sublevels are enough to accommodate all the electrons in the 109 known elements.

The sublevels within a principal energy level correspond to different amounts of energy. The *s* sublevel has lower-energy electrons than the *p*; *p* is lower energy than *d*; and *d* is lower energy than *f* (Figure 3.26). All *s* sublevels contain a maximum of 2 electrons, *p* sublevels hold a maximum of 6, *d* sublevels hold a maximum of 10, and *f* sublevels have a maximum of 14 electrons.

Table 3.7 shows the pattern. Principal energy level 3, for example, can hold 18 (2×3^2) electrons: 2 electrons in the 3*s* sublevel, 6 electrons in the 3*p* sublevel, and 10 electrons in the 3*d* sublevel.

> **EXAMPLE 3.15** List by number and letter the sublevels present in principal energy level (a) 2 and (b) 4.
>
> **SOLUTION** (a) 2*s*, 2*p*; (b) 4*s*, 4*p*, 4*d*, 4*f* (see Table 3.7).

Atomic Orbitals. One level of subdivision remains. Sublevels are each composed of **atomic orbitals,** spaces that have a high probability of containing the electrons. Although sublevels within a principal energy level differ in energy, the orbitals within a sublevel have electrons with the same energy. Since each orbital holds a maximum of two electrons, you can readily calculate how many orbitals are in each type of sublevel (Table 3.7).

Figures 3.27 and 3.28 represent *s* and *p* orbitals. The shapes of *s* and *p* orbitals are the boundaries drawn to enclose the region having a 90% probability of containing the electron at any instant. These shapes do *not* represent an orbit or any other path the electrons follow.

All *s* orbitals have a spherical shape; the 1*s* orbital is simply a smaller sphere than the 2*s*, which in turn is smaller than the 3*s* sphere, and so on. All *p* orbitals consist of two lobes joined at the nucleus. The three *p* orbitals in a subshell are perpendicular to each other in space along the *x, y,* and *z* axes and are designated p_x, p_y, and p_z, respectively. The 2*p* orbitals are smaller than the 3*p*, which in turn are smaller than the 4*p*, and so on. But all *p* orbitals have the same shape. The shapes of *d* and *f* orbitals are more complex, and we don't need to discuss them here.

Figure 3.25 Erwin Schrödinger (1887–1961) received the Nobel prize in physics in 1933 for developing the quantum mechanical model of the atom.

> **EXAMPLE 3.16** Name the sublevels in principal energy level 3, and the number of orbitals in each sublevel.
>
> **SOLUTION** 3*s* (1 orbital), 3*p* (3 orbitals), 3*d* (5 orbitals)

Figure 3.26 Relative energy of electrons in the sublevels of principal energy level 4.

Table 3.7 Arrangements of Electrons in Atoms

Principal Energy Level (*n*)	Maximum Number of Electrons $2(n^2)$	Number of Sublevels	Name of Sublevels	Maximum Number of Electrons and Number of Atomic Orbitals per Type of Sublevel
1	2	1	1*s*	*s*–2*e* (1 orbital)
2	8	2	2*s*, 2*p*	*p*–6*e* (3 orbitals)
3	18	3	3*s*, 3*p*, 3*d*	*d*–10*e* (5 orbitals)
4	32	4	4*s*, 4*p*, 4*d*, 4*f*	*f*–14*e* (7 orbitals)
5	(50)*	(5)	5*s*, 5*p*, 5*d*, 5*f*, (. . .)	
6	(72)	(6)	6*s*, 6*p*, 6*d*, (6*f*, . . .)	
7	(98)	(7)	7*s*, (7*p*, 7*d*, 7*f*, . . .)	

* Values in parentheses not needed (or not fully needed) to account for the electrons in the 109 known elements.

Rule 1 is called the Aufbau principle. Aufbau is a German word that means "building up."

Rule 2 is called Hund's rule after the German physicist Frederick Hund.

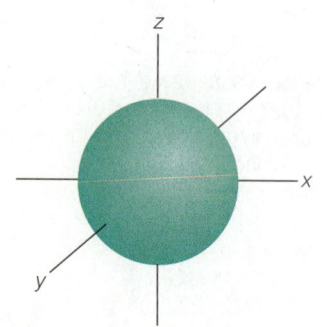

Figure 3.27 Boundary of a 1s orbital that has 90% probability of containing the electron at any instant. The nucleus is at the center of the figure.

When an electron arrangement is represented this way, the sum of the exponents equals the atomic number.

Electron Configurations of Elements. With this background, we can figure out the electron configuration for an atom of any element by following two basic rules.

Rule 1: *Electrons occupy the lowest energy sublevel possible.* How can you know the relative energy of each sublevel? Figure 3.29 shows a way to remember the sublevels in order of increasing energy. Just follow the arrows, starting at the bottom with 1s.

Rule 2: *Electrons in partially filled sublevels spread out to occupy as many orbitals as possible in that sublevel.* If a 4p sublevel, for example, has three electrons, then one electron is in each of the three orbitals (one in the $4p_x$, one in the $4p_y$, and one in the $4p_z$).

Figure 3.30 shows how to represent electrons in sublevels. For example, $4p^6$ means there are 6 electrons in the p sublevel of principal energy level 4. To represent the 4p orbitals, each containing 2 electrons, write: $4p_x^2$, $4p_y^2$, $4p_z^2$.

EXAMPLE 3.17 Atoms of manganese (Mn) have five electrons in their d sublevel of principal energy level 3. Represent that information using the system shown in Figure 3.30.

SOLUTION $3d^5$

Now let's write the electron arrangement for a specific element, fluorine (F). Looking at the periodic table, you see that F has atomic number 9. So F has nine electrons. Assign those nine electrons to sublevels following the sequence in Figure 3.29 and using the symbols shown in Figure 3.30. The electron configuration of fluorine is:

$1s^2 2s^2 2p^5$

Or, showing the details of the partially filled 2p sublevel, you can write:

$1s^2 2s^2 2p_x^2 2p_y^2 2p_z^1$

Another way to show the electron configuration is to use a box for each orbital, filling the box with an arrow for each electron present.[3] Higher boxes

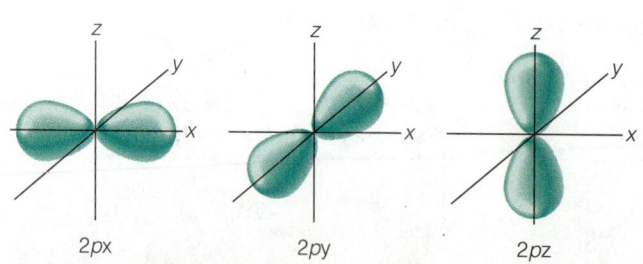

Figure 3.28 Boundaries of the three 2p orbitals; each has a 90% probability of containing the electron at any instant. The three orbitals have the same shape and are at right angles to each other in space. The nucleus is at the center of each figure.

---

[3] Pairs of arrows point in opposite directions because the 2 electrons in an orbital have opposite spins.

represent orbitals of higher energy. For fluorine, the configuration is:

```
  1s    2s    2pₓ   2p_y  2p_z
 [↑↓]  [↑↓]  [↑↓]  [↑↓]  [↑ ]
```

Potassium (K) has atomic number 19. Its electron configuration is:

$1s^2 2s^2 2p^6 3s^2 3p^6 4s^1$

This electron configuration, using boxes, is:

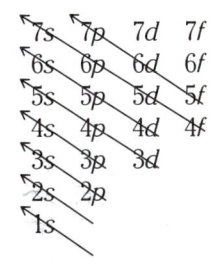

```
  1s    2s    2pₓ   2p_y  2p_z   3s    3pₓ   3p_y  3p_z   4s
 [↑↓]  [↑↓]  [↑↓]  [↑↓]  [↑↓]  [↑↓]  [↑↓]  [↑↓]  [↑↓]  [↑ ]
```

Figure 3.29 The sequence of sublevels in order of increasing energy. The sequence of arrows starts at the bottom.

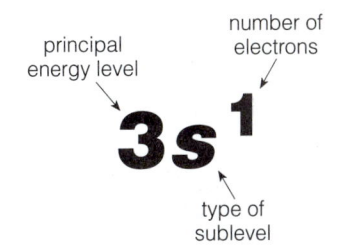

Figure 3.30 How to represent the number of electrons in a sublevel. The notation $3s^1$ means there is one electron in the s sublevel of principal energy level 3. (principal energy level / type of sublevel / number of electrons → $3s^1$)

EXAMPLE 3.18 Use boxes to represent the electron configuration of sulfur (S).

SOLUTION

```
                                            3pₓ   3p_y  3p_z
                                           [↑↓]  [↑ ]  [↑ ]
                                     3s
                                    [↑↓]
                        2pₓ   2p_y  2p_z
                       [↑↓]  [↑↓]  [↑↓]
                 2s
                [↑↓]
           1s
          [↑↓]
```

EXAMPLE 3.19 Use the "$1s^2$ . . ." system to write electron configurations for (a) lithium (Li), (b) phosphorus (P), and (c) iodine (I). For any partially filled p sublevel, list the electrons in each p orbital.

SOLUTION (a) $1s^2 2s^1$
(b) $1s^2 2s^2 2p^6 3s^2 3p_x^1 3p_y^1 3p_z^1$
(c) $1s^2 2s^2 2p^6 3s^2 3p^6 4s^2 3d^{10} 4p^6 5s^2 4d^{10} 5p_x^2 5p_y^2 5p_z^1$

Electron Dot Structures. Electrons in the highest principal energy level are called **valence electrons.** Valence electrons are the ones that interact most when elements form compounds, as you will learn in Chapter 4.

Chemists represent valence electrons by an **electron dot structure,** which consists of the symbol of an element along with a dot for each valence electron. Hydrogen (H), for example, has one valence electron and H· is its electron dot symbol. The maximum number of valence electrons is 8.

For elements with one to four valence electrons, write the dot(s) unpaired on the four sides of the element's symbol; additional valence electrons are paired with one of those four. For example, ·B· is boron (three valence

Electron dot structures are often called Lewis structures because they were devised by Gilbert N. Lewis (1875–1946) at the University of California, Berkeley.

electrons); ·C̈· is carbon (four valence electrons); ·N̈· is nitrogen (five valence electrons); and :Ö· is oxygen (six valence electrons).

Using the "$1s^2$ . . ." system, you can immediately determine the number of valence electrons and then write electron dot structures. For example, fluorine has the configuration $1s^2 2s^2 2p^5$. The highest principal energy level containing electrons is 2. In that energy level are a total of seven electrons (two in the $2s$ sublevel, and five in the $2p$ sublevel). Thus, fluorine has seven valence electrons and :F̈· is its electron dot structure.

With larger atoms, valence electrons aren't always listed together using the $1s^2$ . . . system. For example, tin (Sn) is atomic number 50 and has the configuration $1s^2 2s^2 2p^6 3s^2 3p^6 4s^2 3d^{10} 4p^6 5s^2 4d^{10} 5p^2$. Notice that the highest principal energy level containing electrons is energy level 5. In that level, tin has a total of four valence electrons (two in sublevel $5s$, and two in sublevel $5p$). ·Sn·, then, is the electron dot structure for tin.

EXAMPLE 3.20 Write electron dot structures for (a) oxygen, (b) zinc, (c) bromine, and (d) aluminum.

SOLUTION (a) :Ö·, (b) ·Zn·, (c) :B̈r·, (d) ·Al·

3.6 ELECTRON CONFIGURATION AND THE PERIODIC TABLE

The Connection. Now we return to the question: What do elements in the same group in the periodic table have in common that accounts for their similar chemical and physical properties?

The answer is in Table 3.8, which shows electron dot structures for elements in the A groups. Notice that the number of dots (valence electrons)

Table 3.8 Valence Electrons and Electron Dot Structures of Elements

Group	1/IA	2/IIA	13/IIIA	14/IVA	15/VA	16/VIA	17/VIIA	18/VIIIA
Number of Valence Electrons	1	2	3	4	5	6	7	8 (Except He)
Period				**Electron Dot Structure**				
1	H·							He:
2	Li·	·Be·	·B̈·	·C̈·	·N̈·	:Ö·	:F̈·	:N̈e:
3	Na·	·Mg·	·Al·	·Si·	·P̈·	:S̈·	:C̈l·	:Är:
4	K·	·Ca·	·Ga·	·Ge·	·Äs·	:Se·	:B̈r·	:K̈r:
5	Rb·	·Sr·	·In·	·Sn·	·S̈b·	:Te·	:Ï·	:Xe:
6	Cs·	·Ba·	·Ti·	·Pb·	·Bi·	:Po·	:Ät·	:Rn:
7	Fr·	·Ra·						

within each group matches the group's Roman numeral. Only helium (He) in group VIIIA is an exception. Thus, *elements in the same group have the same number of valence electrons.* Valence electrons, in turn, account for many properties of an element, as you will learn in Chapter 4.

B-group elements (transition elements) are a bit more complicated. If you work out their electron dot structures using the rules described in this chapter, you will conclude that all transition elements have two valence electrons. Most do have two, but exceptions are common because of additional rules beyond the scope of this book. You will learn some of the exceptions in the next chapter.

EXAMPLE 3.21 From their electron dot structures, predict the A-group number for each of the following elements: (a) :R̈n:, (b) Rb·, (c) :S̈e·, (d) ·S̈i·

SOLUTION (a) VIIIA, (b) IA, (c) VIA, (d) IVA

EXAMPLE 3.22 From their group numbers, write electron dot structures for each of the following: (a) calcium, (b) boron, (c) argon, (d) iodine, (e) nitrogen.

SOLUTION (a) ·Ca·, (b) ·Ḃ·, (c) :Är:, (d) :Ï·, (e) ·N̈·

The Scientific Method at Work. In this chapter, you have seen how the scientific method can work. First, scientists made careful observations and used them to formulate the laws of conservation of mass and of definite proportions. Dalton then proposed a hypothesis to explain those laws. His idea worked so well that it became widely accepted and known as atomic theory.

Mendeleev and Meyer discovered how to classify elements by their properties. But they didn't know why elements in the same group in the periodic table have similar properties. The answer had to await later discoveries about atomic structure. Bohr's model was a great advance, but it had to be revised because it didn't fit with new experimental data. An improved version—the quantum mechanical model—was then developed. Along the way, atomic number and atomic mass were related to atomic structure. Finally, scientists discovered that the arrangement of electrons in atoms explains why elements in the same group have similar properties.

This work required insight, imagination, reasoning, and experimentation. In the end, the information fit together in a way both satisfying and beautiful.

I believe there are 15, 747, 724, 136, 275, 002, 577, 605, 653, 961, 181, 555, 468, 044, 717, 914, 527, 116, 709, 366, 231, 425, 076, 185, 631, 031, 296 protons in the universe and the same number of electrons.

Source: Arthur S. Eddington (1882–1944)

SUMMARY

Homogeneous matter consists of pure substances and solutions. Pure substances are either elements or compounds. Compounds can be broken down into elements; elements cannot be broken down into simpler substances. Chemical changes produce different pure substances; physical changes do not.

To explain the law of conservation of mass and the law of definite proportions, Dalton proposed that all elements are composed of atoms. Later research revealed that atoms are composed of protons, neutrons, and electrons. Protons are positively charged, neutrons are uncharged, and electrons are negatively charged. Electrons are outside the nucleus. Protons and neutrons are in the nucleus and provide nearly all of an atom's mass.

The atomic number of an element is the number of protons in its atoms. Elements are composed of isotopes, which vary in their number of neutrons, and thus in their mass. The mass number of an isotope equals the number of protons and neutrons. The atomic mass of an element is an average mass resulting from the masses and relative abundance of its isotopes.

Mendeleev and Meyer independently formulated the periodic table of the elements. Elements with similar properties are in the same group, or vertical column. A horizontal row is a period. Metals are on the left of the staircase line, nonmetals on the right, and metalloids next to the line. Some important groups are the alkali metals, alkaline earth metals, transition elements, halogens, and noble gases.

Bohr proposed that electrons orbit a nucleus in specific energy levels. In the quantum mechanical model, the notion of electrons in fixed orbits was replaced by regions of high probability called orbitals. Electrons are classified by principal energy level, sublevel, and orbital. The electron configuration of an atom of any element is determined by assigning electrons to the lowest available energy sublevels and distributing electrons in partially filled sublevels among the maximum number of orbitals in that sublevel.

Electrons in the highest principal energy level are called valence electrons and can be represented in electron dot structures. The Roman numeral preceding A group elements in the periodic table is the number of valence electrons for those elements (except helium).

KEY TERMS

Alkali metal (3.4)
Alkaline earth metal (3.4)
Atom (3.2)
Atomic mass (3.3)
Atomic number (3.3)
Atomic orbital (3.5)
Atomic theory (3.2)
Chemical change (3.1)
Compound (3.1)
Electron (3.3)
Electron dot structure (3.5)
Element (3.1)
Group (3.4)
Halogen (3.4)
Isotope (3.3)
Law of conservation of mass (3.2)
Law of definite proportions (3.2)
Mass number (3.3)
Metal (3.4)
Metalloid (3.4)
Neutron (3.3)
Noble gas (3.4)
Nonmetal (3.4)
Nucleus (3.3)
Period (3.4)
Periodic table of the elements (3.4)
Physical change (3.1)
Principal energy level (3.5)
Proton (3.3)
Pure substance (3.1)
Quantum mechanical model (3.5)
Sublevel (3.5)
Transition element (3.4)
Valence electron (3.5)

EXERCISES

Even-numbered exercises are answered at the back of this book.

Classifying Matter

1. Classify each of the following as homogeneous or heterogeneous matter: (a) book, (b) butter, (c) chair, (d) sandwich, (e) aluminum, (f) milk, (g) urine, (h) bee, (i) skin.

2. Which of the following represent homogeneous matter?
   (a) Table salt stirred into water.
   (b) Oil stirred into water.
   (c) Sand stirred into water.
   (d) A sugar cube stirred into coffee.

3. Which substances in exercises 1 and 2, if any, are elements?

4. Would the separation of a mixture (such as sugar water) into pure substances (sugar and water) be a physical or a chemical change?

5. Would the breakdown of a compound (such as water) into its elements (hydrogen and oxygen) be a physical or a chemical change?

6. Which of the following are chemical changes? **(a)** taking a shower, **(b)** dough rising, **(c)** shaving, **(d)** stepping on a banana, **(e)** growing hair.

7. Which of the following are chemical changes? **(a)** boiling water, **(b)** hot wax hardening as it cools, **(c)** old milk turning sour, **(d)** cutting a piece of paper, **(e)** burning a piece of paper, **(f)** a wound healing.

8. Look around you. What substances, if any, do you see that are elements rather than compounds?

9. Look at Table 3.5. Which elements, if any, can exist in your body as elements rather than as compounds?

10. If you kept subdividing a bar of gold, would you ever reach something that you couldn't divide further without destroying the gold? Explain.

Conservation of Mass

11. When gasoline burns completely, it disappears. Does this violate the law of conservation of mass? Explain.

12. If 6 g of carbon reacts completely with 16 g of oxygen gas, how many g of carbon dioxide would be produced?

13. Producing 180 kg of water requires the complete reaction between 20 kg of hydrogen and how many kg of oxygen?

14. The mass of water always is 11.1% hydrogen and 88.9% oxygen. According to Dalton's atomic theory, why is this so?

15. Using the information in exercise 14, calculate how many kg of oxygen would be produced by the complete breakdown of 58 kg of water into oxygen and hydrogen.

16. Two grams of hydrogen react completely with 16 g of oxygen to produce 18 g of water. How much water would form if 2 g of hydrogen were allowed to react with 32 g of oxygen?

Atomic Structure and Mass

17. Dalton proposed that all atoms of an element are identical, and different from the atoms of other elements. What later discovery made it necessary to modify this statement?

18. Which subatomic particle has a positive electrical charge?

19. Which subatomic particle contributes almost nothing to atomic mass?

20. Name the elements with the following atomic numbers: **(a)** 1, **(b)** 28, **(c)** 36, **(d)** 56, **(e)** 82.

21. List the atomic numbers for the following elements: **(a)** neon, **(b)** sodium, **(c)** silver, **(d)** chlorine, **(e)** radium.

22. List the name of the element and the mass number of atoms having the following composition: **(a)** 6 protons, 7 neutrons; **(b)** 38 protons, 52 neutrons; **(c)** 53 protons, 78 neutrons.

23. Which isotope of potassium contains 22 neutrons?

24. List the number of protons (p), neutrons (n), and electrons (e) in one atom of each of the following isotopes: **(a)** iron-57, **(b)** sodium-23, **(c)** xenon-132, **(d)** mercury-200, **(e)** $^{60}_{27}Co$, **(f)** $^{235}_{92}U$, **(g)** $^{1}_{1}H$.

25. Rubidium consists of 72.17% $^{85}_{37}Rb$ and 27.83% $^{87}_{37}Rb$. Calculate the atomic mass of Rb, assuming that the mass of each isotope is exactly equal to its mass number.

26. A natural sample of lead (Pb) consists of 1.400% lead-204, 24.10% lead-206, 22.10% lead-207, and 52.40% lead-208. Calculate the atomic mass of Pb, assuming that the mass of each isotope is exactly equal to its mass number.

27. Define "atomic mass unit" (amu).

Periodic Table of the Elements

28. Using the 1–18 group numbering system, list the group number in the periodic table for each of the following: **(a)** any halogen, **(b)** aluminum, **(c)** carbon, **(d)** iron, **(e)** barium.

29. Using the Roman numeral and A/B numbering system, list the group in the periodic table for each of the following: **(a)** any alkaline earth element, **(b)** any noble gas, **(c)** nitrogen, **(d)** bromine, **(e)** platinum.

30. Write the symbols for the following elements: **(a)** iron, **(b)** chlorine, **(c)** sodium, **(d)** potassium, **(e)** magnesium, **(f)** fluorine, **(g)** arsenic.

31. Name the following elements: **(a)** P, **(b)** Cr, **(c)** S, **(d)** C, **(e)** Cu, **(f)** Br, **(g)** Hg.

32. Are most elements metals, nonmetals, or metalloids?

33. Are the transition elements metals, nonmetals, or metalloids?

34. Name the element described: **(a)** alkali metal in period 4; **(b)** group 16, period 2; **(c)** noble gas in period 4; **(d)** group IB, period 5.

35. In what period in the periodic table is uranium (U)?

36. What are the typical characteristics of metals?

Elements in the Body

37. Metals are sometimes classified as macronutrients or micronutrients, depending on whether relatively large or small amounts, respectively, are required in the diet. Which metals listed in Table 3.6 are macronutrients?
38. A deficiency of which three metals can cause a shortage of red blood cells (anemia)?
39. What are the three most abundant elements in the body on the basis of mass?
40. Why does the body need iodine?
41. Why does the body need calcium?

Arrangement of Electrons in Atoms

42. According to Bohr's model of the atom, where are electrons located?
43. According to the quantum mechanical model of the atom, where are electrons located?
44. List the maximum number of electrons in each of the following: **(a)** principal energy level 3, **(b)** a $3d$ sublevel, **(c)** a $4d$ sublevel, **(d)** a $2s$ sublevel, **(e)** a $5f$ sublevel, **(f)** a $4f$ orbital, **(g)** a $3p$ orbital, **(h)** principal energy level 5.
45. Which type of sublevel consists of five orbitals?
46. Which one of the following sublevels has electrons with the lowest energy? **(a)** $4p$, **(b)** $4s$, **(c)** $3d$, **(d)** $3p$
47. Criticize the following statement: A p orbital consists of two lobes because it can contain two electrons.
48. Compare the size and shape of a $3s$ and a $4s$ orbital.
49. Using the "$1s^2$ . . ." system, write the electron configuration for **(a)** aluminum, **(b)** arsenic, **(c)** xenon, **(d)** mercury.
50. Using the "$1s^2$ . . ." system, write the electron configuration for **(a)** carbon, **(b)** magnesium, **(c)** titanium, **(d)** tin.
51. Using boxes to represent orbitals, write the electron configuration for **(a)** oxygen, **(b)** sodium, **(c)** argon.
52. Using boxes to represent orbitals, write the electron configuration for **(a)** calcium, **(b)** nitrogen, **(c)** chlorine.
53. Write electron dot structures for the elements listed in exercise 49.
54. Write electron dot structures for the elements listed in exercise 50.
55. Write electron dot structures for the elements listed in exercise 51.
56. Write electron dot structures for the elements listed in exercise 52.
57. Using the electron dot structure for the A group elements (Al, As, and Xe) in exercise 49, predict the Roman numeral group for these elements.
58. Using the electron dot structure for the A group elements (C, Mg, and Sn) in exercise 50, predict the Roman numeral group for these elements.
59. Suppose a new, superheavy element is synthesized in Group VIA. Its symbol is Q. Write the electron dot structure for this element.
60. From its group number, write electron dot structures for **(a)** neon (Ne), **(b)** rubidium (Rb), **(c)** selenium (Se), **(d)** radium (Ra), **(e)** lead (Pb).
61. Which element has the electron configuration $1s^2 2s^2 2p^6 3s^1$?
62. Which element has the electron configuration $1s^2 2s^2 2p^6 3s^2 3p^6 4s^2 3d^{10} 4p^4$?
63. For the element described in exercise 62, how are the electrons distributed among the $4p$ orbitals?
64. Name an element that has one of its d sublevels half filled.

Electron Arrangement and the Periodic Table

65. Count the number of transition elements in period 4 or 5 of the periodic table. What kind of sublevel do you suppose is being filled? Write out the electron configurations, if necessary, to check your conclusion.
66. Count the number of inner transition elements in either row at the bottom of the periodic table. What kind of sublevel do you suppose is being filled? Write out the electron configurations, if necessary, to check your conclusion.
67. Count the number of groups to the left of the transition elements. What kind of sublevel do you suppose is being filled? Write out the electron configurations, if necessary, to check your conclusion.
68. Count the number of groups to the right of the transition elements. What kind of sublevel do you suppose is being filled? Write out the electron configurations, if necessary, to check your conclusion.
69. For the elements in period 2, what is the highest principal energy level that contains electrons?
70. For the elements in period 4, what is the highest principal energy level that contains electrons?
71. From the group in the periodic table, predict the number of valence electrons for **(a)** fluorine, **(b)** sodium, **(c)** phosphorus, **(d)** calcium, **(e)** cadmium.
72. From the group in the periodic table, predict the number of valence electrons for **(a)** hydrogen, **(b)** sulfur, **(c)** neon, **(d)** potassium, **(e)** boron.

73. Look at groups 13–17 in the periodic table. Does going down within a group increase the tendency for metallic or nonmetallic properties?
74. Strontium-90 is a radioactive material that accumulates in the body. Do you think it tends to replace **(a)** iodine in the thyroid gland, **(b)** iron in red blood cells, **(c)** calcium in bones, or **(d)** sulfur in proteins? [Hint: Look at the groups in the periodic table.]

DISCUSSION EXERCISES

1. Do you believe atoms really exist? Explain. What difference does it make whether they exist or not?
2. Rationalize why oxygen, carbon, hydrogen, nitrogen, calcium, and phosphorus are the most abundant elements in the body.
3. If you could subdivide a piece of gold indefinitely, at what point would it cease to be gold? Explain.
4. What is the best location for hydrogen (H) in the periodic table? What are the advantages and disadvantages of that location?

4

Compounds: Ions and Molecules

1. Why do atoms tend to gain or lose electrons? What is the octet rule?
2. What are ionic compounds, and how do they form? How do you write formulas for ionic compounds and name them?
3. What are alloys? How do they form?
4. What are covalent compounds, and how do they form? How do you write formulas for covalent compounds and name them?
5. What are polar and nonpolar covalent bonds? What are polar and nonpolar molecules? How do you predict the shapes and polarities of molecules?

Table Salt

Chemists use the word *salt* to refer to many different substances called *ionic compounds*. But most people use it to mean table salt, whose chemical name is sodium chloride. Your body needs salt to stay alive. Both sodium and chloride are required in the diet (see Table 3.6), and both are abundant in blood (see Table 4.4). Small amounts of sodium or potassium iodide may also be added to salt (called *iodized salt*). This provides iodide in the diet and helps prevent the thyroid disorder, goiter (see Figure 3.17).

Getting enough dietary salt is so important that in Roman times workers received part of their pay as salt: it was called *salarium*, the root word for *salary*. This also explains the expression that certain people are not "worth their salt."

Farmers and wildlife managers know the importance of adequate amounts of salt in animals' diets. They sometimes put a salt lick—a large cube of sodium chloride (Figure 4.1)—in a pasture for animals to lick.

But many people consume too much salt. Salt is found in many foods, in drinking water and other beverages, and in nonprescription drugs. Over a long time, excess salt intake contributes to hypertension (high blood pressure), a major cause of heart disease and strokes.

Sodium chloride has many uses. Adding salt to frozen highways and sidewalks melts ice by lowering the freezing temperature of water (see Section 7.4). Salt has been used for centuries to preserve meat, fish, and hides. In industry, sodium chloride is involved in producing or processing dyes, soap, cloth, plastics, and many other products.

Figure 4.1 A salt lick provides NaCl for bighorn sheep.

Figure 4.2 The human body is made almost entirely of compounds. Hair and skin, for example, are made of proteins—compounds of carbon, hydrogen, oxygen, nitrogen, and sulfur.

Look around you. Everything you can see and touch is made from elements and their atoms. Even some things that you can't see—like the air—contain atoms.

But you won't see many pure elements. You might spot mercury in a thermometer or see gold, silver, or copper in jewelry. Most jewelry, however, contains other metals to make it harder and more durable.

Nearly everything you see is a combination of elements and their atoms (Figure 4.2). In the natural world, pure elements are rare. The mercury in a thermometer, for example, occurs in nature mostly as cinnabar, a reddish ore composed of mercury and sulfur (Figure 4.3). Most metals occur in the earth combined with other elements such as sulfur and oxygen. Mining and chemical treatments isolate metals from their ores.

Figure 4.3 Mercury occurs naturally combined with sulfur in the ore cinnabar.

4.1 STABLE ELECTRON STRUCTURES: 8 AND 2 AS SPECIAL NUMBERS

Chemical Stability. What makes atoms of different elements combine to form compounds? Atoms do this because they can become more stable, which means they have lower potential energy (Section 2.4). Matter tends toward the lowest state of potential energy it can achieve.

Your body, for example, metabolizes sugar to form carbon dioxide and water. In this process, atoms in sugar (C, H, and O) and atoms in oxygen gas combine to form compounds (carbon dioxide and water) that have lower potential energy. As they reach a lower-energy, more stable state, sugar and oxygen release energy that your body uses (Figure 4.4).

Chemists say that atoms form **chemical bonds,** which hold atoms together in new combinations that are more stable. The three main types of bonding are *ionic*, *metallic*, and *covalent*.

The Octet Rule. How can you recognize a stable state for an atom? The periodic table contains an important clue: The noble gases stand out in terms of their tendency *not* to form compounds.

What can we conclude from this? If atoms react to become more stable, and noble gases don't react, apparently noble gas atoms are very stable. What makes them stable? When atoms interact, their valence, or outermost, electrons

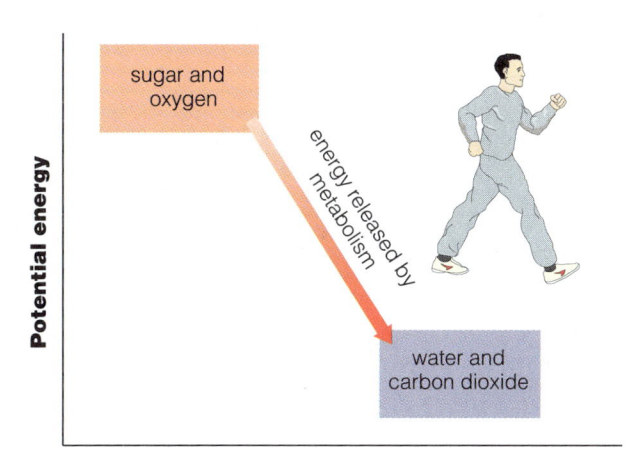

Figure 4.4 Metabolizing sugar and oxygen to products with lower potential energy.

are the first to encounter each other. Could it be that noble gas atoms—with either two (He) or eight (Ne, Ar, Kr, Xe, Rn) valence electrons—don't react because they already have stable electron configurations?

This simple idea is very useful in predicting how atoms form compounds. According to the **octet rule,** *atoms tend to lose, gain, or share electrons to achieve the electron configuration of the nearest noble gas in the periodic table.* Most atoms, then, tend to change electron structures to end up with eight valence electrons. The lightest elements, namely H, Li, and Be, tend to end up with two electrons, like helium.

EXAMPLE 4.1 Name the noble gas that has a total number of electrons nearest to the element listed: (a) oxygen, (b) iodine, (c) hydrogen, (d) calcium, (e) cesium, (f) aluminum.

SOLUTION (a) Ne, (b) Xe, (c) He, (d) Ar, (e) Xe, (f) Ne.

4.2 IONIC COMPOUNDS: FORMING IONS

Ionic Bonds: Metals and Nonmetals. Some atoms gain or lose electrons to get a stable octet of valence electrons. Scientists can measure the amount of energy it takes to remove one electron from an atom; this is called *ionization energy.* Such measurements show that ionization energy increases going from left to right, and from bottom to top, in the periodic table (Figure 4.5). In other words, elements near the top and right of the periodic table (nonmetals) are less likely to lose electrons than those to the bottom or left (metals). So, *in a reaction between a metal and a nonmetal, metal atoms tend to give up electrons to nonmetal atoms.* In the process, both become more stable.

Consider the reaction between sodium and chlorine atoms to form table salt, sodium chloride. Sodium atoms (Group 1/IA) have one valence electron, whereas chlorine atoms (Group 17/VIIA) have seven. A chlorine atom would have eight valence electrons, like argon, if it *gained* one electron (Figure 4.6). If it *lost* one valence electron, a sodium atom would be left with eight valence electrons (in principal energy level 2), like neon (Figure 4.7).

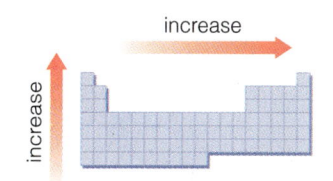

Figure 4.5 Ionization energy increases going from left to right, and from bottom to top, in the periodic table.

64 Part I General Chemistry

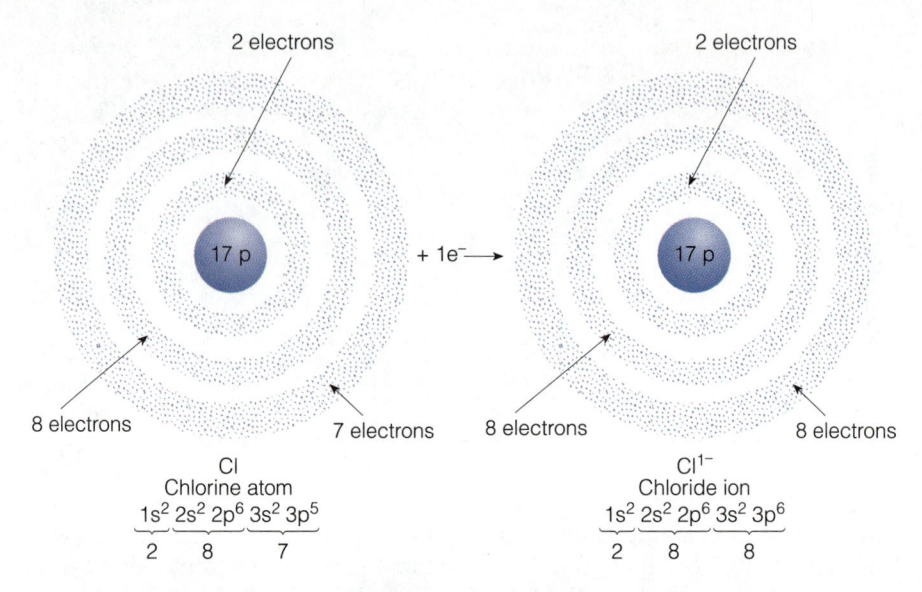

Figure 4.6 A chlorine atom tends to gain one electron to become more stable. Electrons are grouped by principal energy level.

Cl
Chlorine atom
$1s^2\ 2s^2\ 2p^6\ 3s^2\ 3p^5$
2   8   7

Cl^{1-}
Chloride ion
$1s^2\ 2s^2\ 2p^6\ 3s^2\ 3p^6$
2   8   8

Since both become more stable in the process, an electron transfers from sodium to chlorine:

Once atoms gain or lose electrons, they become **ions.** Because they don't have equal numbers of protons and electrons, ions have a net electrical charge. A sodium ion, for example, has 11 protons in its nucleus but only 10 electrons. It has a net charge of 1+. A chloride ion has a charge of 1− because it has 17 protons and 18 electrons.

Oppositely charged sodium and chloride ions attract each other and form a highly ordered, three-dimensional array called an *ionic crystal lattice* (Figure 4.8). The attractive forces between ions are called **ionic bonds,** and compounds formed in this way are called **ionic compounds.** The next time you see a crystal of table salt, sodium chloride, imagine its underlying structure—an ordered array of sodium and chloride ions held together by ionic bonds.

Sodium ion has
11 protons = 11+
10 electrons = 10−
net charge = 1+

Chemists use the word salt *for many ionic compounds, not just sodium chloride.*

Na
Sodium atom
$1s^2\ 2s^2\ 2p^6\ 3s^1$
2   8   1

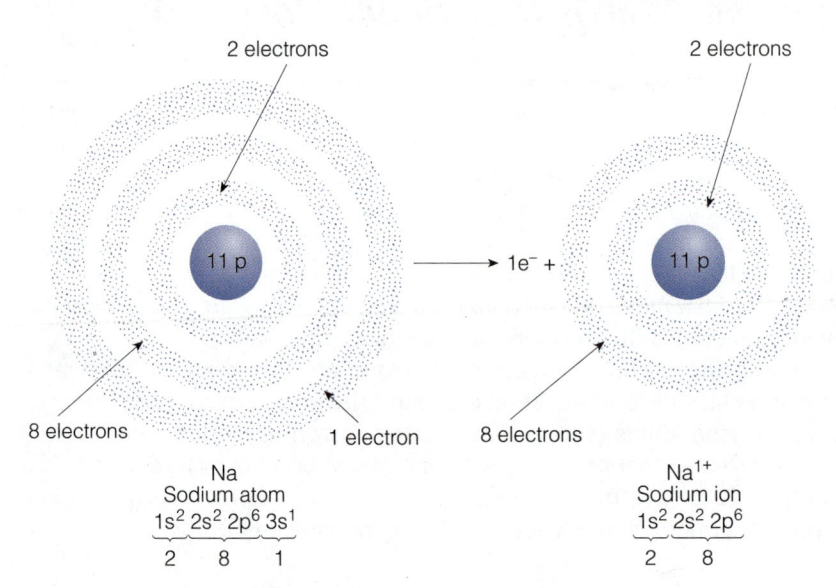

Na^{1+}
Sodium ion
$1s^2\ 2s^2\ 2p^6$
2   8

Figure 4.7 A sodium atom tends to lose one electron to become more stable. Electrons are grouped by principal energy level.

Metals react with nonmetals to form ionic compounds. The pattern is simple. Atoms of metals lose electrons and become positively charged ions; the amount of positive charge equals the number of electrons lost. Atoms of nonmetals gain electrons and become negatively charged ions; the amount of negative charge equals the number of electrons gained.

You can use the periodic table to predict the electrical charges of simple ions. Atoms of group 2/IIA elements, for example, have two valence electrons. These atoms (such as Mg) lose two electrons and form 2+ ions (such as Mg^{2+}). Atoms of group 16/VIA elements have six valence electrons; their atoms (such as S) gain two electrons and form 2− ions (such as S^{2-}). Table 4.1 shows these relationships.

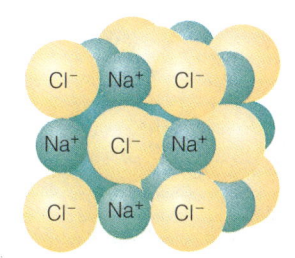

Figure 4.8 Sodium and chloride ions attract each other by ionic bonds and pack into an ionic crystal lattice. For each ionic compound, the pattern in the lattice depends on the relative size and number of positive and negative ions.

EXAMPLE 4.2 Predict the electrical charge on ions of the following elements: (a) oxygen, (b) magnesium, (c) phosphorus, (d) bromine, (e) lithium.

SOLUTION (a) 2−, (b) 2+, (c) 3−, (d) 1−, (e) 1+

Reacting elements don't always gain or lose the same number of electrons. Consider a reaction between sodium and sulfur. A sodium atom readily loses one electron, but each sulfur atom tends to gain two. In order for all the atoms to become stable ions, each sulfur atom has to react with two sodium atoms:

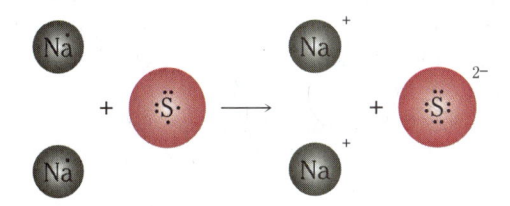

A reaction between aluminum and fluorine, in contrast, requires one aluminum atom and three fluorine atoms:

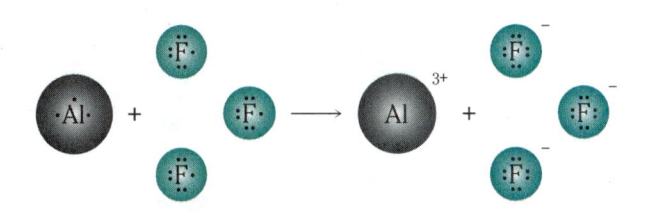

Table 4.1 Group Number and Ionic Charge

Group	Ionic Charge
1/IA	1+
2/IIA	2+
3–12/B	2+*
13/IIIA	3+
14/IVA	**
15/VA	3−
16/VIA	2−
17/VIIA	1−
18/VIIIA	**

* Many exceptions.
** Ions not common.

Writing Formulas for Binary (Two-Element) Ionic Compounds. The formula for an ionic compound is the simplest whole-number ratio of ions in the compound. This combination is called a **formula unit.**

Compounds that have exactly two elements are called binary compounds. To write formulas for binary ionic compounds, write the symbol for the metal element and then the symbol for the nonmetal. After each symbol, write as a subscript the number of those ions in a formula unit. Don't write subscripts of 1. In sodium chloride, for example, the ratio is 1 Na per 1 Cl. So the formula is written as NaCl (not Na_1Cl_1). The ionic compound formed from sodium and sulfur, shown previously, has the formula Na_2S.

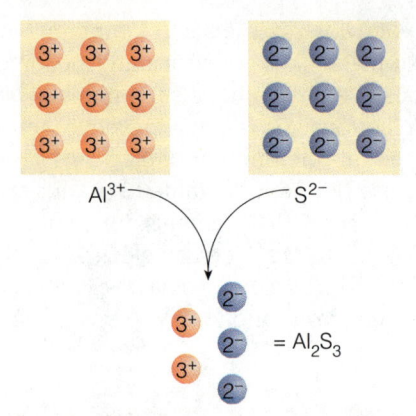

Figure 4.9 Two aluminum "balls" and three sulfide "balls" have a net charge of zero. The formula for aluminum sulfide is Al_2S_3.

EXAMPLE 4.3 Write the formula for the ionic compound formed by the reaction of aluminum with fluorine.

SOLUTION AlF_3

To find the number of each ion in a formula unit, you need to know the electrical charges of the ions (see Table 4.1). Then calculate how many of each ion are needed to give a net charge of zero.

One way to do this is shown in Figure 4.9. Think of each ion as a ball with the appropriate electrical charge. Imagine two boxes, one holding the negatively charged balls and the other holding the positively charged balls. Then figure out the smallest number of balls from each box that give a net charge of zero. Those numbers are the subscripts in the formula. For example, two Al ions (each having a 3+ charge) and three S ions (each having a 2− charge) provide a net charge of zero. The formula, then, is Al_2S_3.

Or look at it this way. You need to find a "common multiple" for the Al^{3+} and S^{2-} ions. We cannot make the charges on both ions add to 3 or 2, but we can make them both 6 (a "common multiple"):

2 Al^{3+} ions = +6
3 S^{2-} ions = −6
net charge = 0

The formula, then, is Al_2S_3.

$Mg^{2+}Br^- \longrightarrow MgBr_2$

$Mg^{2+}O^{2-} \longrightarrow Mg_2O_2$
$\downarrow$
MgO

$K^+S^{2-} \longrightarrow K_2S$

$Al^{3+}Cl^- \longrightarrow AlCl_3$

Figure 4.10 Crisscross method to predict formulas for ionic compounds.

EXAMPLE 4.4 Using the system in Figure 4.9, write formulas for the ionic compounds that form between (a) Li and Br, (b) Zn and F, (c) Ca and P, (d) Na and S, (e) Cd and Se, (f) Al and Cl.

SOLUTION (a) LiBr, (b) ZnF_2, (c) Ca_3P_2, (d) Na_2S, (e) CdSe, (f) $AlCl_3$.

Another method of writing formulas is to "crisscross" the ionic charges so that each number becomes a subscript for the other element (Figure 4.10). For the ionic compound formed from magnesium and bromine, first write the ionic charges: Mg²⁺Br⁻. Then crisscross the numbers as shown in Figure 4.10, remove the + and − signs, and write the formula, Mg₁Br₂, or more correctly, MgBr₂. If both subscripts are the same—as with Mg and O, for example—the formula consists of one ion of each, the simplest ratio.

EXAMPLE 4.5 Write formulas for ionic compounds that form between (a) Ca and Cl, (b) Li and O, (c) Al and F, (d) Sr and S, (e) Al and O, (f) Na and P.

SOLUTION (a) CaCl₂, (b) Li₂O, (c) AlF₃, (d) SrS, (e) Al₂O₃, (f) Na₃P.

Naming Binary Ionic Compounds. To name binary ionic compounds, simply use the name of the metal followed by the stem of the nonmetal plus the suffix *-ide*. Subscripts are not included in the name. NaCl, for example, is sodium chloride. K₂O is potassium oxide. Na₃N is sodium nitride, and MgI₂ is magnesium iodide.

Stems of a few nonmetals are ox-, sulf-, chlor-, nitr-, and phosph-.

EXAMPLE 4.6 Name the ionic compounds listed in Example 4.5.

SOLUTION (a) calcium chloride, (b) lithium oxide, (c) aluminum fluoride, (d) strontium sulfide, (e) aluminum oxide, (f) sodium phosphide.

Although most transition elements have two valence electrons and form ions with a charge of 2+, a few do not. In addition, some transition and other metals form two (or more) different positive ions (Table 4.2). Exceptions occur in transition elements when incompletely filled *d* sublevels alter the normal filling pattern for valence electrons.

Compounds made from elements with two or more possible ionic charges bear the name of the ion listed in Table 4.2. Combining Fe²⁺ with Cl⁻, for

---

Table 4.2 Transition Elements That Have an Ionic Charge Other Than 2+, and Some Elements That Occur in Multiple Ionic Forms

		Name of Ion	
Element	**Ion**	**Systematic**	**Common**
Silver	Ag⁺		silver
Copper	Cu⁺	copper (I)	cuprous
	Cu²⁺	copper (II)	cupric
Mercury	Hg⁺*	mercury (I)	mercurous
	Hg²⁺	mercury (II)	mercuric
Iron	Fe²⁺	iron (II)	ferrous
	Fe³⁺	iron (III)	ferric
Tin	Sn²⁺	tin (II)	stannous
	Sn⁴⁺	tin (IV)	stannic
Lead	Pb²⁺	lead (II)	plumbous
	Pb⁴⁺	lead (IV)	plumbic

* Actually occurs as Hg₂²⁺

Figure 4.11 Iron (II) chloride (left) and iron (III) chloride (right).

example, produces $FeCl_2$, which is named iron (II) chloride or ferrous chloride (Figure 4.11). Fe^{3+} combined with S^{2-} has the formula Fe_2S_3 and is called iron (III) sulfide or ferric sulfide.

> **EXAMPLE 4.7** Write formulas and names of the compounds that consist of (a) Cu^+ and O^{2-}, (b) Pb^{4+} and Cl^-, (c) Ag^+ and Br^-.
>
> **SOLUTION** (a) Cu_2O, copper (I) oxide; (b) $PbCl_4$, lead (IV) chloride; (c) AgBr, silver bromide.

Table 4.3 Common Polyatomic Ions and Their Uses or Effects

Name	Formula	Use or Effect in the Body
Ammonium	NH_4^+	NH_4NO_3 in fertilizer; $(NH_4)_2CO_3$ in smelling salts; NH_4Cl is a diuretic (increases urination)
Hydronium	H_3O^+	Produced by acids; sour taste; caustic
Hydroxide	OH^-	Produced by bases; bitter taste; caustic; NaOH (lye) is cleanser; $Mg(OH)_2$ is milk of magnesia
Permanganate	MnO_4^-	$KMnO_4$ is a disinfectant
Nitrate	NO_3^-	NH_4NO_3 in fertilizer; $NaNO_3$ is a food preservative; $AgNO_3$ used to prevent eye infections in newborns
Nitrite	NO_2^-	$NaNO_2$ is a meat preservative; high levels are toxic and bind to hemoglobin in blood
Cyanide	CN^-	HCN and NaCN are very toxic
Bicarbonate	HCO_3^-	$NaHCO_3$ is baking soda; used as antacid
Carbonate	CO_3^{2-}	$CaCO_3$ is limestone; used as chalk and antacid; Li_2CO_3 is used to treat manic-depression
Sulfate	SO_4^{2-}	$MgSO_4$ is a laxative; one form of $CaSO_4$ is plaster of Paris, used in making casts; $BaSO_4$ is used as a radio-opaque substance for taking X rays of internal organs
Sulfite	SO_3^{2-}	Na_2SO_3 is a food preservative; causes allergic response in some people
Phosphate*	PO_4^{3-}	Major component of teeth and bones; water pollutant from using certain detergents; fertilizer; $AlPO_4$ is an antacid

* Related ions are hydrogen phosphate (HPO_4^{2-}) and dihydrogen phosphate ($H_2PO_4^-$).

Polyatomic Ions. Atoms of two or more different elements (usually nonmetals) can combine to form electrically charged units called **polyatomic ions.** One carbon and three oxygen atoms, for example, combine to form the polyatomic *carbonate ion,* which has a charge of 2− and the formula CO_3^{2-} (read as "C-O-three-two minus").

In Section 4.4, you will learn about another type of bonding (covalent bonding) that holds atoms together in polyatomic ions. Table 4.3 lists the names, formulas, and uses of some common polyatomic ions, and Table 4.4 lists some simple and polyatomic ions in blood.

To write formulas for compounds that contain polyatomic ions, use the same methods as before to achieve a net charge of zero. Figure 4.12 shows how to do this using the crisscross method. Notice that when more than one unit of a particular polyatomic ion is in the formula, the polyatomic ion is enclosed in parentheses and the subscript placed after the parentheses. When the subscript is 1, don't write the number or use parentheses.

$Na^+ \, SO_4^{2-} \longrightarrow Na_2SO_4$

$Sr^{2+} \, NO_3^- \longrightarrow Sr(NO_3)_2$

$Ca^{2+} \, SO_3^{2-} \longrightarrow Ca_2(SO_3)_2$
$\phantom{Ca^{2+} \, SO_3^{2-} \longrightarrow } \downarrow$
$\phantom{Ca^{2+} \, SO_3^{2-} \longrightarrow } CaSO_3$

$NH_4^+ \, PO_4^{3-} \longrightarrow (NH_4)_3PO_4$

Figure 4.12 Crisscross method to predict formulas of compounds with polyatomic ions.

EXAMPLE 4.8 Write formulas for ionic compounds that contain the following ions: (a) sodium and carbonate, (b) calcium and nitrate, (c) ammonium and sulfide, (d) magnesium and sulfate, (e) aluminum and cyanide.

SOLUTION (a) Na_2CO_3, (b) $Ca(NO_3)_2$, (c) $(NH_4)_2S$, (d) $MgSO_4$, (e) $Al(CN)_3$.

To name compounds containing polyatomic ions, use the name of the positively charged ion, then the negative ion. K_2SO_4, for example, is potassium sulfate; NH_4Cl is ammonium chloride; and $Mg_3(PO_4)_2$ is magnesium phosphate.

EXAMPLE 4.9 Name the following compounds: (a) $Mg(OH)_2$, (b) $NaNO_2$, (c) $Ca(HCO_3)_2$, (d) Li_2SO_4, (e) $(NH_4)_2S$.

SOLUTION (a) magnesium hydroxide, (b) sodium nitrite, (c) calcium bicarbonate, (d) lithium sulfate, (e) ammonium sulfide.

Table 4.4 Major Ions in Blood

Ion	Approximate mg/dL Blood	Ion	Approximate mg/dL Blood
Positive Ions		**Negative Ions**	
Sodium (Na^+)	335	Chloride (Cl^-)	350
Potassium (K^+)	20	Bicarbonate (HCO_3^-)	165
Calcium (Ca^{2+})	10	Sulfate (SO_4^{2-})	3
Magnesium (Mg^{2+})	2	Phosphate (PO_4^{3-})	3
Zinc (Zn^{2+})	0.3	Fluoride (F^-)	0.3
Iron (Fe^{2+}, Fe^{3+})	0.01		
Copper (Cu^{2+})	0.01		

CHEMISTRY SPOTLIGHT

Lithium Ion and Manic Depression

Most drugs that alter moods work in one direction: They either stimulate or depress mental activity. Likewise, drugs that treat unwanted moods counteract either anxiety or depression.

People with manic depression, however, have both types of problem. Their moods swing between periods of depression and of manic, uncontrolled outbursts. Unlike most drugs, lithium ion (Li⁺) treats both mania and depression. In clinical tests, ionic lithium compounds such as lithium carbonate (Li_2CO_3) calm hyperactive children, protect people against depression, and control the outbursts of excitement in schizophrenia and manic depression.

Lithium ion acts by inhibiting a system (called the inositol phosphate system) in nerve cells. This action prevents those cells from sending faulty messages to other nerve cells.

4.3 METALLIC BONDING

Metallic Bonds and Metallic Properties. Coins, jewelry, coat hangers, needles, and many other items are made from metals. When all atoms in a substance are of metal elements, which tend to lose valence electrons to become stable, no atoms are present to accept those electrons. How can this work?

In a pure metal, all atoms are of the same element. The metal is solid (except for mercury), with its atoms arranged in a *metal crystal lattice* (Figure 4.13). According to one theory, each metal atom becomes an ion by losing its valence electrons to become more stable. Those valence electrons stay in the lattice and move throughout it. You can think of the arrangement as metal ions floating in a fluid of loosely held valence electrons. **Metallic bonding** is the attraction between this "sea" of mobile electrons and the positive metal ions.

A mixture of elements with metallic properties is called an **alloy.** Brass, for example, is an alloy of copper, zinc, and small amounts of tin, lead, and iron. Steel is a mixture of iron and carbon. Alloys of steel, titanium, or chromium and cobalt serve as artificial hip joints; titanium and other alloys are used for lightweight artificial thigh and other bones. As mixtures (Section 3.1), alloys have varying proportions of elements and don't have specific formulas like ionic compounds do.

Pure metals and alloys conduct electricity because of their mobile electrons. Electrons from electricity enter one end of the metal and leave the other end at the same rate. Metals can be flattened into sheets or foils, or drawn into wires, because layers of atoms slide past each other, with the mobile electrons adjusting to the new shape. The arrangement of electrons at the surface also tends to reflect light, making metals shiny.

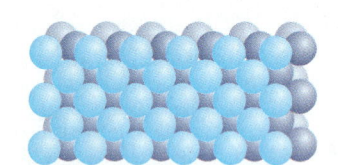

Figure 4.13 Model of a metal crystal lattice, an array of metal atoms packed together. Each sphere represents an atom that becomes an ion by losing its valence electron(s).

Typical composition of two alloys:
sterling silver
    93% Ag, 7% Cu
pewter
    85% Sn, 3% Cu, 2% Zn, 6% Pb, 4% Sb

CHEMISTRY SPOTLIGHT

Mercury: A Dangerous Metal

In his book, *The Periodic Table,* Primo Levi describes mercury (Figure 4.14) as "truly a bizarre substance: it is cold and elusive, always restless, but when it is quite still you can see yourself in it better than in a mirror."

Metallic mercury is toxic when inhaled because it destroys lung tissue. It is less dangerous when swallowed because the body excretes mercury within a few days. Ionic mercury salts at high concentrations are also hazardous. Hatters used mercury (II) nitrate, $Hg(NO_3)_2$, to treat felt to prevent fungi growth. When they chewed on the felt to soften it, they ingested mercury ions. The resulting disease was the basis for the Mad Hatter in Lewis Carroll's *Alice in Wonderland.* Analysis of a hair sample of the famous scientist Isaac Newton indicates he also suffered from exposure to mercury compounds.

The most dangerous form is methyl mercury (CH_3Hg^+), which remains in the body for months and attacks brain tissue, the central nervous system, liver, and kidneys. Near the bottom of acidic bodies of water, where little oxygen is present, anaerobic bacteria convert mercury deposits into methyl mercury. Fish and other aquatic life then get contaminated with mercury and become unsafe to eat.

In the late 1950s, 52 people died and 150 suffered brain and nerve damage from eating mercury-contaminated fish from Japan's Minimata Bay. In addition, several children were born with birth defects linked to mercury poisoning. A chemical plant had discharged mercury into the bay, where it was converted into methyl mercury.

Mercury enters water as fallout from the air and from discharges by mercury-using chemical plants, especially chlor-alkali plants such as the one at Minimata Bay. Large amounts of mercury vaporize naturally from the earth's crust, and huge stores are in bottom sediments in oceans. Thermometer factories and smelters are other sources of mercury.

Mercury damages the nervous system and other tissues by binding to sulfur groups in enzymes (Chapter 22) and other proteins (Chapter 21), preventing them from carrying out critical functions in the body.

Figure 4.14 Mercury.

4.4 COVALENT COMPOUNDS: FORMING MOLECULES

Sharing Electrons. When metals react with nonmetals, their atoms transfer electrons and form ionic compounds. When metal atoms interact, valence electrons become a mobile sea among metal ions, producing metallic bonds. One combination is left. What happens when atoms of nonmetals, which tend to gain electrons, react with each other?

The answer is that such atoms come close together and share electrons. In this way, atoms gain access to more valence electrons and get a valence electron structure more like that of a noble gas. A pair of electrons shared between two atoms is called a **covalent bond.** A group of atoms held together by covalent bonds is called a **molecule.** Substances formed in this way are **covalent compounds.**

One example is hydrogen gas. Hydrogen atoms each have one valence electron. They would be more stable if they had two valence electrons, like the noble gas helium. When two hydrogen atoms come close enough together, their 1s orbitals overlap.

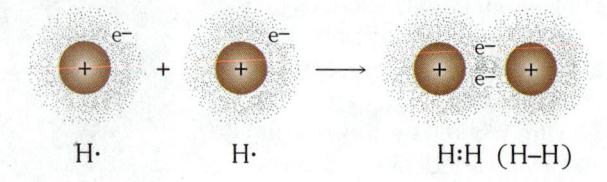

H·   H·   H:H (H–H)

The electron from each atom has a high probability of being in the common, overlapped 1s region for both atoms. There both nuclei attract the two electrons. Although the two nuclei repel each other, at a certain distance the attractive forces exceed the repulsive forces.

This sharing arrangement gives both atoms access to two valence electrons, a situation more stable (lower potential energy) than two individual atoms. So pairs of hydrogen atoms stay together joined by a covalent bond, shown with a dash: H—H. The basic particles in hydrogen gas, then, are molecules having the formula H_2.

A similar arrangement occurs when chlorine atoms interact. Each chlorine atom has seven valence electrons with the configuration $3s^2 3p_x^2 3p_y^2 3p_z^1$ (Section 3.5). These atoms would be more stable if they each gained an electron in their $3p_z$ orbitals (to have eight valence electrons). So two chlorine atoms come close enough together for their partially filled $3p_z$ orbitals to overlap and share a pair of electrons.

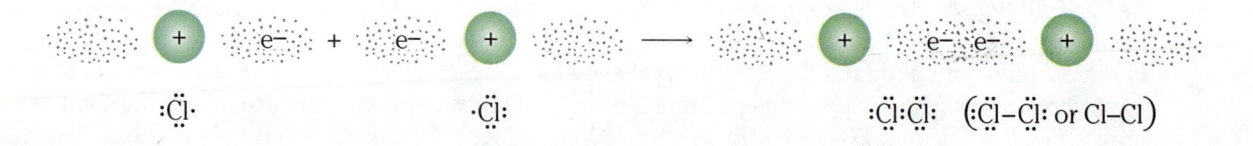

:Cl̈·   ·Cl̈:   :Cl̈:Cl̈:   (:Cl̈–Cl̈: or Cl–Cl)

The resulting chlorine molecule has the formula Cl_2.

When hydrogen and oxygen react, the arrangement is slightly different. Oxygen atoms have six valence electrons and need to gain two to become more stable. Hydrogen atoms, however, only need to gain one electron. In order for each type of atom to become stable, two hydrogen atoms combine with one oxygen atom:

$$H\cdot + \cdot \ddot{O} : \longrightarrow H : \ddot{O} : \atop H \quad \left(H-\ddot{O}: \text{ or } H-O \atop \phantom{H-\ddot{O}:} | \phantom{\text{ or }} | \atop \phantom{H-\ddot{O}:} H \phantom{\text{ or }} H \right)$$

The H—O dash, representing a covalent bond, means the same as H:O.

They form a molecule of water, H_2O.

Notice in these examples that hydrogen and chlorine atoms, which need to gain one electron, each form one covalent bond. Oxygen atoms, which need to gain two electrons, form two covalent bonds. This is part of a pattern. As Table 4.5 shows, the number of covalent bonds that nonmetal atoms tend to form equals the number of valence electrons they need to gain to become stable.

You can predict structures of molecules by combining atoms as simply as possible, connecting them with the appropriate number of covalent bonds. For example, when N and H atoms combine, the simplest arrangement provides each N with three bonds and each H with one:

$$\text{H}-\text{N}-\text{H} \quad (NH_3) \atop | \atop \text{H}$$

From this structure, you can write the formula using subscripts to specify how many atoms of each element are in one molecule of the substance. The element on the left in the periodic table usually comes first in the formula. A few exceptions occur, mostly with H.

In formulas, hydrogen (H) typically follows elements in groups IA to VA (as in NH_3) and precedes elements in groups VIA and VIIA (as in H_2O).

EXAMPLE 4.10 For the following combinations, connect the atoms by the appropriate number of covalent bonds, and then write the formula for the compound: (a) H and Br, (b) S and Cl, (c) C and F.

SOLUTION (a) H—Br, HBr; (b) Cl—S, SCl_2; (c) F—C—F, CF_4
$|$
$Cl F$

Some molecules, especially those of carbon, are more complex. Figure 4.15 shows a few examples. Notice that the same rules of bonding apply: Carbon

Table 4.5 Number of Covalent Bonds Elements Tend to Form

Group	Valence Electrons	Additional Valence Electrons Needed	Number of Covalent Bonds Formed	Example
IA (1) (H only)	1	1	1	H—
IVA (14)	4	4	4	—C—
VA (15)	5	3	3	—N—
VIA (16)	6	2	2	—O—
VIIA (17)	7	1	1	F—

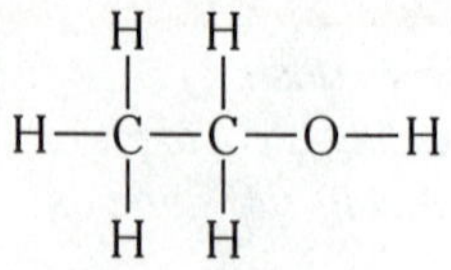

ethanol, C$_2$H$_6$O
(in alcoholic drinks)

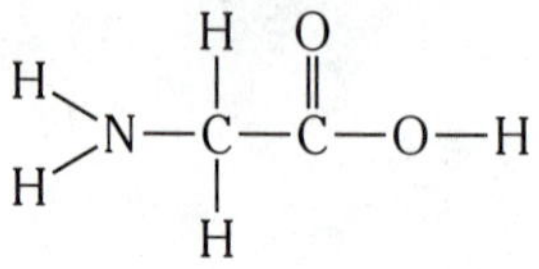

glycine, C$_2$H$_5$NO$_2$
(an amino acid)

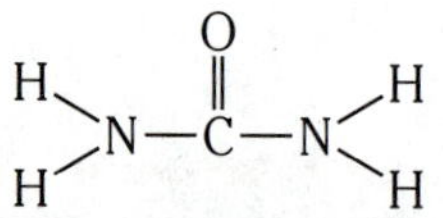

urea, CH$_4$N$_2$O
(in urine)

Figure 4.15 Formulas of three carbon compounds.

atoms have four bonds, hydrogen atoms have one, oxygen atoms have two, and nitrogen atoms have three. You will learn more about carbon compounds beginning in Chapter 11.

Multiple Covalent Bonds. Atoms that form more than one covalent bond sometimes form multiple (double or triple) covalent bonds. A double bond is two pairs of electrons shared between atoms. Figure 4.15 shows two compounds that have a double covalent bond between carbon and oxygen.

Carbon dioxide, CO$_2$, is another example. How can one C atom, which forms four covalent bonds, combine with two O atoms, each of which forms two covalent bonds? The only way is with two double bonds:

$$O=C=O \quad (\ddot{O}::C::\ddot{O} \quad \text{or} \quad \ddot{O}=C=\ddot{O})$$

About 80% of the mass of air is nitrogen gas, N$_2$. Since N atoms form three covalent bonds, an N$_2$ molecule is held together by a triple covalent bond, which is *three pairs* of shared electrons:

$$N{\equiv}N \quad (:N:::N: \quad \text{or} \quad :N{\equiv}N:)$$

Toxic hydrogen cyanide (H—C≡N) also has a triple bond in its molecules.

A Few Exceptions. Although the octet rule and the types of bonding we have discussed work for most compounds, they don't work for every compound. The octet rule can give a good first approximation, but it doesn't apply, for example, to certain compounds of beryllium, boron, nitrogen, phosphorus, and sulfur, which can have 4, 6, 7, 10, and 12 valence electrons, respectively. Two examples are:

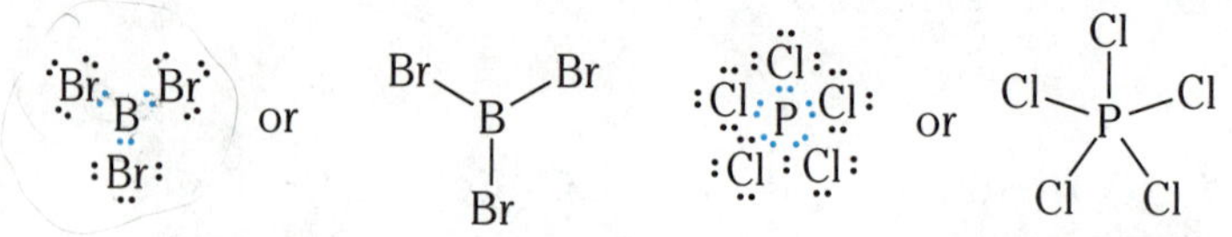

6 valence e   10 valence e

And covalent bonds form in different ways. Sometimes one atom contributes *both* electrons; these are called **coordinate covalent bonds.** For example, when a hydrogen ion (H$^+$, a hydrogen atom that has lost its electron) reacts with NH$_3$, the latter furnishes a pair of electrons that it shares with H$^+$:

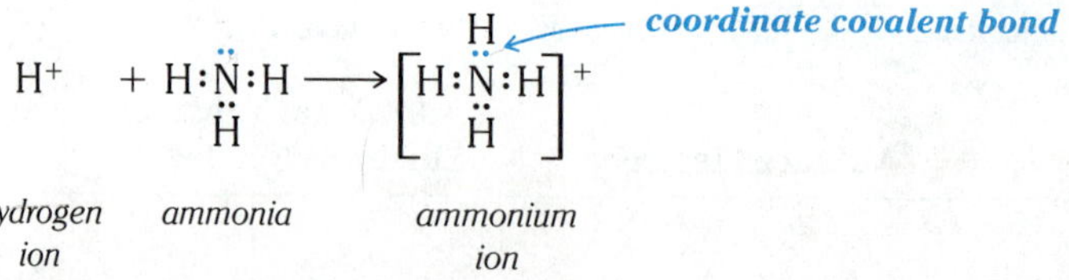

H$^+$ + H:N̈:H ⟶ [H:N:H]$^+$ ← *coordinate covalent bond*

hydrogen   ammonia   ammonium
ion   ion

Once it forms, a coordinate covalent bond is like any other covalent bond.

Polyatomic Ions: Another Look. Polyatomic ions (Table 4.3) are held together by covalent bonds. The ammonium ion (NH$_4^+$), shown in the preceding reaction, is one example. These combinations are ions instead of molecules because they gained or lost electrons to become stable. Ammonium ion (NH$_4^+$), for example, has a total of one less electron than it does protons. A carbonate ion (CO_3^{2-}), in contrast, had to gain two electrons to become a stable unit (Figure 4.16, bottom). Some polyatomic ions have coordinate covalent bonds.

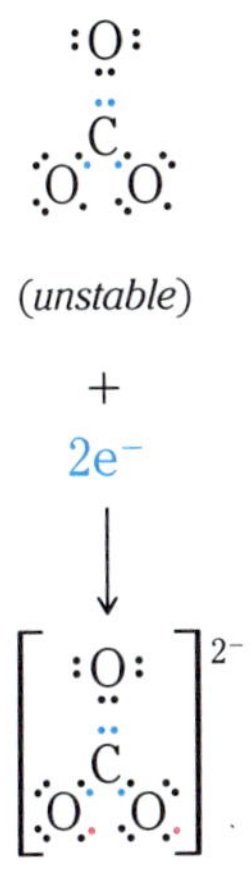

Electron Dot Structures of Molecules and Polyatomic Ions. You can use formulas and a simple procedure to write electron dot (Lewis) structures for molecules and polyatomic ions. The steps are:

1. Add the number of valence electrons from each atom in the formula. For a polyatomic ion, also add (for negative ions) or subtract (for positive ions) the number of electrons in the charge. This is the total number of valence electrons for the electron dot structure.
2. Join atoms in the formula with covalent bonds, providing each atom an appropriate number of bonds (see Table 4.5). Hydrogen and halogen atoms have end positions. The least electronegative atom (besides H) usually is the central atom.
3. Since each bond (step 2) represents a pair of valence electrons, count how many valence electrons those bonds represent. Subtract this number from the total number of valence electrons determined in step 1. Then distribute these remaining electrons in pairs to give each atom eight electrons (except H, which has two).
4. If you have too few electrons to assign in step 3, return to step 2 and insert one or more multiple bonds; then do step 3 again.

Let's follow this procedure to write the electron dot structure for urea (Figure 4.15):

1. Carbon, hydrogen, oxygen, and nitrogen atoms have 4, 1, 6, and 5 valence electrons, respectively. So the total number of valence electrons for urea, CH_4ON_2, is $(1 \times 4) + (4 \times 1) + (1 \times 6) + (2 \times 5) = 24$.

2. Carbon, hydrogen, oxygen, and nitrogen atoms tend to form 4, 1, 2, and 3 covalent bonds, respectively. Draw the structure with covalent bonds, making C the central atom:

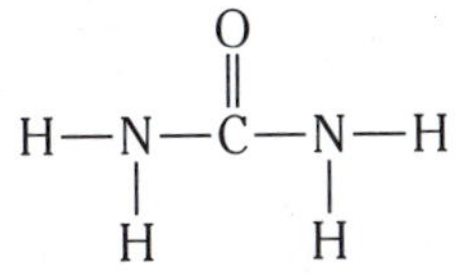

```
       O
       ||
  H—N—C—N—H
     |   |
     H   H
```

3. The 8 bonds in the preceding structure account for $8 \times 2 = 16$ valence electrons. That leaves $24 - 16 = 8$ valence electrons to add to the structure. C and H atoms already have enough electrons in their bonds to make them stable. So distribute the remaining 8 electrons to give the N and O atoms 8 valence electrons each:

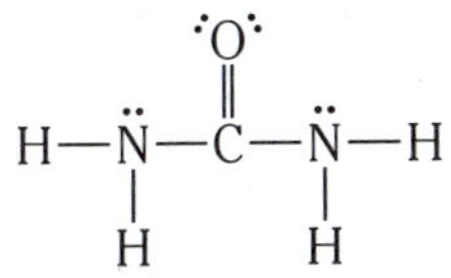

```
       ··
      :O:
       ||
  H—N̈—C—N̈—H
     |   |
     H   H
```

For carbonate ion (CO_3^{2-}) the procedure is:

1. Total valence electrons = $(1 \times 4) + (3 \times 6) + 2 = 24$
2. C is less electronegative than O, so make C the central atom.

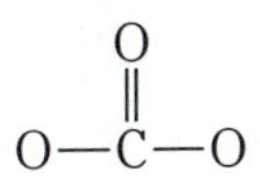

```
      O
      ||
   O—C—O
```

(unstable)

$+$
$2e^-$
$\downarrow$

$$\left[\begin{array}{c} :\ddot{O}: \\ \| \\ :\ddot{O}\!-\!C\!-\!\ddot{O}: \end{array} \right]^{2-}$$

(stable carbonate ion)

Figure 4.16 (Top) Carbon, with its four valence electrons (blue), forms covalent bonds with three oxygen atoms, each with six valence electrons (black dots). (Bottom) Addition of two electrons (magenta) produces a stable carbonate ion (CO_3^{2-}), with a net charge of 2− and each atom having eight valence electrons. The additional two electrons come from a metal (such as Ca) whose atoms give up electrons to become positively charged ions (such as Ca^{2+}).

Add two electrons in step 1 because carbonate ion has a 2− charge.

Structures of polyatomic ions are written in brackets, with the charge written as an exponent outside the brackets.

3. Remaining electrons to distribute = 24 − (4 × 2) = 16

$$\left[:\ddot{\underset{..}{O}}-\underset{\underset{\|}{O}}{C}-\ddot{\underset{..}{O}}: \right]^{2-}$$

For nitrate ion $(NO_3)^-$, the procedure is:

1. (1 × 5) + (3 × 6) + 1 = 24 valence electrons

2.

O—N—O
  |
  O

3. 24 − 6 = 18 electrons to distribute:

:Ö—N—Ö:
     |
    :Ö:

4. not enough electrons for N, so add a double bond and repeat steps 2 and 3:

$$\left[:\ddot{O}=N-\ddot{\underset{..}{O}}: \atop | \atop :\ddot{\underset{..}{O}}: \right]^{-}$$

Table 4.6 Prefixes for Naming Covalent Compounds

Number of Atoms	Prefix
1	mono-
2	di-
3	tri-
4	tetra-
5	penta-
6	hexa-
7	hepta-
8	octa-
9	nona-
10	deca-

Names of covalent compounds with three or more elements are more complex. Many are carbon compounds (Figure 4.16), which you will learn about beginning in Chapter 11.

EXAMPLE 4.11 Write electron dot structures for (a) formaldehyde (CH_2O), (b) hydrogen cyanide (HCN), (c) hydroxide ion (OH^-), (d) NF_3.

SOLUTION (a) $H-\underset{\underset{\|}{O}}{C}-H$, (b) $H-C\equiv N:$, (c) $[:\ddot{\underset{..}{O}}-H]^-$,

(d) :F̈—N̈—F̈:
        |
       :F̈:

Naming Covalent Compounds. To name binary covalent compounds, use the name of the first element in the formula, followed by the name of the second element, which ends in the suffix *-ide*. This is the same as in naming two-element ionic compounds. With covalent compounds, however, we also use prefixes to specify the number of atoms of each element in the formula. Table 4.6 lists the common prefixes.

The prefix *mono-* seldom is used, especially with the first element in the formula. CO_2, for example, is carbon dioxide (not monocarbon dioxide) whereas CO is carbon monoxide. NCl_3 is nitrogen trichloride, and P_4S_7 is tetraphosphorus heptasulfide.

EXAMPLE 4.12 Name the following compounds: (a) CF_4, (b) NO_2, (c) H_2O, (d) S_4N_2, (e) P_2O_5.

SOLUTION (a) carbon tetrafluoride, (b) nitrogen dioxide, (c) dihydrogen monoxide (water), (d) tetrasulfur dinitride, (e) diphosphorus pentoxide.

CHEMISTRY SPOTLIGHT

N₂O: A Chemical of Many Uses

The formal name for N₂O is dinitrogen monoxide, though it is often called nitrous oxide. In 1776 Joseph Priestley (Figure 4.17), an English clergyman and amateur scientist, first prepared this covalent compound by heating ammonium nitrate, NH₄NO₃. In the early 1800s, N₂O became one of the first general anesthetics used in surgery. It is still used, especially in dental and obstetric surgery. N₂O has a sweet taste and pleasant smell, and breathing it produces mild intoxication. You probably know it as *laughing gas*.

Nitrous oxide is a colorless gas that dissolves well in fats. It is used in self-whipping creams sold in spray cans. When this product is sprayed from the can, pressure is released and N₂O bubbles form in the whipped cream.

Animal wastes, fertilizers, and burning plant products put N₂O into the air. It warms the earth by absorbing heat that the earth radiates into space; this is known as the *greenhouse effect*. Because the atmospheric concentrations of gases that have this effect are increasing, the earth may gradually be getting warmer. N₂O accounts for about 6% of the greenhouse effect. Other major contributors are CO₂ (about 50%), CH₄ (15–20%), and chlorofluorocarbons (15–20%).

Figure 4.17 Joseph Priestley (1733–1804).

4.5 SHAPES OF MOLECULES

What Does a Molecule Look Like? Scientists can measure where atoms are located in molecules, and have discovered a variety of shapes. One useful explanation for those shapes is the *valence-shell electron-pair repulsion theory*, or **VSEPR theory**. The name is long, but it summarizes this simple idea: Pairs of negatively charged valence electrons repel each other and get as far away from each other as possible.

Atoms in a molecule typically have four pairs of valence electrons. Some pairs are shared in covalent bonds; others may not be shared. According to

78 Part I General Chemistry

Figure 4.18 According to VSEPR theory, four pairs of valence electrons are about 109° apart around the nucleus.

Figure 4.19 At temperatures colder than −79°C, carbon dioxide (CO_2) is a solid commonly called dry ice. It is used to seed clouds for rain and to store frozen food and biological tissues.

Table 4.7 Shapes of Molecules with Four Separated Electron Pairs

Examples	Shape
H:C̈l:	linear
H:Ö: H	bent
H:N̈:H H	pyramidal
H H:C:H H	tetrahedral

VSEPR theory, the pairs (whether involved in a covalent bond or not) are as far away from each other as possible. Three common situations occur:

1. *Electron pairs in four locations* In three dimensions, four separated electron pairs are about 109° apart (Figure 4.18). This separation can produce several molecular shapes (Table 4.7).

In the tetrahedral shape, such as CH_4, you see the effect of all four pairs bonded to atoms. In the pyramidal shape, such as NH_3, three electron pairs are in covalent bonds to N and one is not. In the bent shape, such as H_2O, two pairs form bonds to O and two do not. In all cases the four electron pairs around the central atom repel each other, keeping each other (and all bonded atoms) about 109° apart.

2. *Electron pairs in three locations* When valence electrons are in three locations, maximum separation is 120°, which produces a flat, triangular-shaped molecule (Table 4.8). One example is BCl_3, in which boron is an exception to the octet rule and has only 3 pairs of valence electrons. The same shape occurs in formaldehyde, $H-\overset{\overset{O}{\|}}{C}-H$, where carbon has a double bond and two single bonds. Here four pairs of valence electrons are arranged in only three locations because two pairs are in the same region to provide the double bond.

3. *Electron pairs in two locations* When valence electrons are in only two locations, maximum separation produces a linear molecule (Table 4.8). Examples include beryllium compounds such as $BeCl_2$, because Be contains only two pairs of valence electrons (another exception to the octet rule). This shape also occurs when carbon contains two double bonds. In CO_2, for example, two pairs of electrons are between the carbon atom and each oxygen atom (Figure 4.19).

EXAMPLE 4.13 Predict the shapes of the following molecules: (a) CCl_4, (b) OCl_2, (c) NI_3, (d) BF_3, (e) N_2, (f) BeF_2.

SOLUTION (a) tetrahedral, (b) bent, (c) pyramidal, (d) triangular, (e) linear, (f) linear.

4.6 POLAR AND NONPOLAR MOLECULES

Types of Bonding: Another Look. We have classified bonding between nonmetal atoms as covalent, bonding between metals as metallic, and bonding between metals and nonmetals as ionic. Using this pattern, you can predict that sulfur (a nonmetal) forms covalent bonds with oxygen (a nonmetal) and ionic bonds with sodium (a metal).

EXAMPLE 4.14 Predict the type of bonding for each of the following compounds: (a) MgO, (b) CH_4, (c) NCl_3, (d) F_2, (e) CaI_2.

SOLUTION (a) ionic, (b) covalent, (c) covalent, (d) covalent, (e) ionic.

But bonding in most substances falls somewhere between pure covalent and pure ionic. For one thing, the distinction between metals and nonmetals isn't definite. So it is more accurate to describe bonding as being principally ionic or covalent. In nonmetal–nonmetal compounds, for example, bonding is principally covalent because the tendency to share electrons usually outweighs the tendency to transfer them.

To understand this partially ionic, partially covalent character of bonding, you need to learn about *bond polarity* and *electronegativity*.

Polar and Nonpolar Covalent Bonds. When two atoms of the same element share electrons, each atom attracts the shared electrons equally. This equal sharing results in a pure covalent bond, called a **nonpolar covalent bond.**

When atoms of two different elements share electrons, one atom usually attracts the shared electrons more strongly. The tendency of an atom to attract shared electrons is called **electronegativity.** As Figure 4.20 shows, electronegativity increases going to the right and up in the periodic table.

What difference does this make? When atoms of different elements share electrons, the electrons are attracted to the atom with the greater electronegativity. Instead of being shared equally, electrons on the average are nearer to the more electronegative atom (pole). This arrangement, called a **polar covalent bond,** lies between an ionic bond (where electrons are completely transferred) and a nonpolar covalent bond (where electrons are shared equally).

HCl is one example of this partial transfer, partial share arrangement. The shared electron pair is pulled toward the more electronegative chlorine atom and away from hydrogen. This produces a slight negative charge on the chlorine atom and a slight positive charge on the hydrogen atom, though the overall charge is zero.

Table 4.8 Shapes of Molecules with Electron Pairs in Two or Three Locations

Examples	Shape
Two Locations	
Cl:Be:Cl	
O::C::O	linear
Three Locations	
Cl:B:Cl (with Cl)	
H:C::O (with H)	triangular

Noble gases are excluded from Figure 4.20. They don't typically have shared electrons to attract.

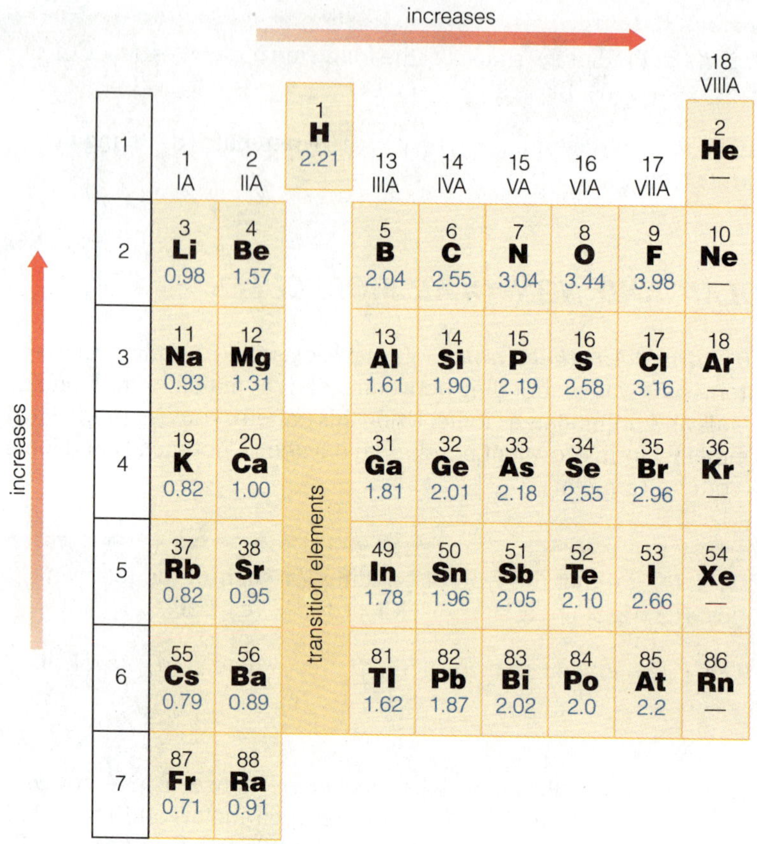

Figure 4.20 Electronegativity values (in color) for A group elements.

	1 IA	2 IIA	13 IIIA	14 IVA	15 VA	16 VIA	17 VIIA	18 VIIIA
1	1 H 2.21							2 He —
2	3 Li 0.98	4 Be 1.57	5 B 2.04	6 C 2.55	7 N 3.04	8 O 3.44	9 F 3.98	10 Ne —
3	11 Na 0.93	12 Mg 1.31	13 Al 1.61	14 Si 1.90	15 P 2.19	16 S 2.58	17 Cl 3.16	18 Ar —
4	19 K 0.82	20 Ca 1.00	31 Ga 1.81	32 Ge 2.01	33 As 2.18	34 Se 2.55	35 Br 2.96	36 Kr —
5	37 Rb 0.82	38 Sr 0.95	49 In 1.78	50 Sn 1.96	51 Sb 2.05	52 Te 2.10	53 I 2.66	54 Xe —
6	55 Cs 0.79	56 Ba 0.89	81 Tl 1.62	82 Pb 1.87	83 Bi 2.02	84 Po 2.0	85 At 2.2	86 Rn —
7	87 Fr 0.71	88 Ra 0.91						

increases →
increases ↑

The amount of charge at each pole is less than 1 (the charge of an electron or proton). Indeed, an electron would have to be transferred completely (as in ionic bonding) to produce an electrical charge as large as 1. Chemists use the Greek letter delta (δ) to represent the partial positive (δ^+) and partial negative (δ^-) charges at the poles:

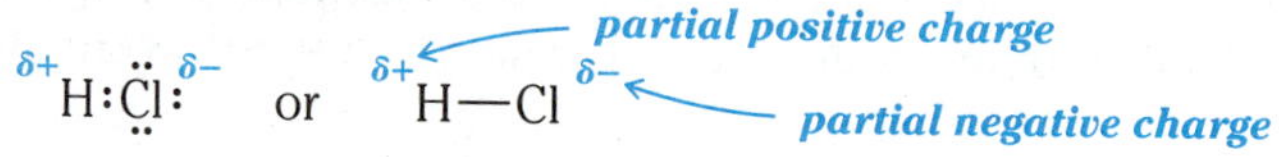

$$\overset{\delta+}{H}:\overset{\delta-}{\ddot{Cl}}: \quad \text{or} \quad \overset{\delta+}{H}—\overset{\delta-}{Cl}$$

← *partial positive charge*
← *partial negative charge*

The more electronegative element is the δ^- pole in a polar covalent bond.

Bonds are classified by the difference in electronegativity between bonding atoms: The greater the difference in electronegativity, the more ionic the bond. Figure 4.21 shows the range of bonding based on differences in electronegativity.

Polar and Nonpolar Molecules. Molecules containing only nonpolar covalent bonds have no partially charged poles. They are called **nonpolar molecules.** Molecules consisting of two atoms of the same element—such as H_2, Cl_2, N_2, O_2, and F_2—are nonpolar.

Another important class is nonpolar molecules consisting only of carbon and hydrogen. These two elements are close enough in electronegativity (a difference of only 0.34; see Figure 4.20) that bonds between C and H are essentially nonpolar. You will learn much more about these compounds beginning in Chapter 11.

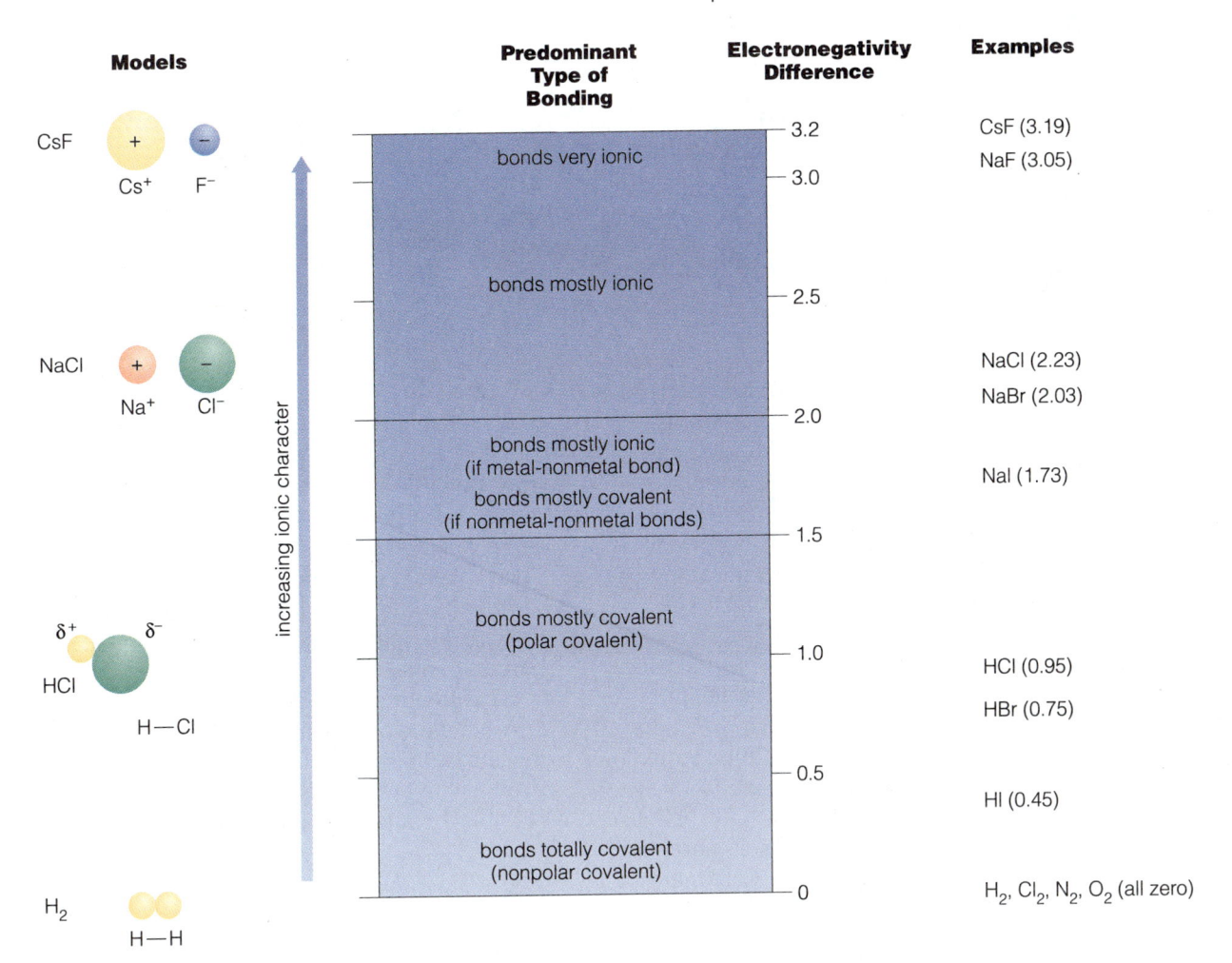

Most molecules containing polar covalent bonds are **polar molecules** or **dipoles;** that is, they have separated centers of partial positive and partial negative charge. Such molecules are said to have a *dipole moment.*

Any diatomic (two-atom) molecule with a polar covalent bond has a dipole moment and is polar. For example, the centers of δ^+ and δ^- are separated in HF (Figure 4.22). Water is also polar because the center of δ^- is at the O atom and separated from the center of δ^+, which is midway between the H atoms (Figures 4.22 and 4.23).

A few molecules have polar covalent bonds but no dipole moment. This happens when two or more polar covalent bonds are equal in polarity and extend in opposite directions in space. Then the centers of δ^+ and δ^- are not separated, and the poles essentially cancel each other.

The most important examples involve Group 14/IVA elements, especially carbon (Figure 4.24). In CF_4, for example, carbon forms four polar covalent bonds of equal strength that extend equally into space at 109° angles. The centers of δ^+ and δ^- coincide at the carbon atom, so CF_4 has no dipole moment and is nonpolar. Likewise, when carbon forms a linear molecule such

Figure 4.21 Bonding can be classified according to differences in electronegativity.

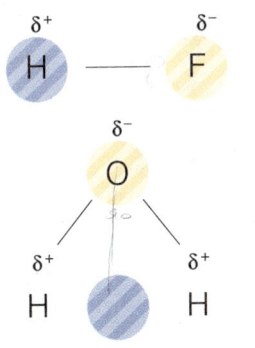

Figure 4.22 HF and H_2O are polar because their centers of partial positive (blue striped sphere) and partial negative (yellow striped sphere) charge are separated.

CHEMISTRY SPOTLIGHT

Carbon Monoxide: A Deadly, Polar Molecule

Carbon monoxide (CO) is a colorless, odorless, poisonous gas produced by burning carbon compounds—such as coal, gasoline, and cigarettes—where insufficient oxygen is present. People who direct traffic or work in enclosed areas such as tunnels or garages may be exposed to high CO levels. Indoor sources include charcoal grills or hibachis, kerosene heaters that aren't properly vented, and tobacco smoke.

Carbon monoxide is a polar molecule with the electron dot structure :C≡O:. Its linear shape (like O_2 molecules) and polarity (unlike O_2, which is nonpolar) enable it to bind to hemoglobin (Hb) in red blood cells more effectively than O_2 does. Hemoglobin bound to CO (COHb) cannot bind O_2, which red blood cells normally carry from the lungs to other cells. Cells that don't receive O_2 die.

Table 4.9 shows the effects of exposure to various levels of CO. Cigarette smoke contains 200–400 parts per million (ppm) of CO and ties up about 20% of the smoker's hemoglobin. Automobile exhaust contains 1000–7000 ppm of CO. Typical concentrations inside automobiles (for nonsmokers) range from 25–115 ppm in downtown traffic and traffic jams, from 10–75 ppm on freeways during rush hour, and 5–20 ppm in urban residential areas.

The treatment for CO poisoning is to replace CO on hemoglobin with O_2. This is done by removing victims from CO-contaminated air and treating them with high concentrations of O_2, if possible, or at least with fresh air.

Table 4.9 Effects of Carbon Monoxide

CO Levels in Air (ppm)*	Exposure	% of Hb as COHb	Symptoms
Less than 100	Indefinite	0–10	None
100–200	Indefinite	10–20	Slight headache, tightness across forehead
200–300	5–6 hours	20–30	Throbbing headache
400–600	4–5 hours	30–40	Severe headache, dizziness, nausea, weakness, impaired vision, collapse
700–1000	3–4 hours	40–50	Increased pulse and breathing, increased tendency to collapse
1000–3000	1–3 hours	50–70	Coma, intermittent convulsions, depressed heart action, possible death at 1600 ppm and higher

* 1 ppm (part per million) is 1 μL of CO per L of air

Compounds: Ions and Molecules   Chapter 4   83

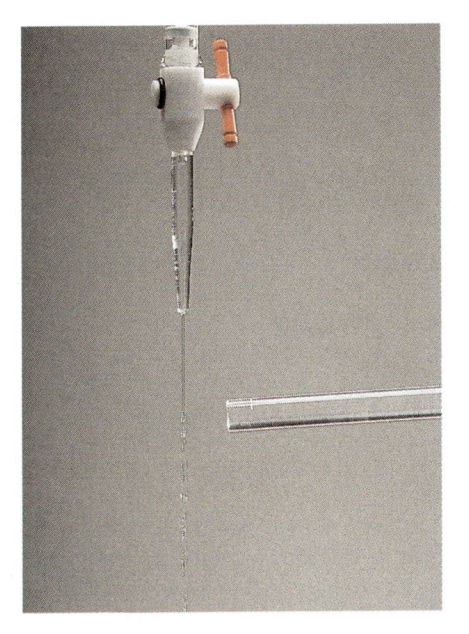

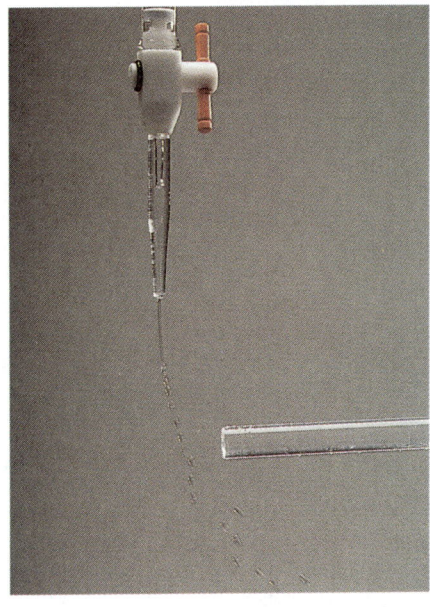

Figure 4.23 (Left) Hexane, C_6H_{14}, is nonpolar and does not respond to a charged rod. (Right) Water is polar, so its molecules have charged poles that do respond to a charged rod.

as CO_2 or CS_2, the two double bonds are of equal polarity and opposite in direction, leaving no dipole moment.

EXAMPLE 4.15
(a) Which molecules depicted in Table 4.7 are polar?
(b) Use δ^+ and δ^- to show the main regions of partial positive and partial negative charge in a molecule of ethanol (see Figure 4.14).

SOLUTION
(a) HCl, H_2O, and NH_3 are polar because they have polar covalent bonds and a dipole moment.

(b)
$$H-\underset{\underset{H}{|}}{\overset{\overset{H}{|}}{C}}-\underset{\underset{H}{|}}{\overset{\overset{H}{|}}{C}}-\overset{\delta^-}{O}-\overset{\delta^+}{H}$$

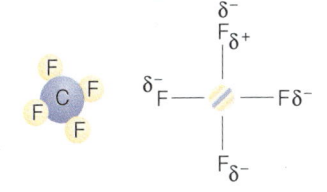

Figure 4.24 Although they have polar covalent bonds, the symmetrical shapes of CF_4 and CO_2 cause the center of the partial negative charge (small striped yellow sphere) to be in the same place as the center of partial positive charge (small striped blue sphere), making the molecules nonpolar.

What practical difference does it make whether a molecule is polar or not? All the difference in the world. If water molecules weren't polar, water would be a gas—and life as we know it couldn't exist on earth. In Chapter 6, you'll find out why water is a liquid, and how polarity helps make substances solids, liquids, or gases.

> All changes we produce consist in separating particles that are in a state of cohesion or combination, and joining those that were previously at a distance.
>
> *John Dalton (1766–1844)*

SUMMARY

Atoms tend to gain, lose, or share electrons to achieve an electron configuration like a noble gas—eight valence electrons (or two for very light elements). Such a state has a lower potential energy and is more stable.

Metals react with nonmetals to form ionic compounds, held in an ionic crystal lattice by ionic bonds. Metal atoms tend to lose electrons to become positively charged ions, whereas nonmetal atoms gain electrons and become negatively charged ions. The electrical charge of an ion depends on the number of electrons the atom gained or lost. Formulas of ionic compounds show as subscripts the simplest ratio of ions of each type to achieve electrical neutrality. Polyatomic ions are charged groups of atoms of two or more elements held together by covalent bonds. Ionic compounds are named using the suffix *-ide* for the negative ion, unless that ion is a polyatomic ion; in that case the name of the polyatomic ion is used.

In a pure metal, or in a metallic mixture of elements (called an alloy), metal atoms exist as ions, with their valence electrons free to move throughout the material.

Nonmetal atoms (except noble gases) share electrons with each other to become more stable. A shared pair of electrons is a covalent bond. A group of atoms held together by covalent bonds is a molecule. The formula for a covalent compound lists as subscripts the number of atoms of each element in a molecule. A pair of electrons shared equally is a nonpolar covalent bond. A pair of electrons shared by atoms of different electronegativities is a polar covalent bond. Electronegativity is a measure of an atom's ability to attract shared electrons. The more electronegative atom in a polar molecule is the partial negative pole.

Molecular shapes are predicted by VSEPR theory, based on the idea that valence electron pairs separate as much as possible. Molecules with polar covalent bonds have a dipole moment and are polar unless their symmetric shape causes the poles to cancel each other in space. Molecules having only nonpolar covalent bonds are nonpolar.

KEY TERMS

Alloy (4.3)
Chemical bond (4.1)
Coordinate covalent bond (4.4)
Covalent bond (4.4)
Covalent compound (4.4)
Dipole (4.6)
Electronegativity (4.6)

Formula unit (4.2)
Ion (4.2)
Ionic bond (4.2)
Ionic compound (4.2)
Metallic bonding (4.3)
Molecule (4.4)
Nonpolar covalent bond (4.6)

Nonpolar molecule (4.6)
Octet rule (4.1)
Polar covalent bond (4.6)
Polar molecule (4.6)
Polyatomic ion (4.2)
VSEPR theory (4.5)

EXERCISES

Even-numbered exercises are answered at the back of this book.

Stable Electron Structures

1. Gasoline burns (reacts with oxygen gas) to produce carbon dioxide and water. Do the reactants (gasoline and oxygen gas) or products (carbon dioxide and water) have lower potential energy? Explain.

2. In photosynthesis, plants use solar energy to produce glucose and oxygen gas. Do the reactants (carbon dioxide and water) or products (glucose and oxygen) have lower potential energy? Explain.

3. Which one of the following elements is the least reactive? **(a)** iron, **(b)** oxygen, **(c)** neon, **(d)** nitrogen, **(e)** hydrogen.

4. Atoms of which element in exercise 3 have the most stable valence electron configuration?

5. Atoms of which element in exercise 3 have as their most stable state a total of two valence electrons?

Ions

6. Which element in exercise 3 commonly exists in two different ionic forms?

7. Which noble gas has the same electron configuration as a calcium ion?

8. What is the total number of electrons in a calcium ion?
9. What is the number of valence electrons in a calcium ion?
10. What noble gas has the same electron configuration as a sulfide ion?
11. What is the total number of electrons in a sulfide ion?
12. What is the number of valence electrons in a sulfide ion?
13. List the ionic charge for elements in the following groups: (a) halogen, (b) IIIA, (c) IIA, (d) VIIIA, (e) alkali metal.
14. List the element in period 3 that typically has the following ionic charge: (a) 3−, (b) 1−, (c) 2+, (d) 1+, (e) 2−.
15. Which one in each of the following pairs of elements has the greater ionization energy? (a) N or Ne, (b) Cl or Br, (c) Na or Cl.
16. Do metals or nonmetals tend to have greater ionization energy?

Ionic Compounds

17. Write formulas for ionic compounds that form by reaction of the following pairs of elements: (a) Na and S, (b) Mg and O, (c) Al and Cl, (d) K and N, (e) Li and I.
18. Write formulas for ionic compounds that form by reaction of the following pairs of elements: (a) O and K, (b) I and Sr, (c) Al and S, (d) Mg and P, (e) Mg and S.
19. Explain what is wrong with each of the following formulas: (a) CaF, (b) Sr_2O_2, (c) Be_2Br, (d) $AlBr_2$.
20. Suppose the symbol for a group IIA element is A and the symbol for a group VIIA element is Z. Write the formula for the ionic compound that forms from these two elements.
21. Name the compound that would form from each combination of elements in exercise 17.
22. Name the compound that would form from each combination of elements in exercise 18.
23. Write formulas for the following ionic compounds: (a) sodium oxide, (b) iron (III) chloride, (c) rubidium nitride, (d) aluminum oxide, (e) potassium iodide.
24. Write formulas for the following ionic compounds: (a) sodium chloride, (b) copper (I) sulfide, (c) lead (II) selenide, (d) cesium phosphide, (e) zinc fluoride.
25. Write the name or formula of the following: (a) potassium sulfate, (b) sodium carbonate, (c) $Mg(NO_3)_2$, (d) $Al(OH)_3$, (e) calcium phosphate.
26. Write the name or formula of the following: (a) sodium cyanide, (b) strontium sulfite, (c) magnesium bicarbonate, (d) $LiNO_2$, (e) $Ba_3(PO_4)_2$, (f) $Al_2(SO_4)_3$.
27. Write formulas for each of the following: (a) calcium nitride, (b) calcium nitrite, (c) calcium nitrate.
28. Write formulas for each of the following: (a) sodium sulfide, (b) sodium sulfite, (c) sodium sulfate.

Pure Metals and Alloys

29. In terms of chemical structure, explain why pure metals and alloys conduct electricity.
30. Is an alloy classified as a pure substance? Does it follow the law of definite proportions?

Covalent Compounds

31. List the number of covalent bonds each of the following elements tends to form: (a) Cl, (b) O, (c) H, (d) C.
32. List the number of covalent bonds each of the following elements tends to form: (a) N, (b) Br, (c) Se, (d) I.
33. Using dashes to represent covalent bonds, draw structural formulas for the following molecules: (a) F_2, (b) PH_3, (c) H_2, (d) CCl_4, (e) H_2S.
34. Using dashes to represent covalent bonds, draw structural formulas for the following molecules: (a) NH_3, (b) N_2, (c) SCl_2, (d) H_2O, (e) Cl_2.
35. Which elements in the compounds in exercise 33 have one or more pairs of valence electrons not shared in a covalent bond?
36. Which elements in the compounds in exercise 34 have one or more pairs of valence electrons not shared in a covalent bond?
37. Using dashes to represent covalent bonds, draw structural formulas for (a) H_2O_2 (hydrogen peroxide), and (b) phosgene, $COCl_2$. [Hint: Phosgene contains a double bond.]
38. Using dashes to represent covalent bonds, draw structural formulas for (a) C_3H_8 (propane), and (b) $CHCl_3$ (chloroform).
39. Ethylene, C_2H_4, is the material used to make polyethylene, a common plastic. Using dashes to represent covalent bonds, draw the structural formula for ethylene. [Hint: This compound has a double bond.]
40. Formic acid, H_2CO_2, is a substance in insect stings. Using dashes to represent covalent bonds, draw the structural formula for formic acid. [Hint: This compound has a double bond.]

Electron Dot (Lewis) Structures

41. Write electron dot structures for each substance listed in exercise 33.
42. Write electron dot structures for each substance listed in exercise 34.
43. Write electron dot structures for both substances in exercise 37.
44. Write electron dot structures for both substances in exercise 38.
45. Write electron dot structures for the substances in exercises 39 and 40.
46. Write electron dot structures for the following polyatomic ions: (a) CN^-, (b) ClO_3^-, (c) NO_3^-.

Naming Covalent Compounds

47. Write the formal name for each of the following: (a) C_2H_6, (b) SO_3, (c) S_4N_4, (d) PH_3.
48. Write the formal name for each of the following: (a) PCl_5, (b) N_2O_3, (c) H_2O_2, (d) S_2F_{10}.
49. Naphthalene is a substance used in one type of mothballs. Write the name for this covalent compound based on its molecular formula, $C_{10}H_8$.

Shapes of Molecules

50. According to VSEPR theory, maximum separation between four pairs of valence electrons leaves them how many degrees apart?
51. Predict the shape of each of the molecules listed in exercise 33.
52. Predict the shape of each of the molecules listed in exercise 34.
53. CS_2, like CO_2, consists of a central carbon atom double bonded to two group VIA atoms. Write the electron dot structure for a CS_2 molecule and predict its shape.

Electronegativity and Type of Bonding

54. Predict the principal type of bonding (ionic or covalent) for each of the following: (a) Na_2S, (b) SO_2, (c) NCl_3, (d) CH_4, (e) KNO_3.
55. Predict the principal type of bonding (ionic or covalent) for each of the following: (a) $(NH_4)SO_4$, (b) HCl, (c) O_2, (d) He, (e) $CaCl_2$.
56. Which atom is the more electronegative in each of the following pairs? (a) N and O, (b) S and Cl, (c) Cl and Br, (d) P and F.

57. What is the most electronegative element? What is the next most electronegative element?

Polar and Nonpolar Molecules

58. Which of the following contain one or more nonpolar covalent bonds? (a) CCl_4, (b) NO_2, (c) N_2, (d) SCl_2, (e) He, (f) H_2.
59. Which of the molecules in exercise 58 contain polar covalent bonds?
60. Why are compounds consisting only of carbon and hydrogen classified as nonpolar?
61. Using dashes to represent covalent bonds, draw structural formulas of the following and use δ^+ and δ^- to represent partial positive and partial negative poles, respectively: (a) H_2O, (b) NI_3, (c) OF_2.
62. Using dashes to represent covalent bonds, draw structural formulas of the following and use δ^+ and δ^- to represent partial positive and partial negative poles, respectively: (a) HF, (b) SCl_2, (c) FCl, (d) NH_3.
63. For which of the following shapes is it possible for a molecule to be nonpolar despite containing polar covalent bonds? (a) tetrahedral, (b) bent, (c) pyramidal, (d) linear.
64. Which of the following are polar molecules? (a) HCl, (b) Cl_2, (c) CCl_4, (d) CO, (e) NH_3.
65. Which of the following are polar molecules? (a) CO_2, (b) O_2, (c) OCl_2, (d) H_2O, (e) H_2, (f) NCl_3.
66. Look at the molecules depicted in Table 4.8. Which are polar and which are nonpolar?
67. Does the shape of a water molecule make a difference in terms of whether it is polar or nonpolar? Explain.

DISCUSSION EXERCISES

1. Since there are many exceptions to the octet rule, what good is it?
2. From the following information, predict whether gases at room temperature are composed of molecules that are (1) relatively heavy or light, and (2) polar or nonpolar. Then predict which of the following are gases at room temperature: (a) C_2H_6, (b) $C_{10}H_{22}$, (c) SCl_2, (d) As_2Se_3, (e) H_3P, (f) CBr_4.

    gases: H_2, O_2, N_2, F_2, Cl_2, CO_2, NH_3, CH_4, HF, HCl, noble gases
    liquids: CS_2, Br_2, CCl_4, C_6H_{14}, PCl_3, H_2O, H_2O_2
    solids: I_2, AsI_3, SiI_4, P_2O_5, Ses, $C_{22}H_{46}$

3. What is the fundamental unit of structure in each of the following? **(a)** pure silver, **(b)** table salt, **(c)** neon gas, **(d)** nitrogen gas, **(e)** dry ice (carbon dioxide), **(f)** baking soda (sodium bicarbonate). How are these fundamental units alike and how are they different? What would be an appropriate formula to represent each substance?

4. Lithium ion, used to treat manic depression (see Chemistry Spotlight in Section 4.2), occurs naturally in many common ionic compounds that cannot be patented. Thus pharmaceutical companies have less financial incentive to market this drug. Should companies be allowed to patent such common mineral compounds? Why or why not?

5

Chemical Reactions

1. How do you write a balanced chemical equation?
2. What are oxidation–reduction reactions? How do you recognize them? What are some of their important uses?
3. What is Avogadro's number? What is a mole? How are the two related, and how do you use them to interconvert moles, grams, and number of particles of a substance?
4. How do you use a balanced chemical equation to calculate the moles, grams, or numbers of particles of reactants and products in a chemical reaction?
5. What types of energy changes occur in chemical reactions?

The Haber Process

Nitrogen gas (N₂) is not very reactive. The only abundant nitrogen compound that occurs naturally is sodium nitrate (NaNO₃), called saltpeter. So how do chemists synthesize huge amounts of nitrogen-containing fertilizers, drugs, dyes, and fabrics such as nylon?

One key step was discovered in 1908 by German scientist Fritz Haber (Figure 5.1). The Haber process is a practical way to make N₂ into ammonia gas (NH₃). The chemical reaction is written as follows:

$$N_2 + 3H_2 \longrightarrow 2NH_3$$

In the Haber process, N₂ gas reacts with H₂ gas at high pressure and temperature (at least 500°C) in the presence of iron and other substances that accelerate the reaction.

Ammonia is much more reactive than N₂ and can be made into urea (see Figure 4.15) or ammonium compounds such as ammonium sulfate, (NH₄)₂SO₄. It can also be converted into nitrates such as NaNO₃. All of these products can be used as fertilizers to provide plants with a usable form of nitrogen. Indeed, the first of three numbers on a fertilizer label (such as 35-10-10) refers to its percentage of nitrogen by mass. (The second and third numbers refer to its phosphorus and potassium content, respectively.) Ammonia is also made into a vast array of other nitrogen compounds (see Chapter 17).

For his work, Haber was awarded the Nobel prize in chemistry in 1918. But that award was clouded in controversy. Haber originally developed his process as a practical way to make explosives (such as NH₄NO₃) for use during World War I. He also helped Germany develop and use gas warfare. In 1933, however, upset over Nazi

Figure 5.1 Fritz Haber (1868–1934).

mistreatment of Jews, he left his native land for England.

In the end, the discovery of the Haber process, though born of dubious motives, turned out to be of great benefit to humankind.

A healthy body is a walking, talking, living bundle of chemical reactions. Your brain is removing glucose from blood and converting it into carbon dioxide and water. These reactions release energy to help your mind work and to keep you warm. Other cells are using energy to join amino acids together to produce the proteins that make up hair, skin, and muscle (Figure 5.2). In your liver and kidneys, ammonia and carbon dioxide are reacting to make urea, which you excrete in urine.

To understand how these reactions keep you alive, or how they can cause harm, you need to know more about chemical reactions in general.

Figure 5.2 The body uses amino acids to synthesize proteins that make muscles.

5.1 WRITING CHEMICAL REACTIONS

Chemical Reactions. Recall that a chemical change (Section 3.1) occurs whenever starting substances (reactants) break down or combine chemically to form different substances (products). A chemical change results from a **chemical reaction.**

Chemical changes are almost always accompanied by physical changes. You can usually conclude that a chemical reaction has taken place when you observe any of the following physical changes (Table 5.1):

- Bubbles of gas form without heating.
- Odors and/or colors change.
- The solution turns cloudy, or a solid insoluble material called a **precipitate** forms (Figure 5.3).

Table 5.1 Examples of Physical Changes That Accompany Chemical Reactions

Chemical Reaction	Physical Changes
Baking soda ($NaHCO_3$) reacts in vinegar ($HC_2H_3O_2$)	Bubbles of gas (CO_2) appear in the liquid
Sunlight causes a chemical reaction in skin	Skin "tans"
Sulfur (S) reacts with oxygen (O_2) in air	Yellow solid sulfur changes into a sharp-smelling gas (SO_2)
Water solutions of sodium chloride (NaCl) and silver nitrate ($AgNO_3$) react (Figure 5.3)	Two clear, colorless solutions form a solid white precipitate (AgCl) that does not dissolve
Gasoline (contains C_8H_{18}) reacts with O_2 (Figure 5.4)	Heat and light are emitted as liquid gasoline changes into gases (CO, CO_2, and H_2O vapor)

- The material burns (Figure 5.4).
- Heat, light, or both are given off (Figure 5.4).

Representing Chemical Reactions. Chemists represent reactions by writing formulas of the reactants and products, separated by an arrow:

$$\text{Reactant(s)} \longrightarrow \text{Product(s)}$$

The arrow means "reacts to form" or "yields." For the reaction of iron (Fe) with oxygen gas (O_2) to form rust, iron (III) oxide (Fe_2O_3; Figure 5.5), you would write:

$$\underset{\text{iron}}{Fe} + \underset{\text{oxygen}}{O_2} \longrightarrow \underset{\text{iron (III) oxide}}{Fe_2O_3}$$

The first step is to *write correct formulas for the reactants and products*. You learned in Chapter 4 how to write formulas for ionic and covalent compounds. If a reactant or product is an element (such as iron in the previous example), simply write the symbol for the element (Fe)—unless the element occurs as molecules. The main elements of this type to remember are H_2, N_2, O_2, and the halogens (F_2, Cl_2, Br_2, and I_2). All of these elements exist as molecules of two like atoms joined by one or more covalent bonds.

EXAMPLE 5.1 Use correct formulas to represent the following reactions:
  (a) sodium + fluorine $\longrightarrow$ sodium fluoride
  (b) hydrogen + oxygen $\longrightarrow$ water
  (c) phosphorus + chlorine $\longrightarrow$ phosphorus trichloride

SOLUTION (a) $Na + F_2 \longrightarrow NaF$; (b) $H_2 + O_2 \longrightarrow H_2O$;
(c) $P + Cl_2 \longrightarrow PCl_3$

Writing Balanced Equations. You may have noticed that the reaction of iron and oxygen, and those in Example 5.1, aren't balanced as written—the reactants and products don't have equal numbers of atoms of each element. In the reaction for iron (III) oxide, for example, the reactants have one atom of iron (Fe) and two atoms of oxygen (O_2), whereas the product (Fe_2O_3) contains two atoms of iron and three of oxygen.

Figure 5.3 Solutions of $AgNO_3$ and NaCl react to form insoluble AgCl, a white precipitate.

Figure 5.4 Oil fields in Kuwait burn, emitting heat and light, as oil reacts with oxygen (O_2).

Figure 5.5 Iron (Fe) reacts with oxygen (O_2) in air to form iron (III) oxide, Fe_2O_3, or rust.

Something more is needed. According to the law of conservation of mass (Section 3.2), products have the same mass as reactants. In a chemical reaction, atoms cannot be created or destroyed. The reaction must be balanced, with reactants and products having the same number of atoms of each element.

To write a balanced chemical reaction (called a **chemical equation**), you need to determine how many particles of each reactant and product participate in the reaction. You do this by placing an appropriate number (called a *coefficient*) in front of the formula for each substance. *Don't change formulas (subscripts) to balance an equation;* an altered formula would no longer represent the actual substance in the reaction.

For the reaction of iron with oxygen, the process goes like this:

$$\_Fe + \_O_2 \longrightarrow \_Fe_2O_3$$

The blanks represent spaces for writing coefficients. A coefficient of 1 is not written.

The first step is to count how many atoms of each element are on each side of the arrow. One way to show the count is:

$$\_Fe + \_O_2 \longrightarrow \_Fe_2O_3$$

| Fe | 1 | 2 |
| O | 2 | 3 |

Next, identify which elements can be balanced by placing a coefficient in front of one of the formulas. In this case, placing a *2* in front of Fe will balance the irons. So do this, and count the atoms on each side again:

$$2Fe + \_O_2 \longrightarrow \_Fe_2O_3$$

| Fe | 2 | 2 |
| O | 2 | 3 |

Oxygen still isn't balanced. The simplest way to balance it is to have six atoms of O on each side of the equation. To do this, place a coefficient of 3 in front of O_2 and a coefficient of 2 in front of the oxygen-containing product (Fe_2O_3). Then count again:

$$2Fe + 3O_2 \longrightarrow 2Fe_2O_3$$

| Fe | 2 | 4 |
| O | 6 | 6 |

After you balance an equation, check the coefficients to see if they all can be divided by a simple number such as 2 or 3. If they can, divide all the coefficients by that number. The equation will stay balanced when all coefficients are divided by the same number.

Notice that *a coefficient in front of a formula multiplies every element in the formula by that amount*. Placing a 2 in front of Fe_2O_3 doubles both the number of Fe and O atoms. The new coefficients have balanced oxygen, but unbalanced iron. The final step is to change the coefficient in front of Fe to *4*:

$$4Fe + 3O_2 \longrightarrow 2Fe_2O_3$$

| Fe | 4 | 4 |
| O | 6 | 6 |

If you stated this balanced equation in words, you would say that four atoms of iron react with three molecules of oxygen to yield two formula units of iron (III) oxide. Figure 5.6 illustrates what the equation represents.

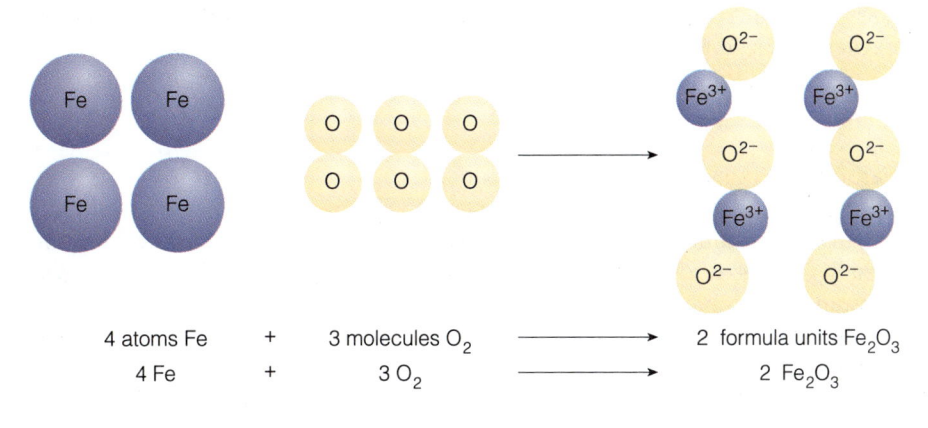

4 atoms Fe	+	3 molecules O₂	⟶	2 formula units Fe₂O₃
4 Fe	+	3 O₂	⟶	2 Fe₂O₃

Figure 5.6 Representing a balanced chemical equation. (Notice also that Fe and O atoms change size as they lose and gain valence electrons, respectively.)

EXAMPLE 5.2 Write balanced equations for the three chemical reactions in Example 5.1.

SOLUTION (a) $2Na + F_2 \longrightarrow 2NaF$; (b) $2H_2 + O_2 \longrightarrow 2H_2O$; (c) $2P + 3Cl_2 \longrightarrow 2PCl_3$

EXAMPLE 5.3 Use spheres, as in Figure 5.6, to represent the reactants and products in the reaction: $2H_2 + O_2 \rightarrow 2H_2O$.

5.2 OXIDATION–REDUCTION REACTIONS

Combustion. When you burn fuels (Figure 5.4), you are using combustion reactions. **Combustion** is the burning of an element or compound with oxygen (O_2) to produce oxygen-containing products such as CO_2, H_2O, and SO_2. One example is burning home heating oil, which contains $C_{15}H_{32}$:

$$C_{15}H_{32} + 23O_2 \longrightarrow 15CO_2 + 16H_2O$$

When your body generates energy by metabolizing carbohydrates or fats to CO_2 and H_2O, the reactions are like combustion reactions except that they don't generate intense light and heat.

Oxidation and Reduction. Combustion reactions belong to a general category of reactions, called *oxidation–reduction reactions*. Oxygen is the most abundant element on earth (Table 5.2) and in the human body (see Table 3.5). It combines with nearly every other element. For these reasons, early chemists

Table 5.2 Elements in the Earth's Crust

Element	Approx. Composition (% by Mass)
Oxygen	49.5
Silicon	25.8
Aluminum	7.5
Iron	4.7
Calcium	3.4
Sodium	2.6
Potassium	2.4
Magnesium	1.9
Hydrogen	0.9
Titanium	0.6
All others	0.7

classified reactions around oxygen. Any element or compound that gained oxygen was said to be *oxidized*.

The equation for the combustion of coal, which is mostly carbon, is:

$$C + O_2 \longrightarrow CO_2$$

Carbon gains oxygen atoms and is oxidized.

Because oxygen is very electronegative (Figure 4.20) and strongly attracts electrons, substances that react with oxygen lose electrons partially or completely. This is a common definition of **oxidation:** the loss of electrons by a substance in a chemical reaction.

In the preceding reaction, each carbon atom ($\cdot \overset{\cdot}{C} \cdot$) in coal has four valence electrons. Once it reacts with oxygen, a carbon atom in CO_2 shares a total of eight valence electrons:

$$\overset{..}{\underset{..}{O}} :: C :: \overset{..}{\underset{..}{O}}$$

Because oxygen is so much more electronegative than carbon, a carbon atom in CO_2 has less than a 50% share of the eight shared electrons. Since carbon in CO_2 has an average of *less* than four valence electrons in the vicinity of its nucleus, the C in coal loses electrons in the reaction with O_2 and is oxidized.

Oxidation is a loss of electrons, but it often doesn't involve oxygen. Zinc (Zn), for example, is oxidized when an atom loses electrons to form zinc ion (Zn^{2+}):

$$Zn \longrightarrow Zn^{2+} + 2e^-$$

This is only half of a reaction, however, and it cannot take place by itself. Zinc atoms must give their electrons to a substance that accepts electrons. Copper (II) ions (Cu^{2+}), for example, can accept electrons from zinc and change into copper atoms:

$$Cu^{2+} + 2e^- \longrightarrow Cu$$

Or, two chlorine atoms in a chlorine molecule (Cl_2) can accept electrons from zinc to become chloride ions:

$$Cl_2 + 2e^- \longrightarrow 2Cl^-$$

The process by which an element gains electrons is called **reduction.** Whenever an element gains electrons, and thus has a more negative charge, it is said to be *reduced* (Figure 5.7).

In an **oxidation–reduction reaction,** at least one substance is oxidized and at least one is reduced. Table 5.3 summarizes three signs you can use to recognize an oxidation–reduction reaction.

Oxidation and reduction often occur in reactions involving hydrogen. Since hydrogen is less electronegative than other reactive nonmetals (Figure 4.18), nonmetals that react with hydrogen (H_2) partially gain the valence electron in each hydrogen atom and are reduced. Hydrogen, then, partially loses an electron and is oxidized. One example is the synthesis of ammonia (NH_3) by the Haber process:

$$N_2 + 3H_2 \longrightarrow 2NH_3$$

reduced oxidized

The reduced substance takes electrons from (oxidizes) another substance and thus is called the **oxidizing agent.** The oxidized substance is called the **reducing agent.** In the preceding reaction, N_2 is the oxidizing agent and H_2 is

This general definition of oxidation also applies to reactions that don't involve oxygen.

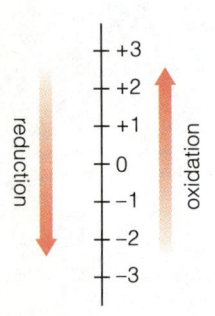

Figure 5.7 Reduction (gaining electrons) produces a more negative charge. Oxidation (losing electrons) produces a more positive charge.

Table 5.3 Three Signs of Oxidation and Reduction

Oxidation	Reduction
lose e^-	gain e^-
gain O	lose O
lose H	gain H

the reducing agent. Since N is more electronegative than H (see Figure 4.18), N attracts the shared electrons in NH_3 more strongly than H does. Thus an H atom in H_2 partially loses its valence electron (and is oxidized) when it shares that electron with N in NH_3.

EXAMPLE 5.4 In the following reactions, identify which reactant is oxidized, which is reduced, which is the oxidizing agent, and which is the reducing agent:
(a) $2H_2 + O_2 \longrightarrow 2H_2O$
(b) $2Br^- + Cl_2 \longrightarrow Br_2 + 2Cl^-$
(c) $C_{57}H_{98}O_6 + 5H_2 \longrightarrow C_{57}H_{108}O_6$

SOLUTION (a) H_2 oxidized (reducing agent), O_2 reduced (oxidizing agent); (b) Br^- oxidized (reducing agent), Cl_2 reduced (oxidizing agent); (c) H_2 oxidized (reducing agent), $C_{57}H_{98}O_6$ reduced (oxidizing agent)

5.3 SOME PRACTICAL OXIDATION–REDUCTION REACTIONS

Bleaches. Household liquid bleach (Figure 5.8) is a small amount of sodium hypochlorite (NaOCl) dissolved in water. Bleach makes clothes whiter because its hypochlorite ions (OCl^-) oxidize certain colored molecules to colorless ones:

colored molecule + OCl^- ⟶ Cl^- + colorless molecule
    oxidized   reduced
(reducing agent) (oxidizing agent)

Figure 5.8 Bleaches use sodium hypochlorite (NaOCl) as an oxidizing agent to decolorize stains.

In this reaction, OCl^- loses O to become Cl^-. Loss of O is a sign that OCl^- is reduced (Table 5.3).

Some people use solutions of hydrogen peroxide (H_2O_2) to bleach their hair. Hydrogen peroxide oxidizes hair pigments to colorless products, leaving hair blond.

Some stain removers use oxidizing or reducing agents in a similar way. Often they don't actually remove the stain—they just make the stain colorless, and therefore less noticeable. Table 5.4 lists some common stain removers.

Table 5.4 Oxidizing and Reducing Agents Used to Remove Stains

Name	Formula	Type of Stains Removed
Oxidizing Agents		
Sodium hypochlorite	NaOCl	most stains on cotton and linen (don't use on wool or silk)
Hydrogen peroxide	H_2O_2	blood stains on cotton and linen
Potassium permanganate	$KMnO_4$	most stains on white fabrics except rayon (then remove $KMnO_4$ with reducing agent such as oxalic acid; see below)
Reducing Agents		
Oxalic acid	$H_2C_2O_4$	rust and $KMnO_4$ stains
Sodium thiosulfate	NaS_2O_3	iodine and silver stains

Oxidizing agents kill bacteria by attacking critical substances that give up electrons. Polyunsaturated fats in cell membranes (see Sections 20.1 and 20.5) are one target.

Disinfectants. Some oxidizing agents are also used as disinfectants. Sodium hypochlorite and bleaching powder (calcium hypochlorite, $Ca(OCl)_2$) both kill bacteria. Adding hypochlorite ions to a swimming pool or chlorinating wastewater (with Cl_2) disinfects the water.

Milder oxidizing agents are used as disinfectants on skin. Two familiar examples are tincture of iodine (a solution of I_2 dissolved in ethyl alcohol) and a weak solution of hydrogen peroxide dissolved in water.

Batteries. Recall that Zn gives up electrons to Cu^{2+} in an oxidation–reduction reaction:

$$Zn + Cu^{2+} \longrightarrow Zn^{2+} + Cu$$

In an electrochemical cell, those electrons travel through a wire to provide electrical current. One or more of a series of electrochemical cells is a *battery*.

Figure 5.9 shows a simple battery. To separate the oxidation and reduction reactions, Zn and Cu^{2+} are placed in different compartments. The compartments are connected by a wire between two metals (the *electrodes*) and separated by a porous barrier that prevents mixing but allows negative ions to flow from one compartment to the other. As Zn^{2+} forms on one side and Cu^{2+} changes into Cu on the other, the flow of ions balances the charges. When either the Zn or Cu^{2+} is depleted, the battery no longer generates a current.

Automobile, flashlight, and other batteries—including those that power hearing aids, electronic calculators, and watches—operate on similar oxidation–reduction reactions.

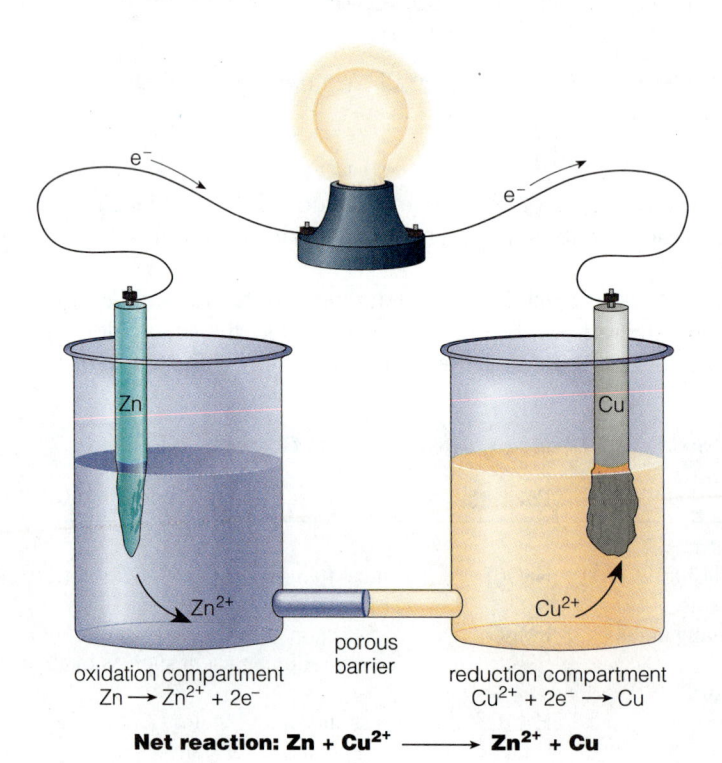

Figure 5.9 A simple battery. The current results from electrons flowing from Zn atoms to Cu^{2+} ions.

oxidation compartment
$Zn \longrightarrow Zn^{2+} + 2e^-$

porous barrier

reduction compartment
$Cu^{2+} + 2e^- \longrightarrow Cu$

Net reaction: $Zn + Cu^{2+} \longrightarrow Zn^{2+} + Cu$

CHEMISTRY SPOTLIGHT

Exercise and Lactic Acid

Figure 5.10 Intense physical activity produces lactic acid in muscles.

Oxygen (O_2) is a strong oxidizing agent that gains electrons in many reactions in the body. In order to complete these oxidation–reduction reactions, however, the body also needs reducing agents.

One good source of electrons is hydride ion [H:]$^-$, a hydrogen atom that has gained one electron and has a negative charge. In the body, hydride ion is often transferred directly between various compounds (AH_2) and a complex substance called NAD$^+$ (see Section 22.3):

$$AH_2 + NAD^+ \rightleftharpoons NADH + H^+ + A$$

NAD$^+$ gains two electrons from hydride ion (a hydrogen nucleus with two valence electrons) and is reduced to NADH. NADH, in turn, can give up hydride ion and be oxidized back to NAD$^+$ in the reverse reaction. The body often uses this system of hydride transfer in oxidation–reduction reactions.

An important example is what happens during strenuous exercise, such as sprinting, wrestling, or swimming (Figure 5.10). At such times, muscles don't receive enough oxygen and go into "oxygen debt."

The body converts carbohydrates into pyruvic acid, which usually is oxidized to CO_2 and H_2O by reaction with O_2. But during oxygen debt, pyruvic acid is reduced to lactic acid by hydride transfer from NADH [plus a proton (H^+)] (see equation). Lactic acid forms in muscle and enters the blood. A rise in blood lactic acid indicates that an athlete is exercising too intensely for his or her oxygen supply and will soon have to slow down.

$$\begin{array}{c} H \\ | \\ H-C-H \\ | \\ C=O \\ | \\ C-O-H \\ \| \\ O \end{array} + NADH + H^+ \rightleftharpoons$$

pyruvic acid

$$\begin{array}{c} H \\ | \\ H-C-H \\ | \\ H-C-O-H \\ | \\ C-O-H \\ \| \\ O \end{array} + NAD^+$$

lactic acid

When exercise slows or stops, oxygen levels return to normal. Various tissues then take up lactic acid from blood and oxidize it back to pyruvic acid using the preceding reaction in the reverse direction. Tissues can oxidize pyruvic acid in the usual way by reaction with O_2, thus repaying the "oxygen debt." Temporary muscle soreness is all that remains of the brief encounter with lactic acid.

EXAMPLE 5.5 Longer-life alkaline batteries typically generate electricity by the reaction:

$$2Zn + 3MnO_2 + 2H_2O \longrightarrow 2Zn(OH)_2 + Mn_3O_4$$

Identify the substance oxidized, the substance reduced, the oxidizing agent, and the reducing agent.

SOLUTION Zn is oxidized and is the reducing agent. As it loses electrons, its charge increases from 0 to 2+ (Zn^{2+} ion in $Zn(OH)_2$). MnO_2 is reduced and is the oxidizing agent. Its electrical charge becomes less positive, forming a product (Mn_3O_4) with less O per Mn atom than MnO_2.

Staying Alive. Your body has to generate enough energy for you to grow new cells, think, dance, and do everything else that makes you alive. That energy comes from oxidizing food. When you metabolize food, such as the carbohydrate glucose ($C_6H_{12}O_6$), you produce energy:

$$C_6H_{12}O_6 + 6O_2 \longrightarrow 6CO_2 + 6H_2O + energy$$

oxidized   reduced

Look at this equation carefully. Notice that the carbon in glucose combines with oxygen to become CO_2; this is a sign that glucose is oxidized (see Table 5.3). Oxygen gas gains hydrogen to become H_2O, a sign that O_2 is reduced.

Oxygen is the major oxidizing agent in the body. The oxygen you breathe is used to oxidize carbohydrates, fats, and proteins—generating the energy you need to stay alive. Your body is a walking, talking factory of oxidation–reduction reactions. The details of these processes are discussed in Part III: Biochemistry.

5.4 MOLES

Avogadro's Number. A balanced equation shows how many atoms, molecules, and formula units of reactants and products participate in a chemical reaction. But a person working in a laboratory or a factory can't weigh out atoms and molecules—they're too small. The periodic table lists the average atomic weights of elements in atomic mass units (amu), and 1 amu is only about 1.66×10^{-24} g.

Atomic mass, however, is also a *relative* mass. Oxygen atoms, for example, have an average mass of 16.00 amu, about 16 times greater than that of hydrogen atoms, (atomic mass of 1.01 amu). Any number of oxygen atoms, then, would have a mass about 16 times greater than that same number of hydrogen atoms. Table 5.5 illustrates how the proportion stays constant as the number of atoms increases.

How many oxygen atoms would it take to have a mass of 16.00 g? The same as the number of hydrogen atoms it would take to have a mass of 1.01 g. In fact, that number of any kind of atom would have a mass in g equal to the atomic mass of that atom.

That number turns out to be 6.02×10^{23} (Table 5.5). Named in honor of the

---

In these calculations, we take atomic masses to two places past the decimal point.

Avogadro was the first to propose that equal volumes of gases under the same conditions contain equal numbers of particles (molecules or, in the case of noble gases, atoms). This explained, in terms of the number of reacting particles, the volume changes when gases react.

Chemical Reactions Chapter 5 99

Table 5.5 Comparing the Masses of Equal Numbers of Hydrogen and Oxygen Atoms

Number of O Atoms	Mass of O Atoms (g)	Number of H Atoms	Mass of H Atoms (g)	Ratio of Masses (O/H)
1	2.66×10^{-23}	1	1.67×10^{-24}	16
2 dozen	6.37×10^{-22}	2 dozen	4.01×10^{-23}	16
100	2.66×10^{-21}	100	1.67×10^{-22}	16
5.00×10^{15}	1.33×10^{-7}	5.00×10^{15}	8.35×10^{-9}	16
6.02×10^{23}	16.00	6.02×10^{23}	1.01	16

Italian physicist Amedeo Avogadro (Figure 5.11), it is known as **Avogadro's number.** Thus, 6.02×10^{23} atoms of oxygen have a mass of 16.00 g, and 6.02×10^{23} atoms of hydrogen have a mass of 1.01 g.

EXAMPLE 5.6 What would be the mass of Avogadro's number of atoms of (a) iron, (b) mercury, and (c) arsenic?

SOLUTION (a) 55.85 g, (b) 200.59 g, (c) 74.92 g

You can calculate the mass of Avogadro's number of molecules (for covalent compounds) or formula units (for ionic compounds) in a similar way. Simply add the atomic masses of each atom in the formula. The sum is called the *formula mass,* or **formula weight.** The mass of Avogadro's number of molecules or formula units is the formula weight in grams.

This calculation for carbon dioxide (CO_2), for example, is:

1 C in CO_2, so $1 \times 12.01 = 12.01$ amu

2 O in CO_2, so $2 \times 16.00 = 32.00$ amu

formula weight of CO_2 = 44.01 amu

mass of Avogadro's number of CO_2 molecules = 44.01 g

When the formula contains a polyatomic ion in parentheses, be especially careful when counting the number of atoms of each element. Figure 5.12 shows how to do this for $Ca(NO_3)_2$.

EXAMPLE 5.7 Count the number of atoms of each element in the following formulas: (a) $(NH_4)_2CO_3$, (b) $Mg_3(PO_4)_2$, (c) $Al(OH)_3$

SOLUTION (a) 2 atoms of N, 8 H, 1 C, 3 O; (b) 3 Mg, 2 P, 8 O; (c) 1 Al, 3 O, 3 H.

EXAMPLE 5.8 Calculate the formula weight, and the mass of Avogadro's number of particles, for each of the following: (a) CCl_4, (b) F_2, (c) Na_3PO_4, (d) $Al_2(SO_4)_3$

Figure 5.11 Amedeo Avogadro (1776–1856).

Formula: $Ca(NO_3)_2$
a. Picture version

$1Ca + 2N + 6O$

b. Calculation version
$1 \times Ca_1 + 2(NO_3) =$
$1Ca + 2(N_1) + 2(O_3) =$
$1Ca + 2N + 6O$

Figure 5.12 Counting the number of atoms (or ions) in a formula.

Avogadro's number is huge. For example, you would have to remove the top 2 m (6 ft) of surface from the entire Sahara desert to get Avogadro's number of sand particles.

> **SOLUTION** (a) 12.01 + 4(35.45) = 153.81 amu; (b) 2(19.00) = 38.00 amu; (c) 3(22.99) + 30.97 + 4(16.00) = 163.94 amu; (d) 2(26.98) + 3(32.07) + 12(16.00) = 342.17 amu. Avogadro's number of particles have a mass of (a) 153.81 g, (b) 38.00 g, (c) 163.94 g, (d) 342.17 g.

Moles. The amount of material containing Avogadro's number of particles (atoms, molecules, or formula units, depending on the substance) is called a **mole.** Chemists use the mole unit to enlarge the scale of the tiny world of atoms to more practical amounts that can be measured using ordinary laboratory equipment (Figure 5.13).

The mass of 1 mole of a substance is the **molar mass.** Thus, the masses listed in Example 5.8 are the molar masses for CCl_4, F_2, Na_3PO_4, and $Al_2(SO_4)_3$, respectively.

Interconverting Moles, Grams, and Number of Particles. Once you can calculate the molar mass of a substance, you can readily interconvert between moles, grams, and number of particles using the factor-label method you learned in Chapter 2. The conversion factors for molar mass are:

$$\frac{1 \text{ mol}}{\text{molar mass (g)}} \quad \text{or} \quad \frac{\text{molar mass (g)}}{1 \text{ mol}}$$

For example:

$$\frac{22.99 \text{ g Na}}{1 \text{ mol Na}} \quad \text{or} \quad \frac{1 \text{ mol Na}}{22.99 \text{ g Na}} \quad \text{and} \quad \frac{28.01 \text{ g N}_2}{1 \text{ mol N}_2} \quad \text{or} \quad \frac{1 \text{ mol N}_2}{28.01 \text{ g N}_2}$$

The conversion factors using Avogadro's number are:

$$\frac{6.02 \times 10^{23} \text{ particles}}{1 \text{ mol}} \quad \text{or} \quad \frac{1 \text{ mol}}{6.02 \times 10^{23} \text{ particles}}$$

In their never-ending quest to save time, scientists use mol *as the abbreviation for* mole.

Figure 5.13 One mole each of bromine, sulfur, carbon, aluminum, copper, mercury, and cobalt (clockwise from top).

where the particles may be atoms, molecules, or formula units, depending on the substance.

For example, to calculate the number of moles in 23.9 g of NaCl, use the conversion factor relating molar mass to moles:

molar mass of NaCl = 22.99 g + 35.45 g = 58.44 g

$$23.9 \text{ g NaCl} \times \frac{1 \text{ mol NaCl}}{58.44 \text{ g NaCl}} = 0.409 \text{ mol NaCl}$$

— molar mass conversion factor

To calculate the g of NaCl in 3.14 mol NaCl, use the inverse of the same conversion factor:

$$3.14 \text{ mol NaCl} \times \frac{58.44 \text{ g NaCl}}{1 \text{ mol NaCl}} = 184 \text{ g NaCl}$$

To calculate the formula units in 0.409 mol NaCl, use the appropriate conversion factor containing Avogadro's number:

— Avogadro's number conversion factor

$$0.409 \text{ mol NaCl} \times \frac{6.02 \times 10^{23} \text{ formula units NaCl}}{1 \text{ mol NaCl}}$$
$$= 2.46 \times 10^{23} \text{ formula units NaCl}$$

You can also calculate, for example, the moles and mass of 5.37×10^{20} molecules of the fuel, propane (C_3H_8). The number of moles is:

Avogadro's number conversion factor

$$5.37 \times 10^{20} \text{ molecules } C_3H_8 \times \frac{1 \text{ mol } C_3H_8}{6.02 \times 10^{23} \text{ molecules } C_3H_8}$$
$$= 8.92 \times 10^{-4} \text{ mol } C_3H_8$$

Then use molar mass to calculate the mass of this many moles of propane:

molar mass of C_3H_8 = (3 × 12.01) + (8 × 1.01) = 44.11 g

— molar mass conversion factor

$$8.92 \times 10^{-4} \text{ mol } C_3H_8 \times \frac{44.11 \text{ g } C_3H_8}{1 \text{ mol } C_3H_8} = 3.93 \times 10^{-2} \text{ g } C_3H_8$$

EXAMPLE 5.9 Calculate the number of moles for each of the following:
(a) 961 g NaCl and (b) 88.7 g C_3H_8.

SOLUTION (a) $961 \text{ g NaCl} \times \dfrac{1 \text{ mol NaCl}}{58.44 \text{ g NaCl}} = 16.4 \text{ mol NaCl}$

(b) $88.7 \text{ g } C_3H_8 \times \dfrac{1 \text{ mol } C_3H_8}{44.11 \text{ g } C_3H_8} = 2.01 \text{ mol } C_3H_8$

> **EXAMPLE 5.10** Calculate the number of molecules in each of the following: (a) 8.048×10^{-6} mol C_3H_8, (b) 1 qt (946 g) of H_2O
>
> **SOLUTION** (a) 8.048×10^{-6} mol $C_3H_8 \times$
>
> $$\frac{6.02 \times 10^{23} \text{ molecules } C_3H_8}{1 \text{ mol } C_3H_8} = 4.84 \times 10^{18} \text{ molecules } C_3H_8$$
>
> (b) molar mass of $H_2O = 2(1.01) + 16.00 = 18.02$ g
>
> $$946 \text{ g } H_2O \times \frac{1 \text{ mol } H_2O}{18.02 \text{ g } H_2O} \times \frac{6.02 \times 10^{23} \text{ molecules } H_2O}{1 \text{ mol } H_2O}$$
> $$= 3.16 \times 10^{25} \text{ molecules } H_2O$$

> **EXAMPLE 5.11** Calculate the mass in g of each of the following: (a) 44 mol C_3H_8, (b) 2.548×10^{-8} mol NaCl
>
> **SOLUTION** (a) $44 \text{ mol } C_3H_8 \times \dfrac{44.11 \text{ g } C_3H_8}{1 \text{ mol } C_3H_8} = 1.9 \times 10^3 \text{ g } C_3H_8$
>
> (b) $2.548 \times 10^{-8} \text{ mol NaCl} \times \dfrac{58.44 \text{ g NaCl}}{1 \text{ mol NaCl}} = 1.489 \times 10^{-6} \text{ g NaCl}$

5.5 STOICHIOMETRY

If you owned a chemical factory, you would want to know how much reactant you would need to produce each kg of product. The chemical factory in your body has similar restrictions. For example, each g of fat you eat produces a specific amount of CO_2 and H_2O when you metabolize it. **Stoichiometry** is the study of the amounts of products and reactants in chemical reactions.

Balanced Equations and Moles. Now we will put together what you have learned in Sections 5.1 and 5.4. The coefficients in a balanced equation are the numbers of atoms, molecules, or formula units of each participant in the reaction. We can increase the scale of a reaction to use more material, but we must keep the same proportions. For the following balanced equation, a few examples are:

2Na	+	Cl_2	$\longrightarrow$	2NaCl
2 atoms		1 molecule		2 formula units
2 dozen atoms		1 dozen molecules		2 dozen formula units
2×10^6 atoms		1×10^6 molecules		2×10^6 formula units
$2 \times (6.02 \times 10^{23})$ atoms		$1 \times (6.02 \times 10^{23})$ molecules		$2 \times (6.02 \times 10^{23})$ formula units
2 moles		1 mole		2 moles

Notice especially the bottom line. If we scale the reaction by Avogadro's number (6.02×10^{23}), the proportion of products and reactants is expressed in moles. In other words, *coefficients in a balanced equation specify the proportion*

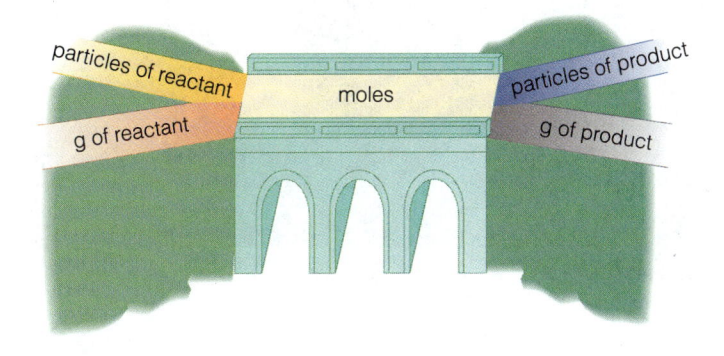

Figure 5.14 Moles are readily interconverted with g or number of particles for all reactants and products in a chemical reaction.

in moles of each substance in the reaction. This proportion is called the **mole ratio.** The mole ratios in the preceding reaction, for example, are:

$$\frac{2 \text{ mol Na}}{1 \text{ mol Cl}_2}, \frac{1 \text{ mol Cl}_2}{2 \text{ mol Na}}, \frac{2 \text{ mol Na}}{2 \text{ mol NaCl}}, \frac{2 \text{ mol NaCl}}{2 \text{ mol Na}}, \frac{1 \text{ mol Cl}_2}{2 \text{ mol NaCl}}, \text{ and}$$

$$\frac{2 \text{ mol NaCl}}{1 \text{ mol Cl}_2}$$

The mole unit turns out to be a bridge which you can cross to other units (Figure 5.14). Recall that for any substance, you can interconvert moles with g (using the molar mass conversion factor) or with number of particles (using Avogadro's number conversion factor) (Section 5.4). Now you can use the mole ratio in a balanced equation to calculate the relative amounts (in g or in number of particles) of any reactants and products in a reaction.

Mole–Mole Calculations. The coefficients in a balanced equation provide conversion factors relating the number of moles of any two substances. For example, when natural gas (CH_4) burns as fuel, the reaction is:

$$CH_4 + O_2 \longrightarrow CO_2 + H_2O$$

Let's calculate how many moles of H_2O are produced when 1.25 mol of CH_4 burns. The first step is to write a balanced equation:

$$CH_4 + 2O_2 \longrightarrow CO_2 + 2H_2O$$

The coefficients in this equation provide the mole ratios. The mole ratios relating CH_4 and H_2O are:

$$\frac{1 \text{ mol CH}_4}{2 \text{ mol H}_2\text{O}} \quad \text{or} \quad \frac{2 \text{ mol H}_2\text{O}}{1 \text{ mol CH}_4}$$

Then, using the factor-label method, begin with the amount given in the question (1.25 mol CH_4) and use the mole ratio term that cancels the starting unit (mol CH_4) and leaves the desired unit (mol H_2O):

$$1.25 \text{ mol CH}_4 \times \underbrace{\frac{2 \text{ mol H}_2\text{O}}{1 \text{ mol CH}_4}}_{\text{mole ratio}} = 2.50 \text{ mol H}_2\text{O}$$

CHEMISTRY SPOTLIGHT

Metabolism and Respiratory Quotient

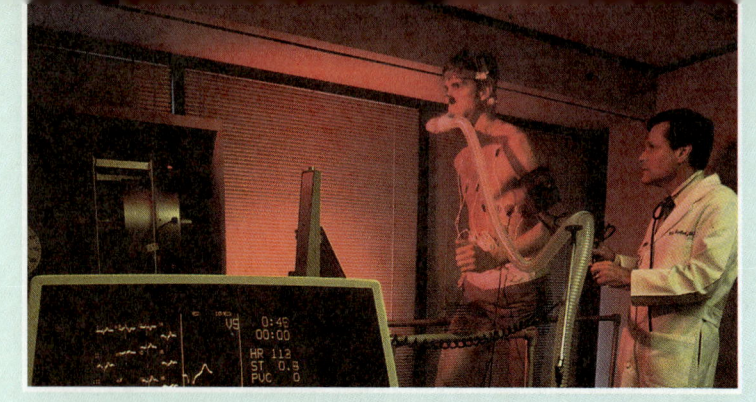

Figure 5.15 Respiratory quotient (RQ) and other values are measured during a treadmill test.

Your body uses carbohydrates, fats, and proteins for energy. After eating a high-calorie meal, you use mostly carbohydrates for energy. After several hours without eating, you use more stored energy—mainly fat. After days of starvation, the body starts breaking down muscle and other tissues, using protein for energy.

How can you tell what fuel your body is using at any one time? One way is to determine your *respiratory quotient (RQ)*, the ratio of CO_2 exhaled to O_2 consumed. You can see the difference in this ratio from these balanced equations:

for carbohydrates, $C_6H_{12}O_6 + 6O_2 \longrightarrow 6CO_2 + 6H_2O$

for a typical fat, $C_{55}H_{100}O_6 + 77O_2 \longrightarrow 55CO_2 + 50H_2O$

From the coefficients you can calculate that the RQ (CO_2/O_2) for carbohydrates is 1.00 (6/6). For a typical fat, the RQ is 0.71 (55/77). The RQ for protein, which isn't normally a major fuel, is about 0.81.

RQ is measured by having the subject breathe into a chamber (Figure 5.15) that collects O_2 and CO_2. The difference between inhaled and exhaled O_2 is the volume of O_2 consumed. The volumes of CO_2 exhaled and O_2 consumed are proportional to their numbers of moles. Thus, RQ equals the volume of CO_2 produced divided by the volume of O_2 consumed.

RQ changes during exercise. Muscles store carbohydrate, which they use for energy. During steady, prolonged exercise, stored fat enters the blood and muscles to supplement carbohydrate fuel. But this fat supply system is too slow to support intense exercise.

Early in a treadmill test (Figure 5.15) a subject's RQ typically is about 0.8, due to metabolism of both fat and carbohydrate. As the treadmill goes progressively faster or gets progressively steeper, the subject becomes exhausted. RQ eventually approaches 1.0 as overworked muscles metabolize stored carbohydrate almost exclusively.

EXAMPLE 5.12 From the preceding equation, calculate how many moles of (a) CH_4 and (b) O_2 it would take to produce 46 moles of CO_2.

SOLUTION (a) $46 \text{ mol CO}_2 \times \dfrac{1 \text{ mol CH}_4}{1 \text{ mol CO}_2} = 46 \text{ mol CH}_4$

(b) $46 \text{ mol CO}_2 \times \dfrac{2 \text{ mol O}_2}{1 \text{ mol CO}_2} = 92 \text{ mol O}_2$

Mole–Mass Calculations. Recall that you use molar mass as a conversion factor to interconvert moles and grams (Section 5.4). You then can calculate, for example, how many g of ethyl alcohol (C_2H_6O) the body metabolizes to produce 0.14 mol of CO_2 by the reaction:

$$C_2H_6O + O_2 \longrightarrow CO_2 + H_2O$$

The first step is to write a balanced equation:

$$C_2H_6O + 3O_2 \longrightarrow 2CO_2 + 3H_2O$$

Next, calculate the molar mass of ethyl alcohol:

molar mass of C_2H_6O = 2(12.01) + 6(1.01) + 16.00 = 46.08 g

Then start with the known (0.14 mol CO_2) and multiply by conversion factors to change to the desired unit (g ethanol). You know conversion factors to go from mol $CO_2 \rightarrow$ mol ethanol $\rightarrow$ g ethanol as follows:

$$0.14 \text{ mol } CO_2 \times \underbrace{\frac{1 \text{ mol } C_2H_6O}{2 \text{ mol } CO_2}}_{\text{mole ratio}} \times \underbrace{\frac{46.08 \text{ g } C_2H_6O}{1 \text{ mol } C_2H_6O}}_{\text{molar mass}} = 3.2 \text{ g } C_2H_6O$$

The best strategy usually is to get into moles, then convert from one substance to another using the mole ratios from the coefficients in the balanced equation. Then convert moles into any other desired unit.

EXAMPLE 5.13 For the preceding reaction, calculate the g O_2 needed to produce 0.14 mol CO_2.

SOLUTION molar mass of O_2 = 2(16.00) = 32.00 g

$$0.14 \text{ mol } CO_2 \times \frac{3 \text{ mol } O_2}{2 \text{ mol } CO_2} \times \frac{32.00 \text{ g } O_2}{1 \text{ mol } O_2} = 6.7 \text{ g } O_2$$

Mass–Mass Calculations. You also calculate the masses of substances in a reaction, using moles as the bridge. For the preceding reaction, for example, you can calculate how many g of O_2 would be needed for the metabolism of 86.8 g of ethanol. To do this, start with the known quantity (86.8 g C_2H_6O), and arrange conversion factors to convert to the desired unit (g O_2):

g ethanol (C_2H_6O) $\longrightarrow$ mol C_2H_6O $\longrightarrow$ mol O_2 $\longrightarrow$ g O_2

$$86.8 \text{ g } C_2H_6O \times \underbrace{\frac{1 \text{ mol } C_2H_6O}{46.08 \text{ g } C_2H_6O}}_{\text{molar mass}} \times \underbrace{\frac{3 \text{ mol } O_2}{1 \text{ mol } C_2H_6O}}_{\text{mole ratio}} \times \underbrace{\frac{32.00 \text{ g } O_2}{1 \text{ mol } O_2}}_{\text{molar mass}} = 181 \text{ g } O_2$$

Notice that mole ratios are conversion factors that relate the number of moles of two different substances in a reaction. Conversion factors relating molar mass or number of particles per mole for a single substance don't depend on the equation.

EXAMPLE 5.14 How many g of CO_2 are produced by the metabolism of 12.3 g of ethanol?

SOLUTION
convert g ethanol (C_2H_6O) $\longrightarrow$ mol C_2H_6O $\longrightarrow$ mol CO_2 $\longrightarrow$ g CO_2
molar mass of CO_2 = 12.01 + 2(16.00) = 44.01 g

$$12.3 \text{ g } C_2H_6O \times \frac{1 \text{ mol } C_2H_6O}{46.08 \text{ g } C_2H_6O} \times \frac{2 \text{ mol } CO_2}{1 \text{ mol } C_2H_6O} \times \frac{44.01 \text{ g } CO_2}{1 \text{ mol } CO_2}$$

= 23.5 g CO_2

Limiting Reactant. In a reaction with two or more reactants, the reactants often are not combined in exact molar ratios. Then one of the reactants limits how much product can form; this is the **limiting reactant** (or *limiting reagent*).

In the preceding mole–mass and mass–mass calculations, for example, we examined the reaction:

$$C_2H_6O + 3O_2 \longrightarrow 2CO_2 + 3H_2O$$

Those calculations were based on a specific amount of ethyl alcohol (C_2H_6O) without specifying the amount of O_2. We *assumed* that plenty of O_2 was available for the reaction. Thus C_2H_6O (not O_2) is the limiting reactant.

In reactions where one reactant isn't clearly in excess, you need to identify the limiting reactant. Then use the amount of that reactant to calculate how much product can form.

One way to do this is to use conversion factors to calculate how much product (either in g or in mol) could be produced from each reactant. *Whichever reactant produces less product is the limiting reactant.*

Suppose, for example, that 20.0 g of ethanol reacted with 35.0 g of O_2 in the preceding reaction. Which is the limiting reactant—C_2H_6O or O_2? To answer this question, calculate how many mol of product each reactant could produce. You can choose either product (CO_2 or H_2O), but use the same one for both calculations. In this example, we will use CO_2. Recall that the molar masses of C_2H_6O, O_2, and CO_2 are 46.08 g, 32.00 g, and 44.01 g, respectively. The calculations are:

$$20.0 \text{ g } C_2H_6O \times \frac{1 \text{ mol } C_2H_6O}{46.08 \text{ g } C_2H_6O} \times \frac{2 \text{ mol } CO_2}{1 \text{ mol } C_2H_6O} = 0.868 \text{ mol } CO_2$$

$$35.0 \text{ g } O_2 \times \frac{1 \text{ mol } O_2}{32.00 \text{ g } O_2} \times \frac{2 \text{ mol } CO_2}{3 \text{ mol } O_2} = 0.729 \text{ mol } CO_2$$

Since the given amount of O_2 produces less product (CO_2), O_2 is the limiting reactant.

You can calculate either g or mol of product to determine the limiting reactant. Calculating mol of product is simpler because it requires one less conversion factor.

EXAMPLE 5.15 What is the limiting reactant if 3.7 mol O_2 reacted with 51.8 g of C_2H_6O in the preceding reaction?

SOLUTION $3.7 \text{ mol } O_2 \times \dfrac{2 \text{ mol } CO_2}{3 \text{ mol } O_2} = 2.47 \text{ mol } CO_2$

$51.8 \text{ g } C_2H_6O \times \dfrac{1 \text{ mol } C_2H_6O}{46.08 \text{ g } C_2H_6O} \times \dfrac{2 \text{ mol } CO_2}{1 \text{ mol } C_2H_6O} = 2.25 \text{ mol } CO_2$

Therefore, C_2H_6O is the limiting reactant.

Percentage Yield. In all the calculations so far, we have assumed that the limiting reactant reacts completely to form the products written in the equation. This rarely happens in actual practice. Usually some of the reactant fails to react, or forms products different from those written. In the real world, it is

Figure 5.16 Liquid mercury and cinnabar (HgS), the ore from which mercury is isolated.

important to know the percentage yield—that is, how much of the product written in the equation actually forms.

$$\text{percentage yield} = \frac{\text{actual yield}}{\text{theoretical yield}} \times 100\%$$

Consider, for example, a smelter that produces mercury metal from cinnabar, or mercury (II) sulfide (HgS) (Figure 5.16). The reaction is:

$$HgS + O_2 \longrightarrow Hg + SO_2$$

Suppose that each 1.00 kg of cinnabar (HgS) reacts with excess O_2 to produce 765 g of mercury (the actual yield). What is the percentage yield for the reaction?

First, make sure the equation is balanced. (It already is.) Then identify the limiting reactant—HgS in this example. Next, calculate how much mercury should form from 1.00 kg HgS according to the equation; this is the theoretical yield. Calculate this as for mass–mass interconversions, starting with the known quantity (1.00 kg HgS). You know conversion factors to convert kg HgS → g HgS → mol HgS → mol Hg → g Hg.

molar mass of HgS = 200.59 + 32.07 = 232.66 g

molar mass of Hg = 200.59 g

$$1.00 \text{ kg HgS} \times \frac{10^3 \text{ g HgS}}{1 \text{ kg HgS}} \times \frac{1 \text{ mol HgS}}{232.66 \text{ g HgS}} \times \frac{1 \text{ mol Hg}}{1 \text{ mol HgS}} \times \frac{200.59 \text{ g Hg}}{1 \text{ mol Hg}}$$
$$= 862 \text{ g Hg (theoretical yield)}$$

Now calculate the percentage yield using the preceding equation:

$$\% \text{ yield} = \frac{\text{actual yield}}{\text{theoretical yield}} \times 100\% = \frac{765 \text{ g Hg}}{862 \text{ g Hg}} \times 100\% = 88.7\%$$

> **EXAMPLE 5.16** For the preceding reaction, calculate how much Hg is produced from 52 g HgS in the presence of excess O_2 if the percentage yield is 73.1%.

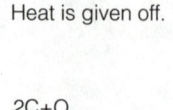

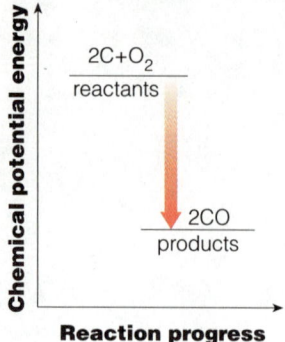

SOLUTION First, calculate the theoretical yield from 52 g HgS:

$$52 \text{ g HgS} \times \frac{1 \text{ mol HgS}}{232.66 \text{ g HgS}} \times \frac{1 \text{ mol Hg}}{1 \text{ mol HgS}} \times \frac{200.59 \text{ g Hg}}{1 \text{ mol Hg}} = \frac{45 \text{ g Hg}}{\text{(theoretical)}}$$

$$73.1\% = \frac{\text{actual yield}}{45 \text{ g Hg}} \times 100\%;$$

$$\text{actual yield} = \frac{73.1\% \times 45 \text{ g Hg}}{100\%} = 33 \text{ g Hg}$$

5.6 CHEMICAL REACTIONS AND ENERGY CHANGES

Chemical reactions either consume or give off energy. Those that give off heat are called **exothermic;** those that take up heat are called **endothermic.** Energy is taken up when the products have more energy than the reactants do; it is given off when the products have less energy than the reactants do (Figure 5.17). The amount of energy released or consumed per mole of reactant is the **heat of reaction.**

Reactions give off or consume energy in forms such as heat, light, and electricity. For example, oxygen reacts with wood, coal, natural gas, or oil to give off heat and light (Figure 5.4). Plants use solar energy to synthesize carbohydrates; those carbohydrates react with O_2 in plant and animal cells, releasing energy. Many manufacturing processes—such as making aluminum metal from its ore—consume large amounts of heat and electricity. Your body uses energy-releasing reactions for such things as pumping blood, staying warm, and synthesizing new cells and tissues.

Recall that heat energy is measured in units of calories (cal) or joules (J) (Section 2.2). Carbohydrate, fat, and protein foods may be called high-calorie. But it is only when those foods are oxidized (metabolized) that they release their energy in a form that can be measured in calories.

A *calorimeter* is used to measure heats of reaction (Figure 5.18). This device is a reaction vessel surrounded by a known volume of water at a particular temperature. Heat released from an exothermic reaction warms both the reaction vessel and the water, and the corresponding temperature increase is measured. Since it takes 1 cal to raise the temperature of 1 mL of H_2O by 1 degree Celsius (Section 2.2), the calories released by the reaction to the water can then be determined. Calories released to the reaction vessel can also be calculated using the heat capacity of the vessel.

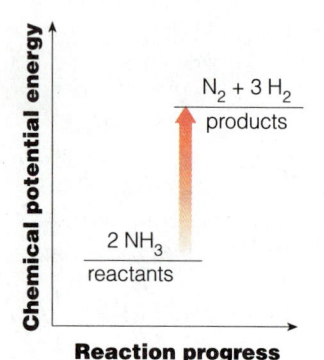

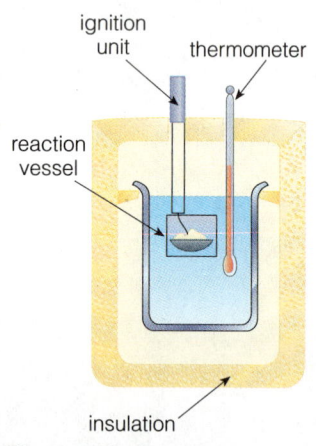

Figure 5.17 Exothermic (top) and endothermic (bottom) reactions.

Figure 5.18 A calorimeter is an insulated water bath containing a reaction vessel and a thermometer to measure the change in water temperature from the reaction.

EXAMPLE 5.17 A 1.310 g sample of glucose ($C_6H_{12}O_6$, molar mass 180.16 g) was burned in O_2 in a calorimeter to produce CO_2 and H_2O. The temperature rose from 20.16°C to 24.74°C. The calorimeter contained 6571 mL of water, and the heat capacity of the reaction vessel was 415 cal/°C. Calculate the energy released in (a) kcal and (b) kcal per mol glucose (the heat of reaction).

CHEMISTRY SPOTLIGHT

Calories and Food

Table 5.6 Typical Calorie and Fat Content in Sandwiches

Sandwich	Calories	Fat (g)	% Calories from Fat
Bacon/lettuce/tomato	356	23	58
Bologna	362	24	60
Egg salad	395	29	66
Ham and cheese	439	21	43
Peanut butter and jelly	320	14	39
Roast beef	315	15	43
Submarine	924	48	58

The caloric value of food comes from three ingredients—carbohydrates, fats, and proteins. When oxidized in the body, carbohydrates and proteins provide 4 Cal/g, and fats produce 9 Cal/g. The reactions of a typical carbohydrate and fat are shown in the Chemistry Spotlight in Section 5.5.

carbohydrate: $C_6H_{12}O_6 + 6O_2 \longrightarrow 6CO_2 + 6H_2O$

fat: $C_{55}H_{100}O_6 + 77O_2 \longrightarrow 55CO_2 + 50H_2O$

Why does the oxidation of fat release more energy (Cal/g) than the oxidation of carbohydrates? The energy released in a reaction depends on the difference in energy between the reactants and products (Figure 5.17). Notice that in both the preceding reactions, O_2 oxidizes the nutrient to the same, low-energy products—CO_2 and H_2O.

Now compare the formula for the carbohydrate with the formula for the fat. The carbohydrate ($C_6H_{12}O_6$) is more oxygen-rich; that is, it is more oxidized than the fat ($C_{55}H_{100}O_6$) to begin with. The reaction with O_2, then, is a more extensive oxidation for fats than for carbohydrates. Thus, the reaction with fats yields more energy.

Food labels now are required to provide information about fat and Calorie content. The labels must specify a realistic serving size and its total Calories, Calories from fat, and g of carbohydrate, protein, total fat, and saturated fat. The American Heart Association recommends that your diet provide less than 30% of its Calories from fat.

That isn't easy. In most sandwiches, for example, fat provides more than 30% of the Calories (see Table 5.6).

You can reduce the fat and Calorie content in sandwiches by using low-fat cheeses and mayonnaise, fewer egg yolks in egg salad, and lean meats—or by replacing meat and eggs with vegetables.

SOLUTION

$$\text{energy released to H}_2\text{O} = 675 \text{ mL H}_2\text{O} \times \frac{1 \text{ cal}}{\text{mL H}_2\text{O} \cdot °C} \times (24.74°C - 20.16°C) = 3090 \text{ cal}$$

$$\text{energy released to vessel} = \frac{415 \text{ cal}}{°C} (24.74°C - 20.16°C) = 1900 \text{ cal}$$

(a) $(3090 \text{ cal} + 1900 \text{ cal}) \times \dfrac{1 \text{ kcal}}{1000 \text{ cal}} = 4.99 \text{ kcal}$

(b) $\dfrac{4.99 \text{ kcal}}{1.310 \text{ g glucose}} \times \dfrac{180.16 \text{ g glucose}}{1 \text{ mol glucose}} = 686 \text{ kcal/mol glucose}$

EXAMPLE 5.18 The heat of reaction of glucose with O_2 is 686 kcal/mol of glucose. When you metabolize 1 mole of glucose, your body saves about 270 kcal in chemical form and releases the rest as heat. What percentage of energy is released as heat?

SOLUTION 686 kcal − 270 kcal = 416 kcal released as heat

$$\% \text{ released as heat} = \frac{416 \text{ kcal}}{686 \text{ kcal}} \times 100\% = 60.6\%$$

We may lay it down as an incontestable axiom that, in all the operations of art and nature, nothing is created; an equal quantity of matter exists before and after the experiment . . .

Antoine Lavoisier (1743–1794)

SUMMARY

A chemical reaction is represented by writing formulas for the reactants and products, separated by an arrow. The equation is balanced by placing coefficients in front of the formulas (without changing the formulas) so that reactants and products have equal numbers of atoms of each element.

In oxidation–reduction reactions, substances that are oxidized lose electrons and are reducing agents; they often gain oxygen atoms or lose hydrogen atoms. Substances that are reduced gain electrons and are oxidizing agents; they often lose oxygen atoms or gain hydrogen atoms. Such reactions are responsible for the action of bleaches, disinfectants, batteries and other products. O_2 is the major oxidizing agent in the body.

The formula weight of a substance is the sum of the atomic masses of each atom in the formula. The formula weight in g is the molar mass of a substance. That amount of material is called a mole (mol), and it contains Avogadro's number (6.02×10^{23}) of particles (atoms, molecules, or formula units, depending on the type of substance).

Coefficients in a balanced equation are the relative numbers of moles of reactants and products in a reaction. For any substance, moles can be readily interconverted with g (by molar mass) and with number of particles (by Avogadro's number). Thus, a balanced equation can be used to calculate the relative amounts of reactants and products in units of moles, grams, or number of particles. The reactant that limits the amount of product formed is the limiting reactant. The percentage yield of a reaction is the actual amount of product divided by the theoretical (calculated) amount of product ($\times 100\%$).

Chemical reactions that give off heat energy are called exothermic; those that require heat energy are called endothermic. The amount of heat released or consumed per mole of reactant is the heat of reaction.

KEY TERMS

Avogadro's number (5.4)
Chemical equation (5.1)
Chemical reaction (5.1)
Combustion (5.2)
Endothermic reaction (5.6)
Exothermic reaction (5.6)
Formula weight (5.4)
Heat of reaction (5.6)

Limiting reactant (5.5)
Molar mass (5.4)
Mole (5.4)
Mole ratio (5.5)
Oxidation (5.2)
Oxidation–reduction reaction (5.2)
Oxidizing agent (5.2)

Percentage yield (5.5)
Precipitate (5.1)
Reducing agent (5.2)
Reduction (5.2)
Stoichiometry (5.5)

EXERCISES

Even-numbered exercises are answered at the back of this book.

Chemical Reactions

1. Which of the following are chemical reactions?
   (a) Wax melts.
   (b) Autumn leaves turn red.
   (c) Spilled acid makes your skin brown.
   (d) Your body changes excess carbohydrate into fat.

2. Which of the following are chemical reactions?
   (a) An ice cube melts in your mouth.
   (b) You produce CO_2 in your lungs and exhale it into the air.
   (c) You trim a toenail.
   (d) You grow hair.

3. Write correct formulas for the following as they would appear in a chemical equation: (a) sodium sulfide, (b) nitrogen gas, (c) helium gas, (d) potassium phosphate, (e) carbon monoxide.

4. Write correct formulas for the following as they would appear in a chemical equation: (a) oxygen gas, (b) carbon tetrachloride, (c) calcium bicarbonate, (d) argon gas, (e) copper (I) oxide.

Writing Balanced Equations

5. Balance the following:
   (a) $HCl + Ca(OH)_2 \longrightarrow H_2O + CaCl_2$
   (b) $Mg + O_2 \longrightarrow MgO$
   (c) $KClO_3 \longrightarrow KCl + O_2$
   (d) $CH_4 + O_2 \longrightarrow CO_2 + H_2O$

6. Balance the following:
   (a) $Fe_2O_3 + Al \longrightarrow Al_2O_3 + Fe$
   (b) $NO + O_2 \longrightarrow NO_2$
   (c) $Li + N_2 \longrightarrow Li_3N$
   (d) $Al + HCl \longrightarrow H_2 + AlCl_3$

7. Balance the following:
   (a) $CO_2 + H_2O \longrightarrow C_6H_{12}O_6 + O_2$
   (b) $Na_2SO_4 + Pb(NO_3)_2 \longrightarrow NaNO_3 + PbSO_4$
   (c) $CH_4 + Cl_2 \longrightarrow CCl_4 + HCl$
   (d) $H_2S + SO_2 \longrightarrow S + H_2O$

8. Balance the following:
   (a) $C_4H_{10} + O_2 \longrightarrow CO_2 + H_2O$
   (b) $H_3PO_4 + NaOH \longrightarrow H_2O + Na_3PO_4$
   (c) $P_2O_5 + C \longrightarrow P + CO$
   (d) $Cu_2O + Cu_2S \longrightarrow Cu + SO_2$

9. Use spheres, as in Figure 5.6, to represent the reactants and products in reactions 7c and 7d.

10. Write balanced equations for the following reactions:
    (a) Sodium metal reacts with chlorine gas to form sodium chloride.
    (b) Silicon dioxide reacts with hydrogen fluoride to form silicon tetrafluoride and water.
    (c) Calcium carbonate reacts to form calcium oxide and carbon dioxide.

11. Write balanced equations for the following reactions:
    (a) Iron (II) oxide reacts with silicon to yield iron and silicon dioxide.
    (b) Nitrogen gas reacts with hydrogen gas to yield ammonia (nitrogen trihydride).
    (c) Combustion reaction using C_5H_{12} as the fuel.

12. Does a balanced equation mean that a reaction necessarily occurs?

13. Consider the equation: $2He + 2NaCl \longrightarrow 2NaHe + Cl_2$
    Do you think this reaction will occur? Why?

Oxidation–Reduction Reactions

14. Write a balanced equation for the combustion of octane (C_8H_{18}).

15. Write a balanced equation for the combustion of natural gas (CH_4).

16. In the equation for exercise 14, identify which substance was oxidized and which was reduced. Identify the oxidizing agent and the reducing agent.

17. In the equation for exercise 15, identify which substance was oxidized and which was reduced. Identify the oxidizing agent and the reducing agent.

18. When iron rusts, it undergoes the following reaction:

    $$4Fe + 3O_2 \longrightarrow 2Fe_2O_3$$

    Identify the substance oxidized, the substance reduced, the oxidizing agent, and the reducing agent.

19. Identify the substance oxidized and the substance reduced in the following reactions:
    (a) $2Ag^+ + 2Br^- \longrightarrow 2Ag + Br_2$
    (b) $Cd + NiO_2 + 2H_2O \longrightarrow Cd(OH)_2 + Ni(OH)_2$
    (c) $WO_3 + 3H_2 \longrightarrow W + 3H_2O$
    (d) $Fe_2O_3 + 2Al \longrightarrow 2Fe + Al_2O_3$

20. Identify the substance oxidized and the substance reduced in the following reactions:
    (a) $6CO_2 + 6H_2O \longrightarrow C_6H_{12}O_6 + 6O_2$
    (b) $Al + 6HCl \longrightarrow 2AlCl_3 + 3H_2$
    (c) $CuO + C \longrightarrow Cu + CO$
    (d) $C_6H_6O_2 + 2Ag^+ \longrightarrow 2Ag + 2H^+ + C_6H_4O_2$

21. Identify the oxidizing agent and the reducing agent in the following reactions:
    (a) $SiO_2 + 2Mg \longrightarrow Si + 2MgO$
    (b) $2BCl_3 + 3H_2 \longrightarrow 2B + 6HCl$

(c) PbS + Fe ⟶ Pb + FeS
(d) 2CH₄O + 3O₂ ⟶ 2CO₂ + 4H₂O

22. Identify the oxidizing agent and the reducing agent in the following reactions:
    (a) CH₂O + Cu(OH)₂ ⟶ CH₂O₂ + Cu + H₂O
    (b) 2Al + 3Cu²⁺ ⟶ 2Al³⁺ + 3Cu
    (c) C₂H₄O₂ + 2O₂ ⟶ 2CO₂ + 2H₂O
    (d) C₅₅H₉₄O₆ + 3H₂ ⟶ C₅₅H₁₀₀O₆

23. Do disinfectants that are oxidizing agents cause bacteria to be (a) oxidized or (b) reduced?

24. In a battery using zinc and copper, is Zn oxidized or reduced?

25. When an apple is peeled and sits out in the air, it becomes brown because a substance in it is oxidized. What do you think is the oxidizing agent?

26. Since oxidation is the loss of electrons, what substance receives the electrons from carbohydrates that are oxidized in your body?

Moles, Grams, and Avogadro's Number

27. Calculate the formula weights of (a) Ca(CN)₂, (b) Al(NO₃)₃, (c) urea (CH₄N₂O), (d) fructose (C₆H₁₂O₆), (e) iron (II) sulfide.

28. Calculate the formula weights of (a) ozone (O₃), (b) NH₄Cl, (c) Al, (d) citric acid (C₆H₈O₆), (e) caffeine (C₈H₁₀N₄O₂).

29. Calculate the mass of 4.13 moles of each substance in exercise 27.

30. Calculate the mass of 2.9 × 10⁻² moles of each substance in exercise 28.

31. For each substance in exercise 27, calculate the number of moles in a 12.5 g sample.

32. For each substance in exercise 28, calculate the number of moles in a 50.0 g sample.

33. Calculate the number of molecules in 46 moles of (a) H₂O and (b) CH₄.

34. Calculate the number of particles in 1.4 × 10⁻⁷ moles of (a) cholesterol (C₂₇H₄₆O) and (b) glucose (C₆H₁₂O₆).

35. Calculate the number of moles in 8.11 × 10²⁴ molecules of H₂O.

36. Calculate the number of moles in 3.71 × 10¹⁹ formula units of CaCl₂.

37. Calculate the number of molecules in 50.0 g of CCl₄.

38. Calculate the number of formula units in 1.00 g of KHCO₃.

39. Calculate the mass in g of one molecule of H₂O.

40. The density of mercury is 13.6 g/mL. How many atoms of mercury are in a thermometer that contains 0.30 mL of mercury?

41. The density of ethyl alcohol (C₂H₆O) is 0.789 g/mL. What is the volume in mL of 8.16 × 10²⁴ molecules of ethyl alcohol?

42. Avogadro's number of Br₂ molecules has a volume of 55 mL. What is the density of Br₂ in g/mL?

Stoichiometry, Limiting Reactant, and Percentage Yield

Use the following equation to complete exercises 43–57:

$$C_6H_{12}O_6 + 6O_2 \longrightarrow 6CO_2 + 6H_2O$$

43. 36.0 moles of glucose (C₆H₁₂O₆) will react with how many moles of O₂?

44. The reaction of 1.2 moles of glucose with excess oxygen will produce how many moles of CO₂?

45. How many moles of O₂ are needed to produce 6.0 × 10³ moles of H₂O?

46. How many g of H₂O are produced by the complete reaction of 62.4 moles of C₆H₁₂O₆?

47. How many moles of CO₂ are produced by the complete reaction of 500.0 g C₆H₁₂O₆?

48. How many g of CO₂ will be produced by the complete reaction of 9.83 × 10²¹ molecules of C₆H₁₂O₆?

49. How many molecules of H₂O will be produced by the complete reaction of 1.00 g C₆H₁₂O₆?

50. When 25.0 g of C₆H₁₂O₆ reacts with 25.0 g of O₂, what is the limiting reactant?

51. How many g of C₆H₁₂O₆ can be oxidized to CO₂ and H₂O by reaction with 3.52 mol of O₂?

52. What is the percentage yield if the reaction of 90.0 g C₆H₁₂O₆ with excess O₂ produces 6.7 g H₂O?

53. What is the percentage yield if the reaction of 120.0 g C₆H₁₂O₆ with excess O₂ produces 20.6 g CO₂?

54. How many g CO₂ would be produced from the reaction of 1.00 g C₆H₁₂O₆ with excess O₂ if the yield is 44.2%?

55. If the yield is 27.4%, how many g H₂O would be produced by the reaction of 50.0 g C₆H₁₂O₆ with excess O₂?

56. How many moles of CO₂ would be produced under the conditions in exercise 54?

57. How many molecules of H₂O would be produced under the conditions in exercise 55?

58. Calculate the Respiratory Quotient (see Chemistry Spotlight in Section 5.5) for the reaction with O₂ of the fat, C₅₇H₁₀₄O₆.

59. Calculate the Respiratory Quotient (see Chemistry Spotlight in Section 5.5) for the reaction with O₂ of the fat, C₅₃H₉₆O₆.

60. Write a balanced equation for the synthesis of ammonia (NH₃) from nitrogen gas and hydrogen gas.

Use the equation for exercise 60 to complete exercises 61–65:

61. Does the synthesis of NH_3 require more g of N_2 or H_2?
62. Does the synthesis of NH_3 require more moles of N_2 or H_2?
63. Using 1.00 g of H_2 and excess N_2, how many g of NH_3 would be produced if the reaction yield is 64%?
64. If the density of NH_3 produced is 0.771 g/L, what would be the volume of NH_3 produced under the conditions in exercise 63?
65. If the density of NH_3 produced is 0.771 g/L, what would be the percentage yield if 1.00 mole of N_2 reacts with excess H_2 to produce 33 L of NH_3?

Chemical Reactions and Energy Changes

66. Identify each of the following as endothermic or exothermic: **(a)** baking a cake, **(b)** metabolizing cake to CO_2 and H_2O, **(c)** making steel, **(d)** burning wood.
67. Is the oxidation of fat in your body an endothermic or exothermic process?
68. When a person goes out in the cold, does the body increase its rate of endothermic or exothermic reactions? Why?
69. When the sugar glucose reacts with O_2 to produce CO_2 and H_2O, the heat of reaction is 686 kcal/mol of glucose. What is the heat of reaction in kJ/mol of glucose? [Review Section 2.2 if necessary.]
70. The heat of reaction of glucose ($C_6H_{12}O_6$) with O_6 is 686 kcal/mol of glucose. Use this information to calculate how many kcal (Cal) are generated when you metabolize 1.00 g of glucose.
71. The heat of reaction of the fat $C_{55}H_{100}O_6$ with O_2 is 7800 kcal/mol of fat. Use this information to calculate how many kcal (Cal) are generated when you metabolize 1.00 g of fat.
72. A typical tuna salad sandwich has 433 Calories and 28 g of fat. What percentage of the Calories comes from fat?
73. A crab quesadilla has 8.0 g of fat, and 32% of its Calories comes from fat. What is the Caloric value of this quesadilla?

DISCUSSION EXERCISES

1. Most metals occur naturally in ores as ionic compounds. A smelter obtains a pure metal from an ore. Does this process require oxidation or reduction of the ore? Why?
2. Estimate the number of water molecules in your body.
3. How far would Avogadro's number of dollar bills, laid end to end, reach? Compare this distance with other distances you know.
4. Suppose an industrial chemical process has a yield of 62%. What types of changes might increase the percentage yield? Why?
5. Should the United States express the energy value of foods in kilojoules (kJ) instead of in Calories (Cal)? Why?
6. Suppose a 2.162 g sample of glucose (molar mass = 180.16 g) reacts completely with O_2 in a calorimeter that contains 1150 mL of H_2O and has an initial temperature of 20.00°C. The heat capacity of the reaction vessel is 612 cal/°C. The heat of reaction is 686 kcal per mol glucose. Calculate the final temperature of the H_2O and reaction vessel.

6

Solids, Liquids, and Gases

1. What is the kinetic molecular theory? How does it explain solids, liquids, or gases?
2. What are the forces of attraction between ions, molecules, and atoms? How do they cause substances to be solids, liquids, or gases?
3. What happens when substances change their physical state? What energy changes are involved?
4. How are the pressure, volume, and temperature of a gas interrelated?
5. How much volume does a given amount of gas occupy? How do you calculate the volume of a gas needed or produced in a chemical reaction?

Earth's Atmosphere and Life: Then and Now

The earth is about 4.6 billion years old. The planet's early atmosphere was mostly hydrogen gas (H_2), with significant amounts of helium (He). Gradually much of these light gases escaped from earth's weak gravitational pull, leaving the atmosphere mostly composed of methane (CH_4), ammonia (NH_3), water (H_2O), nitrogen (N_2), carbon dioxide (CO_2), carbon monoxide (CO), and hydrogen sulfide (H_2S). Little oxygen gas (O_2) was present because oxygen was bound up as compounds in water, rocks, and soil.

Early bacteria consumed simple organic compounds, producing CO_2 and other products. As the levels of CO_2 increased, bacteria and plants emerged that use sunlight, CO_2, and H_2O to carry out photosynthesis, producing O_2 gas:

$$6CO_2 + 6H_2O \xrightarrow{\text{sunlight}} C_6H_{12}O_6 + 6O_2$$

As oxygen levels rose, O_2 reacted to produce ozone (O_3) in the upper atmosphere, making earth's surface habitable for other living things. Oxygen in the air also made life possible for animals that use O_2 to oxidize their food to CO_2 and H_2O:

$$C_6H_{12}O_6 + 6O_2 \longrightarrow 6CO_2 + 6H_2O + \text{energy}$$

A dynamic equilibrium between photosynthesis and O_2 metabolism kept steady levels of CO_2 and O_2 in the atmosphere.

Now human activities—such as burning massive amounts of fossil fuels and clearing the land of plants—are raising the levels of CO_2 in the atmosphere (Figure 6.1). Since CO_2 absorbs heat radiated from the planet's surface, increasing levels of CO_2 cause earth's temperature to rise. This is the *greenhouse effect*. Scientists estimate that CO_2 is responsible for about 50% of the greenhouse effect, which causes global warming.

Figure 6.1 Clearing the land of CO_2-consuming plants raises CO_2 levels in air, making the earth warmer.

Each breath brings oxygen gas (O_2) into your lungs. O_2 enters your blood, binds to hemoglobin, and travels to your cells, where it is used to help build skin, bone, and muscle.

Why is O_2 a gas, blood a liquid, and skin a solid? It isn't magic. We can understand why substances are solids, liquids, and gases in terms of their chemistry.

You know that H_2O isn't always a liquid. You've seen it boil into a gas, and freeze into a solid. We will see how matter changes its physical state, and how energy is required or released when this happens.

You have experienced the properties of gases in many ways: the fizz of a carbonated drink when you remove the cap (Figure 6.2); the rise in tire pressure on a hot day; the "pop" in your ears when you suddenly change altitude. You will learn how the temperature, pressure, volume, and amount of a gas are all interrelated.

6.1 PHYSICAL STATES OF MATTER

Three States of Matter. We can classify materials as solid, liquid, or gas. These three physical states of matter (Section 3.1) are properties of large numbers of particles (atoms, ions, or molecules). It is meaningless to say that a single atom, ion, or molecule is a solid, liquid, or gas. Instead, the physical state depends on interactions between individual particles in a substance.

Kinetic Molecular Theory. The **kinetic molecular theory** pictures particles (atoms, ions, or molecules) in continuous motion. As the temperature increases, so does the average kinetic energy of particles. Moving particles collide with each other but lose no energy. Because they are very far apart compared to their size, particles in a gas occupy an insignificant amount of the total volume.

Figure 6.2 CO_2 gas in a carbonated soft drink.

Table 6.1 Kinetic Molecular Theory and the Three Physical States

Physical State and Property	Kinetic Molecular Theory Explanation
Solid	
definite volume and shape; doesn't flow or mix readily	Attractive forces hold particles in place so they don't move freely.
most dense physical state; not very compressible	Attractive forces hold particles as close together as possible.
Liquid	
indefinite shape; flows and mixes readily; conforms to the shape of its container	Attractive forces allow particles to move about.
definite volume	Attractive forces are strong enough to keep particles from escaping.
more dense than gas; not very compressible	Attractive forces keep particles fairly close together.
Gas	
indefinite volume and shape; readily flows and mixes; conforms to the shape of its container	Attractive forces are so weak that particles move about freely.
least dense physical state; compressible	Particles move freely with much empty space between them.
exerts pressure equally in all directions on surfaces	Particles move rapidly and strike surfaces with equal force.

We can use this theory to explain the properties of solids, liquids, and gases—for example, why gases are the least dense state of matter and have no fixed volume or shape. Read Table 6.1 carefully to see how the attractive forces and kinetic energy of particles account for such properties.

Think of atoms, ions, or molecules as people at a dance. The solid state is like what happens during a very slow number—the dancers sway gently, and stay in about the same place. As the tempo picks up, the action is more like the liquid state. People move energetically about the dance floor, changing locations and neighbors. When the dance is over—and people leave in all directions—the situation is like the gaseous state.

According to kinetic molecular theory, the particles (ions, molecules, or atoms) of a substance move continuously; the warmer the temperature, the greater their average kinetic energy. When their motion is much stronger than the forces attracting them, particles move freely and the substance is a gas (Figure 6.3). When their motion is weaker than the attractive forces, particles stay closer together as a liquid or—in their most dense physical state—as a solid.

Forces Between Molecules. Why is it that—at room temperature—some substances are solids, some are liquids, and some are gases? The difference comes from the strength of attractions between particles.

For example, the attraction between positive and negative ions in a salt such as NaCl is so strong that the ions stay close together in an ionic crystal lattice (Section 4.2). This strong "glue" makes NaCl a solid. It takes a great deal of heat energy to overcome that attraction and change NaCl into a liquid.

Because of their strong attractive forces, ionic compounds are solids at room temperature and have high melting points (Table 6.2). Metals also are solids; their particles stay together because of the strong attraction between positive metal ions and the "sea" of mobile electrons in the metal crystal lattice (Section 4.3). Their melting points, however, aren't as consistently high as those of ionic compounds (Table 6.2).

Noble gases exist as neutral atoms. These atoms have little attraction to each other, so even at very cold temperatures the atoms move freely enough for the substance to be a gas.

EXAMPLE 6.1 Predict the physical state of each of the following at room temperature: (a) $CaCl_2$, (b) He, (c) $Mg(NO_3)_2$, (d) Ra.

SOLUTION (a) solid, (b) gas, (c) solid, (d) solid

With covalent compounds, the situation is more complicated. Forces between molecules typically are weaker than those in metals and ionic compounds, but stronger than those in noble gases. Forces between molecules are classified as: (1) London forces, (2) dipole–dipole interactions, and (3) hydrogen bonds.

Temporary shifts in electron distribution cause *London forces*. Because electrons move constantly, for fleeting instants more electrons are at one side of an atom, ion, or molecule than at the other side. These momentary imbalances of charge create a temporary dipole, with partial positive and negative poles. This dipole can induce a similar dipole in a nearby atom, ion, or molecule. **London forces** are the attractions between temporary dipoles (Figure 6.4).

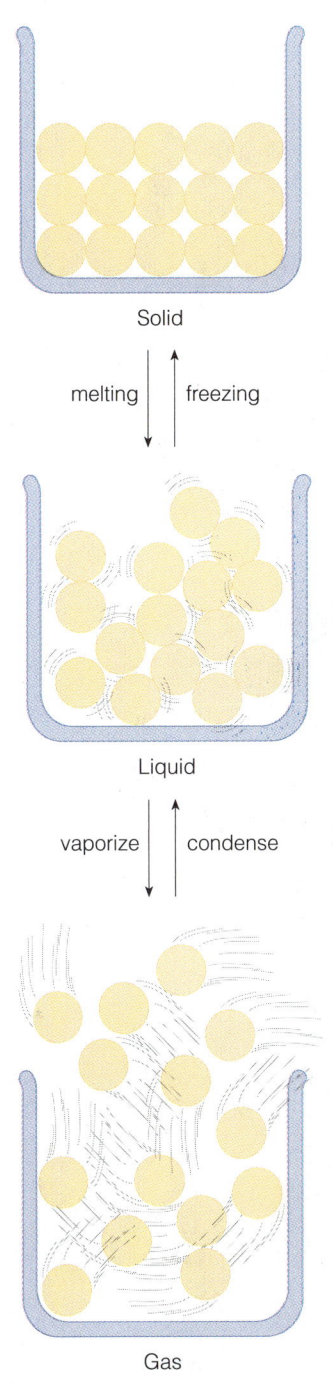

Figure 6.3 Comparison of solids, liquids, and gases.

London forces are named for Fritz London (1900–1954), the German physicist who described them.

Table 6.2 Melting Points of Some Metals and Ionic Chloride Compounds

Substance (Formula)	Melting Point (°C)	Substance (Formula)	Melting Point (°C)
Gold (Au)	1064	Magnesium chloride (MgCl$_2$)	714
Silver (Ag)	961	Mercury (Hg)	−40*
Silver chloride (AgCl)	455	Mercury (II) chloride (HgCl$_2$)	276
Copper (Cu)	1083	Sodium (Na)	98
Copper (II) chloride (CuCl$_2$)	620	Sodium chloride (NaCl)	801
Magnesium (Mg)	649		

*Mercury is the only metal element that is a liquid at room temperature.

London forces occur between all atoms, molecules, and ions. They are the *only* attractions between noble gas atoms and between nonpolar molecules such as H$_2$, CH$_4$, CO$_2$, and Cl$_2$ (Figure 6.4). In most substances, London forces are only about one-thousandth as strong as the typical covalent bond. But larger atoms and molecules have stronger London forces because their larger electron clouds form temporary dipoles more readily. As Figure 6.5 shows, the physical state of different halogens at room temperature changes from gas to liquid to solid as the London forces get stronger.

In addition to London forces, polar molecules have somewhat stronger **dipole–dipole forces** of attraction between their permanent positive and negative poles (Figure 6.6). Because these attractions are between *partial* electrical charges (δ^+ and δ^-), dipole–dipole forces are much weaker than ionic bonds.

A **hydrogen bond** is a special type of dipole–dipole force between a very electronegative atom and a covalently bonded hydrogen atom in the same or a nearby molecule. Although they are the strongest type of intermolecular force, hydrogen bonds are still much weaker than covalent bonds within molecules.

The strongest and most important hydrogen bonds are in molecules having a F—H, —O—H, or —N—H group. Hydrogen covalently bonded to a highly electronegative F, O, or N atom has a relatively large δ^+ charge and is strongly attracted to another F, O, or N (with its relatively large δ^- charge) in the same or another molecule. For example, water (H$_2$O), ammonia (NH$_3$), and hydrogen fluoride (HF) all form hydrogen bonds between their molecules (Figure 6.7).

With small molecules such as those in Figure 6.7, hydrogen bonds form between two different molecules. In Section 7.1, we will discuss how hydrogen bonds give water its distinctive properties. We will also see later that in large

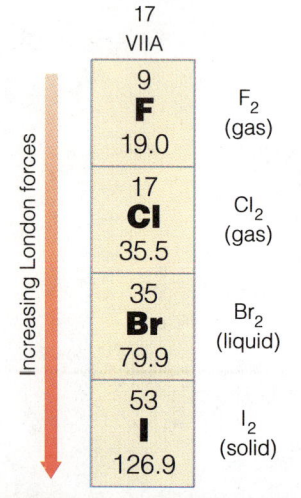

Figure 6.5 Increasing strength of London forces with the weight of molecules affects the physical state at room temperature.

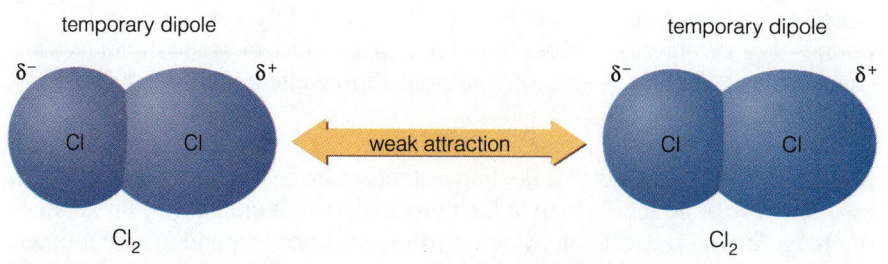

Figure 6.4 London forces of attraction between two nonpolar Cl$_2$ molecules.

Solids, Liquids, and Gases Chapter 6 119

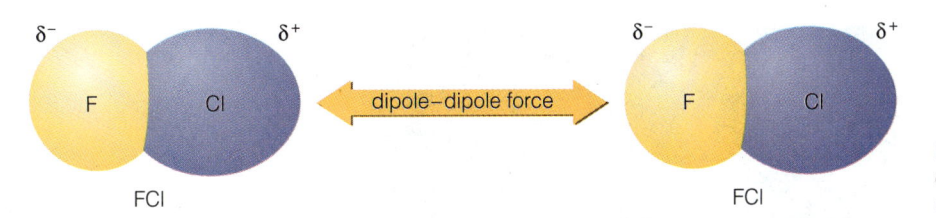

Figure 6.6 Dipole–dipole force of attraction between polar FCl molecules.

molecules—such as proteins (see Chapter 21) and DNA (see Chapter 23)—hydrogen bonds form between two different parts of the same molecule.

EXAMPLE 6.2 List the attractive force(s) between particles of the following: (a) H_2, (b) H_2O, (c) N_2, (d) chloroform ($CHCl_3$), (e) butane (C_4H_{10}).

SOLUTION (a) London forces, (b) London forces and hydrogen bonds, (c) London forces, (d) London forces and dipole–dipole forces, (e) London forces

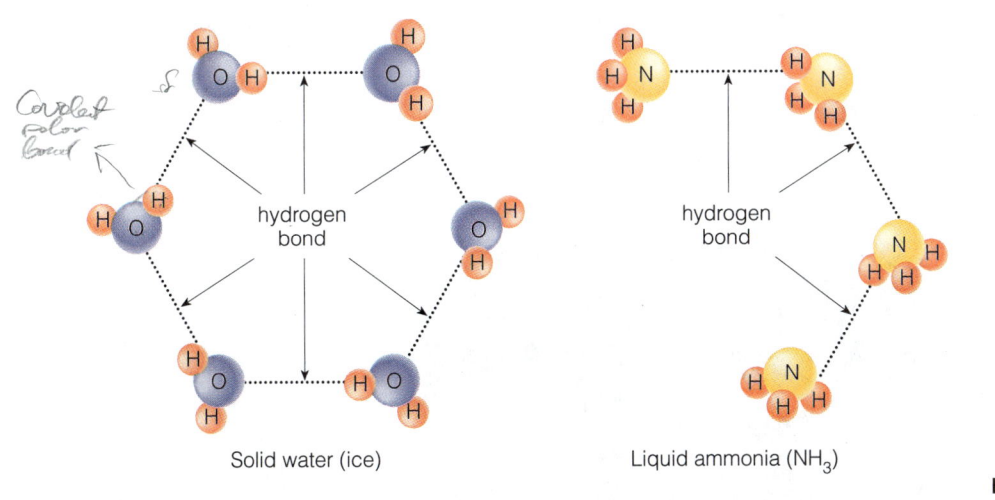

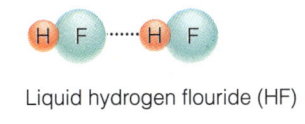

Figure 6.7 Hydrogen bonds (dotted lines) between polar molecules of HF, H_2O, and NH_3.

6.2 CHANGES IN PHYSICAL STATE

Melting and Freezing. When we call something a solid, liquid, or gas, we are usually describing its state at normal pressure and at room temperature. Under different conditions, however, substances change their physical state (Figure 6.8). You can use the kinetic molecular theory and your understanding of attractions between particles to picture what happens when a substance changes physical state.

Particles in a solid stay in place, close together. When heated, they begin to vibrate more. At some temperature, they gain enough energy to overcome the

Figure 6.8 Three states of H₂O. A hot poker changes ice into water and steam. (*Source:* Fritz Goro, *Life* magazine, © 1949 Time Inc.)

attractive forces holding them in place. At this temperature **melting** occurs; a solid changes into a liquid (Figure 6.3). Each pure substance does this at a single, characteristic temperature called the *melting point*.

The change from liquid into solid is **freezing,** or *solidification*. This occurs at the *freezing point*, which for pure substances is the same temperature as the melting point. Above its freezing point a substance stays liquid, but at the freezing point it solidifies.

A few substances change directly from a solid to a gas, a process called **sublimation.** You may have seen dry ice (solid CO_2) do this. Another substance that sublimes is iodine (I_2, Figure 6.9).

Figure 6.9 Sublimation of solid I_2 directly to a gas.

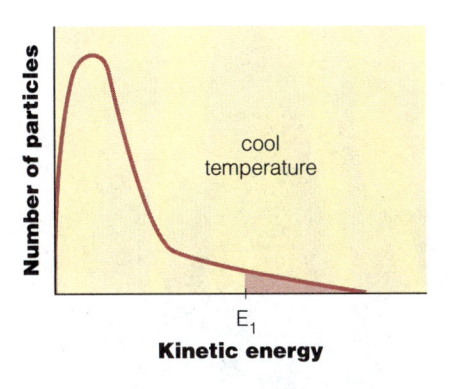

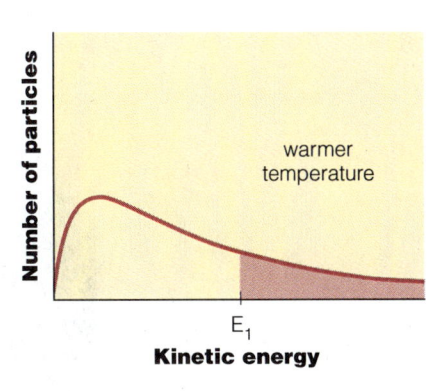

Figure 6.10 Distribution of particles and their kinetic energies at a cool temperature (left) and a warmer temperature (right). The number having E_1 or more energy (in color) is greater at a warmer temperature.

Vaporization and Vapor Pressure. Particles in a liquid vary in energy and speed, but at a given temperature they have a certain average speed (Figure 6.10). If you leave an open container of liquid out long enough, the liquid changes into a gas, a process called **vaporization** (Figures 6.3 and 6.11, top). This happens because particles on the liquid's surface eventually have enough kinetic energy to escape the attractive forces of other particles in the liquid.

If you put a liquid in a covered container, however, particles that change into the gas state can't escape. They eventually return to the liquid state as they are attracted by particles in the liquid. Changing a gas into a liquid is **condensation** (Figure 6.3).

Both vaporization and condensation occur in a closed container. In time, both processes will occur at the same rate, keeping the number of particles in the liquid state and the gas state constant. The balance in the rates of two opposing processes is called a **dynamic equilibrium.** *Dynamic* implies motion. *Equilibrium* implies balance. Although both processes are occurring in a dynamic equilibrium, the number of particles in each state remains the same.

At dynamic equilibrium, the pressure over a liquid in a closed container is the **vapor pressure** of that liquid (Figure 6.11, bottom). The weaker the attractive forces between particles, the more readily a liquid evaporates at a given temperature, and the higher its vapor pressure. For this reason, vapor pressure is a property that can be used to identify a substance.

EXAMPLE 6.3 Pentane and decane are nonpolar liquids with the formulas C_5H_{12} and $C_{10}H_{22}$, respectively. In both compounds, London forces are the only type of intermolecular attraction. Which liquid has the greater vapor pressure?

SOLUTION London forces are stronger in larger molecules (Section 6.1), so decane has stronger London forces. Therefore, pentane vaporizes more easily and has a greater vapor pressure.

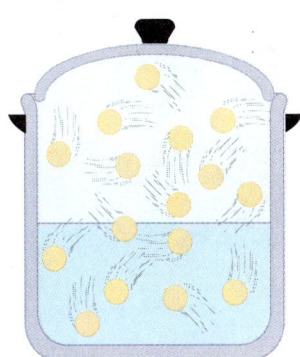

Figure 6.11 (Top) Vaporization occurs in an open container when particles have enough energy to overcome the attractive forces and escape to the gas state. (Bottom) In a closed container, vapor pressure develops as a dynamic equilibrium occurs between vaporization and condensation.

Boiling. When you heat a liquid, its vapor pressure rises and it vaporizes faster (Figure 6.12). At some temperature, a pure substance forms bubbles of vapor in the liquid that escape from an open container (Figure 6.13). This process, known as **boiling,** differs from vaporization in two ways. First, gas forms throughout the liquid, not just at the surface. Second, particles in the gas have

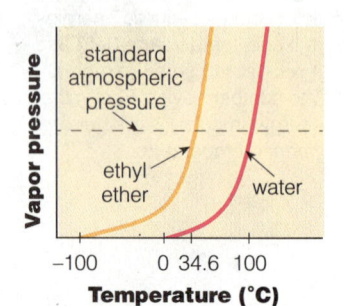

Figure 6.12 Vapor pressure increases with temperature. The boiling points for ethyl ether and water at standard atmospheric pressure are 34.6°C and 100°C, respectively.

Figure 6.13 Gas bubbles form in a boiling liquid.

enough kinetic energy on average to escape from the liquid despite opposing pressure from molecules in the atmosphere above the liquid. The *boiling point* is the temperature at which a liquid's vapor pressure equals the atmospheric pressure above the liquid.

Because atmospheric pressure varies, so does boiling point. For example, water boils at 100°C in a sea-level city such as Boston, but at 95°C in mile-high Denver, where the lower atmospheric pressure makes it easier for water molecules to escape from the liquid (Figure 6.14). The *normal boiling point* of a substance is its boiling point at typical sea-level atmospheric conditions.

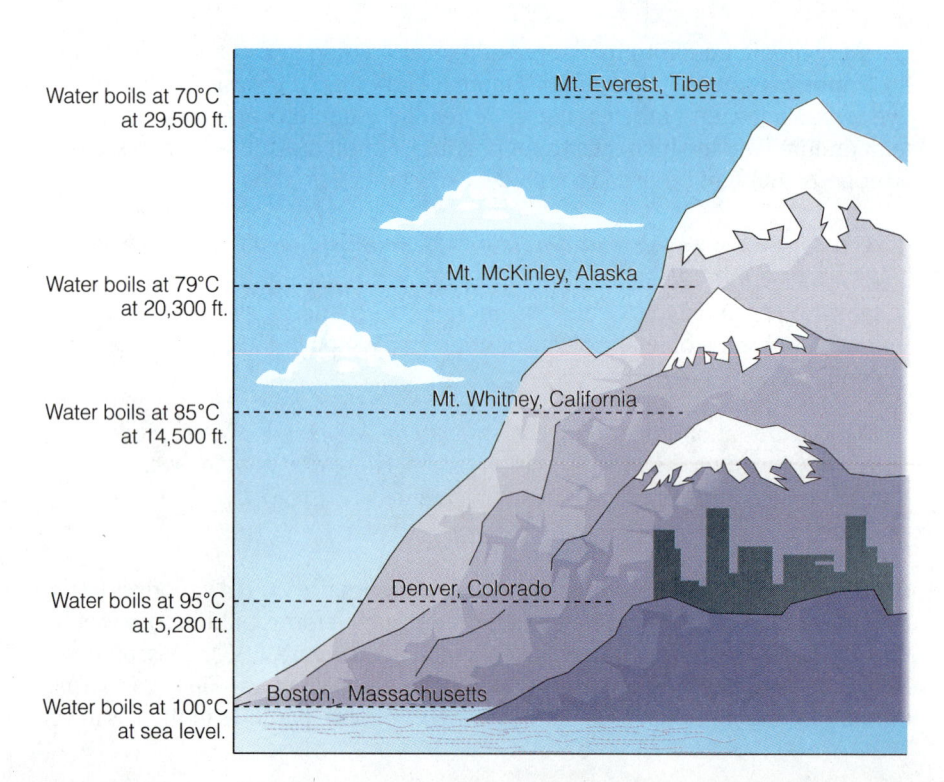

Figure 6.14 Water boils at lower temperatures at higher elevations due to lower atmospheric pressures.

Increasing the pressure on a liquid raises its boiling point (Figure 6.15). The pressure cap on a car radiator keeps water from boiling until it reaches about 120°C. And a pressure cooker cooks food faster because the temperature inside rises above 100°C without water boiling away.

Melting Point, Boiling Point, and Intermolecular Forces. As you would expect, forces of attraction between particles help determine melting and boiling points. The stronger those forces, the more heat energy it takes to overcome them, and the higher the melting and boiling points become (Table 6.3).

Energy Changes. When a substance changes state, extra energy is consumed or given off. For example, think of what happens when you heat water. The

Figure 6.15 Biological and medical equipment can be sterilized in an autoclave. The high pressure inside raises the temperature of steam to 170°C, high enough to kill infectious agents.

Table 6.3 Effect of Increasing Bond Strength on Melting and Boiling Points of Substances of Similar Formula Weights

Substance	Formula Weight (amu)	Strongest Type of Bonding Between Particles	Melting Point (°C)	Boiling Point (°C)
Fluorine (F_2)	38.00	London forces only	−220	−188
Nitric oxide (NO)	30.01	Dipole–dipole forces	−164	−152
Methanol (CH_4O)	32.05	Hydrogen bonds	−94	65
Calcium (Ca)	40.08	Metallic bonding	893	1484
Sodium fluoride (NaF)	41.99	Ionic bonding	993	1695

temperature rises until water reaches its boiling point (100°C). Then the water boils, but its temperature stays at 100°C—despite further heating. At this temperature, heat is consumed as liquid water changes into a gas (steam).

The extra energy required to change a substance at its boiling point from a liquid to a gas is the **heat of vaporization.** The same amount of heat is released when a gas condenses back into a liquid.

Extra energy is also needed to melt a solid at its melting point. This value, the **heat of fusion,** equals the amount of energy given off when a substance freezes.

Table 6.4 lists heats of fusion and vaporization for several substances. Notice that substances with stronger forces of attraction between their particles have higher heats of fusion and vaporization. You can use the heat of fusion (or vaporization) as a conversion factor to calculate the energy for any amount of a substance to melt (or vaporize) at its melting (or boiling) point.

Table 6.4 Heat of Fusion, Heat of Vaporization, and Specific Heat for Some Substances

Substance	Strongest Bonding Forces	Heat of Fusion (cal/g)	Heat of Vaporization (cal/g)	Specific Heat (cal/g °C)
Hydrogen chloride (HCl)	dipole–dipole	13.9	99	0.19
Chloroform (CHCl$_3$)	dipole–dipole	18	59	0.23
Ethanol (C$_2$H$_6$O)	hydrogen	24.9	204	0.58
Water (H$_2$O)	hydrogen	79.8	540	1.00
Aluminum (Al)	metallic	94.4	2500	0.22
Sodium chloride (NaCl)	ionic	124	3130	0.21

EXAMPLE 6.4 Use the information in Table 6.4 to calculate how many calories (cal) and how many joules (J) are needed to melt 50.0 g of ice at 0°C.

SOLUTION $50.0 \text{ g H}_2\text{O} \times \dfrac{79.8 \text{ cal}}{1 \text{ g H}_2\text{O}} = 3990 \text{ cal } (3.99 \times 10^3 \text{ cal})$

$$3990 \text{ cal} \times \dfrac{4.18 \text{ J}}{1 \text{ cal}} = 16{,}700 \text{ J } (1.67 \times 10^4 \text{ J})$$

The energy required to warm 1 g of a substance by 1°C is the **specific heat** of that substance (Table 6.4). The formula for specific heat is:

$$\text{specific heat} = \dfrac{\text{cal (or J)}}{\text{g} \times \text{°C change}}$$

Notice that the specific heat of water is 1.00 calorie per g per °C. This value is very high because extra energy is needed to break water's strong hydrogen bonds.

EXAMPLE 6.5 Using the information in Table 6.4, calculate how many (a) cal and (b) J are needed to heat 465 g of water from 24.0°C to 64.3°C.

SOLUTION Use the specific heat of water as a conversion factor:

(a) $\dfrac{1.00 \text{ cal}}{\text{g H}_2\text{O} \times °\text{C change}} \times 465 \text{ g H}_2\text{O}$

$\times (64.3 - 24.0)°\text{C change} = 18,700 \text{ cal}$
$(1.87 \times 10^4 \text{ cal or } 18.7 \text{ kcal})$

or, you can rearrange the equation:

$$\text{specific heat} = \dfrac{\text{cal}}{\text{g} \cdot °\text{C change}}$$

$\text{cal} = (\text{specific heat})(\text{g})(°\text{C change})$

$= \dfrac{1.00 \text{ cal}}{\text{g H}_2\text{O} \times °\text{C change}} \times 465 \text{ g H}_2\text{O} \times 40.3°\text{C change} = 18,700 \text{ cal}$

(b) $1.87 \times 10^4 \text{ cal} \times \dfrac{4.18 \text{ J}}{1 \text{ cal}} = 7.82 \times 10^4 \text{ J (or 78.2 kJ)}$

EXAMPLE 6.6 Using the information in Table 6.4, calculate how many (a) kcal and (b) kJ are needed to convert 12.4 g of ice at 0.00°C into steam at 100.0°C.

SOLUTION Use heat of fusion for melting ice at 0.00°C:

$12.4 \text{ g ice} \times \dfrac{79.8 \text{ cal}}{\text{g ice}} = 990 \text{ cal}$

Use specific heat to warm liquid water from 0.00°C to 100.0°C:

$12.4 \text{ g H}_2\text{O} \times \dfrac{1.00 \text{ cal}}{\text{g H}_2\text{O} \cdot °\text{C change}} \times 100.0°\text{C change} = 1240 \text{ cal}$

Use heat of vaporization to convert water into steam at 100.0°C:

$12.4 \text{ g H}_2\text{O} \times \dfrac{540 \text{ cal}}{\text{g H}_2\text{O}} = 6700 \text{ cal}$

(a) total energy $= (990 + 1240 + 6700) \text{ cal} \times \dfrac{1 \text{ kcal}}{1000 \text{ cal}} = 8.93 \text{ kcal}$

(b) $8.93 \text{ kcal} \times \dfrac{1000 \text{ cal}}{1 \text{ kcal}} \times \dfrac{4.184 \text{ J}}{1 \text{ cal}} \times \dfrac{1 \text{ kJ}}{1000 \text{ J}} = 37.4 \text{ kJ}$

Because of its high heat of fusion, heat of vaporization, and specific heat, water moderates the earth's temperature. On hot days, water absorbs heat and stays cooler than most materials. Relatively small amounts evaporate, removing large amounts of heat from the earth's surface. At cooler temperatures, water vapor condenses, releasing large amounts of heat to warm the earth.

Your body, which is about 62% water by weight, gets a similar benefit. Water keeps you from overheating on hot days and resists cooling at colder temperatures. Without water, your body couldn't consistently maintain its temperature of 37°C.

CHEMISTRY SPOTLIGHT

Keeping Cool

Figure 6.16 On a hot day, the evaporation of sweat helps cool our bodies.

People exercising on hot days sweat away as much as 1.00 L of water per hour (Figure 6.16). The heat of vaporization of water at 37°C (instead of at its boiling point) is 580 cal/g, and we can use this to calculate how much heat energy the evaporating water removes per hour:

$$1.00 \text{ L H}_2\text{O} \times \frac{1{,}000 \text{ mL}}{1 \text{ L}} \times \frac{1 \text{ g H}_2\text{O}}{1 \text{ mL H}_2\text{O}} \times \frac{580 \text{ cal}}{\text{g H}_2\text{O}}$$
$$= 5.8 \times 10^5 \text{ cal}$$

To appreciate how much heat this is, let's estimate how much the water temperature in a person would rise if this heat *weren't* removed. If we assume that 62% of a 70.0-kg (154-lb) person's weight is water, we can calculate how much water their body contains:

$$70.0 \text{ kg} \times \frac{1{,}000 \text{ g}}{1 \text{ kg}} \times 62\% = 4.3 \times 10^4 \text{ g water}$$

To calculate the warming effect of 5.8×10^5 cal on 4.3×10^4 g of water in the body, we need to use the specific heat of water and then substitute the new information:

$$\text{specific heat of water} = \frac{1.00 \text{ cal}}{1 \text{ g} \times 1°\text{C}}$$
$$= \frac{5.8 \times 10^5 \text{ cal}}{(4.3 \times 10^4 \text{ g})(°\text{C change})}$$

To solve for the unknown (°C change), we need to cross-multiply:

$$(1.00 \text{ cal})(4.3 \times 10^4 \text{ g})(°\text{C change})$$
$$= (1 \text{ g})(1°\text{C})(5.8 \times 10^5 \text{ cal})$$

Rearranging to solve for the °C change:

$$°\text{C change} = \frac{(1 \text{ g})(1°\text{C})(5.8 \times 10^5 \text{ cal})}{(1.00 \text{ cal})(4.3 \times 10^4 \text{ g})}$$
$$= 13°\text{C}$$

Since a Celsius degree is 1.8 times greater than a Fahrenheit degree, a rise of about 13°C is an increase of about 23°F. This would produce a body temperature of 122°F. The highest recorded body temperature from which a person recovered, however, is 115.7°F.

6.3 PHYSICAL PROPERTIES OF GASES

Diffusion and Effusion. The gaseous state is unique in several ways. One liter of a gas weighs much less than one liter of a liquid or solid. It is almost impossible to compress a liquid or a solid, but you can easily squeeze a gas into a smaller volume. And unlike in a liquid or a solid, gas particles collide with the entire container surface, exerting an equal outward push, or *pressure,* on every part of the container.

Two or more gases mix readily, a process called **diffusion** (Figure 6.17). According to kinetic molecular theory (Section 6.1), molecules in a gas (or atoms, in the case of noble gases) move independently and freely, colliding with nearby molecules and mixing with them.

Effusion, in contrast, is the spontaneous movement of gas particles through a tiny opening, going from a region of higher pressure to one of lower pressure. Gas rushing out of a pricked balloon, or escaping from a spray can, are examples of effusion.

Lighter molecules move faster and thus effuse faster than heavier molecules. Thomas Graham (1805–1869), a Scottish chemist, discovered that the rate of effusion is inversely proportional to the molar mass of a gas. This relationship, called **Graham's law,** is stated mathematically as:

$$\frac{\text{rate of effusion of A}}{\text{rate of effusion of B}} = \sqrt{\frac{\text{molar mass of B}}{\text{molar mass of A}}}$$

EXAMPLE 6.7 Does HCl or ammonia (NH_3) effuse faster? How much faster?

SOLUTION The molar masses of HCl and NH_3 are 36.46 g and 17.04 g, respectively, so NH_3, having lighter molecules, effuses faster. The ratio of the rates equals the *square root* of (36.46/17.04), which is 1.46.

Figure 6.17 Molecules of NO_2, a brown gas, diffuse and mix evenly with other molecules in the container.

Pressure, Volume, and Temperature. No matter how a gas behaves chemically, it acts physically like other gases. Its pressure, volume, and temperature are interrelated in a predictable way. As you ride a bicycle or drive a car, air confined to a specific volume inside the tires gets hotter. Since temperature is a measure of the average kinetic energy of particles, air particles inside the tire move faster and hit the tire walls harder and more often, increasing the tire pressure. After you stop, the pressure and temperature of gas inside the tires both drop (you can check the change in pressure with a tire gauge). At constant volume, then, the pressure of a gas is directly related to temperature.

You can enlarge a balloon by putting it in a closed container, then withdrawing air to lower the air pressure. The gas pressure inside the balloon causes it to expand until its internal pressure matches the external pressure (Figure 6.18). If you let air back into the container, the balloon returns to its original,

to vacuum pump

Figure 6.18 The volume of a balloon will increase if the external pressure is lowered (center) and decrease if the temperature is lowered (right).

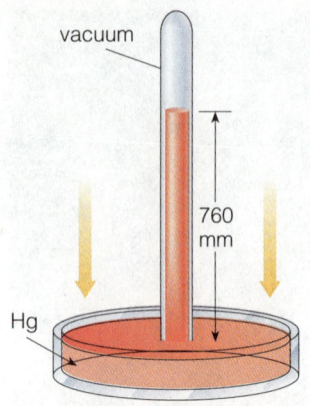

Figure 6.19 A mercury barometer. Normal atmospheric pressure (↓) at sea level pushes Hg to a height of 760 mm (29.9 in) in a tube containing a vacuum. At this height, the downward pressure of the Hg in the tube equals the atmospheric pressure.

smaller size. This illustrates another relationship: at constant temperature, the volume of a gas increases as the external pressure decreases, and *vice versa*.

If you put a balloon in the refrigerator, the balloon shrinks as the air inside it cools. When you take the balloon out, the air gradually expands to its original volume as it warms to room temperature. Volume, like pressure, is directly related to temperature.

Three Important Gas Laws. These physical properties of gases are summarized in three laws:

- *Boyle's law:* At constant temperature, the volume of a fixed amount of gas is inversely proportional to pressure.
- *Charles' law:* At constant pressure, the volume of a fixed amount of gas is directly proportional to temperature.
- *Gay-Lussac's law:* At constant volume, the pressure of a fixed amount of gas is directly proportional to temperature.

These laws are combined into the equation:

$$\frac{P_1 V_1}{T_1} = \frac{P_2 V_2}{T_2}$$

where P_1, V_1, and T_1 are the initial pressure, volume, and temperature of a fixed amount of a gas, and P_2, V_2, and T_2 are the pressure, volume, and temperature of that gas under new conditions.

The units are important. Volumes of gases are usually measured in mL or L. The temperature in this equation *must* be expressed in kelvins.

Several units are used to measure pressure. Average air pressure at sea level, measured by a mercury barometer (Figure 6.19), pushes a column of mercury (Hg) to a height of 760 mm in a tube containing a vacuum. So pressure can be expressed in units of mm Hg. Another unit, the *torr,* equals 1 mm Hg. Yet another unit is the *atmosphere (atm),* which equals 760 torr or 760 mm Hg.

The torr *is named in honor of Italian physicist and mathematician Evangelista Torricelli (1608–1647), who invented the mercury barometer.*

The official SI unit of pressure is the pascal (Pa). 1 atm = 1.01 × 10⁵ Pa (or 1.01 × 10² kPa)

EXAMPLE 6.8 Convert a pressure of 785 mm Hg into (a) torr and (b) atm.

SOLUTION (a) 785 mm Hg × $\frac{1 \text{ torr}}{1 \text{ mm Hg}}$ = 785 torr

(b) 785 mm Hg × $\frac{1 \text{ atm}}{760 \text{ mm Hg}}$ = 1.03 atm

EXAMPLE 6.9 Convert into torr (a) 0.984 atm and (b) 744 mm Hg.

SOLUTION (a) 0.984 atm × $\frac{760 \text{ torr}}{1 \text{ atm}}$ = 748 torr

(b) 744 mm Hg × $\frac{1 \text{ torr}}{1 \text{ mm Hg}}$ = 744 torr

CHEMISTRY SPOTLIGHT

Breathing and the Gas Laws

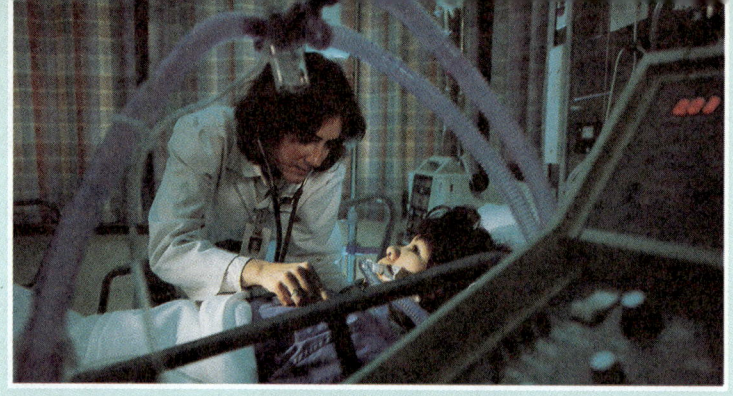

Figure 6.20 A respirator helps people breathe enough oxygen into their lungs.

Each time you breathe, air goes in and out of your lungs in accordance with the gas laws. Inhaling increases the volume of your lungs. According to Boyle's law, this increase in volume reduces the air pressure in your lungs. When this pressure becomes less than the atmospheric pressure outside, air flows into your lungs.

When you exhale you squeeze your lungs, reducing their volume and increasing their pressure. The pressure in your lungs now exceeds the atmospheric pressure, so gases are exhaled.

People get emphysema from smoking and other lung pollutants. Their lungs become less elastic and less able to change volume and pressure. As a result, people with emphysema have difficulty breathing. Some need respirators to help them breathe (Figure 6.20). They also may breathe O_2-enriched gas mixtures to ensure that enough O_2 enters their lungs and blood.

Calculating Changes in Volume, Temperature, and Pressure. Now you are ready to use the equation interrelating pressure, volume, and temperature for a fixed amount of gas. For example, suppose a balloon at 790 torr has a volume of 550 mL at 20°C. What would be the volume at 20°C if the pressure were reduced to 710 mm Hg?

To solve this, first make sure the initial and final pressure, volume, and temperature are each in the same, appropriate units. In the preceding question, express both pressures either in torr or in mm Hg. We'll use torr:

$$\text{final pressure} = 710 \text{ mm Hg} \times \frac{1 \text{ torr}}{1 \text{ mm Hg}} = 710 \text{ torr}$$

Also convert temperatures into kelvins:

20°C + 273 = 293 K

Now substitute in the equation $P_1V_1/T_1 = P_2V_2/T_2$ and solve for the unknown volume (V_2):

$$\frac{(790 \text{ torr})(550 \text{ mL})}{293 \text{ K}} = \frac{(710 \text{ torr})(V_2)}{293 \text{ K}}$$

Since the temperature doesn't change, you can cancel temperature values on both sides of the equation. Then rearrange the equation to solve for V_2:

$$V_2 = \frac{(790 \text{ torr})(550 \text{ mL})}{710 \text{ torr}} = 610 \text{ mL}$$

This answer makes sense. You would expect the volume to increase when the pressure decreases, because gas particles then collide with the container surface less frequently and with the same energy.

CHEMISTRY SPOTLIGHT

Liquid Nitrogen and Cadavers

Making liquid nitrogen from air takes advantage of the gas laws. First, air is passed through materials that remove H_2O and CO_2. The remaining air—mostly N_2 (78%) and O_2 (21%) plus small amounts of Ar, Ne, and He—is compressed to a pressure greater than 100 atm and cooled. When the pressure is suddenly reduced, the gas expands, cooling further. Doing this repeatedly makes the temperature cold enough to liquefy the gases.

Since the liquified gases have different boiling points—N_2 (–195.8°C), O_2 (–183°C), Ar (–189°C), Ne (–246°C), and He (–268.9°C)—they can be separated and purified by evaporation under different conditions.

Liquid nitrogen, the most abundant product, is colorless and very cold (Figure 6.21). It is used to freeze foods and keep them frozen during transit. Physicians sometimes use liquid nitrogen to freeze and remove warts and skin blemishes.

One unusual use of liquid nitrogen is to freeze and store human cadavers. According to one method, the first step is to replace body water, which would expand and rupture cells when frozen. So a protective fluid such as ethylene glycol (the main ingredient in radiator antifreeze) is flushed through the carotid artery and the jugular vein. Next, the body is wrapped in plastic and cooled on crushed ice and salt until its temperature is below 0°C. Then it is wrapped in aluminum foil and stored on dry ice (solid CO_2), at a temperature of –78.5°C. Finally, the cadaver enters a permanent storage facility, essentially a large thermos container chilled with liquid nitrogen.

Figure 6.21 Embryos are frozen and preserved in liquid nitrogen.

Although the public attitude to this has remained cool, several dozen people have arranged to have their bodies preserved frozen until a cure is found for whatever killed them. The plan is then to thaw them out and revive them. In view of the tens of thousands of dollars needed to freeze and maintain these bodies, and the minimal prospects that they will be brought back to life, the clients appear to be getting nothing for something.

EXAMPLE 6.10 Suppose the air in your lungs at 35°C has a volume of 510 mL and a pressure of 0.980 atm.
(a) Calculate the volume in mL if you squeeze your chest cavity so that the pressure becomes 775 mm Hg at 35°C.
(b) If the air in your lungs warmed up to 37°C with the volume staying constant, what would the pressure be in mm Hg?

SOLUTION $0.980 \text{ atm} \times \dfrac{760 \text{ mm Hg}}{1 \text{ atm}} = 745 \text{ mm Hg}$

(a) $\dfrac{(745 \text{ mm Hg})(510 \text{ mL})}{35 + 273 \text{ K}} = \dfrac{(775 \text{ mm Hg})(V_2)}{35 + 273 \text{ K}}$

$V_2 = 490 \text{ mL}$

(b) $\dfrac{(745 \text{ mm Hg})(510 \text{ mL})}{35 + 273 \text{ K}} = \dfrac{(P_2)(510 \text{ mL})}{37 + 273 \text{ K}}$

$P_2 = 750 \text{ mm Hg}$

Partial Pressure. Air is a mixture of gases. The three gas laws, however, apply to mixtures of gases as well as to pure gases. According to kinetic molecular theory, molecules of all gases move about freely, colliding with the walls of their container and exerting pressure. Whether a molecule is of the same or a different substance than its neighboring molecules in a gas sample doesn't matter; each molecule still contributes to the total pressure in the same way.

John Dalton explained this relationship, called the **law of partial pressures** in this way: The total pressure of a gas mixture equals the sum of the partial pressure of each gas in the mixture. This law can be expressed as:

$P_t = P_1 + P_2 + P_3 + \ldots$

where P_t is the total pressure and P_1, P_2, and P_3 are the partial pressures of individual gases in the mixture.

The partial pressure *of a gas is the pressure a gas would exert if it were alone in the container. Its particles collide just as often, with the same force, whether they are alone or mixed with other particles.*

EXAMPLE 6.11 A gas cylinder containing O_2 and CO_2 has a total pressure of 2.18 atm. The partial pressure of O_2 in the tank is 1.89 atm. What is the partial pressure of CO_2?

SOLUTION $2.18 \text{ atm} = 1.89 \text{ atm} + P_{CO_2}$; $P_{CO_2} = 0.29 \text{ atm}$

6.4 THE UNIVERSAL GAS LAW

Avogadro's Law. We have examined the relationship between pressure, volume, and temperature for a fixed sample of gas. But what happens if the amount of gas changes?

Amedeo Avogadro (see Figure 5.11) discovered a relationship between volume and the number of particles in a gas. It is based on the idea that the particles (molecules or atoms) are so tiny that they occupy almost none of the space in a gas sample. According to **Avogadro's law,** *equal volumes of different gases at the same temperature and pressure have equal numbers of molecules.* Since gases with equal numbers of molecules have equal numbers of moles, we can also say that equal volumes of gases at the same temperature and pressure have equal numbers of moles (Figure 6.22).

Scientists find it convenient to specify a standard pressure (1 atm or 760 torr) and temperature (0°C or 273 K) and call these conditions **STP (standard**

Ideally, gases behave exactly this way. In practice, most gases have properties that match closely those of an "ideal" gas.

Figure 6.22 These balloons, one filled with H_2 and the other with air, contain an equal number of molecules. They also contain an equal number of moles.

When pressure is expressed in mm Hg or torr,
$$R = 62.4 \frac{L \cdot mm\ Hg}{mol \cdot K}$$ or
$$62.4 \frac{L \cdot torr}{mol \cdot K}$$, respectively.

temperature and pressure). In accordance with Avogadro's law, it turns out that *one mole of any gas occupies 22.4 L at STP.*

Mathematically, we can say that for one mole of a gas:

$$\frac{P_1 V_1}{T_1} = \frac{P_2 V_2}{T_2} = \frac{(1\ atm)(22.4\ L/mol)}{273\ K} = 0.0821 \frac{L \cdot atm}{mol \cdot K}$$

This value, 0.0821 L · atm/mol · K, is called the *universal gas constant* and has the symbol *R*. For one mole of a substance, then,

$$\frac{PV}{T} = R \quad \text{or} \quad PV = RT$$

For a variable number of moles, the relationships we have discussed about pressure, volume, temperature, and amount of a gas are all summarized mathematically in the **universal gas law**:

$$PV = nRT$$

in which *P* is pressure in atm, *V* is volume in L, *n* is number of moles, *T* is temperature in kelvins, and *R* is the universal gas constant.

You can use this equation to calculate the volume, temperature, or pressure of any amount of gas under any conditions. For example, you can calculate the pressure for 97.0 g of CO_2 gas in a 25.0 L container at 20°C. First convert the information into appropriate units:

$$20°C + 273 = 293\ K$$

$$97.0\ g\ CO_2 \times \frac{1\ mol\ CO_2}{44.01\ g\ CO_2} = 2.20\ mol\ CO_2 \quad \text{(molar mass)}$$

Then use the universal gas law equation to solve for pressure:

$$(P)(25.0\ L) = (2.20\ mol)\left(0.0821 \frac{L \cdot atm}{mol \cdot K}\right)(293\ K)$$

$$P = 2.12\ atm$$

You can also calculate the molar mass and formula weight of a gas if you know the mass of a certain number of moles. For example, suppose 107 g of a gas occupies 16.4 L at 713 torr and 10°C. To calculate the formula weight, first calculate the moles of gas in the 107 g sample.

Convert the information into appropriate units:

$$10°C + 273 = 283\ K$$

$$713\ torr \times \frac{1\ mm\ Hg}{1\ torr} \times \frac{1\ atm}{760\ mm\ Hg} = 0.938\ atm$$

Then use the universal gas law equation:

$$(0.938\ atm)(16.4\ L) = n\ \frac{0.0821\ L \cdot atm}{mol \cdot K}(283\ K)$$

$$n = 1.51\ mol$$

Now use the mass of this sample (107 g) to calculate molar mass:

$$\frac{107\ g}{1.51\ mol} = 70.9\ g/mol = \text{molar mass}$$

formula weight = 70.9 amu

The identity of this gas (from its formula weight) is Cl_2.

EXAMPLE 6.12 A 100.0 L tank contains O_2 gas at a pressure of 2150 torr at 15°C. Calculate the number of (a) moles and (b) molecules of O_2 in the tank.

SOLUTION $2150 \text{ torr} \times \dfrac{1 \text{ atm}}{760 \text{ torr}} = 2.83 \text{ atm}$

(a) $(2.83 \text{ atm})(100.0 \text{ L}) = (n)\left(0.0821 \dfrac{\text{L} \cdot \text{atm}}{\text{mol} \cdot \text{K}}\right)(273 + 15 \text{ K})$

$n = 12.0 \text{ mol}$

(b) $12.0 \text{ mol} \times \dfrac{6.02 \times 10^{23} \text{ molecules}}{1 \text{ mol}} = 7.22 \times 10^{24} \text{ molecules}$

Stoichiometry Involving Gases. When we discussed stoichiometry in Section 5.5, you learned how to calculate the grams or moles of reactants and products in chemical reactions. When a reactant or product is a gas, it is often more practical to measure the volume of gas produced or consumed at a certain temperature and pressure. Now you can use the universal gas law equation to do this.

Let's consider again the reaction that produces mercury from its ore, mercury (II) sulfide (HgS). The balanced equation is:

$$HgS + O_2 \longrightarrow Hg + SO_2$$

You can calculate how many L of SO_2 are produced at 1.00 atm at 450°C for each metric ton (1000 kg or 10^6 g) of ore that reacts with a 75.0% yield.

First, calculate the number of moles of SO_2 produced:

$10^6 \text{ g HgS} \times \underbrace{\dfrac{1 \text{ mol HgS}}{232.66 \text{ g HgS}}}_{\text{molar mass}} \times \underbrace{\dfrac{1 \text{ mol } SO_2}{1 \text{ mol HgS}}}_{\text{mole ratio}} = 4.2981 \times 10^3 \text{ mol } SO_2$
(theoretical yield)

$(4.2981 \times 10^3 \text{ mol})(0.750) = 3.22 \times 10^3 \text{ mol } SO_2$ (actual yield)

Then use the universal gas law to calculate the volume of SO_2 produced at this temperature and pressure:

$(1.00 \text{ atm})(V) = (3.22 \times 10^3 \text{ mol})\left(0.0821 \dfrac{\text{L} \cdot \text{atm}}{\text{mol} \cdot \text{K}}\right)(450 + 273 \text{ K})$

$V = 1.91 \times 10^5 \text{ L } SO_2$

SO_2 is an air pollutant emitted from smelters and from power plants that burn sulfur-containing coal. SO_2 emissions are a major cause of acid rain (see Chemistry Spotlight in Section 9.4).

EXAMPLE 6.13 In the preceding reaction with HgS, calculate the volume of SO_2 produced per metric ton of ore (assuming a 75.0% yield) at 229°C and a pressure of 735 mm Hg.

SOLUTION $735 \text{ mm Hg} \times \dfrac{1 \text{ atm}}{760 \text{ mm Hg}} = 0.967 \text{ atm}$

$(0.967 \text{ atm})(V) = (3.22 \times 10^3 \text{ mol})\left(0.0821 \dfrac{\text{L} \cdot \text{atm}}{\text{mol} \cdot \text{K}}\right)(229 + 273 \text{ K})$

$V = 1.37 \times 10^5 \text{ L } SO_2$

> When water turns ice does it remember one time it was water? When ice turns back into water does it remember it was ice?
>
> *Carl Sandburg (1878–1967)*

SUMMARY

The physical states of a substance are solid, liquid, and gas. The kinetic molecular theory explains typical properties of these states. The state depends on forces of attraction between the particles (ions, molecules, or atoms) and their kinetic energy. Ionic compounds and metals typically have strong attractive forces and are solids at room temperature. Molecules attract each other by London forces, dipole–dipole forces, and hydrogen bonds. Of these, hydrogen bonds usually are the strongest.

Pure substances melt, freeze, boil, and condense at a definite temperature under normal conditions. Heat is absorbed or given off when a substance changes state. H_2O is unusually effective at absorbing heat because of its hydrogen bonds.

Gases readily mix (diffuse), and their rate of effusion is described by Graham's law. According to the law of partial pressures, the pressures of individual gases in a mixture are independent and additive. Relationships of pressure, volume, and temperature of a fixed amount of gas are described by Boyle's law, Charles' law, and Gay-Lussac's law. These three laws are combined with Avogadro's law into the universal gas law, which interrelates pressure, volume, temperature, and moles of a gas: $PV = nRT$. This law is useful in calculating the volumes of gases that participate in chemical reactions.

KEY TERMS

Avogadro's law (6.4)
Boiling (6.2)
Boyle's law (6.3)
Charles' law (6.3)
Condensation (6.2)
Diffusion (6.3)
Dipole–dipole force (6.1)
Dynamic equilibrium (6.2)
Effusion (6.3)
Freezing (6.2)
Gay-Lussac's law (6.3)
Graham's law (6.3)
Heat of fusion (6.2)
Heat of vaporization (6.2)
Hydrogen bond (6.1)
Kinetic molecular theory (6.1)
Law of partial pressures (6.3)
London force (6.1)
Melting (6.2)
Specific heat (6.2)
Standard temperature and pressure (STP) (6.4)
Sublimation (6.2)
Universal gas law (6.4)
Vaporization (6.2)
Vapor pressure (6.2)

EXERCISES

Even-numbered exercises are answered at the back of this book.

Physical States of Matter

1. List the three physical states of matter.
2. According to the kinetic molecular theory, which physical state results when the motion of particles is much weaker than their forces of attraction?
3. Use the kinetic molecular theory to explain why gases condense to become liquids at low temperatures.
4. As temperature increases, is the effect primarily on (a) the forces of attraction between particles or (b) the motion of particles.
5. Use the kinetic molecular theory to explain why solids have a definite shape but liquids and gases do not.
6. According to the kinetic molecular theory, why are gases more compressible than liquids or solids?

7. Is the solid state of a substance typically more dense or less dense than its liquid state? Can you think of a common substance that is an exception?

Attractive Forces and Physical State

8. What is the physical state of ionic compounds at room temperature? Explain why.
9. List the following in order of increasing strength for small molecules: **(a)** dipole–dipole force, **(b)** hydrogen bond, **(c)** London force.
10. Name the strongest intermolecular attraction between each of the following: **(a)** H_2, **(b)** H_2O, **(c)** NCl_3, **(d)** CO_2.
11. Name the strongest intermolecular attraction between each of the following: **(a)** NH_3, **(b)** PCl_3, **(c)** O_2, **(d)** SO_2.
12. Explain why at room temperature water is a liquid while the substances from which it is made—hydrogen and oxygen—are gases.
13. Predict the physical state at room temperature of each of the following: **(a)** Cl_2, **(b)** $MgCl_2$, **(c)** Mg, **(d)** NCl_3, **(e)** C_2H_6.
14. Predict the physical state at room temperature of each of the following: **(a)** K_2O, **(b)** Ar, **(c)** Sr, **(d)** $CHCl_3$ (chloroform), **(e)** C_8H_{18}.
15. CF_4, CCl_4, and CBr_4, are all nonpolar compounds that have no dipole moment.
    (a) What type of intermolecular forces occur between molecules of these compounds?
    (b) Which one of these compounds has the strongest intermolecular attractions?
    (c) Which one of these compounds is the most likely to be a gas at room temperature?
16. Which noble gas has the strongest attraction between its atoms?
17. Predict which noble gas has the lowest boiling point.
18. Which metal apparently has the weakest attractive forces between its atoms?
19. Does Br_2 or $CaBr_2$ have a lower melting point? Explain why.
20. Which noble gas has the highest melting point? Explain why.

Changes in Physical State

21. Two common substances that undergo sublimation are CO_2 and I_2. What kind of intermolecular forces do these substances have?
22. In terms of the kinetic molecular theory, explain why heat causes solids to melt.
23. Diethyl ether, once used widely as an anesthetic, has a boiling point of 34.6°C. Would diethyl ether in an open container evaporate at 20°C? Explain.
24. Identify the term that applies to each of the following physical changes: **(a)** liquid into a solid, **(b)** solid directly into a gas, **(c)** gas into a liquid, **(d)** liquid into a gas, **(e)** solid into a liquid.
25. In a closed container, a dynamic equilibrium exists between the processes of vaporization and condensation. What happens to this equilibrium if the container is opened?
26. Hexane, C_6H_{14}, is a liquid. Predict whether hexane or water has a greater vapor pressure at room temperature.
27. Rationalize why H_2S is a gas at room temperature whereas H_2O is a liquid.
28. At high altitude does it take a longer or a shorter time than normal to **(a)** boil water and **(b)** cook potatoes by boiling them in water?

Energy Changes

29. Explain how water helps warm coastal cities at night and in the winter.
30. How many calories (cal) and how many joules (J) does it take to heat the water in a glass (225 g H_2O) from 20.0°C to 35.0°C on a warm day?
31. How many calories (cal) does it take to warm 85.0 g of H_2O from 20.0°C to its boiling point and then boil the H_2O?
32. Use the information in Table 6.4 to calculate how many calories (cal) it takes to **(a)** warm 50.0 g of ethanol from 20.0°C to its boiling point (78.5°C) and then **(b)** change it into a gas.
33. Use the total number of calories you calculated for exercise 32 to calculate whether this amount of energy would melt 50.0 g of aluminum (Al) at 20°C (see Table 6.4). The melting point of Al is 660°C.
34. Suppose two substances, one polar and one nonpolar, have similar molar masses. Which one would you expect to have the greater heat of fusion? Explain.
35. HCl and F_2 have similar molar masses. Which one would you expect to have the greater heat of fusion? Explain.
36. When one drop each of H_2O and hexane (C_6H_{14}) are placed on the same warm surface, which one evaporates first? Explain.
37. On a hot day, why is a pool of water cooler than the nearby concrete?
38. Assume that 62.0% of the mass of a 67.0 kg person is H_2O. How many extra calories does that H_2O contain when the person's temperature rises from 37.0°C (98.6°F) to 38.5°C (101.3°C)?

Diffusion and Effusion

39. Using the kinetic molecular theory, explain why gases diffuse more readily than solids or liquids.
40. Which two gases effuse the most rapidly?
41. Suppose in the Gas Olympics three gases—acetylene (C_2H_2), hydrogen sulfide (H_2S), and nitrogen (N_2)—are simultaneously released from the same container. Which gas will be the first to cross the finish line 30 m from the container?
42. Calculate the relative rates of effusion of F_2 and Cl_2.
43. What is the molar mass of the gas that effuses four times slower than H_2 does? Identify the gas.

Pressure, Volume, and Temperature of Gases

44. Why does expanding or compressing an accordian cause air to move in or out, producing sound? Which gas law does this illustrate?
45. When you close your mouth over a straw and sip a beverage, what happens to the air pressure in the straw above the liquid? What pushes the liquid into your mouth?
46. Explain why the pressure inside an automobile tire increases on a hot day. What gas law does this illustrate?
47. Look at Figure 6.18. Which gas law does the center balloon illustrate? Which gas law does the balloon on the right illustrate?
48. Convert the following into pressure in torr: **(a)** 650 mm Hg, **(b)** 2.00 atm, **(c)** 74.3 cm Hg, **(d)** 0.88 atm.
49. Convert the following into pressure in atm: **(a)** 875 mm Hg, **(b)** 746 torr, **(c)** 1250 torr, **(d)** 56 mm Hg.
50. If a 2.30 L container of N_2 at STP warms up to 55°C, with the volume remaining constant, what will the pressure be in **(a)** atm and **(b)** torr?
51. Suppose a balloon is partially inflated to contain 75.0 mL of H_2 gas at STP. Then the balloon is put into a chamber from which air is pumped to reduce the pressure. What is the final pressure in torr when the balloon inflates to its maximum volume of 375 mL at 10°C?
52. Suppose you have 265 mL of He in a balloon at STP. If you put the balloon in a freezer at −24°C and 0.976 atm, what will its volume be?

Universal Gas Law

53. How many g of H_2 are in the balloon described in exercise 51?
54. How many moles of H_2 are in the balloon described in exercise 51?
55. How many g of He are in the balloon described in exercise 52?
56. How many moles of He are in the balloon described in exercise 52?
57. Your lungs hold a total of about 6.0 L of air at 37°C and 760 torr. How many **(a)** moles and **(b)** molecules are in that volume of air?
58. At 18.0°C, 70.4 g of a gas has a volume of 36.7 L at 790 torr. What is the molar mass of this gas? Identify the gas.
59. What is the volume of 25.0 g of CH_4 gas at 20.0°C and 0.989 atm?
60. What is the weight in g of 6.17 L of NH_3 at STP?
61. Calculate the density of F_2 gas in g/L at STP.

Partial Pressure

62. N_2, O_2, and H_2O make up about 99.0% of normal air. At 760 torr and a relative humidity of 20%, the partial pressures of O_2 gas and H_2O are 160 torr and 5 torr, respectively. What is the partial pressure of N_2 under these conditions?
63. In exhaled air at 760 torr, the partial pressures of N_2, O_2, and H_2O are about 570 torr, 115 torr, and 50 torr, respectively. Another gas, present only in trace amounts in inhaled air, is abundant in exhaled air. What is its partial pressure? What do you think the gas is?

Stoichiometry

Plants make glucose ($C_6H_{12}O_6$) and O_2 by photosynthesis. The net reaction is:

$$6CO_2 + 6H_2O \longrightarrow C_6H_{12}O_6 + 6O_2$$

Use this balanced equation to answer exercises 64–69.

64. For each 1.00 kg of glucose synthesized, how many g of O_2 are synthesized? How many mol of O_2 are synthesized?
65. For each 1.00 kg of glucose synthesized, how many mol of CO_2 are consumed? How many g of CO_2 are consumed?
66. For exercise 64, what would be the volume at STP of the O_2 synthesized?
67. For exercise 65, what is the volume at STP of the CO_2 consumed?
68. For exercise 64, what would be the volume at 37°C and 0.968 atm of the O_2 synthesized?
69. For exercise 65, what is the volume at 0°C and 725 mm Hg of the CO_2 consumed?
70. What volume would three molecules of O_2 gas occupy at STP?

71. How many molecules would 750 mL of O_2 gas at STP contain?

You metabolize carbohydrates such as glucose ($C_6H_{12}O_6$) by a net reaction that is the reverse of photosynthesis:

$$C_6H_{12}O_6 + 6O_2 \longrightarrow 6CO_2 + 6H_2O$$

Use this balanced equation to answer exercises 72–75.

72. How many mol of O_2 do you consume to metabolize 325 g of glucose?

73. For exercise 72, what volume of O_2 at 1.00 atm and 20°C is consumed?

74. To supply the amount of O_2 you calculated for exercise 72, how many breaths of air would you need if all the O_2 inhaled entered your blood and was used to metabolize glucose. Assume that each breath you take brings into your lungs 520 mL of air at 20°C, and air is 21% O_2 by volume.

75. Do you think the assumption in exercise 74 (that all the inhaled O_2 enters the blood and is used to metabolize glucose) is valid? Explain.

DISCUSSION EXERCISES

1. At STP, do most substances exist as solid, liquid, or gas? Explain why.

2. In terms of the arrangement of H_2O molecules, rationalize why ice floats on liquid water.

3. Some people claim that hot water freezes faster than cold water. Is there any way that could happen? Explain.

4. Explain in terms of the gas laws why it is more difficult to breathe at high altitude.

5. Calculate what gas has a density of 0.17 g/L at 20°C and 745 mm Hg.

7

Solutions

1. What are some important properties of water?
2. What happens when a substance dissolves in water to form a solution? How do solutions differ from suspensions and colloids?
3. How do you predict what substances dissolve in water or other solvents? How do temperature and pressure affect solubility?
4. What are the most common ways to express the concentration of a solution?
5. How do solutions affect the freezing and boiling points of water?
6. What are osmosis and dialysis? How do they affect water balance in the body?

Nitrogen Supersaturation and the Bends

When water plunges over dams or down waterfalls, it carries air deep into the water. At greater depths, the increased pressure drives more air into solution. Since the most abundant gas in air is nitrogen (N_2), extra nitrogen dissolves in the water. Under these conditions, the water is said to be supersaturated with nitrogen.

Nitrogen supersaturation kills fish and amphibians (Figure 7.1). Within an hour or two in supersaturated water, fish absorb extra N_2 in their blood. When these fish swim nearer the surface, the reduced pressure allows N_2 gas to come out of solution in their blood. N_2 bubbles can form in the flesh or in blood vessels, sometimes blocking circulation to vital organs. The eyes, which have a rich blood supply, are a common site of bubble formation, giving the fish a "pop-eyed" appearance. After a month or more of exposure to moderate supersaturation, fish can develop gas bubble disease, which has caused massive fish kills in hatcheries just downstream from dams.

Deep-sea divers face a similar danger. When they are far under water, the increased pressure causes extra amounts of the air they breathe to dissolve in their blood. If they come to the surface too quickly, some of the gas, mostly N_2, comes out of solution in their blood and tissues. The gas bubbles disrupt blood circulation and cause excruciating pain. This condition, known as the *bends,* can be fatal.

To reduce this risk, divers come to the surface gradually. In some situations, they use decompression chambers that slowly reduce the pressure to normal. They also breathe a gas mixture of oxygen and helium. Helium is safer than N_2 because high pressures don't increase the solubility of helium as much as that of N_2. As a result, less gas comes out of solution when divers return to the surface.

Figure 7.1 Nitrogen supersaturation killed this frog. Bubbles of N_2 come out of solution and appear in the flesh.

Figure 7.2 A water molecule is bent and polar.

Covalent bonds occur within water molecules. Hydrogen bonds occur between different water molecules.

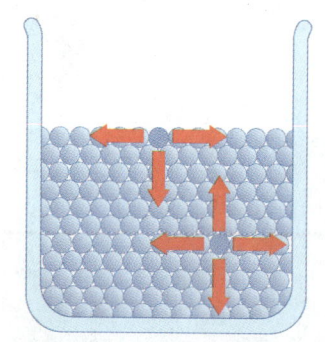

Figure 7.4 Surface water molecules are attracted only sideways and inward by hydrogen bonds. Water molecules in the body of the liquid are attracted equally in all directions.

When you swim, water dissolves the salt and other soluble materials on your skin. Fortunately, water doesn't dissolve your skin itself, or your hair or nails. Inside your body, water dissolves minerals, glucose, waste materials, and hormones, but not cell membranes, fat, or bone. Your life depends on this selective dissolving action of water.

Sometimes the boundary between what dissolves and what doesn't is hazy. For example, water in your blood thoroughly suspends red blood cells, but doesn't actually dissolve them. Blood also carries large lipid–protein (lipoprotein) molecules. This mixture, called a *colloid,* is somewhere between a solution and a suspension.

In solutions, the amount of dissolved material is important. It isn't enough just to know, for example, that your blood carries dissolved glucose. You need to know how much. Too much glucose is a sign of diabetes, and too little can be a sign of malnutrition or hypoglycemia.

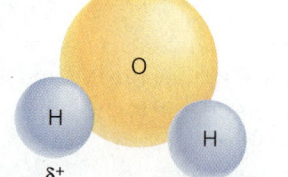

7.1 WATER: A REMARKABLE SUBSTANCE

Important Properties of Water. If you were searching for substances that have extreme or unusual properties, you might expect to find a list of exotic chemicals. But near the top of the list would be water—the most familiar chemical of all.

Recall that water molecules are bent and polar (Section 4.6) (Figure 7.2). They attract each other by hydrogen bonds (see Figure 6.7). The strength of these hydrogen bonds comes from the small size of water molecules and the large partial charges caused by the difference in electronegativity between oxygen and hydrogen. The energy needed to break large numbers of hydrogen bonds gives water its remarkably high specific heat, heat of fusion, and heat of vaporization (Section 6.2). Indeed, hydrogen bonds are what make water a liquid at room temperature, unlike the gases (H_2 and O_2) from which it forms.

You may have seen tiny insects skate across water (Figure 7.3). Water has high **surface tension,** the force that causes the surface of a liquid to contract. Molecules within water are attracted in all directions by hydrogen bonds. But water molecules at the surface of water and air (or anything else) are attracted only by their neighboring water molecules (Figure 7.4). The force of these sideways and inward attractions produces surface tension.

Almost all substances are more dense as solids than as liquids (see Table

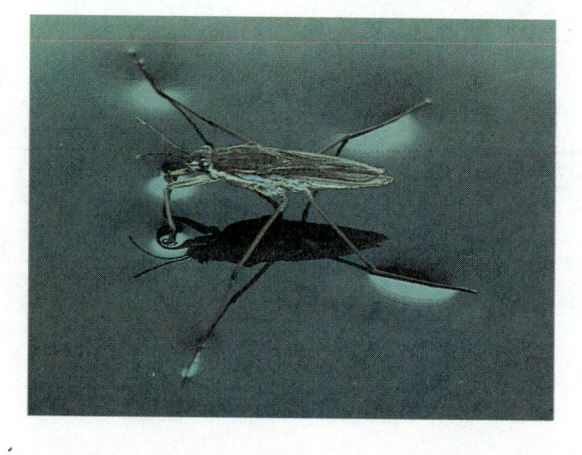

Figure 7.3 The strong surface tension of water supports this water strider.

6.1). But not water. As water cools, its molecules move more slowly and pack more closely together, until it reaches maximum density at 4°C. At colder temperatures, water molecules move even more slowly. At 0°C they begin to form rigid, open, hexagonal networks. The force of hydrogen bonds holds the molecules farther apart than at 4°C (Figure 7.5). In the hexagonal patterns of snowflakes (Figure 7.6), you can see the underlying pattern of water molecules in ice.

In other words, water expands and becomes less dense when it freezes. This has important consequences. The expansion can damage bottles and automobile radiators. But it also makes ice float. When rivers and lakes freeze in winter, a layer of ice forms on top that insulates the water below and keeps it from freezing solid. This helps aquatic organisms stay alive. Few organisms can survive freezing because the expanding ice ruptures cell membranes.

Water also is a superior dissolving agent. In the next section, we examine this property in more detail.

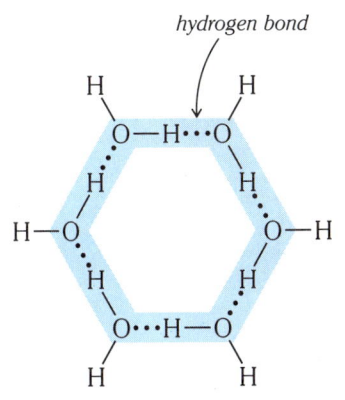

Figure 7.5 Hydrogen bonds (dotted lines) hold water molecules in ice in a rigid, open, hexagonal pattern.

Figure 7.6 The hexagonal (six-sided) shape of snowflakes comes from the underlying geometry of water molecules in ice (see Figure 7.5).

7.2 FORMING SOLUTIONS

Solutions. A **solution** is a homogeneous mixture of two or more substances. We can picture a solution as having the individual atoms, molecules, or ions of its substances evenly interspersed. The substance dissolved, usually present in the lesser amount, is the **solute.** The dissolving agent, present in the greater amount, is the **solvent.** The amount of solute in a given amount of solution is the *concentration* of the solution.

Solutions include gases dissolved in gases (air), solids in solids (alloys),

Table 7.1 Types of Solutions

Type	Examples	Type	Examples
solid in solid	jewelry, coins	liquid in liquid	antifreeze
solid in liquid	salt water	liquid in gas*	mist
solid in gas*	dust in air	gas in solid*	marshmallow
liquid in solid	tooth fillings (silver mercury amalgam)	gas in liquid	carbonated drink
		gas in gas	air

* Are colloids or suspensions rather than true solutions

and other combinations (Table 7.1). Most solutions, however, are solids, liquids, or gases dissolved in a liquid. The most common solvent is water.

The Solution Process. When a solution forms, the forces holding particles together in separate substances must be replaced by forces of attraction between particles of all the substances.

A rough rule of thumb for predicting whether or not substances will form solutions is: *Like dissolves like; unlikes do not.* The two categories are:

- substances with full or partial permanent electrical charges (ionic and polar covalent compounds)
- substances without permanent electrical charges (nonpolar covalent compounds)

Table 7.2 summarizes how this rule works.

Table 7.2 Predicting Whether or Not Solutions Will Form

Type of Solute	Type of Solvent	Solution
ionic or polar covalent	polar covalent	yes
nonpolar covalent	polar covalent	no
ionic or polar covalent	nonpolar covalent	no
nonpolar covalent	nonpolar covalent	yes

The solubility rule predicts that table salt (NaCl) dissolves in water because both substances are in the same category (water is polar covalent and NaCl is ionic). Figure 7.7 shows what happens when grains of NaCl are put in water. Water molecules collide with surface Na^+ and Cl^- ions. Polar ends of the water molecules attract oppositely charged ions strongly enough to pull them away from their ionic crystal lattice. The freed ions stay in solution surrounded by water molecules.

Although many ionic compounds dissolve in water, some have such strong attractions between their ions that water molecules cannot pull them away from their neighboring ions. Table 7.3 lists the most common ionic compounds that dissolve in water.

EXAMPLE 7.1 Which of the following dissolve in water: (a) $CaCl_2$, (b) AgCl, (c) Na_2SO_4, (d) $AlPO_4$.

SOLUTION (a) and (c) (see Table 7.3)

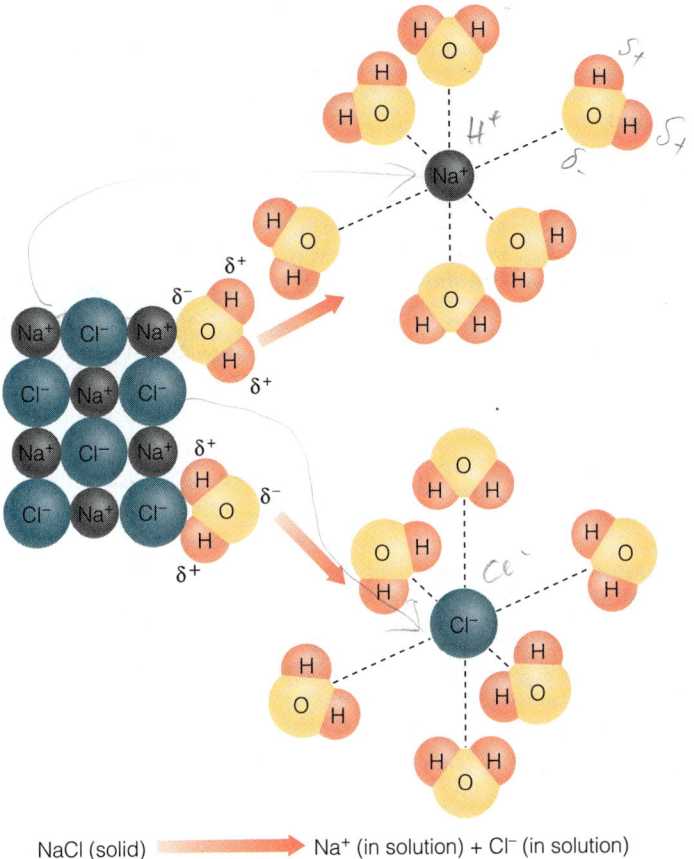

Figure 7.7 Dissolving an ionic solid (sodium chloride) in a polar solvent (water). Negative (O) poles of water molecules attract Na⁺ ions, while positive (H) poles of water attract Cl⁻ ions.

NaCl (solid) ⟶ Na⁺ (in solution) + Cl⁻ (in solution)

    The solubility rule also predicts that table sugar ($C_{12}H_{22}O_{11}$)—because it is polar—dissolves in water. As you will learn in Chapter 19, sugar molecules have many polar —O—H groups. Polar water molecules form hydrogen bonds with these polar groups in sugar molecules to form a stable solution.

    The second part of the rule predicts that oil, a mixture of nonpolar hydrocarbons, doesn't dissolve in water. Since nonpolar molecules have no charge, water molecules aren't attracted to them. Water molecules form hydrogen bonds with other water molecules instead, so oil and water don't mix (Figure 7.8). Because oil is less dense than water, it forms a layer on top.

    You can, however, dissolve oil in nonpolar solvents such as gasoline or kerosene. Here both the solvent and solute molecules are nonpolar, with only weak London forces of attraction to their neighbors. Stirring the mixture intersperses the molecules, which then have no tendency to form separate layers.

Table 7.3 Ionic Compounds that Dissolve in Water*

Compounds containing alkali metals (Group 1A elements)
Compounds containing ammonium (NH_4^+), nitrate (NO_3^-), chlorate (ClO_3^-), perchlorate (ClO_4^-), and acetate ($C_2H_3O_2^-$)
Compounds containing halides (Group VIIA elements), except those of Ag, Hg, and Pb
Compounds containing sulfate (SO_4^{2-}), except those of Ca, Sr, Ba, Ag, Hg, and Pb
Oxides (O^{2-}) and hydroxides (OH^-) of Ba and Sr

* Most other ionic compounds are insoluble in water.

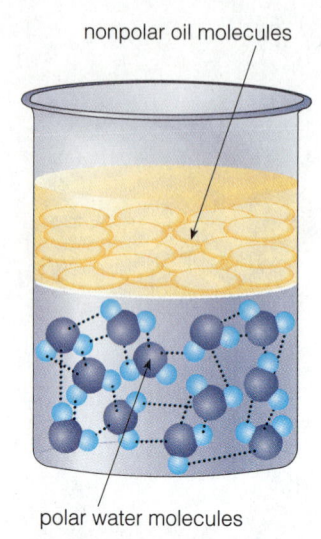

Figure 7.8 Oil and water don't mix because hydrogen bonds (dotted lines) between water molecules are stronger than the weak attraction between nonpolar oil molecules and polar water molecules.

EXAMPLE 7.2 Benzene is a nonpolar solvent with the formula C_6H_6. Which of the following dissolve in benzene? (a) H_2O, (b) Br_2, (c) $CaBr_2$, (d) hexane (C_6H_{14}).

SOLUTION (b) and (d) because they are nonpolar

Effects of Temperature and Pressure. If you want a sugar cube to dissolve faster in water, you can do several things: (a) stir, (b) break up the cube into smaller particles, or (c) heat the water.

Stirring increases the interaction between water molecules and sugar molecules on the surface of the cube. Breaking the cube exposes many more sugar molecules to water, so the sugar dissolves faster. In contrast, a block of salt set out in a pasture lasts through many rainstorms before dissolving completely. Even though NaCl dissolves in water, most Na^+ and Cl^- ions are inside a large block of salt—they can't dissolve because they don't interact with water molecules at the surface.

At warmer temperatures, the average kinetic energy of particles is greater. This not only causes solutions to form faster, but also changes how much substance can be dissolved. The amount of solute that dissolves in a given amount of solvent is the *solubility* of the substance.

Table 7.4 lists a few examples of how temperature affects solubility in water. Warmer temperatures typically make solids and liquids more soluble in water, and gases less soluble. For example, you can see bubbles of CO_2 gas escaping from a carbonated beverage as it warms (Figure 7.9). Your tongue also detects less dissolved CO_2, and the beverage tastes "flat."

Increasing pressure, however, makes gases (but not liquids or solids) more soluble in liquids. This is why carbonated beverages are bottled or canned under pressure. The pressure makes it harder for CO_2 molecules to escape from solution.

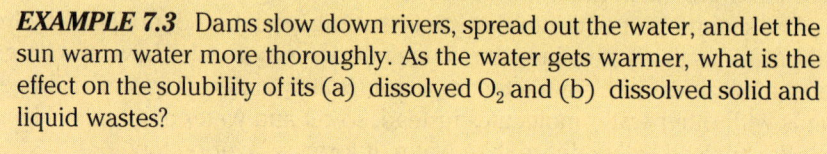

Figure 7.9 The solubility of CO_2 gas decreases as this soft drink gets warmer.

EXAMPLE 7.3 Dams slow down rivers, spread out the water, and let the sun warm water more thoroughly. As the water gets warmer, what is the effect on the solubility of its (a) dissolved O_2 and (b) dissolved solid and liquid wastes?

SOLUTION (a) decreased solubility, (b) increased solubility

Table 7.4 Effect of Temperature on Solubility of Some Substances

Substance	Solubility in H_2O (g/100 mL Solution)	Temp (°C)
sodium chloride (NaCl)	35.8	10
	37.3	60
sodium acetate ($NaC_2H_3O_2$)	40.8	10
	139	60
oxygen (O_2)	.0054	10
	.0023	60
carbon dioxide (CO_2)	.23	10
	.058	60

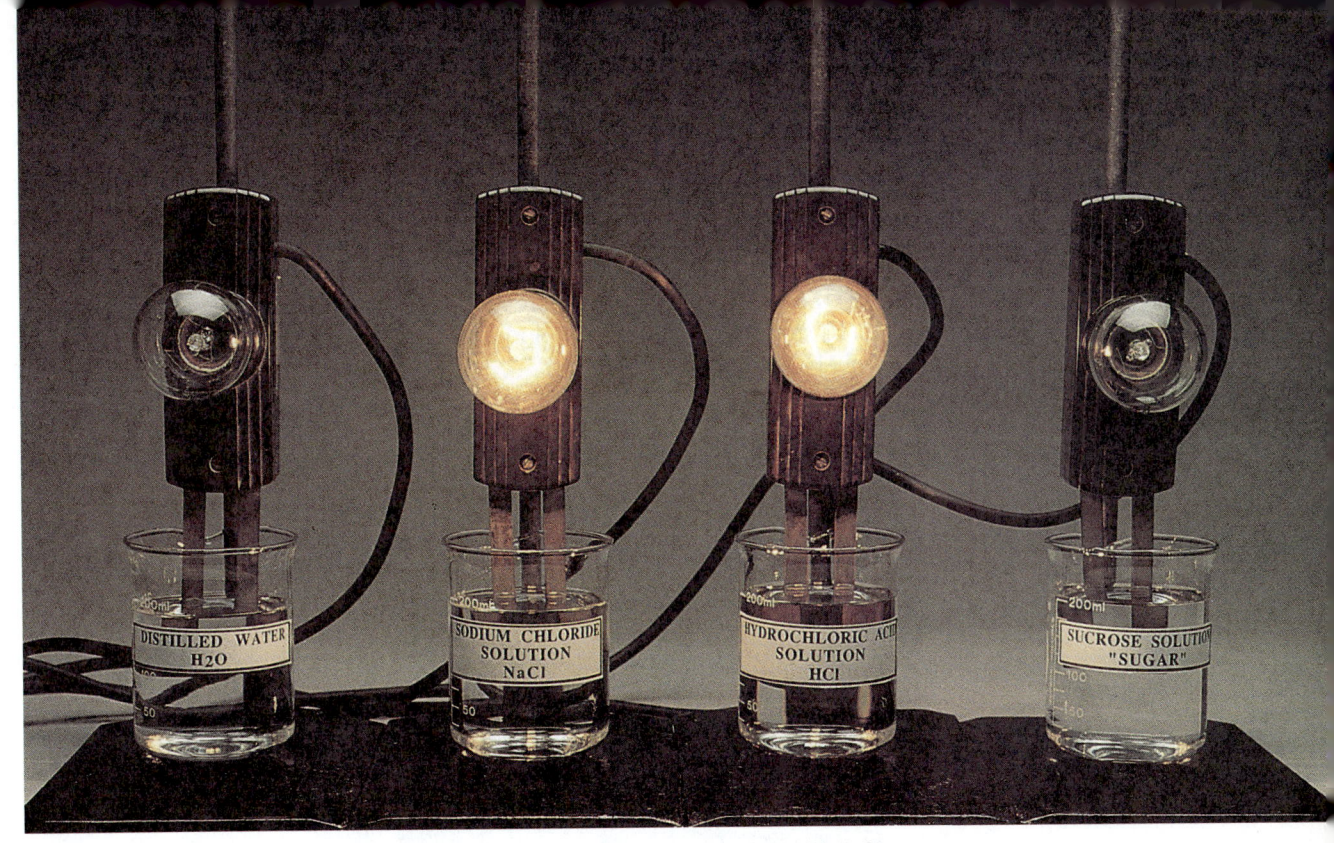

Figure 7.10 (a) Distilled water doesn't conduct enough electricity to light the bulb. (b) NaCl and (c) HCl solutions are good conductors, but (d) a sugar solution is not. NaCl and HCl are electrolytes; sugar is a nonelectrolyte.

Electrolytes. You've heard warnings to keep electrical appliances away from water. Because water conducts electricity, you could get a serious shock. Yet if you test pure, distilled water, you would find that it is a very poor conductor (Figure 7.10). Most water, however, contains dissolved ionic compounds that make the solution a good conductor.

Electrolytes are substances that conduct electricity in the liquid state or when dissolved in water. Electrolytes provide mobile ions that carry electrical current from one electrode to another, completing a circuit.

Solid table salt (NaCl) doesn't conduct electricity because Na^+ and Cl^- ions are held rigidly in place. But at 801°C, NaCl melts. Once its ions can move about, it becomes a conductor. NaCl also provides mobile Na^+ and Cl^- ions when it dissolves in water (Figure 7.7), so a solution of NaCl conducts electricity (Figure 7.10).

Ionic compounds, then, are electrolytes. When they dissolve in water, their ions separate. A few covalent compounds also are electrolytes, but they provide mobile ions in a different way: they react with water in solution to form ions. Hydrogen chloride, for example, reacts with water to form *hydrochloric acid*:

$$HCl + H_2O \longrightarrow H_3O^+ + Cl^-$$

The solution conducts electricity because of its mobile H_3O^+ (hydronium) and Cl^- ions (Figure 7.10).

The main electrolytes in your body are ionic compounds. Table 4.4 and Table 7.5 list the major ions in blood. Na^+ and Cl^- are by far the most abundant, both in blood and in other extracellular fluids. Inside cells, however, K^+ and HPO_4^{2-} (hydrogen phosphate) are the major ions.

Table 7.5 Typical Concentrations of Ions in Blood Plasma

Ion	meq/L
Cations	
Na^+	142
K^+	5
Ca^{2+}	5
Mg^{2+}	3
Total	155
Anions	
HCO_3^-	27
Cl^-	103
HPO_4^{2-}	2
SO_4^{2-}	1
Organic acids	6
Proteins	16
Total	155

You will learn in Chapter 9 that acids are defined as compounds that provide hydronium ions (H_3O^+) in water solution.

Figure 7.11 Light passes through sugar water (left), a true solution, but not through starch water (right), a colloid.

The Tyndall effect was described by John Tyndall (1820–1893), a British scientist.

Figure 7.12 Filter paper doesn't separate out the particles in milk, a colloid.

EXAMPLE 7.4 Which of the following are electrolytes? (a) $CaCl_2$, (b) table sugar ($C_{12}H_{22}O_{11}$), (c) $Zn(NO_3)_2$.

SOLUTION (a) and (c) are ionic compounds and are electrolytes; (b) is covalent and does not ionize, so it is a nonelectrolyte (see Figure 7.11).

7.3 COLLOIDS AND SUSPENSIONS

Colloids. When you stir a sugar cube in a glass of water, sugar dissolves and becomes invisible. If you shine light through the solution, you will see no evidence that it contains sugar. But light does not pass straight through a glass of milk or a starch solution—although both are homogeneous and mostly water (Figure 7.11). The milk and starch particles must be "dissolved" in a different way.

Solute particles in solution are small, with a diameter less than 1 nm (10^{-9} m). In contrast, a **colloid** (or *colloidal dispersion*) is a homogeneous mixture containing particles with larger diameters, typically 1 to 1000 nm. Substances with molar masses of several thousand and above have particles this size. Ordinary filter paper cannot separate out these particles (Figure 7.12), but more specialized filters can. Milk is a colloid because it contains proteins and other large molecules dispersed in water.

Large particles in a colloid reflect and scatter light, called the *Tyndall effect*. For example, sunbeams occur when tiny dust particles in air scatter sunlight (Figure 7.13). Smog, fog, butter, mayonnaise, milk, and whipped cream are all examples of colloids.

Several factors keep large particles dispersed. In water-based colloids, water molecules collide with dispersed particles, keeping them in motion. Water molecules also attract and surround the large particles. In addition, the dispersed particles typically have the same electrical charge and thus repel each other. The last two factors keep particles from combining into larger units that would settle out.

Suspensions. Some mixtures eventually settle out or can be separated by ordinary filtration; these are called **suspensions.** Here the particles are too large to stay dispersed. Muddy water is one example; the large particles eventu-

Table 7.6 Comparison of Solutions, Colloids, and Suspensions

Characteristic	Solution	Colloid	Suspension
particle size	<1 nm	1–1000 nm	>1000 nm
scatters and reflects light	no	yes	yes
settles on standing	no	no	yes
separates with ordinary filter paper	no	no	yes
separates with dialysis membranes (see Section 7.5)	no	yes	yes

Figure 7.13 The Tyndall effect. Suspended dust particles in a forest scatter light.

ally settle out. Blood, however, contains dissolved solutes, colloidal protein and lipoprotein material, and suspended red blood cells. But only the red blood cells can be separated out by gravity or filters. The motion of blood in the body keeps its cells suspended.

Table 7.6 compares solutions, colloids, and suspensions.

7.4 CONCENTRATIONS OF SOLUTIONS

Saturated and Unsaturated Solutions. When you add sugar to sweeten a glass of lemonade, you eventually reach a limit of how much sugar will dissolve. If you add still more, some solid sugar will remain undissolved in the glass. The solid sugar appears unchanged, but the appearance is deceiving. The amount stays constant because of a dynamic equilibrium (Section 6.2) between two

147

CHEMISTRY SPOTLIGHT

Detergents

Water doesn't dissolve nonpolar grease, oil, or dirt. To remove these materials from your skin or clothing, you use a *detergent*. A detergent is a cleaning agent that produces a colloid or suspension between unlike materials such as oil and water. A colloid of two liquids is called an *emulsion*.

Detergents suspend nonpolar materials in polar solvents such as water. They have two chemical features: a nonpolar region (typically made of carbon and hydrogen) and a polar or ionic region. In a mixture of oil and water, the nonpolar end of a detergent dissolves in oil and the polar or ionic end dissolves in water (Figure 7.14). As a result, a detergent is a kind of chemical diplomat that brings together two unlike substances, one nonpolar and the other polar or ionic.

Your body uses detergents to digest fat. The liver makes bile salts, detergents that pass through the bile duct into the intestine. There they suspend dietary fat so it can be absorbed. If your liver fails to make bile salts, or your bile duct is blocked, you cannot digest lipids and excrete them instead.

Nonpolar lipids don't dissolve in blood either. There proteins work as detergents, combining with cholesterol and other lipids to form a colloid of *lipoproteins*. But some cholesterol compounds still deposit in blood vessels. Excess deposits cause *artherosclerosis*, a condition in which blood vessels narrow and raise the blood pressure. This increases the risk of a clot blocking a vessel to the heart (causing a heart attack) or brain (causing a stroke).

Detergents sometimes are used to treat gallstones that form in the gallbladder or bile duct. Gallstones are made of cholesterol, calcium salts, and other materials. Detergents can help suspend and remove those that are mostly cholesterol.

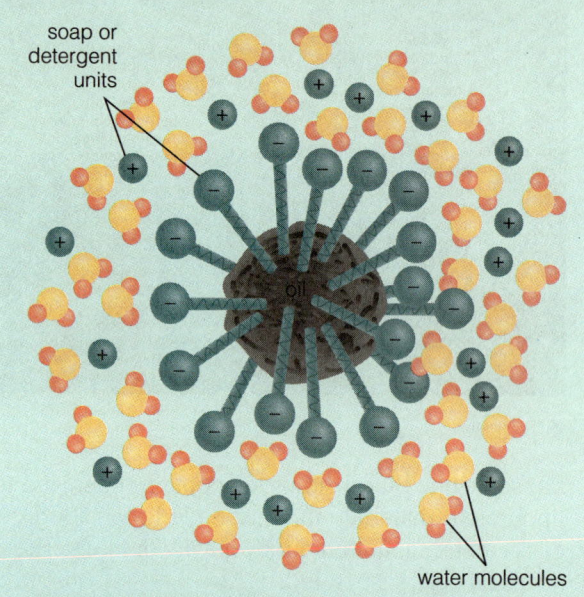

Figure 7.14 Ionic ends of a surfactant dissolve in polar water; nonpolar ends dissolve in nonpolar dirt or grease. This action suspends dirt and grease particles in water.

processes: while solid sugar is dissolving, dissolved sugar is coming out of solution at the same rate.

A solution that holds the maximum amount of dissolved solute at a given temperature is a **saturated solution** (Figure 7.15). The lemonade described previously is saturated; it can't hold any more sugar in solution. An **unsaturated solution,** in contrast, contains less than the maximum amount of solute.

A **supersaturated solution** contains more than the maximum amount of solute. It seems strange that this could happen. But remember that the solubility of a solid or liquid in a liquid typically increases with temperature (Section 7.2). If you heat a saturated solution and add solute, more solute will dissolve. When you cool the solution, the additional solute may stay dissolved, forming a supersaturated solution. The extra solute precipitates out of solution when it can bind to a crystal of solute or some other suitable surface.

Figure 7.15 A saturated solution of copper (II) sulfate, $CuSO_4$.

EXAMPLE 7.5 Examine Table 7.4. Which substance most readily forms a supersaturated solution when a saturated solution cools?

SOLUTION sodium acetate, which is much more soluble at warmer temperatures

Concentrations of Solutions. *Saturated* and *unsaturated solution* aren't precise terms. In a saturated solution, the amount of solute depends on its solubility and the temperature. For example, one liter of a saturated solution of NaCl at room temperature contains less solute than one liter of a saturated solution of sodium acetate ($NaC_2H_3O_2$) because NaCl is less soluble in water than sodium acetate (Table 7.4). And an unsaturated solution of sodium chloride or sodium acetate may contain *any* amount of solute less than the maximum.

Other common terms are also vague. A *concentrated solution* contains a large amount of solute; it may be saturated, or nearly so. A *dilute solution,* in contrast, contains relatively little solute.

Scientists use more precise terms to specify the **concentration** of a solution, the amount of solute in a given amount of solution.

Molarity. The **molarity (*M*)** of a solution is the number of moles of solute per liter of solution (Figure 7.16):

$$\text{molarity } (M) = \frac{\text{mol solute}}{\text{L solution}}$$

Since one mole contains Avogadro's number (6.02×10^{23}) of particles (formula units for ionic compounds; molecules for covalent compounds), molarity also expresses the number of solute particles per liter of solution.

To calculate molar concentrations, remember that the units are mol/L. You can use the factor-label method to calculate, for example, the molarity of 50.0 g $CaCl_2$ dissolved in 675 mL of solution. To calculate molarity, begin with the known information and use appropriate conversion factors to change into the desired units (mol/L):

molar mass of $CaCl_2$ = 40.08 + 2(35.45) = 110.98 g

$$\frac{50.0 \text{ g } CaCl_2}{675 \text{ mL}} \times \frac{1 \text{ mol } CaCl_2}{110.98 \text{ g } CaCl_2} \times \frac{1000 \text{ mL}}{1 \text{ L}} = 0.667 \text{ mol/L} = 0.667 \text{ } M$$

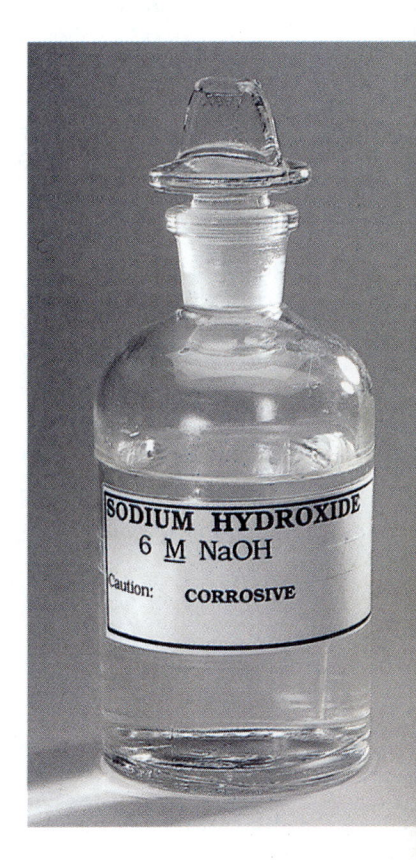

Figure 7.16 This 6 *M* solution of sodium hydroxide contains 6 moles of NaOH per liter.

To calculate how many g of CaCl$_2$ you would need to make 80.0 mL of a 3.12 M solution, begin with the known information and use molarity as a conversion factor to change into the desired units (g CaCl$_2$):

$$80.0 \text{ mL} \times \frac{1 \text{ L}}{1000 \text{ mL}} \times \frac{3.12 \text{ mol CaCl}_2}{1 \text{ L}} \times \frac{110.98 \text{ g CaCl}_2}{1 \text{ mol CaCl}_2} = 27.7 \text{ g CaCl}_2$$

To make up the solution, then, you weigh out 27.7 g of CaCl$_2$ and dissolve it in enough water to make a final volume of 80.0 mL.

EXAMPLE 7.6 What is the molarity (mol/L) of the following solutions? (a) 1.56 g CaCl$_2$ in 1.5 L of solution, (b) 1.56 g CaCl$_2$ in 750.0 mL of solution.

SOLUTION (a) $\dfrac{1.56 \text{ g CaCl}_2}{1.5 \text{ L}} \times \dfrac{1 \text{ mol CaCl}_2}{110.98 \text{ g CaCl}_2} = 9.4 \times 10^{-3} \, M$

(b) $\dfrac{1.56 \text{ g CaCl}_2}{750.0 \text{ mL}} \times \dfrac{1000 \text{ mL}}{1 \text{ L}} \times \dfrac{1 \text{ mol CaCl}_2}{110.98 \text{ g CaCl}_2} = 1.87 \times 10^{-2} \, M$

EXAMPLE 7.7 Calculate how many g of NaCl are needed to make (a) 80.0 mL of 3.12 M solution, (b) 6.33 L of 4.5×10^{-2} M solution.

SOLUTION (a) $80.0 \text{ mL} \times \dfrac{3.12 \text{ mol NaCl}}{1 \text{ L}} \times \dfrac{1 \text{ L}}{1000 \text{ mL}} \times \dfrac{58.44 \text{ g NaCl}}{1 \text{ mol NaCl}}$
$= 14.6 \text{ g NaCl}$

(b) $6.33 \text{ L} \times \dfrac{4.5 \times 10^{-2} \text{ mol NaCl}}{\text{L}} \times \dfrac{58.44 \text{ g NaCl}}{\text{mol NaCl}} = 17 \text{ g NaCl}$

Percent Concentrations. Percent refers to parts per hundred. Just as an exam score of 85% means 85 points out of a total of 100, a 10% solution means 10 parts of solute per 100 units of solution. Three different **percent concentrations** are used.

A common method, designated *w/v*, is the *weight* of solute per *volume* of solution:

$$\% \, (w/v) = \frac{\text{g solute}}{100 \text{ mL solution}}$$

Notice that you don't need to know molar mass to calculate percent concentrations. For example, a 3.14% (w/v) solution of any solute has 3.14 g of solute per 100 mL of solution.

A physiological saline (NaCl) solution in a hospital, for example, is 0.9% (*w/v*) (Figure 7.17). This means that each 100 mL of solution contains 0.9 g NaCl.

You can use % (*w/v*) as a conversion factor. For example, you can calculate the mass of NaCl you would need to make up 6.00 L of a 2.15% (*w/v*) solution:

$$6.00 \text{ L solution} \times \frac{1000 \text{ mL solution}}{1 \text{ L solution}} \times \frac{2.15 \text{ g NaCl}}{100 \text{ mL solution}} = 129 \text{ g NaCl}$$

To make up the solution, you weigh out 129 g NaCl and add enough water to bring the total volume to 6.00 L.

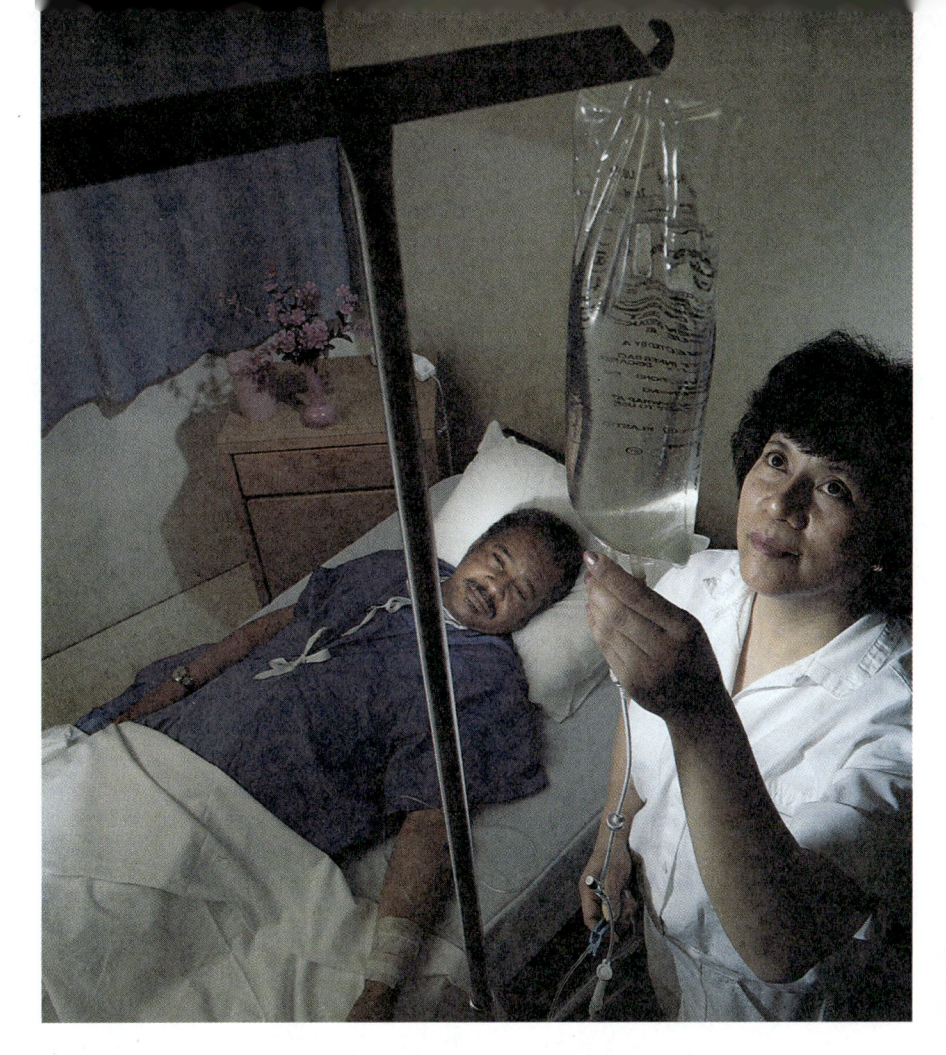

Figure 7.17 This patient is receiving a 0.9% (w/v) saline (NaCl) solution.

EXAMPLE 7.8 How many g of solute do you need to make (a) 865 mL of 5.10% (w/v) glucose solution, and (b) 15 mL of 0.90% (w/v) NaCl solution?

SOLUTION (a) $865 \text{ mL solution} \times \dfrac{5.10 \text{ g glucose}}{100 \text{ mL solution}} = 44.1 \text{ g glucose}$

(b) $15 \text{ mL solution} \times \dfrac{0.90 \text{ g NaCl}}{100 \text{ mL solution}} = 0.14 \text{ g NaCl}$

A similar system is used in clinical chemistry. Concentrations in blood or urine are often expressed as the mass of solute per deciliter (dL). Blood glucose levels, for example, typically are 60–100 mg/dL and Zn^{2+} ion levels are 75–160 μg/dL.

EXAMPLE 7.9 Calculate the g of hemoglobin in your blood if your body contains 5.0 L of blood with a concentration of 14.2% (w/v) hemoglobin (Hb).

SOLUTION $5.0 \text{ L blood} \times \dfrac{1000 \text{ mL blood}}{1 \text{ L blood}} \times \dfrac{14.2 \text{ g Hb}}{100 \text{ mL blood}} = 710 \text{ g Hb}$

Proof was first used to express the strength of alcoholic beverages in the 17th century. Dealers tested whiskey by pouring it on gunpowder and igniting it. If the gunpowder exploded after the whiskey burned off, this was the "proof" of its high ethanol (and low water) content.

The unit mg/L is parts per million because 1 L of a dilute solution has a mass of 1000 g, which is 1,000,000 mg. So mg/L amounts to mg solute per 1,000,000 mg solution.

A second method, designated *w/w*, is the *weight* of solute per *weight* of solution:

$$\% \ (w/w) = \frac{\text{g solute}}{100 \text{ g solution}}$$

To make up a 10.0% (*w/w*) glucose solution, for example, you weigh out 10.0 g of glucose and add water until the total mass of the solution is 100 g.

The third method, designated *v/v*, is the *volume* of solute per *volume* of solution:

$$\% \ (v/v) = \frac{\text{mL solute}}{100 \text{ mL solution}}$$

This is used mostly for solutions of liquids in liquids, especially ethanol in water. For example, table wine with a concentration of 12.0% (*v/v*) has 12.0 mL of ethanol in each 100 mL of wine (Figure 7.18). The *proof* of alcoholic beverages is twice the percent (*v/v*) of ethanol. The aforementioned table wine, for example, would be 24.0 proof.

Parts Per Million. Concentrations of very dilute solutions are sometimes expressed as **parts per million (ppm)**. This is the same as mg solute/L of solution:

$$\text{ppm} = \frac{\text{mg solute}}{\text{L solution}}$$

A concentration of 2.4 ppm Pb, for example, means 2.4 mg Pb per liter of solution.

You can use ppm as a conversion factor to calculate, for example, how many g of mercury (II) chloride, $HgCl_2$, is needed to make 24.0 L of 1.16 ppm $HgCl_2$:

$$24.0 \ \text{L} \times \frac{1.16 \ \text{mg } HgCl_2}{\text{L}} \times \frac{1 \ \text{g } HgCl_2}{10^3 \ \text{mg } HgCl_2} = 2.78 \times 10^{-2} \ \text{g } HgCl_2$$

EXAMPLE 7.10 7500 L water at 20°C contains 8.1 ppm of dissolved oxygen. How many g of O_2 are dissolved in this water?

SOLUTION $7500 \ \text{L water} \times \dfrac{8.1 \ \text{mg } O_2}{1 \ \text{L water}} = 61{,}000 \ \text{mg } O_2 \ (6.1 \times 10^4 \ \text{mg } O_2$ or 61 g O_2)

Equivalents. Scientists sometimes express concentrations in terms of *equivalents* per liter. One **equivalent (eq)** of an ion is the number of grams that provides one mole (Avogadro's number) of electrical charge in solution. One mole (22.99 g) of Na^+, for example, provides Avogadro's number of positive charge in solution since each Na^+ ion carries a +1 charge. In contrast, one mole (40.08 g) of Ca^{2+} provides *two times* Avogadro's number of positive charge in solution, or two equivalents.

The **equivalent weight** of an ion is its formula weight divided by its electrical charge; it represents the g of ion that provides Avogadro's number of electrical charge in solution, or one equivalent. The equivalent weight of Na^+ is

Figure 7.18 This wine contains 12.0% (*v/v*) ethanol in water.

22.99 g/1 = 22.99 g. The equivalent weight of Ca^{2+} is 40.08 g/2 = 20.04 g. A solution containing 30.0 g of Ca^{2+}/L has a concentration of 1.50 eq/L.

EXAMPLE 7.11 Calculate the equivalent weights of the following: (a) S^{2-}, (b) NH_4^+, (c) PO_4^{3-}, (d) Cu^{2+}.

SOLUTION (a) $\dfrac{32.06 \text{ g}}{2} = 16.03$ g; (b) $\dfrac{18.05 \text{ g}}{1} = 18.05$ g; (c) $\dfrac{94.97 \text{ g}}{3} = 31.66$ g; (d) $\dfrac{63.55 \text{ g}}{2} = 31.78$ g

---

You can use equivalent weight as a conversion factor. For example, calculate how many g $MgCl_2$ is needed to make 1.00 L of a solution that is 2.08 eq Mg^{2+}/L. Since Mg^{2+} has a 2+ charge, its equivalent weight = (24.31 g)/2 = 12.16 g. Then calculate the g Mg needed:

$$1.00 \text{ L} \times \frac{2.08 \text{ eq Mg}^{2+}}{\text{L}} \times \frac{12.16 \text{ g Mg}^{2+}}{\text{eq Mg}^{2+}} = 25.3 \text{ g Mg}^{2+}$$

To supply 25.3 g Mg^{2+} in the form of $MgCl_2$, we determine how much of the mass of $MgCl_2$ is Mg^{2+}:

molar mass of $MgCl_2$ = 24.31 g + 2(35.45 g) = 95.21 g

Thus there are 24.31 g Mg^{2+} in every 95.21 g $MgCl_2$. We can use this relationship to determine how many g $MgCl_2$ are needed to furnish 25.3 g Mg^{2+}:

$$25.3 \text{ g Mg}^{2+} \times \frac{95.21 \text{ g MgCl}_2}{24.31 \text{ g Mg}^{2+}} = 99.1 \text{ g MgCl}_2$$

So dissolving 99.1 g $MgCl_2$ in water to make 1.00 L of solution makes a concentration of 2.08 eq Mg^{2+}/L.

EXAMPLE 7.12 How many g NaCl are dissolved in 2.00 L of a solution that contains 1.18×10^{-3} eq Cl^- per liter?

SOLUTION

$$2.00 \text{ L} \times \frac{1.18 \times 10^{-3} \text{ eq Cl}^-}{\text{L}} \times \frac{35.45 \text{ g Cl}^-}{\text{eq Cl}^-} \times \frac{58.44 \text{ g NaCl}}{35.45 \text{ g Cl}^-}$$
$$= 0.138 \text{ g NaCl}$$

The advantage of equivalents, as the name suggests, is that they put all ions on the same basis in terms of amount of charge. That is, 0.14 eq of *any* ion provides the same amount of charge (0.14 times Avogadro's number) as 0.14 eq of any other ion. Concentrations of ions are often expressed as equivalents or milliequivalents (meq, or 10^{-3} eq) per L or deciliter (dL) of blood (Table 7.5).

Diluting Solutions. Suppose you need several concentrations of the same solution. You could prepare each one separately. But often it's more convenient to prepare one concentrated solution and use it to prepare other, lower concentrations by adding appropriate amounts of water.

Adding water to a solution to reduce its concentration is called **dilution.** The more water you add, the more the concentration is reduced. A simple relationship is:

$$C_i \times V_i = C_f \times V_f$$

where C_i and C_f are the initial and final concentrations and V_i and V_f are the initial and final volumes, respectively. The equation works as long as you use the same units for initial and final conditions. Volume is usually measured in mL or L, whereas concentrations can be in molarity, percent, ppm, or eq/L.

For example, you can calculate the new concentration after adding 50.0 mL of water to 38.7 mL of 0.367 M Na$_3$PO$_4$ solution:

$$(0.367\ M)(38.7\ \text{mL}) = C_f \times (38.7\ \text{mL} + 50.0\ \text{mL})$$

Notice that V_f, the final volume, is the initial volume plus the volume of water added.

$$C_f = \frac{(0.367\ M)(38.7\ \text{mL})}{88.7\ \text{mL}} = 0.160\ M$$

You can also calculate how much water to add to dilute a solution to the desired concentration. For example, how much water would you add to 1.79 L of 2.86% (w/v) CuSO$_4$ solution to make a 1.00% (w/v) solution?

$$(2.86\%)(1.79\ \text{L}) = (1.00\%) \times V_f$$

$$V_f = \frac{(2.86\%)(1.79\ \text{L})}{1.00\%} = 5.12\ \text{L}$$

Since the final volume is 5.12 L and you begin with 1.79 L, you would add a total of 3.32 L (5.12 L − 1.79 L) of water.

EXAMPLE 7.13 Calculate the concentration that would result if you add 80.0 mL of water to each of the following solutions: (a) 50.0 mL of 2.17 M glucose, (b) 1.20 L of 4.22% (w/v) KHCO$_3$, (c) 619 mL of 1.24 ppm PbCl$_2$, (d) 145 mL of 1.63 meq K$^+$/L.

SOLUTION (a) 0.835 M, (b) 3.96% (w/v), (c) 1.10 ppm, (d) 1.05 meq K$^+$/L

EXAMPLE 7.14 Suppose you drink 400.0 mL of orange juice that is a 0.045% (w/v) vitamin C solution and excrete all the vitamin C in your urine in 24 hours. If the average concentration of vitamin C in that urine is 0.019% (w/v), how much urine did you excrete?

SOLUTION $(0.045\%)(400.0\ \text{mL}) = (0.019\%)(V_f)$; $V_f = 950$ mL

7.5 COLLIGATIVE PROPERTIES AND WATER BALANCE IN THE BODY

Effects on Freezing and Boiling Points. In winter, people sometimes put salt on sidewalks to melt the ice. They also put antifreeze in car radiators to

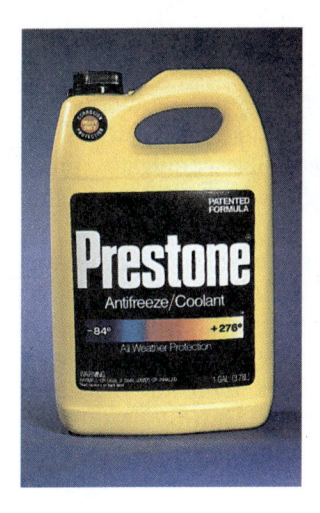

Figure 7.19 (Left) Antifreeze lowers the freezing point of water and thus protects radiators against damage from ice. (Right) Water and a water–antifreeze solution, both at −16°C.

prevent freezing (Figure 7.19). Both materials lower the freezing point of water, keeping it from freezing until the temperature gets colder than 0°C. Antifreeze also raises the boiling point of water, so in hot summer driving the radiator solution is less likely to boil off.

Solutes lower the freezing point of water because their interspersed particles keep water molecules from forming the hexagonal pattern of ice (Figure 7.5). Solutes also raise the boiling point because their interspersed particles reduce the tendency of water molecules to escape into the vapor state; in other words, solutes lower the vapor pressure of solvents.

The effect on freezing and boiling points depends only on the number—not the type—of particles in solution. Effects that depend only on the number of solute particles are called **colligative properties.**

One mole of particles in 1000 g of water raises the boiling point by 0.51°C and lowers the freezing point by 1.86°C. Ionic compounds typically have a greater effect than covalent compounds because their ions separate in solution, increasing the number of particles. One mole of NaCl, for example, provides *two* moles of particles in solution—one mole of Na$^+$ ions and one mole of Cl$^-$ ions; this doubles the colligative effect. In contrast, one mole of most covalent compounds, such as ethanol (C_2H_6O) or sucrose ($C_{12}H_{22}O_{11}$), provides *one* mole of particles in solution.

EXAMPLE 7.15 Predict the freezing point of solutions containing the following in 1000 g of water: (a) 1 mole sucrose ($C_{12}H_{22}O_{11}$), (b) 2 moles NaCl, (c) 2 moles sucrose, (d) 1 mole K_3PO_4.

SOLUTION (a) −1.86°C, (b) −7.44°C, (c) −3.72°C, (d) −7.44°C

Some animals have antifreeze molecules that keep them from freezing. Antarctic fish, for example, have special carbohydrate–protein (glycoprotein) molecules that help them stay alive even in very cold water in winter.

Osmotic Pressure. Figure 7.20 shows what happens if you separate water from a water solution using a membrane that lets only water molecules pass

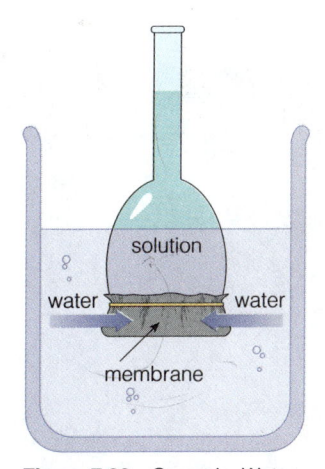

Figure 7.20 Osmosis. Water flows through the membrane into the solution, causing the volume inside to increase and push liquid up the tube. The height of liquid in the tube is a measure of the osmotic pressure of the solution.

through. The effect, called **osmosis,** is a net movement of water molecules from pure water or a dilute solution to a more concentrated solution.

The incoming water (Figure 7.20) increases the volume of the solution. The solution rises in the tube until the downward pressure in the tube matches the upward pressure, called the **osmotic pressure.** Because it is a colligative property, osmotic pressure increases as the number (not type) of solute particles increases.

Osmotic pressure is important in the body. Although cell membranes are very selective about what molecules can enter and leave the cell, they allow water molecules to pass through freely. Thus cells need to have the same osmotic pressure as surrounding fluids to prevent excessive water gain or loss.

Blood has an osmotic pressure about equal to a 0.9% (*w/v*) NaCl solution. Two solutions having the same osmotic pressure are called **isotonic.** A solution having a higher osmotic pressure is **hypertonic;** a solution with a lower osmotic pressure is **hypotonic.**

When red blood cells are taken from the equivalent of a 0.9% NaCl solution and placed in a hypertonic solution, such as 10% (*w/v*) sucrose solution, water leaves the cells for the more concentrated solution outside. When cells lose water they shrivel and shrink, a process called *crenation*.

Putting red blood cells in a hypotonic solution, such as pure water, has the opposite effect. Water rushes into the cells, increasing the volume so much that they burst. This is called *hemolysis*.

> **EXAMPLE 7.16** Should a solution injected intravenously into a patient be hypertonic, isotonic, or hypotonic to blood?
>
> **SOLUTION**   isotonic, to prevent crenation or hemolysis

Dialysis.   Most biological membranes allow certain other small molecules (besides water) and ions to pass through, but not large molecules and colloidal material (see Table 7.6 and Figure 7.21). This process, analogous to osmosis, is called **dialysis.** Blood, for example, undergoes dialysis as it passes through the kidneys. The kidneys remove water and waste products from blood and excrete them in urine. During this process, however, large protein and lipoprotein molecules remain in the blood.

Damaged and diseased kidneys don't dialyze blood effectively, so small ions and molecules that are toxic accumulate in blood. If not treated, this can be fatal. Kidney dialysis machines provide temporary help (see Chemistry Spotlight, Section 28.6). Blood circulates through cellophanelike membranes that remove toxic waste materials. The best long-term solution for such patients is a kidney transplant.

Water Balance in the Body.   About 62% of your weight is water, and water is necessary for you to stay alive. You need water to maintain body temperature (Section 6.2), remove waste products, and transport substances throughout your body.

The body loses 2.0 to 2.5 L of water a day through urine, perspiration, feces, and exhaled air. You need to take in that much water each day, except for about 0.4 L of water that you produce by metabolism.

Figure 7.21   A dialyzing membrane is permeable to water but impermeable to the colloidal material in milk.

CHEMISTRY SPOTLIGHT

Edema

Blood is hypertonic to nearby extracellular fluids, mainly because of the large protein molecules, called albumins, that stay inside blood vessels. As a result, water is attracted from the surrounding tissues into the blood. Excess water in blood is excreted when blood passes through the kidneys.

People who have inadequate protein in their diets have less albumin in their blood. Low albumin also results from damaged kidneys (which then excrete albumin) or cirrhosis of the liver (which reduces albumin synthesis).

Low albumin levels reduce the flow of water from surrounding tissues into blood. Water then stays in tissues and collects between cells. People with this condition, known as *edema,* often have swollen feet, abdominal distention, and mild dehydration. The solution is to treat the underlying cause of edema.

When you don't have enough water, your pituitary gland secretes antidiuretic hormone (ADH), which causes the kidneys to retain water in the blood instead of excreting it. You notice the effect because you urinate less. The urine is a darker color because it contains a higher than normal concentration of pigments and other solutes. Patients who have had their pituitary glands removed for medical reasons lack this water-retaining action; as a result, they excrete (and thus have to consume) as much as 30 L of water a day.

> Only this water in solidified form are this earth, the atmosphere, the heavens, the mountains, plants and trees, wild animals, even to worms, flies and ants—they are all only this water in solidified state.
>
> *The Chandogya Upanishad (Eighth century* B.C.*)*

SUMMARY

Water has high surface tension and expands when it freezes because strong hydrogen bonds form between its molecules. Since water is polar, it tends to dissolve polar and ionic compounds but not nonpolar compounds. The general solubility rule is "like dissolves like."

A homogeneous mixture of two or more substances is a solution. The solvent is the dissolving agent and is present in the greater amount. The substance dissolved, and present in the lesser amount, is the solute. Increasing temperature increases the rate of dissolving and the solubility of most solids and liquids in liquids. It decreases the solubility of gases in liquids. The solubility of gases in liquids increases with increasing pressure. Substances that supply ions in solution are called electrolytes. Their solutions conduct electricity.

Homogeneous mixtures are colloids if their "solute" particles are large enough to scatter and reflect light. If the particles are so large that they settle out or can be filtered, the mixture is called a suspension.

Concentrations of solutions are described by general terms such as saturated, unsaturated, supersaturated, concentrated, and dilute. Specific concentrations are expressed as molarity (M, mol/L), percent (w/v, w/w, or v/v), parts per million (ppm, mg/L), or equivalents/L. Dilutions of solutions are calculated by:

$$C_i \times V_i = C_f \times V_f.$$

Colligative properties depend on the number, but not the type, of particles in solution. Such properties include osmosis, raising boiling points, and lowering freezing points. Osmotic pressure helps maintain water balance in the body. Dialysis—the passage of water and certain other particles through membranes—is used to remove toxic materials from blood and to maintain water balance in the body.

KEY TERMS

Colligative property (7.5)
Colloid (7.3)
Concentration (7.4)
Dialysis (7.5)
Dilution (7.4)
Electrolyte (7.2)
Equivalent (eq) (7.4)
Equivalent weight (7.4)

Hypertonic solution (7.5)
Hypotonic solution (7.5)
Isotonic solution (7.5)
Molarity (7.4)
Osmosis (7.5)
Osmotic pressure (7.5)
Parts per million (ppm) (7.4)
Percent concentration (7.4)

Saturated solution (7.4)
Solute (7.2)
Solution (7.2)
Solvent (7.2)
Supersaturated solution (7.4)
Surface tension (7.1)
Suspension (7.3)
Unsaturated solution (7.4)

EXERCISES

Even-numbered exercises are answered at the back of this book.

Properties of Water

1. What forces are responsible for the surface tension of water?

2. Detergent particles tend to accumulate on the surface of water, getting in between water molecules. What effect do you think this has on the surface tension of water?

3. Are water molecules closer or farther apart in ice than in liquid water?

4. When salt water partially freezes, do you think the frozen part is (a) more salty, (b) less salty, or (c) equally salty compared with the unfrozen liquid? Explain.

5. Does water first freeze at the top or the bottom? Explain why.

6. Why do snowflakes have hexagonal shapes?

Forming Solutions

7. List three internal body parts that do not dissolve in water. Do you think their surfaces are mostly polar or nonpolar?

8. Water has sometimes been called the universal solvent. Is it?

9. An antifreeze manufacturer recommends a mixture of 5 L of antifreeze per 3 L water for maximum protection. In such a solution, which is the solvent and which is the solute? (Consider antifreeze to be a single substance.)

10. In 86-proof whiskey, ethyl alcohol is 43% of the beverage by volume and nearly all the remaining volume is water. Which is the solvent and which is the solute?

11. Predict which of the following dissolve effectively in water: (a) K_2SO_4, (b) Br_2, (c) KCl, (d) CCl_4, (e) HCl.

12. Predict which of the following dissolve effectively in water: (a) I_2, (b) $Mg(NO_3)_2$, (c) $C_{10}H_{22}$, (d) $(NH_4)_2CO_3$, (e) $CuCl_2$.

13. Predict which of the substances in exercise 11 dissolve effectively in gasoline, which is a mixture of nonpolar compounds.
14. Predict which of the substances in exercise 12 dissolve effectively in hexane (C_6H_{14}), a nonpolar solvent.
15. According to the "like dissolves like" rule, should rust (Fe_2O_3) dissolve in water? Does it? What can you conclude about the attractive forces between the Fe^{3+} and O^{2-} ions in rust?
16. Grease is a nonpolar material. Which is the better solvent to remove a grease stain from clothing: **(a)** ethyl alcohol (C_2H_6O) or **(b)** octane (C_8H_{18})?
17. The B vitamins and vitamin C are classified as "water-soluble." Vitamins A, D, E, and K are called "fat-soluble." Which class of vitamins is **(a)** less polar and **(b)** excreted more rapidly from the body?
18. Does stirring a sugar cube in a cup of coffee increase the solubility of sugar? Explain.
19. Does heating a cup of coffee increase the solubility of sugar in the coffee? Explain.
20. Do deep-sea divers have more or less than a normal amount of gases dissolved in their blood?

Electrolytes

21. Glucose ($C_6H_{12}O_6$) is a polar substance that dissolves in water but does not form ions. Would a solution of glucose in water be a good conductor of electricity?
22. Suppose you stir NaCl into hexane (C_6H_{14}), a nonpolar liquid. Would the resulting mixture conduct electricity? Explain.
23. Which of the substances in exercise 11 are electrolytes?
24. Which of the substances in exercise 12 are electrolytes?
25. Does melted table salt (NaCl) conduct electricity? Explain.
26. Does solid table salt (NaCl) conduct electricity? Explain.
27. Do acids tend to be electrolytes or nonelectrolytes? Explain.

Colloids and Suspensions

28. Why do you shake or stir paint before using it? Is paint a solution?
29. List two ways to distinguish between a solution and a colloid.
30. Why isn't milk transparent to light?
31. What keeps the dispersed particles in milk from settling out?
32. If you stirred dirt in water, would the muddy water be a solution, colloid, or suspension?
33. Is blood a solution, colloid, suspension, or combination? Explain.
34. Identify which one of the following is an emulsion: **(a)** butter, **(b)** dust in air, **(c)** peanut butter, **(d)** oil and vinegar salad dressing, **(e)** wine.
35. How is cholesterol, a nonpolar substance, carried in blood, which is polar?

Concentrations of Solutions

36. What could you do to test whether or not a NaCl solution is saturated?
37. If you made a saturated solution of sodium acetate ($NaC_2H_3O_2$) and then heated it, would it still be saturated? Explain. (See Table 7.4.)
38. Is a dilute solution likely to be saturated or unsaturated?
39. Explain how you would prepare a supersaturated solution.
40. Calculate the g of solute needed to make up the following solutions: **(a)** 1.00 L of 0.412 M NaOH, **(b)** 1.00 L of 2.12 M $CuSO_4$, **(c)** 250 mL of 1.00 M $CaCl_2$, **(d)** 6.44 L of 0.00130 M NaCl.
41. Calculate the g of solute needed to make up the following solutions: **(a)** 1.00 L of 3.04 M KOH, **(b)** 2.50 L of 0.0335 M CsCl, **(c)** 475 mL of 1.78 M K_2CO_3, **(d)** 15.0 mL of 0.032 M sodium acetate ($NaC_2H_3O_2$).
42. How many mol of solute are needed to make up each solution in exercise 40?
43. How many mol of solute are needed to make up each solution in exercise 41?
44. How many mol of ions are in each solution in exercise 40?
45. How many mol of ions are in each solution in exercise 41?
46. Identify the molarity of each of the following: **(a)** 65 g NaCl in 850 mL solution, **(b)** 3.12 g $CoCl_2$ in 1.12 L solution, **(c)** 45.0 g $Mg(HCO_3)_2$ in 1.00 L solution, **(d)** 10.0 mg caffeine ($C_8H_{10}N_4O_2$) in 3.00 L solution.
47. Identify the molarity of each of the following: **(a)** 16.4 g NH_4Cl in 1.00 L solution, **(b)** 5.50 g Na_3PO_4 in 750.0 mL solution, **(c)** 1.00 kg NaCl in 1.25 × 10³ L solution, **(d)** 135 mg cholesterol ($C_{27}H_{46}O$) in 2.00 L solution.

48. What is the percent (w/v) of the solutions in exercise 46?

49. What is the percent (w/v) of the solutions in exercise 47?

50. Calculate the g of solute you would need to make up the following solutions: **(a)** 100.0 mL of 0.17% (w/v) glucose, **(b)** 1.00 L of 3.00% (w/v) NaCl, **(c)** 450 g of 5.0 (w/w) sucrose, **(d)** 500.0 g of 12.0% (w/w) $CaCl_2$.

51. Calculate the g of solute you would need to make up the following solutions: **(a)** 1.00 L of 0.002345 (w/v) KCl, **(b)** 25.0 mL of 2.50% (w/v) fructose, **(c)** 3.50 kg of 0.250% (w/w) NaOH, **(d)** 800.0 g of 5.37% (w/w) NH_4NO_3.

52. Calculate how many mL of solute you would need to make up the following solutions: **(a)** 375 mL of 12.0% (v/v) ethanol, **(b)** 81.0 mL of 2.63% (v/v) ethanol, **(c)** 8.0 L of 45% (v/v) ethylene glycol (the main ingredient in radiator antifreeze).

53. What is the molar concentration of 0.89% (w/v) NaCl solution?

54. What is the percent (w/v) of 0.714 M glucose ($C_6H_{12}O_6$) solution?

55. What is the percent (w/v) of 145 ppm $FeCl_2$ solution.

56. A normal concentration of calcium ion in blood is 9.5 mg/dL. What is this concentration in ppm?

57. What is the concentration of calcium ion in blood listed in exercise 56 in units of meq/L?

58. A typical concentration of sulfate ion (SO_4^{2-}) in blood plasma is 1.0 meq/L (Table 7.6). How many g of SO_4^{2-} does each liter of blood plasma contain?

59. What is the equivalent weight of **(a)** PO_4^{3-}, **(b)** Fe^{3+}, **(c)** Fe^{2+}, **(d)** Al^{3+}, **(e)** CO_3^{2-}?

60. How many g of copper (I) carbonate, Cu_2CO_3, are dissolved in 685 mL of a solution that contains 2.00 meq Cu^+ per liter?

Diluting Solutions

61. Identify the volume of water needed to dilute each of the following to make the desired concentration: **(a)** 1.00 L of 2.15 M Na_2SO_4 to make a 1.50 M solution, **(b)** 35.6 mL of 6.00 M HCl to make a 2.50 M solution, **(c)** 745 mL of 1.25% (w/v) sucrose to make a 1.00% (w/v) solution, **(d)** 186 mL of 0.450% (v/v) isopropyl (rubbing) alcohol to make a 0.108% (v/v) solution, **(e)** 1.50 L of 1.42 eq Na^+/L to make a 1.00 eq Na^+/L solution.

62. Identify the volume of water needed to dilute each of the following to make the desired concentration: **(a)** 100.0 mL of 6.00 M KOH to make a 0.480 M solution, **(b)** 3.10 L of 0.130 M $Cu(NO_3)_2$ to make a 3.1×10^{-2} M solution, **(c)** 50.0 mL of 10.0% sucrose (w/v) to make a 3.25% (w/v) solution, **(d)** 18.0 mL of 2.00% (v/v) ethyl alcohol to make a 0.465% (v/v) solution, **(e)** 188 mL of 16.1 meq NO_3^-/L to 3.50 meq NO_3^-/L solution.

63. Find the concentration of the following solutions: **(a)** 55.0 mL of 2.12 M Na_2HPO_4 diluted by adding 100.0 mL water, **(b)** 1.00 L of 3.00 M HCl diluted by adding 250 mL of water, **(c)** 125 mL of 1.48% (w/v) Li_2CO_3 diluted by adding 38.0 mL of water, **(d)** 3.12 L of 10.0% (v/v) wood (methyl) alcohol (CH_4O) diluted by adding 1.000 L of water, **(e)** 165 mL of 0.14 eq Cl^-/L diluted by adding 685 mL of water.

64. Find the concentration of the following solutions: **(a)** 18.0 mL of 4.12×10^{-7} M cholesterol diluted by adding 100.0 mL of water, **(b)** 625 mL of 1.00 M $AgNO_3$ diluted by adding 375 mL of water, **(c)** 147 mL of 1.35% (w/v) $CuSO_4$ diluted by adding 350.0 mL of water, **(d)** 200.0 mL of 12.0% (v/v) ethyl alcohol diluted by adding 1.00 L of water, **(e)** 1.000 L of 4.27 meq Ca^{2+}/L diluted by adding 425 mL of water.

Colligative Properties

65. For each of the following, predict which solution has the higher boiling point: **(a)** 1.0 M NaCl or 2.0 M NaCl, **(b)** 1.0 M NaCl or 1.0 M glucose ($C_6H_{12}O_6$), **(c)** 1.0 M NaCl or 1.0 M $CaCl_2$, **(d)** 1.0 M NaCl or 0.8 M Na_3PO_4.

66. For each pair of substances in exercise 65, predict which one has the higher melting point.

67. A 50–50 mixture of water and antifreeze freezes at a lower temperature than either water or antifreeze alone. Provide a chemical rationale for this.

68. For each pair of substances in exercise 65, predict which one exerts the greater osmotic pressure.

69. Do red blood cells shrink when placed in a **(a)** hypertonic, **(b)** hypotonic, or **(c)** isotonic solution?

70. A solution of 0.9% (w/v) NaCl is isotonic to blood. Is a solution of 0.9% KCl (w/v) also isotonic? Explain.

71. Suppose solution A is separated from solution B, which is more dilute, by a membrane that only lets water molecules pass through. Into which solution, A or B, will there be a net flow of water?

72. List three common colligative properties.

73. Explain how osmosis enables groundwater to rise in a plant.

74. If you dehydrate on a warm day, does the osmotic pressure in your blood increase or decrease?

75. Why is the action of the kidneys on blood called dialysis and not osmosis?

76. Caffeine and ethanol inhibit antidiuretic hormone (ADH). How does this affect the volume of urine?

DISCUSSION EXERCISES

1. How would the world be different if water molecules were linear and nonpolar instead of bent and polar?
2. Suppose you make solutions of glucose ($C_6H_{12}O_6$) and of $CaCl_2$ that have identical concentrations. In which of the following concentration units would these two solutions have the same number of g solute per L of solution? **(a)** *M*, **(b)** % (*w/v*), **(c)** % (*w/w*), **(d)** eq/L.
3. Gargling with salt water can help relieve a sore throat. Meat is preserved by "curing" it, treating it with salt. Explain how exposure to salt can kill bacteria. Does the salt have to be NaCl?

8

Reaction Rates and Equilibrium

1. What are the first and second laws of thermodynamics? How do they limit energy changes in chemical reactions?
2. What determines the speed of chemical reactions? What factors change this rate?
3. What is dynamic chemical equilibrium? What are equilibrium constants, and how do you calculate them?
4. What is Le Châtelier's principle? What conditions change the equilibrium of a chemical reaction?

Treating Patients with Oxygen

Doctors prescribe pure oxygen or oxygen-enriched air for several reasons. Patients with emphysema cannot inhale normal amounts of air, so they need more O_2 in each breath. Premature babies have a similar problem; they have trouble breathing because their lungs aren't fully developed (Figure 8.1). Doctors treat patients for carbon monoxide (CO) poisoning by administering oxygen-rich air to drive CO out of the blood and replace it with O_2 (see the Spotlight in Section 4.6).

Though it doesn't burn by itself, oxygen (O_2) reacts rapidly with many organic (carbon-containing) compounds and other substances. In these materials, carbon reacts with O_2 to form CO_2:

$$C + O_2 \longrightarrow CO_2 + energy$$

Since oxygen is a very reactive element (see Figure 8.10) and its molecules are in constant motion in the gaseous state, O_2 can react rapidly and release large amounts of energy.

This makes high concentrations of oxygen very dangerous. All it takes is one spark or flame to ignite an explosion. In 1967, for example, three U.S. astronauts died when faulty wiring caused an explosive reaction in their pure oxygen atmosphere. Hospital equipment that might spark is kept away from oxygen-rich areas. And people working in such areas avoid wearing materials such as nylon and rubber that can produce sparks.

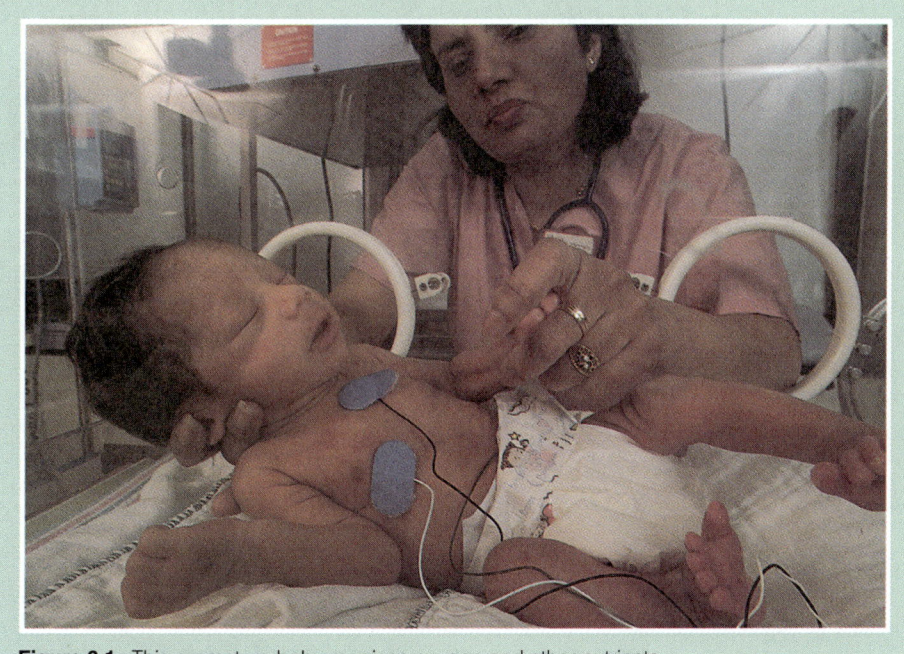

Figure 8.1 This premature baby receives oxygen and other nutrients.

Some chemical reactions are very fast and some are slow. A spark ignites an explosive reaction between gasoline and air in an automobile engine (Figure 8.2). Yet it takes many years for that car to rust completely.

The simple occurrence of a reaction isn't always enough to be useful. Your liver, for example, converts toxic substances such as ethanol and barbiturates into less toxic products. But those reactions need to be fast enough to prevent serious harm to you.

Figure 8.2 Rapid combustion reactions between gasoline and O_2 power this automobile.

Many factors affect the direction of a chemical reaction. In your lungs, for example, oxygen gas reacts with hemoglobin in blood to form an oxygen–hemoglobin complex. The opposite reaction occurs in your tissues: O_2 leaves hemoglobin and enters your cells for metabolism. What causes the reaction to go in one direction in your lungs and the opposite direction in your other tissues?

8.1 FIRST AND SECOND LAWS OF THERMODYNAMICS

First Law. Recall that chemical reactions give off or consume energy (Section 5.6). Those that give off heat are *exothermic,* and those that take up heat are *endothermic.* Changes in physical state also consume or give off energy (Section 6.2).

Thermodynamics is the interconversion of heat and other forms of energy. In studying physical and chemical changes, scientists consistently observe changes in the type, but not the amount, of energy. According to the **first law of thermodynamics** (also called the *law of conservation of energy*), energy is not created or destroyed, but it can change forms.

Energy is conserved if we take into account a system and its surroundings. Burning a lump of coal, for example, changes the coal's chemical potential energy into heat and light, which flow out to warm and illuminate the surroundings (Figure 8.3). Energy isn't created or destroyed, but it changes form.

Another way to word the first law of thermodynamics is: You can't get something (energy) for nothing. Yet another is: You can't win; you can only break even.

Second Law. Because the total amount of energy stays constant, you might think that we could never run out of energy. But after you burn a lump of coal for heat, that energy source is spent. If the energy itself isn't gone, what *is* lost?

According to the **second law of thermodynamics,** a system and its surroundings tend to change to a condition of greater entropy. **Entropy** is a measure of the randomness, or disorder, of a system. When water evaporates, for example, water molecules go from a more ordered state (a liquid) to a less ordered, more random state (a gas). The gaseous state has greater entropy (disorder) than the liquid state.

> **EXAMPLE 8.1** Which of the following has greater entropy? (a) ice or liquid water, (b) your room *before* or *after* you clean it up, (c) a sugar cube in coffee *before* or *after* it dissolves, (d) a falling egg *before* or *after* it hits a sidewalk.
>
> **SOLUTION** (a) liquid water, (b) *before*, (c) *after*, (d) *after*.

When coal burns, its highly ordered carbon atoms (fixed in a solid lump) react to form gaseous CO_2, which spreads out into the atmosphere. Entropy increases. As this happens, potential energy in coal changes into heat and light, which spread out into the universe. The energy still exists, but it can do little useful work.

Energy often changes into heat, which radiates into the environment (Figure 8.4). A light bulb, for example, converts only about 5% of electrical energy into useful light; the other 95% changes into heat. When you metabolize food, your body converts most of the food's potential energy into heat. The heat keeps you warm for a while, but eventually is lost to the atmosphere.

According to the first law of thermodynamics, we will never run out of energy. According to the second law, however, energy changes into less useful forms (mainly heat) during energy conversions. So conversions aren't 100% efficient in terms of useful energy.

Figure 8.3 Chemical potential energy changes into heat and light as this building burns.

Another way to put the second law (regarding useful energy) is: You can't even break even.

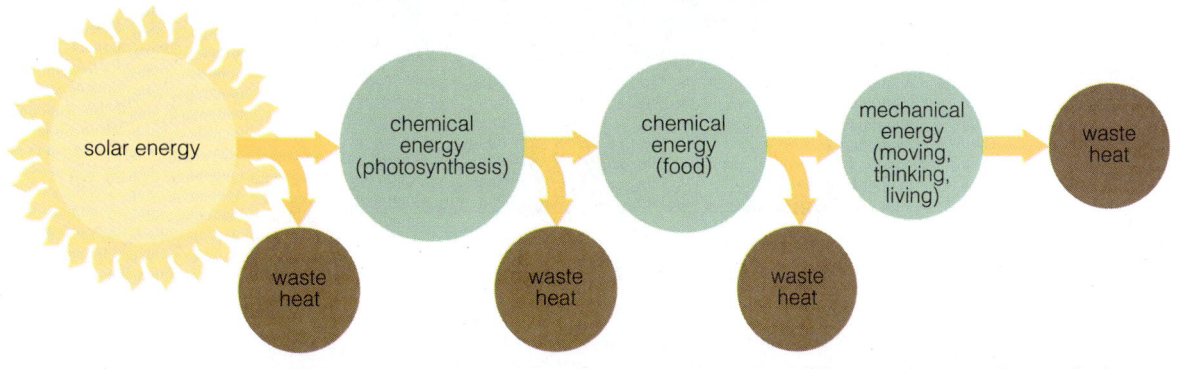

Figure 8.4 The second law of thermodynamics. When it changes into other forms, some energy becomes low-quality heat, which goes into the environment.

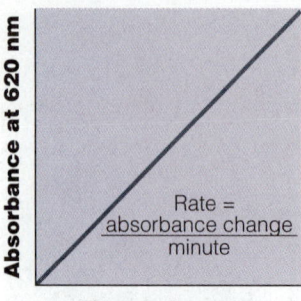

8.2 REACTION RATES

Reaction rate is the speed of a chemical reaction. It is the speed at which product forms or reactant disappears.

When a reactant or product is colored, we can determine the reaction rate by measuring the change in color over a period of time. For example, glucose reacts with a substance (*o*-toluidine) to form a blue-green product:

$$\text{glucose} + o\text{-toluidine} \longrightarrow \text{colored complex}$$
$$\text{colorless} \qquad\qquad\qquad \text{blue-green}$$

You can determine the reaction rate by measuring how much blue-green product forms per minute (Figure 8.5). To do this, you could use an instrument that measures absorbance (due to the blue-green color) at a wavelength of 620 nm.

If the increase in color in 10 minutes, for example, shows an increase in product concentration from 0.000 M to 0.013 M, the average reaction rate is $\frac{0.013 \text{ mol/L} - 0.000 \text{ mol/L}}{10 \text{ minutes}}$ = 0.0013 mol/L per minute.

Figure 8.5 Measuring the rate of reaction between glucose and *o*-toluidine to form a blue-green product.

EXAMPLE 8.2 What is the average rate of a reaction in mol/L per minute if the concentration of reactant changes from 0.532 M to 0.388 M in 1.00 hour?

SOLUTION rate = $\dfrac{0.532\ M - 0.388\ M}{60.0 \text{ min}} = \dfrac{2.40 \times 10^{-3} \text{ mol/L}}{\text{min}}$

What Determines the Rate? Hydrogen gas (H_2) reacts explosively with oxygen gas (Figure 8.6). Other reactions, such as the gradual discoloration of house paint, take years.

Figure 8.6 Explosion of the German dirigible *Hindenberg* in 1937 as its hydrogen (H_2) gas rapidly reacts with oxygen (O_2) gas.

One reason reaction rates vary so much is that reactant particles (atoms, ions, or molecules) have to collide with each other in order to react. The more frequently they collide, the faster the reaction is. But not all collisions produce reactions. Particles must collide with enough energy and with the proper alignment for a reaction to occur (Figure 8.7).

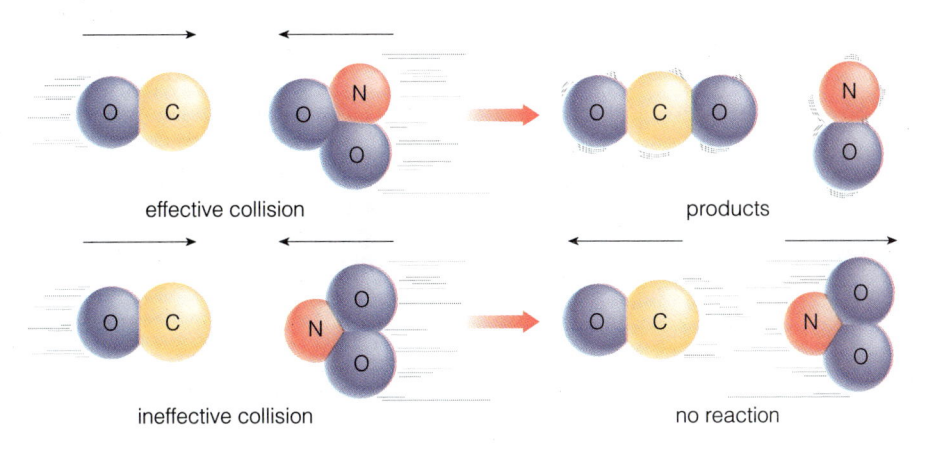

Figure 8.7 For the reaction, CO + NO$_2$ → CO$_2$ + NO, to occur, reactant molecules must collide with the proper orientation.

The minimum energy required for a reaction to occur is called the **activation energy** (or *energy of activation*). Just as a boulder may have to be pushed over a hill before it can roll down, activation energy is a barrier that reactants must overcome before they can react (Figure 8.8).

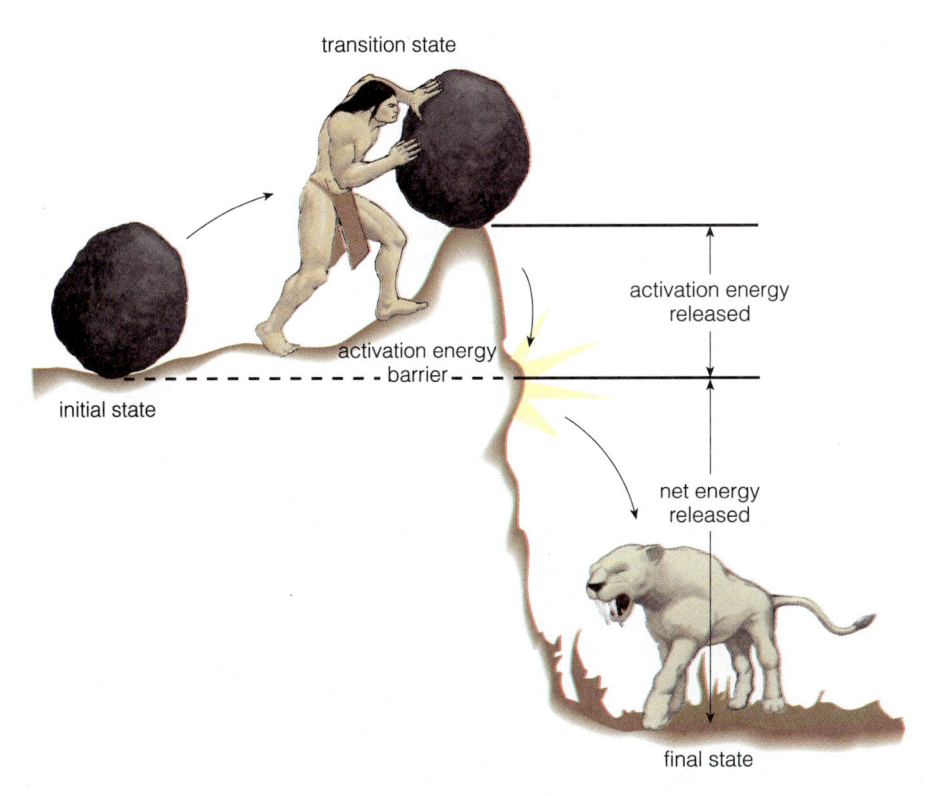

Figure 8.8 An analogy for activation energy. If enough energy is supplied to push the boulder over the hill (activation energy barrier), the boulder will roll down the mountain, releasing energy as it goes to a state of lower energy. Boulders are pushed over the cliff faster when the activation energy is small.

Activation energy affects the reaction rate. When the activation energy is high, few molecules have enough energy to react. A lump of coal, for example, could come in contact with oxygen molecules in air for many years but not react until a flame supplies the activation energy. A high activation energy means a slower reaction, because fewer reactant molecules collide with enough energy to form products.

Figure 8.9 shows a way to represent energy changes in a reaction. Notice that the height of the hill is the activation energy. Also notice that the difference between the energy of reactants and products is the net energy change from the

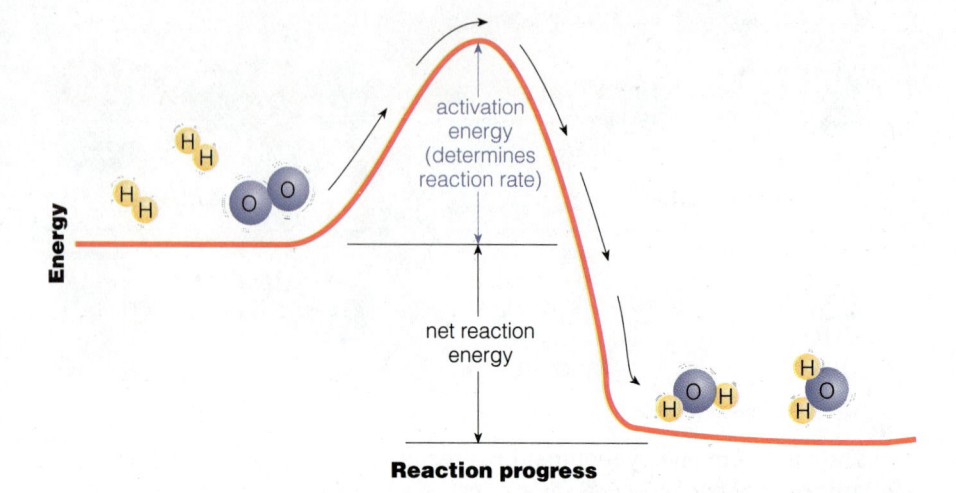

Figure 8.9 A simplified energy diagram for the reaction between H_2 and O_2 to form H_2O. The activation energy is so high that the reaction is extremely slow at 25°C.

reaction. When the products have lower energy, as in Figure 8.8, the reaction gives off energy and is exothermic.

EXAMPLE 8.3 The combustion of coal is an exothermic reaction, like the reaction represented in Figure 8.9. Once a burning match gives reactant particles enough energy to reach the activation energy, the reaction begins. What sustains the reaction after the burning match is consumed?

SOLUTION The net energy released by the reaction activates more reactant particles to react.

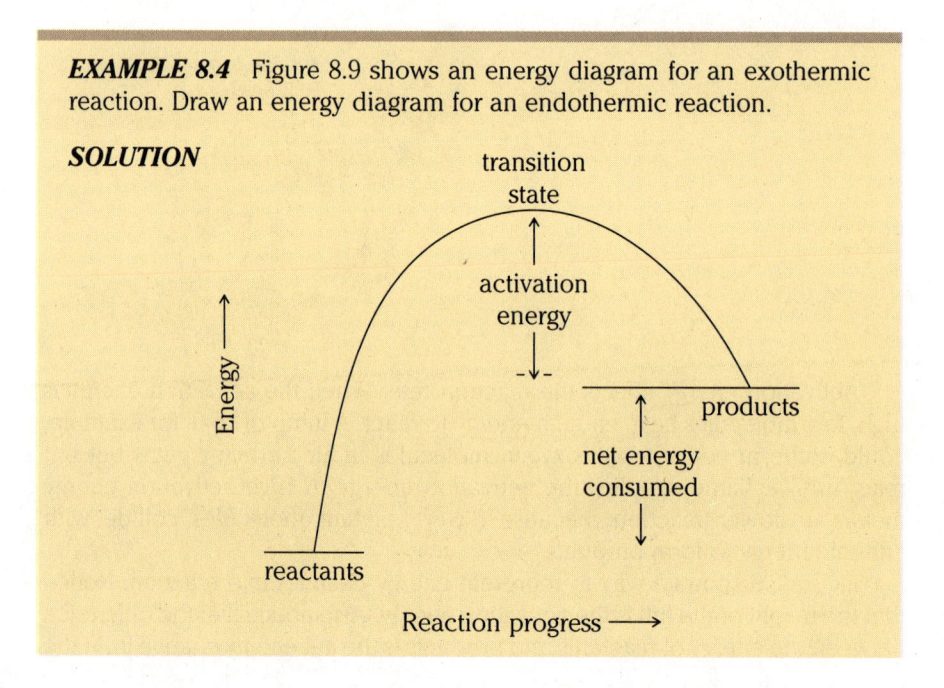

EXAMPLE 8.4 Figure 8.9 shows an energy diagram for an exothermic reaction. Draw an energy diagram for an endothermic reaction.

SOLUTION

8.3 CHANGING REACTION RATES

Why don't all chemical reactions occur at the same rate? In this section, we examine five factors that affect the speed of a reaction: (1) chemical reactivity of the reactants, (2) physical subdivision of reactants, (3) concentration of reactants, (4) temperature, and (5) catalysts.

Chemical Reactivity. Substances react to become more stable. Recall that the octet rule (Section 4.1) describes stable electron arrangements for many substances. We can use the periodic table to predict the tendency of elements to gain or lose electrons.

Recall that *ionization energy* (the energy to remove an electron from an atom) generally increases going from left to right, and from bottom to top, in the periodic table (see Figure 4.5). *Electronegativity* (the attraction for shared electrons) follows a similar pattern (see Figure 4.21). High electronegativity corresponds to high reactivity for nonmetal elements, which tend to attract electrons in a reaction. The opposite pattern applies to metal elements, which tend to lose electrons in a reaction.

Figure 8.10 summarizes these patterns. The reactivity of metals tends to increase going down and to the left in the periodic table (though a few exceptions occur, especially among transition elements). Nonmetal elements (except noble gases, which rarely react) increase in reactivity going up and to the right in the periodic table. The reactivities of alkali metals and halogens, for example, fit this pattern (see Table 3.4); cesium is the most reactive common metal and fluorine is the most reactive nonmetal.

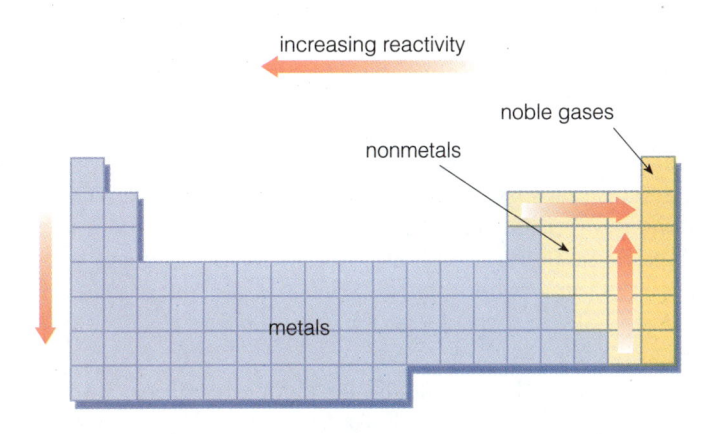

Figure 8.10 Chemical reactivity for metals increases going to the left and down in the periodic table. Reactivity for nonmetals (except noble gases) increases going to the right and up.

EXAMPLE 8.5 In each of the following pairs of elements, which one reacts more rapidly with oxygen gas (O_2): (a) K or Fe, (b) Li or Rb?

SOLUTION (a) K, (b) Rb

Physical Subdivision. Recall that a sugar cube dissolves faster in water if you break it into small pieces (Section 7.2). The greater surface area increases interactions between water and sugar molecules and thus hastens the dissolving action.

Increasing the surface area also speeds up chemical reactions between a solid reactant and a liquid or gas, and for the same reason: more reactant particles interact with each other in a given amount of time.

Increasing the surface area greatly accelerates reactions with solids, whose particles are held rigidly in place. Dried kernels of grain in a silo don't pose much of a fire hazard, but those same kernels ground into flour have so much more surface area that fire becomes a real danger. Dust explosions at grain elevators are a constant threat for the same reason (Figure 8.11).

Figure 8.11 An explosive reaction in this grain elevator in New Orleans, Louisiana, killed 35 people.

EXAMPLE 8.6 Which burns faster in a fireplace: equal amounts of wood in the form of (a) one log, (b) 10 pieces, or (c) many splinters.

SOLUTION (c) because of the increased surface area

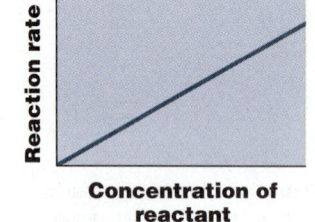

Figure 8.12 For many (but not all) reactions, the rate is directly proportional to the reactant concentration.

Concentration. Reactions go faster at higher reactant concentrations. Putting more reactant particles in a space is like adding dancers to a dance floor. The increased concentration of reactants (or dancers) means that more will collide with each other. Some of the extra collisions will form products.

Figure 8.12 shows a common pattern for reactions in which increased concentrations produce faster rates.

Temperature. Warmer temperatures typically accelerate reactions. A roast cooks faster in an oven at 375°F than at 300°F. And you slow down reactions that spoil food by keeping food cold.

At higher temperatures, the reacting particles have a greater average kinetic energy, and more particles have a particular amount of energy, such as the activation energy (see Figure 6.10). As a result, more particles collide with enough energy (the activation energy) to form products. A very rough rule of thumb is that reaction rates double for each 10°C rise in temperature.

Heat given off in very exothermic reactions can accelerate the reaction. Once a forest fire starts, for example, the heat produced ignites other trees, spreading the fire. Sometimes runaway reactions occur that, once started, produce a ball of flame or an explosion. When a spark starts a reaction between hydrogen and oxygen gas, for example, the energy released makes the reaction go even faster, producing a deafening boom and a fireball (Figure 8.6).

CHEMISTRY SPOTLIGHT

Temperature and Aging

Studies of a wide variety of animals show that those with slower metabolic rates generally live longer. Metabolic reactions can be slowed by lowering body temperature. One experiment, for example, showed that flies kept at low temperatures lived much longer (Table 8.1).

Might cold temperatures help us live longer? Two problems suggest the answer is no. First, experiments with warm-blooded animals (unlike flies, which are cold-blooded) show that cold temperatures produce shorter—not longer—life spans. In the cold, such animals develop faster metabolic rates to generate enough heat to keep warm. We have an internal mechanism, like a thermostat, that resists lowering our body temperature.

And trying to reset our internal thermostats to a lower temperature also may be a bad idea, even if we could do it. At a lower temperature our internal chemical reactions would slow down. We might live longer, but go through life in s. . . l. . . o. . . w motion.

Table 8.1 Effects of Temperature on the Life Span of Fruit Flies

Temperature (°C)	Average Life Span Male (days)	Female (days)
15	162	171
20	102	112
25	40	58
30	22	29

Data from J. Loeb and J.H. Northrop, *J. Biol. Chem.* 32 (1917): 103–121.

Catalysts. **Catalysts** are substances that accelerate a reaction without being chemically changed. They alter the *reaction mechanism,* the exact way in which reactants form chemical intermediates in the process of changing into products. Catalysts enable reactants to form intermediates that require less energy. As a result, they lower the activation energy for a reaction.

For example, if we add finely divided platinum (Pt) metal to a mixture of hydrogen and oxygen gas at room temperature, the reaction goes rapidly:

$$2H_2 + O_2 \xrightarrow{Pt} 2H_2O$$

Platinum, like other catalysts, lowers the activation energy for the reaction (Figure 8.13 and Table 8.2) by altering the reaction mechanism. Less energy is needed because Pt binds H_2 molecules on its surface and weakens H—H covalent bonds; O_2 molecules then interact with H_2 more effectively to form H_2O (Figure 8.14).

Catalysts have many uses. The petroleum industry uses catalysts to refine oil into gasoline, kerosene, and other useful products. Catalytic converters in automobiles oxidize CO and unburned fuel in exhaust gases to less harmful

Table 8.2 Activation Energies for the Reaction: $2H_2O_2 \rightarrow 2H_2O + O_2$

Condition	Activation Energy (kcal/mol)
Uncatalyzed	18
Pt catalyst	13
Enzyme catalyst	7

Figure 8.13 A catalyst accelerates a reaction by lowering the activation energy so that more reactant molecules collide with enough energy to react.

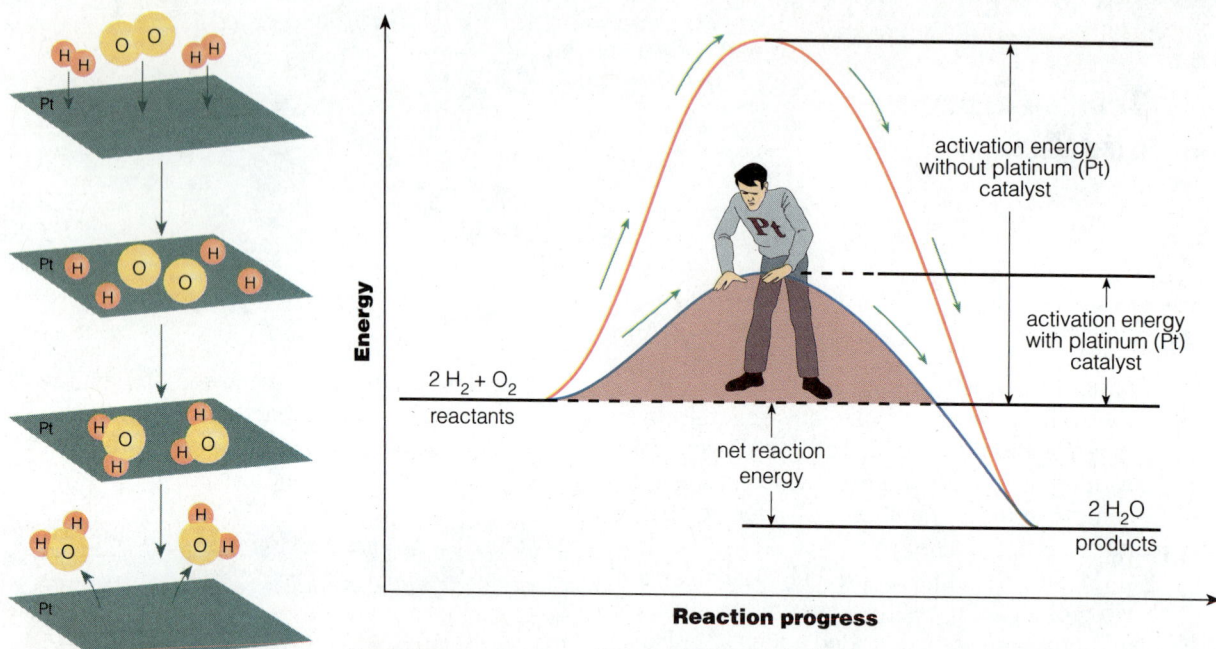

Figure 8.14 Platinum (Pt) works as a catalyst by providing a surface for hydrogen (H₂) and oxygen (O₂) molecules to separate into individual atoms, which recombine to form water (H₂O) molecules.

products, CO_2 and H_2O. Self-cleaning ovens have catalysts that break down smoke particles.

Almost all reactions in your body are accelerated by *enzymes*. Like other catalysts, enzymes lower the activation energy of reactions (Table 8.2). As a result enzymes make reactions go as much as a million times faster than uncatalyzed reactions. And in the body, enzymes allow reactions to go rapidly even at a temperature of only 37°C.

8.4 DYNAMIC CHEMICAL EQUILIBRIUM

Dynamic Equilibrium. Many reactions are reversible; they go in both directions. Such reactions—in a closed container at constant temperature and pressure—don't go to 100% completion. Instead, they reach a state of *dynamic equilibrium* (Section 6.2). Reactants form products, while products form reactants. At equilibrium the forward and reverse reactions occur at the same rate, keeping the concentrations of reactants and products constant (Figure 8.15).

We have already discussed how physical processes reach equilibrium. Recall, for example, that in a closed container a liquid such as water vaporizes while water vapor condenses at the same rate (Section 6.2); the amount of liquid water stays constant, as does the amount of water vapor, even though vaporization and condensation both continue. The system is at equilibrium.

Equilibrium also occurs in a saturated solution containing undissolved solute. If you look at a saturated NaCl solution containing undissolved NaCl, it looks like nothing is happening. But it is. Solid NaCl continues to dissolve while dissolved Na^+ and Cl^- ions come out of solution and deposit on the solid NaCl. There is no *net* change because the two processes occur at the same rate and are at dynamic equilibrium.

Equilibrium also occurs in chemical reactions. Suppose, for example, we put a mixture of sulfur dioxide (SO_2) gas and oxygen gas in a closed container at room temperature and pressure. If we analyze the contents later, we find that

Recall that reactants rarely react completely to form the products written in an equation (Section 5.3).

the container holds not only SO_2 and O_2 gases but also sulfur trioxide (SO_3) gas. SO_3 forms by the *forward reaction* between SO_2 and O_2:

$$2SO_2 + O_2 \longrightarrow 2SO_3$$

SO_2 and O_2 form by the *reverse reaction* of SO_3:

$$2SO_3 \longrightarrow 2SO_2 + O_2$$

At dynamic equilibrium the forward and reverse reactions occur at the same rate, and the concentrations of SO_2, O_2, and SO_3 stay constant. A double arrow ($\rightleftarrows$) shows that the reaction goes in both directions:

$$2SO_2 + O_2 \rightleftarrows 2SO_3$$

If we put just SO_2 and O_2 in a container, the forward reaction is faster at first; SO_2 and O_2 react to form SO_3. Once SO_3 appears, the reverse reaction begins. As the concentration of SO_3 rises, the reverse reaction goes faster; at equilibrium it goes at the same rate as the forward reaction (Figure 8.16).

The equilibrium is *dynamic,* with constant movement. It is like a football game in a large stadium, with a fixed number of spectators. If 500 people leave the stands every minute for refreshments, and 500 others return each minute, the number of people in the stands stays constant. But they aren't exactly the same people. The continuous, steady turnover is a dynamic equilibrium.

Equilibrium Constants. If you put any concentrations of SO_2 and O_2—with or without any amount of SO_3—in a closed container, the system will still reach equilibrium. At equilibrium, the forward and reverse reactions occur at the same rate. But that doesn't mean the concentrations of reactants and products are equal (Figure 8.16). In fact, the concentrations are rarely equal.

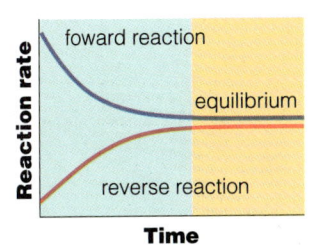

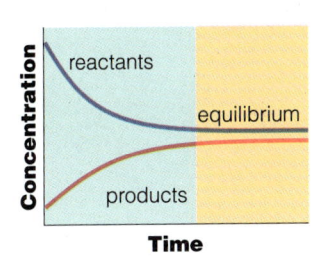

Figure 8.15 At equilibrium the rates of the forward and reverse reactions are equal (top); the concentrations of reactants and products are constant, but not necessarily equal (bottom).

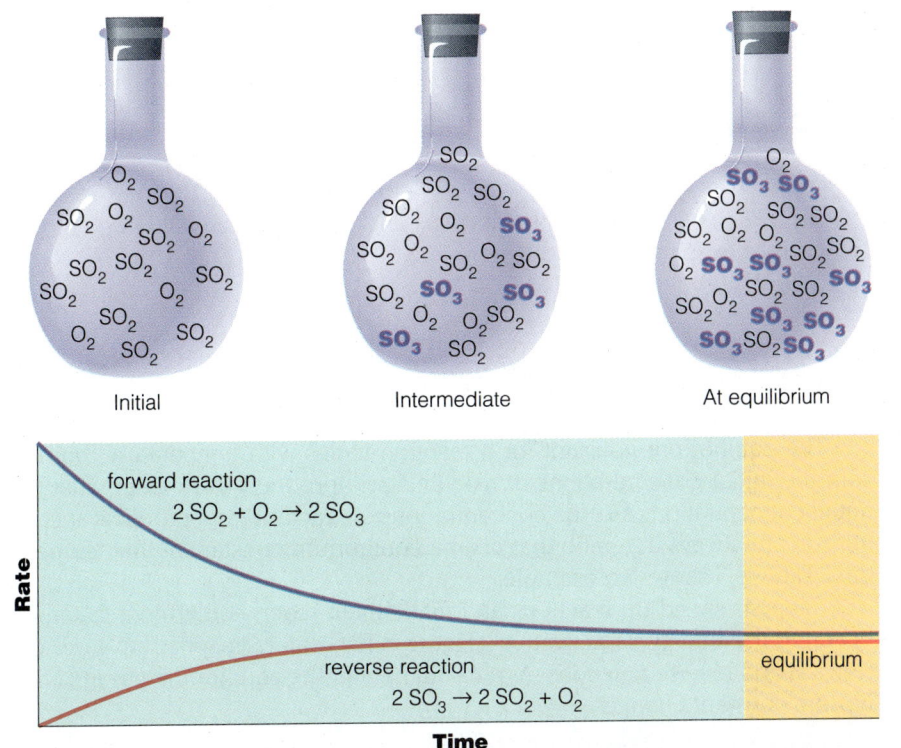

Figure 8.16 The forward reaction goes faster at first. As SO_3 forms, the reverse reaction goes faster. At equilibrium the rates are equal.

No matter what the initial concentrations are in a reaction, reactants and products occur in the same *proportion* at equilibrium. Chemists use an **equilibrium constant (K)** to represent the ratio of products to reactants at equilibrium. For the general reaction

$$aA + bB \rightleftharpoons cC + dD$$

the equilibrium constant is:

$$K = \frac{[C]^c[D]^d}{[A]^a[B]^b}$$

where [ ] represents the concentration in mol/L of each reactant (A and B) and product (C and D), and a, b, c, and d are the coefficients (Section 5.1) in the balanced equation. For the reaction, $2SO_2 + O_2 \rightleftharpoons 2SO_3$,

$$K = \frac{[SO_3]^2}{[SO_2]^2[O_2]}$$

> **EXAMPLE 8.7** Write equations for the equilibrium constants for the following reactions: (a) $N_2 + 3H_2 \rightleftharpoons 2NH_3$, (b) $HCl + H_2O \rightleftharpoons H_3O^+ + Cl^-$, (c) $2N_2O_5 \rightleftharpoons 4NO_2 + O_2$.
>
> **SOLUTION** (a) $K = \dfrac{[NH_3]^2}{[N_2][H_2]^3}$, (b) $K = \dfrac{[H_3O^+][Cl^-]}{[HCl][H_2O]}$,
>
> (c) $K = \dfrac{[NO_2]^4[O_2]}{[N_2O_5]^2}$

K is written with no units. Units for K may be different for different reactions, so we usually omit the units.

To calculate the equilibrium constant for a reaction, you need to know the balanced equation and the concentrations of reactants and products at equilibrium. Suppose in the reaction, $2SO_2 + O_2 \rightleftharpoons 2SO_3$, the concentration at equilibrium of SO_2 is 0.010 mol/L; of O_2 is 0.20 mol/L; and of SO_3 is 0.10 mol/L. You can calculate the equilibrium constant:

$$K = \frac{[SO_3]^2}{[SO_2]^2[O_2]} = \frac{(0.10)^2}{(0.010)^2(0.20)} = 5.0 \times 10^2$$

> **EXAMPLE 8.8** For the reaction, $N_2 + 3H_2 \rightleftharpoons 2NH_3$, calculate the equilibrium constant if the concentrations at equilibrium are 0.082 M N_2, 2.14 M H_2, and 0.26 M NH_3.
>
> **SOLUTION** $K = \dfrac{[NH_3]^2}{[N_2][H_2]^3} = \dfrac{(0.26)^2}{(0.082)(2.14)^3} = 8.4 \times 10^{-2}$

The equilibrium constant for a reaction varies with temperature, but not with the initial concentrations of reactants and products. No matter what the initial concentrations are, the concentrations of reactants and products at equilibrium are always in a ratio that fits the equilibrium constant at that temperature. Table 8.3 shows an example.

Catalysts speed up reactions and make them reach equilibrium faster. By lowering the energy of activation (Figure 8.13), catalysts speed up both the forward and reverse reactions. As a result, neither the equilibrium nor the equilibrium constant changes.

Table 8.3 Effect of Initial Reactant and Product Concentrations on Concentrations at Equilibrium for the Reaction: $H_2 + I_2 \rightleftarrows 2HI$

Initial Concentration (mol/L)			Concentration at Equilibrium (mol/L)			Equilibrium Constant $K = \dfrac{[HI]^2}{[H_2][I_2]}$
H_2	I_2	HI	H_2	I_2	HI	
0	0	0.0150	0.00160	0.00160	0.0118	54.4
0.00932	0.00805	0	0.00257	0.00130	0.0135	54.5
0.00104	0	0.0145	0.00224	0.00120	0.0121	54.5
0.00375	0.00375	0.00375	0.00120	0.00120	0.00886	54.5

Very large or small equilibrium constants indicate which direction predominates. Since product concentrations are in the numerator and reactant concentrations are in the denominator, very large K values indicate a higher concentration of products than reactants. Those reactions go mostly to the right as written, in the forward direction. Reactions having very small equilibrium constants go mostly to the left as written, in the reverse direction. Values greater than about 10^6 or less than 10^{-6} indicate reactions that go almost completely to the right or left, respectively.

EXAMPLE 8.9 For each of the following equilibrium constants, does the reaction go mostly to the left or to the right? (a) 1.1×10^{16}, (b) 2.14×10^{-1}, (c) 8.7×10^{-9}.

SOLUTION (a) right, (b) reactions occur at significant rates in both directions, (c) left.

8.5 LE CHÂTELIER'S PRINCIPLE

Making a Reaction Go Farther. Scientists usually want to get as high a yield of products as possible. For industrial processes, high yield is essential. Three things can help: (a) increasing the concentration of reactants, (b) removing products as they form, and (c) adjusting the temperature and pressure to form more product at equilibrium.

French chemist Henri Le Châtelier (Figure 8.17) proposed a way to predict the shift that occurs in an equilibrium mixture when the conditions change. According to **Le Châtelier's principle:** When a system at dynamic equilibrium is subjected to a stress, the system changes, if possible, to relieve the stress.

We can use this principle to predict whether a new equilibrium mixture contains more or less product.

Concentration of Reactants. For the reaction, $2SO_2 + O_2 \rightleftarrows 2SO_3$, suppose the reactants and products have reached equilibrium in a closed container. What happens if we upset the equilibrium by adding O_2?

According to Le Châtelier's principle, the system shifts to relieve the stress. In this case, the system will use up some excess O_2. To do this, the reaction must

Figure 8.17 Henri Le Châtelier (1850–1936).

CHEMISTRY SPOTLIGHT

Getting Oxygen to Your Cells

Your cells and tissues need oxygen gas to stay alive. Hemoglobin in the blood binds O_2 in your lungs, carries it to your cells, and then releases it. One reason this happens is that changes in O_2 concentrations in the body shift the equilibrium for the reaction that binds oxygen to hemoglobin.

We can represent the binding of hemoglobin (Hb) to oxygen as: $Hb + O_2 \rightleftharpoons HbO_2$. The air you breathe provides a high concentration of O_2 in your lungs; the partial pressure of O_2 is about 100 torr (Figure 8.18). As blood passes by your lungs, high O_2 levels shift the equilibrium position to the right:

$$Hb + O_2 \longrightarrow HbO_2$$

Under these conditions, hemoglobin is almost completely saturated with oxygen.

Oxygenated hemoglobin (HbO_2) then travels in blood from the lungs to other tissues. There the O_2 concentration is much lower because cells consume oxygen. In capillaries of active muscles, for example, the partial pressure of O_2 is about 20 torr (Figure 8.18). Such low levels of O_2 shift the equilibrium position to the left:

$$Hb + O_2 \longleftarrow HbO_2$$

As O_2 is released, then used by cells, hemoglobin becomes less saturated.

Other factors, including CO_2 and acidity levels, also help load oxygen onto hemoglobin at the lungs and unload it at the cells (see Section 9.7). They work the same way as the change in O_2 concentration, causing a shift in equilibrium position for the reaction of Hb with O_2.

Because this system works so well, we are able to work, play, and discuss this nice application of Le Châtelier's principle.

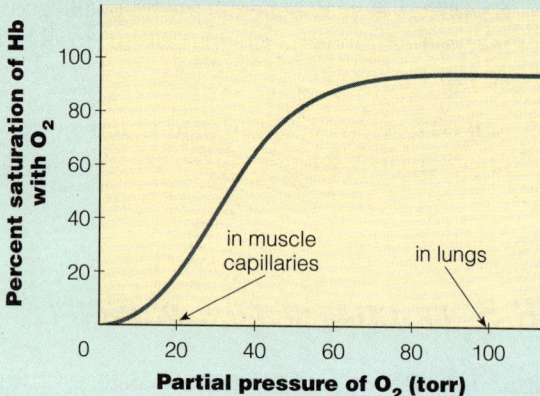

Figure 8.18 Effect of partial pressure of O_2 on the extent of oxygenation of hemoglobin (Hb).

go in the forward direction to form more SO_3. Thus, by increasing the concentration of reactant, we shift the equilibrium (as written) to the right.

If instead of adding excess O_2 we added SO_2, the equilibrium still shifts to the right to relieve the stress. If we increased the concentration of product (SO_3) instead, the equilibrium shifts to the left to remove excess SO_3. In effect, adding reactant or product to a system at equilibrium is like pushing on one side of a seesaw; the equilibrium mixture moves in the opposite direction.

CHEMISTRY SPOTLIGHT

Equilibrium and Steady State

Removing product is an effective way to "pull" a reaction to the right. Your body uses this principle to carry out the set of reactions known collectively as *metabolism*.

You metabolize glucose ($C_6H_{12}O_6$), for example, by the net reaction:

$$C_6H_{12}O_6 + 6O_2 \rightleftharpoons 6CO_2 + 6H_2O$$

But it actually takes more than a dozen reactions to accomplish this net reaction. Each reaction is reversible and has an equilibrium constant. We can represent the reaction sequence as:

$$\text{glucose} \rightleftharpoons A \rightleftharpoons B \rightleftharpoons C \rightleftharpoons D \rightleftharpoons \ldots \rightleftharpoons CO_2 + H_2O$$

The product of each reaction is consumed immediately because it is the reactant for the next reaction. The constant removal of product keeps pulling each previous reaction to the right to replenish product. Even the final products, CO_2 and H_2O, are removed. You exhale CO_2 and eliminate H_2O in your breath, perspiration, and urine. Thus, the preceding net reaction is often written with an arrow going only to the right.

You will learn the details of metabolism in Chapters 24–26. This constant shifting of reactions in one direction by removing product is called a *steady state* condition. It's like water that flows from a central source through a water main to homes—although the water could go in either direction, the flow essentially moves only away from the source, because homes constantly remove and use water.

The reactions in your body keep trying to reach equilibrium. But they aren't able to because products keep getting removed. You are at a steady state, not equilibrium. When your body finally reaches equilibrium, you won't be in any condition to celebrate it.

EXAMPLE 8.10 For the reaction $2SO_2 + O_2 \rightleftharpoons 2SO_3$, what is the effect on the equilibrium position if (a) SO_2 or (b) O_2 is removed?

SOLUTION In both cases the equilibrium position shifts to the left to replenish SO_2 or O_2.

Removing Product. Le Châtelier's principle applies in the same way to removing product from an equilibrium mixture. Consider again the reaction between SO_2 and O_2 to form SO_3:

$$2SO_2 + O_2 \rightleftharpoons 2SO_3$$

Suppose at equilibrium we remove some SO_3. The equilibrium position shifts to the right to replenish the product (SO_3).

Pressure. Pressure changes can shift the equilibrium of reactions involving gases. Consider once again the reaction involving three gases:

$$2SO_2 + O_2 \rightleftharpoons 2SO_3$$

Suppose we reduce the volume of the container, increasing the pressure inside after the system reaches equilibrium. According to Le Châtelier's principle, the system will change to relieve the stress. But how can this system relieve the increase in pressure?

Recall the universal gas law, $PV = nRT$ (Section 6.4). At a given temperature (T) and volume (V), pressure (P) is proportional to n, the number of moles (or particles) of a gas. Increasing the number of molecules increases the pressure, and reducing the number of molecules reduces the pressure.

How can the equilibrium position of the preceding reaction shift to reduce the pressure? Notice that there are three molecules of reactants per two molecules of product. A shift to the right reduces the number of molecules and thus reduces pressure. So increased pressure shifts the equilibrium to the right to relieve the change in pressure.

Figure 8.19 shows an example of how pressure affects a system at equilibrium.

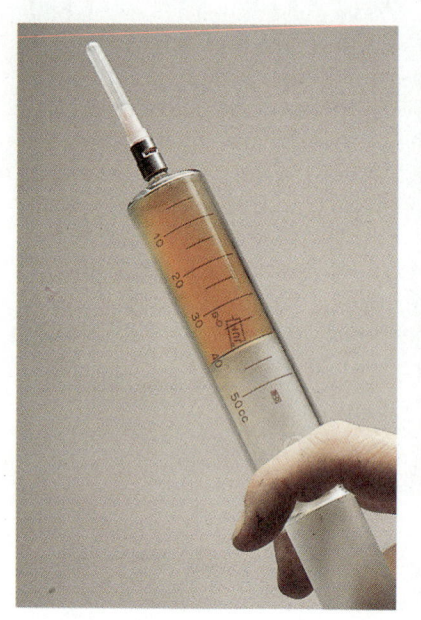

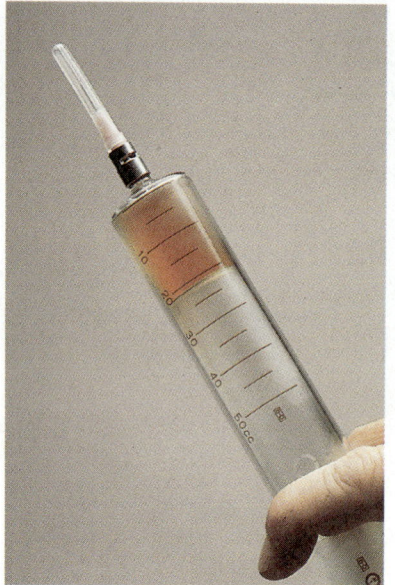

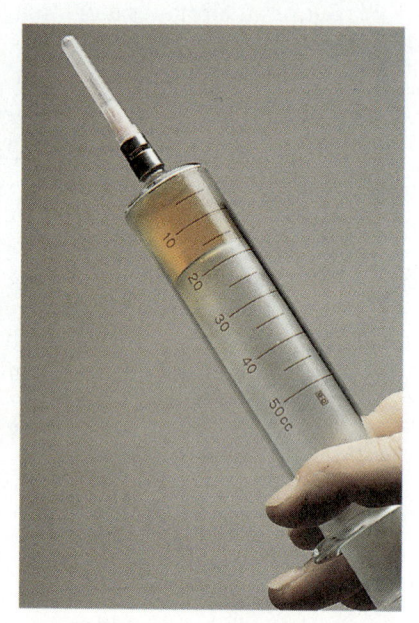

Figure 8.19 Effect of pressure on equilibrium of the reaction, $2NO_2 \rightleftharpoons N_2O_4$. (Left) The syringe contains an equilibrium mixture of NO_2 (red-brown) and N_2O_4. (Center) Increasing the pressure increases the concentration of both substances, so the color due to NO_2 is more intense. This color eventually decreases (right) because higher pressure shifts the equilibrium position to the right, reducing the concentration of colored NO_2.

EXAMPLE 8.11 A decrease in pressure shifts the equilibrium position in which direction in the following reactions: (a) $2H_2 + O_2 \rightleftharpoons 2H_2O$, (b) $2H_2O \rightleftharpoons 2H_2 + O_2$, (c) $N_2 + 3H_2 \rightleftharpoons 2NH_3$, (d) $H_2 + I_2 \rightleftharpoons 2HI$.

SOLUTION (a) left, (b) right, (c) left, (d) no shift (reactants and products have equal numbers of molecules).

Temperature. Temperature changes an equilibrium constant. But if you don't know the equilibrium constant, you can still use Le Châtelier's principle to predict whether a temperature change shifts a reaction to the left or right.

To do this, you only need to know whether a reaction is endothermic or exothermic.

Let's consider two reactions, one endothermic (requiring heat) and one exothermic (giving off heat):

Endothermic: $N_2 + O_2 + \text{heat} \rightleftharpoons 2NO$
Exothermic: $2NO_2 \rightleftharpoons N_2O_4 + \text{heat}$

If we heat a system at equilibrium, how does the system respond? According to Le Châtelier's principle, the reaction shifts in the direction that uses up the excess heat. The preceding endothermic reaction, for example, shifts to the right to use up some excess heat. The preceding exothermic reaction, in contrast, shifts to the left to use up excess heat (Figure 8.20).

You can think of heat as a component in the reaction. Then predict the shift for a change in temperature in the same way as you would for a change in concentration of a reactant or product. Increasing temperature shifts the reaction in the direction that consumes heat; cooling produces a shift in the opposite direction.

In classifying reversible reactions as endothermic or exothermic, we are considering the forward reaction (reading the equation from left to right as written). The reverse reaction has the opposite classification.

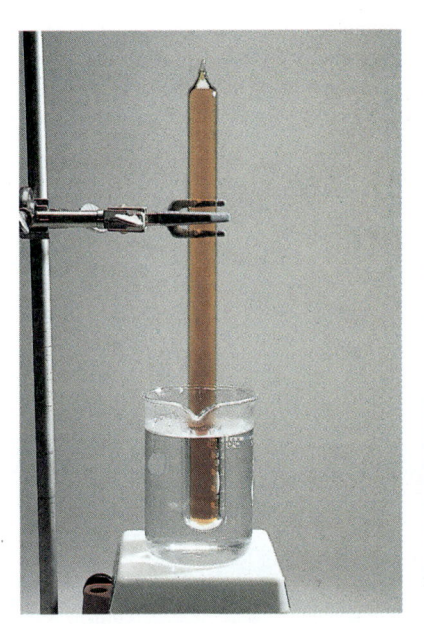

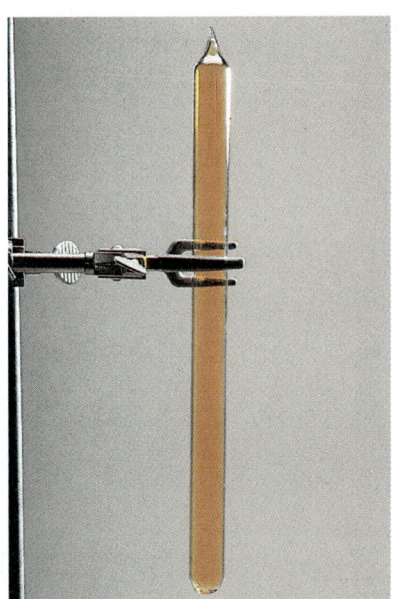

EXAMPLE 8.12 What is the effect of cooling on the position of equilibrium in the two preceding reactions?

SOLUTION The first reaction ($N_2 + O_2 + \text{heat} \rightleftharpoons 2NO$) shifts to the left. The second reaction ($2NO_2 \rightleftharpoons N_2O_4 + \text{heat}$) shifts to the right (Figure 8.20).

Figure 8.20 Effect of temperature on equilibrium of the reaction, $2NO_2 \rightleftharpoons N_2O_4 + \text{heat}$. (Center) An equilibrium mixture at room temperature contains NO_2 (red-brown) and N_2O_4 (colorless). Warming the mixture (left) shifts the equilibrium position to the left, increasing the amount of NO_2 and color. Cooling the mixture (right) shifts the reaction to the right, lessening the amount of NO_2 and color.

If your theory is found to be against the second law of thermodynamics, I can give you no hope; there is nothing to do but collapse in deepest humiliation.

Arthur S. Eddington (1882–1944)

SUMMARY

According to the first law of thermodynamics, energy is not created or destroyed, but it can change forms. The second law states that a system and its surroundings tend to change to a condition of greater entropy (disorder). When energy is used to do work, or changes into other forms, some of it changes into less useful forms, often heat.

In a chemical reaction, reaction rate is the speed of forming products or consuming reactants. To form products, the reactants must collide with enough energy and the proper alignment. The rate is faster for reactions that have a low activation energy. Factors that affect the reaction rate are temperature, the presence of catalysts, and the chemical reactivity, physical subdivision, and concentration of reactants. Catalysts accelerate reactions without being chemically changed; they lower the activation energy of a reaction by enabling reactants to form different intermediates while changing into products.

Reversible reactions reach dynamic equilibrium in a closed container. At equilibrium, forward and reverse reactions occur at the same rate, leaving the concentrations of reactants and products constant. The equilibrium constant K is the ratio of concentrations of products to reactants. For the reaction, $aA + bB \rightleftarrows cC + dD$:

$$K = \frac{[C]^c[D]^d}{[A]^a[B]^b}$$

The value of K indicates the extent to which a reaction goes to the left or right. It varies with temperature but is independent of the initial concentrations of reactants and products. Catalysts make a system reach equilibrium faster but don't change K.

According to Le Châtelier's principle, when a system at dynamic equilibrium is subjected to a stress, it changes, if possible, to relieve the stress. The equilibrium position can be shifted by changing the temperature or the concentration of reactants or products at equilibrium. Pressure changes shift the equilibria of some reactions involving gases.

KEY TERMS

Activation energy   (8.2)
Catalyst   (8.3)
Entropy   (8.1)
Equilibrium constant (K)   (8.4)
First law of thermodynamics   (8.1)
Le Châtelier's principle   (8.5)
Reaction rate   (8.2)
Second law of thermodynamics   (8.1)

EXERCISES

Even-numbered exercises are answered at the back of this book.

First and Second Laws of Thermodynamics

1. State the first law of thermodynamics.
2. When you burn gasoline in an automobile, are you producing energy? Explain your answer in terms of the first law of thermodynamics.
3. According to the first law of thermodynamics, are we in danger of running out of energy?
4. State the second law of thermodynamics.
5. According to the second law of thermodynamics, are we in danger of running out of energy?
6. Explain, in terms of the second law of thermodynamics, what is happening to energy when you burn gasoline in an automobile.
7. Explain, in terms of the second law of thermodynamics, what is happening to energy when you eat a cookie and metabolize it.
8. Is the chemical process described in exercise 7 endothermic or exothermic?
9. Identify which of the following has the greater entropy: **(a)** 1 mol of H_2 gas in a small container or in a

larger container; **(b)** 2 kg of sewage dumped in a river or the same sewage after being carried 2 km downstream; **(c)** gasoline before or after it is burned; **(d)** your hair before or after you comb it; **(e)** a slice of bread before or after you eat it.

Reaction Rates

10. Suppose bleach oxidizes a dye to a colorless material. The general reaction is:

    dye + bleach ⟶ oxidized dye + reduced bleach
    (*colored*)   (*colorless*)

    What is a convenient way to measure the rate of this reaction?

11. Ethyl ether (C_2H_6O), a liquid that vaporizes readily at room temperature, reacts with oxygen (O_2) gas to form CO_2 and H_2O. Ethyl ether reacts faster as a gas than as a liquid. Rationalize why.

12. Does a high activation energy determine whether a reaction is exothermic or endothermic? Explain.

13. Does a high activation energy affect reaction rate? Explain.

14. What is the effect of a catalyst on the activation energy of a reaction?

15. Nitrogen (N_2) and oxygen (O_2) gas stay mixed in the atmosphere indefinitely without reacting. Yet an intense source of energy—lightning—causes the gases to react. Explain why.

16. If the concentration of product in a reaction increases from 0.0042 *M* to 0.0090 *M* in 16 min, what is the average reaction rate in mol/L per min?

17. If the rate of a reaction is 2.4×10^{-3} mol/L per min, how many minutes does it take for the concentration of reactant to change from 0.632 *M* to 0.512 *M*?

18. Which alkaline earth metal is the most reactive?

19. Which alkali metal is the most reactive?

20. Which halogen is the most reactive?

21. Which will burn faster: a bound catalog or the same pages individually crumpled? Explain.

22. Would 10 kg of iron (Fe) react faster with oxygen (O_2) gas to form rust (Fe_2O_3) when the iron is in the form of **(a)** a smooth ball or **(b)** a flat brick 1 cm high? Explain why.

23. For the reaction, $2Mg + O_2 \rightarrow 2MgO$, would the rate be faster if two moles of Mg and one mole of O_2 reacted in a volume of **(a)** 20 L or **(b)** 40 L? Explain.

24. Look at Figure 8.12. How would the graph be different if a catalyst were added to the reaction mixture?

25. In a plot of reaction rate versus concentration of reactants, the rate eventually slows down. Why does this happen?

26. Why don't all collisions between reactants form products?

27. List two characteristics of collisions between reactants that account for higher reaction rates at higher temperatures.

28. In the presence of certain catalysts, such as enzymes, the initial reaction rate increases with increasing temperature. But then the rate slows down considerably. A graph might look like this:

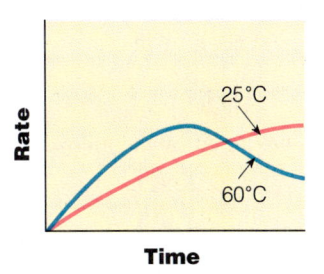

    List two reasons why the rate might eventually slow down.

29. How does a catalyst affect the net energy consumed or given off in a chemical reaction?

30. For the reaction, $2H_2 + O_2 \rightarrow 2H_2O$, platinum (Pt) is a catalyst. Suggest a reason why finely divided Pt is more effective than an equal amount of Pt in the form of a cube.

Dynamic Chemical Equilibrium

31. Why is chemical equilibrium frequently called *dynamic* chemical equilibrium?

32. The reaction, $2SO_2 + O_2 = 2SO_3$, will reach dynamic chemical equilibrium if the substances are in a closed container. Why does the container need to be closed?

33. For the reaction in exercise 32, suppose SO_2 and O_2 are put in a closed container. How do the rates of the forward and reverse reactions change as the system approaches equilibrium?

34. For the reaction in exercise 32, are equal concentrations of SO_2 and SO_3 necessarily present at equilibrium? Explain.

35. Write equations for the equilibrium constants for the following reactions: **(a)** $CO + H_2O \rightleftarrows CO_2 + H_2$, **(b)** $C_2H_4 + H_2 \rightleftarrows C_2H_6$, **(c)** $CH_4 + 2O_2 \rightleftarrows CO_2 + 2H_2O$, **(d)** $NaCl \rightleftarrows Na^+ + Cl^-$.

36. Write equations for the equilibrium constants for the following reactions: **(a)** $2H_2 + O_2 \rightleftarrows 2H_2O$,

(b) $H_2SO_4 + H_2O \rightleftharpoons H_3O^+ + HSO_4^-$, (c) $2NO + O_2 \rightleftharpoons 2NO_2$, (d) $2Al + 6HCl \rightleftharpoons 3H_2 + 2AlCl_3$.

37. Write each reaction in exercise 35 in the reverse direction. Then write equations for the equilibrium constants for each reaction.

38. Write each reaction in exercise 36 in the reverse direction. Then write equations for the equilibrium constants for each reaction.

39. Compare the equilibrium equation for exercise 36a with your answer for exercise 38a. If you know the equilibrium constant for a reaction, what is a simple way to calculate the equilibrium constant for that reaction written in the reverse direction?

40. Write the equation described by the following equilibrium constant: $K = \dfrac{[KCl]^2[O_2]^3}{[KClO_3]^2}$

41. Write the equation described by the following equilibrium constant: $K = \dfrac{[CO_2]^3[H_2O]^4}{[C_3H_8][O_2]^5}$

Use the reaction, $2H_2 + O_2 \rightleftharpoons 2H_2O$, to answer exercises 42–44.

42. At equilibrium are there necessarily twice as many moles of H_2 as O_2? Explain.

43. At equilibrium are there necessarily 1½ times as many moles of reactant as product? Explain.

44. If the reaction in a closed container begins with an equal number of molecules of H_2 and O_2, can the system still reach equilibrium? Explain.

45. For the reaction, $2NO_2 \rightleftharpoons N_2O_4$, calculate the equilibrium constant at 20°C if the system at equilibrium contains 2.05×10^{-2} mol/L N_2O_4 and 8.45×10^{-2} mol/L NO_2.

46. For the reaction, $2NOCl \rightleftharpoons 2NO + Cl_2$, calculate the equilibrium constant at 20°C if the system at equilibrium contains 6.84×10^{-3} mol/L NOCl, 7.35×10^{-4} mol/L NO, and 3.68×10^{-4} mol/L Cl_2.

47. For the reaction, $N_2 + O_2 \rightleftharpoons 2NO$, $K = 1.1 \times 10^{-2}$ at 20°C. Calculate the concentration of NO at equilibrium if the concentrations of N_2 and O_2 are 7.35×10^{-3} mol/L and 5.18×10^{-3} mol/L, respectively.

48. For the reaction, $PCl_5 \rightleftharpoons PCl_3 + Cl_2$, $K = 4.30 \times 10^{-2}$ at 20°C. Calculate the concentration of PCl_5 at equilibrium if the concentrations of PCl_3 and Cl_2 are 6.50×10^{-2} mol/L and 8.80×10^{-3} mol/L, respectively.

49. Does a reaction in which $K = 4.12 \times 10^{14}$ go mostly to the left or to the right as written?

50. Does a reaction in which $K = 9.53 \times 10^{-20}$ go mostly to the left or to the right as written?

Le Châtelier's Principle

51. Are the concentrations of all reactants and products necessarily the same at equilibrium when $K = 1$? Explain.

52. For the reaction in exercise 45, does the equilibrium position shift to the left or to the right when, at equilibrium, (a) NO_2 is added, (b) N_2O_4 is removed, (c) NO_2 is removed, or (d) pressure is reduced?

53. For the reaction in exercise 46, does the equilibrium position shift to the left or to the right when, at equilibrium (a) Cl_2 is added, (b) NO is removed, (c) NOCl is added, or (d) pressure is increased?

54. For the reaction in exercise 47, do you need to add or remove NO at equilibrium to shift the equilibrium position to the left?

55. Consider the reaction, $Hb + O_2 \rightleftharpoons HbO_2$, where Hb is the symbol for hemoglobin. When you are at high altitude, where the concentration of oxygen (O_2) gas is reduced, in which direction is the equilibrium position shifted? What is the effect on you?

56. One way to make nylon is to place the reactants in two liquid layers, one on top of the other. Solid nylon forms at the boundary where the liquids meet. If you used tweezers to pull out nylon as it forms, how would this affect the amount of nylon produced?

57. Cooling shifts the equilibrium position of exothermic reactions to the _____ (left or right) and endothermic reactions to the _____ (left or right).

58. Ammonia (NH_3) is made commercially by the Haber process by the reaction: $N_2 + 3H_2 \rightleftharpoons 2NH_3$ (see Chapter 5 Opener). High temperature and pressure are needed to form large amounts of product. Why does high pressure favor the formation of NH_3?

59. Is the following reaction endothermic or exothermic?

$H_2 + I_2 + \text{heat} \rightleftharpoons 2HI$

In which direction does the equilibrium position shift when the temperature is increased?

60. Is the following reaction endothermic or exothermic?

$CH_4 + 2O_2 \rightleftharpoons CO_2 + 2H_2O + \text{heat}$

In which direction does the equilibrium position shift when the temperature is increased?

DISCUSSION EXERCISES

1. Patent offices will not consider patent applications for perpetual motion machines, which produce more energy than they use. Assess this policy.

2. Do organisms carry out mostly exothermic or endothermic reactions? Explain why.

3. "The existence of a human body goes against the natural tendency for increasing entropy (disorder), and it violates the second law of thermodynamics." Do you agree or disagree with this statement? Explain why.

4. The rates of virtually all reactions in the body are controlled by catalysts called enzymes (see Chapter 22). Compare the use of catalysts with other factors that alter reaction rates. How practical are those other factors in controlling reaction rates in the body?

5. Why is it advantageous for an organism *not* to reach equilibrium?

9

Acids and Bases

1. What are the definitions of an acid and a base? What are the names and properties of acids and bases?
2. What are strong and weak acids and bases? What are acid and base ionization (dissociation) constants, and how do they indicate the strength of an acid or base?
3. How is acidity measured by the pH scale?
4. What are common reactions of acids? How do you use neutralization reactions to determine by titration the concentrations of acidic and basic solutions?
5. What are buffers? How do they work?
6. How does the body maintain the right level of acidity? What are the causes and effects of too much or too little acid?

Antacids

Table 9.1 Some Antacids and Their Ingredients

Product	Ingredients
Alka-Seltzer	$NaHCO_3$
Di-Gel, Maalox and Mylanta	$Al(OH)_3$ and $Mg(OH)_2$
Milk of Magnesia	$Mg(OH)_2$
Riopan	$AlMg(OH)_5$
Rolaids	$AlNa(OH)_2CO_3$
Tums	$CaCO_3$

Your stomach normally produces hydrochloric acid (HCl) to help digest food. But stress, overeating, and certain bacteria can cause the stomach to produce too much HCl. When this condition persists, excess acid destroys tissues lining the stomach or intestine, causing ulcers. Often you can get temporary relief by taking an antacid.

Antacids typically are metal hydroxides, carbonates, or bicarbonates (Table 9.1). They all neutralize acids. Carbonate and bicarbonate antacids produce CO_2 gas when they react with acid in the digestive tract (Figure 9.1).

Each type of antacid has advantages and disadvantages. Those with $Al(OH)_3$, for example, are effective but tend to cause constipation. Antacids with carbonates or bicarbonates provide prompt relief but may provoke the body to secrete extra HCl (called acid rebound), thus aggravating the problem.

Because it is mostly calcium carbonate ($CaCO_3$), Tums brand antacid is also advertised as a source of calcium ion (Ca^{2+}). Relatively few U.S. residents get the Recommended Daily Allowance of calcium—800 mg per day for adults over age 25 (1500 mg for women after menopause). Many physicians consider Ca^{2+} intake essential to prevent osteoporosis, a condition in which bones lose mass and become fragile. Tums contains 200 mg Ca^{2+} per tablet.

Figure 9.1 This antacid tablet is neutralizing an acid.

186 Part I General Chemistry

Figure 9.2 Svante Arrhenius (1859–1927) was a graduate student in 1884 when he proposed his new ideas about acids, bases, and electrolytes. The professors on his committee were not impressed. They gave him the lowest possible passing score for his Ph.D. thesis. Less than 20 years later—in 1903—Arrhenius won the Nobel prize in chemistry for this work.

*Another version of the Arrhenius definition is that an **acid** provides hydronium ions (H_3O^+) in water solutions.*

When you hear the word *acid,* you might think of an old science-fiction movie in which thick fumes rise as acid eats away everything it touches. Yet you probably drank several acids today, and your body is busy making others right now.

Sometimes acid in your stomach gives you "heartburn." One solution is to neutralize it by taking an antacid (Figure 9.1). An antacid is a *base,* and it converts acids into ionic compounds, or *salts.*

Your body carefully regulates its level of acidity. When that balance is upset, you become ill. Extra acidity in blood and urine, for example, are signs of diabetes and other disorders. But when you get excited and breathe very fast, you feel ill because your body has a shortage of acid. Breathing into a paper sack restores the balance.

9.1 ACIDS AND BASES: DEFINITIONS, PROPERTIES, AND NAMES

Arrhenius Definition. Recall that acids are electrolytes (Section 7.2); they produce ions in water that let electrical current flow through the acid solution (Figure 7.10).

Svante Arrhenius (Figure 9.2), a Swedish chemist, defined an **acid** as a substance that provides hydrogen ions (H^+) in water solutions. Hydrochloric acid (HCl), for example, is a covalent compound that forms ions in water:

$$H:\ddot{Cl}: \longrightarrow H^+ + :\ddot{Cl}:^-$$

When HCl forms ions, the pair of electrons shared in the covalent bond stays with Cl, which is more electronegative than H. Cl thus gains an electron and becomes Cl^- ion. The electron Cl gains came from an H atom. This ionization leaves H with no electrons and a nucleus consisting of one proton. So H^+ ion is simply a proton.

When H^+ ions form in water, they immediately react with water molecules to form **hydronium ions (H_3O^+)**. H^+ becomes more stable when it shares a pair of electrons with oxygen in a water molecule:

$$H^+ + :\ddot{O}: \longrightarrow :\ddot{O}:^+\!\!-\!H \quad \text{(hydronium ion)}$$
$$H\ H H\ H$$

Arrhenius defined a **base** as a substance that provides hydroxide ions (OH^-) in water solutions. Many bases are ionic compounds that contain hydroxide ion (OH^-). Those ions separate (dissociate) when bases dissolve in water. Sodium hydroxide (NaOH), for example, dissolves in water to produce hydrated Na^+ and OH^- ions:

$$NaOH \longrightarrow Na^+ + OH^-$$

EXAMPLE 9.1 Are bases electrolytes? Explain.

SOLUTION Bases are electrolytes because they provide ions in water solutions; those solutions conduct electricity.

Brønsted–Lowry Definition. In 1923 Johannes Brønsted and Thomas Lowry defined an **acid** as a proton donor and a **base** as a proton acceptor in a chemical reaction.

The Brønsted–Lowry definitions include all substances Arrhenius defined as an acid or base, but they go even farther. For one thing, they don't restrict acids and bases to water solutions. In addition, these definitions emphasize **acid–base reactions,** in which substances transfer protons to each other.

For example, when gaseous ammonia (NH_3) dissolves in water the reaction is:

$$NH_3 + H_2O \rightleftharpoons NH_4^+ + OH^-$$

Since NH_3 gains (accepts) a proton (H^+) from H_2O in the forward reaction, NH_3 is a base and H_2O is an acid (proton donor). In the reverse reaction, however, NH_4^+ donates a proton (H^+) to hydroxide ion (OH^-); here NH_4^+ is an acid and OH^- is a base.

In reversible reactions, the products are called *conjugate* acids and bases because they correspond to (are formed from) the reactant acids and bases. In the reaction with ammonia, for example, we can write:

$$NH_3 + H_2O \rightleftharpoons NH_4^+ + OH^-$$
base   acid   conjugate acid   conjugate base

EXAMPLE 9.2 Identify the acid, base, conjugate acid, and conjugate base in the reaction, $HCl + H_2O \rightleftharpoons H_3O^+ + Cl^-$.

SOLUTION $HCl + H_2O \rightleftharpoons H_3O^+ + Cl^-$
                acid   base   conjugate acid   conjugate base

---

Properties of Acids and Bases. Table 9.2 summarizes some properties of acids and bases. Two properties—taste and feel—are too dangerous to use for general identification. The sour taste of vinegar, however, tells you it is an acid, whereas unsweetened chocolate and many herbs that contain bases taste bitter. Bases feel slippery because they break down fats in skin (Section 20.1), making skin cells flat and able to slide over each other.

Dipping a strip of litmus paper into a water solution is a safe and simple way to identify acids or bases. Litmus paper contains a vegetable dye from certain lichens. Acids turn litmus red; bases turn it blue (Figure 9.3).

Litmus is one of a group of substances, called *indicators,* that change color when the acidity or basicity of a solution changes. Another indicator is phenolphthalein, which is crimson in basic solutions and colorless in acidic solutions (Figure 9.3).

Naming Acids and Bases. Besides ammonia (NH_3), the only bases you need to recognize and name contain hydroxide ion (OH^-). Their formulas end with OH (Table 9.2), and you name them the same way you name other ionic compounds (Section 4.2). $Ca(OH)_2$ and KOH, for example, are calcium hydroxide and potassium hydroxide, respectively.

Formulas of acids typically begin with H (Table 9.2). To name an acid, first identify from the formula whether or not it contains oxygen. For acids that don't contain oxygen, use the prefix *hydro-* followed by the name of the nonhydrogen element with the suffix *-ic;* then add the word *acid.* Solutions of HCl and H_2S, for example, are hydrochloric acid and hydrosulfuric acid, respectively.

188 Part I General Chemistry

Figure 9.3 The acid solutions, vinegar (left) and 7-Up (center) turn litmus paper red and show no color with the indicator phenolphthalein. Ammonia solution (right), a base, turns litmus paper blue and phenolphthalein deep pink.

Table 9.2 Properties of Acids and Bases

Property	Acids	Bases
Ions produced in water	H_3O^+	OH^-
Proton (H^+) transfer	Proton donor	Proton acceptor
Formula	Usually starts with H	Usually ends with OH
Taste	Sour or tart	Bitter
Feel	—	Slippery
Litmus paper	Red	Blue
Other indicators	Colors vary	Colors vary
pH	Less than 7	More than 7
Neutralization reaction	Neutralize bases	Neutralize acids

Name acids that don't contain oxygen hydro____ic acid.

Name oxygen-containing
        ate
acids ____ ic acid
        ite
or ____ ous acid.

Formulas of oxygen-containing acids end with a polyatomic ion (Table 4.3). The number of hydrogens in the formula matches the negative charge in the ion. The acid that contains sulfate ion (SO_4^{2-}), for example, has the formula H_2SO_4.

To name oxygen-containing acids, you first identify the polyatomic ion following H in the formula. For example, notice the carbonate ion (CO_3^{2-}) in the formula H_2CO_3. Name the polyatomic ion, changing a suffix of *-ate* to *-ic*, or *-ite* to *-ous;* then add the word *acid*. Don't use the prefix *hydro-*. H_2CO_3, then, is called carbonic acid. Using this system, you would name H_2SO_4 and H_2SO_3 sulfuric acid and sulfurous acid, respectively.

EXAMPLE 9.3 Identify the following as acids or bases and name them: (a) HBr, (b) HNO_3, (c) NaOH.

SOLUTION (a) hydrobromic acid, (b) nitric acid, (c) sodium hydroxide (a base).

EXAMPLE 9.4 Write formulas for the following: (a) magnesium hydroxide, (b) sulfuric acid, (c) nitrous acid.

SOLUTION (a) $Mg(OH)_2$, (b) H_2SO_4, (c) HNO_2.

Acids having one H in the formula, such as nitric acid (HNO_3) and hydrochloric acid (HCl), are called *monoprotic* (one proton per molecule) acids. Sulfuric acid (H_2SO_4) and carbonic acid (H_2CO_3) are *diprotic* acids, and phosphoric acid (H_3PO_4) is a *triprotic* acid.

9.2 STRENGTH OF ACIDS AND BASES

Strong and Weak Acids. Acids are classified as strong or weak. **Strong acids,** such as hydrochloric acid (HCl), react completely with water—or nearly so—to produce hydronium ions. Placing a double reaction arrow ($\rightleftharpoons$) between the reactants and products, with the longer arrow pointing to the right, indicates that the forward reaction predominates:

Strong Acid: $HCl + H_2O \rightleftharpoons H_3O^+ + Cl^-$

Weak acids, such as the poisonous gas hydrogen cyanide (HCN), give only a fraction of their potential hydronium ions when dissolved in water. Here the longer arrow points to the left, showing that the reverse reaction predominates:

Weak Acid: $HCN + H_2O \rightleftharpoons H_3O^+ + CN^-$

The reaction goes in both directions, but at equilibrium the concentration of HCN and H_2O exceeds that of hydronium (H_3O^+) and cyanide (CN^-) ions.

Table 9.3 lists some strong and weak acids and their uses. Although strong acids can eat holes in your clothes and burn your skin, notice that some weak acids (such as ascorbic, benzoic, and boric acid) help you stay healthy (Figure 9.4).

Table 9.3 Some Common Acids

Name	Formula	Uses
Strong Acids		
Hydrochloric acid	HCl	Digesting food; cleaning floors, toilet bowls, bricks, and metals
Nitric acid	HNO_3	Manufacturing fertilizer, explosives, dyes, and plastics
Sulfuric acid	H_2SO_4	Automobile batteries, drain cleaner; manufacturing plastics and paper
Weak Acids		
Acetic acid	$HC_2H_3O_2$	Vinegar
Ascorbic acid	$HC_6H_7O_6$	Vitamin C
Benzoic acid	$HC_7H_5O_2$	Food preservative
Boric acid	H_3BO_3	Eyewash
Carbonic acid	H_2CO_3	Carbonated beverages
Hydrocyanic acid	HCN	Fumigant
Phosphoric acid	H_3PO_4	Manufacturing fertilizers

Strong and Weak Bases. Like acids, bases are classified as strong or weak (Table 9.4). Ammonia (NH₃) is a **weak base** because it reacts with water to provide only a small fraction of the potential hydroxide (OH⁻) ions:

$$\text{Weak Base: } NH_3 + H_2O \rightleftharpoons NH_4^+ + OH^-$$

The arrows show that, at equilibrium, NH₃ and H₂O are present in higher concentrations than ammonium (NH₄⁺) and hydroxide (OH⁻) ions.

A **strong base** such as sodium hydroxide (NaOH), in contrast, provides virtually all of its hydroxide ions in water:

$$\text{Strong Base: } NaOH \rightleftharpoons Na^+ + OH^-$$

Measuring Strengths of Acids and Bases. Writing equations with arrows pointing mainly to the right or left doesn't tell us *how* strong or weak an acid or base is. To assess the strength of an acid, for example, we need to know at equilibrium how much of an acid produces hydronium ions.

Recall that an equilibrium constant (K) shows the ratio of products to reactants at equilibrium (Section 8.4). If we represent an acid as HA, the reaction with water is:

$$HA + H_2O \rightleftharpoons H_3O^+ + A^-$$

The equilibrium constant is:

$$K = \frac{[H_3O^+][A^-]}{[HA][H_2O]}$$

In dilute solutions, the concentration of water is essentially constant and is combined with the equilibrium constant. We then rearrange the equation:

$$K[H_2O] = \frac{[H_3O^+][A^-]}{[HA]} = K_a$$

The new constant, K_a, is called the **acid ionization constant** or **acid dissociation constant**.

Figure 9.4 Vinegar is a water solution of acetic acid (HC₂H₃O₂), a weak acid that reacts as follows:

$$HC_2H_3O_2 + H_2O \rightleftharpoons H_3O^+ + C_2H_3O_2^-$$

Table 9.4 Some Common Bases

Name	Formula	Uses
Strong Bases		
Sodium hydroxide	NaOH	Refining petroleum and vegetable oils; drain and oven cleaners
Potassium hydroxide	KOH	Producing liquid soaps, detergents, and fertilizers
*Calcium hydroxide	Ca(OH)₂	Neutralizing soil acidity; producing mortar, bleaching powder, and paper
*Magnesium hydroxide	Mg(OH)₂	Antacids and laxatives (milk of magnesia)
Weak Bases		
Aluminum hydroxide	Al(OH)₃	Antacids; purifying water; fixing dyes
Ammonia	NH₃	Household cleaner; producing fertilizer, explosives, and plastics

*Not very soluble in water, so solutions have low concentrations of OH⁻. The material that does dissolve, however, dissociates nearly 100%.

CHEMISTRY SPOTLIGHT

Corrosive Acids and Bases

Strong acids and bases (Tables 9.3 and 9.4) destroy tissue on contact. The damage depends on the contact time and the volume, concentration, temperature, and strength of the acid or base.

If you spill acid or base on your skin, you should immediately flush the skin with water to dilute the concentration. Likewise, people who swallow corrosive acids or bases should immediately drink water or milk to dilute the solution. They should not, however, try to vomit because this would again expose their throat and nose tissues to the acid or base. For additional treatment, a health professional may neutralize the acid or base with a weak base or acid, respectively (see Section 9.4).

Strong acids destroy tissues in several ways. They often dehydrate tissues by reacting with water. The reaction of water with sulfuric acid, for example, is:

$$H_2SO_4 + H_2O \longrightarrow H_3O^+ + HSO_4^-$$

Phosgene ($COCl_2$), a deadly gas used in World War I, reacts with moisture in the lungs to form hydrochloric acid (HCl), a strong acid:

$$COCl_2 + H_2O \longrightarrow 2HCl + CO_2$$

HCl then attracts water into the lungs, suffocating the victim.

Strong acids and bases also destroy proteins that provide the structure of skin, hair, nails, and connective tissue (see Chapter 21). In addition, strong bases break down fats in skin and in cell membranes, killing the cells.

EXAMPLE 9.5 Write equations for the reactions of (a) HCN and (b) $HC_2H_3O_2$ with water. Then write equations for the acid ionization constants for these weak acids.

SOLUTION (a) $HCN + H_2O \rightleftharpoons H_3O^+ + CN^-$; $K_a = \dfrac{[H_3O^+][CN^-]}{[HCN]}$

(b) $HC_2H_3O_2 + H_2O \rightleftharpoons H_3O^+ + C_2H_3O_2^-$; $K_a = \dfrac{[H_3O^+][C_2H_3O_2^-]}{[HC_2H_3O_2]}$

Recall that large K values mean a reaction goes mostly to the right as written; small K values mean a reaction goes mostly to the left (Section 8.4). The same is true of K_a values. A large K_a, for example, indicates a strong acid, which is mostly ionized.

Table 9.5 lists acid ionization constants for several acids. Notice two things. First, strong acids have larger K_a values than weak acids. Second, di- and triprotic acids have two and three K_a values, respectively, because they participate in more than one reaction that produces hydronium ion. Each molecule of phosphoric acid (H_3PO_4), for example, can produce three hydronium ions by

Table 9.5 K_a Values of Some Acids

Acid	K_a
HCl	*Very large
HNO$_3$	*Very large
HC$_2$H$_3$O$_2$	1.8×10^{-5}
HCN	4.9×10^{-10}
H$_2$SO$_4$	*Very large
HSO$_4^-$	1.3×10^{-2}
H$_2$CO$_3$	4.2×10^{-7}
HCO$_3^-$	4.8×10^{-11}
H$_3$PO$_4$	7.5×10^{-3}
H$_2$PO$_4^-$	6.2×10^{-8}
HPO$_4^{2-}$	4.8×10^{-13}

*Ionizes nearly 100%.

successive reactions:

$$H_3PO_4 + H_2O \rightleftharpoons H_3O^+ + H_2PO_4^-$$
$$H_2PO_4^- + H_2O \rightleftharpoons H_3O^+ + HPO_4^{2-}$$
$$HPO_4^{2-} + H_2O \rightleftharpoons H_3O^+ + PO_4^{3-}$$

Each reaction has a characteristic K_a value.

EXAMPLE 9.6 From the K_a values in Table 9.5, rank the following in order of decreasing strength as acids: (a) H$_3$PO$_4$, (b) H$_2$PO$_4^-$, (c) HPO$_4^{2-}$.

SOLUTION H$_3$PO$_4$ > H$_2$PO$_4^-$ > HPO$_4^{2-}$

The strength of bases is measured in a similar way. Ammonia (NH$_3$), for example, is a weak base that in water provides only a small percentage of the possible hydroxide ions:

$$NH_3 + H_2O \rightleftharpoons NH_4^+ + OH^-$$

The equilibrium constant is called the **base ionization constant** (or **base dissociation constant**) and has the symbol K_b. For the preceding reaction,

$$K[H_2O] = \frac{[NH_4^+][OH^-]}{[NH_3]} = K_b$$

Strong bases have large K_b values and weak bases have small ones.

EXAMPLE 9.7 Write the equation for the dissociation of KOH (a strong base) in water. Then write the equation for the K_b value.

SOLUTION $KOH \longrightarrow K^+ + OH^-$; $K_b = \dfrac{[K^+][OH^-]}{[KOH]}$

Ionization of Water. Notice in the preceding reactions that water is a base (proton acceptor) when it reacts with acids, such as HCl and HCN:

$$HCl + H_2O \rightleftharpoons H_3O^+ + Cl^-$$
acid   base

But in the reaction with the base ammonia (NH$_3$), water is an acid (proton donor):

$$NH_3 + H_2O \rightleftharpoons NH_4^+ + OH^-$$
base   acid

Water molecules also react with each other both as proton donors and acceptors:

$$H_2O + H_2O \rightleftharpoons H_3O^+ + OH^-$$
acid   base

Pure water contains equal concentrations of hydronium (H_3O^+) and hydroxide (OH^-) ions. At room temperature, their concentrations are 1×10^{-7} M.

The equilibrium constant for this reaction is:

$$K = \frac{[H_3O^+][OH^-]}{[H_2O]^2}$$

Just as the H_2O term is combined with K to make acid and base ionization constants, here we combine the H_2O term, $[H_2O]^2$, with K to make an equilibrium constant called the **ion-product constant of water,** which has the symbol K_w:

$$K[H_2O]^2 = [H_3O^+][OH^-] = K_w$$

At room temperature, the concentrations of H_3O^+ and OH^- in pure water are both 1×10^{-7} M. Thus $K_w = (1 \times 10^{-7})(1 \times 10^{-7}) = 1 \times 10^{-14}$.

K_w equals 1×10^{-14} at room temperature in acidic and basic solutions, too. So if you know the concentration of H_3O^+ (or OH^-), you can use K_w to calculate the concentration of OH^- (or H_3O^+). For example, if $[H_3O^+] = 1 \times 10^{-4}$ M, you can calculate $[OH^-]$:

$$K_w = [H_3O^+][OH^-]$$
$$1 \times 10^{-14} = (1 \times 10^{-4})[OH^-]$$
$$[OH^-] = \frac{1 \times 10^{-14}}{1 \times 10^{-4}} = 1 \times 10^{-10} \ M$$

Table 9.6 lists other examples of this relationship between $[H_3O^+]$ and $[OH^-]$.

Like other equilibrium constants, K_w varies with temperature but is otherwise constant.

EXAMPLE 9.8 Calculate the concentration of H_3O^+ in a solution at room temperature containing 5×10^{-9} M OH^-.

SOLUTION $[H_3O^+] = \dfrac{1 \times 10^{-14}}{5 \times 10^{-9}} = 2 \times 10^{-6} M$

Table 9.6 Relationship between pH and $[H_3O^+]$ and $[OH^-]$

	Concentration of H_3O^+ (M)		Concentration of OH^- (M)	
pH	Exponential	Decimal	Exponential	Decimal
1	10^{-1}	0.1	10^{-13}	0.0000000000001
2	10^{-2}	0.01	10^{-12}	0.000000000001
4	10^{-4}	0.0001	10^{-10}	0.0000000001
4.7	$10^{-4.7} = 2.0 \times 10^{-5}$	0.00002	$10^{-9.3} = 5.0 \times 10^{-10}$	0.0000000005
5.8	$10^{-5.8} = 1.6 \times 10^{-6}$	0.0000016	$10^{-8.2} = 6.3 \times 10^{-9}$	0.0000000063
7.0	10^{-7}	0.0000001	10^{-7}	0.0000001
9.0	10^{-9}	0.000000001	10^{-5}	0.00001
9.5	$10^{-9.5} = 3.2 \times 10^{-10}$	0.00000000032	$10^{-4.5} = 3.2 \times 10^{-5}$	0.000032
12	10^{-12}	0.000000000001	10^{-2}	0.01

9.3 pH

The pH Scale. To measure the acidity of water solutions, chemists use the *pH scale*. The **pH** of a solution is a measure of the concentration of hydronium ions in mol/L.

For most solutions, pH values range from 0 to 14 (Figure 9.5). Pure water has equal amounts of hydronium (H_3O^+) and hydroxide (OH^-) ions. It is *neutral*—neither acidic nor basic—and has a pH of 7.0. Solutions with excess hydronium ions or hydroxide ions are *acidic* or *basic*, respectively. Basic solutions also are called *alkaline*.

Solutions with pH values below 7 are acidic. The farther below 7, the more acidic the solution is. Alkaline solutions have pH values above 7. Higher pH values indicate greater alkalinity.

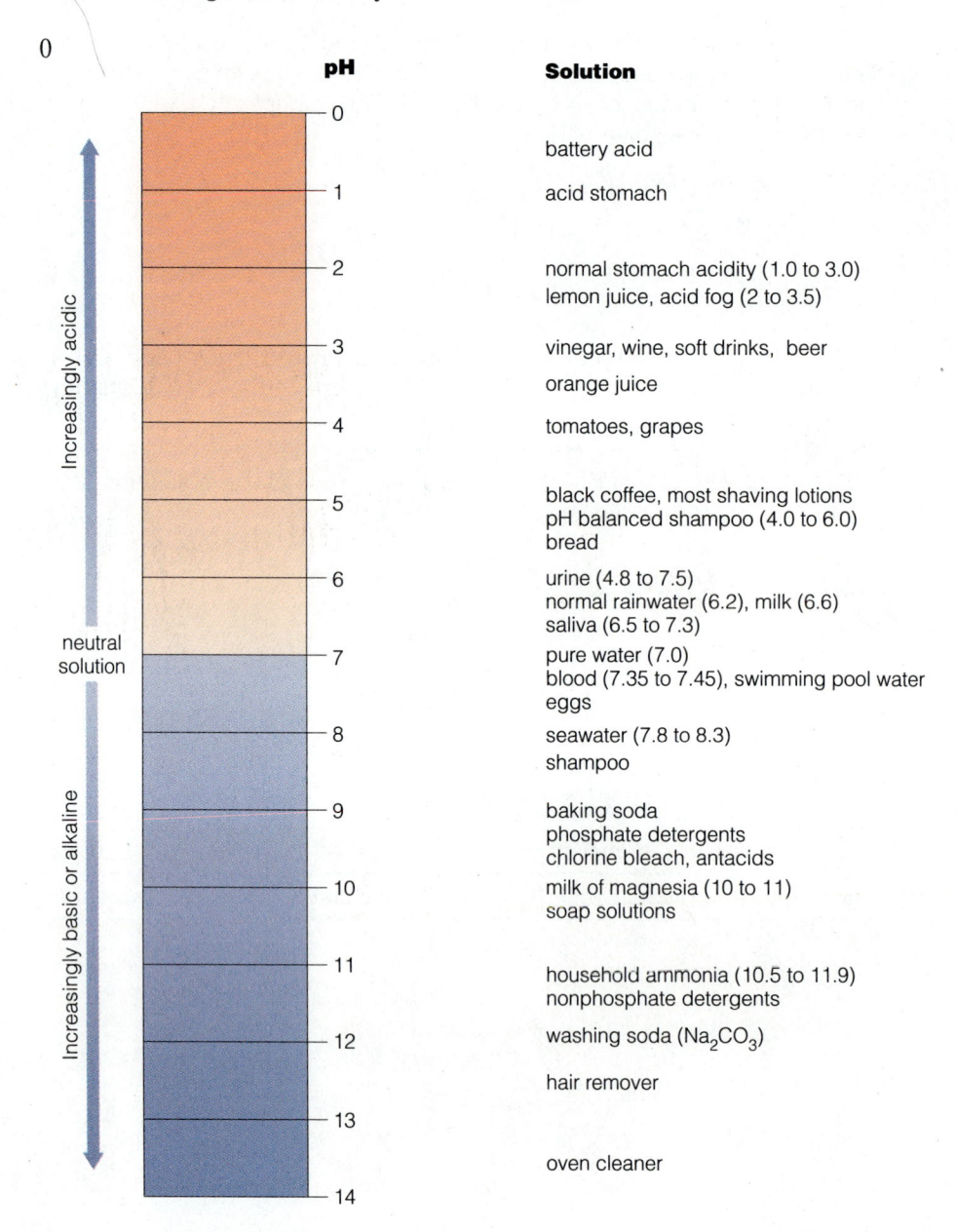

pH	Solution
0	battery acid
1	acid stomach
2	normal stomach acidity (1.0 to 3.0) lemon juice, acid fog (2 to 3.5)
3	vinegar, wine, soft drinks, beer orange juice
4	tomatoes, grapes
5	black coffee, most shaving lotions pH balanced shampoo (4.0 to 6.0) bread
6	urine (4.8 to 7.5) normal rainwater (6.2), milk (6.6) saliva (6.5 to 7.3)
7	pure water (7.0) blood (7.35 to 7.45), swimming pool water eggs
8	seawater (7.8 to 8.3) shampoo
9	baking soda phosphate detergents chlorine bleach, antacids
10	milk of magnesia (10 to 11) soap solutions
11	household ammonia (10.5 to 11.9) nonphosphate detergents
12	washing soda (Na_2CO_3)
13	hair remover
14	oven cleaner

Increasingly acidic ↑ / neutral solution / Increasingly basic or alkaline ↓

Figure 9.5 The pH scale. Values are approximate.

Each change of 1 pH unit corresponds to a tenfold change in hydronium ion concentration (Table 9.6). For example, vinegar (pH 3) is 10 times more acidic and has a concentration of hydronium ions 10 times greater than grape juice (pH 4). Lemon juice (pH 2) is 10 times more acidic than vinegar and 100 (10 × 10) times more acidic than grape juice.

The mathematical definition of pH is:

$$pH = -\log[H_3O^+]$$

where $[H_3O^+]$ is the concentration of hydronium ion in mol/L (M). Log is the abbreviation for *logarithm* and is simply the exponent of 10 in the molar concentration of H_3O^+. For a solution with a H_3O^+ concentration of 10^{-4} M, $\log[H_3O^+]$ is -4. Substituting this in the preceding equation, pH = $-(-4)$ = 4. So $[H_3O^+] = 10^{-pH}$.

EXAMPLE 9.9 What is the pH of solutions with the following concentrations of H_3O^+: (a) 10^{-11} M, (b) 10^{-1} M, (c) 10^{-7} M. Identify each solution as acidic, alkaline, or neutral.

SOLUTION (a) 11 (alkaline), (b) 1 (acidic), (c) 7 (neutral).

EXAMPLE 9.10 Find the concentrations in mol/L of H_3O^+ in solutions with the following pH values: (a) 9, (b) 13, (c) 6. Identify each solution as acidic, basic, or neutral.

SOLUTION (a) 10^{-9} M (basic), (b) 10^{-13} M (basic), (c) 10^{-6} M (acidic).

Suppose rainwater has a hydronium ion concentration of 8.4×10^{-7} M. What is its pH? To answer this question, you convert the entire concentration term into a logarithm (exponent of 10). On most calculators with logarithm functions, you simply enter the concentration and press the *log* button. In this example, you press 8.4, the *EE* or EXP (exponent) button, 7, the +/− button, and the *log* button. The display is −6.0757. (This means 8.4×10^{-7} M hydronium ion is equal to $10^{-6.0757}$ M.) Round this number to the appropriate number of significant figures (2, in this example) and reverse the negative sign. The pH of this solution is 6.1.

EXAMPLE 9.11 Calculate the pH of solutions with the following concentrations of H_3O^+: (a) 7.43×10^{-4} M, (b) 5.7×10^{-10} M, (c) 2.85×10^{-6} M. Identify each solution as acidic, alkaline, or neutral.

SOLUTION (a) 3.13 (acidic), (b) 9.2 (alkaline), (c) 5.55 (acidic).

For Part (a), press 7.43, EE or EXP, 4, +/−, log on your calculator.

You also can calculate the hydronium ion concentration from pH. For a solution with pH 8.76, you need to convert the logarithm (−8.76) into a corresponding number called the antilogarithm. To do this on a calculator, use the *inv* (for inverse) button and then the *log* button. Simply press 8.76, the +/− button, the *inv* button, and the *log* button. When you do this, the calculator displays 1.7378-09. The last part (−09) means the exponent of 10 is −9. Round off the first part of the display to the appropriate number of significant figures (3, in this example). Then write the hydronium ion concentration: 1.74×10^{-9} M.

For Part (a), press 6.338, +/−, inv, log on your calculator.

> **EXAMPLE 9.12** Calculate the hydronium ion concentration of solutions with the following pH values: (a) 6.338, (b) 9.9, (c) 1.40. Identify each solution as acidic, basic, or neutral.
>
> **SOLUTION** (a) 4.592×10^{-7} M (acidic), (b) 1.3×10^{-10} M (basic), (c) 3.98×10^{-2} M (acidic).

Measuring pH. A common way to estimate pH in the laboratory is to use pH paper, which contains a mixture of indicators (Section 9.1). A piece of pH paper dipped into a solution will turn different colors, depending on the pH. By comparing the color to a color scale, you can estimate the pH (Figure 9.6).

You get more accurate and precise results by using a *pH meter*. When an electrode is dipped into a solution, the meter displays the pH value of the solution (Figure 9.7).

pH of Salt Solutions. Solutions of different ionic compounds have different pH values. For example, a water solution of NaCl is neutral (pH = 7.0), a solution of Na_2CO_3 is alkaline, and a solution of NH_4Cl is acidic (Figure 9.8).

Why do these salt solutions differ in pH? Some salts must react with water to produce H_3O^+ or OH^- ions, making a solution acidic or basic, respectively. The pattern is:

- Salts from a strong acid and a strong base produce neutral solutions.
- Salts from a strong acid and a weak base produce acidic solutions.
- Salts from a weak acid and a strong base produce basic (alkaline) solutions.
- Salts from a weak acid and a weak base can produce acidic, neutral, or basic solutions.

Let's consider a salt of a strong acid and a strong base. NaCl is one example; its parent acid and base are HCl and NaOH, respectively. Strong acids and bases

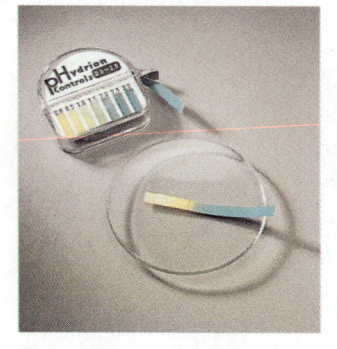

Figure 9.6 A piece of pH paper changes color depending on the pH of the solution.

Figure 9.7 A pH meter. This solution has a pH of 3.

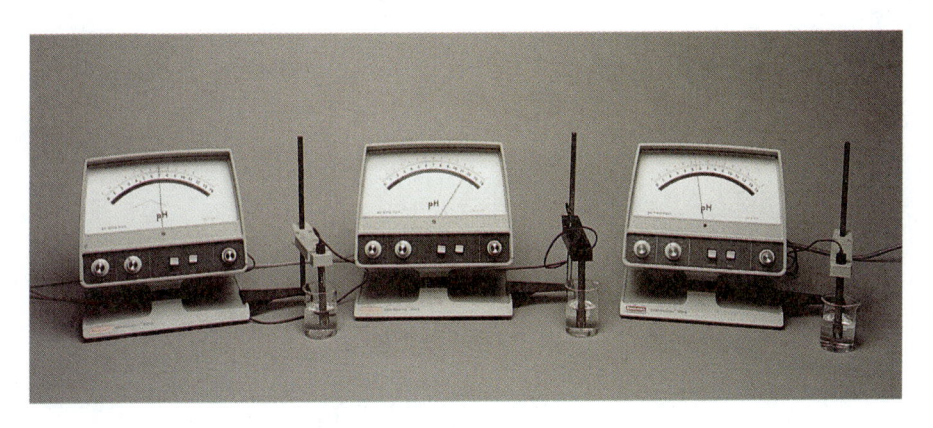

Figure 9.8 Solutions of NaCl (left), Na$_2$CO$_3$ (center), and NH$_4$Cl (right) are neutral, alkaline, and acidic, respectively.

exist almost entirely as ions in water. Neither Na$^+$ (from NaOH) nor Cl$^-$ (from HCl) reacts with water, so the solution is neutral.

NH$_4$Cl, however, is the salt of a strong acid (HCl) and a weak base (NH$_3$). As in NaCl, Cl$^-$ doesn't react with water. But ammonium ion (NH$_4^+$) does react:

$$NH_4^+ + H_2O \rightleftharpoons NH_3 + H_3O^+$$

The production of hydronium ion (H$_3$O$^+$) makes the solution acidic.

Na$_2$CO$_3$ is the salt of a weak acid (H$_2$CO$_3$) and a strong base (NaOH). Na$^+$ doesn't react with water, but carbonate ion (CO$_3^{2-}$) does:

$$CO_3^{2-} + H_2O \rightleftharpoons HCO_3^- + OH^-$$

The production of OH$^-$ makes Na$_2$CO$_3$ solution alkaline.

We will not try to predict pH for solutions of salts of a weak acid and a weak base.

EXAMPLE 9.13 Predict whether solutions of the following compounds are acidic, neutral, or alkaline: (a) NaC$_2$H$_3$O$_2$, (b) KNO$_3$, (c) NH$_4$NO$_3$, (d) LiCl.

SOLUTION (a) alkaline, (b) neutral, (c) acidic, (d) neutral.

9.4 REACTIONS OF ACIDS

Reactions with Active Metals. Acids react with certain metals to produce hydrogen (H$_2$) gas and a salt (ionic compound). Hydrochloric acid, for example, reacts with aluminum to form H$_2$ gas and aluminum chloride (Figure 9.9):

$$2Al + 6HCl \longrightarrow 3H_2 + 2AlCl_3$$

Reactions with Metal Hydroxides and Oxides. Recall that in an *acid–base reaction* an acid donates a proton to a base (Section 9.1). The general reaction is:

$$HX + MOH \longrightarrow H_2O + MX$$

   acid   base   salt

Because the resulting solution is neither acidic nor basic, this is also called a **neutralization** reaction.

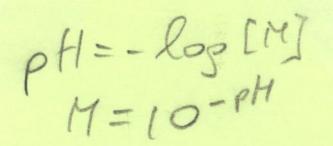

Figure 9.9 Hydrochloric acid (HCl) reacts with aluminum foil to produce hydrogen gas (H$_2$), which inflates the balloon.

CHEMISTRY SPOTLIGHT

Acid Deposition

Natural rainwater is slightly acidic (about pH 5.6) because it carries dissolved CO_2 that forms carbonic acid (H_2CO_3), a weak acid:

$$CO_2 + H_2O \rightleftharpoons H_2CO_3$$

Rain, snow, and fog, however, sometimes have pH values as low as 4 when they also carry nitric acid and sulfuric acid (Figure 9.10). This precipitation is known as *acid deposition,* commonly called *acid rain.*

Sulfuric acid in air comes mostly from combustion of sulfur-containing fossil fuels, especially coal. Sulfur reacts with oxygen to form sulfur dioxide (SO_2) (see following equation).

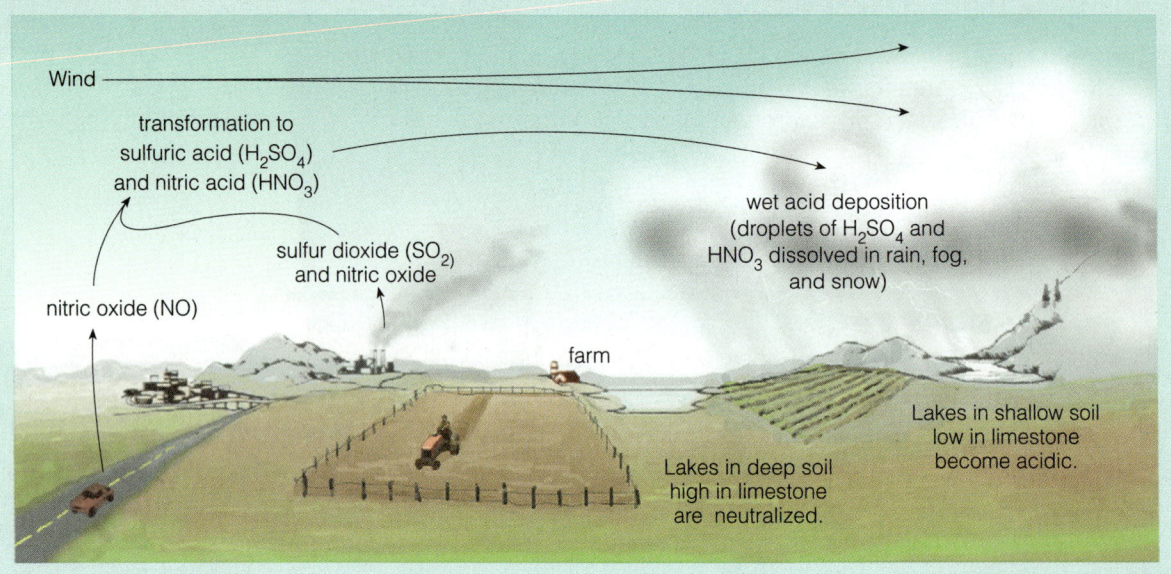

Figure 9.10 Acid deposition.

When hydrochloric acid (a strong acid) reacts with sodium hydroxide (a strong base), a proton transfers from the acid to the base to form water and sodium chloride:

$$HCl + NaOH \longrightarrow H_2O + NaCl$$

A similar reaction with a weak acid, acetic acid ($HC_2H_3O_2$), makes oven cleaning safer and easier. Most oven cleaners contain NaOH, which breaks down grease spattered on oven walls. But NaOH also irritates the hands wiping

$$S + O_2 \longrightarrow SO_2$$

SO_2 reacts in the air to form sulfur trioxide (SO_3) and sulfuric acid (H_2SO_4):

$$2SO_2 + O_2 \longrightarrow 2SO_3$$
$$SO_3 + H_2O \longrightarrow H_2SO_4$$

A major source of nitric acid is automobile exhaust. At high temperature and pressure in internal combustion engines, N_2 and O_2 in air react to form nitric oxide (NO):

$$N_2 + O_2 \rightleftharpoons 2NO$$

Nitric oxide emitted in automobile exhaust then reacts in air to form nitric acid (HNO_3):

$$2NO + O_2 \longrightarrow 2NO_2$$
$$3NO_2 + H_2O \longrightarrow 2HNO_3 + NO$$

Regions rich in alkaline minerals such as limestone, $CaCO_3$, and dolomite, $CaMg(CO_3)_2$, resist acidification because their minerals neutralize acids like antacids do (see the Chapter Opener). Lakes and thin soils in some areas, however, are vulnerable to acid deposition because they lack these alkaline materials (Figure 9.11).

Acids from acid deposition react with and corrode materials such as marble, limestone, mortar, and nylon. Many buildings and works of art have been damaged in this way. Acid rain aggravates human respiratory diseases and harms trees, crops, and aquatic life. For example, acidity increases the solubility of toxic metals such as lead and mercury stored in lake sediments. Organisms in such lakes (and predators of those organisms) then are exposed to higher concentrations of the toxic substances.

The most effective remedy for acid deposition is the prevention of SO_2 and NO emissions. Scientists are trying to reduce NO emissions from automobiles by developing catalytic converters that change NO in exhaust back into N_2 and O_2.

Power plants that use coal to generate electricity now control SO_2 emissions by using *scrubbers*. Stack gases pass through a chamber containing limestone ($CaCO_3$) that is heated to produce lime (CaO). Lime then reacts with SO_2 in the stack to produce solid calcium sulfite ($CaSO_3$) or calcium sulfate ($CaSO_4$), which is removed:

$$CaO + SO_2 \longrightarrow CaSO_3$$
$$CaO + SO_3 \longrightarrow CaSO_4$$

Lime is also used to neutralize excess acidity in lakes or soil. The reaction with sulfuric acid is:

$$H_2SO_4 + CaO \longrightarrow CaSO_4 + H_2O$$

Although the reaction neutralizes acids, it leaves *salinity* (saltiness) in lakes or soil because of products such as $CaSO_4$.

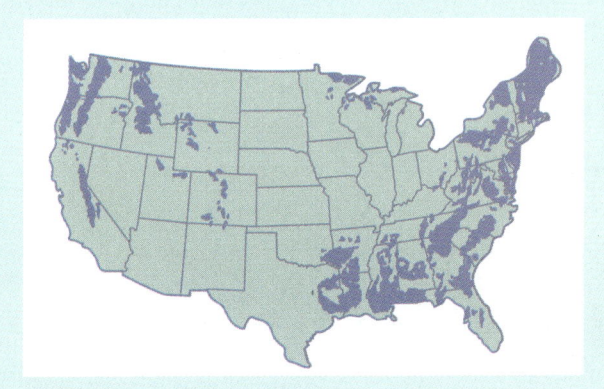

Figure 9.11 Areas in the United States where lakes and streams are especially vulnerable to acid deposition because they have low concentrations of alkaline substances that can neutralize acids.

the oven. Spraying the oven with vinegar (a solution of acetic acid, $HC_2H_3O_2$) neutralizes excess NaOH:

$$HC_2H_3O_2 + NaOH \xrightarrow{\text{proton transfer}} H_2O + NaC_2H_3O_2$$

The neutral mixture of water and sodium acetate ($NaC_2H_3O_2$) can be safely wiped away with a rag.

Acids react to form the same products with metal oxides as with metal hydroxides. Calcium oxide (CaO) and calcium hydroxide [$Ca(OH)_2$], for example, neutralize hydrochloric acid (HCl) to form the same products:

$$2HCl + CaO \longrightarrow H_2O + CaCl_2$$

$$2HCl + Ca(OH)_2 \longrightarrow 2H_2O + CaCl_2$$

EXAMPLE 9.14 Write balanced equations for the neutralization of nitric acid (HNO_3) by (a) $Mg(OH)_2$ and (b) MgO.

SOLUTION (a) $2HNO_3 + Mg(OH)_2 \longrightarrow 2H_2O + Mg(NO_3)_2$
(b) $2HNO_3 + MgO \longrightarrow H_2O + Mg(NO_3)_2$

Reactions with Metal Carbonates and Bicarbonates. Acids donate protons to salts containing carbonate (CO_3^{2-}) or bicarbonate (HCO_3^-) ions to produce carbonic acid (H_2CO_3), which decomposes into CO_2 and H_2O. The other product is a salt. For example, if you pour vinegar (acetic acid, $HC_2H_3O_2$) on baking soda (sodium bicarbonate, $NaHCO_3$), you see gas (CO_2) bubbles form (Figure 9.12). The reaction is:

$$HC_2H_3O_2 + NaHCO_3 \longrightarrow H_2O + CO_2 + NaC_2H_3O_2$$

Acids react with carbonates to form the same products. Sulfuric acid (H_2SO_4), for example, reacts with washing soda (sodium carbonate, Na_2CO_3) to form sodium sulfate (Na_2SO_4):

$$H_2SO_4 + Na_2CO_3 \longrightarrow H_2O + CO_2 + Na_2SO_4$$

Figure 9.12 Vinegar ($HC_2H_3O_2$) reacts with baking soda ($NaHCO_3$) to produce bubbles of CO_2 gas. The same reaction occurs in Figure 9.1.

EXAMPLE 9.15 Write balanced equations for the reaction of hydrochloric acid (HCl) with (a) $CaCO_3$ and (b) $Ca(HCO_3)_2$.

SOLUTION (a) $2HCl + CaCO_3 \longrightarrow H_2O + CO_2 + CaCl_2$
(b) $2HCl + Ca(HCO_3)_2 \longrightarrow 2H_2O + 2CO_2 + CaCl_2$

Reactions with Ammonia. Acids donate protons to ammonia (NH_3) to form ammonium (NH_4^+) salts. Hydrochloric acid (HCl) reacts with ammonia to form ammonium chloride (NH_4Cl):

$$HCl + NH_3 \longrightarrow NH_4Cl$$

As we discuss in Section 9.7, the kidneys use this reaction to remove excess acidity from the blood and excrete it in urine as ammonium ion (NH_4^+).

9.5 ACID–BASE TITRATIONS

Titration. Workers in laboratories often need to determine the concentrations of solutions. They do this by **titration,** a process of measuring the volume of a solution of known concentration (called a *standard solution*) that reacts completely with a second solution. From this measurement, the concentration of the second solution can be calculated.

Figure 9.13 Titrating a solution of HCl with a standard NaOH solution. To measure the volume (in mL) of NaOH used, the solution is delivered from a *buret*. Phenolphthalein, an indicator, is added to the HCl solution to identify the end point of the titration. When HCl is neutralized, the pH suddenly changes from acidic to basic and phenolphthalein changes from colorless (left) to deep pink (right).

To determine the concentration of an acid or base solution, you use a standard solution that neutralizes the acid or base. If you have an acid solution of unknown concentration, for example, you would titrate it with a standard base solution such as NaOH. If you put an appropriate indicator in the acid solution you are titrating, you can see when the solution is exactly neutralized; the sudden rise in pH at that point (called the *end point*) causes the indicator to change color (Figure 9.13). When you reach the end point, you stop the titration and measure the volume of standard solution needed to neutralize the unknown solution.

Calculating Molarity of Solutions. Suppose you titrate 69.1 mL of a sulfuric acid (H_2SO_4) solution and it takes 44.9 mL of 0.374 M NaOH to reach the end point. To calculate the concentration of H_2SO_4, first write a balanced equation for the reaction:

$$2NaOH + H_2SO_4 \longrightarrow 2H_2O + Na_2SO_4$$

Then calculate how many moles of NaOH reacted. Remember that the units of M are mol/L. We can use this as a conversion factor:

$$44.9 \text{ mL NaOH} \times \frac{1 \text{ L NaOH}}{1,000 \text{ mL NaOH}} \times \frac{0.374 \text{ mol NaOH}}{1 \text{ L NaOH}} = 0.0168 \text{ mol NaOH}$$

Now use the mole ratio from the balanced equation to calculate how many mol of H_2SO_4 react with 0.0168 mol NaOH:

$$0.0168 \text{ mol NaOH} \times \frac{1 \text{ mol } H_2SO_4}{2 \text{ mol NaOH}} = 0.00840 \text{ mol } H_2SO_4$$

Thus, 69.1 mL of solution contained 0.00840 mol of H_2SO_4. The concentration in M (mol/L), then, is:

$$\frac{0.00840 \text{ mol } H_2SO_4}{69.1 \text{ mL}} \times \frac{1,000 \text{ mL}}{1 \text{ L}} = \frac{0.122 \text{ mol } H_2SO_4}{1 \text{ L}} = 0.122 \text{ } M \text{ } H_2SO_4$$

EXAMPLE 9.16 Calculate the molar concentration of H_2SO_4 if it takes 14.3 mL of 1.25 M NaOH solution to titrate (neutralize) 48.1 mL of H_2SO_4 solution.

SOLUTION

$$14.3 \text{ mL NaOH} \times \frac{1 \text{ L NaOH}}{1{,}000 \text{ mL NaOH}} \times \frac{1.25 \text{ mol NaOH}}{1 \text{ L NaOH}}$$

$$\times \frac{1 \text{ mol } H_2SO_4}{2 \text{ mol NaOH}} = 0.00894 \text{ mol } H_2SO_4$$

$$\frac{0.00894 \text{ mol } H_2SO_4}{48.1 \text{ mL}} \times \frac{1{,}000 \text{ mL}}{1 \text{ L}} = \frac{0.186 \text{ mol } H_2SO_4}{1 \text{ L}} = 0.186 \ M \ H_2SO_4$$

EXAMPLE 9.17 Calculate the molar concentration of HCl solution if it takes 63.0 mL of 0.219 M NaOH solution to neutralize 265 mL of HCl solution.

SOLUTION The balanced equation for the reaction is:

$$\text{NaOH} + \text{HCl} \longrightarrow \text{NaCl} + H_2O$$

$$63.0 \text{ mL NaOH} \times \frac{1 \text{ L NaOH}}{1000 \text{ mL NaOH}} \times \frac{0.219 \text{ mol NaOH}}{1 \text{ L NaOH}} \times \frac{1 \text{ mol HCl}}{1 \text{ mol NaOH}}$$

$$= 0.0138 \text{ mol HCl}$$

$$\frac{0.0138 \text{ mol HCl}}{265 \text{ mL}} \times \frac{1000 \text{ mL}}{1 \text{ L}} = 0.0521 \text{ mol HCl/L} = 0.0521 \ M \ \text{HCl}$$

Calculating Normality of Solutions. Scientists sometimes express concentrations of acids and bases in *normality* (N) instead of molarity (M) to simplify calculations.

Normality (N) is the number of equivalents/L. Since an acid or base provides H_3O^+ or OH^- ions, which both have a charge of 1, the equivalent weight is simply the formula weight divided by the number of H or OH in the formula. The formula weight of sulfuric acid (H_2SO_4), for example, is 98 g. Since H_2SO_4 has two H in its formula, its equivalent weight is 98 g/2 = 49 g.

Recall that equivalents are the number of grams (equivalent weight) that provides one mole (Avogadro's number) of electrical charge in solution (Section 7.4).

EXAMPLE 9.18 Calculate the equivalent weights of the following: (a) HCl (formula weight is 36.5 g), (b) H_3PO_4 (FW = 98.0 g), (c) $Mg(OH)_2$ (FW = 58.3 g), (d) LiOH (FW = 23.9 g).

SOLUTION (a) 36.5 g, (b) 32.7 g, (c) 29.2 g, (d) 23.9 g.

Now let's compare molarity (mol/L) with normality (eq/L). The formula weight is the mass of 1 mole; the formula weight divided by 1, 2, or 3 is the equivalent weight. So you convert molarity into normality by multiplying molarity by 1, 2, or 3, depending on the number of H or OH in the formula. A 0.37 M solution of H_2SO_4, for example, is $0.37 \times 2 = 0.74 \ N$.

EXAMPLE 9.19 What is the concentration in normality (N) of the following solutions: (a) $1.30\ M$ HCl, (b) $0.104\ M\ H_3PO_4$, (c) $2.56\ M$ KOH, (d) $3.4 \times 10^{-4}\ M\ Ca(OH)_2$.

SOLUTION (a) $1.30\ N$, (b) $0.312\ N$, (c) $2.56\ N$, (d) $6.8 \times 10^{-4}\ N$.

When solution concentrations are expressed as *normality* (N), titration calculations are simpler because one equivalent of any acid neutralizes one equivalent of any base. You don't need to balance equations and use coefficients and mole ratios in the calculations. Instead, you can use the equation:

$$V_{acid} \times N_{acid} = V_{base} \times N_{base}$$

where V and N are volume and normality, respectively.

Now let's use normality to solve the same question we did before: Suppose you titrate 69.1 mL of sulfuric acid (H_2SO_4) solution and it takes 44.9 mL of $0.374\ M$ NaOH to reach the end point. To calculate the concentration of H_2SO_4 using the preceding equation, convert the concentration of NaOH into normality (N). Since there is one OH in the formula, $0.374\ M$ NaOH is $0.374\ N$. Now solve the equation:

$$(69.1\ \text{mL})(N\ \text{of}\ H_2SO_4) = (44.9\ \text{mL})(0.374\ N)$$

$$N\ \text{of}\ H_2SO_4 = \frac{(44.9\ \text{mL})(0.374\ N)}{(69.1\ \text{mL})} = 0.243\ N\ H_2SO_4$$

The normality of H_2SO_4 is two times its molarity, so $0.243\ N\ H_2SO_4$ is $0.243/2 = 0.122\ M\ H_2SO_4$. The answer is the same as we calculated previously for this problem, but the calculation was easier using normality.

EXAMPLE 9.20 Calculate the molar concentration of H_2SO_4 solution if it takes 14.3 mL of $1.25\ M$ NaOH solution to titrate (neutralize) 48.1 mL of H_2SO_4 solution. (Notice that this is the same question as in Example 9.16. This time, do the calculation using normality.)

SOLUTION $(48.1\ \text{mL})(N\ \text{of}\ H_2SO_4) = (14.3\ \text{mL})(1.25\ N)$

$$N\ \text{of}\ H_2SO_4 = \frac{(14.3\ \text{mL})(1.25\ N)}{48.1\ \text{mL}} = 0.372\ N$$

$$0.372\ N\ H_2SO_4 = 0.186\ M\ H_2SO_4$$

EXAMPLE 9.21 Calculate the molar concentration of HCl solution if it takes 63.0 mL of $0.219\ M$ NaOH solution to neutralize 265 mL of HCl solution. (This is a repeat of Example 9.17. Calculate it this time using normality.)

SOLUTION $(63.0\ \text{mL})(0.219\ N) = (265\ \text{mL})(N\ \text{of}\ HCl)$

$$N\ \text{of}\ HCl = \frac{(63.0\ \text{mL})(0.219\ N)}{(265\ \text{mL})} = 0.0521\ N\ HCl = 0.0521\ M\ HCl$$

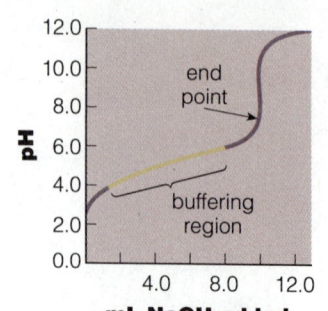

Figure 9.14 Effect on pH of adding 0.100 M NaOH to 10.0 mL of 0.100 M HC$_2$H$_3$O$_2$. The most effective buffering is near pH 5.

9.6 BUFFERS

Buffers. If you titrate acetic acid (HC$_2$H$_3$O$_2$) with a standard solution of NaOH and measure the change in pH, you get a pattern like that shown in Figure 9.14. Notice that adding a given amount of base doesn't always change pH by the same amount. In some pH regions adding base produces a large pH change; in other regions, adding base makes only a small change in pH.

A **buffer** is a substance that allows only small changes in pH when acid or base is added. Looking at Figure 9.14, you can see that acetic acid buffers best at about pH 5. At this pH, adding base (or acid) produces only a small increase (or decrease) in pH.

The most effective buffers are mixtures of a weak acid and its conjugate base or a weak base and its conjugate acid. For acetic acid the equation is:

$$HC_2H_3O_2 + H_2O \rightleftharpoons H_3O^+ + C_2H_3O_2^-$$
acid   *conjugate base*

Because it is a weak acid, acetic acid at equilibrium exists both as the acid and as the conjugate base, acetate ion (C$_2$H$_3$O$_2^-$). This mixture neutralizes additions of either acids or bases. If you add HCl, for example, the H$_3$O$^+$ it produces is neutralized by reaction with acetate ion:

$$C_2H_3O_2^- + H_3O^+ \rightleftharpoons H_2O + HC_2H_3O_2$$

Notice that adding acid shifts the equilibrium position to consume hydronium ion (H$_3$O$^+$) and increase the amount of acetic acid. When base (such as NaOH) is added, some acetic acid neutralizes the OH$^-$ ions and forms acetate ion (C$_2$H$_3$O$_2^-$):

$$HC_2H_3O_2 + OH^- \rightleftharpoons H_2O + C_2H_3O_2^-$$

Figure 9.15 shows the effect on pH of adding the same volume of the same HCl solution to water at pH 7 (which doesn't contain a buffer) and to a water solution at pH 7 containing a mixture of H$_2$PO$_4^-$ and HPO$_4^{2-}$ (a buffer solution). The drop in pH is much less in the buffer solution.

The most effective buffering occurs when an acid (or base) and its conjugate base (or acid) are present at equal concentrations. Having equal amounts of both forms provides the maximum capacity to neutralize additions of both acids and bases; this minimizes changes in pH.

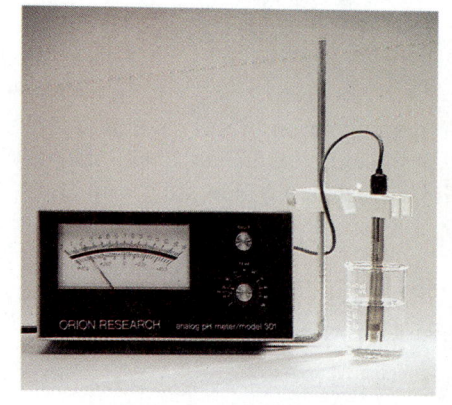

Figure 9.15 Addition of 5.0 mL of 1 M HCl produces a larger drop in the pH of water (left) than in a buffered solution of H$_2$PO$_4^-$ and HPO$_4^{2-}$ (right). Both were pH 7.0 before the addition of HCl solution.

pK_a Values. How can you tell the pH range in which a substance is a buffer? Chemists use **pK_a** values to do this. Just as pH is defined mathematically as $-\log[H_3O^+]$, pK_a is defined mathematically:

$$pK_a = -\log K_a$$

You calculate pK_a values the same way you calculate pH from H_3O^+ concentrations. For example, the acid ionization constant (K_a) for acetic acid is 1.8×10^{-5} (Table 9.5). To calculate its pK_a:

$$pK_a = -\log(1.8 \times 10^{-5})$$
$$K_a = 4.7$$

On your calculator, press 1.8, EE or EXP, 5, +/−, log. The display is −4.7447, so the answer is 4.7.

Therefore, acetic acid/acetate ion buffers best at pH 4.7. The range of effective buffering is about 1 pH unit above and below pK_a, so acetic acid/acetate ion buffers in the pH range 3.7–5.7 (Figure 9.14).

EXAMPLE 9.22 Calculate the pK_a value for hydrocyanic acid (HCN), which has a K_a value of 4.9×10^{-10}. List the effective buffering range for HCN and its conjugate base (CN^-).

SOLUTION pK_a = 9.3; buffering range is pH 8.3–10.3.

A useful equation (derived from the equation for K_a in Section 9.2) is:

$$pH = pK_a + \log \frac{[A^-]}{[HA]}$$

This equation is commonly known as the Henderson–Hasselbalch equation.

where [HA] and [A^-] are the molar concentrations of the acid and its conjugate base, respectively. Notice mathematically that when pH = pK_a, log ([A^-]/[HA]) equals zero; thus [A^-]/[HA] = 1. Stated in words, pK_a is the pH at which there are equal concentrations of an acid and its conjugate base (their ratio equals 1).

Since buffers are most effective when the acid and conjugate base are at equal concentrations, *a buffer is most effective at a pH matching its* pK_a *value*. Once large amounts of acid or base are added, the [A^-]/[HA] ratio changes so much that the mixture no longer buffers effectively.

Table 9.5 lists K_a values for several acids. Notice that polyprotic acids have multiple K_a values. This is because in solution more than one equilibrium exists. The reactions for carbonic acid (H_2CO_3) are:

$$H_2CO_3 + H_2O \rightleftharpoons H_3O^+ + HCO_3^- \quad K_a = 4.2 \times 10^{-7}, pK_a = 6.4$$
$$HCO_3^- + H_2O \rightleftharpoons H_3O^+ + CO_3^{2-} \quad K_a = 4.8 \times 10^{-11}, pK_a = 10.3$$

Each reaction has a K_a, and thus a pK_a, value. Carbonic acid, then, buffers best near pH 6.4 and pH 10.3.

9.7 pH IN THE BODY

Different parts of your body vary in pH. Your stomach is the most acidic, about pH 2. In contrast, your pancreas sends alkaline secretions (about pH 8) into your intestine, where they neutralize acidity from the stomach. Most cells and

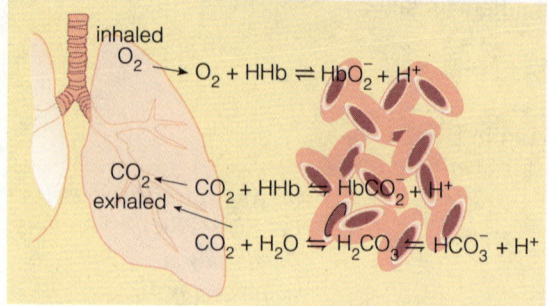

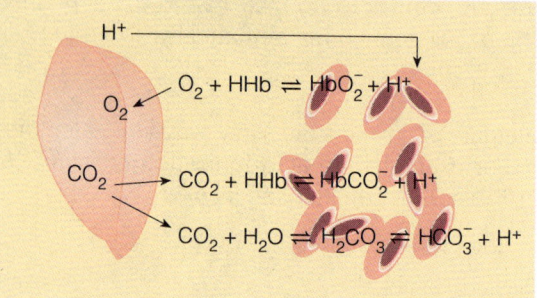

Figure 9.16 Summary of reactions in blood as it passes by lung (left) and other tissues (right). At the lungs, inhaled O_2 is bound to hemoglobin (HHb), acidity is neutralized, and CO_2 is exhaled. At other tissues, CO_2 and H^+ are removed, acidity is neutralized, and O_2 is released from hemoglobin in the blood.

organs are slightly acidic, about pH 6–7. Saliva is about pH 6.5–7.3. The pH of urine normally varies from about 4.8 to 7.5, with an average value of about 6.

Blood pH Levels. Your blood is slightly alkaline, having a pH of 7.35–7.45 (Figure 9.5). The main buffer system is carbonic acid (H_2CO_3) and bicarbonate ion (HCO_3^-). Since pK_a for this buffer is 6.4, a blood pH of 7.4 is at the upper edge of its buffering capacity. Here a H_2CO_3/HCO_3^- mixture is more effective at neutralizing excess acid than excess base. Buffering is also provided by proteins (Chapter 21) and by a mixture of hydrogen phosphate ion (HPO_4^{2-}) and dihydrogen phosphate ion ($H_2PO_4^-$), which has a pK_a of 7.2.

Hemoglobin (Hb), a complex protein that carries oxygen, helps maintain blood pH. Blood pH, in turn, helps load O_2 onto Hb at the lungs and unload O_2 at other tissues. It also helps you exhale waste carbon dioxide (CO_2) produced by metabolism.

Figure 9.16 shows how this works. Here we represent hemoglobin (Hb) as HHb because Hb functions as an acid and donates a proton (H^+) when it binds O_2.

Recall that as blood passes by the lungs, high O_2 levels shift the equilibrium to the right to form more oxygenated hemoglobin (HbO_2) (Section 8.5, Spotlight). We can write the equation:

$$HHb + O_2 \rightleftharpoons HbO_2^- + H^+$$

The H^+ released is immediately neutralized by two reactions. It combines with bicarbonate ion (HCO_3^-) in blood, shifting to the right the equilibrium of the reaction:

$$H^+ + HCO_3^- \rightleftharpoons H_2CO_3 \rightleftharpoons H_2O + CO_2$$

H^+ also is neutralized by Hb that carries CO_2 (represented as $HbCO_2$):

$$H^+ + HbCO_2^- \rightleftharpoons HHb + CO_2$$

CO_2 produced in both reactions is exhaled from the lungs.

When oxygenated blood goes from the lungs to other cells, these reactions go in the opposite direction. Cells consume O_2 and produce CO_2 and acids when they metabolize carbohydrates, fats, and proteins (see Chapters 24–27). The lower concentration of O_2 and higher concentrations of CO_2 and H^+ shift the equilibria of the reactions in the blood:

$$HHb + O_2 \rightleftharpoons HbO_2^- + H^+$$

$$H^+ + HCO_3^- \rightleftharpoons H_2CO_3 \rightleftharpoons H_2O + CO_2$$

$$H^+ + HbCO_2^- \rightleftharpoons HHb + CO_2$$

This interplay between chemical reactions in the lungs and other tissues is an excellent example of Le Châtelier's principle (Section 8.5)—and of how chemistry works in your body to keep you alive.

CHEMISTRY SPOTLIGHT

Acidosis and Alkalosis

The pH of blood normally stays close to 7.40. Too much acidity (pH 7.30 or lower) or too much alkalinity (pH 7.50 or higher) are called **acidosis** or **alkalosis**, respectively. Both conditions indicate illness, and blood pH values below 6.8 or above 7.8 can be fatal. Acidosis makes you feel light-headed or disoriented, and in severe cases puts you in a coma. Alkalosis can bring headaches, cramps, and even convulsions.

Both acidosis and alkalosis can be caused by abnormal breathing rates. Recall the equation:

$$H^+ + HCO_3^- \rightleftharpoons H_2CO_3 \rightleftharpoons H_2O + CO_2$$

Rapid breathing (called *hyperventilation*) removes CO_2 and shifts the equilibrium to the right, which removes acidity (H^+). This condition, called *respiratory alkalosis,* makes blood more alkaline than normal. Very slow breathing (*hypoventilation*), in contrast, fails to remove acidity at the normal rate and leaves blood more acidic than normal. This is *respiratory acidosis*.

Other causes of pH imbalance are called *metabolic acidosis* and *metabolic alkalosis*. Here the respiration rate changes to help correct the imbalance. Table 9.7 summarizes causes, effects, and treatments for both types of acidosis and alkalosis.

Table 9.7 Causes, Effects, and Treatment of Acidosis and Alkalosis

Causes	Effects	Treatment
Respiratory acidosis hypoventilation due to emphysema, other lung disorders, asthma, depressants	blood pH <7.30, high HCO_3^- in blood	treat underlying cause, neutralize with intravenous $NaHCO_3$ solution
Respiratory alkalosis hyperventilation due to high altitude, fever, crying, excitement	blood pH >7.50, low blood HCO_3^-	treat underlying cause, breathe into a paper bag to retain CO_2
Metabolic acidosis diabetes, diarrhea with loss of HCO_3^-, low carbohydrate diet, or starvation	blood pH <7.30, low blood HCO_3^-, hyperventilation, increased urination	treat underlying cause, neutralize with intravenous $NaHCO_3$ solution
Metabolic alkalosis vomiting with loss of HCl due to many causes, excess use of antacids	blood pH >7.50, high blood HCO_3^-, hypoventilation	treat underlying cause, neutralize with intravenous NH_4Cl, which is mildly acidic

The last two reactions convert CO_2 into HCO_3^- and $HbCO_2^-$, which travel to the lungs where CO_2 is exhaled. Acidity produced by these reactions of CO_2 is neutralized by HbO_2^- in the first reaction. This not only neutralizes the acidity, but also releases O_2 for cells to use.

The kidneys also help maintain blood pH within narrow limits. Kidneys remove excess acidity from blood by reacting H_3O^+ with ammonia (NH_3) to produce ammonium ion (Section 9.4):

$$H_3O^+ + NH_3 \rightleftharpoons H_2O + NH_4^+$$

Ammonium ions then are excreted in urine.

> ... after having taken in one breath of [hydrochloric acid] you expel from your nose two short plumes of white smoke ... and you feel your teeth turn sour in your mouth, as when you have bitten into a lemon.
> *Source: The Periodic Table* by Primo Levi (1919–1987)

SUMMARY

Acids are proton (H^+) donors that provide hydronium ions (H_3O^+) in water. Bases are proton acceptors and provide hydroxide ions (OH^-) in water. Acids taste sour, turn litmus red, neutralize bases, and have pH values below 7. Bases taste bitter, feel slippery, turn litmus blue, neutralize acids, and have pH values above 7. Acids that don't contain oxygen are named using the prefix *hydro-* followed by the name of the nonhydrogen element with the suffix *-ic* and the word *acid*. Those containing oxygen are named by changing the suffix of the polyatomic ion from *-ate* to *-ic*, or *-ite* to *-ous*, and adding *acid*. Acids that provide one, two, or three protons per molecule are called monoprotic, diprotic, or triprotic acids, respectively.

Strong acids provide nearly all of their protons to produce H_3O^+ ions in water; weak acids provide relatively few. Strong bases provide virtually all their OH^- ions in water; weak bases provide only a small percentage of their potential OH^- ions. The acid ionization (dissociation) constant (K_a) for an acid (HA) is:

$$K_a = \frac{[H_3O^+][A^-]}{[HA]}$$

K_a indicates the strength of an acid. K_b values, defined in a similar way, indicate the strengths of bases. The ion-product constant of water (K_w) equals $[H_3O^+][OH^-]$ and is 1×10^{-14} at room temperature.

The pH of a solution is a measure of H_3O^+ concentration in mol/L. Mathematically, $pH = -\log [H_3O^+]$. Solutions with pH values below 7 are acidic; those above pH 7 are basic (alkaline); those at pH 7.0 are neutral. Salts of weak acids or weak bases produce basic or acidic solutions, respectively.

Acids react with certain metals, ammonia (NH_3) and metal hydroxides, oxides, carbonates, and bicarbonates. Acids react with bases to produce water and a salt; these are neutralization reactions. Titration of an acid with a standard base solution (or *vice versa*) is used to determine the concentration of the acid (or base) solution. When concentrations are expressed as normality (N), $V_{acid} \times N_{acid} = V_{base} \times N_{base}$. Normality has the units equivalents/L and equals 1, 2, or 3 times molarity (M), depending on the number of H or OH in the formula of the acid or base.

A buffer allows only small changes in pH when acid or base is added to a solution. A mixture of a weak acid (or base) and its conjugate base (or acid) is an effective buffer, especially at a pH near the substance's pK_a value. pK_a equals $-\log K_a$. The pH of blood is normally maintained very close to 7.40 by protein, phosphate, and bicarbonate buffers. Excess acidity or alkalinity in blood is called acidosis or alkalosis, respectively.

KEY TERMS

Acid (9.1)
Acid–base reaction (9.1)
Acid ionization (dissociation) constant (K_a) (9.2)
Acidosis (9.7)
Alkalosis (9.7)
Base (9.1)
Base ionization (dissociation) constant (K_b) (9.2)
Buffer (9.6)
Hydronium ion (H_3O^+) (9.1)
Ion-product constant of water (K_w) (9.2)
Metabolic acidosis (9.7)
Metabolic alkalosis (9.7)
Neutralization (9.4)
Normality (N) (9.5)
pH (9.3)
pK_a (9.6)
Respiratory acidosis (9.7)
Respiratory alkalosis (9.7)
Strong acid (9.2)
Strong base (9.2)
Titration (9.5)
Weak acid (9.2)
Weak base (9.2)

EXERCISES

Even-numbered exercises are answered at the back of this book.

Definitions and Properties of Acids and Bases

1. Why does hydronium ion have a positive charge?
2. How did Arrhenius define *acid*?
3. How did Arrhenius define *base*?
4. Identify which of the following are acids, according to Arrhenius: **(a)** NaOH, **(b)** NaCl, **(c)** HCl, **(d)** KNO_3, **(e)** HNO_3, **(f)** KOH.
5. Which of the substances in exercise 4 are bases, according to Arrhenius?
6. Which of the substances in exercise 4 are bases, according to the Brønsted–Lowry definition?
7. In terms of the Brønsted–Lowry definition of acids and bases, water can function either as an acid or a base. Explain.
8. When water functions as an acid, what is its conjugate base?
9. When water functions as a base, what is its conjugate acid?
10. What is the conjugate base of nitric acid (HNO_3) in water solution?
11. Methyl red is red below pH 5 and yellow above that pH. A substance that changes color at a particular pH is called a(n) _____.
12. List three simple tests you could do to determine whether a water solution is acidic or basic.

Names and Formulas of Acids and Bases

13. Name the following compounds: **(a)** H_2SO_4, **(b)** $Ca(OH)_2$, **(c)** HI, **(d)** $HC_2H_3O_2$.
14. Name the following compounds: **(a)** H_3PO_4, **(b)** H_2CO_3, **(c)** KOH, **(d)** HNO_3.
15. Identify each acid in exercises 13 and 14 as monoprotic, diprotic, or triprotic.
16. Write formulas for the following: **(a)** hydrocyanic acid, **(b)** magnesium hydroxide, **(c)** nitric acid, **(d)** nitrous acid.
17. Write formulas for the following: **(a)** sodium hydroxide, **(b)** hydrochloric acid, **(c)** sulfurous acid, **(d)** acetic acid.
18. Write the names and formulas of three strong acids.
19. Write the names and formulas of three weak acids.
20. Write the names and formulas of two strong bases and one weak base.

Strength of Acids and Bases

21. Suppose you had separate 0.1 M solutions of HCl (a strong acid) and H_3BO_3 (a weak acid). Would the two solutions have the same pH? Explain.
22. Suppose you had separate 0.1 M solutions of NH_3 (a weak base) and KOH (a strong base). Would the two solutions have the same pH? Explain.
23. Write reactions for the production of OH^- in water solutions from **(a)** NH_3 and **(b)** KOH. Write the longer arrow pointing in the direction of the reaction that predominates.
24. Write equations for the acid ionization constants (K_a) for **(a)** H_2CO_3 and **(b)** $HC_7H_5O_2$ (benzoic acid).
25. Write equations for the base ionization constants (K_b) for **(a)** NH_3 and **(b)** $Ca(OH)_2$.
26. K_a for carbonic acid (H_2CO_3) is 4.2×10^{-7}. K_a for boric acid (H_3BO_3) is 6.4×10^{-10}. Which is the stronger acid?
27. At room temperature, what is $[OH^-]$ in a water solution that contains $2.0 \times 10^{-4}\ M\ H_3O^+$?
28. At room temperature, what is $[H_3O^+]$ in a water solution that contains $8.3 \times 10^{-6}\ M\ OH^-$?

pH

29. From the following pH values classify each solution as strongly acidic, acidic, neutral, alkaline, or strongly alkaline: **(a)** 7.9, **(b)** 2.1, **(c)** 13.0, **(d)** 6.2, **(e)** 7.2.

30. Does a solution at pH 5.4 contain more H_3O^+ or OH^- ions?

31. Identify the pH of solutions having the following hydronium ion (H_3O^+) concentrations: (a) 1.00×10^{-12} M, (b) 6.33×10^{-6} M, (c) 9.40×10^{-8} M, (d) 2.58×10^{-3} M.

32. Identify the pH of solutions having the following hydronium ion (H_3O^+) concentrations: (a) 4.44×10^{-2} M, (b) 7.72×10^{-9} M, (c) 1.00×10^{-7} M, (d) 0.000047 M.

33. Classify each of the solutions in exercise 32 as acidic, basic, or neutral.

34. Calculate the concentrations of H_3O^+ in solutions with the following pH values: (a) 7.45, (b) 3.37, (c) 12.71, (d) 5.80.

35. Calculate the concentrations of H_3O^+ in solutions with the following pH values: (a) 3.00, (b) 10.75, (c) 6.94, (d) 7.06.

36. What is the concentration of H_3O^+ in blood at pH 7.40?

37. How could you estimate the pH of a solution if you didn't have litmus paper or a pH meter?

38. Predict whether water solutions of the following are acidic, basic, or neutral: (a) KNO_3, (b) Na_3PO_4, (c) $AlCl_3$, (d) $LiCN$.

39. Predict whether water solutions of the following are acidic, basic, or neutral: (a) NH_4NO_3, (b) $CaCO_3$, (c) $KC_2H_3O_2$, (d) KCl.

Reactions of Acids

40. Hydrochloric acid reacts with magnesium metal to liberate hydrogen gas and a salt. Write a balanced equation for the reaction.

41. Complete the following reactions and balance them:
    (a) $Mg(OH)_2 + HCl \longrightarrow$
    (b) $MgO + H_2SO_4 \longrightarrow$
    (c) $NaOH + H_2SO_4 \longrightarrow$
    (d) $Ca(OH)_2 + HNO_3 \longrightarrow$

42. Complete the following reactions and balance them:
    (a) $H_3PO_4 + KOH \longrightarrow$
    (b) $HC_2H_3O_2 + CaO \longrightarrow$
    (c) $H_2CO_3 + NaOH \longrightarrow$
    (d) $HBr + Al(OH)_3 \longrightarrow$

43. Lemon juice has a pH of about 2.3. Identify which of the following would *not* neutralize lemon juice:
    (a) $NaCl$, (b) NH_3, (c) KOH, (d) H_3BO_3, (e) $Al(OH)_3$.

44. What is the pH of a solution made by mixing exactly 15.0 mL of 0.100 M HCl with 15.0 mL of 0.100 M NaOH?

45. Complete the following reactions and balance them:
    (a) $MgCO_3 + HCl \longrightarrow$
    (b) $Mg(HCO_3)_2 + HCl \longrightarrow$
    (c) $K_2CO_3 + HNO_3 \longrightarrow$
    (d) $NH_3 + HBr \longrightarrow$

46. Complete the following reactions and balance them:
    (a) $Na_2CO_3 + H_3PO_4 \longrightarrow$
    (b) $CaCO_3 + HC_2H_3O_2 \longrightarrow$
    (c) $NaHCO_3 + H_2SO_4 \longrightarrow$
    (d) $NH_3 + HC_2H_3O_2 \longrightarrow$

47. If you take an antacid that contains carbonate or bicarbonate ion (Table 9.1), you may experience gas when the antacid reacts with HCl in your stomach. What gas is produced?

48. Does acidosis result in increased or decreased excretion of ammonium ion (NH_4^+) in the urine?

Acid–Base Titrations

49. In titrating an acid solution with a standard base solution, what could you use besides an appropriate indicator to show you when you reach the end point?

50. Identify which of the following standard solutions you could use to titrate a solution of HCl to determine its concentration: (a) H_2SO_4, (b) KOH, (c) distilled H_2O, (d) $NaCl$, (e) $NaOH$.

51. Calculate the equivalent weights of the following: (a) $LiOH$, (b) H_2SO_4, (c) $Mg(OH)_2$, (d) $HC_2H_3O_2$.

52. Calculate the equivalent weights of the following: (a) H_2CO_3, (b) $Al(OH)_3$, (c) HNO_3, (d) $NaOH$.

53. Calculate the normality (N) of the following solutions: (a) 0.128 M H_2SO_4, (b) 6.00 M HCl, (c) 3.41×10^{-7} M $Ca(OH)_2$, (d) 1.00 M H_3PO_4.

54. Calculate the normality (N) of the following solutions: (a) 3.2×10^{-8} M $Al(OH)_3$, (b) 0.148 M HNO_3, (c) 6.12×10^{-2} M H_2SO_4, (d) 3.50 M KOH.

55. If it takes 38.8 mL of 0.245 M NaOH to neutralize 135.2 mL of an H_2SO_4 solution, what is the molar concentration of H_2SO_4?

56. If it takes 1.4 mL of 0.100 M HCl to neutralize 87.4 mL of an $Al(OH)_3$ solution, what is the molar concentration of $Al(OH)_3$?

57. What volume of 0.854 M NaOH solution is needed to neutralize 65.0 mL of 0.200 M H_2SO_4 solution?

58. What volume of 1.25 M HCl solution is needed to neutralize 20.0 mL of 2.19 M NaOH solution?

59. If you were titrating 50.0 mL of a solution of HCl that you suspected was about 10^{-3} M, identify which concentration of standard NaOH solution would be the most practical to use: (a) 1.00 M, (b) 1.00×10^{-3} M, (c) 1.00×10^{-6} M. Explain why.

Buffers

60. Calculate pK_a values for substances having the following K_a values: **(a)** 1.7×10^{-4}, **(b)** 5.35×10^{-11}, **(c)** 4.89×10^{-7}, **(d)** 4.75×10^{-2}.

61. Calculate pK_a values for substances having the following K_a values: **(a)** 5.60×10^{-12}, **(b)** 3.44×10^{-6}, **(c)** 9.87×10^{-3}, **(d)** 8.16×10^{-9}.

62. Which substance in exercise 60 is the most effective buffer at pH 7.40?

63. Which substance in exercise 61 is the most effective buffer at pH 8.0?

64. Which substance, with its conjugate acid or base, is the most effective buffer (at an appropriate pH): **(a)** HCl, **(b)** $HC_2H_3O_2$, **(c)** NaOH, **(d)** NaCl, **(e)** KOH.

65. Lactic acid has a K_a value of 1.4×10^{-4}. In what pH range is lactic acid most effective as a buffer?

66. What is the pH of a solution containing 0.10 M NaH_2PO_4 and 0.10 M Na_2HPO_4? (See Table 9.5.)

67. What is the pH of a solution containing 0.014 M Na_2SO_4 and 0.14 M $NaHSO_4$? (See Table 9.5.)

68. Does a solution containing 0.15 M NaH_2PO_4 and 0.05 M Na_2HPO_4 have a higher or lower pH value than the solution described in exercise 66?

69. Does a solution containing 0.020 M Na_2SO_4 and 0.008 M $NaHSO_4$ have a higher or lower pH value than the solution described in exercise 67?

pH in the Body

70. Most cells and organs are slightly acidic, about pH 6–7. Guess why.

71. List four substances that buffer blood to keep its pH near 7.40.

72. When blood passes by the lungs, in which direction does the following equilibrium position shift: $HHb + O_2 \rightleftarrows HbO_2^- + H^+$? What causes this shift?

73. For the reactions, $H^+ + HCO_3^- \rightleftarrows H_2CO_3 \rightleftarrows H_2O + CO_2$, in which direction does the equilibrium shift in the lungs? What factors cause this shift?

74. For the reaction in exercise 73, in which direction does the equilibrium shift as blood passes by muscle cells? What factors cause this shift?

75. What factors cause O_2 to come off hemoglobin (Hb) as blood passes by muscle cells?

76. How does metabolic acidosis affect breathing rate? Explain why.

77. Why do people at high altitude sometimes get relief by breathing into a paper sack?

78. Are blood levels of bicarbonate ion (HCO_3^-) high, normal, or low in people with respiratory alkalosis?

DISCUSSION EXERCISES

1. If you spill a strong acid on your skin, you should flush the acid thoroughly with water to dilute it. What solution could you then use to neutralize any remaining acid? How would the treatment differ if you spilled a strong base on your skin?

2. Dough used to make cakes and biscuits contains a weak acid and baking soda, $NaHCO_3$. What makes the dough rise?

3. What is the pH of 1.00 M HCl solution?

4. Limestone, marble, and roofing slate contain calcium carbonate, $CaCO_3$. Why are these materials damaged by acid deposition containing H_2SO_4? Write an equation for the reaction between H_2SO_4 and $CaCO_3$.

5. Is water an acid, a base, or neither? Is HSO_4^- an acid, a base, or neither? What is the criterion for classifying these substances?

6. Some chemists prefer not to use normality (N) as a concentration unit. What are the arguments for and against using normality? Which arguments do you favor?

10

Nuclear Chemistry

1. What are the main types of radioactivity from natural sources? What are their properties?
2. How do you write equations to represent nuclear reactions?
3. How long do substances stay radioactive?
4. How is radioactivity measured?
5. How does radioactivity affect the body?
6. What are some beneficial uses of radioactivity, particularly in medicine?
7. What are nuclear fission and fusion, and how are they used to produce energy?

Marie Curie

Scientists learned the hard way how radioactivity harms living things. When Marie Curie (Figure 10.1) worked with her husband Pierre in their Paris laboratory—an abandoned shed that was hot and smelly in summer and cold and damp in winter—little was known about the potential dangers. Of this time near the beginning of the 20th century, their younger daughter, Eve, later wrote (in the biography, *Madame Curie*):

> Marie continued to treat, kilogram by kilogram, the tons of [radioactive]pitchblende residue which were sent from St. Joachimsthal. With her terrible patience, she was able to be, every day for four years, a physicist, a chemist, a specialized worker, an engineer, and a laboring man all at once.

From that pitchblende residue, Marie isolated a few tenths of a gram of two new elements, both radioactive. She named one polonium (Po) for her homeland, Poland, and the other radium (Ra) from the Latin for *ray*. Radium, which has a phosphorescent bluish glow, has been used on watch dials to make them glow in the dark. It is also used to treat cancer (see Section 10.5).

Marie Curie was the first woman to receive a Nobel prize, sharing the 1903 award in physics with Pierre Curie and Henri Becquerel for their fundamental work in radioactivity. Marie also became the first person to receive a second Nobel prize (in chemistry in 1911) for discovering polonium and radium.

Little did she know, however, what her work was doing to her. Marie suffered from anemia, a common symptom of radiation sickness, and had one miscarriage. Another daughter, Irène, also worked with radioactive materials and won the Nobel prize in chemistry in 1935. That award came one year after Marie died from leukemia, no doubt caused by her long exposure to radioactivity. Irène also died from leukemia, probably for the same reason.

Figure 10.1 Marie Curie (1867–1934).

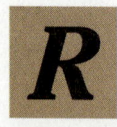

R adiation is all around you. Solar radiation streams to earth; television sets beam sound and light waves; radio waves pass invisibly through the air; heat leaves earth. Our air, food, and water all give off some level of radioactive emissions (Figure 10.2).

Figure 10.2 These tasty treats contain small amounts of radioactivity.

Figure 10.3 shows the range of electromagnetic radiation measured by wavelength, the distance between each wave crest. Although you see only a tiny fraction—the visible spectrum—other radiation still surrounds you.

In this chapter we focus mainly on *radioactivity*, which comes from the nuclei of atoms. Radioactivity includes gamma (γ) rays, a type of high-energy electromagnetic radiation (Figure 10.3). It also includes alpha and beta emissions—nuclear particles that stream out from atoms. Radioactivity and other radiation can harm us, but we also use it to diagnose and treat diseases.

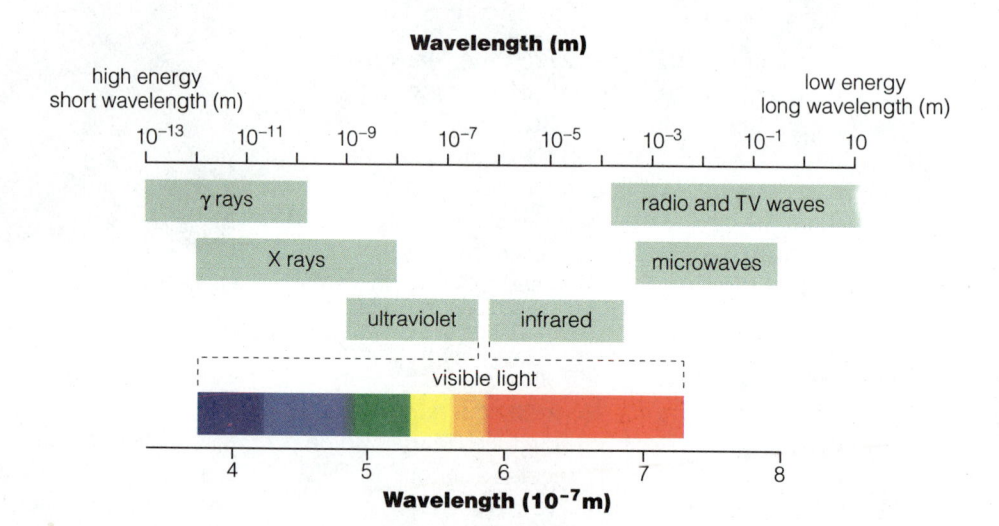

Figure 10.3 The spectrum of electromagnetic radiation.

10.1 TYPES OF RADIOACTIVITY

Radioactivity. Recall that almost all of an atom's mass is crammed into its tiny nucleus (Section 3.3). The nucleus contains protons and neutrons. Often these particles stay together in a stable arrangement. But sometimes they part company. Nuclei that have an unstable combination of protons and neutrons shoot out bits of mass and energy, an action called **radioactivity.**

CHEMISTRY SPOTLIGHT

The Discovery of Radioactivity

In 1896, French physicist Henri Becquerel (Figure 10.4) was studying chemicals that glow (fluoresce) after exposure to sunlight. He wondered whether fluorescence contained X rays, which had been discovered a year earlier.

Becquerel planned an experiment. He would expose some rocks to sunlight, then see if they emitted radiation that would develop photographic film the way X rays do. The Paris skies were cloudy that day, so he put the rocks away in his desk drawer and left for the weekend. When he returned, he noticed that one of the rocks in the drawer had been next to an unexposed photographic plate. The plate, though wrapped, bore a foggy image of the rock, which happened to contain uranium.

Becquerel concluded that the rock had spontaneously emitted energetic rays that penetrated the wrapping and exposed the photographic plate. Further research by Becquerel and others revealed three types of radiation; these were called alpha (α), beta (β), and gamma (γ) radiation. Marie Curie (see the Chapter Opener), who also worked in Paris, suggested that this phenomenon be called *radioactivity*.

Figure 10.4 Henri Becquerel (1852–1908) received the 1903 Nobel prize in physics for discovering radioactivity.

Because their nuclear compositions differ, some isotopes are stable and some are not. A radioactive isotope—in other words, one with an unstable nucleus—is called a **radioisotope.**

Types of Radioactivity. In studying natural radioisotopes, scientists have observed several types of radioactive emissions. Three common types are named after the first three letters of the Greek alphabet—*alpha* (α), *beta* (β), and *gamma* (γ). Table 10.1 summarizes their features.

An **alpha particle (α)** is a package of two neutrons and two protons. It is identical to the nucleus of a helium-4 atom, and its two protons provide a 2+ electrical charge. Because alpha particles have the largest charge and size of the radioactive emissions, they are the least penetrating form of radiation. They

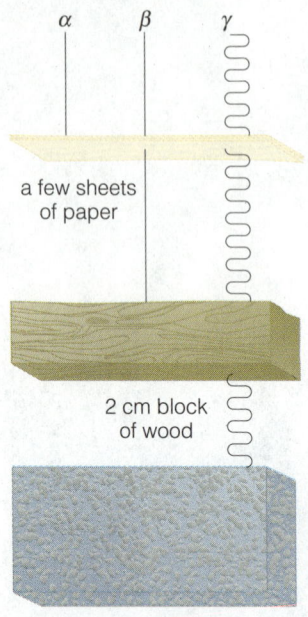

Figure 10.5 The three major types of natural radioactivity vary considerably in their penetrating power.

Table 10.1 Characteristics of Alpha, Beta, and Gamma Emissions

Symbol	Identity	Charge	Mass (amu)
Alpha $^{4}_{2}$He, $^{4}_{2}\alpha$	Helium nucleus	2+	4
Beta $^{0}_{-1}$e, $^{0}_{-1}\beta$	Electron	1−	1/1840
Gamma γ	Similar to X rays	0	0

cannot pass through skin, a thin sheet of aluminum, or several sheets of paper (Figure 10.5).

Beta particles (β) are fast-moving electrons with a charge of 1−. Because of their smaller charge and higher speeds, they are more penetrating than alpha particles. It takes an 0.3-cm aluminum plate or a 2-cm block of wood to stop them (Figure 10.5).

Gamma rays (γ) are not particles but a form of high-energy electromagnetic radiation similar to X rays (Figure 10.3). Traveling at the speed of light, they are so penetrating that only thick layers of lead or concrete can stop them.

10.2 WRITING NUCLEAR EQUATIONS

When a radioisotope emits radiation, the composition of its nucleus changes. This sometimes makes it a different element. Just as we write equations for chemical reactions, we write equations for nuclear reactions.

A *nuclear equation* represents each reactant and product in a nuclear reaction. The rule for balancing a nuclear equation is: The sum of the atomic numbers (subscripts) must be the same on both sides of the equation, as must the sum of the mass numbers (superscripts).

Uranium-238 ($^{238}_{92}$U), for example, emits alpha radiation. Since an alpha particle has an atomic number of 2 and a mass number of 4, we represent it as $^{4}_{2}$He or $^{4}_{2}\alpha$ (Table 10.1). We begin to write the nuclear equation like this:

$$^{238}_{92}\text{U} \longrightarrow {}^{4}_{2}\text{He} + \underline{\quad}$$

To fill in the blank, calculate the mass number and atomic number of the product. The total mass number on the left of the arrow is 238; the total on the right, then, is $4 + x = 238$, so the mass number of the product (x) is 234. The total atomic number on the left is 92; the total on the right is $2 + y = 92$, so the atomic number (y) of the product is 90. Now write the equation:

$$^{238}_{92}\text{U} \longrightarrow {}^{4}_{2}\text{He} + {}^{234}_{90}\underline{\quad}$$

To complete the equation, write the symbol in the periodic table that matches the atomic number (90, in this example). The complete equation is:

$$^{238}_{92}\text{U} \longrightarrow {}^{4}_{2}\text{He} + {}^{234}_{90}\text{Th}$$

The calculated mass number specifies the isotope formed in the reaction (thorium-234 in the preceding reaction). That mass number doesn't necessarily match the atomic mass listed for the element in the periodic table (such as 232.04 for thorium), which is the weighted *average* of naturally occurring isotopes of that element (see Section 3.3).

Notice that the sum of the superscripts (mass numbers) is 238 on each side of the arrow, and the sum of the subscripts (atomic numbers) is 92 on each side.

Nuclear Chemistry Chapter 10 217

> **EXAMPLE 10.1** Write a nuclear equation for the emission of alpha radiation by (a) radium-226 and (b) samarium (Sm)-147.
>
> **SOLUTION** (a) $^{226}_{88}Ra \longrightarrow ^{4}_{2}He + ^{222}_{86}Rn$, (b) $^{147}_{62}Sm \longrightarrow ^{4}_{2}He + ^{143}_{60}Nd$

Beta emissions are a little more complicated. Since electrons are found only outside the nucleus, how can a nuclear reaction emit a beta particle, which is an electron? It turns out that an electron is produced in the nucleus (and then emitted) when a neutron changes into a proton (Figure 10.6). This change doesn't alter the mass number (because protons and neutrons both count as 1), but it does increase the atomic number (number of protons) by 1.

Now, using one of the symbols for a beta particle (Table 10.1), let's write a nuclear equation for the emission of a beta particle by carbon-14. The total mass number on the left is 14, so the mass number of the product (x) must be 14 since $0 + x = 14$. The atomic number of the product added to -1 must equal 6; thus the number is 7. The equation, then, is:

$$^{14}_{6}C \longrightarrow ^{0}_{-1}e + ^{14}_{7}\underline{}$$

To complete the equation, fill in the symbol of the element that has atomic number 7:

$$^{14}_{6}C \longrightarrow ^{0}_{-1}e + ^{14}_{7}N$$

Emitting a beta particle changes a radioisotope into an element with the next higher atomic number.

The mass numbers on each side of the equation equal 14, and the atomic numbers equal 6.

> **EXAMPLE 10.2** Write a nuclear equation for the emission of a beta particle by (a) iodine-131 and (b) nickel-63.
>
> **SOLUTION** (a) $^{131}_{53}I \longrightarrow ^{0}_{-1}e + ^{131}_{54}Xe$, (b) $^{63}_{28}Ni \longrightarrow ^{0}_{-1}e + ^{63}_{29}Cu$

Since gamma rays are not particles, they have no charge or mass. When a radioisotope emits only gamma rays, its atomic number and mass number don't change. We don't need to write equations for these reactions.

Many radioisotopes emit gamma rays along with alpha or beta radiation. Gamma rays remove excess energy from unstable nuclei. Uranium-234, for example, emits both alpha and gamma radiation. We can write the equation to show both:

$$^{234}_{92}U \longrightarrow ^{4}_{2}He + \gamma + ^{230}_{90}Th$$

Two other types of radioactivity are *positron emission* and *electron capture*. A *positron* is a positively charged electron and has the symbol $^{0}_{+1}\beta$ or $^{0}_{+1}e$ (see Section 10.5). Electron capture occurs when an electron, usually in the 1s energy sub-level, enters and remains in the nucleus. Both positron emission and electron capture convert a proton into a neutron (the opposite of what happens in beta emission), so the atomic number decreases by 1. An example of each is:

positron emission $^{39}_{19}K \longrightarrow ^{0}_{+1}\beta + ^{39}_{18}Ar$

electron capture $^{55}_{26}Fe + ^{0}_{-1}e \longrightarrow ^{55}_{25}Mn$

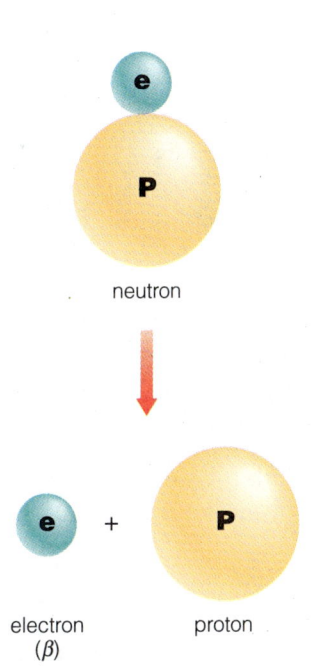

Figure 10.6 When an electron (β particle) is emitted, a neutron changes into a proton. Picturing a neutron as a proton and electron fused together may help you remember this change.

You can't tell from an isotope's atomic number or mass number whether or not it is radioactive or, if it is, what type of radiation it emits. But if you know what a radioisotope emits, you can write a nuclear equation and use simple arithmetic to figure out the products of the reaction.

10.3 HALF-LIFE

Rates of Radioactivity. When a radioisotope emits radiation, it may form a product that is also radioactive. A sequence of reactions continues until a stable isotope forms. Naturally occurring uranium-238, for example, goes through a long series of nuclear reactions and finally becomes a stable isotope, lead-206 (Figure 10.7). While some of the reactions are fast, others are very slow. The entire process takes billions of years to complete.

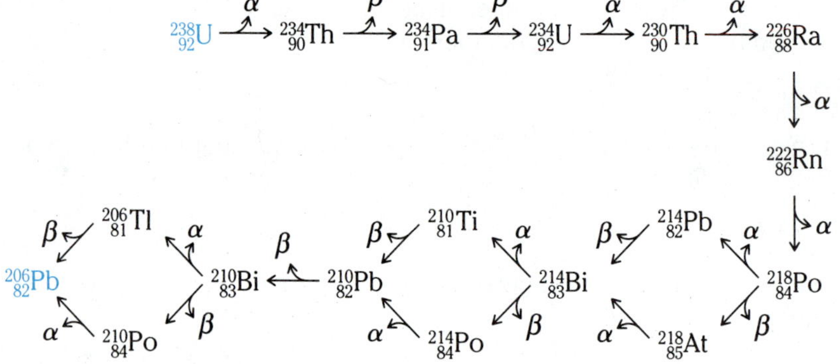

$$^{238}_{92}U \xrightarrow{\alpha} {}^{234}_{90}Th \xrightarrow{\beta} {}^{234}_{91}Pa \xrightarrow{\beta} {}^{234}_{92}U \xrightarrow{\alpha} {}^{230}_{90}Th \xrightarrow{\alpha} {}^{226}_{88}Ra \xrightarrow{\alpha} {}^{222}_{86}Rn$$

Figure 10.7 Conversion of uranium-238 into lead-206, which is not radioactive. Each reaction has a characteristic half-life. The first reaction has the longest half-life, 4.5 billion years.

Each radioisotope has a characteristic rate of emitting radiation. **Half-life** is the time it takes for half of the nuclei in a sample of a radioisotope to decay (give off their emissions). Table 10.2 lists a few radioisotopes and their half-lives. Notice, for example, how long it takes for uranium-238 to decay to thorium-234. In contrast, radioisotopes such as iodine-131 have less stable nuclei that emit their radiation more quickly.

Mathematically, half-life is like traveling to a destination 100 miles away and each day going half of the remaining distance. For a radioisotope with a half-life of 1 day, during each 24-hour period, half of the remaining nuclei would emit their radiation. For example, if you start with 64 g of material, 32 g of that radioisotope remains after 1 day, 16 g after 2 days, 8 g after 3 days, and so on (Figure 10.8). The rest of the original 64 g of mass becomes the product(s) of the decay reaction.

Table 10.2 Half-Lives of Some Radioisotopes

Isotope	Radiation	Half-life
$^{3}_{1}H$	β	12.3 yr
$^{14}_{6}C$	β	5730 yr
$^{24}_{11}Na$	β, γ	15 hr
$^{60}_{27}Co$	β, γ	10.5 mo
$^{90}_{38}Sr$	β	28 yr
$^{99}_{43}Tc$	β, γ	6.0 hr
$^{131}_{53}I$	β, γ	8.1 day
$^{235}_{92}U$	α	7.1×10^8 yr
$^{238}_{92}U$	α, γ	4.5×10^9 yr
$^{239}_{94}Pu$	α, γ	2.4×10^4 yr

EXAMPLE 10.3 If you have 24.0 mg of technetium-99, which has a half-life of 6 hours (Table 10.2), how many mg of technetium-99 remains after 24 hours?

SOLUTION 1.5 mg (12.0 mg after 6 hrs, 6.0 mg after 12 hrs, and 3.0 mg after 18 hrs)

Estimating the Age of Objects. The half-life of a radioisotope is constant. It is a kind of clock that keeps ticking at the same rate regardless of heat, pressure,

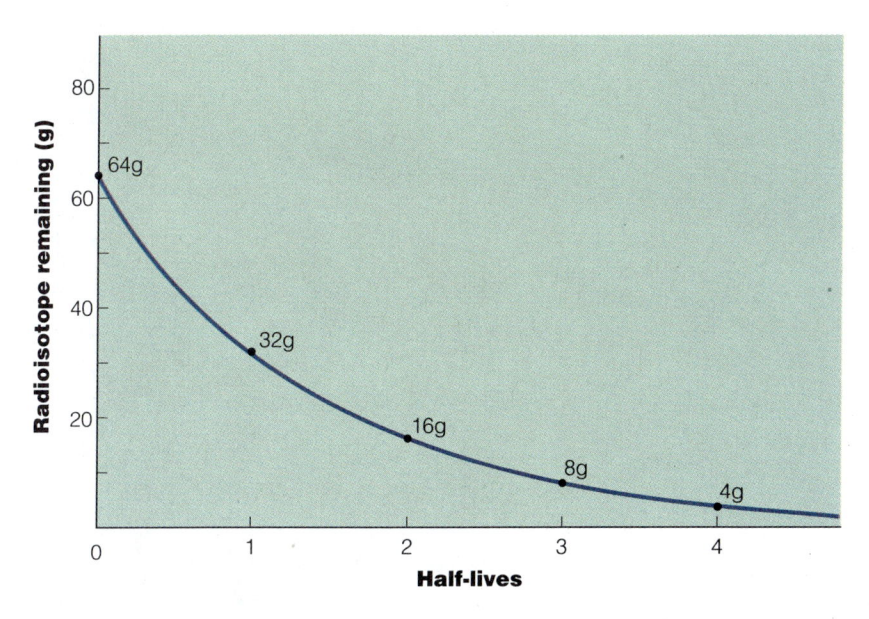

Figure 10.8 Pattern of decay for 64 g of any radioisotope. During each half-life, the amount of radioisotope decreases by one-half.

or chemical reactions. This makes radioisotopes useful in estimating the age of objects.

Scientists measure the relative amounts of a radioisotope and its products in an object. They assume all of the product comes from decay of the radioisotope. If equal amounts of radioisotope and its product are present the radioisotope has gone through one half-life.

Consider, for example, the conversion of uranium-238 into lead-206 (Figure 10.7). When 1 mole (238 g) of uranium-238 goes through one half-life (4.5 billion years; see Table 10.2), it produces 0.5 mole (103 g) of lead-206 while 0.5 mole (119 g) of uranium-238 remains. The ratio of lead-206 to uranium-238 at this time is 103 g/119 g = 0.866. If an ancient rock has this ratio of the two isotopes, the rock's estimated age is 4.5 billion years. A lower ratio of lead-206 to uranium-238 in the rock would correspond to an age younger than the half-life of uranium-238. If most of the radioisotope hasn't yet decayed, the object is much younger than the half-life.

One common method, *radiocarbon dating,* uses radioactive carbon-14 to estimate the age of plants, wood, teeth, bone fossils, and other carbon-containing materials. Carbon-14 forms continuously in the atmosphere, and is incorporated at a steady rate into plant and animal tissues as long as they are alive (Figure 10.9). After a plant or animal dies, its carbon-14 slowly gives off beta particles and changes into nitrogen-14:

$$^{14}_{6}\text{C} \longrightarrow \, ^{0}_{-1}\text{e} + \, ^{14}_{7}\text{N}$$

Most of an organism's carbon, however, is nonradioactive carbon-12, which doesn't change. Because the half-life of carbon-14 is 5730 years (Table 10.2), the amount of carbon-14 decreases by one-half during that time. Thus, measuring the ratio of radioactive carbon-14 in an object's total carbon gives an estimate of the object's age.

Although the conversion of uranium-238 into lead-206 takes many reactions, the first reaction (see Figure 10.6) has by far the longest half-life and thus determines the rate of lead-206 formation.

EXAMPLE 10.4 At death, a human body contains enough ^{14}C to provide 16 counts/min of radioactivity per g of carbon. Suppose that in the year 2000, some human remains are found with a radioactivity of 4 counts/min per g of carbon. Estimate the age of the remains.

Figure 10.9 How carbon-14 can be used to estimate the age of archaeological artifacts that were once living.

Cosmic rays collide with atmosphere, producing high-energy neutrons.

Neutrons hit nitrogen atoms, producing radioactive carbon-14.

Carbon-14 and oxygen combine, forming radioactive carbon dioxide.

Plants take in normal and radioactive carbon dioxide to build tissues.

Animals consume plant tissues containing carbon-14.

Humans consume plant and animal tissues containing carbon-14.

Radioactive carbon –14 in dead plant and animal tissues decays to nitrogen –14, without being replenished. Because this happens at a fixed rate, comparison of the radioactivity in fossil and living specimens gives the elapsed time since death.

SOLUTION ^{14}C has gone through two half-lives to reduce the radioactivity from 16 counts/min → 8 counts/min → 4 counts/min. Two half-lives = 2 × 5730 yr = 11,460 yr. The remains, then, date back to 2000 − 11,460 = 9460 B.C.

Radioisotope dating can give accurate results when the object's age is no more than ten times larger or smaller than the radioisotope's half-life. Radiocarbon dating, for example, is limited to objects younger than about 50,000 years—because older objects have too little ^{14}C left to measure accurately. The ages of objects older than 50,000 years are estimated using radioisotopes with longer half-lives. For example, uranium-238 (Table 10.2) dating of meteorites and other very old rocks indicates our planet is about 4.6 billion years old.

10.4 EFFECTS OF RADIOACTIVITY

Measuring Radioactivity. One basic unit of radioactivity is the **curie (Ci)**, which equals 3.7×10^{10} emissions per second, or the amount of radiation emitted by 1 g of radium. Smaller units are the millicurie (mCi) and microcurie (μCi).

The effects of radioactivity depend not only on the number of emissions, but also on their energy and type. Radioactive emissions typically strike electrons in surrounding materials and produce ions. The **roentgen** is a unit based on the number of ions produced as emissions pass through air.

Another unit, the **rad** (for *r*adiation *a*bsorbed *d*ose), is defined as 1×10^{-5} J of energy per g of tissue irradiated. Since 1 rad doesn't produce the same effect in each part of the body, the rad is often converted into the **rem** (for *r*oentgen *e*quivalent for *m*an), which is a measure of the effect of radiation on the body. Our exposure to radioactive emissions often is measured in rems or millirems (mrem)(Table 10.3).

The curie is named after Marie and Pierre Curie (see the Chapter Opener). The roentgen is named after Wilhelm Roentgen, who discovered X rays in 1895.

SI units are: 1 becquerel (Bq) = 1 emission/sec; 1 sievert (Sv) = 100 rem; 1 gray (Gy) = 100 rad.

Table 10.3 Average Annual Exposure to Radioactivity in the United States

Source	Approximate Dose (mrem)
Natural Sources (background radioactivity)	
cosmic rays from space	20–50
radioactive minerals in rocks and soil	25–30
radioactivity in the body from air, water, and food	40
radon-222 in the air	200
Sources from Human Activities	
medical and dental X rays and nuclear medicine	50–75
air travel	5
smoking (40 mrem for a pack-a-day smoker)	—
consumer products (such as TV sets)	10
Total	**360**

Source: National Council on Radiation Protection, Report 93, 1987.

People who work near radioactive materials often wear radiation badges to measure their exposure (Figure 10.10). The badge contains photographic film that fogs when exposed to radiation.

Another measuring device is the Geiger counter (Figure 10.11). Radiation entering the tube ionizes argon gas, which causes a pulse of current to flow between two electrodes. The number of pulses measures the radioactivity entering the tube. Because they often don't penetrate the window of a Geiger counter, alpha particles aren't counted as effectively as beta and gamma emissions.

Effects on the Body. The effects of radioactivity on the body depend on the amount and frequency of exposure, the type of radiation, and whether the radioactive emissions come from outside or inside the body.

From the outside, alpha emissions are the least dangerous because they don't penetrate skin (Figure 10.5). Beta emissions don't penetrate skin well,

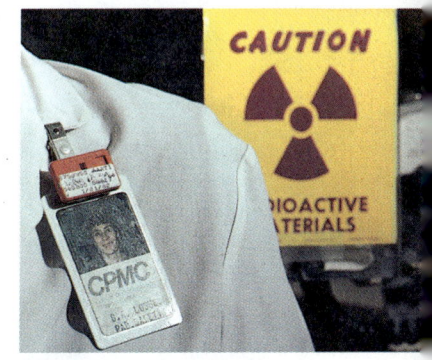

Figure 10.10 A radiation badge measures this person's exposure to radioactivity.

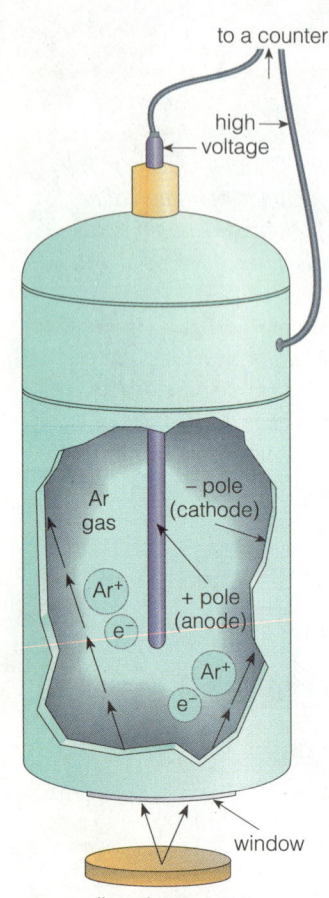

either. Gamma rays are the most dangerous because they easily enter the body. Alpha emitters, however, are the most dangerous radioisotopes inside the body. Alpha particles don't travel as far as gamma or beta emissions, but their large size and charge causes considerable damage to nearby cells.

You can't escape being exposed to radioactivity. You eat, drink, and breathe small amounts of radioactive materials every day. Natural sources make up *background radioactivity* (Table 10.3).

A major source is radon-222, which forms in rocks and soil from the decay of uranium-238 (Figure 10.7). Radon-222 gas seeps through cracks and drains into buildings and can accumulate to high levels, especially in energy-efficient buildings that don't bring in much fresh air. Breathing such air draws radon-222 into the lungs, where its alpha emissions can cause serious damage.

Table 10.4 lists the effects of high doses of radioactivity over a short time. In comparison, the average annual exposure of about 360 mrem (0.360 rem; Table 10.3) is a small dose, and is spread out over a year. Because the body repairs some damage, small doses of radioactivity over a long time cause less damage than the same total dose all at once. The National Research Council estimates that radon-222 gas causes up to 20,000 lung cancer deaths a year; typical radiation exposure from other sources causes about 1% of all fatal cancers and 5% to 6% of the genetic disorders in the United States.

Radioactivity causes materials in cells to ionize. A major victim is DNA, the hereditary material in chromosomes. In DNA, radioactivity causes chemical changes called *mutations*. Egg and sperm cells that carry mutations can transmit genetic disorders to children. Mutations also can trigger the cancer process.

When cells divide (Figure 10.12), their DNA is especially vulnerable to radiation. Since blood cells and cells lining the digestive tract divide rapidly, many of them are damaged during exposure to radiation. This is why radioactivity causes nausea, vomiting, diarrhea, a shortage of blood cells, and fatigue (Table 10.4).

One tragic example of the dangers of radioactivity occurred in the 1920s, when women worked in factories painting radium onto the dials of watches to make them glow in the dark. When they put the tips of their brushes on their tongues to make a finer point for painting, the workers ingested small amounts of radium, an alpha and gamma emitter. Many died from anemia (a shortage of red blood cells) and bone cancer.

Figure 10.11 Diagram of a Geiger counter. Radioactive emissions enter and ionize argon atoms. This produces an electrical pulse that is counted.

Table 10.4 Effects on Humans of Whole-Body Exposure to Radioactivity over a Short Period of Time

Dose (rems)	Effects
0–50	No consistent symptoms
50–200	Decreased white blood cells, nausea, vomiting; about 10% die within months at 200 rems
200–400	Loss of blood cells, fever, hemorrhage, hair loss, nausea, vomiting, diarrhea, fatigue, skin blotches; about 20% die within months
400–500	Same symptoms as 200–400 rems but more severe, increased infections due to lack of white blood cells; 50% die within months at 450 rems
500–1000	Severe gastrointestinal and central nervous system damage, cardiovascular collapse; doses above 700 rems fatal within a few weeks
10,000	Death in hours

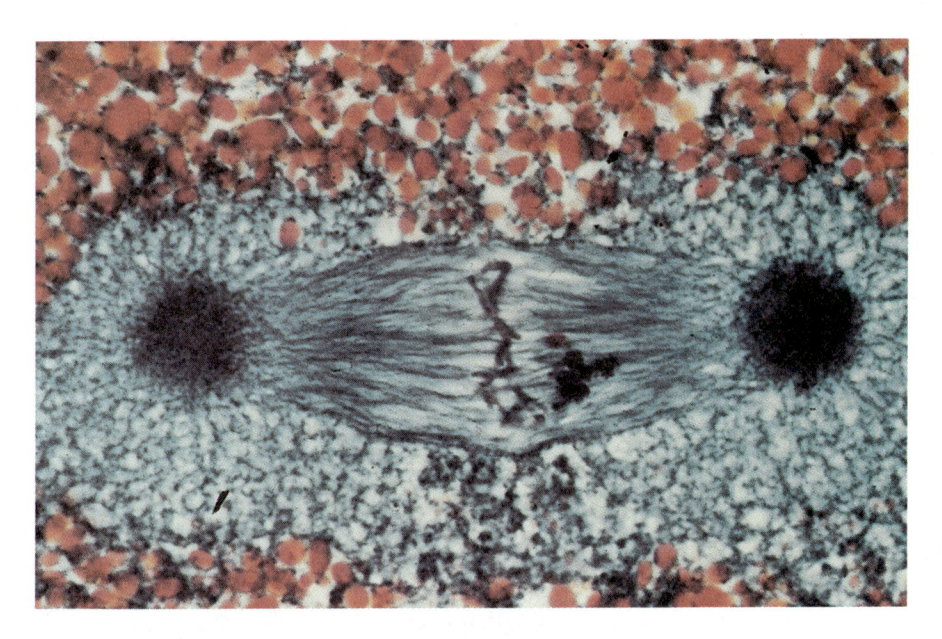

Figure 10.12 Cells are especially susceptible to radiation damage when they are dividing.

10.5 MEDICAL USES OF RADIOACTIVITY

Treating Disease. Because they damage DNA and other cell materials, radioisotopes can be used to destroy harmful cells. People with overactive thyroid glands (hyperthyroidism), for example, are given radioactive iodine-131. The radioisotope collects in the thyroid gland and destroys some of its tissue.

Radioactivity can cause cancer, but it also is used to treat cancer. Because cells and their DNA are especially vulnerable to radioactivity when they divide, radiation is an effective way to kill rapidly dividing cancer cells. But the treatment also kills some normal cells. This is why people treated with radiation often experience side effects such as nausea, diarrhea, and low blood-cell counts (Table 10.4).

Most treatments use an external source of radiation to penetrate the skin and reach the tumor (Figure 10.13). Common sources are gamma emitters such

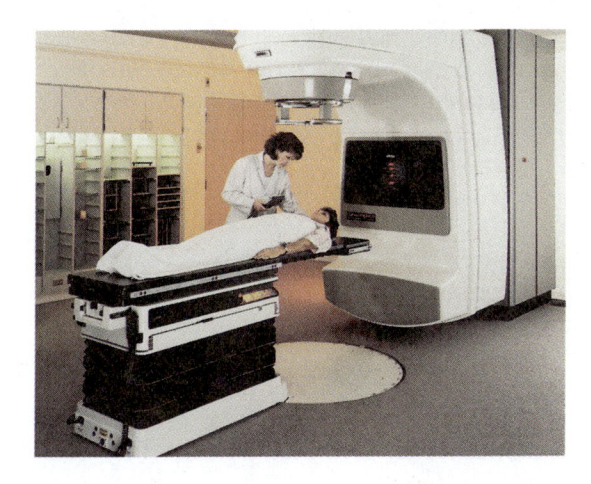

Figure 10.13 Internal cancer tumors can be treated with gamma rays emitted by cobalt-60.

as cobalt-60 and cesium-137, or X rays. A narrow beam of radiation is focused precisely on the tumor area to minimize damage to other cells. Lead shields block radiation to nearby parts of the body.

Very high doses are used to kill cells. Cancer patients typically receive 150–200 rems per treatment. Some receive 6000 or more rems during a two- to three-month period. By comparing this dose with the information in Table 10.4, you can see why this treatment kills cells but must be confined to the tumor area.

A new, experimental approach treats cancer by attaching radioactive atoms to antibodies that selectively bind to cancer cells. Once the antibodies bind, their radioactive cargo kills the cancer cells.

Diagnosing Diseases. To diagnose certain illnesses, a radioisotope is ingested or injected into the body. This poses a risk that must be weighed against the benefits of doing the diagnosis. Gamma or beta emitters with short half-lives are used to minimize the risk.

To diagnose malfunctions of the thyroid gland, for example, a patient drinks a solution containing iodine-123, a gamma emitter. Then the physician scans the thyroid area with a radioactivity detector to see if iodine accumulates there at the normal rate (Figure 10.14). Sodium-24 is used to detect constrictions or obstructions in blood vessels. Injected into the bloodstream, its radioactivity traces the flow of blood in veins and arteries.

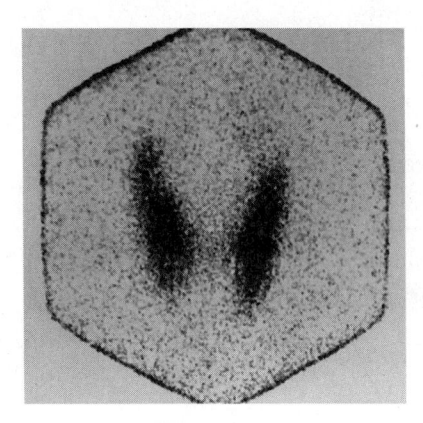

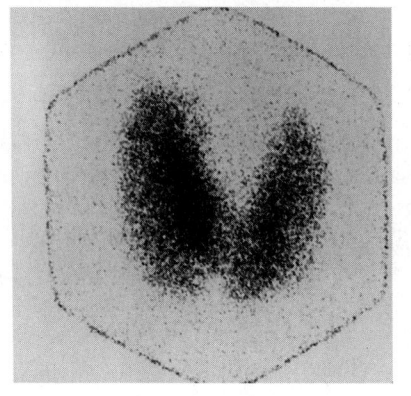

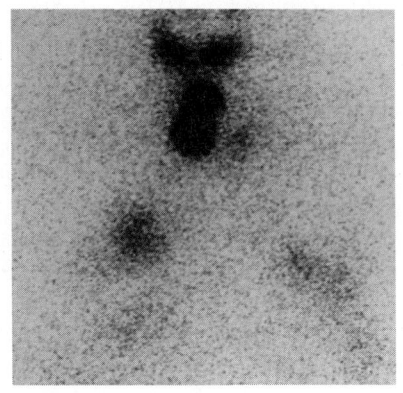

Figure 10.14 Scans showing the uptake of radioactive iodine in a normal (left), overactive (center), and cancerous (right) thyroid gland.

A similar approach is used to diagnose and precisely locate cancer. Because cancer tumors grow very rapidly, they tend to accumulate materials faster than most tissues. Brain tumors, for example, rapidly absorb large amounts of copper-64, technetium-99, and indium-111. Different isotopes are used to detect different types of cancer.

Table 10.5 lists a few radioisotopes used in medicine.

Imaging Techniques. Radioisotopes inside the body can be used to obtain an image of the areas where they accumulate. Iodine-123, for example, gives an image of the brain. Technetium-99 provides images of the heart, liver, and lungs.

Ever since Wilhelm Roentgen discovered X rays in 1895, physicians have used radiation to "see" inside the body and diagnose diseases. **X rays** are high-energy radiations (Figure 10.3) but aren't radioactivity since they don't come from the nucleus.

CHEMISTRY SPOTLIGHT

Synthesizing Isotopes

For maximum safety, radioisotopes taken internally must have short half-lives. Technetium-99, used to detect brain cancer and bone fractures, has a half-life of only 6 hours (Table 10.2). Such radioisotopes can't be stored for long. They have to be produced when needed.

Some isotopes are made using a *cyclotron* or *linear accelerator*. These machines accelerate charged particles through a series of tubes with alternating electrical charge. As particles are repelled out of each tube and attracted to the next, they go faster. When accelerated particles enter the nuclei of target materials, they produce new isotopes. Accelerated alpha particles striking bismuth-209 nuclei, for example, produce the heaviest halogen, astatine (At), and two neutrons (^1_0n):

$$^{209}_{83}Bi + ^4_2He \longrightarrow ^{211}_{85}At + 2\,^1_0n$$

Alpha and other positive particles have to be accelerated to overcome the repulsion of positively charged nuclei that they enter. But a neutron, having no charge, easily enters a nucleus without being accelerated. Neutrons are used to make molybdenum-98 into an isotope that changes into technetium-99 by beta emission. The reactions are:

$$^{98}_{42}Mo + ^1_0n \longrightarrow ^{99}_{42}Mo$$

$$^{99}_{42}Mo \longrightarrow ^{99}_{43}Tc + ^{\ \ 0}_{-1}e$$

The cyclotron was invented by Ernest Lawrence at the University of California, Berkeley. He received the Nobel prize in physics in 1939. Scientists have used cyclotrons to synthesize elements with atomic numbers greater than 92 (called *transuranium elements*), which don't occur in nature. Lawrence is honored in element 103, lawrencium (Lw).

Table 10.5 Radioisotopes Used in Medicine

Radioisotope	Medical Use
carbon-11	PET scans, especially for brain function
sodium-24	detect obstructions in blood vessels
phosphorus-32	treat leukemia
cobalt-60	treat cancers
gallium-67	treat lymphomas
strontium-85	bone scans
technetium-99	image the brain, heart, lungs, and liver
iodine-123	image the brain; diagnose thyroid malfunction
iodine-131	diagnose thyroid malfunction; treat hyperthyroidism
xenon-133	detect lung malfunction

Because of their high energy (Figure 10.3) and penetrating ability, X rays are stopped in the body only by dense materials such as bone. When used in combination with dyes or other dense materials, however, they

Figure 10.15 X ray of intestinal tract containing BaSO$_4$.

can also produce pictures of soft tissues. X rays of patients given a BaSO$_4$ mixture, for example, show details of the gastrointestinal tract (Figure 10.15).

CT (Computerized Tomography) **scans** use X rays to give finer, three-dimensional detail. As a machine rotates around the patient, brief X rays are taken at various angles. A computer then reconstructs the internal images. CT scans are used to diagnose brain and other tumors, tiny fractures, and many other disorders (Figure 10.16).

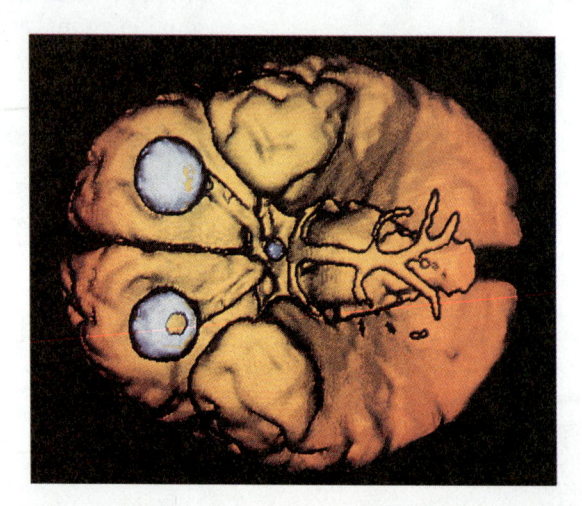

Figure 10.16 CT scan of the brain.

PET (Positron Emission Tomography) **scans** detect substances containing synthetic radioisotopes that emit positrons. The main isotopes used are carbon-11, nitrogen-13, and oxygen-15. Since a positron immediately collides with an electron to produce a pair of gamma rays that travel in opposite directions, gamma detectors are used to locate positron emissions. PET scans can show metabolism in the brain of glucose labeled with oxygen-15. They also show abnormal patterns in disorders such as manic depression, schizophrenia, and Alzheimer's disease (Figure 10.17).

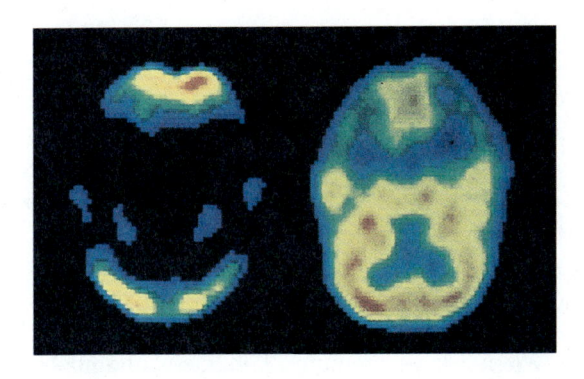

Figure 10.17 PET head scans of a normal person (right) and a person with Alzheimer's disease (left).

MRI (Magnetic Resonance Imaging) **scans** use low-energy radio waves (Figure 10.3) to cause the nuclei of atoms, often hydrogen, to absorb and emit specific amounts of energy. The pattern is measured and converted by computer into an image of the region. Because water and fat are abundant sources of hydrogen, the image largely shows their distributions. Different tissues absorb and emit energy at different rates, so MRI scans show details in soft tissues.

Unlike X rays, CT scans, and PET scans, MRI uses no radiation known to be harmful. Though expensive (a knee exam can cost $1200), MRI is used to diagnose brain, spinal cord, joint, and other soft-tissue disorders.

10.6 NUCLEAR FISSION AND FUSION

Nuclear Fission. In 1938, German radiochemist Otto Hahn and his student Fritz Strassman bombarded uranium with neutrons. Among the reaction products, they discovered tiny amounts of barium (Ba), cerium (Ce), and lanthanum (La). How could barium, which has an atomic number of 56, come from uranium, which has an atomic number of 92? It was as if the uranium nucleus had been cut nearly in half.

Hahn and Strassman had discovered **nuclear fission,** a process in which a large nucleus breaks into two or more smaller nuclei. This happens when a neutron (^1_0n) enters certain nuclei and makes them unstable. Figure 10.18 shows a fission reaction with uranium-235.

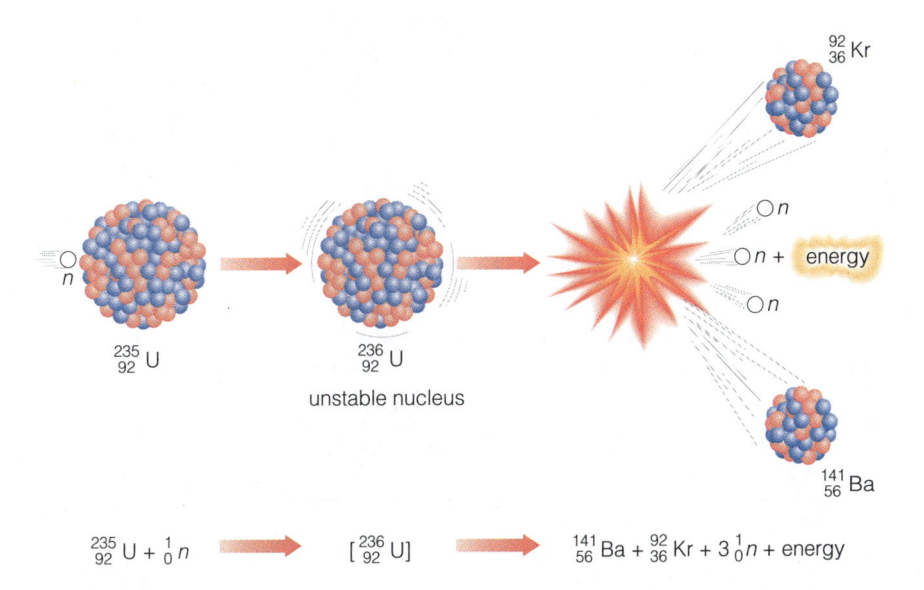

$$^{235}_{92}U + ^1_0n \longrightarrow [^{236}_{92}U] \longrightarrow ^{141}_{56}Ba + ^{92}_{36}Kr + 3\,^1_0n + \text{energy}$$

Figure 10.18 Typical products of the fission of a uranium-235 nucleus by a slow-moving neutron.

Fission releases huge amounts of energy when mass in the nucleus changes into energy. In 1905 Albert Einstein proposed, in his special theory of relativity, that energy and mass are two aspects of the same thing. He expressed this relationship in the famous equation:

$$E = mc^2$$

where E represents energy, m represents mass, and c represents the speed of light. Because c is a very large number, small amounts of mass (multiplied by c^2) change into very large amounts of energy.[1]

Nuclei about the size of iron (Fe) are the most stable and have the least internal energy (Figure 10.19). When a large uranium-235 nucleus undergoes fission, it releases energy as it changes into smaller, lower-energy nuclei.

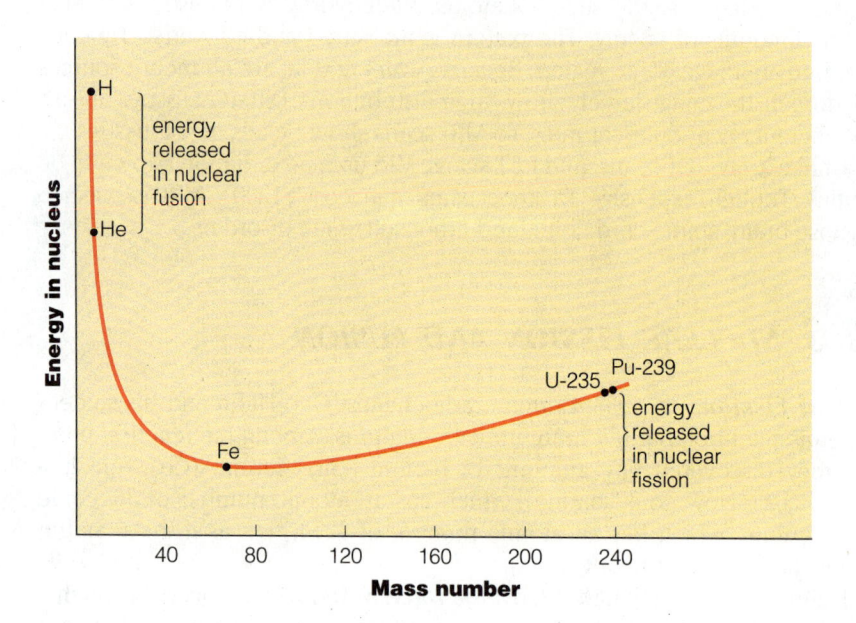

Figure 10.19 Effect of nuclear size on nuclear energy and stability.

Nuclear Energy. Notice in Figure 10.18 that when a neutron causes uranium-235 to undergo fission, the reaction releases more neutrons. Those neutrons, in turn, can cause other uranium-235 nuclei to split. When this happens, a *chain reaction* occurs (Figure 10.20), with many nuclei splitting almost simultaneously. The minimum amount of fissionable fuel needed to keep a chain reaction going is called the *critical mass*.

An atomic bomb releases a vast amount of energy all at once. On August 6, 1945, a U.S. airplane dropped a uranium-235 bomb on Hiroshima, Japan (Figure 10.21). Three days later a plutonium-239 bomb was dropped on Nagasaki. These bombs instantly killed 110,000 people. Within several months, another 100,000 people died from injuries and radiation exposure. A few days after the bombs were dropped, Japan surrendered to end World War II.

Nuclear power plants use controlled fission to generate electricity (Figure 10.22). The fuel—uranium-235 or plutonium-239—is packed into fuel rods and

---

[1] This discovery forced scientists to modify the laws of conservation of mass (Section 3.2) and conservation of energy (Section 8.1). The new version, the *law of conservation of mass–energy*, states that the total amount of mass and energy—taken together—stays constant even though mass and energy may change forms and be interconverted. For ordinary chemical reactions, however, the law of conservation of mass still works.

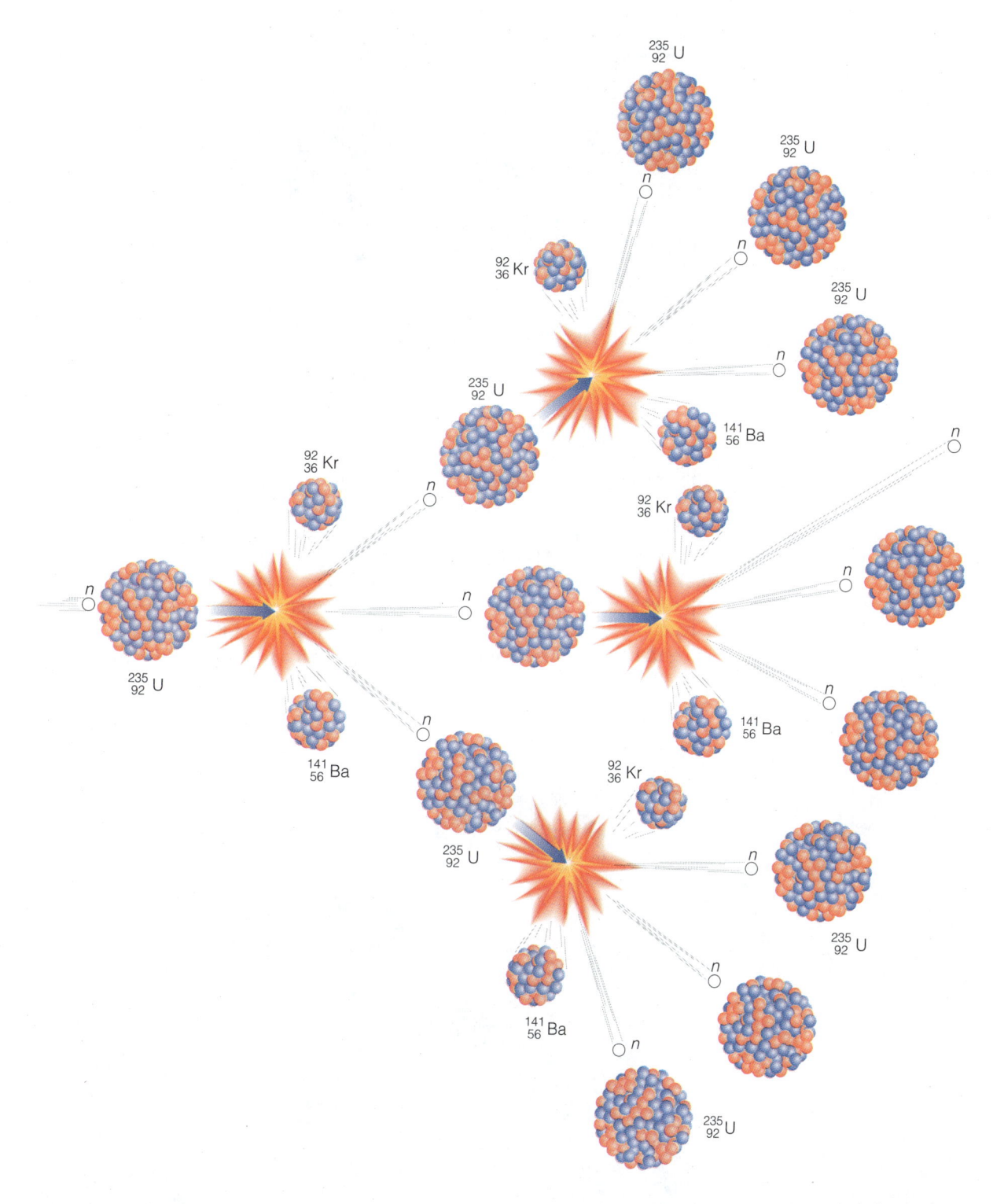

surrounded by control rods made of neutron-absorbing materials. Moving control rods in and out of the reactor regulates the fission rate and thus the amount of power produced. The fuel isn't concentrated enough to have the critical mass for a nuclear explosion.

Figure 10.20 A chain reaction started by one neutron triggering fission in a single uranium-235 nucleus.

230 Part I General Chemistry

Figure 10.21 A nuclear fission bomb of the type exploded over Hiroshima, Japan. The bomb is 3 m long, weighs 4000 kg, and has the explosive power equal to 17 million kg (20,000 tons) of TNT.

Water functions as a moderator, slowing neutrons and increasing their chances of producing fission when they enter fuel nuclei. Water also cools fuel rods and carries heat to a heat exchanger. The heat makes steam, which spins the blades of a turbine, which in turn runs a generator to produce electricity.

Fission supplies a higher percentage of energy in Europe (particularly in France and Belgium), Japan, and Taiwan than in the United States. The major concerns include safety, storage of radioactive wastes, and cost.

A nuclear reactor cannot blow up like an atomic bomb because it doesn't have enough fissionable fuel in the right configuration to allow a runaway chain reaction. But a reactor core can overheat because of mechanical errors, causing radioactive material to escape into the air and melt into the ground. Accidents at the Three Mile Island facility in Pennsylvania and at Chernobyl in Russia have increased these concerns about safety.

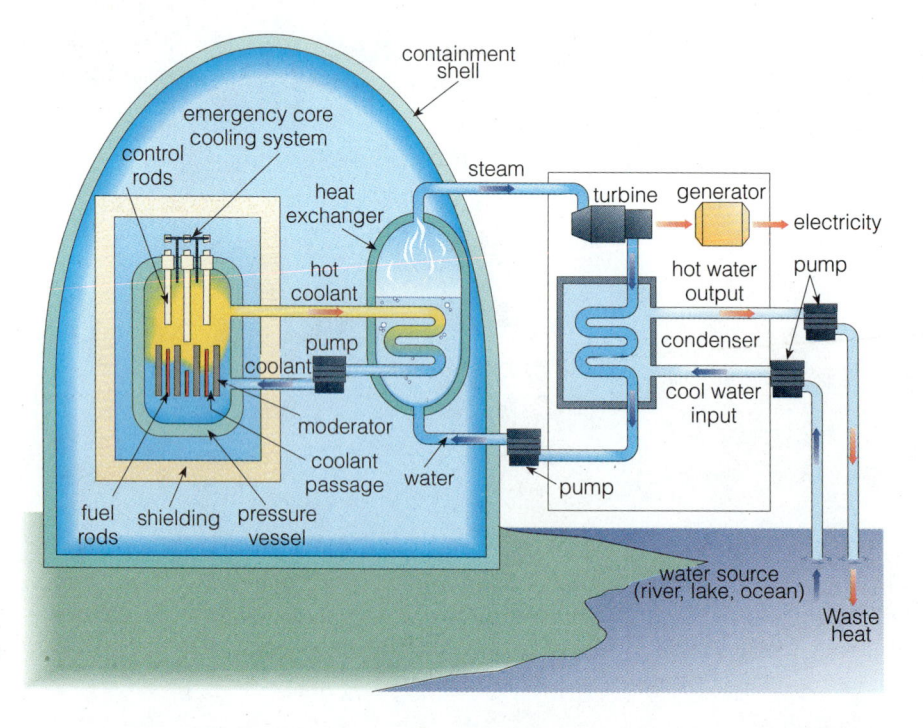

Figure 10.22 A nuclear power plant with a light-water reactor.

CHEMISTRY SPOTLIGHT

J. Robert Oppenheimer and the Manhattan Project

At age 38, J. Robert Oppenheimer (Figure 10.23) was one of the world's most brilliant physicists. During World War II, the U.S. government asked him to leave his teaching job at the University of California at Berkeley to direct the Los Alamos (New Mexico) Laboratory. His job was to develop the first atomic bomb. The code name for this top-secret operation was the Manhattan Project.

Under Oppenheimer's direction, the project was completed in less than three years. The first atomic bomb, detonated at a test site near Alamogordo, New Mexico, filled Oppenheimer and other onlookers with awe. The massive outburst of heat and light reminded him of a line from the Hindu *Bhagavad-Gita:* "I am become death, the shatterer of worlds." Less than a month later, atomic bombs were dropped on Hiroshima and Nagasaki, Japan, bringing the war to an end.

After the war, Oppenheimer became chairman of the General Advisory Committee to the Atomic Energy Commission (AEC). He had reservations, however, about embarking on another crash project to develop even more powerful nuclear fusion (hydrogen) bombs. This view made him politically unpopular.

Anti-Communist fever was running high in the early 1950s when a security hearing was held to assess Oppenheimer's loyalty. His prewar associations with Communists and Communist sympathizers were linked to his reluctance to embrace new nuclear weapons. In 1954 the AEC voted four to one to deny him security clearance. The person most responsible for helping the United States develop the atomic bomb died in 1967, mistrusted by his country.

Figure 10.23 J. Robert Oppenheimer (1904–1967).

Fission produces many radioactive products, some with long half-lives. Finding a safe method and a secure site to store these wastes for hundreds of years is another unresolved issue. The cost of storage—plus the costs of constructing and maintaining nuclear plants that meet stringent safety standards—make nuclear energy expensive.

Nuclear Fusion. Scientists also have learned how to get small nuclei (usually isotopes of hydrogen) to join, or fuse, into a larger nucleus; this process is **nuclear fusion.**

Nuclear fusion is the major source of energy in the universe. Each second about 4–5 billion kg of the sun's mass, which is mostly hydrogen, changes into energy that streams to earth and other parts of our solar system. The reaction is:

$$4\,{}^{1}_{1}H \longrightarrow {}^{4}_{2}He + 2\,{}^{0}_{+1}e + energy$$

The two particles represented as ${}^{0}_{+1}e$ are positrons.

Scientists are trying to use nuclear fusion to produce electricity. A major problem is that the nuclei have to be very hot—10–100 million °C—to overcome their electrical repulsion and fuse. No known materials can contain fuel that hot. One solution is to use powerful electromagnetic fields to contain the fuel in a vacuum—a sort of invisible "bottle."

Because fusion has advantages as a potential energy source, research continues despite these difficulties. For one thing, fusion releases far more energy per gram of fuel than does fission (Figure 10.191). The fuel supply (hydrogen-2 and hydrogen-3 in water) is abundant. And fusion produces less radioactive waste than does fission. Most energy experts, however, don't expect fusion to be a significant energy source until at least 2050.

> In some sort of crude sense which no vulgarity, no humor, no overstatement can quite extinguish, the physicists have known sin; and this is a knowledge which they cannot lose.
>
> *J. Robert Oppenheimer (1904–1967)*

SUMMARY

Unstable nuclei emit bits of mass and energy, an action called radioactivity. Natural radioisotopes emit alpha, beta, or gamma radiation. Alpha particles consist of two protons and two neutrons; beta particles are electrons; gamma rays have no mass and are the most penetrating.

Nuclear equations represent radioactive processes. The sums of the superscripts (mass numbers) on both sides of the equation are equal, as are the sums of the subscripts (atomic numbers).

Half-life is the time required for half the nuclei of a radioisotope to emit their radiation. Half-life is constant and used to estimate the age of objects.

Units of radioactivity include the roentgen, curie (Ci), rad, and rem. Radioactivity harms the body by disrupting electron arrangements in normal substances and producing ions. Damage to DNA produces mutations that can lead to cancer or birth defects. U.S. residents are exposed to an average of about 360 millirems per year from background and other radiation.

Gamma rays and X rays are used to treat cancer; the intense radiation destroys rapidly dividing cells. Nuclear medicine also includes the use of radiation to diagnose diseases, often by obtaining images of internal body parts. Techniques include X rays, CT scans, PET scans, and MRI scans.

Uranium-235 and plutonium-239 undergo nuclear fission, in which nuclei fragment into much smaller pieces. At very high temperatures, light isotopes—mainly those of hydrogen—fuse to make larger nuclei. Both fission and fusion release large amounts of energy as nuclear mass is converted into energy. Both processes have been used in bombs, and fission is used in nuclear power plants.

Nuclear Chemistry Chapter 10 233

KEY TERMS

Alpha particle (α) (10.1)
Beta particle (β) (10.1)
CT scan (10.5)
Curie (Ci) (10.4)
Gamma ray (γ) (10.1)
Half-life (10.3)

MRI scan (10.5)
Nuclear fission (10.6)
Nuclear fusion (10.6)
PET scan (10.5)
Rad (10.4)
Radioactivity (10.1)

Radioisotope (10.1)
Rem (10.4)
Roentgen (10.4)
X ray (10.5)

EXERCISES

Even-numbered exercises are answered at the back of this book.

Types of Radioactivity

1. Is your body radioactive? Explain.
2. What part of the atom is responsible for radioactivity?
3. Iodine-128 emits beta and gamma radiation. What, if anything, can you conclude about the radioactivity of iodine-127?
4. For many elements, some isotopes are radioactive and some are not. Explain why.
5. What is the electrical charge of (a) an alpha particle, (b) a beta particle, and (c) a gamma ray?
6. List alpha, beta, and gamma emissions in order of increasing mass.
7. Do you think radioisotopes that emit alpha particles tend to have large or small nuclei? Explain.
8. Which type of radioactivity is similar to X rays?
9. Which type of radioactive emission—alpha, beta, or gamma—is the most surprising to come out of the nucleus? Explain why.
10. If a strong magnet has no effect on the direction of a radioactive emission, what type of emission is it? Explain.
11. What is the relationship between the wavelength of radiation and its energy?
12. Arrange the following in order of increasing energy: (a) radio waves, (b) ultraviolet (UV) rays, (c) visible light, (d) gamma rays, (e) X rays.

Nuclear Equations

13. Complete the following nuclear equations:
    (a) $^{200}_{79}\text{Au} \longrightarrow ^{0}_{-1}e\, +$ _____
    (b) $^{228}_{91}\text{Pa} \longrightarrow ^{4}_{2}\text{He}\, +$ _____
    (c) $^{140}_{56}\text{Ba} \longrightarrow ^{0}_{-1}\beta + \gamma +$ _____
    (d) $^{197}_{83}\text{Bi} \longrightarrow ^{4}_{2}\alpha\, +$ _____

14. Complete the following nuclear equations:
    (a) $^{12}_{5}\text{B} \longrightarrow ^{0}_{-1}\beta\, +$ _____
    (b) $^{242}_{94}\text{Pu} \longrightarrow ^{4}_{2}\alpha\, +$ _____
    (c) $^{27}_{12}\text{Mg} \longrightarrow ^{0}_{-1}e + \gamma +$ _____
    (d) $^{210}_{84}\text{Po} \longrightarrow ^{4}_{2}\text{He} + \gamma +$ _____

15. Complete the nuclear equations below to identify which new transuranium element was synthesized in this way:
    (a) $^{238}_{92}\text{U} + ^{1}_{0}\text{n} \longrightarrow ^{0}_{-1}\beta\, +$ _____
    (b) $^{253}_{99}\text{Es} + ^{4}_{2}\text{He} \longrightarrow ^{1}_{0}n\, +$ _____
    (c) $^{246}_{96}\text{Cm} + ^{12}_{6}\text{C} \longrightarrow 4\, ^{1}_{0}n\, +$ _____
    (d) $^{241}_{95}\text{Am} + ^{4}_{2}\alpha \longrightarrow 2\, ^{1}_{0}n\, +$ _____

16. Write a nuclear equation to determine the substance that reacts with bismuth-209 to produce unnilseptium-262 and a neutron.

17. Write a nuclear equation to identify the radioisotope that by beta emission produces gold-197.

18. If scientists tried to synthesize element 114 from californium (Cf), they would need to accelerate nuclei of which element to react with Cf?

19. Carbon-14 is produced in the atmosphere when a neutron reacts with a nitrogen-14 nucleus. Write a nuclear equation for the reaction and identify what other particle (besides carbon-14) is produced.

Half-life

20. What happens to the half-life of a radioisotope when those atoms react with atoms of a different element to form molecules?

21. Which type of nucleus is more stable: carbon-14 or tritium (hydrogen-3)? (Hint: See Table 10.2.)

22. Would you use radiocarbon dating to estimate the age of a leg bone suspected of being 800,000 years old? Explain.

23. Protactinium-234 has a half-life of 1 minute. Suppose you start with 8.0 g of this isotope.
    (a) How long does it take until 1.0 g of protactinium-234 remains?

(b) When 1.0 g of protactinium-234 remains, what happened to the other 7.0 g of material you started with?

24. Each g of carbon in your body has enough radioactive carbon-14 to emit 16 counts of radioactivity per minute while you are alive. If you die in the year 2050, in what year would your remains emit 2 counts per minute per g of carbon?

25. Why doesn't a radioisotope with a half-life of 1.3 billion years give an accurate estimate of the age of an object several hundred years old?

26. For 64.0 g radioisotope with a half-life of 10 hours, calculate the amount of radioisotope present at 10, 20, 30, 40, and 50 hours. Then plot g radioisotope *versus* hours and draw a curve through the points from your calculations.

27. Look at the shape of the curve for exercise 26. How long would it take for 25% of that radioisotope to emit its radioactivity: **(a)** 5 hours, **(b)** more than 5 hours, or **(c)** less than 5 hours?

28. What is the half-life of a radioisotope if after 10.0 days a 10.0 g sample decays to 2.5 g of original radioisotope?

29. What percentage of a sample of radioisotope remains after four half-lives?

Units of Radioactivity

30. One curie (Ci) of radioactivity equals 3.7×10^{10} emissions per second. How many emissions per second are given off by 2.0 microcuries (μCi) of a radioisotope?

31. If a radioactive sample gives off 6.1×10^8 emissions per second, how many millicuries (mCi) does it contain?

32. Which unit of radioactivity is defined in terms of the number of ions produced when the radiation passes through air?

33. Which unit of radioactivity is a measure of the effect of radiation on the body?

Effects of Radioactivity on the Body

34. Which type of radioactivity—alpha, beta, or gamma—is the most dangerous from outside the body? Explain why.

35. Which type of radioactivity—alpha, beta, or gamma—is the most dangerous from inside the body? Explain why.

36. Indoor air in some homes has hazardous levels of radon-222, which is inhaled into the lungs. Exposure to this radioisotope over 20–30 years may cause lung cancer. Predict the type of radioactive emission from radon-222.

37. For most people, what is the largest source of background radioactivity?

38. Is a total dose of 300 mrems of radioactivity likely to cause more harm when it occurs in one exposure or in multiple exposures of smaller doses? Explain.

39. Why is bone a common place for radioactive strontium-90 to accumulate? [Hint: Look at the elements in the same group as Sr in the periodic table.]

40. What substance in the nuclei of cells is especially vulnerable to damage by radioactivity?

41. The watch-dial painters suffered from the ingestion of radium, an alpha emitter. Would it have made a difference if radium were a beta emitter instead? Explain.

42. Why are cancer cells especially likely to be destroyed during a short exposure to a high dose of radiation?

Medical Uses of Radioactivity and Other Radiation

43. Examine the effects of whole-body exposure to 50–200 rems of radioactivity over a short period of time (Table 10.4). Why does radioactivity cause these particular effects?

44. What kinds of emitters—alpha, beta, or gamma—are used to irradiate internal cancer tumors? Explain why.

45. Compare the dosage of radioactivity used to treat cancer with the dosages that harm the body (Table 10.4). Why doesn't the treatment kill cancer patients?

46. Should radioisotopes taken internally for medical diagnosis have long or short half-lives? Explain why.

47. Would a radioisotope with a half-life of 10 seconds be suitable to be taken internally for medical diagnosis? Explain why.

48. Sodium-24 is taken internally to detect constrictions or obstructions in blood vessels. Predict the type of radioactive emissions this radioisotope gives off.

49. Identify which one of the following could *not* be accelerated by a cyclotron: **(a)** proton, **(b)** neutron, **(c)** alpha particle, **(d)** oxygen nucleus.

50. Which particle in exercise 49 could most easily enter a nucleus without first being accelerated? Explain why.

51. Identify which of the following is probably the safest: **(a)** X rays, **(b)** CT scans, **(c)** PET scans, **(d)** MRI scans.

52. Of the techniques listed in exercise 51, which one requires that the patient ingest radioactive material?

53. What are some advantages of CT scans compared to conventional X rays?

Nuclear Fission and Fusion

54. Where does the energy come from in nuclear fission?
55. Where does the energy come from in nuclear fusion?
56. What size nucleus has the least energy and is the most stable?
57. What particle triggers fission when it enters the nucleus of an appropriate radioisotope?
58. Which radioisotopes are the main fuels for nuclear fission?
59. Why does an amount of fissionable fuel smaller than the critical mass not sustain a chain reaction?
60. What is the function of control rods in a nuclear reactor?
61. Where does nuclear fusion occur naturally?
62. List three advantages and one disadvantage of nuclear fusion as a potential energy source compared with nuclear fission.

DISCUSSION EXERCISES

1. Would the world be a better place in which to live if radioactivity didn't exist? Explain why.
2. Do you think the earth is more radioactive or less radioactive today than it was 10 million years ago? Explain why.
3. What are some possible sources of error in using half-life to estimate the ages of ancient bones?
4. What are the main sources of radioactivity you are exposed to?

II

ORGANIC CHEMISTRY

11

Introduction to Organic Chemistry

1. Why is carbon unique? Why does it form so many different compounds?
2. What are the general shapes of organic molecules? How do you use VSEPR theory and hybrid orbitals to predict those shapes? How are the shapes written in two dimensions?
3. What are the three major types of elemental carbon?
4. What are functional groups? Why are they the basis for classifying organic compounds?

The Vital Force

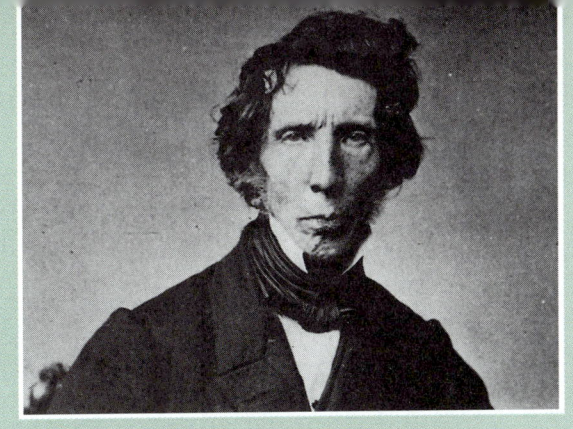

Figure 11.1 Friedrich Wöhler (1800–1882).

Living things are exquisite chemical systems, and we don't know exactly what makes them alive. No one yet has mixed some chemicals in a test tube and produced something that crawled out.

In our experience, new life comes only from something that's already alive.

Why is that so? For a long time, scientists believed that part of the answer was in the chemicals. Perhaps the chemicals in living things had unique, special qualities that helped make organisms alive. Two centuries ago, this life-giving quality was called the "vital force."

In 1828 Friedrich Wöhler, a German chemist (Figure 11.1), produced urea by heating ammonium cyanate, a mineral material that presumably lacked the "vital force." The reaction was:

$$NH_4OCN \xrightarrow{heat} H_2NCONH_2$$

ammonium cyanate   *urea*

He wrote in a letter to a friend, "I can no longer, as it were, hold back my chemical urine; and I have to let out that I can make urea without needing a kidney, whether of man or dog."

How could this happen? Urea is in urine, so it presumably possessed the "vital force." How could it form simply by heating something else that lacked the "vital force"? Wöhler's experiment, and many others that followed, eventually convinced scientists that there is no such thing as a "vital force" that inhabits the chemicals in organisms.

The chemistry of living organisms is now called biochemistry (see Chapter 19).

We usually think of living things—such as kangaroos, caterpillars, roses, or skunks—in terms of what they look like and what they do. We're less likely to think of them as chemical systems. But what is it that accounts for a kangaroo's pouch, a caterpillar's cocoon, the waxy petals on a rose, and a skunk's special perfume? These things all come from the organism's chemistry.

The study of chemicals in organisms was once called organic chemistry. Besides being part of the life process, however, those chemicals are special in another way: most of them are compounds of carbon. Indeed, the crucial molecules of life—carbohydrates, fats, proteins, nucleic acids, and vitamins—all are carbon compounds. Now **organic chemistry** is defined as the study of carbon-containing compounds, regardless of their occurrence in organisms.

That takes in a lot of territory, for more than 10 million organic substances are known, and this list grows by about 100 per week. There seems to be almost no limit to the number that can exist.

In order to understand the chemistry of living things, you first need to learn about organic chemistry. Organic compounds are not only essential to life as we know it, but they also provide many useful products, such as clothing, drugs, dyes, cosmetics, food additives, structural materials, and fuel (Figure 11.2).

Figure 11.2 Coal is made mostly of carbon.

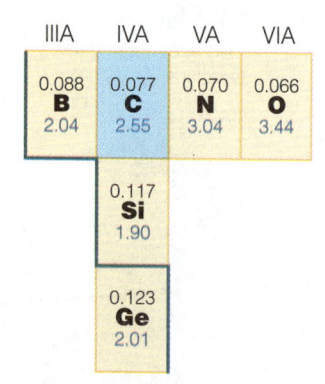

Figure 11.3 Comparison of carbon with other nearby elements in the periodic table. The average atomic radius (top number) and electronegativity (bottom number) are listed for each element.

11.1 CARBON: A UNIQUE ELEMENT

What's so special about carbon? Why is an entire branch of chemistry devoted to compounds of this one element, while another branch (inorganic chemistry) deals with the other 108 elements?

Carbon is unique because it forms stable covalent bonds with other carbon atoms; this allows carbon to form a remarkable variety of molecules. Carbon atoms link together to form chains, branched chains, and rings of many different sizes. No other element can bond to other atoms of the same element to make such large and diverse molecular networks.

Carbon has just the right combination of intermediate electronegativity (Figure 11.3) and relatively small size to form strong, stable covalent bonds with other carbon atoms. You might expect that silicon, which is in the same group in the periodic table, would have similar properties. Silicon does form some large molecules (see Chapter 18), but its lower electronegativity and greater size make it less successful than carbon in this regard.

Table 11.1 Comparison of Organic and Inorganic Carbon Compounds

Compound (Formula)	Melting Point (°C)	Boiling Point (°C)	Forces Between Particles	Conductivity in Water Solution
Organic	low	low	various	no**
acetic acid (CH_3CO_2H)	17	118	hydrogen bonds	yes
methyl cyanide (CH_3CN)	−46	82	dipole–dipole	no
nicotine ($C_{10}H_{14}N_2$)	−79	247	dipole–dipole	no
octane (C_8H_{18})	−57	126	London	no
Inorganic	high	high	ionic bonds	yes
lithium carbonate (Li_2CO_3)	723	1310*	ionic bonds	yes
potassium cyanate (KOCN)	700–900*		ionic bonds	yes
sodium cyanide (NaCN)	564	1496	ionic bonds	yes

* Decomposes at this temperature.
** Except organic acids and bases; furthermore, many organic compounds aren't polar enough to dissolve effectively in H_2O.

A few carbon compounds aren't usually considered part of organic chemistry. They include carbon dioxide (CO_2), carbon monoxide (CO), carbon disulfide (CS_2), and compounds in which carbon is part of a polyatomic ion, especially carbonate (CO_3^{2-}), bicarbonate (HCO_3^-), carbide (C_2^{2-}), cyanide (CN^-), and the material Wöhler used—cyanate (OCN^-). But almost all other carbon compounds are included in organic chemistry.

Ionic compounds containing carbon have very different properties than molecular carbon compounds (Table 11.1). For example, ionic carbon-containing compounds melt and boil only at temperatures higher than 400°C because of their strong ionic bonds. Most organic substances, however, have weaker attractions between molecules, so they melt and boil at much lower temperatures.

11.2 SHAPES OF ORGANIC MOLECULES

The shape of an organic molecule helps determine its properties. For example, the symmetric shape of a CCl_4 molecule makes it nonpolar despite having four polar C—Cl bonds. Since CCl_4 is nonpolar, it has only weak London forces between its molecules; this makes CCl_4 a liquid instead of a solid at room temperature.

Organic molecules get their shapes mostly from the way atoms bond to carbon. Carbon has four valence electrons ($\cdot \dot{C} \cdot$), so it needs four more electrons to become stable, and forms four covalent bonds to do so. We discussed earlier how to use valence-shell electron-pair repulsion (VSEPR) theory to predict the shapes of molecules (Section 4.5). According to this idea, pairs of valence electrons, whether involved in a covalent bond or not, repel each other and spread as far apart from each other as possible.

Using VSEPR theory, you would predict that carbon bonded to four other atoms would take the shape of a tetrahedron, with carbon in the center and the other four atoms at the corners (Figure 11.4). Table 11.2 summarizes the shapes of carbon compounds predicted by VSEPR theory.

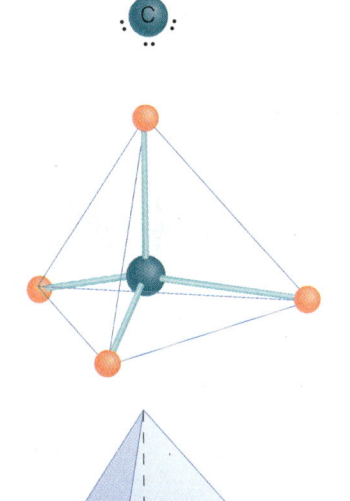

Figure 11.4 Valence-shell electron pairs of carbon are as far apart from each other as possible. As a result the four atoms bonding to carbon form the shape of a tetrahedron.

Table 11.2 Shapes of Carbon Compounds

Bonds Around Carbon	Structural Formula	Electron Dot Structure	Space-Filling Model	Shape Around Carbon
4 single bonds	H—C(H)(H)—H	H:C:H with H above and below	(CH₄ model)	tetrahedral (109.5° angles)
2 single bonds, 1 double bond	H₂C=O	H:C(H):O (with lone pairs on O)	(H₂CO model)	triangular (120° angles)
1 single bond, 1 triple bond	H—C≡N	H:C:::N:	(HCN model)	linear (180° angles)

Hybrid Orbitals. VSEPR theory correctly predicts the shapes of organic molecules. But it doesn't explain why compounds with carbon-to-carbon double bonds are more reactive than those having only single bonds. The theory also doesn't fit well with the electron configuration for carbon:

$$1s^2 2s^2 2p_x^1 2p_y^1$$

Two valence electrons (in principal energy level 2) are in the 2s orbital, which is spherical (Section 3.5); one valence electron is in each of two 2p orbitals, which are double lobes at right angles to each other. Yet experiments with carbon compounds show that all four bonding atoms are located equidistant around carbon with angles of 109.5° to each other.

How can we account for this discrepancy between experimental results and theory? The answer is to revise the theory to fit the results. A useful idea is to picture the 2s orbital and three 2p orbitals of carbon blending together to produce four new orbitals with a new, identical shape. This process is called *orbital hybridization,* and the new orbitals are called **hybrid orbitals.** Since they form from one s and three p orbitals, each of the four new orbitals is called an *sp³* hybrid orbital.

Figure 11.5 shows the formation and shapes of *sp³* orbitals. Like *p* orbitals, they contain two lobes. Unlike *p* orbitals, one lobe is much larger than the other. Each *sp³* orbital in carbon contains one valence electron and forms a covalent bond by overlapping an orbital of the other bonding atom that contains one electron. The *sp³* orbitals are arranged at 109.5° angles from the nucleus, which matches the way four bonded atoms actually fit around carbon (Table 11.2).

Single covalent bonds of carbon atoms are called **sigma (σ) bonds.** Atoms joined by sigma bonds rotate freely around the axis of the bond (Figure 11.6).

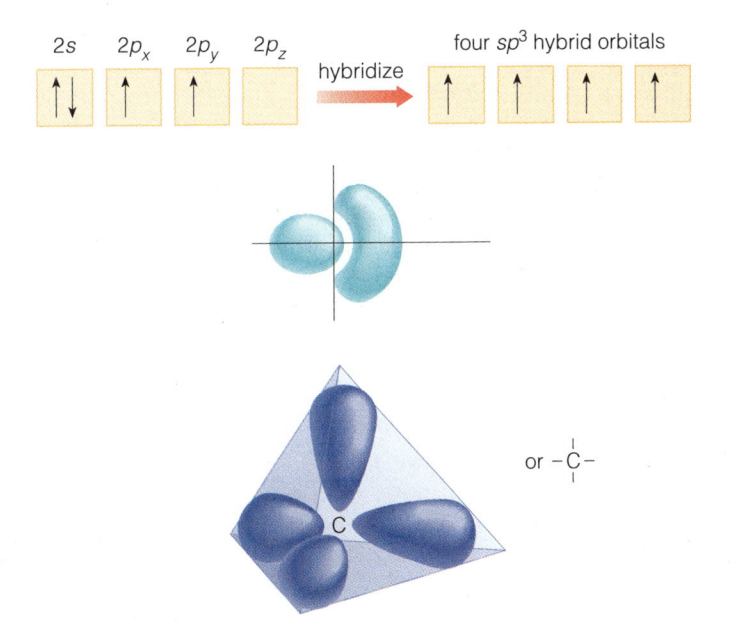

Introduction to Organic Chemistry   Chapter 11   **243**

Figure 11.5 The 2s and three 2p orbitals of carbon hybridize to form four sp^3 hybrid orbitals (top), whose boundaries each consist of two lobes (center) with the larger lobes projecting out from the nucleus to form the shape of a tetrahedron (bottom).

EXAMPLE 11.1 Predict the arrangements around carbon atoms in each of the following molecules:

(a) H—C(H)(H)—H, (b) Cl—C(H)(Cl)—Cl, (c) H—C(H)(H)—C(H)(H)—H

SOLUTION In all three compounds, carbon atoms form single (sigma) bonds with four other atoms, so around each carbon is a tetrahedral arrangement with bond angles of 109.5°.

When carbon atoms form multiple (double or triple) bonds, the arrangement changes. Carbon joined to another atom by a double bond is bonded to only three atoms. According to VSEPR theory, the three bonded atoms are in the same plane as carbon and separated by angles of 120° (Table 11.2). Measurements show that this is indeed how such atoms are arranged in space.

How can we account for this shape in terms of carbon's orbitals? Since s and p orbitals don't produce 120° angles, and neither do sp^3 hybrid orbitals, some other arrangement is necessary.

Double-bonded carbon atoms form a different type of hybrid orbital. The $2s$ orbital in carbon blends with *two* (not all three) of the $2p$ orbitals to produce three hybrid orbitals, called sp^2 orbitals (Figure 11.7). The three sp^2 orbitals, each containing one electron, are arranged at 120° angles from the nucleus and in the same plane. The two lobes of the remaining, unhybridized $2p$ orbital (also containing one electron) are perpendicular to this plane.

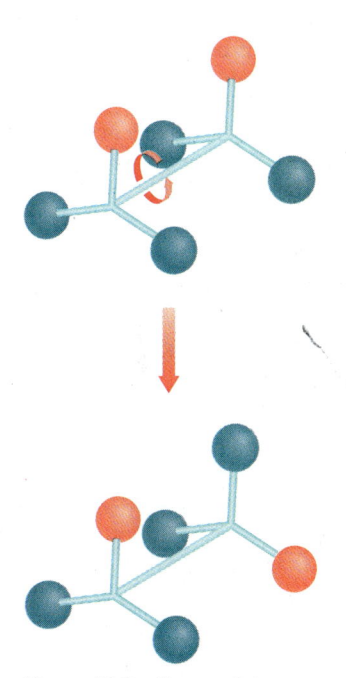

Figure 11.6 Atoms rotate freely around a single (sigma) bond.

244 Part II Organic Chemistry

In order to have side-to-side overlap of p orbitals, double-bonded carbon atoms have to be closer together than single-bonded atoms. Measurements confirm that this occurs.

Now picture what happens when carbon forms a double bond with another carbon (or some other atom) and single bonds to two other atoms. Carbon uses an *sp²* hybrid orbital to bond with each of the three other atoms; these are all sigma (single) bonds, with free rotation and bond angles of 120°. The second bond in the double bond comes from side-to-side overlap using the remaining *p* orbital that has one electron (Figure 11.7). Atoms cannot rotate around the axis of the double bond, for that would break the alignment necessary for side-to-side sharing between *p* orbitals.

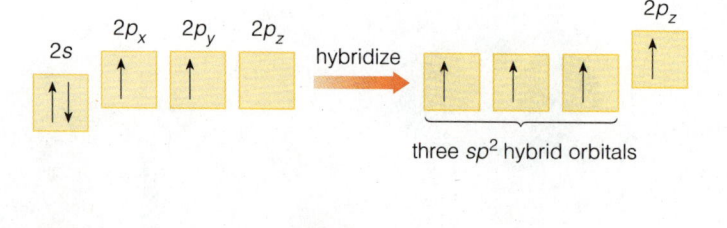

Figure 11.7 The 2s and two 2p orbitals hybridize to form three *sp²* hybrid orbitals, leaving one *p* orbital unhybridized. The shapes of the orbitals (lower left) account for the shapes of molecules that form by overlap of orbitals (lower right).

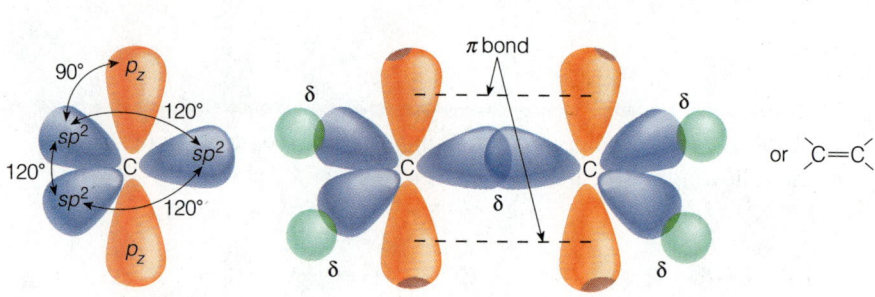

This side-to-side sharing, where no rotation occurs, is called a **pi bond (π bond)**, and the shared electrons are called **pi (π) electrons.** A double bond, then, consists of one sigma bond (from the direct overlap of *sp²* hybrid orbitals) and one pi bond (from side-to-side sharing between *p* orbitals).

Triple-bonded carbon atoms (and carbons with two double bonds) bond to only two atoms. According to VSEPR theory, the valence electrons are in only two regions, and maximum separation puts them on opposite sides of the carbon atom. This produces a bond angle of 180° and a linear arrangement (Table 11.2). Measurements show that this is the actual alignment.

We also can explain this shape by picturing each carbon forming two *sp* hybrid orbitals, with the two remaining *p* orbitals at right angles (90°) to the *sp* hybrid orbitals (Figure 11.8). Carbon uses its *sp* hybrid orbitals to form single (sigma) bonds with two other atoms. The remaining two bonds form by side-to-side overlap of both *p* orbitals. A triple bond, then, consists of one sigma bond (from the overlap of *sp* orbitals) and two pi bonds. Atoms cannot rotate around a triple bond because that would break the pi bonds.

Notice that you can use both VSEPR theory and hybrid orbitals to predict the shapes of organic molecules. Both give the same, correct answer.

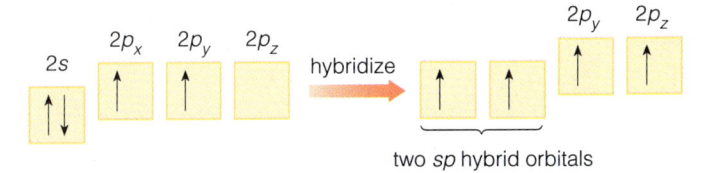

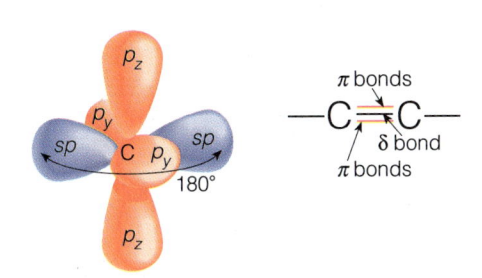

Figure 11.8 The 2s and one 2p orbital of carbon hybridize to form two sp hybrid orbitals, leaving two p orbitals unhybridized. The sp orbitals are 180° apart (lower left) and produce linear molecules (lower right).

EXAMPLE 11.2 Predict the bond angles around carbon atoms in the following compounds:

(a) H₂C=O, (b) H—C≡N, (c) H—C≡C—H, (d) H₂C=CH₂

SOLUTION (a) 120°, (b) 180°, (c) 180°, (d) 120°

11.3 REPRESENTING CARBON CHAINS AND RINGS

In most textbooks, including this one, you will see carbon chains shown in two dimensions with bonds at right angles (90°):

But remember that carbon bonded to four atoms actually has a tetrahedral shape with 109.5° angles. Figure 11.9 shows how single-bonded carbon chains look as represented by textbooks and by ball-and-stick models. Carbon also bonds to a variety of other atoms, as we discuss in Section 11.5.

Atoms joined by single (sigma) bonds rotate freely and can twist into a variety of shapes. A linear, five-carbon chain, for example, can have any of the

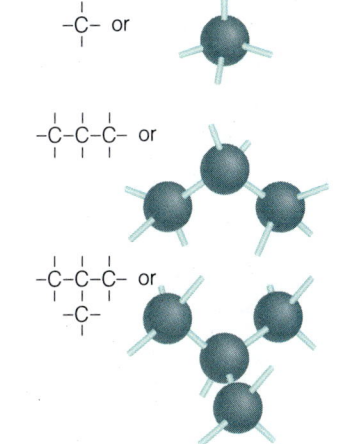

Figure 11.9 Representing carbon chains with two-dimensional structures (left) and with ball-and-stick models, in which sticks represent bonds to other atoms.

shapes shown in Figure 11.10. Other shapes occur, too. You could write the chain in several different ways. Some of them are:

$$-\overset{|}{\underset{|}{C}}-\overset{|}{\underset{|}{C}}-\overset{|}{\underset{|}{C}}-\overset{|}{\underset{|}{C}}-\overset{|}{\underset{|}{C}}- \quad \text{or} \quad -\overset{|}{\underset{|}{C}}-\overset{|}{\underset{|}{C}}-\overset{|}{\underset{|}{C}}-\overset{|}{\underset{|}{C}}- \\ \quad\quad\quad\quad\quad\quad\quad\quad\quad\quad\quad\quad\quad\quad -\overset{|}{\underset{|}{C}}- \quad \text{or}$$

$$\begin{array}{c} -\overset{|}{\underset{|}{C}}- \\ -\overset{|}{\underset{|}{C}}-\overset{|}{\underset{|}{C}}-\overset{|}{\underset{|}{C}}- \\ -\overset{|}{\underset{|}{C}}- \end{array} \quad \text{or} \quad \begin{array}{c} \quad\quad -\overset{|}{\underset{|}{C}}-\overset{|}{\underset{|}{C}}- \\ -\overset{|}{\underset{|}{C}}-\overset{|}{\underset{|}{C}}-\overset{|}{\underset{|}{C}}- \end{array}$$

If you can draw one continuous line through all the carbon atoms, the chain is linear.

These structures all represent the same chain as it actually exists in space. It is the same chain because, in each case, there is a sequence of five carbon atoms bonded to each other, with each carbon atom having its four bonded atoms attached in a tetrahedral arrangement.

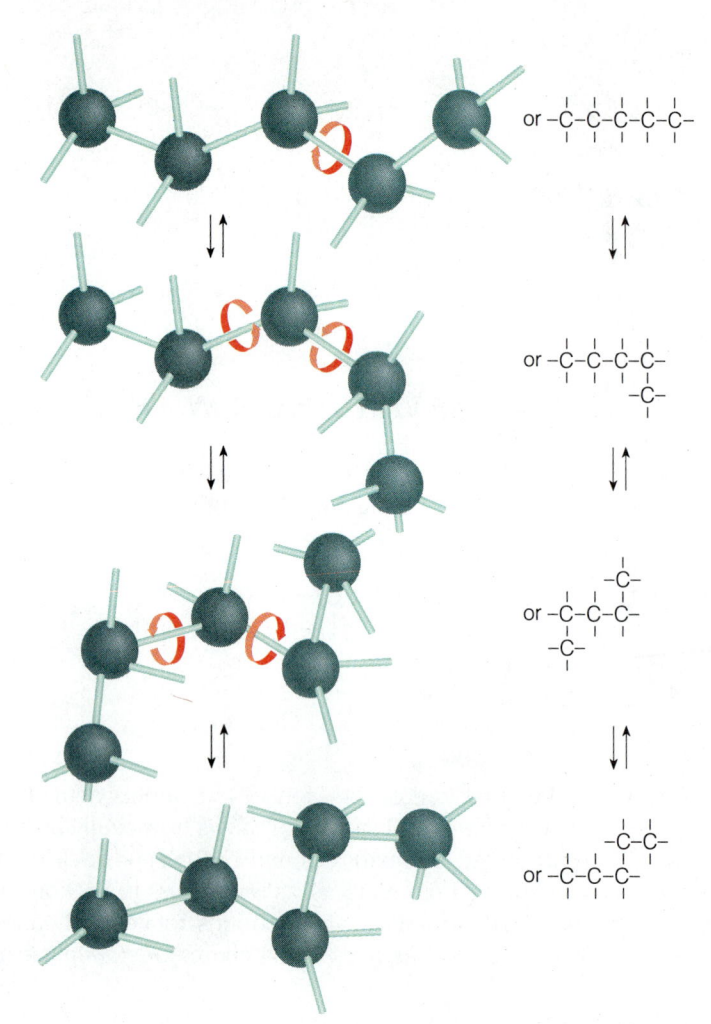

Figure 11.10 Different shapes of the same, linear, five-carbon chain.

With branched carbon chains, you also need to recognize what structures represent the same molecules. For example, you can write several different structures for this same, branched, six-carbon chain:

In all of these structures, you can draw a continuous line through five carbons. The sixth carbon is bonded to a next-to-end carbon in the linear part of the chain.

EXAMPLE 11.3 Identify which of the following represent the same chains:

(a), (b), (c), (d), (e)

SOLUTION Structures (c) and (e) are the same; both are linear chains of five carbon atoms. Structures (a), (b), and (d) are the same; they are four-carbon chains with a carbon branch at a next-to-end carbon atom.

Structures for carbon rings are usually written as regular polygons. Rings containing six carbon atoms are the most common; they are written as hexagons (Figure 11.11).

Again, the two-dimensional representation doesn't accurately show the actual shape of the molecule. In a regular hexagon, the angles are 120°, which doesn't match the 109.5° angles for tetrahedral carbon atoms. Six-membered carbon rings containing single bonds are not really flat; instead, they have a "puckered" arrangement to allow for the tetrahedral bond angles. Figure 11.11 shows the two most common conformations, the "chair" and the "boat."

Carbon atoms that form four bonds have as close to a tetrahedral arrangement as possible. So when you see flat, two-dimensional figures, keep in mind the shapes of these molecules as they actually exist. The shape of a molecule makes a crucial difference in how it functions. For example, enzymes that

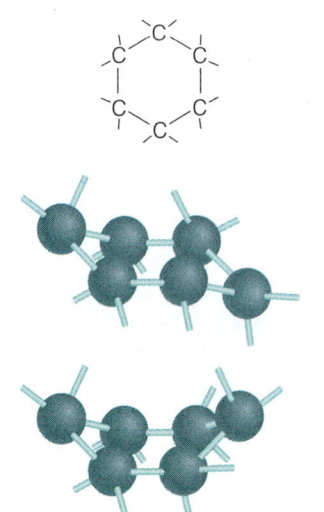

Figure 11.11 Although six-carbon ring compounds containing only single bonds are often shown as planar hexagons (top), they actually have tetrahedral (109.5°) bond angles and exist in the chair form (center) or, less commonly, in the boat form (bottom).

CHEMISTRY SPOTLIGHT

The Right Shape

Molecules must have just the right shape to function in your body. One example is growth hormone (GH), secreted by your pituitary gland. GH stimulates muscle, bone, and cartilage to grow. People who don't make enough GH often are short and small.

Human GH is very specific; it stimulates growth in humans, but not in other animals. Furthermore, growth hormones from other animals don't work in humans.

Why is human growth hormone so specific? First, it is a large protein molecule (Chapter 21) that folds into a complex, specific shape. In order to stimulate muscle, bone, and cartilage growth, GH must fit with certain molecules (called receptors) on the cells' surfaces. Human GH has just the right three-dimensional shape to bind to those receptors (Figure 11.12).

This elaborate interlocking of GH and receptors stimulates cells to grow. Growth hormones from other animals have slightly different shapes, so they don't fit with the receptors in human cells. Indeed, only human cells that have the correct receptor molecules bind and respond to human growth hormone.

Your body is full of chemical reactions that occur only because the reacting molecules have exactly the right shapes.

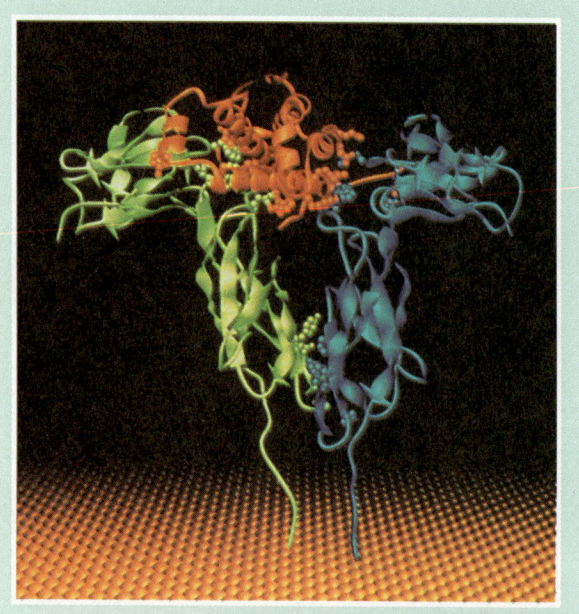

Figure 11.12 Model of human growth hormone (red) and two receptor molecules (green and blue) on the surface of muscle, bone, and cartilage cells. The hormone has to fit exactly to stimulate cell metabolism and growth.

convert food into energy, hormones that regulate our growth (see the Chemistry Spotlight), antibiotic drugs that kill infectious bacteria, and many other organic compounds work only when they have the right shape.

11.4 FORMS OF THE ELEMENT CARBON

Substances that contain only carbon usually aren't considered part of organic chemistry, but they do show how carbon bonds help determine a substance's

Introduction to Organic Chemistry   Chapter 11   249

properties. Carbon itself exists in three main forms. One is *graphite* (Figure 11.13, a slippery black material used in lubricants, "lead" pencils, shafts of expensive golf clubs, and artificial body parts (Figure 11.14).

Graphite is an array of flat, hexagonal carbon rings joined together with carbon atoms bonded to three—not four—other carbon atoms. These carbon atoms bond with three sp^2 hybrid orbitals. The unhybridized *p* orbital forms pi bonds. The pi electrons move freely within a ring, making graphite a good conductor of electricity. Clusters of joined rings act like tiny sheets, with only weak attractions between sheets. These sheets readily slide past each other, making graphite slippery.

Another form of carbon—*diamond*—could hardly be more different (Figure 11.13. Here each carbon atom bonds with sp^3 hybrid orbitals to four other carbon atoms in a tetrahedral arrangement, forming a huge, interconnected network of carbon atoms. Diamonds are hard, shiny, and clear. Unlike graphite, diamonds do not conduct electricity. In fact, the stable electron structure of their carbon atoms makes diamonds good electrical insulators.

Figure 11.13 Two forms of carbon, graphite and diamond.

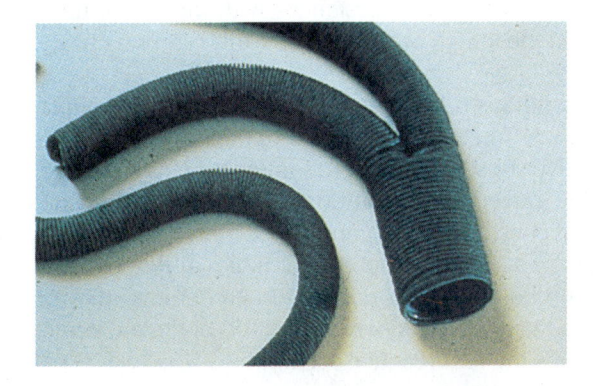

Figure 11.14 Artificial blood vessel segments coated with a thin layer of carbon, which reduces the risk of blood clots.

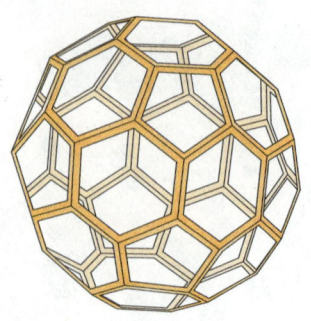

Figure 11.15 Structure of buckminsterfullerene.

In 1985 scientists discovered that carbon also exists as large, spherical molecules they named after Buckminster Fuller (1895–1983), the imaginative designer of the geodesic dome, which resembles this molecule. The most common form has 60 carbon atoms and is called buckminsterfullerene or buckyballs. With 12 five-carbon rings (shown as pentagons) interlaced with 20 six-carbon rings (hexagons), the structure looks like a hollow soccer ball (Figure 11.15). Carbon–carbon bonds are a mixture of single and double bonds.

The properties of fullerenes are just beginning to be explored. They may prove useful in manufacturing lubricants and electrical superconductors, and in forming diamond films.

11.5 TYPES OF ORGANIC COMPOUNDS: FUNCTIONAL GROUPS

Compounds of Carbon. Several factors account for the astonishing number of organic compounds. First and foremost, carbon forms a virtually unlimited variety of chains and branched chains. Such chains can contain thousands of carbon atoms. Carbon also forms single and multiple ring systems—and chains of various sizes and branching patterns can be attached to these rings.

In addition, carbon forms multiple covalent bonds, especially with other carbon atoms (double and triple bonds) and oxygen atoms (double bonds only). Besides oxygen (O), carbon forms covalent bonds with nonmetals such as hydrogen (H), nitrogen (N), sulfur (S), fluorine (F), chlorine (Cl), and, to a lesser extent, with bromine (Br), iodine (I), phosphorus (P), silicon (Si), and boron (B).

You can easily see, then, why there are millions of different organic compounds. Building organic molecules is like taking hundreds, or even thousands, of Tinker Toys or blocks and putting them together in all kinds of chains and rings of varying sizes and shapes. You can attach neighboring units by one, two, or even three links (bonds), and you also can add several other kinds of pieces (elements) in different places on each ring or chain.

Functional Groups. We can simplify things a bit by focusing on the most reactive parts of organic molecules, called **functional groups.** Most organic molecules contain regions of carbon and hydrogen atoms joined by single covalent bonds. This part of the molecule isn't very reactive. But other parts (the functional groups) are much more likely to react; they give organic molecules their distinctive properties. Indeed, the art and science of organic chemistry is based on understanding the chemistry of functional groups.

This makes it easier to classify organic molecules. Even if a molecule has tens, hundreds, or thousands of carbon atoms, you only need to find its key functional groups to know how it will react. Substances with the same functional group, even if they differ in the length or branching of their carbon chain, react similarly.

Chapters 12–17 on organic chemistry are organized by functional groups. We begin with compounds of carbon and hydrogen containing only single bonds. These are the least reactive substances. The action picks up as we move on to compounds with multiple carbon-to-carbon bonds, and then to organic compounds containing elements besides carbon and hydrogen.

Table 11.3 shows the structures of some important functional groups. The symbol, *R*, represents a carbon chain (with bonded hydrogen atoms) of any length, branched or not. You can usually think of *R* as the "rest of the molecule you can ignore," since it contains the less reactive section.

R comes from the German word Radikal, *which means "group."*

Table 11.3 Some Important Functional Groups

Group Name	Structure	Example
halide	*R—X (X=F, Cl, Br, I)	CHCl₃ chloroform
alkene	C=C	H₂C=CH₂ ethylene
alkyne	—C≡C—	H—C≡C—H acetylene
alcohol	R—O—H	CH₃—CH₂—O—H ethyl alcohol
ether	R—O—R'	CH₃—CH₂—O—CH₂—CH₃ diethyl ether
aldehyde	R—C(=O)—H	CH₃—C(=O)—H acetaldehyde
ketone	R—C(=O)—R'	CH₃—C(=O)—CH₃ acetone
carboxylic acid	R—C(=O)—O—H	CH₃—C(=O)—O—H acetic acid
ester	R—C(=O)—O—R'	CH₃—C(=O)—O—CH₃ methyl acetate
thiol (mercaptan)	R—S—H	CH₃—CH₂—S—H ethanethiol
amine	R—N(H)—H	CH₃—CH₂—N(H)—H ethyl amine
amide	R—C(=O)—N(H)—H	CH₃—C(=O)—N(H)—H acetamide

*R and R' represent a group of one or more carbons with bonded hydrogens.

EXAMPLE 11.4 Using the information in Table 11.2, identify and name the functional groups in the following (in color):

(a) glutamine (an amino acid)

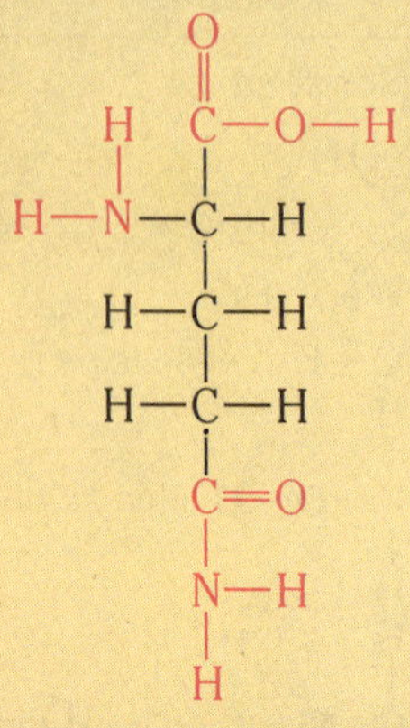

(b) glucose (a simple sugar)

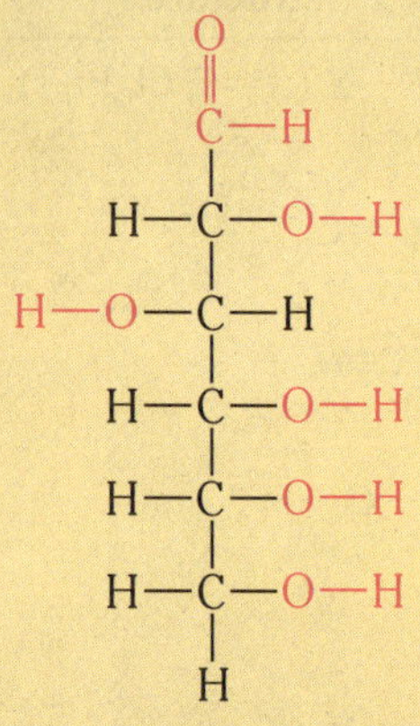

(c) geraniol (in rose oil)

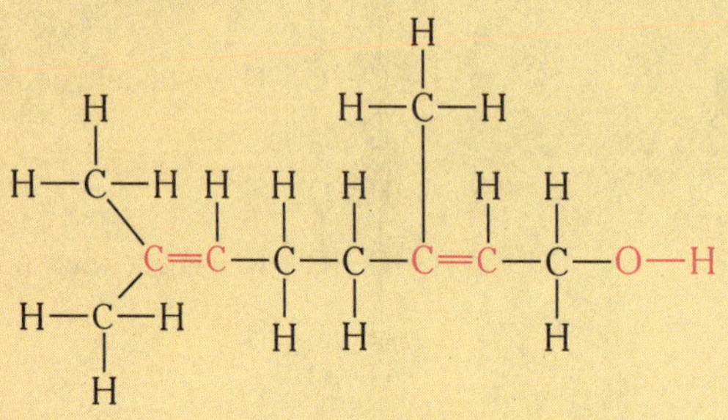

(d) halothane (an anesthetic)

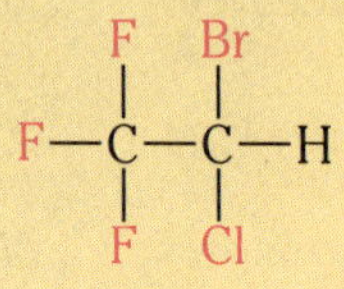

SOLUTION (a) carboxylic acid, amine, and amide groups
(b) one aldehyde and five alcohol groups
(c) one alcohol and two alkene groups
(d) F, Cl, and Br halide groups

> The most important hypothesis in all of biology . . . is that everything that animals do, atoms do. In other words, there is nothing that living things do that cannot be understood from the point of view that they are made of atoms acting according to the laws of physics.
>
> *Richard Feynman (1918–1988)*

SUMMARY

Organic chemistry is the study of carbon-containing compounds. Carbon is unique because it forms strong, stable covalent bonds with other carbon atoms.

Shapes of organic molecules are predicted by VSEPR theory (Section 4.5) or by the concept of hybrid orbitals of carbon formed between its 2s orbital and one, two, or three of its 2p orbitals. Carbon bonded to four other atoms has four sp^3 hybrid orbitals, a tetrahedral shape, and 109.5° angles between bonded atoms; all four bonds are single (sigma) bonds. Carbon double-bonded to one atom and single-bonded to two other atoms has three sp^2 hybrid orbitals and one p orbital, a flat, triangular shape, and 120° bond angles; the double bond consists of one sigma and one pi bond. Carbon triple-bonded to another atom has two sp hybrid orbitals and two p orbitals, a linear shape, and 180° bond angles; the triple bond consists of one sigma and two pi bonds.

Linear and branched structures are often written in two dimensions with carbon atoms bonded at 90° angles. Cyclic compounds are written as regular polygons. These structures don't show the actual bond angles present.

Elemental carbon occurs in three main forms: graphite, diamond, and fullerenes.

Organic compounds have a wide variety of carbon chains, rings, and atoms of other nonmetal elements. Functional groups are distinctive parts of organic molecules that provide most of the substance's physical and chemical properties.

KEY TERMS

Functional group (11.5)
Hybrid orbital (11.2)
Organic chemistry (11.1)
Pi (π) bond (11.2)
Pi (π) electron (11.2)
Sigma (σ) bond (11.2)

EXERCISES

Even-numbered exercises are answered at the back of this book.

Organic Chemistry

1. Does a molecule of urea produced in a cell differ in any way in its physical and chemical properties from a molecule of urea synthesized in a laboratory? Explain.
2. Can your body distinguish between a molecule of vitamin C in orange juice and a molecule of vitamin C in a tablet? Explain.
3. Why do "organic" chemistry and "organism" come from the same root word?
4. Distinguish between biochemistry and organic chemistry. Why is it necessary to study organic chemistry before studying biochemistry?
5. How is carbon unique among the elements?

Properties of Organic Compounds

6. Why do ionic compounds containing carbon have higher melting points than organic compounds? [Review Section 6.1 if necessary.]
7. Of the elements that commonly form covalent bonds with carbon (H, B, N, O, F, Si, P, S, Cl, Br, and I), which ones are more electronegative than carbon? [Review Section 4.6 if necessary.]
8. Of the elements that commonly form covalent bonds with carbon (see exercise 7), which ones are less electronegative than carbon?

Bonding and Shapes of Molecules

9. Is CF_4 polar or nonpolar? Explain. [Review Section 4.6 if necessary.]
10. Is CH_3F polar or nonpolar? Explain. [Review Section 4.6 if necessary.]
11. Identify which of the following are hybrid orbitals: (a) sp, (b) s, (c) sp^2, (d) p, (e) sp^3.
12. How does orbital hybridization "explain" the observed shapes of organic molecules containing only single bonds?
13. How does the shape of organic molecules with single bonds as predicted by hybrid orbitals (answer to exercise 12) fit with the shapes predicted by VSEPR theory?
14. What is the shape (tetrahedral, triangular, or linear) around the carbon atom in each of the following molecules?

(a)
$$H-\overset{\overset{H}{|}}{\underset{\underset{Cl}{|}}{C}}-Cl$$

(b)
$$H-\overset{\overset{H}{|}}{\underset{\underset{H}{|}}{C}}-\overset{\overset{H}{|}}{\underset{\underset{H}{|}}{C}}-\overset{\overset{H}{|}}{\underset{\underset{H}{|}}{C}}-H$$

(c)
$$\overset{H}{\underset{H}{\diagdown}}C=C\overset{H}{\underset{H}{\diagup}}$$

(d)
$$\overset{O}{\underset{Cl\quad Cl}{\overset{\|}{C}}}$$

15. What are the bond angles around each of the carbon atoms in exercise 14?
16. What type of hybrid orbitals occur around each of the carbon atoms in exercise 14?

17. List the number of (a) sigma bonds and (b) pi bonds formed by each carbon atom in each compound in exercise 14.
18. Write structural formulas (like those shown in exercise 14) for the following organic compounds: (a) CCl_4, (b) CH_2O, (c) CH_5N, (d) C_2H_4.
19. From the structural formulas for the compounds in exercise 18, predict the shape and bond angles around each carbon atom.
20. Which compound—C_2H_4 or C_2H_6—contains one pi bond?
21. Do the orbitals overlap more completely in (a) sigma bonds or (b) pi bonds?
22. Do atoms rotate freely about sigma bonds? Do they rotate about pi bonds? Explain both answers.
23. Which of the following do *not* represent a linear carbon chain?

(a) —C—C—C—
         |
        —C—

(b) —C—C—C—
         |
        —C—

(c) —C—C—
    |
   —C—
    |
   —C—

(d) —C—C—
   —C—C—

(e) —C—
    |
    C
   / \
  C   C

(f) —C—
    |
   —C—C—
    |
   —C—

24. In exercise 23, which structures are different forms of the same, linear four-carbon chain? How many other kinds of four-carbon chains are shown?
25. Would a sample of butane,

H H H H
| | | |
H—C—C—C—C—H,
| | | |
H H H H

actually have molecules with each of the linear carbon-chain shapes shown in exercise 23? Would they have any other shapes?
26. Why is a ring of six carbon atoms, joined by single covalent bonds, not flat?
27. Use an appropriate geometric figure to represent C_5H_{10}, which consists of a ring of five carbon atoms, each bonded to two hydrogen atoms.
28. An equilateral triangle has angles of 60°, so a ring of three carbon atoms has angles of 60°. Do you expect such a ring to be more stable or less stable than a five-carbon ring, which has angles of 108°? Why?

29. Both graphite and buckminsterfullerene contain carbon-to-carbon double bonds. What kind of hybrid orbitals do such carbon atoms have?

Forms of the Element Carbon

30. We can think of a diamond as being a single, huge molecule. Does this also apply to a piece of graphite?
31. Which of the three forms of carbon contains five-carbon rings?
32. Which of the three forms of carbon contains atoms bonded only at 109.5° angles using only sigma bonds?

Functional Groups

33. Of the elements listed in Section 11.5 as commonly forming covalent bonds with carbon, which ones form more than one covalent bond to become stable? How many bonds do each of these elements form?
34. Examine the functional groups in Table 11.3. Which groups form hydrogen bonds with each other? What effect does this have on their melting points, boiling points, and solubility in water? [Review Sections 6.1, 6.2, and 7.2 if necessary.]
35. In addition to carbon and hydrogen, amine functional groups contain the element _____, mercaptan (thiol) functional groups contain the element _____, and alcohol functional groups contain the element _____.
36. Why are functional groups so important in organic chemistry?
37. In Table 11.3, what does the symbol R (or R') represent?
38. Use R to represent the structures of the following in simplified form:

(a)
H H
| |
H—C—C—Cl
| |
H H

(b)
H H H
| | |
H—C—O—C—C—H
| | |
H H H

(c)
H H H O H H
| | | || | |
H—C—C—C—C—O—C—C—H
| | | | |
H H H H H

(d)
H O
| ||
H—C—C—O—H
|
H

39. Use Table 11.3 to name all the functional groups in the structures listed in exercise 38.

40. Identify and name all the functional groups in the following:

(a) CHCl₃
chloroform

(b)
$$\begin{array}{c} O \\ \parallel \\ C-H \\ | \\ H-C-H \\ | \\ H-C-O-H \\ | \\ H-C-O-H \\ | \\ H-C-O-H \\ | \\ H \end{array}$$
deoxyribose (in DNA)

(c)
$$\begin{array}{c} O \\ \parallel \\ H-N-C-N-H \\ | \quad\quad | \\ H \quad\quad H \end{array}$$
urea

(d)
$$\begin{array}{c} H \quad\quad\quad\quad\quad\quad\quad H \\ \diagdown \quad\quad O \quad\quad | \\ C=C-\overset{\parallel}{C}-O-C-H \\ \diagup \quad | \quad\quad\quad\quad | \\ H \quad H-C-H \quad\quad H \\ \quad\quad | \\ \quad\quad H \end{array}$$
methyl methacrylate (used to make Lucite and Plexiglas)

DISCUSSION EXERCISES

1. Do you think it will ever be possible to synthesize life by combining the appropriate chemicals? Explain.

2. For each of the functional groups in Table 11.3, list the strongest type of force between like molecules. [Recall that the three types of forces between molecules are London forces, dipole–dipole interactions, and hydrogen bonds (Section 6.1).]

12

Alkanes and Alkyl Halides

1. What are the names and uses of some alkanes?
2. How do you write condensed structural formulas for organic compounds?
3. What is structural isomerism?
4. What are the physical properties and typical chemical reactions of alkanes?
5. What are the main natural sources of alkanes?
6. What are the names, properties, and uses of some alkyl halides?

Hydrocarbons and Skin Care

Skin is the largest organ in the body, weighing 4–5 kg and having a thickness of 5–11 mm. The top layer is the *stratum corneum* (Figure 12.1), a protective covering of dead cells. Sebaceous glands exude oily sebum, which coats the stratum corneum and helps maintain moisture.

Skin gets dirty, dry, cracked, irritated, or diseased when you don't give it some care. One way to help is to use cleansing creams, which are mixtures of hydrocarbons (compounds composed only of carbon and hydrogen). Two common materials in cleansing creams are mineral oil and petroleum jelly (one brand is Vaseline). Cold cream is a mixture of borax and beeswax, which has the structure:

$$CH_3(CH_2)_{14}-\overset{\overset{\displaystyle O}{\|}}{C}-O-(CH_2)_{29}CH_3$$

Although beeswax contains an ester functional group (see Table 11.3), 45 of its 46 carbons are hydrocarbon material; this makes beeswax nonpolar. Unlike water, nonpolar cleansing creams dissolve oil and grease and are very effective in removing heavy, oil-based makeup. Once you wipe off the resulting solution with a tissue, your skin is clean.

Moisturizing products supplement the natural action of sebum to help skin stay swollen with water. *Emollients,* for example, coat skin with a waterproof layer that prevents water from escaping. Emollients can contain a wide variety of hydrocarbons and other nonpolar compounds, including petroleum products (mineral oil, petroleum jelly), animal oils (mink, lanolin), and plant oils (avocado, sesame, almond, jojoba). All serve as barriers to prevent drying and to make skin feel smooth and supple.

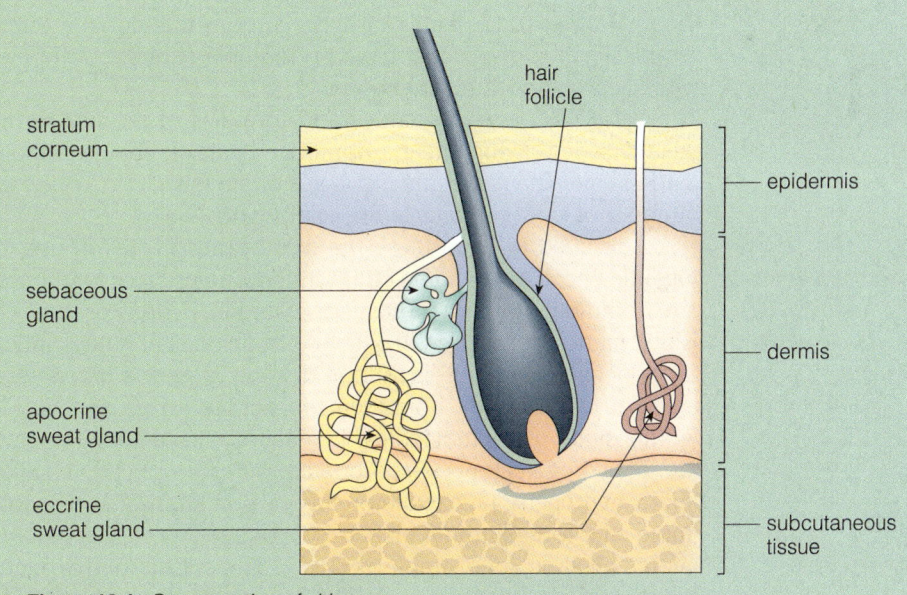

Figure 12.1 Cross section of skin.

What do the gasoline in your car (Figure 12.2), a dab of Vaseline, and a puff of natural gas have in common? Certainly not their physical states; one is a solid, one is a liquid, and one is a gas. It turns out that all three are *hydrocarbons,* compounds containing only hydrogen and carbon.

As you will see, several different types of hydrocarbons exist. The three materials above contain only single covalent bonds. This makes them members of a group of hydrocarbons called *alkanes.*

Figure 12.2 Crude oil is the source of gasoline.

12.1 ALKANES

Hydrocarbons are organic compounds that contain only hydrogen and carbon. **Alkanes** are a family of hydrocarbons that contain only single covalent bonds. Hydrocarbons that have no multiple (double or triple) bonds also are called **saturated hydrocarbons.**

Table 12.1 lists the names and formulas of the first ten members of the alkane series having linear, unbranched carbon chains. Notice that the names end in *-ane.* The rest of the name is a prefix (shown in color) that indicates the number of carbon atoms in the compound.

You need to learn these prefixes because you will use them throughout organic chemistry to name compounds. The prefixes for numbers 5 through 10 are commonly used in words. For example, *pent-* is used in pentagon (a five-sided figure), *oct-* is used in octave (a span of eight notes) and octet (a group of eight people), and *dec-* is used in decade (a period of ten years).

You can also see from their **molecular formulas** (which show only the composition of the molecules) that alkanes have a regular pattern in the number of hydrogen atoms per carbon atom (Table 12.1). Each alkane has twice as many hydrogens, plus two more, as carbon atoms. In other words, the general molecular formula of an alkane is C_nH_{2n+2}, where *n* is the number of carbon atoms. Hexane, for example, is C_6H_{14}. The reason for this pattern is easy to see in linear alkanes, where each interior carbon atom is bonded to two hydrogen atoms and both end carbons have one extra hydrogen (a total of three hydrogens).

For numbers 5 through 10, these are the same prefixes you use in naming covalent compounds (Section 4.4; Table 4.6).

Table 12.1 Ten Linear Alkanes

Name	Molecular Formula	Structural Formula	Melting Point (°C)	Boiling Point (°C)	State at 25°C
methane*	CH_4	H—C(H)(H)—H	−183	−162	gas
ethane	C_2H_6	H—C(H)(H)—C(H)(H)—H	−183	−89	gas
propane	C_3H_8	H—C—C—C—H (with H's)	−190	−42	gas
butane	C_4H_{10}	H—C—C—C—C—H (with H's)	−138	0	gas
pentane	C_5H_{12}	H—C—C—C—C—C—H (with H's)	−130	36	liquid
hexane	C_6H_{14}	H—C—C—C—C—C—C—H (with H's)	−95	69	liquid
heptane	C_7H_{16}	H—C—C—C—C—C—C—C—H (with H's)	−91	98	liquid
octane	C_8H_{18}	H—C—C—C—C—C—C—C—C—H (with H's)	−57	126	liquid
nonane	C_9H_{20}	H—C—C—C—C—C—C—C—C—C—H (with H's)	−51	151	liquid
decane	$C_{10}H_{22}$	H—C—C—C—C—C—C—C—C—C—C—H (with H's)	−30	174	liquid

* Alkanes are named by the appropriate prefix (in color) and the suffix -ane.

Cyclic alkanes, however, have two fewer hydrogens per carbon. Their general molecular formula is C_nH_{2n}. Ring carbons in a cyclic alkane each bond to two hydrogens; unlike linear alkanes, they have no "end" carbons bonded to three hydrogens. Figure 12.3 compares formulas of the linear and cyclic six-carbon alkanes.

hexane (C₆H₁₄)

cyclohexane (C₆H₁₂)

Figure 12.3 Formulas of a six-carbon linear alkane (top) and a cyclic alkane (bottom).

Unlike molecular formulas, **structural formulas** show the arrangements of atoms and their covalent bonds. Table 12.1 shows structural formulas for ten alkanes.

EXAMPLE 12.1 Predict the molecular formula for (a) a linear hydrocarbon and (b) a cyclic hydrocarbon, each having 20 carbon atoms.

SOLUTION (a) $C_{20}H_{42}$, (b) $C_{20}H_{40}$

12.2 STRUCTURAL ISOMERISM

Alkanes with four or more carbon atoms exist in both linear and branched forms. **Structural isomers** are two or more compounds that have the same molecular formula but different structural formulas. They are not just different shapes of the same molecule (Section 11.3); they are different molecules that cannot be interconverted by rotation around single bonds.

The simplest example is butane, which has two isomers (Figure 12.4). Both have the molecular formula C_4H_{10}, but they differ in their bonding structures. Notice the tetrahedral arrangement around each carbon atom; the 109.5° bond angles in alkanes result from the single carbon bonds using sp^3 hybrid orbitals (Section 11.2).

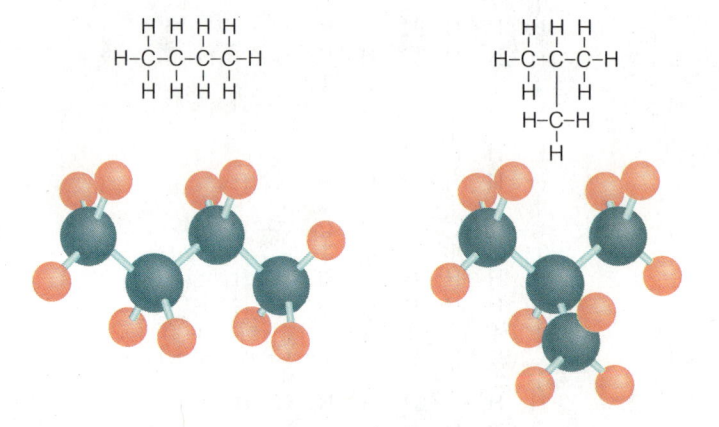

Figure 12.4 Structural formulas (top) and ball-and-stick models (bottom) of the two structural isomers of butane, C_4H_{10}.

You can write other structural formulas for butane that have different shapes because atoms rotate about single bonds. Although other forms may look different from the two in Figure 12.4, each is one of the two isomers of butane.

Because they have different structures, structural isomers have slightly different properties. Branched alkanes, for example, typically have lower melting and boiling points than their linear counterparts because branched molecules cannot pack as closely together. As a result, they have weaker London forces between molecules and thus vaporize and melt more readily than linear isomers of the same molecular weight.

Alkanes and Alkyl Halides Chapter 12   261

EXAMPLE 12.2 Identify each of the following conformations as the linear or the branched isomer of butane:

(a)
```
    H H H
    | | |
H—C—C—C—H
    | | |
    H H
    |
H—C—H
    |
    H
```

(b)
```
        H
        |
    H—C—H
    H | H
    | | |
H—C—C—C—H
    | | |
    H H H
```

(c)
```
        H
        |
    H—C—H
    H |
    | |
H—C—C—H
    | |
    H |
    H—C—H
        |
        H
```

(d)
```
        H
        |
    H—C—H
    H |
    | |
H—C—C—H
    | |
    H |
    H—C—H
        |
        H
```

SOLUTION Structures (a) and (d) are linear; (b) and (c) are the branched isomer.

EXAMPLE 12.3 The branched isomer of butane has melting and boiling points of −160°C and −12°C, respectively. Predict whether these values are higher or lower than those of linear butane.

SOLUTION They are lower (see Table 12.1).

EXAMPLE 12.4 Write molecular and structural formulas for the three structural isomers of pentane.

SOLUTION All have the molecular formula C_5H_{12}. The structural formulas are:

```
    H H H H H
    | | | | |
H—C—C—C—C—C—H
    | | | | |
    H H H H H
```

```
    H H H H
    | | | |
H—C—C—C—C—H
    | | | |
    H | H H
        |
    H—C—H
        |
        H
```

continued

SOLUTION, continued

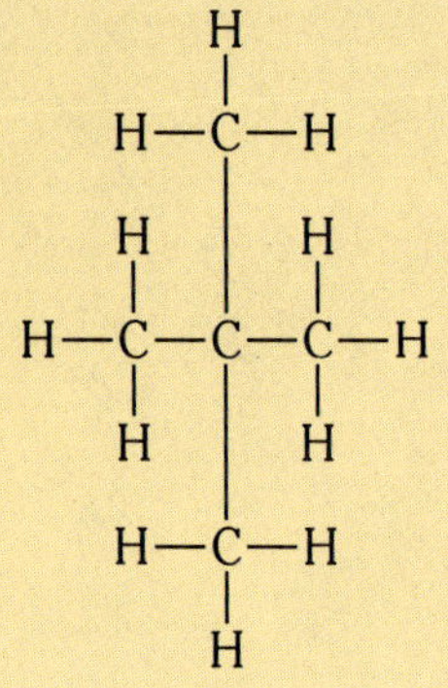

```
      H
      |
    H—C—H
   H  |  H
   |  |  |
 H—C——C——C—H
   |  |  |
   H  |  H
    H—C—H
      |
      H
```

The number of possible structural isomers increases rapidly as the number of carbon atoms in a molecule increases (Table 12.2). Structural isomerism is one reason so many different organic compounds exist.

Carbon atoms are classified according to how many other carbons are bonded to them:

primary carbon atom—bonded to one (or no) other carbon
secondary carbon atom—bonded to two carbon atoms
tertiary carbon atom—bonded to three carbons
quaternary carbon atom—bonded to four other carbons

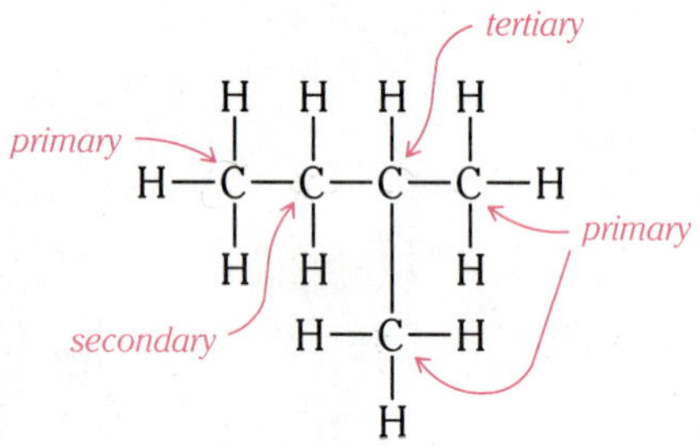

| Table 12.2 Number of Possible Isomers of Alkanes ||
Molecular Formula	Total Isomers Possible
CH_4	1
C_2H_6	1
C_3H_8	1
C_4H_{10}	2
C_5H_{12}	3
C_6H_{14}	5
C_7H_{16}	9
C_8H_{18}	18
C_9H_{20}	35
$C_{10}H_{22}$	75
$C_{15}H_{32}$	4347
$C_{20}H_{42}$	366,319
$C_{30}H_{62}$	4.11×10^9

The following isomer of pentane, for example, has three different types of carbon atoms:

```
              tertiary
     H  H  H  H
     |  |  |  |
primary→ H—C—C—C—C—H
     |  |  |  |
     H  H  |  H  ← primary
secondary→    H—C—H
              |
              H
```

EXAMPLE 12.5 How many carbons of each type occur in the compound below:

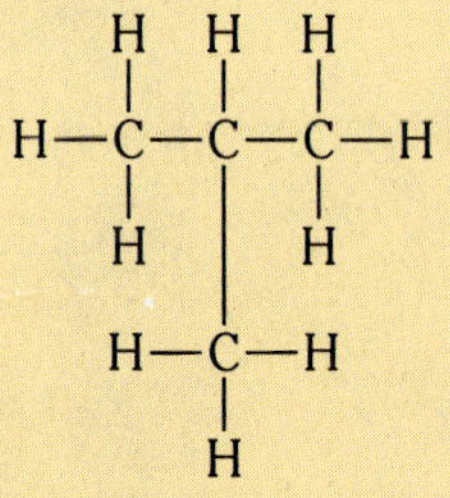

```
   H  H  H
   |  |  |
 H—C—C—C—H
   |  |  |
   H  |  H
    H—C—H
      |
      H
```

SOLUTION Three primary carbons and one tertiary carbon (near the center).

12.3 CONDENSED STRUCTURAL FORMULAS

Chemists use a kind of shorthand, called **condensed structural formulas,** to write structures of organic compounds. With straight-chain (linear) and branched compounds, these condensed formulas show each carbon atom together with its bonded hydrogen (or other) atoms. To save time and space, bonds to hydrogen and single bonds to carbon atoms in the unbranched chain aren't shown. But bonds to carbon at branches in the chain often are written as vertical bonds. Double and triple bonds also are written. Table 12.3 shows how this works.

In long linear chains, the structural formula is sometimes condensed even more by enclosing consecutive —CH_2— groups in parentheses, with a subscript to show the number of —CH_2— groups. For example, you can write hexane as $CH_3CH_2CH_2CH_2CH_2CH_3$ or condense it to $CH_3(CH_2)_4CH_3$. Condensed formulas for branched chains sometimes collect —CH_3 groups in a similar way. For example, you can condense $CH_3CHCH_2CHCH_3$ (with CH_3 branches) to $(CH_3)_2CHCH_2CH(CH_3)_2$.

Condensed structural formulas are simplified even more for ring compounds. An appropriate geometric figure (usually a triangle, square, pentagon, or hexagon) represents carbon atoms in the ring, and hydrogen atoms bonded to ring carbons are not shown. If an atom other than hydrogen is bonded to a ring carbon, however, it is shown in the formula (Table 12.3).

Table 12.3 Writing Condensed Structural Formulas

Complete Structural Formula	Condensed Structural Formula	Complete Structural Formula	Condensed Structural Formula
Six-carbon linear hydrocarbon (hexane)	$CH_3CH_2CH_2CH_2CH_2CH_3$ or $CH_3(CH_2)_4CH_3$	Six-carbon ring hydrocarbon (cyclohexane)	(hexagon)
Four-carbon linear hydrocarbon containing a double bond	$CH_2=CHCH_2CH_3$	Five-carbon ring containing bromine and a double bond	(cyclopentene with Br)
Five-carbon branched chain containing chlorine	$CH_3CHCH_2CH_2Cl$ (with CH_3 branch) or $(CH_3)_2CHCH_2CH_2Cl$		

You need to learn how to write and read condensed structural formulas, for they appear throughout this and other chemistry books.

EXAMPLE 12.6 Write condensed structural formulas for the following hydrocarbons:

(a) 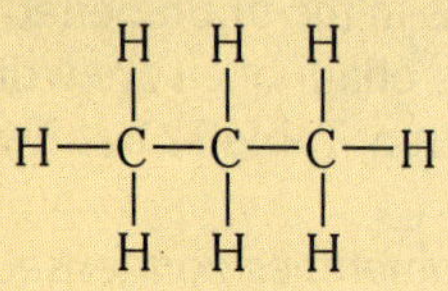 propane structure shown

(b) 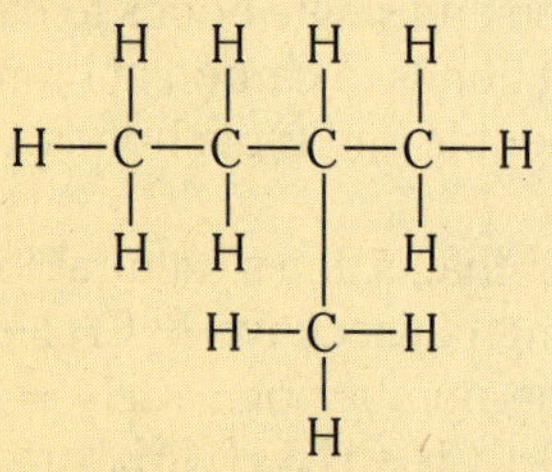 isobutane structure shown

(c) 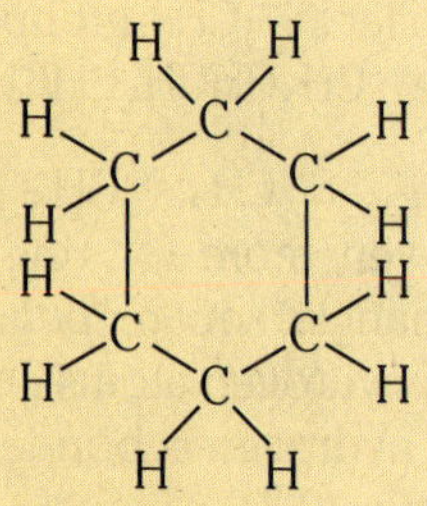 cyclohexane structure shown

(d) 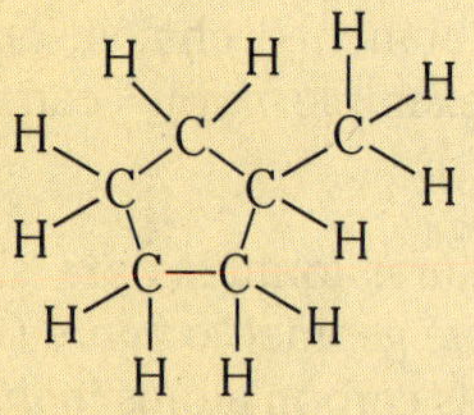 methylcyclopentane structure shown

SOLUTION (a) $CH_3CH_2CH_3$ (b) 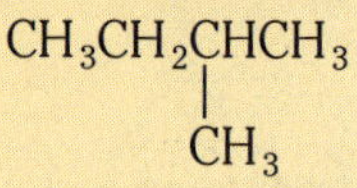 $CH_3CH_2CHCH_3$ or $CH_3CH_2CH(CH_3)_2$
    with CH_3 branch

(c) 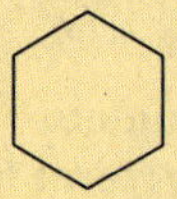 cyclohexane (hexagon)   (d) 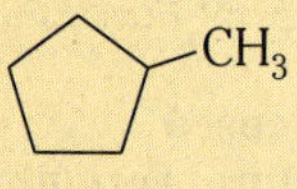 cyclopentane with CH_3

12.4 NAMING ALKANES

IUPAC Names. Many organic compounds, such as urea, are known by the common names given when they were discovered. Common names are still widely used, but this method couldn't handle the growing number of known organic compounds (such as the enormous number of isomers in large alkanes). In response, the International Union of Pure and Applied Chemistry (IUPAC) adopted a standard system to name organic compounds.

To name alkanes by the IUPAC system, use the suffix *-ane* along with the appropriate prefixes, depending on the number of carbon atoms. For linear chains, use the names in Table 12.1. For branched alkanes, such as the one below, use the following rules:

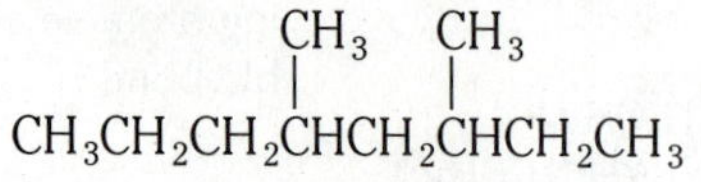

$$CH_3CH_2CH_2CHCH_2CHCH_2CH_3$$
with two CH_3 branches

Rule 1. *Find the longest continuous carbon chain and use the name for the alkane with that number of carbon atoms.*

Here the longest continuous chain (in color) has eight carbon atoms, so use the base name *octane*.

$$\text{CH}_3\text{CH}_2\text{CH}_2\overset{\underset{|}{\text{CH}_3}}{\text{CH}}\text{CH}_2\overset{\underset{|}{\text{CH}_3}}{\text{CH}}\text{CH}_2\text{CH}_3$$

Rule 2. *Number the carbons in the longest continuous carbon chain from the end nearer the first substituent group outside the chain.*

$$\underset{1\ \ 2\ \ \ 3\ \ \ 4\ \ 5\ \ \ 6\ \ 7\ \ 8}{\text{CH}_3\text{CH}_2\text{CH}_2\overset{\underset{|}{\text{CH}_3}}{\text{CH}}\text{CH}_2\overset{\underset{|}{\text{CH}_3}}{\text{CH}}\text{CH}_2\text{CH}_3} \quad \text{or} \quad \underset{8\ \ 7\ \ \ 6\ \ \ 5\ \ 4\ \ \ 3\ \ 2\ \ 1}{\text{CH}_3\text{CH}_2\text{CH}_2\overset{\underset{|}{\text{CH}_3}}{\text{CH}}\text{CH}_2\overset{\underset{|}{\text{CH}_3}}{\text{CH}}\text{CH}_2\text{CH}_3}$$

Numbering from the left leaves the first substituent group (CH$_3$) outside the chain attached to carbon 4. Numbering from the right, the first group is bonded to carbon 3. Therefore, number the carbon chain from the right.

Rule 3. *Identify substituent groups outside the chain by name.*

$$\underset{8\ \ 7\ \ \ 6\ \ \ 5\ \ 4\ \ \ 3\ \ 2\ \ 1}{\text{CH}_3\text{CH}_2\text{CH}_2\overset{\underset{|}{\text{CH}_3}}{\text{CH}}\text{CH}_2\overset{\underset{|}{\text{CH}_3}}{\text{CH}}\text{CH}_2\text{CH}_3}$$

Hydrocarbon substituent groups containing only single bonds are called **alkyl groups.** Table 12.4 shows the names and structures of common alkyl groups. You need to learn these, for you will use them throughout organic chemistry.

Alkyl groups are named by replacing *-ane* in the alkane name with *-yl*. Both substituent groups in the compound above (in color) are one-carbon alkyl groups, called *methyl* groups. Methyl groups, like all linear alkyl groups (Table 12.4), have a primary carbon atom that bonds to the carbon chain.

You will need practice to recognize branched alkyl groups, or alkyl groups in which an interior carbon atom bonds to a carbon chain (Table 12.4). These are named by using prefixes. *Iso-* refers to a branched carbon chain having two

Table 12.4 Names and Structures of Common Alkyl Groups

Name	Structure	Name	Structure
Linear Alkyl Groups		**Branched Alkyl Groups**	
methyl	CH$_3$—		
ethyl	CH$_3$CH$_2$—	isopropyl	CH$_3$CH— with CH$_3$ above
propyl	CH$_3$CH$_2$CH$_2$—		
butyl	CH$_3$CH$_2$CH$_2$CH$_2$—	isobutyl	CH$_3$CHCH$_2$— with CH$_3$ above
		sec-butyl	CH$_3$CH$_2$CH— with CH$_3$ above
		tert-butyl	CH$_3$C— with CH$_3$ above and CH$_3$ below

methyl groups bonded to a next-to-end carbon atom. *Sec-* refers to an alkyl group attached by a secondary carbon atom. *Tert-* (or *t-*) is used for an alkyl group attached by a tertiary carbon atom. In each case, the name following the prefix (such as butyl or propyl) depends on the *total* number of carbons in the substituent group, branched or not.

> **Rule 4.** *Identify by number the carbon atom in the chain to which each substituent bonds. If the same substituent occurs more than once, identify each carbon number for that substituent, separate the numbers by commas, and use the appropriate prefix (such as di-, tri, tetra-) to specify how many of that substituent occurs in the compound.*

In this example, two methyl groups occur, bonded to carbons 3 and 5 in the chain. So write *3,5-dimethyl*.

> **Rule 5.** *Name the compound by listing the substituents preceded by their numbers and a hyphen, followed by the base name for the carbon chain. If more than one type of substituent occurs, arrange them in alphabetical order, ignoring all prefixes except iso- and cyclo-. Connect all numbers and prefixes by hyphens.*

Substituents may also be listed in order of increasing size and complexity instead of in alphabetical order. In this text we will use alphabetical order.

In our example, the full IUPAC name is 3,5-dimethyloctane.
Another example is shown below with its IUPAC name:

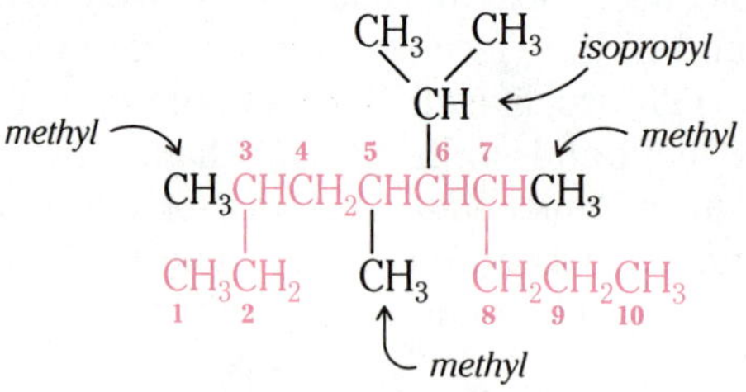

$$\underset{\text{6-isopropyl-3,5,7-trimethyldecane}}{\text{structure}}$$

EXAMPLE 12.7 Write IUPAC names for the following:

(a) 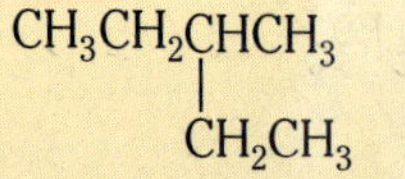 CH₃CH₂CHCH₃
          |
          CH₂CH₃

(b) 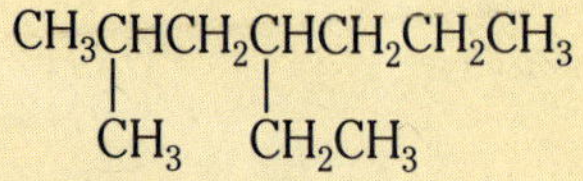 CH₃CHCH₂CHCH₂CH₂CH₃
         | |
         CH₃   CH₂CH₃

(c) 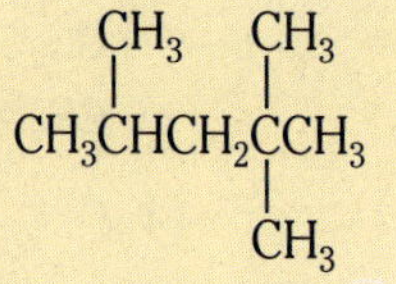 CH₃CHCH₂CCH₃
         | |
         CH₃ CH₃
              |
              CH₃

(d) 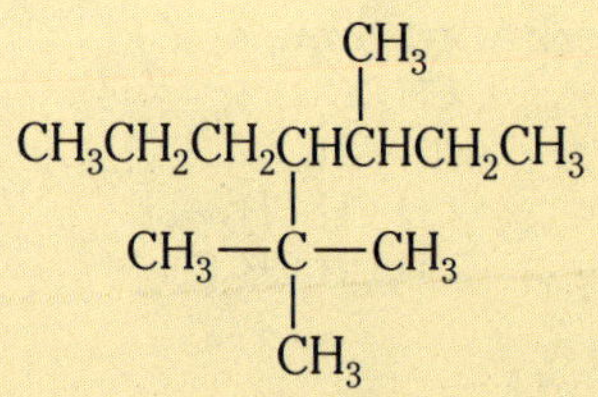 CH₃CH₂CH₂CHCHCH₂CH₃
                  | |
                  CH₃

      CH₃—C—CH₃
           |
           CH₃

(e) 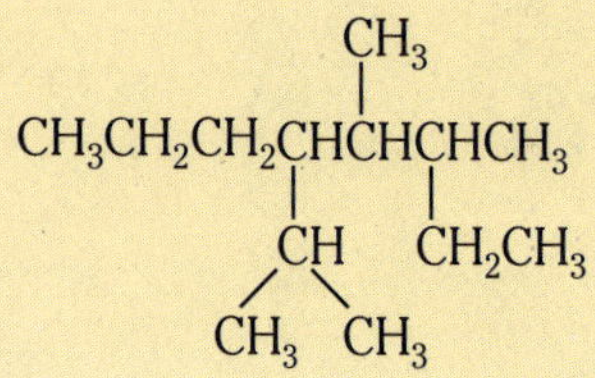 CH₃CH₂CH₂CHCHCHCH₃
                 | |
                 CH₃ CH₂CH₃

                 CH
                / \
              CH₃ CH₃

SOLUTION
(a) 3-methylpentane (The longest continuous carbon chain has five carbon atoms.)
(b) 4-ethyl-2-methylheptane
(c) 2,2,4-trimethylpentane (Numbering from the right results in lower numbers in the name. Use a separate number for each methyl group, even though two are attached to the same carbon atom.)
(d) 4-*t*-butyl-3-methylheptane (*t*-butyl precedes methyl alphabetically because you ignore the prefix *t*-.)
(e) 5-isopropyl-3,4-dimethyloctane [The longest chain has eight carbons; alphabetically, isopropyl (count the *iso*- prefix) precedes dimethyl (don't count the *di*- prefix).]

EXAMPLE 12.8 Write condensed structural formulas for the following: (a) 3,3-dimethylhexane, (b) 4-ethyl-2,3-dimethylheptane, (c) 6-*sec*-butyl-4-isopropyldecane, and (d) 4,4-diethyl-2-methyloctane.

SOLUTION

(a)
$$\begin{array}{c} \text{CH}_3 \\ | \\ \text{CH}_3\text{CH}_2\text{CCH}_2\text{CH}_3 \\ | \\ \text{CH}_3 \end{array}$$

(b)
$$\begin{array}{c} \text{CH}_3 \ \ \text{CH}_2\text{CH}_3 \\ | \ \ \ \ \ \ | \\ \text{CH}_3\text{CHCHCHCH}_2\text{CH}_3 \\ | \\ \text{CH}_3 \end{array}$$

(c)
$$\begin{array}{c} \text{CH}_3 \ \ \text{CH}_3 \\ \ \ \ \diagdown \diagup \\ \text{CH} \\ | \\ \text{CH}_3\text{CH}_2\text{CH}_2\text{CHCH}_2\text{CHCH}_2\text{CH}_2\text{CH}_3 \\ | \\ \text{CH}_3\text{CHCH}_2\text{CH}_3 \end{array}$$

(d)
$$\begin{array}{c} \text{CH}_3 \ \ \text{CH}_2\text{CH}_3 \\ | \ \ \ \ \ \ | \\ \text{CH}_3\text{CHCH}_2\text{CCH}_2\text{CH}_2\text{CH}_2\text{CH}_3 \\ | \\ \text{CH}_2\text{CH}_3 \end{array}$$

Study the solutions to these examples carefully to see if you understand how to name alkanes. Exercises at the end of this chapter will give you more practice in identifying alkyl groups, naming alkanes, and drawing structural formulas from the names.

To name cyclic alkanes, simply put the prefix *cyclo-* before the alkane name. A six-carbon cyclic alkane, for example, is called cyclohexane. Single substituent groups are included in the name without a number. If the ring has two or more substituent groups, number the ring carbons so that a substituent group is at carbon 1 and in the direction of the next closest substituent.

> **EXAMPLE 12.9** Name the following cycloalkanes:
>
> (a) △ (b) cyclopentane with CH₂CH₃ substituent (c) cyclohexane with CH₃ groups at 1, 3, 5 positions
>
> **SOLUTION**
> (a) cyclopropane
> (b) ethylcyclopentane
> (c) 1,3,5-trimethylcyclohexane (Notice how the ring is numbered to give the lowest numbers to the substituents.)

Common Names. Common names of alkanes differ from IUPAC names in two ways. First, common names use the prefix *n*- for unbranched alkanes. For example, butane (IUPAC) may be called *n*-butane (common) and octane (IUPAC) is *n*-octane (common).

Another difference occurs in alkanes that are linear except for a methyl group substituent on a next-to-end carbon. Common names for these alkanes have the prefix *iso-* followed by the alkane name matching the *total* number of carbon atoms *including* the branched methyl group. Two examples are:

$$CH_3CHCH_2CH_2CH_3 \qquad CH_3CH_2CH_2CH_2CH_2CH_2CHCH_3$$
$$\;\;\;\;|\;|$$
$$CH_3 \qquad\qquad\qquad\qquad\qquad\qquad CH_3$$

2-methylpentane (IUPAC)   2-methyloctane (IUPAC)

isohexane (common name)   *isononane* (common name)
(not isopentane)   (not isooctane)

Notice that hydrocarbons with a common name beginning with iso- *have an IUPAC name beginning with* 2-methyl.

> **EXAMPLE 12.10**
> (a) Write a condensed structural formula for isopentane.
> (b) Write IUPAC and common names for $CH_3(CH_2)_3CHCH_3$
> $\;|$
> $\;CH_3$
>
> **SOLUTION**
>
> $\qquad\qquad CH_3$
> $\qquad\qquad\;\;|$
> (a) $CH_3CH_2CHCH_3$
> (b) 2-methylhexane (IUPAC); isoheptane (common)

12.5 PHYSICAL PROPERTIES AND CHEMICAL REACTIONS OF ALKANES

Physical Properties. Think of alkanes you're familiar with. They include natural gas (mostly methane), propane, butane, and octane for fuels, kerosene, motor oil, heating oil, Vaseline, and wax (Figure 12.2). All are gases, liquids, or soft solids.

Figure 12.5 The oil-polluted Cuyahoga River in Cleveland, Ohio, on fire in the 1960s.

Why is that so? Alkanes are nonpolar because the electronegativities of carbon and hydrogen are similar enough (Section 4.6) for their bonds to be essentially nonpolar. Being nonpolar, hydrocarbon molecules attract each other only with weak London forces (Section 6.1); thus they tend to be gases and liquids at room temperature. Since larger molecules have stronger London forces, alkanes go from being gases to liquids to soft solids as they increase in size (Table 12.1).

Remember that *like dissolves like* (Section 7.2). Since alkanes are nonpolar, they dissolve in nonpolar solvents, but not in polar substances such as water. Oil spills (Figure 12.5) are a reminder that oil (a mixture of hydrocarbons) forms a separate layer when mixed with water; since alkanes are less dense than water, they float on top.

You can use alkanes to dissolve other nonpolar materials. Hexane or kerosene (a mixture of alkanes), for example, dissolve away grease from clothing. And gasoline or soft wax both remove bits of tar from automobiles.

This principle of *like dissolves like* is also important in the body. Parts of the body in contact with water—such as hair, skin, cell membranes, and blood vessels—all contain large amounts of nonpolar, hydrocarbon material that keeps them from dissolving in water. On the other hand, inhaled alkanes can cause considerable damage by dissolving nonpolar materials in delicate membranes in the lungs. And coatings of oil destroy the natural insulating properties of feathers, making birds (especially diving birds) unable to stay warm (Figure 12.6).

Cyclopropane is used as an inhalation anesthetic (Table 12.5). The anesthetic gas mixture also contains helium to prevent an explosive reaction between cyclopropane and oxygen. Because it has low toxicity, acts rapidly, and is quickly eliminated from the body, cyclopropane is often used for surgery on babies and small children.

Chemical Reactions. Alkanes are not very reactive. At room temperature, for example, they don't react with acids, bases, or oxidizing or reducing agents. In fact, the *R* that represents the "rest of the molecule you can ignore" in terms of reactivity (Section 11.3) is an alkyl group. Under the right conditions, however, alkanes undergo two types of reactions.

1. Combustion In the most familiar reaction, called *combustion*, (Section 5.2), alkanes react (burn) with oxygen gas (O_2) to produce carbon dioxide (CO_2), water, and energy (Figures 5.4 and 12.5). You use this reaction when

Figure 12.6 A seabird coated with crude oil from an oil spill. Such birds will die unless the oil is removed.

Table 12.5 Some Inhalation Anesthetics

Name	Formula	Uses and Characteristics
Gaseous Inhalants		
cyclopropane	CH_2–CH_2 with CH_2 on top (triangle) or △	Flammable, nonirritating, fast acting; often used in surgery on small children
nitrous oxide (laughing gas)	N_2O	Nonflammable, nonirritating; used in dentistry and obstetrics, often with more powerful anesthetics and muscle relaxants
Volatile Liquid Inhalants		
halothane (Fluothane)	F–C(F)(F)–C(Cl)(Br)–H	Nonflammable, nonirritating, and powerful; rapid recovery; often used with nitrous oxide and muscle relaxants

you burn fuels for heat or to run automobiles and other engines. Two typical fuels are natural gas (methane, CH_4) and heating oil (which contains $C_{16}H_{34}$):

$$CH_4 + 2O_2 \longrightarrow CO_2 + 2H_2O + energy$$

$$2C_{16}H_{34} + 49O_2 \longrightarrow 32CO_2 + 34H_2O + energy$$

When combustion isn't complete, the reaction produces some carbon monoxide (CO) instead of carbon dioxide. For example, if enough oxygen is present in an automobile engine, gasoline (represented here by one component, octane, C_8H_{18}) would burn completely to CO_2 and H_2O:

$$2C_8H_{18} + 25O_2 \longrightarrow 16CO_2 + 18H_2O$$

If there isn't enough oxygen to do this, the reaction can be:

$$2C_8H_{18} + 22O_2 \longrightarrow 10CO_2 + 6CO + 18H_2O$$

Recall that CO is toxic (Table 4.9) because it binds to hemoglobin in red blood cells, preventing cells from getting the O_2 they need (Section 4.6 Spotlight).

EXAMPLE 12.11 If the amount of oxygen available to burn octane were reduced even more than in the preceding reaction, what effect would it have on the amount of carbon monoxide (CO) produced? Balance the equation below and compare the CO produced with the preceding reaction:

$$2C_8H_{18} + 19O_2 \longrightarrow \_\_CO_2 + \_\_CO + 18H_2O$$

SOLUTION $2C_8H_{18} + 19O_2 \longrightarrow 4CO_2 + 12CO + 18H_2O$
When less oxygen is available, less CO_2 and more carbon monoxide (CO) is produced.

2. Halogenation Alkanes react with halogens in the presence of heat or ultraviolet (UV) light to form **alkyl halides** (see Table 11.3), which have one or more halogen atoms bonded to carbon. *Halogenation* is a **substitution reaction,** in which an atom (in this case, chlorine) or group replaces a different kind of atom (in this case, hydrogen) in a molecule.

Fluorine (F_2) reacts explosively with alkanes, bromine (Br_2) reacts slowly, and iodine (I_2) doesn't react. As a result, chlorine (Cl_2) is the most common halogen used. Heat or UV light first breaks the covalent bond in the halogen molecule. The reaction with chlorine is:

$$:\ddot{Cl}:\ddot{Cl}: \xrightarrow{\text{UV light or heat}} :\ddot{Cl}\cdot + \cdot\ddot{Cl}:$$

chlorine molecule → chlorine atoms

Each chlorine atom has an unpaired electron. Such atoms, or molecular fragments, are called **free radicals.** Free radicals are unstable and seek another valence electron to form a stable octet. A chlorine atom thus attacks an alkane molecule, such as methane, to gain an electron and become stable:

$$H:\overset{H}{\underset{H}{\ddot{C}}}:H + \cdot\ddot{Cl}: \longrightarrow H:\overset{H}{\underset{H}{\ddot{C}}}\cdot + H:\ddot{Cl}:$$

methane — chlorine atom — methyl free radical — hydrogen chloride

The methyl free radical, in turn, is unstable and can react with another chlorine atom to form a stable product:

$$H:\overset{H}{\underset{H}{\ddot{C}}}\cdot + \cdot\ddot{Cl}: \longrightarrow H:\overset{H}{\underset{H}{\ddot{C}}}:\ddot{Cl}:$$

The product can be halogenated further by reaction with additional chlorine atoms. Halogenation reactions produce a mixture of products with different numbers of halogen atoms. For example, methane reacts with chlorine in the presence of heat or ultraviolet light to form a mixture of alkyl halides (see Section 12.7):

$$CH_4 + Cl_2 \xrightarrow{\text{UV light}} CH_3Cl + CH_2Cl_2 + CHCl_3 + CCl_4$$

12.6 FOSSIL FUELS AND PETROCHEMICALS

Fossil fuels—coal, natural gas, and petroleum—formed from plants and animals that died and decayed millions of years ago. Buried deep in the earth, these remains slowly changed into coal (which is mostly carbon), natural gas (mostly methane), and petroleum, a mixture of many hydrocarbons.

Fossil fuels provide most of the energy used in the United States and the world (Figure 12.7). But petroleum also is our main raw material for making organic products such as plastics, drugs, pesticides, fertilizers, and synthetic fabrics. These products, known as **petrochemicals,** account for most manufactured organic compounds.

Much of the coal that isn't used directly for fuel is heated in the absence of air at 1000–1300°C as follows:

$$\text{coal} \xrightarrow[\text{heat}]{\text{no air}} \text{coke + coal tar + coal gas}$$

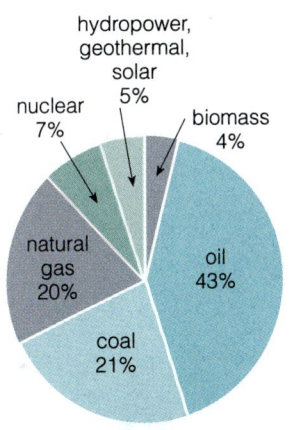

United States

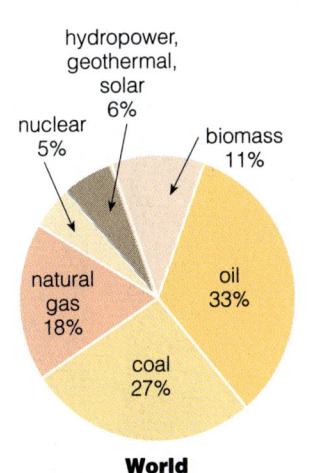

- hydropower, geothermal, solar 5%
- nuclear 7%
- biomass 4%
- oil 43%
- coal 21%
- natural gas 20%

World

- hydropower, geothermal, solar 6%
- nuclear 5%
- biomass 11%
- oil 33%
- coal 27%
- natural gas 18%

Figure 12.7 Commercial energy use in the United States (top) and the world (bottom).

272 Part II Organic Chemistry

Coke, the solid material that remains, is used to make iron and steel. Coal tar is a liquid that contains many cyclic organic compounds used to make dyes, plastics, medicines, explosives, and pesticides. Coal gas is used as a fuel.

Petroleum, not coal, is the largest source of cyclic organic compounds.

Petroleum Refining. Crude oil has too many different substances to be very useful by itself. So oil refineries separate petroleum into fractions that are more homogeneous in the size of their molecules.

Refineries separate hydrocarbons by **fractional distillation** in huge fractionating columns (Figure 12.8). A furnace at the bottom heats crude petroleum to about 315°C in the absence of oxygen. Nonpolar hydrocarbons, with only weak London forces between their molecules, have low boiling points that depend mostly on the size of the molecule (Table 12.1). Smaller, lower-boiling

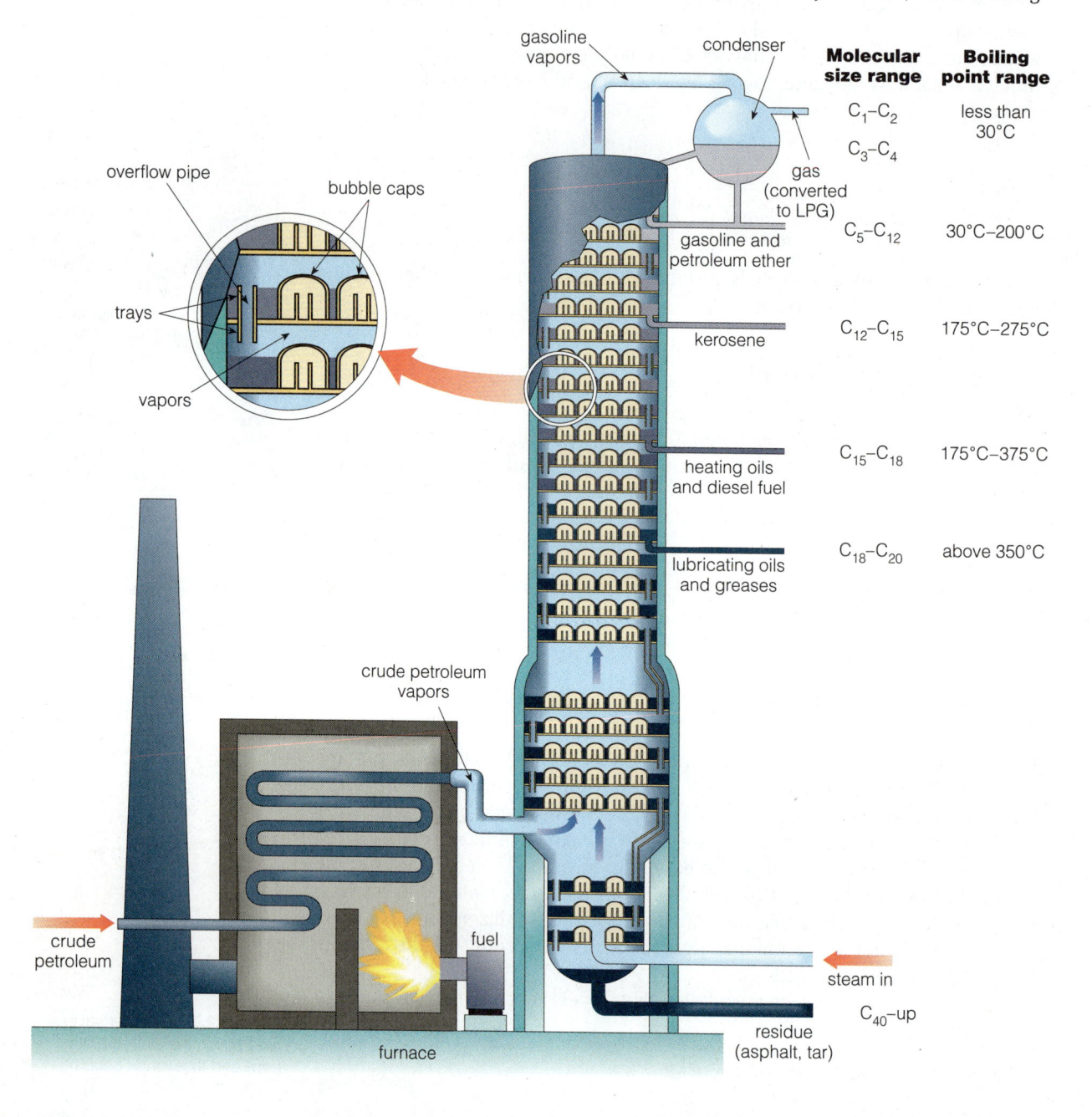

Figure 12.8 Fractional distillation separates crude petroleum into fractions based on their boiling points. Actual columns reach as high as 30 m (100 ft).

Molecular size range	Boiling point range
C_1–C_2	less than 30°C
C_3–C_4	
C_5–C_{12}	30°C–200°C
C_{12}–C_{15}	175°C–275°C
C_{15}–C_{18}	175°C–375°C
C_{18}–C_{20}	above 350°C
C_{40}–up	

substances that vaporize and rise to the top of the column are collected as gases. As larger molecules rise in the column, they cool enough to liquefy. As a result, various fractions separate and condense at various heights in the fractionating column, where they then can be isolated.

Many fractions are useful fuels. Propane and butane are gases at standard temperature and pressure, but are often supplied in liquid form for use in camp stoves, cigarette lighters, and automobiles. Other liquid hydrocarbons serve as solvents or fuels for automobiles, trucks, and furnaces. Fractions containing larger molecules (C_{16} and above) are used as lubricants, waxes, and road materials (asphalt and tar).

The fraction in greatest demand is gasoline, a mixture of hydrocarbons containing about five to nine carbon atoms. Since too much distilled petroleum goes into other fractions—and the gasoline fraction that does form is a poor fuel for high-compression engines—refineries have to (a) convert other fractions into gasoline, and (b) upgrade the gasoline fraction.

At high temperatures and pressures, in the presence of special catalysts and in the absence of oxygen, alkanes break into smaller hydrocarbons; this is called *cracking*. Cracking produces an alkane plus a hydrocarbon with a carbon-to-carbon double bond. A typical reaction is:

$$CH_3CH_2CH_2CH_2CH_2CH_2CH_2CH_2CH_2CH_2CH_2CH_2CH_3 \quad (C_{14}H_{30})$$

$$\downarrow$$

$$CH_3CH_2CH_2CH_2CH_2CH_2CH_2CH_3 \quad (C_8H_{18})$$
$$+ CH_2{=}CHCH_2CH_2CH_2CH_3 \quad (C_6H_{12})$$

Cracking is also an important industrial process for producing hydrocarbons with carbon-to-carbon double bonds, called alkenes (see Chapter 13).

Refineries crack heating oil into gasoline, especially in early spring when the demand for heating oil drops and the demand for gasoline increases.

Refineries also use catalysts, such as aluminum chloride ($AlCl_3$), to make alkanes branched or cyclic. This process is called *catalytic reforming*. A typical reaction is:

$$CH_3CH_2CH_2CH_2CH_2CH_2CH_2CH_3 \xrightarrow{AlCl_3} CH_3\underset{}{\overset{CH_3}{\underset{}{C}H}}CH_2\underset{CH_3}{\overset{CH_3}{\underset{}{C}}}CH_3$$

octane   2,2,4-trimethylpentane ("isooctane")

Catalytic reforming improves the quality of gasoline because branched and cyclic hydrocarbons have less tendency to ignite prematurely. Premature ignition upsets the timing between ignition and the moving pistons, causing an engine to "knock."

Octane number indicates the antiknock quality of gasoline; the higher the number, the lower the tendency to cause engine knock. Gasoline is tested in standard engines and assigned an octane number based on its antiknock properties. One isomer of octane—2,2,4-trimethylpentane (see the preceding reaction)—has excellent antiknock qualities and is assigned an octane number of 100. Pure heptane, in contrast, performs very poorly and is assigned a value of 0. On this scale, gasoline with an octane number of 85 performs the same as a mixture of 85% 2,2,4-trimethylpentane and 15% heptane. Most high-compression engines require fuel with octane values of about 90 (Figure 12.9).

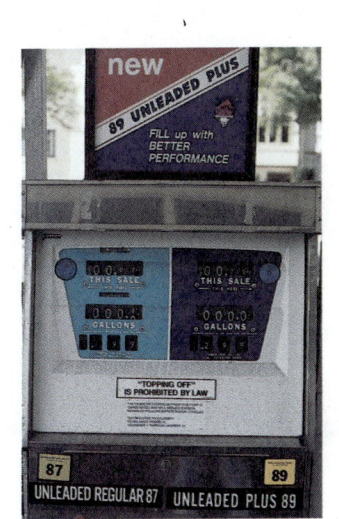

Figure 12.9 The octane numbers of this gasoline are displayed on the pump.

12.7 ALKYL HALIDES

One of the few reactions alkanes undergo is halogenation by a free-radical process (Section 12.5). You can name the products, called *alkyl halides*, either by IUPAC or common names. For IUPAC names, consider the halogens as substituent groups on a carbon chain or ring, and change their *-ine* suffix to *-o*. Then identify by number (when necessary) the carbon atoms to which fluoro-, chloro-, bromo-, or iodo- groups are bonded. Three examples are:

$$\text{CH}_3\text{CHCH}_3 \qquad \text{CH}_3\text{CH}_2\text{CH}_2\text{F} \qquad \text{CH}_3\text{CH}_2\text{Cl}$$
$$\quad\;\; |$$
$$\quad\;\; \text{Br}$$

2-bromopropane   1-fluoropropane   chloroethane
(no number is needed)

To write a common name, simply name the alkyl group (Table 12.4) as a substituent, followed by the name of the halide, using the *-ide* suffix. For example, the preceding compounds have the common names isopropyl bromide, propyl fluoride, and ethyl chloride.

EXAMPLE 12.12 Write IUPAC and common names for the following alkyl halides:

(a) $\text{CH}_3\text{CH}_2\text{I}$ (b) $\text{CH}_3\overset{\overset{\text{CH}_3}{|}}{\underset{\underset{\text{Cl}}{|}}{\text{C}}}\text{CH}_3$ (c) $\text{CH}_3\text{CH}_2\text{CHClCH}_3$ (d) cyclopentyl—Br

SOLUTION
(a) iodoethane (no number is needed); ethyl iodide (common name)
(b) 2-chloro-2-methylpropane; *t*-butyl chloride (common)
(c) 2-chlorobutane; *sec*-butyl chloride (common)
(d) bromocyclopentane; cyclopentyl bromide (common)

Properties and Uses. Organic halides have many uses. Athletic trainers often spray ethyl chloride on a bruised shin or forearm to provide temporary relief from minor pain (Figure 12.10). Ethyl chloride vaporizes quickly and takes with it heat from the skin, leaving the bruised area feeling a bit numb or "frozen."

Although chlorine, iodine, and bromine are more electronegative than carbon (see Figure 4.21), the differences are small enough so that organic compounds with these halides are not highly polar. Methylene chloride (CH_2Cl_2), chloroform (CHCl_3), and carbon tetrachloride (CCl_4) have been used as cleaning solvents to remove nonpolar materials, and chloroform was used as a general anesthetic. We know now that these chlorinated compounds shouldn't be inhaled; chloroform and CCl_4, for example, cause liver damage and are potential carcinogens (cancer-causing agents). Safer anesthetics, some of

---

Recall that CCl_4 is nonpolar because the symmetric shape of the molecule produces no net dipole moment (Section 4.6).

which are also organic halides (Table 12.5), have replaced chloroform in the operating room (Figure 12.11).

Organic compounds with most or all of their hydrogens replaced by fluorine are inert, nonreactive substances. One example is Teflon, a fluorocarbon used in nonstick cookware and tubing (see Section 18.2). Another example is perfluorodecalin (trade name Fluosol; Figure 12.12), a clear liquid that dissolves large amounts of oxygen (Figure 12.13) but is not metabolized in the body. It is used as a temporary blood substitute in emergencies when the needed blood, or blood type, isn't available. It also avoids the risk of transmitting infectious diseases such as AIDS.

Figure 12.10 Ethyl chloride evaporates quickly from the skin, relieving soreness.

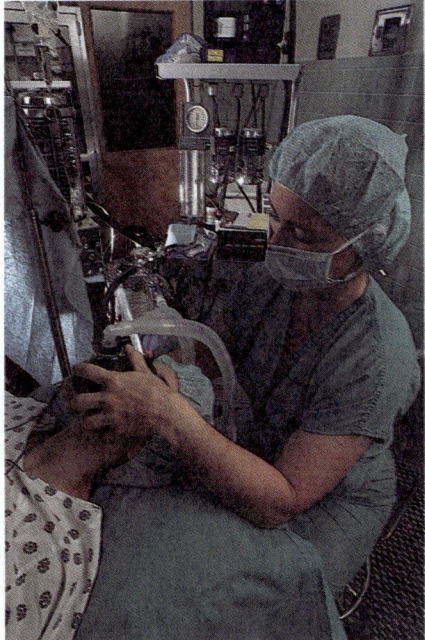

Figure 12.11 Some anesthetics are organic halides.

Figure 12.12 Perfluorodecalin (Fluosol).

Organic halides are also effective pesticides. Methyl bromide, for example, is a fungicide used to fumigate hazelnuts (filberts) and many other crops. DDT, lindane, and aldrin are common examples of organochlorine pesticides. They are nonpolar and not very reactive, so they persist in the environment (typically for several years) and accumulate in the body in nonpolar tissues such as liver and fat cells. This has led to a ban on some organochlorine pesticides—such as DDT—for general use in the United States.

Figure 12.13 This rat suspended in a container of perfluorodecalin continues to breathe the oxygen dissolved in that liquid.

> Life takes the atoms and molecules and crystals; but, instead of making a mess of them like the stone, it combines them into new and more elaborate patterns of its own.
>
> *Aldous Huxley (1894–1963)*

CHEMISTRY SPOTLIGHT

Chlorofluorocarbons and the Ozone Layer

Up in the stratosphere, 20–50 km (10–30 miles) above earth, a protective layer of ozone (O₃) absorbs most of the sun's harmful ultraviolet (UV) radiation. But our ozone layer is being depleted, and chlorofluorocarbons (CFCs)—a group of organic halides commonly known as Freons—are one of the major culprits.

The United States and some other countries no longer allow CFCs in spray cans, but Freons are still widely used in refrigeration and air-conditioning units (which keep CFCs sealed inside cooling coils). CFCs also are used as foaming agents in polyurethane and other materials.

CFCs are not very reactive. Once in the air, many CCl₂F₂ (Freon-12) and other CFC molecules don't react until they encounter high-energy UV radiation in the stratosphere. The encounter generates free radicals such as Cl atoms. One example is:

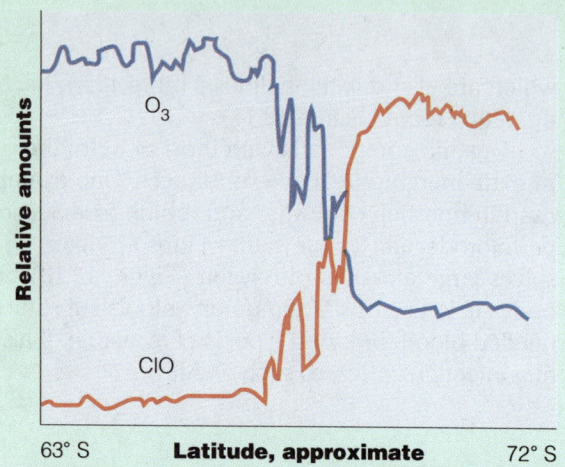

$$\text{Cl}-\underset{\underset{F}{|}}{\overset{\overset{F}{|}}{C}}:\text{Cl} \xrightarrow{\text{UV radiation}} \text{Cl}-\underset{\underset{F}{|}}{\overset{\overset{F}{|}}{C}}\cdot + \text{Cl}\cdot$$

Freon - 12   CClF₂   chlorine atom

Figure 12.14 A NASA airplane flying toward Antarctica measured concentrations of ozone (O₃) and ClO. As the plane reached the ozone hole, O₃ levels decreased sharply as ClO levels increased.

Because it needs another valence electron to become stable, a chlorine atom readily reacts with ozone:

$$\text{Cl}\cdot + \text{O}_3 \longrightarrow \text{ClO} + \text{O}_2$$

Figure 12.14 shows how the presence of ClO correlates with the disappearance of O₃ in the "ozone hole" over Antarctica.

According to the National Academy of Sciences, each 1% loss of ozone increases by 2% the amount of UV radiation reaching earth. More UV radiation means more skin cancer and cataracts in humans, as well as more intense photochemical smog, damage to many animals, and lower crop yields.

The major industrialized nations have signed a treaty to ban their use of CFCs by the end of this century. As that deadline approaches, chemical companies are working to develop less harmful substitutes.

SUMMARY

Alkanes are saturated hydrocarbons containing only single bonds. Their molecular formulas are C_nH_{2n+2} for chain compounds and C_nH_{2n} for cyclic ones. Compounds with the same molecular formula but different structural formulas are structural isomers; such isomers have different properties. Condensed structural formulas for linear and branched compounds show carbon atoms together with their bonded hydrogen atoms and omit single bonds in the linear chain. Condensed formulas for cyclic compounds are geometric figures that may omit ring carbons and their bonded hydrogens.

The names of some linear alkanes are in Table 12.1. To name other compounds, find the longest continuous carbon chain and number from the end nearer the first substituent. Then designate substituents by number and list them alphabetically in front of the alkane name matching the longest carbon chain. Some common alkyl groups are shown in Table 12.4. Use the prefix *cyclo-* to name cyclic alkanes. Common names use the prefix *n-* for linear (straight-chain) alkanes and *iso-* for alkanes having a methyl (CH_3—) group bonded to a next-to-end carbon in the chain.

Alkanes are nonpolar, don't dissolve in water, and are gases, liquids, or soft solids at room temperature, depending on their molecular size. They are not very reactive, but under appropriate conditions they burn in combustion reactions and undergo halogenation by a free-radical process.

Our main sources of petrochemicals are petroleum, coal, and natural gas. Crude oil is refined by fractional distillation, cracking, and catalytic reforming to produce gasoline and other useful products.

Alkyl halides are organic compounds containing one or more halogen atoms. In IUPAC names, the halogens are substituent groups. Common names list the alkyl group (Table 12.4) followed by the name of the halide, using the suffix *-ide*. Alkyl halides have many uses. Chlorofluorocarbons (CFCs), also known as Freons, are a major culprit in depleting the earth's protective ozone layer.

KEY TERMS

Alkane (12.1)
Alkyl group (12.3)
Alkyl halide (12.5)
Condensed structural formula (12.3)
Fossil fuel (12.6)
Fractional distillation (12.6)

Free radical (12.5)
Hydrocarbon (12.1)
Molecular formula (12.1)
Octane number (12.6)
Petrochemical (12.6)
Primary carbon atom (12.2)
Quaternary carbon atom (12.2)

Saturated hydrocarbon (12.1)
Secondary carbon atom (12.2)
Structural formula (12.1)
Structural isomer (12.2)
Substitution reaction (12.5)
Tertiary carbon atom (12.2)

EXERCISES

Even-numbered exercises are answered at the back of this book.

Alkanes

1. To be an alkane, what structural characteristics must a molecule have?
2. Identify which of the following are alkanes: **(a)** CH_3Cl, **(b)** C_2H_2, **(c)** CH_4, **(d)** C_2H_4, **(e)** $C_{12}H_{26}$.
3. What are the bond angles and shapes around the carbon atoms in each compound in exercise 2? [Review Section 11.2 if necessary.]

Formulas of Alkanes

4. Write structural formulas for each compound in exercise 2.
5. Explain the difference between a molecular formula and a structural formula.
6. What hybrid orbitals do carbon atoms have in each compound in exercise 2? [Review Section 11.2 if necessary.]
7. Write molecular and structural formulas for **(a)** pentane and **(b)** heptane.
8. Many camp stoves use propane for fuel. Write the molecular and structural formulas for propane.
9. Identify which of the following could be a molecular formula for a saturated hydrocarbon: **(a)** $C_{10}H_{22}$, **(b)** C_2H_8, **(c)** C_6H_{12}, **(d)** C_6H_{14}, **(e)** $C_{164}H_{330}$.
10. Which compound in exercise 9 could be a cyclic alkane?

Structural Isomerism

11. Identify in which one of the following ways two structural isomers are the same: **(a)** melting point, **(b)** structural formula, **(c)** density, **(d)** molecular formula, **(e)** boiling point.
12. Write the molecular formula of the alkane that has two structural isomers.
13. Write structural formulas for the structural isomers of the compound in exercise 12.
14. Write structural formulas and IUPAC names for the five different isomers of C_6H_{14}.

15. Is cyclobutane a structural isomer of butane? Explain.
16. Is isohexane a structural isomer of hexane? Explain.

Condensed Structural Formulas

17. Write a condensed structural formula for a saturated hydrocarbon with the molecular formula C_3H_6.
18. Write condensed structural formulas for the following:

    (a) H—C—C—H with H's (ethane-like, 2 carbons)

    (b) H—C—C—Cl (2 carbons with Cl)

    (c) 5-carbon straight chain

    (d) 5-carbon chain with a CH branch

    (e) 4-carbon chain with F substituent

    (f) 7-carbon chain with multiple methyl branches

19. In the structure in exercise 18f, identify each carbon as a primary, secondary, or tertiary carbon atom.
20. Which compounds in exercise 18 contain a tertiary carbon atom?
21. Write a condensed structural formula for a structural isomer of the compound in exercise 18f.
22. Write a condensed structural formula for a structural isomer of the compound in exercise 18d.
23. Write a condensed structural formula for a structural isomer of the compound in exercise 18e.
24. Write condensed structural formulas for (a) cyclobutane and (b) cyclohexane.
25. Write condensed structural formulas for the following: (a) isohexane, (b) isobutyl chloride, (c) cyclohexyl bromide, (d) chloroform, (e) 4-t-butyl-2,2,5-trimethylheptane, (f) 1,2,3,4,5,6-hexachlorocyclohexane (the insecticide lindane).
26. Write condensed structural formulas for the following: (a) isopropyl chloride, (b) 4-t-butyl-2-methylheptane, (c) isobutyl iodide, (d) 4-ethyl-2,2,5-trimethyldecane, (e) 1,3-dimethylcyclohexane.

Naming Alkanes

27. Write extended and condensed structural formulas for 2-ethylpentane. Why is this not the correct name for this compound? What is the correct IUPAC name?
28. Write IUPAC and common names for the substances in exercise 18.
29. Suppose you discover that two compounds have the same IUPAC name. Are they (a) different structural isomers, or (b) different forms of the same structural isomer?
30. Write IUPAC names for compounds a–d in exercise 25.
31. Write IUPAC names for the following compounds:

    (a) CH_3CH_2 $CH_3CHCH_2CH_3$
                 | Br |
    $CH_3CHCH_2CCH_2CHCH_2CH_3$
                    |
                    CH_3CH
                      |
                      CH_3

    (b) $CH_3CHCH_2CH(CH_2)_3CH_3$
         CH_3CCH_3 $CH_2CH_2CH_3$
            |
          CH_2Cl

32. Write IUPAC names for the following compounds:

    (a) $CH_3CHCH_2CHCHCH_2CH_2CH_2CH_3$ with CH_3, CH, $CHCH_3$, H_3C, CH_3, CH_2CH_3 substituents

    (b) $CH_3CH_2CHClCHCH_2CH(CH_2)_3CH_3$ with $CH(CH_3)_2$ and $C(CH_3)_3$ substituents

Physical Properties of Alkanes

33. Predict whether the following are solids, liquids, or gases at room temperature and standard pressure: (a) $C_{10}H_{22}$, (b) C_2H_6, (c) $C_{25}H_{52}$, (d) C_5H_{12}.
34. Which compounds in exercise 33 are more soluble in hexane than in water?
35. Some gasolines are formulated for cold climates by adding pentanes and other small molecules to help

the gasoline vaporize better and ignite. Why does pentane vaporize more readily than octane?

36. Water does not dissolve well in gasoline, yet gasoline spilled on the skin dries out the skin. Explain how this happens.

Chemical Reactions of Alkanes

37. Write a chemical equation for the complete combustion of butane in a cigarette lighter.
38. Write a chemical equation for the complete combustion of cyclopentane.
39. What is the role of heat or UV light in the halogenation of an alkane?
40. For the following reactions, write condensed structural formulas and IUPAC names for all the different monochloride products that could form:
    (a) $CH_3CH_2CH_2CH_3 + Cl_2 \xrightarrow{UV}$
    (b) $CH_3\overset{\overset{\displaystyle CH_3}{|}}{C}HCH_3 + Cl_2 \xrightarrow{UV}$
41. Write common names for each monochloride product that could form in the reactions in exercise 40.
42. For the reactions in exercise 40, what would happen in the absence of UV light?

Fossil Fuels and Petrochemicals

43. Classify each of the following as a combustion, halogenation, cracking, or reforming reaction:
    (a) converting decane into C_6 and C_4 hydrocarbons
    (b) converting decane into 3,4,6-trimethylheptane
    (c) making lindane (see exercise 25f) from cyclohexane
    (d) an airplane burning jet fuel
    (e) converting home heating oil into gasoline
44. Explain, in terms of intermolecular attractions, why pentane rises higher than decane in a distillation column before condensing from a gas into a liquid.
45. What does an octane number of 90 mean?
46. Less than 5% of our fossil fuels are used for petrochemicals. What are most of our fossil fuels used for?

Alkyl Halides

47. Write IUPAC names for (a) methylene chloride (CH_2Cl_2), and (b) chloroform ($CHCl_3$).
48. Leaded gasoline contains the additive 1,2-dibromoethane to convert lead into $PbBr_2$, which is emitted in the exhaust. Write (a) the structural formula for this additive and (b) the IUPAC name for the structural isomer of this additive.
49. What alkyl halides are thought to be the most responsible for depleting the ozone layer?
50. Write condensed structural formulas for all eight structural isomers having the molecular formula $C_5H_{11}Cl$.
51. Write IUPAC names for all of the compounds in exercise 50.
52. Iodoform is a yellow powder occasionally used on the skin as an antiseptic. From the formula for chloroform ($CHCl_3$), predict the formula for iodoform. Write its IUPAC name.
53. Name the alkyl halides in exercise 18.

DISCUSSION EXERCISES

1. Do you think the percentage of petroleum used for petrochemicals will increase or decrease in the 21st century as world petroleum supplies are depleted? Why?
2. Write IUPAC names and condensed structural formulas for the 18 structural isomers of octane (C_8H_{18}). How does this number (18) compare with the number of structural isomers with the molecular formula $C_8H_{17}F$?
3. Carbon tetrachloride (CCl_4) and hexane (C_6H_{14}) are liquids that have densities of 1.60 g/mL and 0.66 g/mL, respectively. Predict the number of layers and the composition of each layer in the following mixtures: (a) CCl_4 and H_2O, (b) C_6H_{14} and H_2O, (c) CCl_4 and C_6H_{14}, (d) CCl_4, C_6H_{14}, and H_2O.

13

Unsaturated and Aromatic Hydrocarbons

1. What are the names and structures of some unsaturated hydrocarbons called alkenes and alkynes?
2. What is geometric isomerism?
3. What are some physical properties, uses, and chemical reactions of alkenes and alkynes?
4. What are the names and structures of some aromatic hydrocarbons?
5. What are some physical properties, uses and chemical reactions of aromatic compounds?

The Birth of Birth Control Pills

In the 1930s physicians used a female sex hormone called progesterone (Figure 13.1) to treat menstrual disorders and miscarriages. Because it was inactivated in the digestive tract, progesterone—like all sex hormones—had to be given by injection.

Two chemical discoveries led to the synthetic sex hormones now used in birth control pills. In the 1940s German scientists found that attaching an acetylene group (—C≡CH) to the five-carbon ring in natural hormones protected the compounds from inactivation in the digestive tract. This meant the hormones could be taken orally and remain active. Then a group of American scientists discovered a way to make hormones more active by removing the methyl group bonded to carbon at the juncture of two rings (Figure 13.1). In the early 1960s, two drug companies (Syntex and G.D. Searle) began marketing oral birth control pills containing synthetic estrogen and progesterone compounds.

These synthetic hormones prevent ovulation by creating a "false pregnancy." They also cause changes in the cervical mucus, in the lining of the uterus, and in the action of the fallopian tubes that further reduce the chances of conception and implantation. Side effects include nausea, weight gain, headaches, reduced menstrual flow, depression, and an increased risk of clots forming in blood vessels.

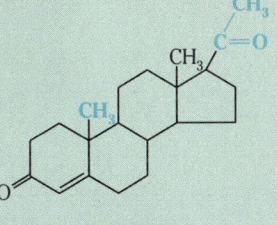

Figure 13.1 A natural female sex hormone (progesterone) and a synthetic hormone (norethindrone) used in birth control pills. Structural differences are shown in color.

In 1990 the U.S. Food and Drug Administration approved the first injectable contraceptive, a synthetic progesterone encased in plastic tubes. Six matchstick-size tubes (above) implanted under the skin of a woman's arm slowly release hormone and prevent conception for up to five years.

As you read this sentence, your eyes see letters and words because of some incredibly fast chemical reactions involving carbon-to-carbon double bonds in vitamin A (Figure 13.2). Vitamin A is just one of many important compounds in your body that have covalent double bonds between carbon atoms. We call such compounds *unsaturated*.

Figure 13.2 As you look at this picture, chemical reactions involving unsaturated vitamin A are occurring in your eyes.

Other organic compounds, called *aromatic* compounds, contain one or more six-carbon rings (or something similar) with a special blending of single and double bonds between ring atoms. A few examples are estrogens (female sex hormones), insulin, sunscreen compounds, Styrofoam, and vitamin K.

13.1 ALKENES: NAMES AND STRUCTURES

Alkenes are a family of hydrocarbons that have one or more double bonds between carbon atoms. Because carbon atoms joined by a double bond cannot bond to as many hydrogen atoms as those joined only by single bonds, alkenes are also called **unsaturated hydrocarbons.**

Compare the structures of an alkene and an alkane with the same number of carbon atoms:

an alkane (C_3H_8)   an alkene (C_3H_6)

Notice that the three-carbon alkane (propane, C_3H_8) has two more hydrogens than the three-carbon alkene. Since carbon atoms form only four covalent bonds, double-bonded carbon atoms in an alkene have fewer bonds left to bond to hydrogen. Each double bond in a hydrocarbon results in two fewer hydrogen atoms. So while alkanes have the general formula C_nH_{2n+2}, alkenes containing one double bond have the general formula, C_nH_{2n} (Table 13.1).

Cyclic alkenes with one double bond have the formula C_nH_{2n-2} (see cyclohexene in Table 13.1).

Naming Alkenes. Once you can name alkanes, it is easy to name alkenes. Consider the following compound as an example:

$$CH_3CHCHCH_2CHCH_2CH_2CH_3$$
with Cl on C2, CH_3 on C3, and $CH=CH_2$ on C5

Table 13.1 Some Alkenes and Alkynes

Name	Molecular Formula	Condensed Structural Formula	Melting Point (°C)	Boiling Point (°C)	State at 25°C
Alkenes					
ethene (ethylene)	C_2H_4	$CH_2{=}CH_2$	−169	−104	gas
propene (propylene)	C_3H_6	$CH_2{=}CHCH_3$	−185	−47	gas
1-butene	C_4H_8	$CH_2{=}CHCH_2CH_3$	−185	−6	gas
cis-2-butene	C_4H_8	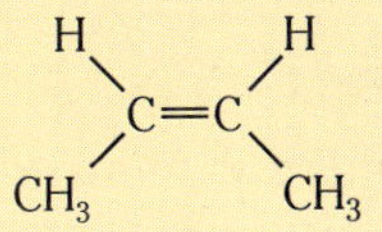 (H,H on same side; CH₃,CH₃ on same side of C=C)	−139	4	gas
trans-2-butene	C_4H_8	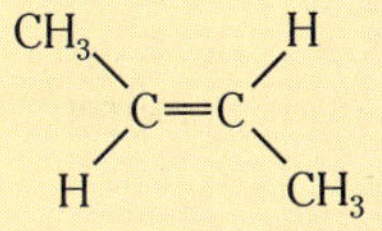 (CH₃,H / H,CH₃ across C=C)	−105	1	gas
1,3-butadiene	C_4H_6	$CH_2{=}CHCH{=}CH_2$	−109	−4	gas
1-pentene	C_5H_{10}	$CH_2{=}CH(CH_2)_2CH_3$	−138	30	liquid
1-hexene	C_6H_{12}	$CH_2{=}CH(CH_2)_3CH_3$	−140	63	liquid
cyclohexene	C_6H_{10}	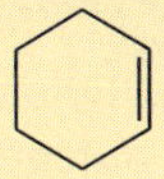 (cyclohexene ring)	−103	83	liquid
1-decene	$C_{10}H_{20}$	$CH_2{=}CH(CH_2)_7CH_3$	−66	171	liquid
Alkynes					
ethyne (acetylene)	C_2H_2	$CH{\equiv}CH$	−81	−84	gas
1-butyne	C_4H_6	$CH{\equiv}CCH_2CH_3$	−126	8	gas
1-hexyne	C_6H_{10}	$CH{\equiv}C(CH_2)_3CH_3$	−132	71	liquid

In the IUPAC system, alkenes are named in the same way as alkanes, except for three changes:

1. *Find the longest continuous carbon chain that contains the double bond(s), even if it isn't the longest carbon chain,* and *number from the end nearer the double bond.* In this example, the longest carbon chain *containing the double bond* is 7. Number from the right:

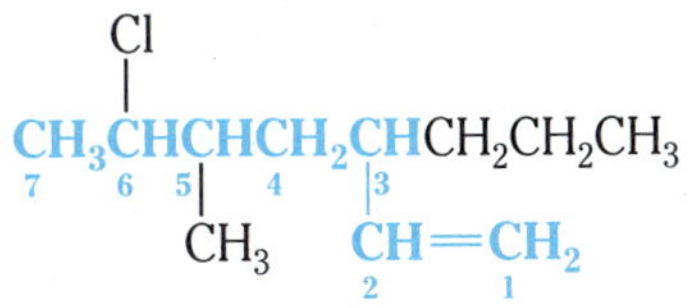

$$\underset{7}{CH_3}\underset{6}{CH}\underset{5}{\underset{|}{CH}}\underset{4}{CH_2}\underset{3}{\underset{|}{CH}}CH_2CH_3$$
$$\quad\quad\;\; CH_3 \quad\; \underset{2}{CH}{=}\underset{1}{CH_2}$$

(with Cl on C-6)

2. *Write the alkane name corresponding to the longest carbon chain, then change the -ane suffix to -ene.* In this example the name of the seven-carbon chain is *heptene*.

3. *Identify by number the lower-numbered carbon atom containing a double bond, followed by a hyphen directly before the alkene name; precede this by naming substituents in the usual way. If two or more double bonds occur, specify each double bond by number and use an appropriate prefix (such as di- or tri-) before -ene to indicate the number of double bonds.* Since in this example the double bond is between carbons 1 and 2, show this with the number *1-heptene*. Then collect prefixes for substituent groups and arrange them in alphabetical order to make the name for this compound: *6-chloro-5-methyl-3-propyl-1-heptene*.

The simplest alkenes (Table 13.1) are ethene and propene. No number (such as 1-propene) is used to locate the double bond because you would always number the chain to put the double bond between carbons 1 and 2. A four-carbon alkene, however, can have two different forms:

$$\begin{array}{c} H\ H\ H\ H \\ |\ |\ |\ | \\ H-C-C-C=C-H \\ |\ | \\ H\ H \end{array} \qquad \begin{array}{c} H\ H\ H\ H \\ |\ |\ |\ | \\ H-C-C=C-C-H \\ |\ \ \ \ \ \ \ \ \ | \\ H\ \ \ \ \ \ \ \ \ H \end{array}$$

1-butene (C_4H_8)   2-butene (C_4H_8)

Notice how this name fits with Rule 3.

Compounds with more than one multiple carbon-to-carbon bond are **polyunsaturated.** The most familiar examples are vegetable oils (Figure 13.3). These compounds have long hydrocarbon chains (see Chapter 20). A simpler example of a polyunsaturated compound is:

$$\begin{array}{c} H\ \ H\ \ H\ \ H \\ |\ \ \ |\ \ \ |\ \ \ | \\ H-C=C-C=C-H \end{array}$$

1,3-butadiene (C_4H_6)

Condensed structural formulas for alkenes show all double bonds. The formulas for 1-butene, 2-butene, and 1,3-butadiene are, respectively, CH_2=$CHCH_2CH_3$, CH_3CH=$CHCH_3$, and CH_2=$CHCH$=CH_2.

Figure 13.3 Sunflower oil, like many vegetable oils, is a polyunsaturated compound. It is also an ester (see Section 16.5).

EXAMPLE 13.1 Name the following alkenes:
(a) $CH_3CH_2CH_2CH$=$CHCH_3$ (b) $CH_3CHCH_2CH_2CH$=CH_2
                                                                                |
                                                                                CH_2CH_3

(c) [3-chlorocyclopentene structure]   (d) CH_2=$CH(CH_2)_4CH$=CH_2

SOLUTION
(a) 2-hexene (number from the right)
(b) 5-methyl-1-heptene (number from the right; longest chain with the double bond has 7 carbon atoms; list substituent groups before the location of the double bond)
(c) 3-chlorocyclopentene (no number to locate the double bond; it is always between C_1 and C_2 in cylic compounds)
(d) 1,7-octadiene (like 1,3-butadiene above)

Common names are rare for alkenes. The main ones are ethylene (for ethene) and propylene (for propene). The common name for 2-methylpropene is isobutylene, and the group, CH_2=CH—, is called the *vinyl* group. Two examples are:

CH_2=$CHCl$ $\begin{array}{c} CH_3 \\ | \\ CH_3-C=CH_2 \end{array}$

chloroethene (IUPAC)   2-methylpropene (IUPAC)
vinyl chloride (common)   *isobutylene* (common)

EXAMPLE 13.2 Write condensed structural formulas for the following: (a) 3-hexene, (b) vinyl bromide, (c) 2-chloro-1-butene.

SOLUTION (a) $CH_3CH_2CH= CHCH_2CH_3$, (b) $CH_2= CHBr$, (c) $CH_2= CClCH_2CH_3$

13.2 ALKYNES: NAMES AND STRUCTURES

Hydrocarbons that contain a triple bond between carbon atoms are called **alkynes.** The alkyne of greatest importance is the simplest one, called ethyne or acetylene (Table 13.1). Because they produce a very hot flame, acetylene torches are used to cut or weld metals such as steel (Figure 13.4).

Name alkynes in the same way as alkenes, using the suffix *-yne* instead of *-ene.* Acetylene is the only common name.

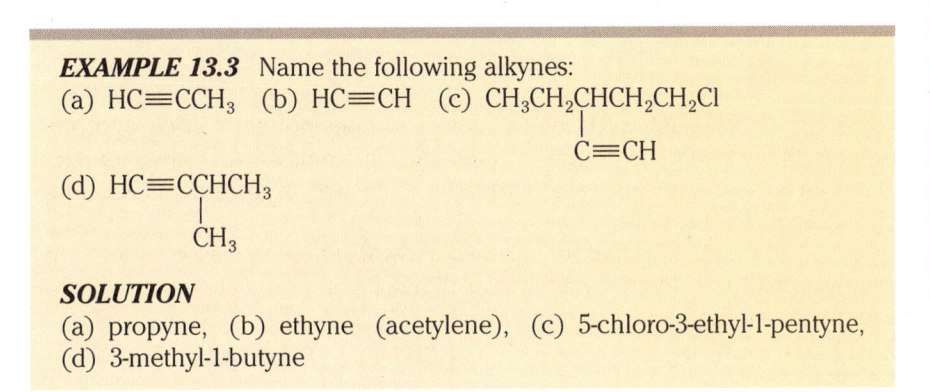

EXAMPLE 13.3 Name the following alkynes:
(a) $HC\equiv CCH_3$ (b) $HC\equiv CH$ (c) $CH_3CH_2CHCH_2CH_2Cl$
$\qquad\qquad\qquad\qquad\qquad\qquad\qquad\quad |$
$\qquad\qquad\qquad\qquad\qquad\qquad\quad C\equiv CH$

(d) $HC\equiv CCHCH_3$
$\qquad\quad |$
$\qquad\ CH_3$

SOLUTION
(a) propyne, (b) ethyne (acetylene), (c) 5-chloro-3-ethyl-1-pentyne, (d) 3-methyl-1-butyne

Figure 13.4 An acetylene torch is used for welding.

13.3 GEOMETRIC ISOMERS

Recall that double-bonded carbon atoms use sp^2 hybrid orbitals (Section 11.2); the three bonded atoms are in the same plane as carbon and separated by 120° bond angles (Figure 11.7). The arrangement is flat and triangular.

Since alkanes have only single bonds, free rotation around those bonds makes the arrangements of the bonded atoms in space interchangeable. The only kind of isomers alkanes have are structural isomers, which differ in the order in which the atoms are bonded to one another. But alkenes differ in a more subtle way. The double bond prevents rotation and fixes atoms in space. As a result, molecules with the same atoms bonded to each other may differ in how their atoms are arranged in space. These are called **geometric isomers.**

The simplest example is 2-butene. Each double-bonded carbon bonds to a hydrogen and a methyl group. The methyl groups and hydrogens on carbons 2 and 3 can be arranged in two ways: on the same side or on opposite sides (Table 13.1; Figure 13.5). Because of restricted rotation around the double bond, these two forms are not interconvertible. They are different isomers of 2-butene that have slightly different physical and chemical properties. Nonhydrogen groups are on opposite sides in the ***trans* isomer;** those groups are on the same side in the ***cis* isomer.**

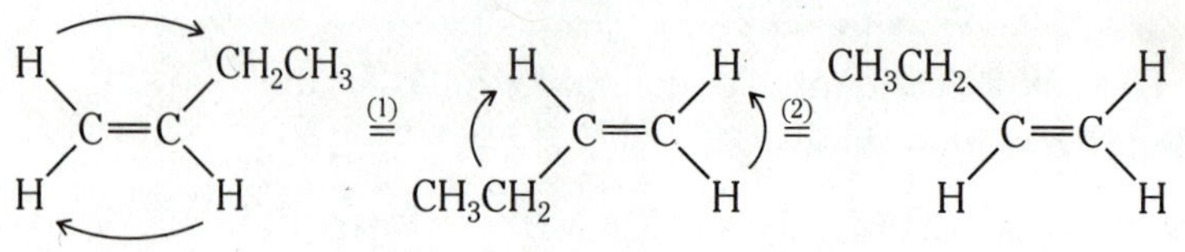

Figure 13.5 Rotating a molecule (1) in the plane of this paper (bringing groups on the right to the left and *vice versa*) and then (2) through the plane of this paper (bringing bottom groups to the top and *vice versa*) shows different forms of the *same* molecule. 1-Butene exists in only one form, which can be written in different ways (top row across). 2-Butene, however, occurs as *cis* and *trans* isomers (each can be written in two ways) that aren't interconvertible because they cannot rotate about the double bond.

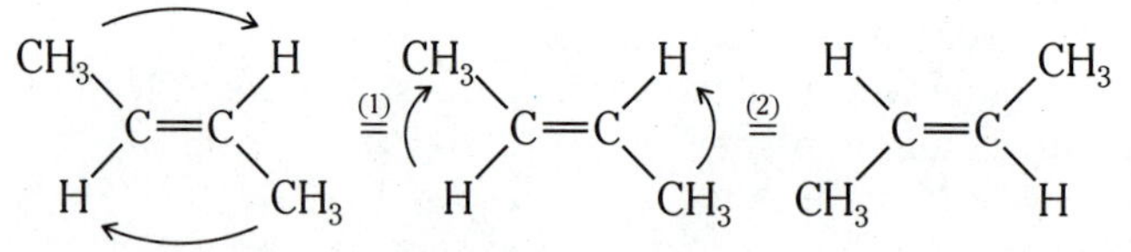

1-butene

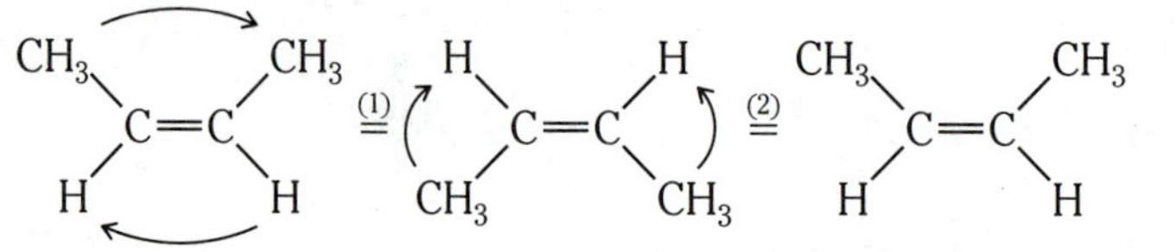

cis-2-butene

trans-2-butene

Alkenes occur as cis *and* trans *geometric isomers when neither double-bonded carbon bonds to two identical groups.*

A compound such as 1-butene, however, does not exist in *cis* and *trans* isomers. Since one of its double-bonded carbon atoms bonds to two identical atoms or groups (in this case, hydrogen), the two arrangements are interchangeable (Figure 13.5).

The structural formulas of alkenes shown earlier in this chapter don't clearly identify a structure as *cis* or *trans*. When it is important to do so, write the structures as follows:

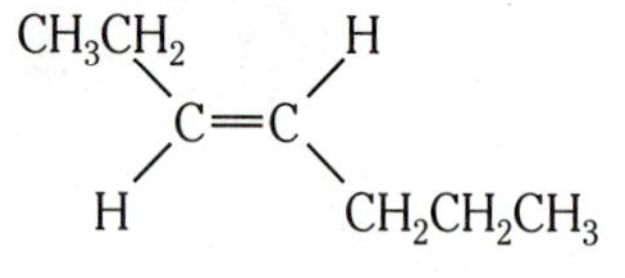

trans-3-heptene 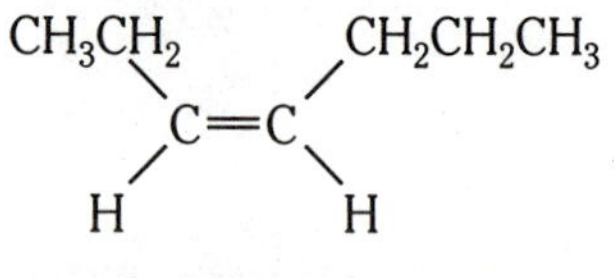   cis-3-heptene

EXAMPLE 13.4

(a) Which of the following exist as geometric isomers in the *cis* and *trans* forms:

(1) $CH_3CH_2CH_2CH=C(CH_3)_2$ (2) $CH_3CH_2CH_2CH=CHCH_3$
(3) $CH_3CH=CHCH_3$

(b) Write structural formulas to show the *trans* isomer of each.

SOLUTION (a) Compound (1) doesn't have *cis* and *trans* forms because one double-bonded carbon bonds to two identical (methyl) groups; (2) and (3) have *cis* and *trans* forms. The *trans* forms are:

(b) (2)   (3)

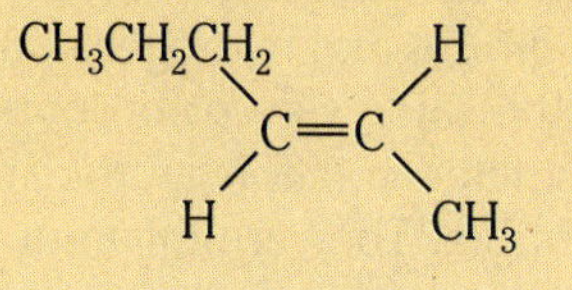

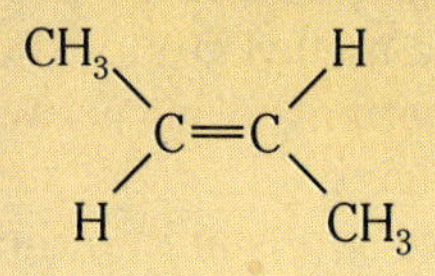

What practical difference does it make whether a compound is *cis* or *trans*? An important example is occurring in your body right now. As you read this,

CHEMISTRY SPOTLIGHT

Vitamin A and Scientific Knowledge

Figure 13.6 George Wald (1906–   ).

George Wald (Figure 13.6), a biochemist at Harvard University, discovered how vitamin A reacts in the eye when vision occurs. For this work he received the Nobel prize in physiology or medicine in 1967.

But scientific knowledge can be used in many ways, as Wald soon learned. In the late 1960s, the U.S. Department of Defense asked him to help develop chemical blinding agents. The very knowledge he had discovered—the chemistry of vision—had a potential use in chemical warfare. Wald angrily refused the request.

photons of light are changing millions of molecules of a vitamin A derivative in your eyes from the *cis* to the *trans* form (see Chemistry Spotlight). Special catalysts (enzymes called isomerases) then convert the *trans* isomers back into the *cis* form so you can use them again (Figure 13.7).

cis isomer

↓ light

trans isomer

Figure 13.7 A pulse (photon) of light converts the *cis* isomer of a vitamin A derivative (retinal) into the *trans* isomer to initiate the events that produce vision. The change is shown in color.

13.4 PHYSICAL PROPERTIES AND CHEMICAL REACTIONS OF ALKENES AND ALKYNES

Physical Properties and Uses. Like alkanes, alkenes and alkynes are nonpolar hydrocarbons that don't dissolve in water and other polar solvents. Being less dense than water, they float. Their melting and boiling points are about the

Figure 13.8 A miner's acetylene lamp. Acetylene (ethyne) gas is lit by a spark from a sealed internal light mechanism.

In addition reactions, unlike substitution reactions (Section 12.5), all atoms in the organic reactant remain in the organic product.

same (or slightly lower) as those of comparable alkanes, so small alkenes and alkynes (up to 4 carbons) are gases, while those with 5–17 carbons are liquids at room temperature (Table 13.1).

Alkenes and alkynes have many uses. Synthetic materials such as polyethylene, Teflon, polyvinyl chloride (PVC), and Plexiglas (see Chapter 18) are all made from alkenes. Many natural products—cholesterol, insect sex attractants, vitamins, and vegetable oils, to name a few—contain alkene groups. As fruits and vegetables ripen, they produce ethylene (ethene) gas, which helps them ripen even more. Food processors take advantage of this effect by treating produce with ethylene just before bringing it to market.

Besides producing very hot flames for welding, acetylene illuminates miners' lamps (Figure 13.8). Attaching acetylene groups to natural hormones also was a key to developing oral birth control pills (see Chapter Opener).

Chemical Reactions. Multiple bonds are functional groups (Table 11.3) that make alkenes and alkynes more reactive than alkanes. Recall that multiple bonds contain pi bonds from the side-to-side overlap of *p* orbitals (Section 11.2). Pi bonds are relatively weak because this orbital overlap is less extensive than in sigma bonds. As a result, pi electrons are vulnerable to attack by substances that seek electrons to become more stable.

Unsaturated compounds like alkenes and alkynes participate in **addition reactions** to form saturated products. In an addition reaction, a pair of pi electrons is used to form two sigma bonds. The general reaction with an alkene is:

$$R-CH=CH_2 + X-Y \longrightarrow R-\underset{X}{\underset{|}{CH}}-\underset{Y}{\underset{|}{CH_2}} \text{ or } R-\underset{Y}{\underset{|}{CH}}-\underset{X}{\underset{|}{CH_2}}$$

Alkynes add *two* moles of reagent per mole of alkyne because two pi bonds in each molecule are broken. One pattern is:

$$R-C\equiv C-H + X-Y \longrightarrow R-\underset{X}{\underset{|}{C}}=\underset{Y}{\underset{|}{C}}-H$$

$$R-\underset{X}{\underset{|}{C}}=\underset{Y}{\underset{|}{C}}-H + X-Y \longrightarrow R-\underset{X}{\overset{X}{\underset{|}{\overset{|}{C}}}}-\underset{Y}{\overset{Y}{\underset{|}{\overset{|}{C}}}}-H$$

Figure 13.9 summarizes the main addition reactions for alkenes. They react with hydrogen gas (H_2), halogens (especially Cl_2 and Br_2), inorganic acids such as HCl and HBr, and water.

1. Catalytic hydrogenation The reaction with hydrogen gas is called **catalytic hydrogenation** because it requires a metal catalyst such as platinum, nickel, or palladium. Hydrogen adds to the two carbons joined by a double bond to form an alkane. One example is:

$$CH_3CH=CHCH_3 + H_2 \xrightarrow{Ni} CH_3\underset{H}{\underset{|}{CH}}\underset{H}{\underset{|}{CH}}CH_3 \quad (CH_3CH_2CH_2CH_3)$$

2-butene
(cis or trans)

butane

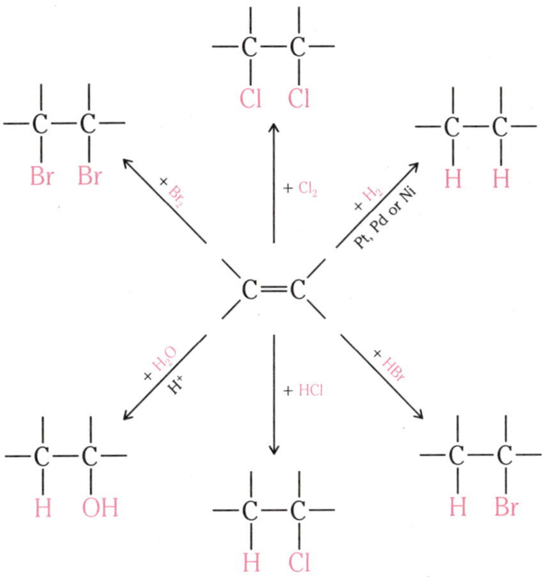

Figure 13.9 Addition reactions of alkenes. Reactions with HCl, HBr, and H₂O follow Markovnikov's rule.

EXAMPLE 13.5 Write structural formulas and names for the products of the following reactions:

(a) $CH_2=CHCl + H_2 \xrightarrow{Pt}$

(b) 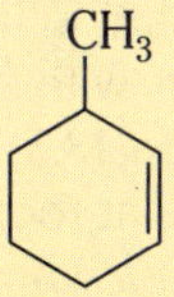 [methylcyclohexene] $+ H_2 \xrightarrow{Ni}$

(c) $CH_3C≡CH + 2 H_2 \xrightarrow{Pd}$

SOLUTION (a) CH_3CH_2Cl, chloroethane (ethyl chloride)

(b) 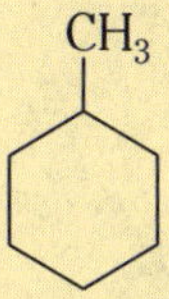 [methylcyclohexane structure], methylcyclohexane

(c) $CH_3CH_2CH_3$, propane

Vegetable oil molecules typically have three long hydrocarbon chains (C_{12}–C_{20}) containing one or more double bonds (Section 20.1). Food companies use catalytic hydrogenation to make vegetable oils into soft solids such as shortening (Figure 13.10), which provides the desired texture for cooking certain foods.

2. Halogenation Chlorine (Cl_2) and bromine (Br_2) add readily to alkenes without the need for a catalyst. The products have a chlorine (or bromine) atom bonded to each carbon in the original double bond. For example:

$$CH_3CH=CH_2 + Br_2 \longrightarrow CH_3CH(Br)-CH_2(Br)$$

  propene   *1,2-dibromopropane*

$$\begin{array}{l} CH_2-O-\overset{\overset{O}{\|}}{C}-(CH_2)_7-CH=CHCH_2\,CH=CHCH_2CH=CHCH_2CH_3 \\ | \quad\quad\quad O \\ \quad\quad\quad \| \\ CH-O-C-R \\ | \quad\quad\quad O \\ \quad\quad\quad \| \\ CH_2-O-C-R \end{array}$$

vegetable oil

$+ H_2 \downarrow Pt$

$$\begin{array}{l} CH_2-O-\overset{\overset{O}{\|}}{C}-(CH_2)_7-CH=CHCH_2\,CH_2CH_2CH_2CH_2CH_2CH_2CH_3 \\ | \quad\quad\quad O \\ \quad\quad\quad \| \\ CH-O-C-R \\ | \quad\quad\quad O \\ \quad\quad\quad \| \\ CH_2-O-C-R \end{array}$$

shortening

Figure 13.10 Partial hydrogenation of vegetable oil produces shortening. Changes are shown in color using simplified structures.

Recall that Br_2 reacts with alkanes only in the presence of UV light or heat (Section 12.5).

This addition reaction is a useful way to distinguish between alkanes and unsaturated hydrocarbons (Figure 13.11). Typically used as a solution in CCl_4, bromine has a reddish-brown color that disappears when it reacts with alkenes and alkynes (but not alkanes or aromatic compounds; see Section 13.5). If you add a small amount of Br_2 solution to an unknown hydrocarbon at room temperature, you can identify it as saturated or unsaturated by seeing whether or not the color disappears. The saturated dibromo products are colorless.

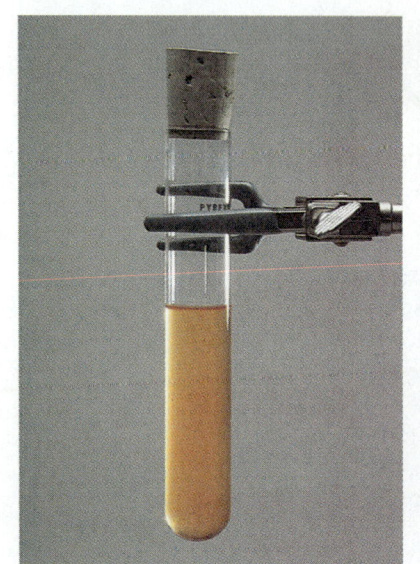

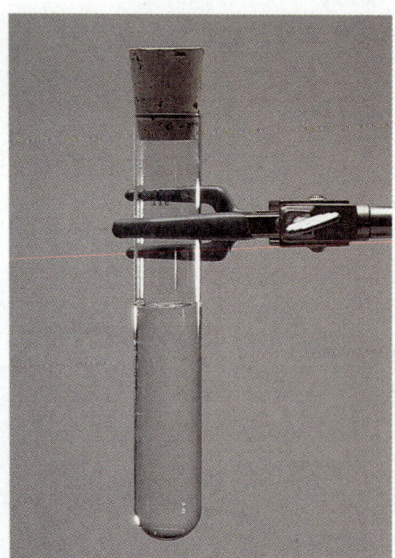

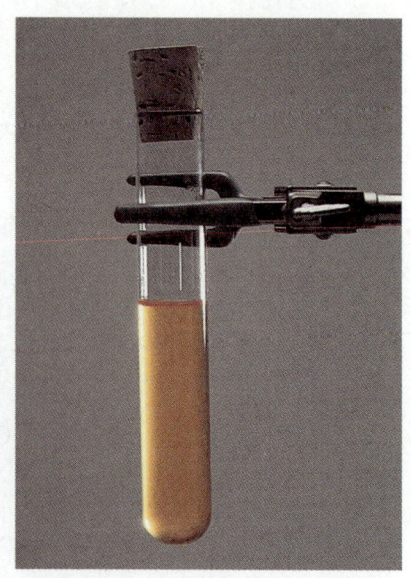

Figure 13.11 When a solution of Br_2 in CCl_4 is added to an alkane, hexane (left), the color from Br_2 remains. Adding Br_2 solution to an alkene, 1-hexene (center), causes the color to disappear. The color remains when Br_2 is added to benzene (right; see Section 13.5).

3. Addition of HCl and HBr The inorganic acids HCl and HBr readily add to alkenes and alkynes. Hydrogen bonds to one of the double- (or triple-) bonded carbons and the halogen (Cl or Br) bonds to the other. If we represent the acid as H—X, the general reaction is:

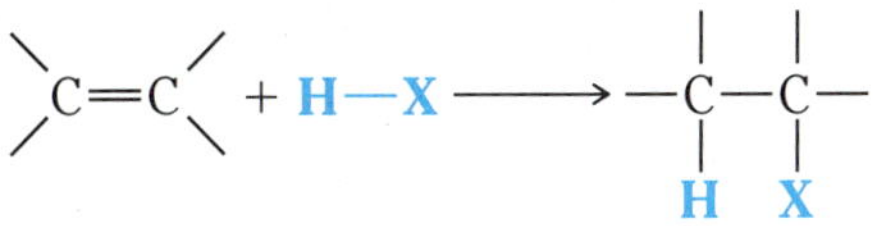

$$\begin{array}{c}\diagdown\\ /\end{array}C=C\begin{array}{c}\diagup\\ \diagdown\end{array} + H-X \longrightarrow \begin{array}{c}|\;\;\;|\\-C-C-\\|\;\;\;|\\H\;\;X\end{array}$$

But H—X substances—unlike H_2, Cl_2 and Br_2—contain two different groups. How do you know which part (H or X) adds to which carbon in the double bond?

In a reaction with ethene, for example, the order doesn't matter. The same product forms either way:

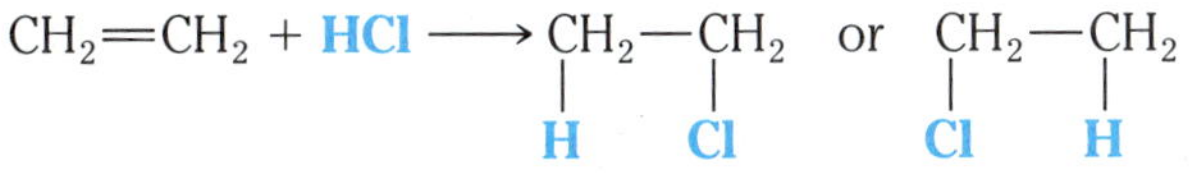

$$CH_2{=}CH_2 + HCl \longrightarrow \underset{\underset{H\;\;\;Cl}{|\;\;\;|}}{CH_2{-}CH_2} \text{ or } \underset{\underset{Cl\;\;\;H}{|\;\;\;|}}{CH_2{-}CH_2}$$

ethene   both are *chloroethane* (*ethyl chloride*)

With propene, though, two products are possible:

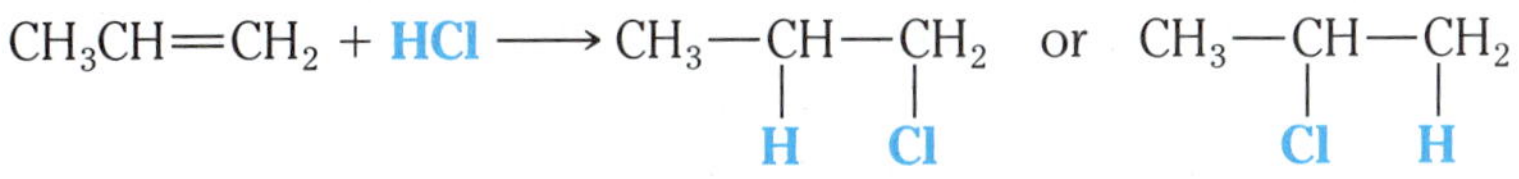

$$CH_3CH{=}CH_2 + HCl \longrightarrow \underset{\underset{H\;\;\;Cl}{|\;\;\;|}}{CH_3{-}CH{-}CH_2} \text{ or } \underset{\underset{Cl\;\;\;H}{|\;\;\;|}}{CH_3{-}CH{-}CH_2}$$

propene   *1-chloropropane*   *2-chloropropane*

More than a century ago, a Russian chemist named Vladimir Markovnikov studied many reactions like this and discovered a pattern, known as **Markovnikov's rule** (see Chemistry Spotlight): *When H—X adds to an alkene or alkyne, H adds predominantly to the carbon in the multiple bond that already has the greater number of hydrogens; X adds to the other carbon.*

In the preceding reaction, carbon 2 in propene is bonded to one hydrogen while carbon 1 is bonded to two hydrogens. According to Markovnikov's rule, the H in HCl should add predominantly to carbon 1 while Cl adds to carbon 2 to produce 2-chloropropane. This is indeed the major product in the reaction.

When H—X substances add to an alkyne, Markovnikov's rule still applies, but you need to look at the reaction in two steps. For example:

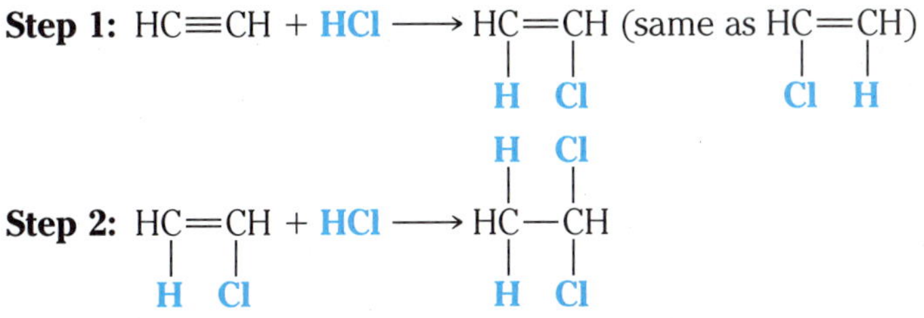

Step 1: $HC{\equiv}CH + HCl \longrightarrow \underset{\underset{H\;\;\;Cl}{|\;\;\;|}}{HC{=}CH} \text{ (same as } \underset{\underset{Cl\;\;\;H}{|\;\;\;|}}{HC{=}CH}\text{)}$

Step 2: $\underset{\underset{H\;\;\;Cl}{|\;\;\;|}}{HC{=}CH} + HCl \longrightarrow \underset{\underset{H\;\;\;Cl}{|\;\;\;|}}{\overset{\overset{H\;\;\;Cl}{|\;\;\;|}}{HC{-}CH}}$

1,1-dichloroethane

In the first step, H and Cl could add either way since both triple-bonded carbons are bonded to the same number of hydrogens (one). But in step 2, Markovnikov's rule applies and H adds to the carbon that received H in the first step. The product, then, is 1,1-dichloroethane, not 1,2-dichloroethane.

CHEMISTRY SPOTLIGHT

The Last Word on Reaction Mechanisms

Why does Markovnikov's rule on the addition of HX compounds to alkenes work? It isn't by magic. The details of the reaction, called the *reaction mechanism,* can help you understand why certain products form and others don't.

Consider the addition reaction between propene and HCl. Acids are proton donors (Section 9.1). Since HCl is a strong acid (Section 9.2), it separates almost completely into H⁺ and Cl⁻ ions. H⁺ has no valence electrons; it would be more stable if it had 2. One way to achieve the more stable structure is to react with an unshared pair of electrons on an oxygen atom in H₂O to form H₃O⁺ (hydronium ion). H⁺ can also react with the weakly shared pi electrons in propene, forming a covalent bond and breaking the double bond:

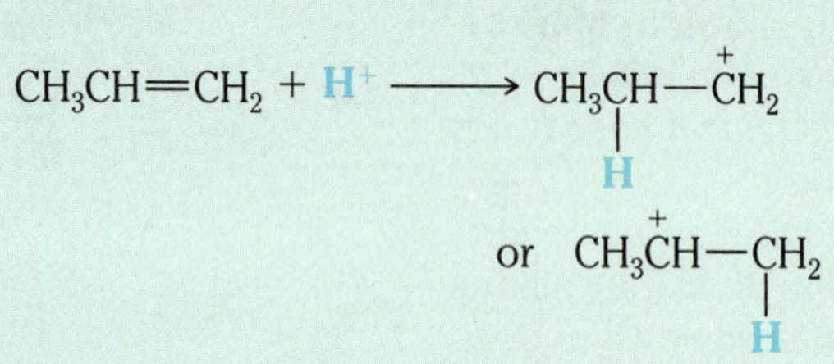

$$CH_3CH=CH_2 + H^+ \longrightarrow CH_3\overset{+}{C}H-CH_2 \atop |\ \ \ H$$

$$\text{or}\quad CH_3CH-\overset{+}{C}H_2 \atop |\ \ \ H$$

This reaction produces a positively charged carbon, called a *carbocation.*

Notice that two carbocations could form: the one with the charge on a primary carbon is a *primary carbocation;* the one with the positive charge on a secondary carbon atom is a *secondary carbocation.* Much evidence shows that primary carbocations are less stable than secondary carbocations (which in turn are less stable than tertiary carbocations). In the preceding reaction the secondary carbocation is the main product because it is more stable.

Carbocations react quickly with substances that provide an electron to stabilize them. The secondary carbocation, then, reacts immediately with Cl⁻ ion (formed from HCl) to form the main product, 2-chloropropane:

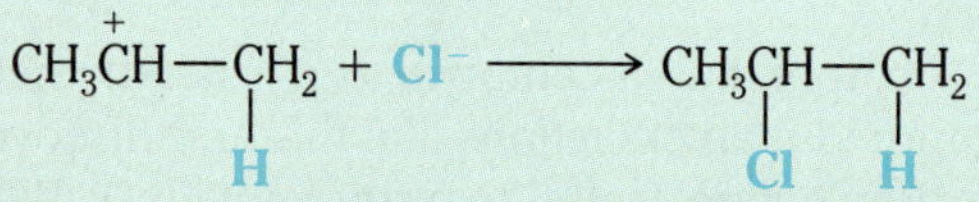

$$CH_3\overset{+}{C}H-CH_2 + Cl^- \longrightarrow CH_3CH-CH_2 \atop |\quad\quad\quad\quad\quad\quad\quad\quad\ \ |\ \ \ \ | \atop H\quad\quad\quad\quad\quad\quad\quad\quad\quad\ Cl\ H$$

It is the relative stability of carbocation intermediates that leads to the products predicted by Markovnikov's rule. Markovnikov, however, had no knowledge of this reaction mechanism.

Chemists use reaction mechanisms to understand a wide variety of organic reactions, including those in the following chapters. Although the details of these mechanisms are useful, they are beyond the scope of this book.

EXAMPLE 13.6 Write structural formulas for the major products in the following reactions:

(a) 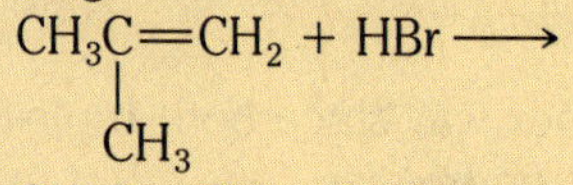 $CH_3C=CH_2 + HBr \longrightarrow$
   $\ \ \ \ \ \ |$
   $\ \ \ \ CH_3$

(b) $CH_3C\equiv CH + 2\ HBr \longrightarrow$

(c) $CH_3C\equiv CCH_3 + 2\ HCl \longrightarrow$

SOLUTION (a) CH₃C(CH₃)(Br)CH₃, (b) CH₃C(Br)₂CH₃, (c) CH₃C(Cl)₂CH₂CH₃

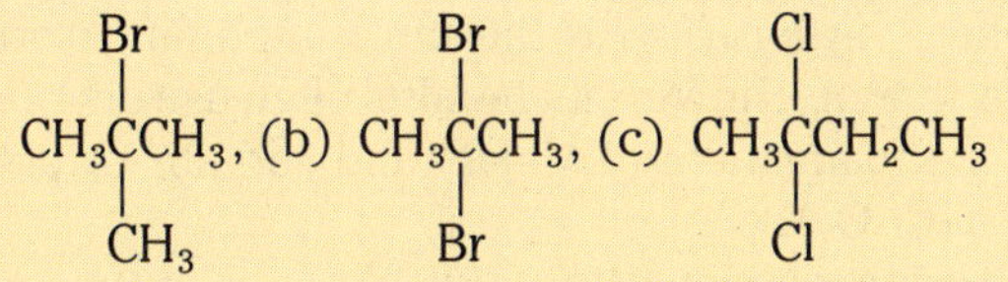

4. Hydration In the presence of a strong acid catalyst such as H_2SO_4 (sometimes represented simply as H^+), water adds to multiple bonds in a **hydration** reaction. If we represent water as H—OH, the general reaction is:

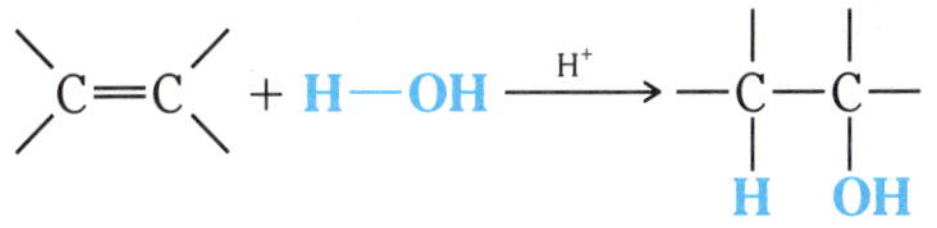

$$\ce{>C=C< + H-OH ->[H^+] -C(H)-C(OH)-}$$

The reaction mechanism is like that for the addition of HCl or HBr to an alkene. As a result, Markovnikov's rule applies in the same way (and for the same reason) to predict the major product when more than one can form. One example of a hydration reaction is:

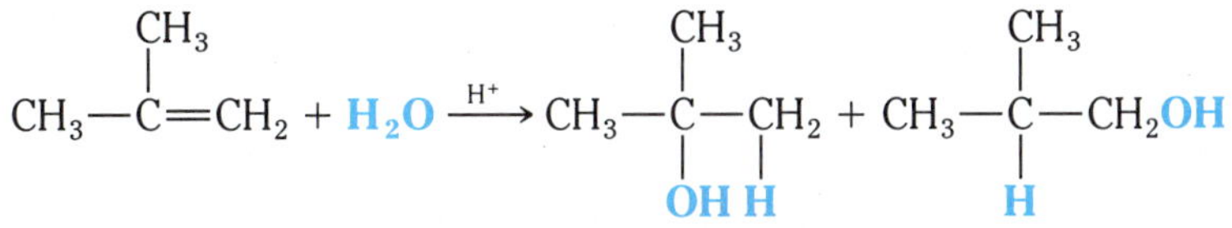

$$\ce{CH_3-C(CH_3)=CH_2 + H_2O ->[H^+] CH_3-C(CH_3)(OH)-CH_3 + CH_3-C(CH_3)(H)-CH_2OH}$$

major product

Hydration reactions are an important method to synthesize alcohols (see Chapter 14).

EXAMPLE 13.7 Write structural formulas for the major product in each of the following reactions:

(a) $\ce{CH_3CH_2CH_2CH=CH_2 + H_2O ->[H^+]}$

(b) 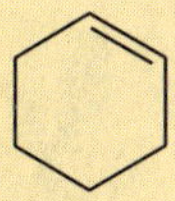 cyclohexene $+ H_2O \xrightarrow{H_2SO_4}$

SOLUTION (a) 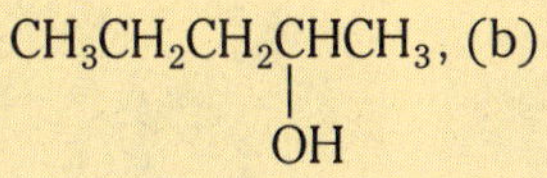 $\ce{CH_3CH_2CH_2CH(OH)CH_3}$, (b) 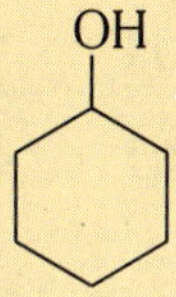 cyclohexanol (cyclohexane with OH)

---

5. Oxidation Remember that oxidation occurs when a substance loses electrons, and reduction occurs when a substance gains electrons (Section 5.2). These definitions can get complicated when we apply them to organic reactions, so we will use a related, but slightly different, version: *oxidation is the gain of oxygen atoms or loss of hydrogen atoms; reduction is the loss of oxygen or gain of hydrogen atoms* (Table 5.3).

Like all hydrocarbons, alkenes and alkynes are oxidized by O_2 in combustion reactions to form CO_2 and H_2O. If you burn 1-butene, for example, the reaction is:

$$C_4H_8 + 6O_2 \longrightarrow 4CO_2 + 4H_2O$$

You can tell from the equation that this is an oxidation reaction because both the carbon and hydrogen in 1-butene gain oxygen atoms to form CO_2 and H_2O.

294 Part II Organic Chemistry

Alkenes, with their pi electrons, are a ready target for oxidizing (electron-seeking) agents. Oxygen (O_2) is the most important oxidizing agent in your body. In the laboratory, however, chemists often use other effective agents such as potassium permanganate ($KMnO_4$) and sodium or potassium dichromate ($Na_2Cr_2O_7$ or $K_2Cr_2O_7$).

In the following reactions, the oxidizing agent is represented as (O), and no attempt is made to write a balanced equation. This is a common practice for two reasons: (1) we are interested mostly in the organic reactants and products, and (2) oxidizing agents often form inorganic products that make it complicated to write balanced equations.

Alkenes are oxidized under mild conditions as follows:

Notice from the structure of the product that this reaction fits the definition of oxidation (the gaining of oxygen atoms).

$$R-\underset{H}{\overset{H}{C}}=\underset{H}{\overset{H}{C}}-R \xrightarrow{(O)} R-\underset{OH}{\overset{H}{\underset{|}{C}}}-\underset{OH}{\overset{H}{\underset{|}{C}}}-R$$

If the oxidizing agent were $KMnO_4$, you could literally see that oxidation occurred (Figure 13.12) because purple $KMnO_4$ is reduced to brown manganese dioxide (MnO_2). This color change is another simple chemical test (besides the decoloration of Br_2) to distinguish between an alkane and an alkene or alkyne.

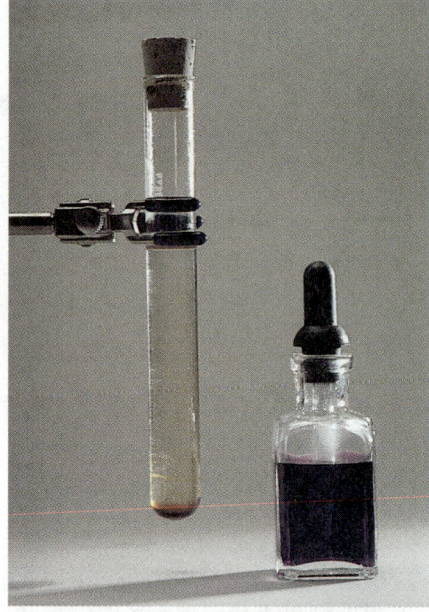

Figure 13.12 Addition of purple $KMnO_4$ to the alkene 1-hexene (left) changes the color to brown MnO_2 (right).

Some oxidizing agents break alkenes at the double bond to form two organic products:

$$R-\underset{H}{\overset{H}{C}}=\underset{H}{\overset{H}{C}}-R \xrightarrow{(O)} R-\overset{H}{\underset{|}{C}}=O + O=\overset{H}{\underset{|}{C}}-R$$

You may have noticed that vegetable oils and other fatty materials eventually turn rancid. This happens because oxygen in air oxidizes unsaturated fats to form foul-smelling products.

CHEMISTRY SPOTLIGHT

August Kekulé and the Whirling Snakes

Figure 13.13 August Kekulé (1829–1896).

According to legend, one evening in 1865 August Kekulé (Figure 13.13), was sitting half asleep in front of his fireplace. The German chemist had long puzzled over the structure of benzene. As his mind wandered, he imagined atoms dancing like snakes in the fire. Suddenly a snake seized its tail and whirled around in a circle. Kekulé awoke and spent the rest of the night working out what he had seen. He proposed that benzene is a ring of six carbon atoms, with alternating single and double bonds, and each carbon bonded to one hydrogen atom:

Kekulé wrote of his experience, "Let us learn to dream, gentlemen, then perhaps we shall find the truth. But let us beware of publishing our dreams till they have been tested by the waking understanding."

13.5 AROMATIC HYDROCARBONS

The structure of an organic compound called benzene was a mystery for a long time (see Chemistry Spotlight). Its formula is C_6H_6, and the only way a molecule with six carbon atoms could have so few hydrogens is if it is highly unsaturated. But benzene doesn't act like an alkene or alkyne. When you add Br_2 in CCl_4 to this compound, for example, the bromine color does not disappear (Figure 13.11).

Eventually, chemists concluded that benzene molecules are six-carbon rings with each carbon bonded to one hydrogen. But its ring carbons cannot be joined by alternating single and double bonds as proposed by Kekulé because benzene doesn't participate in typical addition reactions. Later measurements also showed that each ring carbon is an equal distance from its two neighboring carbons. This couldn't occur if each carbon were double-bonded to one carbon and single-bonded to the other, because double-bonded carbons are closer together (to share the pi electrons) than single-bonded carbons.

Instead, the six ring carbons equally share the three extra pairs of pi electrons that would have provided the three double bonds. It's as if each carbon joins its neighbor in the ring with an average of one and one-half bonds, with the six pi electrons moving freely about the ring.

You can picture how this works using the same sp^2 hybrid orbitals that occur in alkenes (Section 11.2). Each ring carbon in benzene bonds to two neighboring carbons and one hydrogen by sigma bonds using its three sp^2 orbitals. This puts the entire carbon ring and the six attached hydrogens in the same plane, with angles of 120° throughout the molecule.

Each carbon has one remaining *p* orbital with one electron. These six *p* orbitals and their six electrons form a doughnut-shaped space above and below the plane of the ring (Figure 13.14). They are called pi electrons, though they don't form specific pi bonds between any two adjacent carbons. The modern condensed structural formula for benzene is often written with a circle in the center to represent those mobile pi electrons (Figure 13.15).

Any compound that contains a benzene or benzenelike ring is called an **aromatic compound.** Substances with benzene rings fused together are called **polycyclic aromatic compounds.** The simplest is naphthalene, $C_{10}H_8$, with two benzene rings joined together (Table 13.2). Naphthalene is used as an insecticide in one kind of mothballs. Compounds with more fused rings occur in small amounts in cigarette smoke and automobile exhaust; many can cause cancer.

Figure 13.14 Benzene carbons with *p* orbitals (blue) perpendicular to the ring (left) form doughnut-shaped clouds (yellow) of pi electrons above and below the ring (center). A space-filling model (right) shows the flat ring and 120° angles between all atoms.

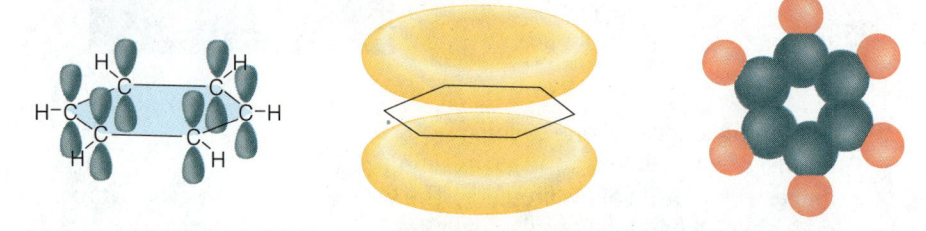

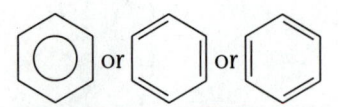

Figure 13.15 The condensed structural formula for benzene is often written with a circle, but the other two formulas (center and right) also are commonly used.

The term aromatic compound *originated because many of these substances have characteristic aromas.*

Table 13.2 Some Polycyclic Aromatic Hydrocarbons

Name	Condensed Structural Formula	Some Uses or Effects
naphthalene		Used in some mothballs
anthracene		Used in dyes
phenanthrene		Used in dyes, explosives, and synthesis of drugs
benzo[*a*]pyrene		Carcinogen in cigarette and wood smoke

Naming Aromatic Compounds. To name substances in which one hydrogen in benzene is replaced by a different atom or group, name the substituent group followed by *benzene,* unless a common name applies (Figure 13.16).

chlorobenzene   nitrobenzene   ethylbenzene

toluene   phenol   aniline

Figure 13.16 Some monosubstituted aromatic compounds. Those in the bottom row are named with their common names.

You need to learn these common names.

When benzene has two substituent groups, indicate their locations by special prefixes. Use the prefix *o-* (for *ortho*) for groups on adjacent carbons (the numbering would be 1,2), *m-* (*meta*) for substituents on positions 1 and 3, and *p-* (*para*) for 1,4 substituents. Figure 13.17 shows how this works.

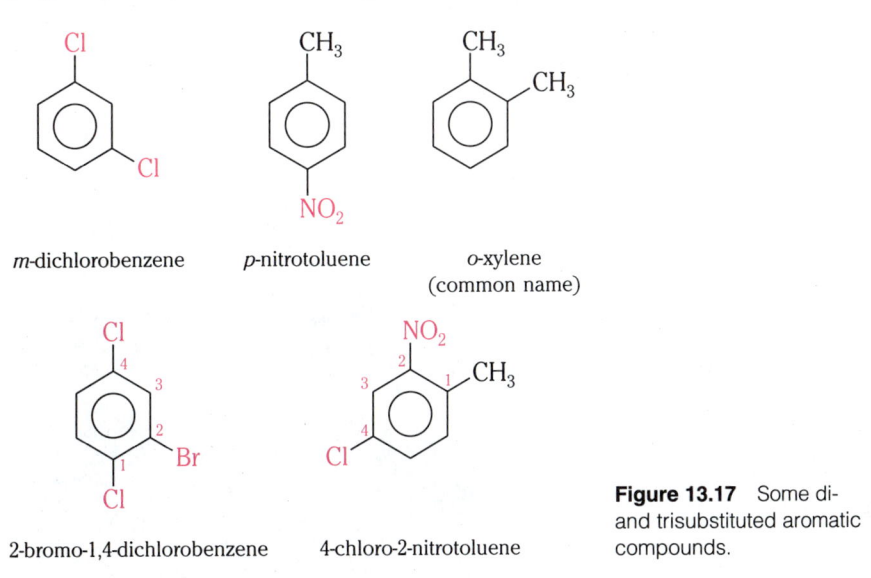

m-dichlorobenzene   *p*-nitrotoluene   *o*-xylene (common name)

2-bromo-1,4-dichlorobenzene   4-chloro-2-nitrotoluene

Figure 13.17 Some di- and trisubstituted aromatic compounds.

If a common name includes one of the substituent groups, use it in naming the compound. For example,

includes the structure for toluene (in color), so name it *p*-chlorotoluene (instead of *p*-chloromethylbenzene). In the case of xylene (Figure 13.17), the name includes *both* substituent methyl groups. If two different substituent groups both need to be named, arrange them in alphabetical order.

If the ring has three or more substituent groups, use numbers to locate substituents (Figure 13.17). Number ring carbons to give the lowest numbers for the substituents. If one substituent can be named within a common name (for example, the methyl group in toluene), make that group carbon 1 and use the common name.

EXAMPLE 13.8 Draw condensed structural formulas for (a), (b), and (c). Write IUPAC or common names (where appropriate) for the rest of the compounds. (a) *m*-chloroaniline, (b) 2,4,6-trinitrotoluene (TNT), (c) *p*-xylene,

SOLUTION

(d) *p*-bromophenol, (e) *m*-dichlorobenzene, (f) 1,3,5-trichlorobenzene, (g) *o*-ethyliodobenzene

---

dichlorodiphenyltrichloroethane (DDT)

Figure 13.18 DDT, an organochlorine pesticide containing two phenyl groups.

Sometimes it is more convenient to consider benzene as a substituent group. Then benzene is called a *phenyl* group. For example,

CH$_3$CHCH$_2$CH$_2$CH$_3$

is 2-phenylhexane. *Phenyl* also appears in the name for DDT (Figure 13.18). Phenylalanine is an amino acid (a component of proteins) that contains a phenyl group.

13.6 PHYSICAL PROPERTIES AND CHEMICAL REACTIONS OF AROMATIC COMPOUNDS

Physical Properties and Uses. Like all hydrocarbons, benzene, toluene, and other aromatic hydrocarbons are nonpolar. As a result, they have low melting and boiling points and are liquids at room temperature (except for polycyclic or other large molecules, which have strong enough London forces to be solids). Since they are nonpolar, aromatic hydrocarbons don't dissolve in water or other polar solvents but do dissolve in nonpolar solvents such as hexane.

Attaching polar groups (such as —OH, —NH$_2$, or —NO$_2$) to the benzene ring modifies these properties a bit. The melting and boiling points increase a little, as does the solubility in water, because such molecules have stronger dipole–dipole forces of attraction or hydrogen bonds (Section 6.1). The electronegativities of chlorine and bromine are close enough to carbon so that aromatic halides aren't highly polar. This is why a pesticide such as DDT (Figure 13.18) is insoluble in water and accumulates in nonpolar fatty tissues in the body.

Aromatic compounds have many important uses. You will learn in later chapters that nucleic acids (DNA and RNA), proteins, and certain vitamins are aromatic. An aromatic hydrocarbon called styrene is used to make polystyrene (Styrofoam) for insulation, packing material, or flotation (Section 18.2). Some polycyclic aromatic compounds are dyes. Aromatic drugs include aspirin, penicillin, cocaine, tetracycline, heroin, and diazepam (Valium). In addition to DDT, some other aromatic halides are toxic enough to use as pesticides or herbicides (2,4,5-T, also called "Agent Orange"; Figure 13.19).

Chemical Reactions. Although it contains pi electrons, benzene doesn't have pi bonds like alkenes or alkynes. As a result, the benzene ring doesn't participate in addition or oxidation reactions except under very vigorous conditions. Instead, ring carbons react to replace a bonded hydrogen with a different atom or group in a *substitution reaction* (see Section 12.5).

It's easy to recognize the difference between a substitution reaction and an addition reaction:

Substitution C$_6$H$_5$H + X$_2$ → C$_6$H$_5$X + HX

Addition C$_6$H$_6$ + X$_2$ → C$_6$H$_6$X$_2$

With aromatic compounds, substitution reactions are favored because they don't disrupt the stable arrangement of pi electrons above and below the plane of the benzene ring.

Figure 13.19 The aromatic herbicides 2,4,5-T ("Agent Orange") and 2,4-D were used as defoliants in the Vietnam War (left).

2,4-D (2,4-dichlorophenoxyacetic acid)

2,4,5-T (2,4,5-trichlorophenoxyacetic acid)

The main reactions are with halogens (especially Cl_2 and Br_2), nitric acid (HNO_3), sulfuric acid (H_2SO_4), and alkyl halides (mainly *R*—Cl and *R*—Br):

1. **Halogenation**

   $$C_6H_6 + Cl_2 \xrightarrow{AlCl_3 \text{ or Fe}} C_6H_5Cl + HCl$$

   $$C_6H_6 + Br_2 \xrightarrow{AlBr_3 \text{ or Fe}} C_6H_5Br + HBr$$

2. **Nitration**

   $$C_6H_6 + HO-NO_2 \xrightarrow{H_2SO_4} C_6H_5NO_2 + H_2O$$
   (HNO$_3$)

3. **Sulfonation**

   $$C_6H_6 + HO-SO_3H \longrightarrow C_6H_5SO_3H + H_2O$$
   (H$_2$SO$_4$)

4. **Alkylation**

   $$C_6H_6 + R-Cl \text{ (or } R-Br\text{)} \xrightarrow{AlCl_3 \text{ (or AlBr}_3\text{)}} C_6H_5R + HCl \text{ (or HBr)}$$

Notice that each reaction requires a catalyst. The catalyst makes the substituent into a positively charged ion that needs an electron to become stable and thus attacks a ring carbon, displacing hydrogen. We won't, however, consider the details of the reaction mechanisms.

The product that forms in reaction 3 is called benzenesulfonic acid. Sulfonic acids are strong acids. A derivative of benzenesulfonic acid is sulfanilamide, a sulfa drug (see Section 22.4).

Reaction Conditions. The conditions necessary for reactions to occur are important. Indeed, reaction conditions determine whether or not a reaction will

> **EXAMPLE 13.9**
> (a) Name the aromatic products formed in the preceding reactions 1 and 2.
> (b) Write reactions for the formation of toluene and isopropyl benzene from benzene using the appropriate preceding reaction.
>
> **SOLUTION**
> (a) chlorobenzene and bromobenzene (reaction 1); nitrobenzene (reaction 2)
>
> (b) toluene: C$_6$H$_6$ + CH$_3$Cl $\xrightarrow{AlCl_3}$ C$_6$H$_5$CH$_3$ + HCl
>
> isopropyl benzene: C$_6$H$_6$ + CH$_3$—CHCl—CH$_3$ $\xrightarrow{AlCl_3}$ C$_6$H$_5$—CH(CH$_3$)$_2$ + HCl

proceed and, if several reactions are possible, which one will occur. Consider, for example, the possibilities when the following two reactants are combined:

C$_6$H$_5$—CH=CHCH$_2$CH$_3$ + Cl$_2$ ⟶

Reaction conditions make all the difference here. If no heat, UV light, or catalyst is present, the reaction is an addition of Cl$_2$ to the double bond (Section 13.4). In the presence of AlCl$_3$ or Fe, chlorine substitutes for one or more hydrogens bonded to the benzene ring (Section 13.6). And if UV light or heat is supplied, chlorine atoms replace hydrogens in the saturated part of the molecule (Section 12.5).

You need to pay attention to reaction conditions and the catalysts used. By controlling these conditions, organic chemists can tailor a reaction to produce as much of the desired product as possible.

> Chemistry without catalysis would be a sword without a handle, a light without brilliance, a bell without sound.
>
> *Alwyn Mittasch (1869–1953)*

SUMMARY

Alkenes are unsaturated hydrocarbons having carbon-to-carbon double bonds; alkynes have carbon-to-carbon triple bonds. Compounds having more than one multiple bond are polyunsaturated. To name alkenes and alkynes, use the endings *-ene* and *-yne*, respectively; select the longest carbon chain containing the multiple bond and number it from the end nearer the multiple bond. Some alkenes occur as *cis* and *trans* geometric isomers: different structures in space that cannot be interconverted because they cannot rotate about the double bond.

Alkenes and alkynes are nonpolar, insoluble in water, and have melting and boiling points similar to alkanes. They participate in addition reactions with H_2, halogens (usually Cl_2 and Br_2), HCl, HBr, and H_2O. Markovnikov's rule predicts the main product from the latter three reactants. Alkenes and alkynes also react with oxidizing agents.

Aromatic compounds contain benzene or benzene-like rings. Benzene is stabilized by pi electrons above and below the plane of the ring. To preserve this stability, aromatic compounds undergo substitution—not addition—reactions with halogens (Cl_2 and Br_2), nitric acid (HNO_3), sulfuric acid (H_2SO_4), and alkyl halides in the presence of appropriate catalysts.

Several substituted benzene compounds have common names; others are named by writing the substituent name and then *benzene*. Disubstituted compounds use *o-* (*ortho*), *m-* (*meta*), and *p-* (*para*) to locate ring substituents. Locations of three or more substituents are specified by numbers. Aromatic hydrocarbons are nonpolar, insoluble in water, and have melting and boiling points like other hydrocarbons of a similar molecular size. Polar substituents increase their melting and boiling points and solubility in water.

KEY TERMS

Addition reaction (13.4)
Alkene (13.1)
Alkyne (13.2)
Aromatic compound (13.5)
Catalytic hydrogenation (13.4)
Cis isomer (13.3)

Geometric isomer (13.3)
Hydration reaction (13.4)
Markovnikov's rule (13.4)
Polycyclic aromatic compound (13.5)
Polyunsaturated compound (13.1)

Trans isomer (13.3)
Unsaturated hydrocarbon (13.1)

EXERCISES

Even-numbered exercises are answered at the back of this book.

Alkenes and Alkynes

1. What is the effect of a carbon-to-carbon double bond on the number of hydrogen atoms a carbon atom can bind?

2. Identify which of the following could be alkenes: (a) CH_4, (b) C_6H_{12}, (c) C_3H_8, (d) C_2H_5Cl, (e) $C_{20}H_{40}$.

3. Which of the compounds in exercise 2 could be linear or branched alkanes?

4. Which of the compounds in exercise 2 could be cyclic alkanes?

5. Which of the compounds in exercise 2 could be alkynes?

6. Write molecular formulas for (a) 1-butene, (b) 2-butene, (c) 2-octene, (d) 1-hexene, (e) cyclohexene.

7. Write condensed structural formulas for the compounds in exercise 6.

8. Write condensed structural formulas for the following: (a) 2-chloro-1-pentene, (b) 1,3-butadiene, (c) 3-methylcyclohexene, (d) trichloroethene.

9. Write condensed structural formulas for the following: (a) isobutylene, (b) 1-chloro-2-butyne, (c) acetylene, (d) vinyl bromide.

10. Try drawing structural formulas for (a) 2,2-dimethyl-1-heptene and (b) 3,3-dichloro-2-butyne. Why can't these compounds exist?

11. Name the following: (a) CH$_3$CH=CHCH$_2$CH$_3$,
(b) CH$_2$=CCH$_2$CHClCH$_3$ with CH$_2$CH$_3$ group on the =C,
(c) CH$_3$CHCH=CHCHCH$_3$ with CH$_2$CH$_3$ and CH$_3$ substituents,
(d) [cyclohexene with Cl substituent], (e) CH$_2$=CHCH$_2$CH=CH$_2$,
(f) CH≡CCl

12. Name the following: (a) CH$_3$CH$_2$C≡CH,
(b) CH$_3$CHCH=CH$_2$ with CH$_3$ substituent,
(c) [methylcyclopentene with CH$_3$],
(d) CH$_3$CHCHCH$_3$ with CH$_3$ and CH=CH$_2$ substituents,
(e) CH$_3$CH$_2$CH$_2$C=CH$_2$ with CH(CH$_3$)$_2$ substituent,
(f) CHBr=CHCH(CH$_3$)$_2$

Geometric Isomers

13. Distinguish between a structural isomer and a geometric isomer.
14. Identify which of the following may occur as geometric isomers: (a) alkanes, (b) alkenes, (c) alkynes, (d) benzene, (e) ethyl chloride.
15. Which of the compounds in exercise 6 can occur as geometric isomers?
16. Which of the compounds in exercise 8 can occur as geometric isomers?
17. Which of the compounds in exercise 11 can occur as geometric isomers?
18. Which of the compounds in exercise 12 can occur as geometric isomers?
19. Draw structural formulas for the following: (a) *cis*-2-pentene, (b) *trans*-2,3-dichloro-2-hexene, (c) *cis*-5-methyl-3-octene, (d) *trans*-5-chloro-2-pentene.
20. Draw structural formulas for the following: (a) *cis*-3-heptene, (b) *trans*-4-methyl-2-hexene, (c) *trans*-1,1-dichloro-2-butene, (d) *cis*-4,6-diisopropyl-2-decene.
21. Is it the sigma bond or the pi bond in an alkene double bond that prevents free rotation about the double bond?
22. In 2-butene, how many sigma bonds and how many pi bonds are formed by (a) carbon 1 and (b) carbon 2?
23. Describe the geometry and bond angles around a double-bonded carbon atom in an alkene.
24. What hybrid orbitals does a multiple-bonded carbon atom use to form sigma bonds in (a) an alkene and (b) an alkyne?
25. Alkynes cannot rotate freely about the triple bond, but (unlike alkenes) do not occur as geometric isomers. Rationalize why.

Physical Properties of Alkenes and Alkynes

26. For alkenes and alkynes, which provides the main attraction *between* molecules: (a) ionic bonds, (b) covalent bonds, (c) London forces, (d) dipole–dipole forces, or (e) hydrogen bonds? [Review Section 6.1 if necessary.]
27. Is 2-methyl-2-butene more soluble in water or in benzene? Explain.
28. Does 2-methyl-2-butene have a higher or lower melting point than heptane? Explain.
29. Identify which of the following is the most soluble in water: (a) isopentane, (b) ethene, (c) toluene, (d) *cis*-2-butene, (e) phenol.
30. Arrange the compounds in exercise 29 in order of increasing boiling point. Explain why.

Chemical Reactions of Alkenes and Alkynes

31. Which substance(s) in exercise 29 decolorize a solution of bromine (Br$_2$) in CCl$_4$?
32. Suppose you have a liquid alkane or alkene that boils at about 66°C. Consult Tables 12.1 and 13.1 to determine which alkene or alkane it might be. What simple chemical test could you do to determine which one it is? Explain what you would do and what you would observe.
33. Identify which of the following participate in addition reactions with Cl$_2$: (a) alkanes, (b) alkenes, (c) alkynes, (d) aromatic compounds.
34. Using molecular formulas, write a balanced chemical equation for the complete combustion of cyclohexene with oxygen.
35. Using condensed structural formulas, write a balanced chemical equation for the catalytic hydrogenation of 3-hexene.

36. Write condensed structural formulas for the major products formed in each of the reactions below. Write "no reaction" where appropriate.
    (a) $CH_3CH_2CH=CH_2 + H_2O \xrightarrow{H^+}$
    (b) C₆H₅—CH=CH₂ + Br₂ ⟶
    (c) $CH_3CH=CHCH_2CH_3 + H_2 \longrightarrow$

37. Write condensed structural formulas for the major products formed in each of the reactions below. Write "no reaction" where appropriate.
    (a) $CH_3CH=CHCH_3 + HBr \longrightarrow$
    (b) $CH_3C{\equiv}CCH_3 + 2HBr \longrightarrow$
    (c) $CH_3CH_2CH_2CH=CH_2 + HCl \longrightarrow$

38. In which reactions in exercise 36 does Markovnikov's rule apply?

39. In which reactions in exercise 37 does Markovnikov's rule apply?

40. Write a balanced chemical equation to show how each of the substances listed below could be produced. Use as your starting material any alkene or alkyne you want and show what it reacts with. Include any reaction conditions or catalysts that may be necessary.
    (a) 2-chloropentane
    (b) 2,3-dichloropentane
    (c) 2,2-dichloropentane

41. Which of the organic compounds in exercises 36 and 37 would react with $KMnO_4$? What would you observe if a reaction occurred?

42. What group bonded to female sex hormones protects them against inactivation in the digestive tract (see Chapter Opener)?

43. List two simple chemical reactions in which a color change (or absence of a change) can distinguish between butane and 1-butene.

44. Suppose a substance with the formula C_5H_{10} reacts readily with $KMnO_4$. Is the substance a cyclic or a chain compound? Explain.

Aromatic Hydrocarbons

45. What kind of hybrid orbitals do ring carbons in benzene use?

46. What chemical evidence indicates that benzene does not contain normal carbon-to-carbon double bonds?

47. Write condensed structural formulas for the following: (a) aniline, (b) *m*-xylene, (c) 3,4-diiodotoluene, (d) nitrobenzene, (e) naphthalene, (f) 1,3-dibromo-5-chlorobenzene.

48. Write condensed structural formulas for the following: (a) phenol, (b) 2,4,6-trinitrotoluene (TNT), (c) *p*-dichlorobenzene, (d) 2-phenyloctane, (e) ethylbenzene, (f) *o*-chlorophenol.

49. Name the following:
    (a) naphthalene, (b) 4-nitrophenol, (c) 2,4-dichlorophenol, (d) C₆H₅—CH₂CH₂CH=CH(CH₂)₂CH₃, (e) 1,2-dibromobenzene

50. Name the following:
    (a) aniline (C₆H₅NH₂), (b) 3-nitrotoluene, (c) 2,5-dimethyltoluene (p-xylene with extra CH₃), (d) iodobenzene, (e) 1,3,5-tribromobenzene

51. Is benzene more soluble in toluene or in water? Explain.

52. Arrange the following in order of increasing boiling point: (a) toluene, (b) benzene, (c) nitrobenzene. Explain why.

53. Identify which of the following participate in substitution reactions with Cl_2 under appropriate reaction conditions: (a) alkanes, (b) alkenes, (c) alkynes, (d) aromatic compounds.

54. Why does the benzene ring participate in substitution instead of addition reactions?

55. Write condensed structural formulas for the major organic products formed in the reactions below. Where appropriate, write "no reaction."
    (a) C₆H₆ + Cl₂ $\xrightarrow{UV\ light}$
    (b) C₆H₆ + $CH_3CHClCH_3$ $\xrightarrow{AlCl_3}$

(c) 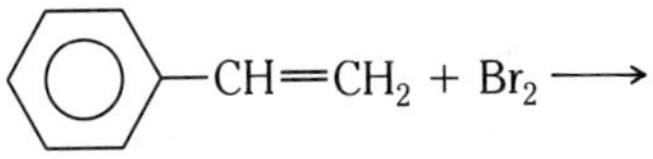 C₆H₅—CH=CH₂ + Br₂ ⟶

(d) 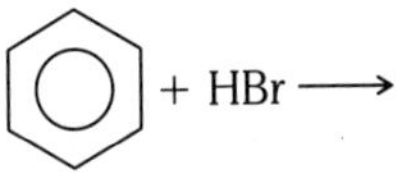 C₆H₆ + HBr ⟶

56. Write condensed structural formulas for the major organic products formed in the reactions below. Where appropriate, write "no reaction."

(a) 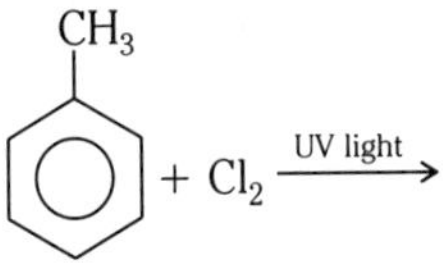 C₆H₅—CH₃ + Cl₂ —UV light→

(b) 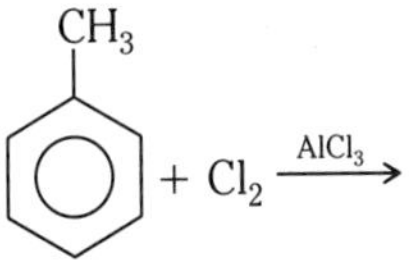 C₆H₅—CH₃ + Cl₂ —AlCl₃→

(c) 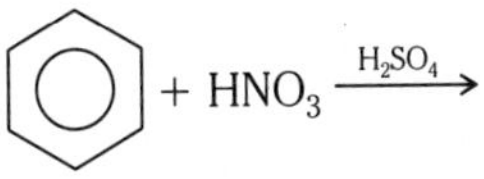 C₆H₆ + HNO₃ —H₂SO₄→

(d) 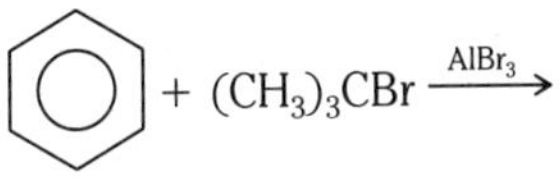 C₆H₆ + (CH₃)₃CBr —AlBr₃→

57. You have learned several halogenation reactions in which a hydrocarbon reacts with Cl₂. Fill in the table below to distinguish between these reactions:

Reactant	Products	Type of Reaction	Reaction Conditions
alkane	___	substitution	___
aromatic	___	___	Fe or AlCl₃
___	alkyl dihalides	___	no special conditions

DISCUSSION EXERCISES

1. Read the Chemistry Spotlight about George Wald (Section 13.3). To what extent is a scientist responsible for applications of knowledge she or he discovers? Explain.

2. *Cis*-2-butene and *trans*-2-butene have boiling points of −139°C and −105°C, respectively. What structural features and intermolecular forces are responsible for this difference?

3. Are aromatic hydrocarbons polyunsaturated compounds? Why?

4. Predict which one of the aromatic compounds in Figure 13.16 is a solid at room temperature. Justify your prediction.

14

Alcohols, Phenols, Thiols, and Ethers

1. What are the names and structures of some alcohols? What are their major uses and chemical reactions?
2. What are the names, structures, and uses of some phenols?
3. How are thiols (mercaptans) named and what are they used for?
4. What are ethers and how are they named? What are their properties and uses?

Thiol Groups and Toxic Metals

Figure 14.1 BAL and a BAL–lead complex.

Certain metals and metal compounds are poisons. Mercury, lead, and arsenic (a metalloid; Section 3.4) inactivate proteins by binding to thiol (—SH) groups (see Table 11.3) in the amino acid cysteine. Lead, for example, blocks the action of a protein enzyme necessary for making hemoglobin. As a result, people with lead poisoning often have anemia, a shortage of red blood cells.

One way to counteract poisoning by metals is to attract them to other thiol groups. During World War I, British scientists searched for an antidote to lewisite, an arsenic-containing gas developed for gas warfare. The most effective compound they found had two —SH groups and was called *British Anti-Lewisite (BAL)*. The thiol groups in BAL tie up the metal in a BAL–metal complex (Figure 14.1), keeping the metal from causing damage elsewhere in the body. Today BAL is used in hospitals to treat poisoning by arsenic, mercury, lead, nickel, antimony, and manganese.

In an emergency, drinking egg whites and milk, followed by vomiting, helps treat mercury, lead, and other metal poisons. Egg whites and milk work by providing protein —SH groups to bind metal ions and prevent them from doing additional damage in the body.

Arsenic.

As many as 2% of all babies born in the world's developed nations have neurological and other damage because their mothers drank ethyl alcohol (ethanol) during pregnancy. Children with this disorder (called **fetal alcohol syndrome**) have abnormal brain development, impaired coordination, and poor learning, thinking, and language skills (Figure 14.2). Ethanol is the leading cause of mental retardation in the developed world.

Ethanol is one of a group of organic substances, called *alcohols,* that we examine in this chapter. Three other groups we will study are *phenols, thiols* (or *mercaptans*), and *ethers.* Compounds with these functional groups help give a skunk its special perfume, protect car radiators from freezing in the winter, and help keep your skin soft and moist.

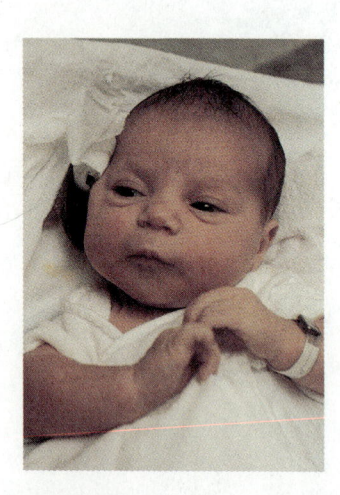

Figure 14.2 This baby was born with fetal alcohol syndrome.

14.1 Alcohols: Names and Structures

We can think of the functional groups discussed in this chapter as derivatives of H_2O. The relationship is:

water
- replace one H with *R* → R–O–H alcohol
- replace one H with benzene → C$_6$H$_5$–O–H phenol
- replace O with S and H with *R* → R–S–H thiol
- replace both H with *R* groups → R–O–R′ ether

We begin with organic compounds that have an alkyl group bonded to the functional group —O—H; these are **alcohols** (see Table 11.3). Many simple alcohols have common names consisting of the name of the alkyl group plus the word *alcohol.* Table 14.1 shows some examples.

IUPAC names for alcohols are like IUPAC names for alkanes except for two additional rules:

1. *Change the final -e to -ol.* For example, CH_3—OH is methanol, and CH_3CH_2—OH is ethanol.

2. *Select the longest carbon chain containing the* —OH *group and number from the end nearer the* —OH *group. Specify the carbon(s) with the* —OH *by number (followed by a hyphen) immediately before the name of the alcohol and after any other substituents.*

$CH_3CH_2CH_2$—OH

1-propanol

$CH_3CHCH_2CH_2CH_3$ with OH on C2

2-pentanol

Table 14.1 Some Alcohols and Phenols

Name	Molecular Formula	Condensed Structural Formula	Melting Point (°C)	Boiling Point (°C)	State at 25°C
Alcohols					
methanol (methyl alcohol)	CH_4O	CH_3OH	−94	65	liquid
ethanol (ethyl alcohol)	C_2H_6O	CH_3CH_2OH	−117	78	liquid
1,2-ethanediol (ethylene glycol)	$C_2H_6O_2$	$HOCH_2CH_2OH$	−11	198	liquid
1-propanol (*n*-propyl alcohol)	C_3H_8O	$CH_3CH_2CH_2OH$	−126	97	liquid
2-propanol (isopropyl alcohol)	C_3H_8O	CH_3CHCH_3 | OH	−89	82	liquid
1-hexanol	$C_6H_{14}O$	$CH_3(CH_2)_4CH_2OH$	−47	158	liquid
1-decanol	$C_{10}H_{22}O$	$CH_3(CH_2)_7CH_2OH$	7	229	liquid
Phenols					
phenol	C_6H_6O	(benzene ring with OH)	43	182	solid
o-cresol	C_7H_8O	(benzene ring with OH and CH₃ ortho)	31	191	solid
m-cresol	C_7H_8O	(benzene ring with OH and CH₃ meta)	11	202	liquid
p-cresol	C_7H_8O	(benzene ring with OH and CH₃ para)	35	202	solid

Now try using these rules to name a more complicated alcohol:

$$\underset{1}{CH_3}-\underset{2}{CH}-\underset{3}{CH}-\underset{}{CH_2CH_2CH_3}$$
$$\quad\;\;\;\; | \qquad\;\; |$$
$$\quad\;\;\;\;OH \quad CH_2CHCH_2CH_2Cl$$
$$\qquad\qquad\;\; \underset{4}{}\;\underset{5}{|}\;\underset{6}{}\;\underset{7}{}$$
$$\qquad\qquad\qquad CH_3$$

The longest continuous carbon chain containing the OH group has seven carbon atoms. Number from the left so the OH group is bonded to carbon 2, making the name (with the *-ol* suffix) 2-heptanol. Collect other substituent groups and arrange them in alphabetical order to make the complete name: 7-chloro-5-methyl-3-propyl-2-heptanol.

Compounds with two or three —OH groups are called *diols* and *triols*, respectively. Their names start with the numbers of the carbons bonded to

—OH groups, followed by the alkane name and the suffix *-diol* or *-triol*. For example, HOCH$_2$CH$_2$CH$_2$CH$_2$OH is 1,4-butanediol.

> **EXAMPLE 14.1** Write IUPAC and common (where appropriate) names for a–d. Write condensed structural formulas for e–h.
> (a) CH$_3$CH$_2$CH$_2$CH$_2$OH (b) CH$_3$CH$_2$CHCH$_3$ (c) CH$_3$CHCH$_2$CH$_2$OH
> | |
>   OH   CH$_2$CH$_2$Br
>
> (d) C$_6$H$_{11}$—OH (e) *t*-butyl alcohol, (f) isopropyl alcohol,
>
> (g) 1,2-propanediol, (h) 3,3-dimethyl-1-pentanol
>
> **SOLUTION** (a) 1-butanol (*n*-butyl alcohol), (b) 2-butanol (*sec*-butyl alcohol), (c) 5-bromo-3-methyl-1-pentanol, (d) cyclohexanol (cyclohexyl alcohol),
>
> (e) CH$_3$—C(CH$_3$)(CH$_3$)—OH (f) CH$_3$CHCH$_3$ (g) CH$_3$—CH—CH$_2$—OH
> | |
>   OH   OH
>
> (h) CH$_2$CH$_2$CCH$_2$CH$_3$
> | |
>   OH   CH$_3$ (with CH$_3$ branch)

Alcohols are classified as *primary, secondary,* or *tertiary alcohols* depending on whether the —OH group bonds to a primary, secondary, or tertiary carbon atom (Section 12.2). For example, CH$_3$CH$_2$CH$_2$OH (1-propanol) is a primary alcohol because carbon 1 bonds to one other carbon; CH$_3$CHCH$_3$ with OH (2-propanol) is a secondary alcohol because carbon 2 bonds to two other carbons; and (CH$_3$)$_2$CCH$_2$CH$_3$ with OH (2-methyl-2-butanol) is a tertiary alcohol.

> **EXAMPLE 14.2** Identify the compounds in example 14.1 as primary, secondary, or tertiary alcohols.
>
> **SOLUTION** Compounds (a), (c), and (h) are primary alcohols; (b), (d), and (f) are secondary alcohols; (g) has both a primary and a secondary alcohol group; and (e) is a tertiary alcohol.

14.2 Phenols: Names and Structures

Compounds having —OH bonded to a benzene ring are called **phenols.** The simplest phenol has the common name phenol (Figure 13.16 and Table 14.1). A phenol with one methyl group bonded to the benzene ring is called cresol.

Three isomers of cresol exist: *o*-, *m*-, and *p*-cresol (Table 14.1). Name other derivatives using the rules for aromatic compounds discussed in Section 13.5. Three examples are:

p-nitrophenol   *m*-chlorophenol   2,4-dibromophenol

14.3 PHYSICAL PROPERTIES AND USES OF ALCOHOLS AND PHENOLS

Physical Properties. Table 14.1 shows the melting and boiling points of a few alcohols and phenols. Because of their polar —OH group, alcohols and phenols form hydrogen bonds (Figure 14.3); this makes it harder for these substances to vaporize (more heat is required). So alcohols and phenols have higher melting and boiling points than hydrocarbons of similar molecular weights (Figure 14.4).

Figure 14.3 Hydrogen bonds (in color) form between alcohol (or phenol) molecules.

Figure 14.4 Effect of molecular weight on the boiling point of linear alkanes (C_1–C_{10}), primary alcohols (C_1–C_{10}), and phenol.

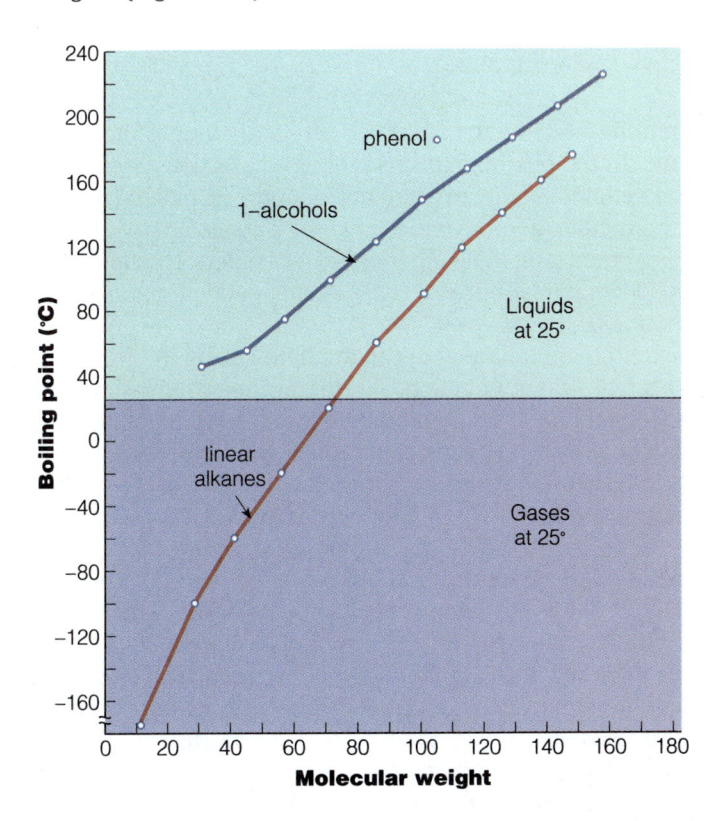

The polar —OH group also makes alcohols and phenols much more soluble in water than hydrocarbons with similar molecular weights (Figure 14.5). As alcohols and phenols dissolve, their polar groups form hydrogen bonds with

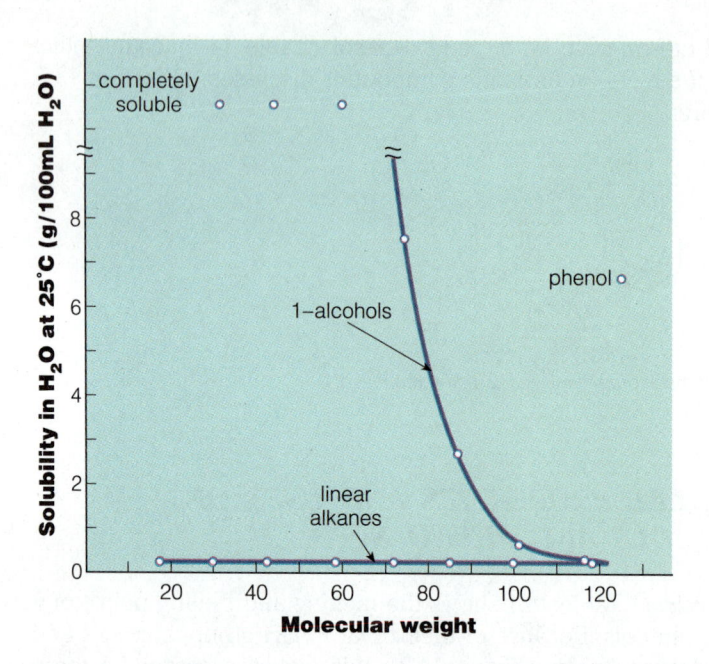

Figure 14.5 Effect of molecular weight on the solubility in water of linear alkanes (C$_1$–C$_8$), primary alcohols (C$_1$–C$_7$), and phenol.

Figure 14.6 Hydrogen bonds (in color) form between water and alcohol (or phenol) molecules.

water molecules (Figure 14.6). Ethanol and water, for example, dissolve completely in each other.

As the length of the nonpolar hydrocarbon chain in an alcohol increases, the effect of the polar —OH group lessens, the molecule becomes less polar, and the solubility in water decreases (Figure 14.5). 1-Pentanol, for example, is only slightly soluble in water, while 1-propanol dissolves completely.

Uses. *Ethanol*, commonly called ethyl alcohol or grain alcohol, is the active ingredient in alcoholic drinks. Microorganisms such as yeasts convert sugar in fruits or vegetables into ethanol in the absence of air, a process called **fermentation.** Apples yield hard cider; grapes yield wine. Beer and ale are made from malt, which is dried, sprouted grain. The ethanol content of fermentation products varies from a few percent to about 15% (*v/v*). At that concentration, yeasts die and fermentation stops.

Drinks with a higher ethanol content are made by distilling the fermentation mixture to make a high-ethanol fraction. Distilling wine produces brandy. Whiskey and vodka are made by distilling fermented grains, while tequila and rum come from distilling fermented agave plant and sugar cane, respectively. The color and flavor of these products come from minor contaminants distilled with the ethanol or introduced while the liquor ages in wooden casks (Figure 14.7).

Ethanol concentration is often expressed in terms of "proof," which is twice the percentage (by volume) of ethanol. So 80-proof whiskey contains 40% (*v/v*) ethanol.

Ethanol is one of the most widely used and abused drugs in the world. It is a depressant, and at high doses causes slurred speech and vision, poor coordination, coma, and in rare instances, death. Long-term, extensive use of this drug increases the risk of liver damage and heart disease.

A 70% solution of ethanol in water is useful as a disinfectant. A mixture of gasoline with a 10% or higher concentration of ethanol is the fuel *gasohol*. Ethanol is used in industry as a solvent in drugs, lotions, and other products. Tincture of iodine, for example, is a solution of iodine (I$_2$) in ethanol that

Figure 14.7 Wine aging in barrels.

disinfects cuts or skin abrasions. *Denatured alcohol*, which isn't drinkable, is ethanol mixed with small amounts of hard-to-remove organic substances that are toxic, foul-smelling, or foul-tasting. This allows ethanol to be produced for industrial purposes without the high taxes levied on drinkable ethanol.

Methanol, often called methyl alcohol or wood alcohol, is burned as a fuel in camp stoves and may someday be used in gasoline. It is made commercially by reaction of carbon monoxide (CO) with hydrogen gas (H_2) at high temperature in the presence of a catalyst:

$$CO + 2H_2 \xrightarrow[\text{catalyst}]{\text{heat}} CH_3-OH$$

Methanol is toxic to drink and occasionally occurs in improperly made moonshine whiskey. Because methanol can damage the fatty sheath around the optic nerve and cause permanent blindness, it gives a literal meaning to the phrase "blind drunk."

Another important alcohol, *isopropyl alcohol* (2-propanol; Table 14.1), is widely used as rubbing alcohol and a disinfectant. It also is toxic, but induces vomiting so effectively that it rarely stays in the body long enough to do severe damage. It is made by hydration of the alkene, propene (see Section 13.4):

$$CH_2=CHCH_3 + H_2O \xrightarrow{H^+} CH_3\overset{\overset{\displaystyle OH}{|}}{C}HCH_3$$

Ethylene glycol (1,2-ethanediol; Table 14.1) is the main ingredient in antifreeze. Its two polar —OH groups make it very soluble in water, so you can mix antifreeze and water in any proportion. Its sweet taste has lured animals and children to drink this poisonous substance.

Another sweet (but nontoxic) substance with multiple —OH groups is *glycerol* (1,2,3-propanetriol), sometimes called glycerin. Glycerol is obtained from fats and oils (see Chapter 20). With its three —OH groups, glycerol forms strong hydrogen bonds with water. Because its water-binding qualities keep skin soft and moist, it is used in many cosmetic preparations.

In the 19th century Joseph Lister, an English surgeon (Figure 14.8), began using phenol to disinfect surgical equipment. Although it burns skin and is toxic when taken internally, phenol was helpful in preventing infections from operations.

The antiseptic action of phenol derivatives is still widely used. One compound, 4-hexylresorcinol, is a common mouthwash and throat lozenge ingredient that kills oral bacteria (Figure 14.9). Another derivative, hexachlorophene

Figure 14.8 Joseph Lister (1827–1912).

As predicted by Markovnikov's rule (see Section 13.4), the main product is 2-propanol, not 1-propanol.

Figure 14.9 4-Hexylresorcinol.

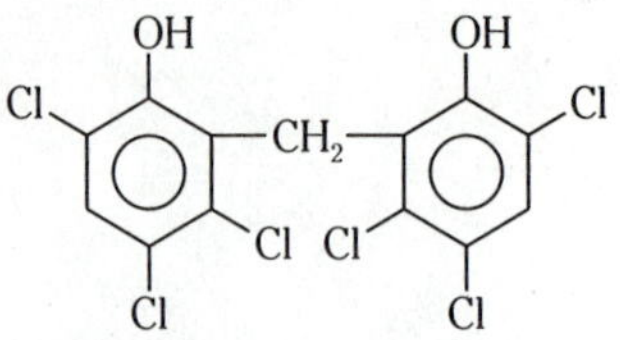

Figure 14.10 Hexachlorophene is used in hospitals as a disinfectant. It is commonly sold as the brand name Phisohex (right).

(Figure 14.10), is used in hospitals to kill infectious bacteria. And cresol, a methylated version of phenol (Table 14.1), is a key component in creosote, a wood preservative. Wooden posts, for example, are commonly soaked with creosote before being put in the ground.

14.4 CHEMICAL REACTIONS OF ALCOHOLS AND PHENOLS

Alcohols participate in two types of reactions—*dehydration* and *oxidation*. In contrast, phenols don't undergo dehydration and are oxidized in a different way from alcohols.

1. Dehydration In a **dehydration reaction,** water is removed from the reactant. In the presence of heat and a strong acid (usually concentrated sulfuric acid, H_2SO_4), alcohols are dehydrated to form alkenes. Removal of an —OH group from one carbon and H from a neighboring carbon produces H_2O. The general reaction is:

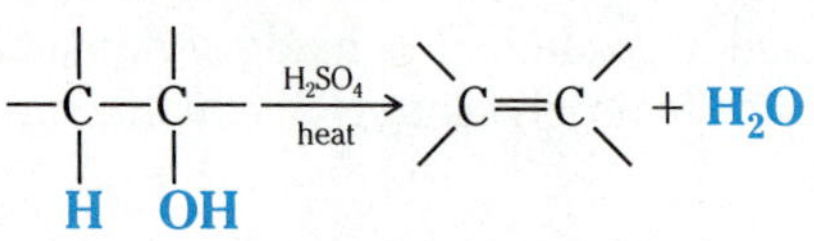

This reaction looks like the reverse of the addition reaction of H_2O to an alkene to form an alcohol (see Section 13.4).

$$-\underset{H}{\overset{|}{C}}-\underset{OH}{\overset{|}{C}}- \xrightarrow[\text{heat}]{H_2SO_4} \;\; \diagup\!\!C=C\!\!\diagdown + H_2O$$

Primary alcohols are dehydrated to produce 1-alkenes. But dehydrating secondary and tertiary alcohols sometimes produces two products. For example, 2-pentanol could dehydrate in two ways:

$$CH_3CH_2CH_2\underset{OH}{\overset{|}{C}}HCH_3 \xrightarrow[180°C]{H_2SO_4}$$

$CH_3CH_2CH_2CH=CH_2 + H_2O$
1-pentene

or

$CH_3CH_2CH=CHCH_3 + H_2O$
2-pentene

This rule, like Markovnikov's rule (Section 13.4), results from a reaction mechanism involving carbocation intermediates. The predominant product in both cases is the one that is more stable.

The general rule is that *the main product is the one that has the most alkyl groups bonded to the double-bonded carbons.*

Now apply this rule in the preceding example. The double-bonded carbons in 1-pentene bond to a total of *one* alkyl group (carbon 2 bonds to a propyl group and a hydrogen; carbon 1 bonds to two hydrogens). But the double-bonded carbons in 2-pentene bond to a total of *two* alkyl groups (carbon 2 bonds to a methyl group and hydrogen while carbon 3 bonds to an ethyl group and hydrogen). According to the rule, then, you predict that 2-pentene is the major product. When the reaction actually occurs, the main product is indeed 2-pentene.

EXAMPLE 14.3 Write a condensed structural formula and IUPAC name for the major product formed in each of the following dehydration reactions:

(a) $CH_3(CH_2)_3CH_2OH \xrightarrow{H_2SO_4,\;180°C}$

(b) $CH_3\underset{OH}{\overset{|}{C}}HCH_2CH_3 \xrightarrow{H_2SO_4,\;180°C}$

(c) $CH_3CHCHCH_3$ with OH on C2 and CH_3 on C3 $\xrightarrow{H_2SO_4,\ 180°C}$

(d) $(CH_3)_3C-OH \xrightarrow{H_2SO_4,\ 180°C}$

SOLUTION (a) $CH_3(CH_2)_2CH=CH_2$ (*1-pentene*),

(b) $CH_3CH=CHCH_3$ (*2-butene*),

(c) $CH_3C(CH_3)=CHCH_3$ (*2-methyl-2-butene*), (d) $(CH_3)_2C=CH_2$ (*2-methylpropene*)

At temperatures lower than those required for dehydration, some alcohols (especially primary ones) react with each other to eliminate a molecule of H_2O and form an ether. This removal of H_2O between *two* molecules is called a *condensation reaction*. The reaction with ethanol is:

$$CH_3CH_2-O-H + H-O-CH_2CH_3 \xrightarrow[H_2SO_4]{140°C} CH_3CH_2-O-CH_2CH_3 + H_2O$$

an ether

Dehydration of one molecule of an alcohol produces an alkene. Dehydration between two molecules of an alcohol (condensation) produces an ether.

This reaction is sometimes used to synthesize ethers (Section 14.6).

The reaction conditions determine the main product. At 180°C and with excess H_2SO_4, alcohols mostly dehydrate to form alkenes. At 140°C with an excess amount of a primary alcohol, the main product is an ether.

EXAMPLE 14.4 Write the structural formula for the main product that forms from the reaction of 1-butanol in the presence of H_2SO_4 at (a) 180°C and (b) 140°C.

— sulphuric acid

SOLUTION (a) $CH_3CH_2CH=CH_2$,
(b) $CH_3CH_2CH_2CH_2-O-CH_2CH_2CH_2CH_3$

Phenols don't dehydrate because removing —OH plus hydrogen from a neighboring carbon to produce a double bond would disrupt the stable arrangement of six pi electrons in the benzene ring.

2. Oxidation The general reaction for the oxidation of an alcohol is:

$$-\underset{H}{\underset{|}{\overset{OH}{\overset{|}{C}}}}- + (O) \longrightarrow -\overset{O}{\overset{\|}{C}}- + H_2O$$

Recall that we can define oxidation as the gain of oxygen atoms or loss of hydrogen atoms (Section 13.4). Notice in the preceding reaction that the alcohol loses two hydrogen atoms to the oxidizing agent, represented as (O). The product contains a **carbonyl group** ($>C=O$), which is a carbon atom double-bonded to an oxygen. A carbonyl group at one end of a carbon chain makes the *aldehyde* functional group, whereas a carbonyl group within a carbon chain makes a *ketone* functional group (Table 11.3). We discuss these functional groups in more detail in the next chapter.

Using these definitions, though, you should be able to predict that a primary alcohol is oxidized to an aldehyde. One example is:

$$CH_3CH_2OH \xrightarrow{(O)} CH_3\overset{\overset{\displaystyle O}{\|}}{C}-H$$

ethanol (a primary alcohol)   an aldehyde

Unless it is removed as soon as it forms, the aldehyde is oxidized further to form a *carboxylic acid* functional group (Table 11.3):

$$CH_3-\overset{\overset{\displaystyle H}{|}}{\underset{\underset{\displaystyle H}{|}}{C}}-OH \xrightarrow{(O)} CH_3-\overset{\overset{\displaystyle O}{\|}}{C}-H \xrightarrow{(O)} CH_3-\overset{\overset{\displaystyle O}{\|}}{C}-OH$$

ethanol   an aldehyde   a carboxylic acid

You can recognize the first reaction as oxidation because ethanol loses two hydrogens. The second reaction is oxidation because the aldehyde gains one oxygen.

This carboxylic acid is called acetic acid. Household vinegar is a weak solution of acetic acid in water. When wine or hard cider is exposed to air, its ethanol reacts with oxygen to form acetic acid, which gives the beverage a vinegary taste. This is why wineries ferment and age wine in airtight containers.

The preceding reaction also is used to measure the amount of ethanol in the breath of drunk-driving suspects. A Breathalyzer test (Figure 14.11) uses a solution of potassium dichromate ($K_2Cr_2O_7$) as the oxidizing agent. A measured

Figure 14.11 This person is being given a breathalyzer test.

volume of the suspect's breath is bubbled into the solution; any ethanol is oxidized to acetic acid while the yellow-orange dichromate ion ($Cr_2O_7^{2-}$) is reduced to green chromium (Cr^{3+}) ion. The amount of color change shows the amount of ethanol in the suspect's breath.

Your body uses protein catalysts (enzymes; Chapter 22) to oxidize ethanol to an aldehyde (acetaldehyde), then to acetic acid, and finally to CO_2 and H_2O, generating considerable energy in the process. In fact, you get just as much energy per gram (4.0 Cal or 17 kJ) from ethanol as from carbohydrates.

Secondary alcohols are oxidized to form ketones. One example is:

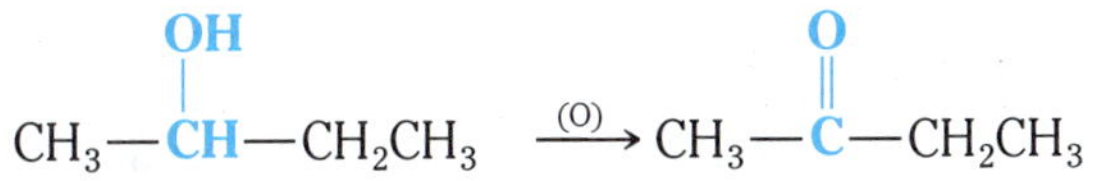

$$CH_3-\underset{\underset{\text{OH}}{|}}{CH}-CH_2CH_3 \xrightarrow{(O)} CH_3-\underset{\underset{\text{O}}{\|}}{C}-CH_2CH_3$$

2-butanol (a secondary alcohol)   a ketone

The ketone product, unlike an aldehyde, is not oxidized further.

Tertiary alcohols aren't oxidized under these conditions because a tertiary carbon doesn't have enough bonds to form an additional double bond to oxygen.

The pattern is:

primary alcohol $\xrightarrow{(O)}$ *aldehyde*

secondary alcohol $\xrightarrow{(O)}$ *ketone*

tertiary alcohol $\xrightarrow{(O)}$ *no reaction*

EXAMPLE 14.5 Write structural formulas for the products that form in the reactions below. If no reaction occurs, write "no reaction."

(a) $CH_3CH_2CH_2CH_2CH_2CH_2OH \xrightarrow{KMnO_4}$

(b) 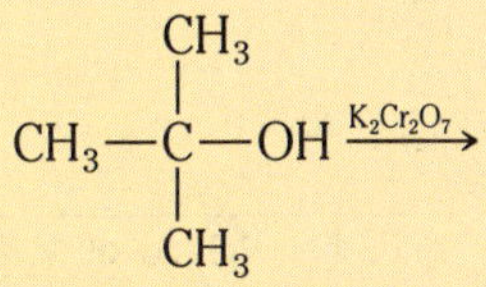 $CH_3-\underset{\underset{CH_3}{|}}{\overset{\overset{CH_3}{|}}{C}}-OH \xrightarrow{K_2Cr_2O_7}$

(c) $CH_3(CH_2)_2\underset{\underset{OH}{|}}{CH}CH_2\underset{\underset{CH_3}{|}}{CH}CH_3 \xrightarrow{(O)}$

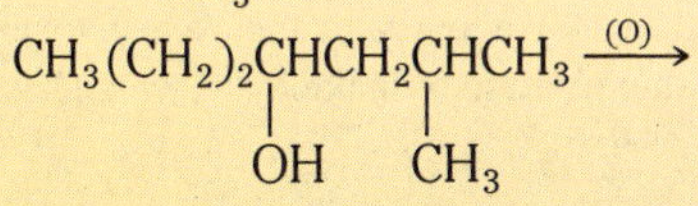

SOLUTION (a) $CH_3(CH_2)_4\overset{\overset{O}{\|}}{C}-H$ (an aldehyde),

(b) no reaction, (c) 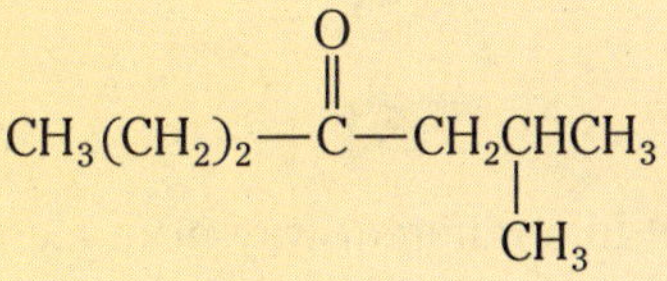 $CH_3(CH_2)_2-\overset{\overset{O}{\|}}{C}-CH_2\underset{\underset{CH_3}{|}}{CH}CH_3$ (a ketone)

BHT and BHA (Figure 14.12) are two examples of phenols.

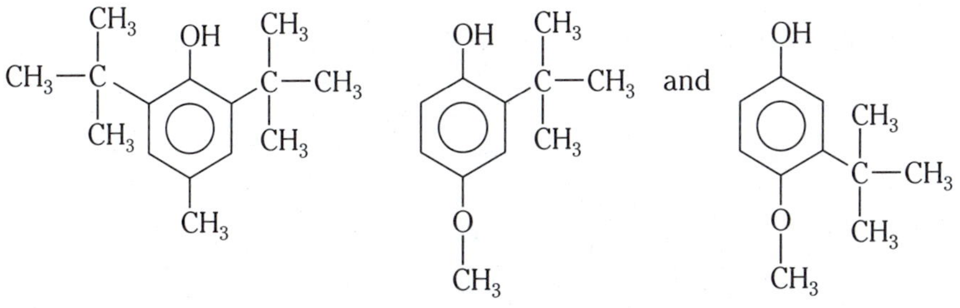

BHT (butylated hydroxytoluene)   BHA (butylated hydroxyanisole)

Figure 14.12 Two food additives used as antioxidants. BHA is a mixture of two isomers.

CHEMISTRY SPOTLIGHT

Treating Alcohol Poisoning

Although methanol and ethylene glycol are toxic (Table 14.2), alcoholics occasionally drink them—as well as isopropyl alcohol (rubbing alcohol)—as cheap substitutes for ethanol. Children may accidentally drink these alcohols, too—some being attracted to ethylene glycol by its sweet taste.

The body metabolizes these alcohols the same way as ethanol, oxidizing them to aldehydes and carboxylic acids. Methanol is oxidized to formaldehyde, a very toxic substance that destroys proteins in the body. Ethylene glycol is metabolized to oxalic acid, which combines with calcium ion (Ca^{2+}) to form insoluble calcium oxalate; these crystals deposit in kidneys and can lead to kidney failure.

Because it is oxidized by the same system, ethanol is an antidote for both methanol and ethylene glycol poisoning. Once the victim receives ethanol, usually by intravenous injection, the body begins to metabolize ethanol instead of methanol or ethylene glycol. This slows the production of toxic oxidation products and gives the body time to excrete the toxic alcohol. In effect, the antidote substitutes a less toxic alcohol (ethanol) for a more toxic one.

Table 14.2 Toxicity of Some Alcohols

Alcohol	LD_{50}*
methanol	**
ethanol	7.06
isopropyl alcohol (2-propanol)	5.8
ethylene glycol	8.5

* Lethal dose in g/kg body weight to kill 50% of test animals (rats); a low value means high toxicity

** Metabolized to formaldehyde, which has LD_{50} value of 0.07

Phenols are oxidized in a different way from alcohols and form complex products. Food manufacturers take advantage of this by using certain phenols, such as BHA and BHT, as additives to protect fatty foods from oxidation. These phenols react with oxidizing agents to prevent foods containing unsaturated and polyunsaturated fats from turning rancid (Section 13.4).

Acidity of Phenols. Phenols are weak acids that neutralize a base such as NaOH:

phenol + NaOH ⟶ phenoxide ion + H_2O + Na^+

With a K_a value of about 10^{-10}, phenol is a much weaker acid than acetic acid, which has a K_a value of 1.8×10^{-5} (see Table 9.5). Phenols are more acidic than alcohols, however, because pi electrons in the benzene ring spread out the negative charge of phenoxide ion, making the ion more stable. As a result, phenols donate protons (to form phenoxide ion) more readily than alcohols do.

Because it is acidic, phenol was once known as carbolic acid.

14.5 THIOLS

The prefix *thio-* refers to sulfur. A **thiol** is like an alcohol, except that oxygen in the alcohol group (—OH) is replaced by sulfur, forming a —SH group. The general formula of a thiol, also called a **mercaptan**, is R—SH (Table 11.3).

Thiols are easy to name. Use the alkane name for the R group and add the suffix *-thiol*. For example, CH_3CH_2—SH is ethanethiol. When thiols have three or more carbons, use a number to identify the carbon bonded to the —SH group. For example, $CH_3CH_2CH_2CHCH_3$ is 2-pentanethiol.
                                              |
                                              SH

The most distinctive property of many thiols is their strong, unpleasant odor. When cut, onions get your attention by releasing 1-propanethiol. And skunks use several thiols, including 3-methyl-1-butanethiol, to greet unwanted visitors.

One chemical reaction of thiols is important in biochemistry. Two thiol groups are oxidized (lose hydrogens) to form a **disulfide** compound (containing —S—S—) as follows:

$$R-SH + R'-SH \xrightarrow{(O)} R-S-S-R'$$

two thiol molecules   *a disulfide*

A disulfide, in turn, is reduced (gains hydrogens) to produce two thiol molecules:

$$R-S-S-R' \xrightarrow{reduction} R-SH + R'-SH$$

These reactions are important in your body. One amino acid, cysteine, has a thiol functional group. The proteins in your body are large molecules made from amino acids, and nearly all your proteins have —SH groups because they contain cysteine (see Chapter Opener). Oxidizing agents such as O_2 can cause neighboring —SH groups in a protein (or even in two different proteins) to form a disulfide bond (Figure 14.13). This may change the protein's shape and ability to function.

Your hair, for example, is a protein material that has large amounts of cysteine. The curl of your hair depends on the arrangement of disulfide groups, which hold each hair strand in a certain shape. But, if you don't like your hair's natural curl, you can buy curly hair at the beauty shop or in a home permanent kit. The first step is to treat your hair with reducing agents to break the disulfide bonds. Next, the hair is forced into the desired shape with curlers. Finally, an oxidizing agent is added to form new disulfide bonds that help hold your hair in its new shape (Figure 14.14).

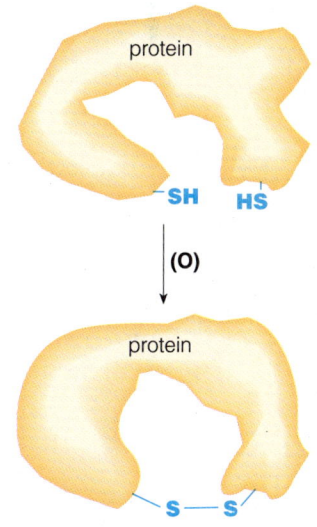

Figure 14.13 Oxidation of thiol (—SH) groups in a protein to form a disulfide bond changes the shape of the protein.

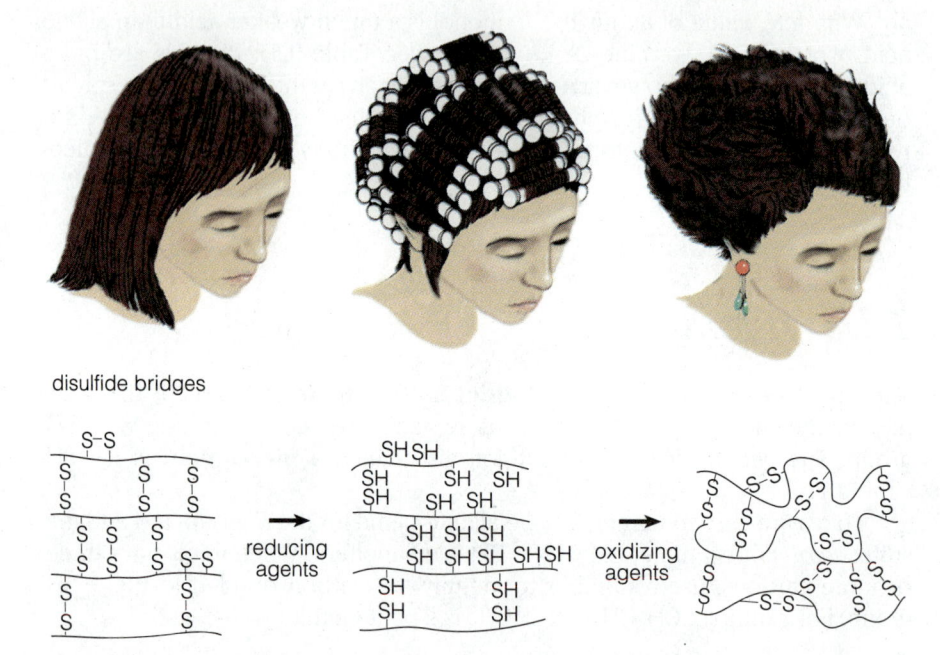

Figure 14.14 Hair permanents involve reduction of disulfide bonds followed by oxidation of —SH groups.

14.6 ETHERS

Ethers have an oxygen atom bonded between two carbon atoms. They are like an alcohol or phenol except that H in the —OH group is replaced by carbon. The general formula of an ether is R—O—R', where R and R' are the same or different alkyl groups (Table 11.3). Ethers also can have oxygen bonded to aromatic groups.

Naming ethers is simple. List the common names of the alkyl or aromatic groups (in alphabetical order) followed by the word *ether*. If both groups are the same, use the prefix *di-*. In simple ethers, however, the prefix *di-* sometimes is omitted. Three examples are:

CH$_3$CH$_2$—O—CH$_2$CH$_3$ CH$_3$CH$_2$CH$_2$—O—CH$_3$
diethyl ether   *methyl propyl ether*

diphenyl ether

EXAMPLE 14.6 Name the following ethers:

(a) CH$_3$(CH$_2$)$_4$—O—CH$_2$CH$_3$, (b) C$_6$H$_5$—O—CH$_3$,

(c) CH$_3$CH(CH$_3$)—O—CH$_2$(CH$_2$)$_2$CH$_3$

SOLUTION (a) ethyl pentyl ether, (b) methyl phenyl ether (common name is anisole; see BHA in Figure 14.12), (c) butyl isopropyl ether.

Properties, Uses, and Reactions. The best known ether is diethyl ether, commonly called ethyl ether or ether. In 1846 William Morton, a Boston dentist, publicly demonstrated that ether could be used as an anesthetic for surgery (Figure 14.15). This launched a new era in medicine, making possible medical and dental surgery that previously caused too much pain to be performed.

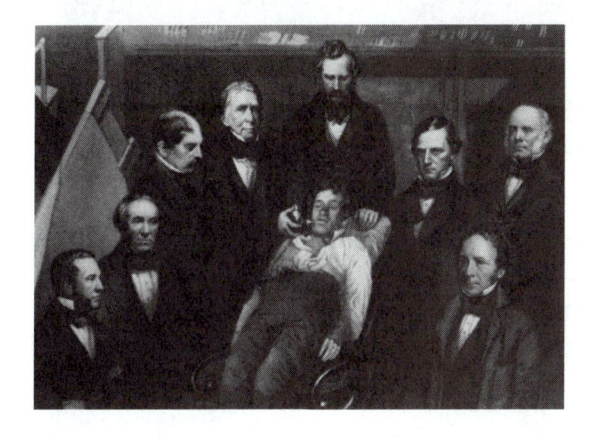

Figure 14.15 William Morton making the first public demonstration of ethyl ether as an anesthetic in 1846. Note the lack of gowns, masks, and gloves. The germ theory of disease had not yet been established.

Figure 14.16 Two inhalation anesthetics that are ethers.

Diethyl ether is rarely used now because it is slow-acting and causes nausea and vomiting in many patients. It is also highly flammable and explosive. When exposed to air for a long time, ethers form explosive compounds called peroxides. A lit match, a spark from an electrical appliance or from brushing against clothing, or even a hot piece of metal near fumes of diethyl ether can cause an explosive combustion reaction. Halogenated ethers such as enflurane and methoxyflurane (Figure 14.16) are safer to use in operating rooms and have helped replace diethyl ether as an anesthetic.

Ether molecules are slightly polar because oxygen is more electronegative than carbon. Like alcohols, ethers form hydrogen bonds with water molecules (Figure 14.17). As a result, ethers are more soluble in water than hydrocarbons of similar molecular weights but less soluble than comparable alcohols (Table 14.3).

Since ethers cannot form hydrogen bonds with other ether molecules, the forces of attraction between ether molecules are much weaker than those be-

Figure 14.17 Ethers accept hydrogen bonds from water (top) but cannot form hydrogen bonds with other ether molecules (bottom) because they lack a hydrogen bonded to a very electronegative atom such as oxygen.

Table 14.3 Comparison of Linear Alkanes, Primary-alcohols, and Ethers of Similar Molecular Weights

	Molecular Weight	Melting Point (°C)	Boiling Point (°C)	State at 25°C	Solubility in Water
propane	44	−190	−42	gas	insoluble
dimethyl ether	46	−138	−25	gas	soluble
ethanol	46	−117	78	liquid	very soluble
butane	58	−138	0	gas	insoluble
ethyl methyl ether	60	—*	11	gas	soluble
1-propanol	60	−126	97	liquid	very soluble
pentane	72	−130	36	liquid	insoluble
diethyl ether	74	−116	35	liquid	slightly soluble
methyl propyl ether	74	—*	39	liquid	slightly soluble
1-butanol	74	−89	117	liquid	very soluble

*Not determined

CHEMISTRY SPOTLIGHT

Sex Secrets of the Gypsy Moth

Figure 14.19 The female gypsy moth (bottom) has an abdomen filled with eggs and less developed antennae than the male gypsy moth (top), which can detect very small amounts of disparlure in the air.

Insects often emit chemical substances, called *pheromones*, to communicate with each other. Many female insects give off highly specific pheromones to attract males for mating.

One such insect is the gypsy moth, which was brought to Massachusetts from Europe in 1869 to start a silk-producing industry. Some moths escaped, and their population increased dramatically in their new, enemy-free environment. Their larvae eat the leaves of trees, killing many trees and defoliating forests. Pesticides such as DDT are effective against gypsy moths, but DDT may no longer be used widely in the United States.

Pheromones are a relatively new weapon against insects. The gypsy moth pheromone, called disparlure (Figure 14.18), is a cyclic ether. Females carry about 400 eggs and are too heavy to fly, so they emit disparlure into the air. Male gypsy moths, with their highly developed antennae (Figure 14.19), follow the scent upwind to the female to fertilize the eggs.

To combat gypsy moth infestations, scientists have synthesized disparlure and used it to bait toxic traps, thus luring male moths to their deaths. As little as 10^{-9} gram of disparlure in a trap (an amount too small to see without a microscope) is enough to keep evaporating for three months, providing a fatal attraction for males. Another tactic is to spray disparlure into the air to prevent the confused (though presumably happy) males from finding females.

$$CH_3(CH_2)_9CH\overset{\overset{O}{\diagup\diagdown}}{}CH(CH_2)_4\overset{\overset{CH_3}{|}}{C}HCH_3$$

Figure 14.18 Disparlure, a cyclic ether used as a sex attractant by female gypsy moths.

---

tween alcohol molecules. Thus, ethers have slightly higher melting and boiling points than hydrocarbons of similar molecular weights, but lower melting and boiling points than comparable alcohols (Table 14.3).

Despite their high flammability, diethyl ether and other ethers are not very reactive. Ethers burn in combustion reactions and can be produced by the condensation of alcohols in the presence of H_2SO_4 (see Section 14.4). They participate in only a few other reactions, none of which we need to discuss here.

> ... the blank whirlwind of emotion, the horror of great darkness [that] swept through my mind and overwhelmed my heart, I can never forget, however gladly I would do so.
>
> *A 19th-century physician describing his self-amputation of a limb without using an anesthetic*

SUMMARY

Alcohols consist of an alkyl group bonded to —O—H. To name them, find the longest carbon chain containing the —OH group and change the final -e in the alkane name to -ol; then precede the name with a number to locate the —OH group. Diols and triols have two and three —OH groups, respectively. Alcohols are classified as primary, secondary, or tertiary. Phenols have —OH bonded directly to a benzene ring. Name them like other aromatic compounds.

Because their —OH group forms hydrogen bonds, alcohols and phenols are more soluble in water and have higher melting and boiling points than alkanes of similar molecular weights. As the size of its hydrocarbon chain increases, an alcohol becomes less polar. Ethanol is in alcoholic beverages, isopropyl alcohol is rubbing alcohol, and ethylene glycol is used as antifreeze. Some phenols are disinfectants.

Alcohols (but not phenols) are dehydrated to alkenes in the presence of heat and concentrated H_2SO_4; at lower temperatures primary alcohols undergo a condensation reaction to form ethers. Primary alcohols are oxidized to aldehydes, and then to carboxylic acids; secondary alcohols are oxidized to ketones; tertiary alcohols aren't oxidized. Phenols are weak acids.

Thiols (mercaptans) have the functional group —S—H. Their names consist of an alkane name plus -thiol. Thiols are oxidized to disulfides; disulfides, in turn, are reduced to thiols.

Ethers have an oxygen atom bonded between two carbon atoms. Their names consist of the hydrocarbon groups followed by *ether*. Ethers are slightly more soluble in water and have similar melting and boiling points compared with alkanes of similar molecular weights. Except for combustion reactions, ethers are not very reactive.

KEY TERMS

Alcohol (14.1)
Carbonyl group (14.4)
Dehydration reaction (14.4)
Disulfide (14.5)

Ether (14.6)
Fermentation (14.3)
Fetal alcohol syndrome (14.1)

Mercaptan (14.5)
Phenol (14.2)
Thiol (14.5)

EXERCISES

Even-numbered exercises are answered at the back of this book.

Alcohols and Phenols

1. Which element in an alcohol is the most electronegative?
2. Use δ^+ and δ^- to indicate the region of partial positive and partial negative charge, respectively, in an ethanol molecule.
3. Use VSEPR theory (Section 4.5) to predict whether the —C—O—H atoms in an alcohol molecule are in a linear or bent arrangement. How does this compare with the arrangement in a water (H—O—H) molecule?
4. Explain, on a chemical basis, why ethanol is a liquid at room temperature while alkanes of similar molecular weights are gases.
5. Arrange the following in order of increasing boiling point: **(a)** 1-butanol, **(b)** pentane, **(c)** 1-pentanol.
6. Arrange the compounds in exercise 5 in order of increasing solubility in water.
7. Notice the solubility in water of 1-pentanol shown in Figure 14.5. Would the solubility of 1,4-pentanediol be greater or less than this? Explain why.
8. Are the melting and boiling points of ethylene glycol higher or lower than those of ethanol? Explain why. (See Table 14.1).
9. Does methanol or 1-decanol dissolve better in ethanol? Explain why.
10. Write IUPAC names for the following:

    (a) 4-methylphenol structure (benzene ring with CH₃ and OH in para positions)

    (b) $HOCH_2CH(CH_2)_3CHCH_2Cl$ with CH_2CH_3 and CH_3 substituents

(c) $CH_3-\underset{\underset{CH_3}{|}}{\overset{\overset{CH_3}{|}}{C}}-OH$ (d) $CH_3CH(CH_2)_5CH_3$ with OH on the CH

11. Write IUPAC names for the following:

(a) $CH_3\underset{\underset{}{}}{\overset{\overset{CH_3}{|}}{CH}}CH_2OH$,

(b) phenol with OH and Br (meta-bromophenol structure)

(c) $CH_3CH_2CH_2CH_2OH$

(d) $CH_3\underset{\underset{OH}{|}}{\overset{\overset{CH_3}{|}}{CH}}CHCH_2CH_3$

12. Write the common name for compound (c) in exercise 10.

13. Write the common names for compounds (a) and (c) in exercise 11.

14. Write condensed structural formulas for the following: **(a)** 2-propanol, **(b)** isopropyl alcohol, **(c)** *o*-chlorophenol.

15. Write condensed structural formulas for the following: **(a)** 3-phenyl-1-butanol, **(b)** *m*-cresol, **(c)** 1,4-pentanediol.

16. Classify each substance in exercise 10 as a primary, secondary, or tertiary alcohol.

17. Classify each substance in exercise 11 as a primary, secondary, or tertiary alcohol.

18. Is rubbing alcohol a primary, secondary, or tertiary alcohol?

19. A 70% (*v/v*) solution of ethanol in water works as a disinfectant. What is the "proof" of this solution?

20. Why do table wines typically have about 12% (*v/v*) ethanol?

21. Methanol is no longer used as an antifreeze in automobiles because it vaporizes so quickly that the antifreeze protection is soon lost. Explain chemically why the substance that replaced methanol in antifreeze—ethylene glycol—vaporizes more slowly.

22. Explain why glycerin (1,2,3-propanetriol) on the skin binds water to keep skin moist.

23. Thymol, which could be called 2-isopropyl-5-methylphenol, is a natural ingredient in the herb thyme used to kill fungi and hookworms. Write a condensed structural formula for this compound.

Reactions of Alcohols and Phenols

24. Write a structural formula for the major organic product of a dehydration reaction with concentrated sulfuric acid and heat for each of the substances in exercise 10. Write "no reaction" where appropriate.

25. Write a structural formula for the major organic product of a dehydration reaction with concentrated sulfuric acid and heat for each of the substances in exercise 11. Write "no reaction" where appropriate.

26. Write a structural formula for the major organic product of a dehydration reaction with concentrated sulfuric acid and heat for each of the substances in exercise 14. Write "no reaction" where appropriate.

27. Write a structural formula for the major organic product of a dehydration reaction with concentrated sulfuric acid and heat for substances (a) and (b) in exercise 15. Write "no reaction" where appropriate.

28. Write a reaction for the synthesis of dipropyl ether from 1-propanol.

29. Name the possible ethers that could form by reaction at 140°C of a mixture of ethanol and 1-butanol in the presence of H_2SO_4.

30. Write structural formulas for the major organic product formed by the oxidation of each compound in exercise 10.

31. Write structural formulas for the major organic product formed by the oxidation of each compound in exercise 11.

32. Write structural formulas for the major organic product formed by the oxidation of each compound in exercise 14.

33. Write structural formulas for the major organic product formed by the oxidation of each compound in exercise 15.

34. What types of substances, besides ethanol, might give a positive reaction with a Breathalyzer test?

Thiols

35. Write condensed structural formulas for the following: **(a)** 2-hexanethiol, **(b)** ethanethiol, **(c)** 4-methyl-2-pentanethiol.

36. Write condensed structural formulas for the following: **(a)** 3-methyl-1-butanethiol, **(b)** 2-isobutyl-1-hexanethiol, **(c)** 1,2-dichloro-3-methyl-4-heptanethiol.

37. Name the following:

(a) phenyl—SH, (b) $CH_3CH_2CH_2CH_2SH$,

(c) $(CH_3)_2CHCH_2SH$.

38. Name the following:
  (a) CH$_3$CHCCl$_2$CH$_3$,
       |
       SH

       CH$_3$
       |
  (b) CH$_3$CHCH(CH$_2$)$_2$CH$_3$ (c) CH$_3$SH
       |
       CH$_2$SH

39. Insulin consists of two protein chains held together by a pair of disulfide bonds. Would insulin have to be oxidized or reduced in order to make it into two separate chains?

40. Represent with a condensed structural formula the product formed from the oxidation of each compound in exercise 37.

41. Represent with a condensed structural formula the product formed from the oxidation of each compound in exercise 38.

Ethers

42. Write condensed structural formulas for the following: (a) diphenyl ether, (b) diisopropyl ether, (c) ethylmethyl ether.

43. Write condensed structural formulas for the following: (a) hexyl methyl ether, (b) *sec*-butyl ethyl ether, (c) ethyl ether

44. Name the following:
  (a) C$_6$H$_5$—O—CH$_2$CH$_3$,
  (b) CH$_3$CH$_2$CH$_2$—O—CH$_3$,
  (c) CH$_3$CH$_2$—O—CH$_2$CH$_3$

45. Name the following: (a) (CH$_3$)$_3$C—O—CH$_3$, (b) CH$_3$—O—CH$_3$, (c) (CH$_3$)$_2$CH—O—(CH$_2$)$_2$CH$_3$

46. Ethanol and dimethyl ether both have the molecular formula C$_2$H$_6$O but have different structural formulas. What term describes this relationship?

47. Since ethanol and dimethyl ether have the same molecular formula, would you expect them to have the same physical properties? Why? Compare, and give a chemical explanation for, the melting and boiling points of these two compounds in Table 14.3.

48. If you spilled diethyl ether on your skin, would it be more likely to dry your skin by (a) removing water or (b) removing natural oils? Explain why.

49. Write a balanced chemical equation for the complete combustion of diethyl ether (C$_4$H$_{10}$O) to form CO$_2$ and H$_2$O.

More Reactions

50. Write a reaction (using those from previous chapters when necessary) to produce each of the following as a major product: (a) CH$_3$CH$_2$—S—S—CH$_2$CH$_3$, (b) 1-chloropropane, (c) toluene, (d) isopropyl alcohol.

51. Starting with any unsaturated hydrocarbon of your choice, write reactions to synthesize each of the following:
  $$\text{(a) } CH_3\overset{O}{\overset{\|}{C}}CH_3, \text{ (b) } CH_3\overset{O}{\overset{\|}{C}}-OH.$$

DISCUSSION EXERCISES

1. Predict how ethanol (CH$_3$CH$_2$OH) compares with ethanethiol (CH$_3$CH$_2$SH) in terms of (a) melting point, (b) boiling point, and (c) solubility in water. Justify your predictions.

2. Arrange the following in order of increasing boiling point: (a) glycerol (1,2,3-propanetriol), (b) propane, (c) propylene glycol (1,2-propanediol), (d) 1-propanethiol, and (e) 1-propanol. Justify your prediction.

3. If a person's breath contained no ethanol but did contain another alcohol or an aldehyde, would it give a positive Breathalyzer test? Explain.

15

cadaveric — tropi

Aldehydes and Ketones

1. What are the names and structures of some aldehydes and ketones?
2. What are the physical properties (such as melting and boiling points and solubility in water) of aldehydes and ketones?
3. What are some practical uses of aldehydes and ketones?
4. How are aldehydes and ketones synthesized from alcohols, and reduced to alcohols?
5. What are some other chemical reactions of aldehydes and ketones?

Understanding Undertaking

You are probably familiar wth the use of formaldehyde to preserve biological specimens. Formaldehyde has two properties that make it an effective preservative. First, it is toxic; it prevents bacterial decay of dead tissue, often for years. Second, formaldehyde reacts with proteins to form chemical cross-links that stabilize proteins; this helps hold tissue structures together. Formaldehyde is just one of the organic compounds undertakers use to preserve cadavers.

The usual procedure is to drain blood from a vein and add embalming fluid through an artery. The fluid typically contains compounds such as ethanol, formaldehyde, phenol, and glycerol. Because ethanol is so soluble in water, it carries the other materials through the blood vessels and into cells, replacing much of the body's water. Formaldehyde prevents tissues from deteriorating. Phenol sterilizes tissues and protects against fungus infections. And the water-attracting properties of glycerol (with its three —OH groups) prevent drying and make the body more pliable.

Natural fluids are also removed from body cavities and hollow organs. They are replaced with a preservative solution stronger than embalming fluid. In addition, undertakers use dyes and cosmetics to make the corpse look as natural and lifelike as possible.

Figure 15.1 The body of Russian leader Vladimir Lenin has been preserved since his death in 1924.

Diabetes is a disease that affects the metabolism of carbohydrates. It is often diagnosed by analyzing glucose concentrations in blood or urine. Glucose—the most common simple sugar in the body—is an *aldehyde* that also has several alcohol (—OH) groups (Figure 15.2).

The blood and urine of diabetics contains unusually high levels of glucose, as well as increased concentrations of another group of compounds called *ketones*. Since diabetics do not metabolize fats completely, they produce more of certain acids and ketones. These products can accumulate to dangerous levels in a condition called *ketosis* or *ketoacidosis*.

You may know of some aldehydes and ketones already. Formaldehyde, for example, is a strong-smelling substance used to preserve biological specimens (see Chapter Opener). The word *ketone* appears in phenylketonuria (PKU), a genetic disease for which nearly all babies in the United States are tested at birth. People with PKU have excess aromatic (phenyl) ketones in their urine.

In this chapter, we examine aldehydes and ketones. In later chapters we look at carbohydrates, diabetes, and PKU.

Figure 15.2 Some fruits contain glucose, a simple sugar that is also an aldehyde.

15.1 ALDEHYDES: NAMES AND STRUCTURES

Aldehydes contain the $-\overset{\overset{\displaystyle O}{\|}}{C}-H$ functional group (Table 11.3). The $>C=O$ arrangement, called a *carbonyl group* (Section 14.4), occurs both in aldehydes and ketones.

Recall the arrangement around double-bonded carbon atoms: carbon forms three sp^2 hybrid orbitals and one p orbital (Figure 11.7, lower left). In aldehydes, carbon forms sigma bonds with hydrogen and another atom using two of its sp^2 hybrid orbitals. In ketones, carbon uses two sp^2 orbitals to form sigma bonds with two other carbon atoms. In both aldehydes and ketones, carbon uses its third sp^2 orbital to form a sigma bond with oxygen and its p orbital to form the second (pi) bond to oxygen. This gives a flat, triangular arrangement with 120° bond angles around the carbonyl carbon atom (Figure 15.3).

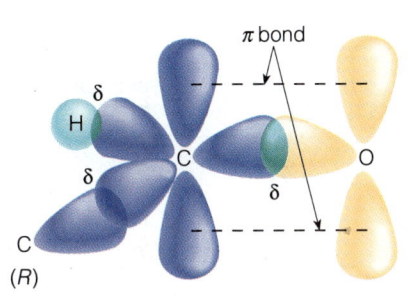

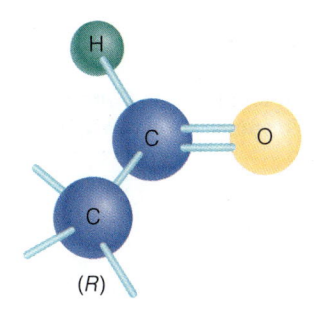

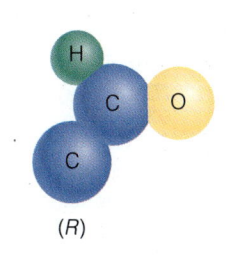

Figure 15.3 (Left) A carbonyl carbon forms three sigma bonds with its sp^2 orbitals, and one pi bond to oxygen with side-to-side overlap of p orbitals. The flat, triangular shape can be represented by a ball-and-stick model (center) or space-filling model (right). An aldehyde is shown. In ketones the H in the figures above is replaced by C from an R group.

Naming Aldehydes. IUPAC rules for naming aldehydes are simple. The only new rule is to use the suffix *-al*:

Find the longest continuous carbon chain containing the aldehyde functional group. To name it, change the final -e in the alkane name to -al. The carbonyl carbon is carbon 1.

$$CH_3CH_2\overset{O}{\overset{\|}{C}}-H \qquad CH_3CHClCH_2\overset{O}{\overset{\|}{C}}-H$$

propanal   *3-chlorobutanal*

Now try to name the following aldehyde:

$$\underset{7}{CH_3}\underset{6}{CH_2}\underset{5}{|}\underset{4}{CHCH_2}\underset{3}{|}\underset{2}{CHCH_2}\overset{O}{\overset{\|}{\underset{1}{C}}}-H$$

with CH_3CH_2 at position 5 and Br at position 3.

The longest continuous carbon chain containing the aldehyde group has seven (not six) carbons. The *-al* suffix makes the name heptanal. Collect substituent groups and number from the aldehyde end to get the complete name: 3-bromo-5-methylheptanal.

Common names are often used with aldehydes. These names are derived from the common names of carboxylic acids with the same numbers of carbons (see Table 16.1), and consist of a prefix related to the number of carbon atoms, followed by the suffix *-aldehyde*. The most common prefixes are:

    C_1 *form-*
    C_2 *acet-*
    C_3 *propion-*
    C_4 *butyr-*
    C_5 *valer-*

Learn these prefixes. In the next chapter you will use them again to name carboxylic acids.

The simplest aldehyde containing a benzene ring is named benzaldehyde.

Table 15.1 gives structures and names of some aldehydes. For simplicity, the aldehyde group sometimes is written as —CHO.

Table 15.1 Some Aldehydes and Ketones

Name IUPAC (Common)	Molecular Formula	Condensed Structural Formula	Boiling Point (°C)	Physical State at 25°C
Aldehydes				
methanal (formaldehyde)	CH_2O	HCHO	−21	gas
ethanal (acetaldehyde)	C_2H_4O	CH_3CHO	21	gas
propanal (propionaldehyde)	C_3H_6O	CH_3CH_2CHO	49	liquid
butanal (butyraldehyde)	C_4H_8O	$CH_3CH_2CH_2$CHO	76	liquid
benzaldehyde (benzaldehyde)	C_7H_6O	Ph—CHO	178	liquid
Ketones				
propanone (acetone) (dimethylketone)	C_3H_6O	$CH_3\overset{O}{\overset{\|}{C}}CH_3$	56	liquid
butanone (methyl ethyl ketone)	C_4H_8O	$CH_3\overset{O}{\overset{\|}{C}}CH_2CH_3$	80	liquid
2-pentanone (methyl propyl ketone)	$C_5H_{10}O$	$CH_3\overset{O}{\overset{\|}{C}}CH_2CH_2CH_3$	102	liquid
cyclohexanone (cyclohexanone)	$C_6H_{10}O$	cyclohexyl=O	156	liquid
benzophenone (diphenyl ketone)	$C_{13}H_{10}O$	Ph-C(=O)-Ph	306	solid

---

EXAMPLE 15.1 Write common names for a–d and IUPAC names for a–f. Write condensed structural formulas for g–j.

(a) $CH_3CH_2CH_2CHO$ (b) HCHO (c) Ph—CHO

(d) 3-bromophenyl—CHO (Br on ring)

(e) $CH_3CH_2CH_2\underset{Cl}{\overset{}{C}H}CH_2\underset{CHO}{\overset{}{C}H}CH_2CH_3$ (f) $ICH_2\underset{CH_3}{\overset{CH_3}{\overset{\|}{C}}}CH_2CH_2\underset{CH_2CH_2CH_3}{\overset{}{C}H}CHO$

(g) acetaldehyde (h) 2,4-dibromohexanal (i) propanal
(j) *p*-ethylbenzaldehyde

SOLUTION
(a) butyraldehyde (common), butanal (IUPAC)
(b) formaldehyde, methanal
(c) benzaldehyde (common and IUPAC)
(d) 3-bromobenzaldehyde (*m*-bromobenzaldehyde)
(e) 4-chloro-2-ethylheptanal
(f) 6-iodo-5,5-dimethyl-2-propylhexanal
(g) CH₃CHO, (h) CH₃CH₂CHBrCH₂CHBrCHO, (i) CH₃CH₂CHO

(j) CH₃CH₂—⟨◯⟩—CHO

15.2 KETONES: NAMES AND STRUCTURES

Ketones have the general formula $R-\overset{\overset{O}{\|}}{C}-R'$ (Table 11.3), where the carbonyl carbon ($>C=O$) bonds to two other carbon atoms. This kind of carbonyl group is often called a **keto group**. Ketones, like aldehydes, have a flat, triangular shape around the carbonyl carbon, with the double-bonded oxygen and both single-bonded carbons lying in the same plane at 120° angles (Figure 15.3).

Naming Ketones. Table 15.1 gives structures and names of some simple ketones. For IUPAC names, use the same rules as for naming aldehydes except for two differences:

1. Use the suffix *-one* to designate a ketone.
2. Select the longest continuous carbon chain containing the keto group. Number from the end nearer the keto group. Place a number and hyphen in front of the ketone name to identify the keto carbon. Name substituents in the usual way.

$$CH_3\overset{\overset{O}{\|}}{C}CH_3$$
propanone
(no number is needed)

$$\underset{1\ \ 2\ 3\ \ \ 4\ \ \ \ 5\ \ 6}{CH_3\overset{\overset{O}{\|}}{C}CH_2CH_2\overset{\overset{CH_3}{|}}{C}HCH_3}$$
5-methyl-2-hexanone

Now let's name the following ketone:

$$\underset{1\ \ \ \ \ 2\ \ \ \ 3\ 4\ \ \ \ 5}{ClCH_2CH_2\overset{\overset{O}{\|}}{C}\overset{\overset{\overset{6\ \ \ \ 7\ \ \ \ 8}{CH_2CH_2CH_3}}{|}}{C}HCH_2CH_3}$$

The longest continuous carbon chain containing the keto group has eight carbon atoms, so the base name is octanone. Number from the left because that is nearer the keto carbon, which is number 3. The base name, then, is 3-octanone. Collect substituent groups to arrive at the complete name: 1-chloro-5-ethyl-3-octanone.

The IUPAC system uses the suffix *-phenone* for a phenyl group bonded to a keto group:

$$-\overset{O}{\underset{\|}{C}}-\phi \quad \text{phenone}$$

Notice in Table 15.1 that when another phenyl group bonds to the phenone group, the name becomes benzophenone. Another common example is acetophenone:

$$CH_3-\overset{O}{\underset{\|}{C}}-\phi$$

Simple ketones have common names. To use them, name each *R* group bonded to the keto (carbonyl) carbon, followed by *ketone*. Consider, for example, the compound below:

ethyl ↘ $\overset{O}{\underset{\|}{}}$ ↙ propyl
$$CH_3CH_2\overset{O}{\underset{\|}{C}}CH_2CH_2CH_3$$

The IUPAC name is 3-hexanone. To determine the common name, identify the ethyl and propyl groups bonded to the keto carbon. The common name, then, is ethyl propyl ketone.

One common name you should memorize is acetone

$$CH_3-\overset{O}{\underset{\|}{C}}-CH_3,$$ *which is nearly always used instead of the names dimethyl ketone or propanone (Table 15.1).*

EXAMPLE 15.2 Write common names for a and b, IUPAC names for a–e, and structural formulas for f–h.

(a) $\phi-\overset{O}{\underset{\|}{C}}-\phi$ (b) $CH_3CH_2\overset{O}{\underset{\|}{C}}CH_2CH_3$ (c) $CH_3\overset{O}{\underset{\|}{C}}CH\overset{CH_3}{\underset{|}{C}}HCH_2Br$

(d) cyclohexyl=O (e) $(CH_3CH_2)_2CHCH_2\overset{O}{\underset{\|}{C}}CH_2CH(CH_3)_2$

(f) acetone (g) methyl phenyl ketone (also called acetophenone)
(h) chloropropanone

SOLUTION (a) diphenyl ketone, benzophenone (IUPAC)
(b) diethyl ketone, 3-pentanone (IUPAC)
(c) 5-bromo-3,4-dimethyl-2-pentanone
(d) cyclohexanone
(e) 6-ethyl-2-methyl-4-octanone

(f) $CH_3\overset{O}{\underset{\|}{C}}CH_3$ (g) $\phi-\overset{O}{\underset{\|}{C}}CH_3$ (h) $CH_3\overset{O}{\underset{\|}{C}}CH_2Cl$

15.3 PHYSICAL PROPERTIES AND USES OF ALDEHYDES AND KETONES

Physical Properties. Because oxygen is much more electronegative than carbon, the carbonyl group in aldehydes and ketones is polar. Oxygen's stronger attraction for shared electrons gives it a partial negative charge and leaves carbon with a partial positive charge (Figure 15.4). The resulting dipole–

Figure 15.4 The polar carbonyl group in aldehydes and ketones (left) produces dipole–dipole forces between like molecules (right).

dipole attractions between aldehyde and ketone molecules gives these compounds higher melting and boiling points than hydrocarbons (and the less polar ethers) of similar molecular weights (Figure 15.5). Since aldehydes and ketones lack the strong hydrogen bonds that occur between alcohol molecules, however, they have lower melting and boiling points than alcohols of similar molecular weights (Figure 15.5).

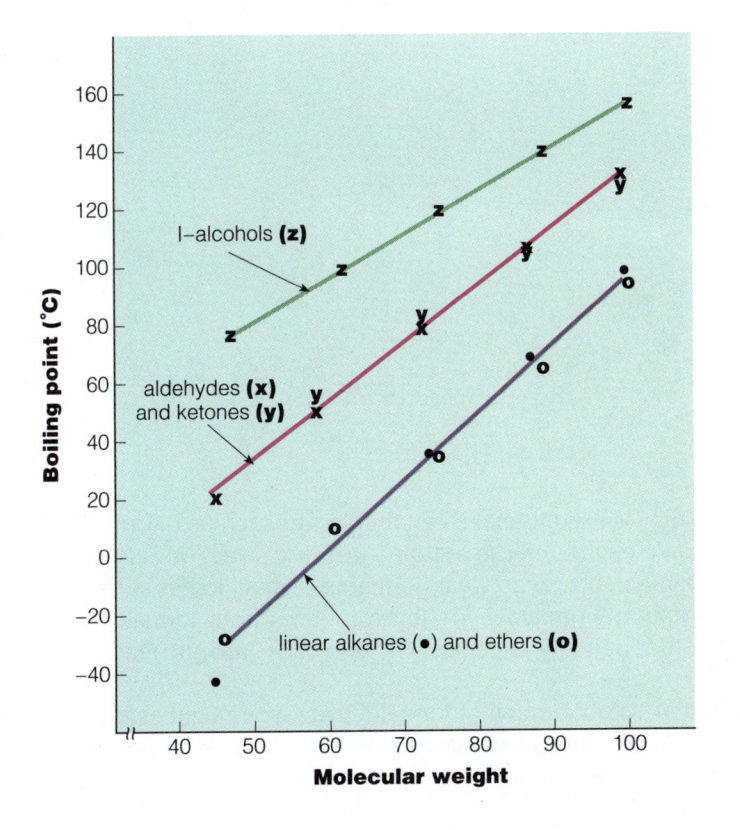

Figure 15.5 Aldehydes and ketones have boiling (and melting) points lower than alcohols and higher than hydrocarbons and ethers of similar molecular weights.

Figure 15.6 Aldehydes and ketones accept hydrogen bonds from water.

Aldehydes and ketones accept hydrogen bonds from water molecules (Figure 15.6), so they are more soluble in water than hydrocarbons and ethers of similar molecular weights, but less soluble than alcohols. As with other organic compounds, aldehydes and ketones have higher melting and boiling points (Figure 15.5) and lower solubility in water as their molecular weights increase.

> **EXAMPLE 15.3** Arrange the following in order of increasing solubility in water: (a) butanone, (b) hexane, (c) propionaldehyde, (d) 2-propanol, (e) 3-hexanone.
>
> **SOLUTION** (b) < (e) < (a) < (c) < (d) The least soluble, (b), is a nonpolar hydrocarbon; (e), (a), and (c) have carbonyl groups and increasing solubility as their hydrocarbon content (and molecular weight) decreases; (d), a low molecular weight alcohol, is the most soluble.

Uses. The aldehyde with the lowest molecular weight, and therefore the highest solubility in water, is formaldehyde (methanal). A solution of formaldehyde gas (about 37%) in water is called *formalin*. A 10% solution of formalin (about 4% formaldehyde) diluted in water or ethyl alcohol has long been used to fix and preserve biological specimens.

Because of recent evidence that it causes cancer and other health problems, however, the use of formaldehyde as a preservative is decreasing. For example, glutaraldehyde (Figure 15.7) has replaced formaldehyde for preserving certain types of tissues because it is a better cross-linking agent.

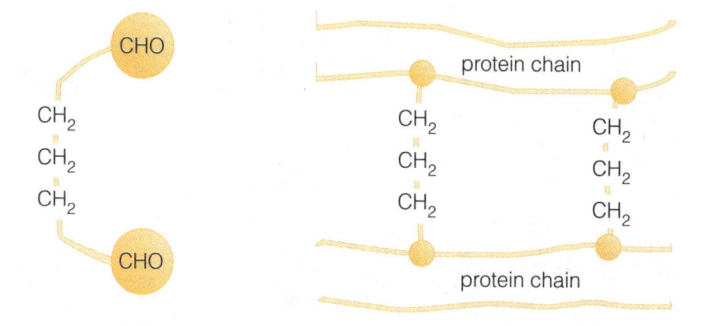

Figure 15.7 Glutaraldehyde (left) has two aldehyde groups that react with functional groups in protein chains to cross-link them (right).

Formaldehyde also is used to make certain plastics such as Bakelite (Section 18.1) and to make clothes crease-resistant. Because sweat frees formaldehyde from fabric, some people who wear permanent-press clothes suffer from rashes and skin irritation. Formaldehyde that vaporizes from formaldehyde–urea insulation or other building materials may also cause allergic reactions in some people.

If you are familiar with the smell of formaldehyde, you might assume that all aldehydes have strong, unpleasant aromas. Some aldehydes, however—especially at low concentrations—are part of the pleasing smells and tastes of

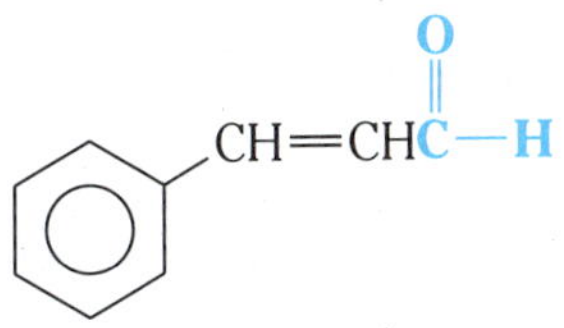

cinnamaldehyde (cinnamon)

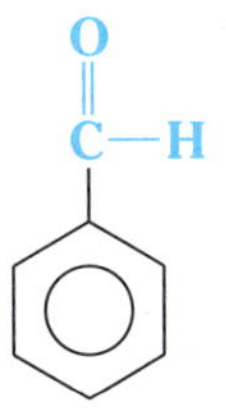

benzaldehyde (almonds)

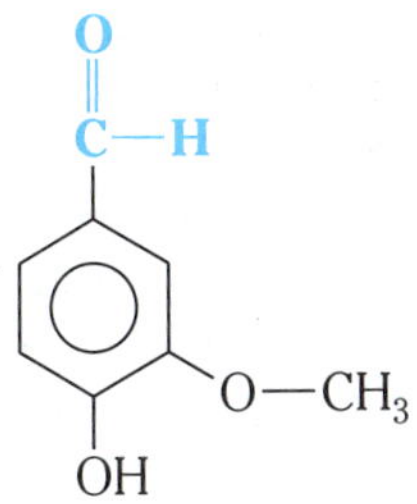

vanillin (vanilla)

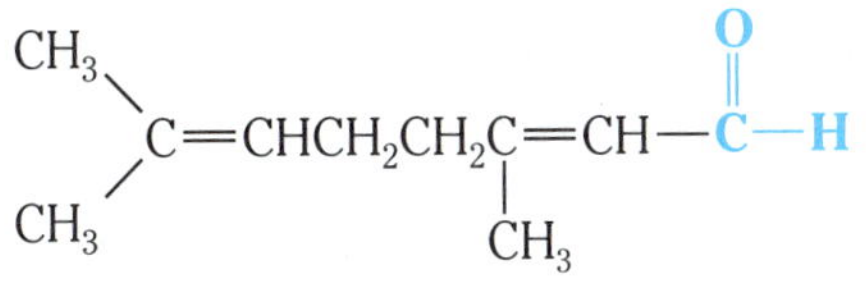

citral (lemon)

Figure 15.8 Aldehydes used in perfumes or as flavoring agents.

natural products. Figure 15.8 shows some aldehydes used in perfumes or as flavoring agents in foods.

Certain aldehydes and ketones that contain multiple alcohol (—OH) groups are sugars. One familiar example is glucose, an aldehyde that is the most common simple sugar in your body. Another is fructose, a ketone found in honey and other natural products. In Chapter 19, we examine sugars in more detail.

The simplest ketone is acetone (Table 15.1). A person in a diabetic coma has *ketosis*, a condition marked by ketones in the urine and blood, plus the sweetish odor of acetone on the breath. If you recognize this odor on an unconscious person, seek medical attention immediately—you could help save a life.

Acetone and other ketones are solvents that remove varnish, paint, and fingernail polish. Small amounts also are used to make plastics and perfumes (Figure 15.9). Familiar ketones in your body (recognizable by the suffix *-one*) are cortisone and the sex hormones, progesterone and testosterone (Section 20.3).

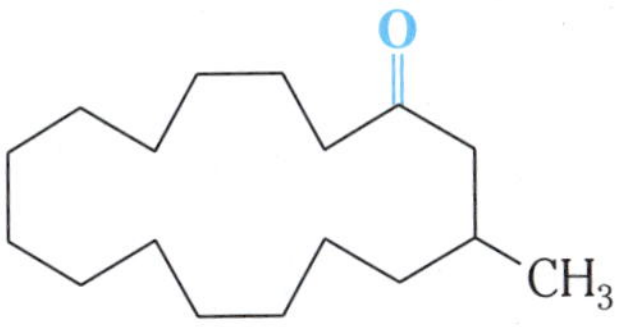

muscone (musk essence)

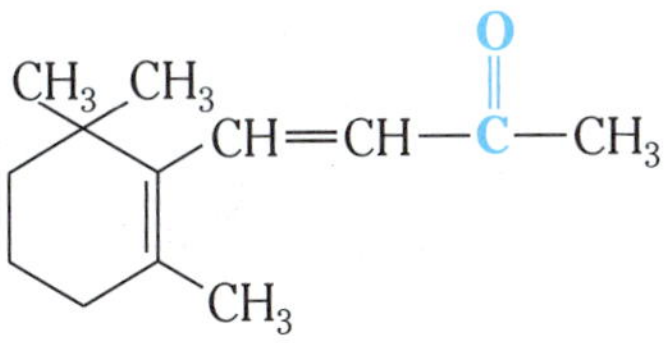

β-ionone (violets essence)

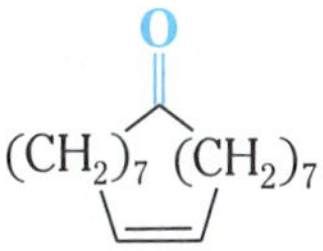

civetone (civet essence)

Figure 15.9 Ketones used in perfumes.

15.4 CHEMICAL REACTIONS OF ALDEHYDES AND KETONES

Carbonyl groups in aldehydes and ketones, like double-bonded carbons in alkenes, have double bonds consisting of one sigma bond and one pi bond. The double bond is a rich source of electrons for electron-seeking (oxidizing) agents. It also makes a compound tend to participate in addition reactions to share the pi electrons with other atoms or groups.

However, aldehydes and ketones differ from alkenes in an important way. Unlike a carbon-to-carbon double bond, the bonds between carbon and oxygen in a carbonyl group are polar. Because it is more electronegative than carbon, oxygen in a carbonyl group carries a partial negative charge (see Figure 15.4). This polarity causes aldehydes and ketones to react somewhat differently than alkenes.

The two main types of reactions we examine here are oxidation and addition.

Oxidation Reactions. In Section 13.4, you learned that alkenes react with oxidizing agents such as potassium permanganate ($KMnO_4$) and sodium (or potassium) dichromate ($Na_2Cr_2O_7$). Recall also that these oxidizing agents oxidize alcohols to aldehydes and ketones (Section 14.4); primary alcohols produce aldehydes while secondary alcohols produce ketones (Figure 15.10).

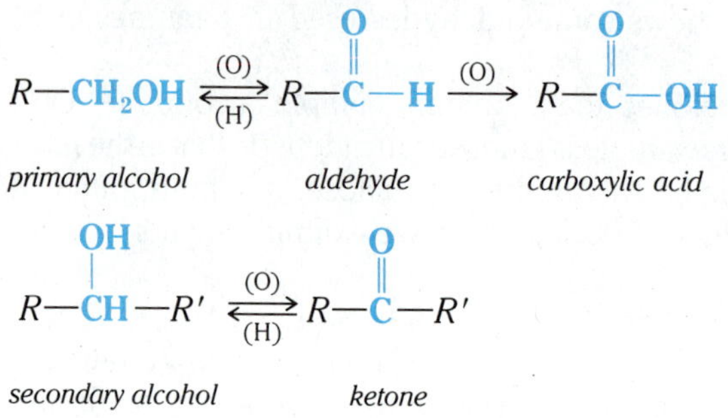

$$R-CH_2OH \underset{(H)}{\overset{(O)}{\rightleftharpoons}} R-\overset{\overset{O}{\|}}{C}-H \xrightarrow{(O)} R-\overset{\overset{O}{\|}}{C}-OH$$

primary alcohol   *aldehyde*   *carboxylic acid*

$$R-\overset{\overset{OH}{|}}{CH}-R' \underset{(H)}{\overset{(O)}{\rightleftharpoons}} R-\overset{\overset{O}{\|}}{C}-R'$$

secondary alcohol   *ketone*

Figure 15.10 Oxidation (O) and reduction (H) reactions of aldehydes and ketones.

In addition, aldehydes themselves are readily oxidized to form carboxylic acids. Ethanol, for example, is oxidized to acetaldehyde, which is oxidized to acetic acid:

$$CH_3-\overset{\overset{H}{|}}{\underset{\underset{H}{|}}{C}}-OH \xrightarrow{(O)} CH_3-\overset{\overset{O}{\|}}{C}-H \xrightarrow{(O)} CH_3-\overset{\overset{O}{\|}}{C}-O-H$$

ethanol   *acetaldehyde (ethanal)*   *acetic acid*

Notice in the first reaction that two hydrogen atoms are lost. In the second reaction, an oxygen atom is gained. Both are signs of oxidation reactions.

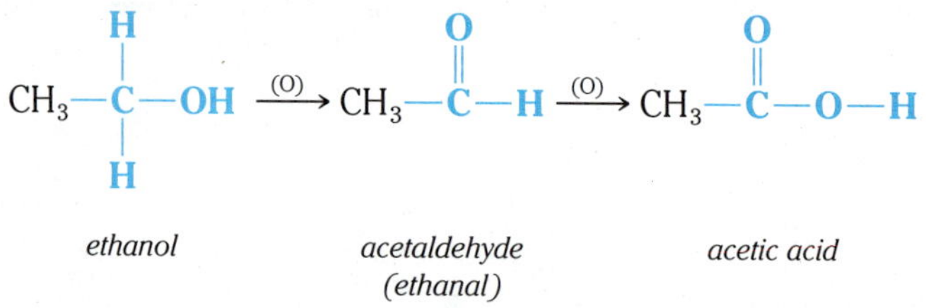

Since aldehydes are easily oxidized to carboxylic acids whereas ketones are not, you can use oxidation reactions to distinguish between the two. The basic idea is to use an oxidizing agent that is itself reduced to form a visible product. Three common procedures—the Tollens test, the Benedict test, and the Fehling test—use metal ions (weak oxidizing agents) that are reduced to colored products.

1. Tollens Test The **Tollens test** uses a reagent consisting of silver nitrate ($AgNO_3$) and ammonia (NH_3) in water. Under these conditions, the oxidizing agent $Ag(NH_3)_2^+$ forms. This ion is reduced to metallic silver, Ag, which you can easily see. The ion reacts with aldehydes but not with ketones:

$$R-\overset{\overset{O}{\|}}{C}-H + 2\ Ag(NH_3)_2^+ + 3\ OH^- \longrightarrow$$

aldehyde   *colorless*

$$R-\overset{\overset{O}{\|}}{C}-O^- + 2\ Ag + 4\ NH_3 + 2\ H_2O$$

carboxylic acid   *silver*
(as an ion)   *mirror*

Metallic silver deposits on the walls of a clean test tube, giving it a mirrored appearance (Figure 15.11).

2. Benedict and Fehling Tests These tests use blue copper (II) ion (Cu^{2+}), a mild oxidizing agent. Cu^{2+} oxidizes aldehydes but not most ketones and is reduced (gains an electron) to Cu^+ ion, which forms red, insoluble copper (I) oxide (Cu_2O):

$$R-\overset{\overset{O}{\|}}{C}-H + 2\ Cu^{2+} + 5\ OH^- \longrightarrow R-\overset{\overset{O}{\|}}{C}-O^- + 3\ H_2O + Cu_2O$$

aldehyde   *blue*   *carboxylic acid*   *red solid*
                         (as an ion)

The color changes from blue to green to yellow to orange to brick red, depending on the amount of oxidizable material present (Figure 15.12).

The only difference in the tests is that the **Benedict test** uses sodium citrate to keep copper (II) ion in solution, whereas the **Fehling test** uses sodium tartrate for that purpose. Because its reagents are more stable, the Benedict test is more commonly used.

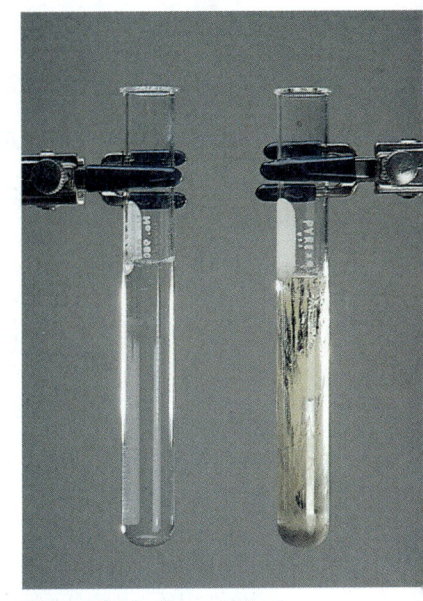

Figure 15.11 Formaldehyde solution gives a positive Tollens test (right), but acetone (a ketone) does not react (left).

EXAMPLE 15.4 Write structural formulas for the major organic product in each of the following reactions (write "no reaction" where appropriate):

(a) $CH_3\overset{\overset{O}{\|}}{C}CH_3 \xrightarrow{(O)}$

(b) $CH_3CH_2CH_2CHO \xrightarrow{(O)}$

(c) $\text{C}_6\text{H}_5-CHO + Ag(NH_3)_2^+ + 3\ OH^- \longrightarrow$

(d) $HCHO \xrightarrow{\text{Benedict test}}$

SOLUTION (a) no reaction, (b) $CH_3CH_2CH_2\overset{\overset{O}{\|}}{C}-OH$,

(c) $\text{C}_6\text{H}_5-\overset{\overset{O}{\|}}{C}-OH$, (d) $H\overset{\overset{O}{\|}}{C}-OH$

CHEMISTRY SPOTLIGHT

Diabetes and Glucose Testing

Simple sugars such as glucose are the main substances in your body that give positive Benedict and Fehling tests. Normal urine contains very little glucose or other sugars. But people with diabetes excrete glucose in their urine. Urine can be tested by the Benedict test (Figure 15.12). The color change (from green to yellow to orange to red) shows the amount of glucose present.

A more useful approach is to measure glucose concentrations directly in blood. Some insulin-dependent diabetics do this as often as three to four times a day to determine the amount of insulin to inject. They use a specific test for glucose based on two reactions. First, an enzyme (glucose oxidase) oxidizes glucose (an aldehyde) to a carboxylic acid and produces hydrogen peroxide (H_2O_2). Then a second enzyme (peroxidase) reacts H_2O_2 with a dye to form a colored product.

$$X-\overset{\overset{O}{\|}}{C}-H + O_2 \longrightarrow X-\overset{\overset{O}{\|}}{C}-O-H + H_2O_2$$

    glucose   gluconic acid

H_2O_2 + dye (colored) $\longrightarrow$ H_2O + oxidized dye (different color)

A drop of blood from a finger prick is placed on the tip of a plastic strip containing the enzyme and dye. The amount of color that forms is proportional to the blood glucose concentration. The color is measured by comparison with a color chart or by an electronic monitor (Figure 15.13).

Figure 15.12 Glucose gives a positive Benedict test, changing the solution from blue (left) to brick red (right).

Figure 15.13 Equipment for home blood glucose monitoring.

Addition Reactions. Recall that alkenes, with their carbon-to-carbon double bonds, participate in a wide variety of addition reactions (Section 13.4). Aldehydes and ketones, which have a carbon-to-oxygen double bond, also participate in addition reactions (Figure 15.14). Like alkenes, their double bonds

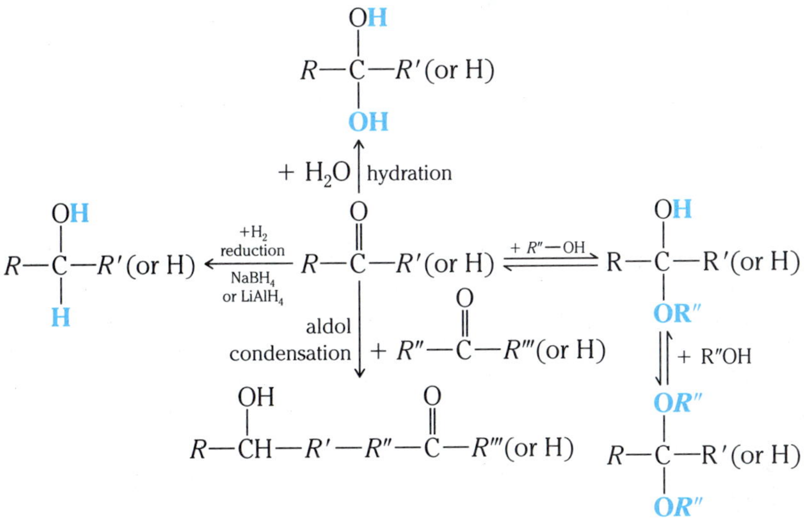

Figure 15.14 Addition reactions of aldehydes and ketones.

(consisting of one sigma and one pi bond) are a rich source of electrons for electron-seeking substances.

1. Addition of H$_2$ (Reduction) Aldehydes and ketones are reduced to alcohols by reagents such as lithium aluminum hydride (LiAlH$_4$) and sodium borohydride (NaBH$_4$). The general reaction is:

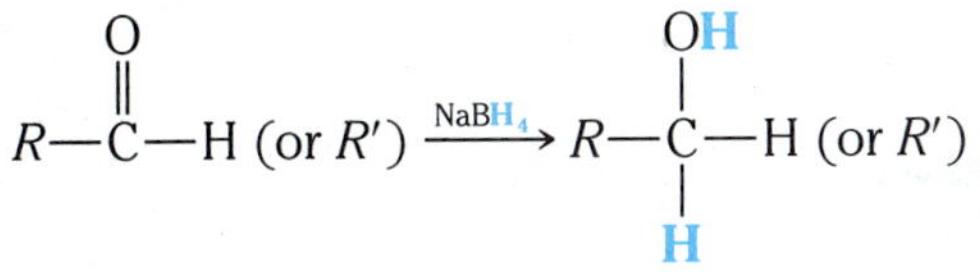

$$R-\overset{O}{\underset{\|}{C}}-H \text{ (or } R') \xrightarrow{NaBH_4} R-\overset{OH}{\underset{H}{\overset{|}{C}}}-H \text{ (or } R')$$

The addition of two hydrogens to a carbonyl group makes this a reduction reaction. Reduction of an aldehyde produces a primary alcohol. One example is:

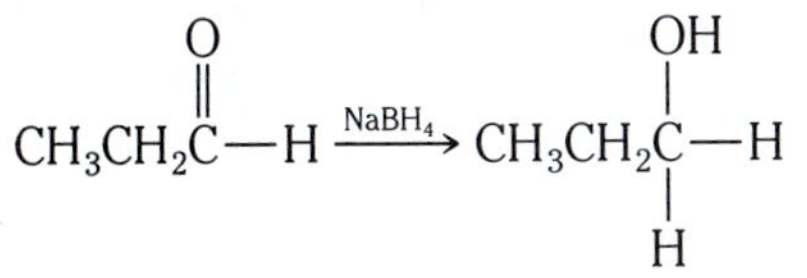

$$CH_3CH_2\overset{O}{\underset{\|}{C}}-H \xrightarrow{NaBH_4} CH_3CH_2\overset{OH}{\underset{H}{\overset{|}{C}}}-H$$

propanal   *1-propanol* (a primary alcohol)

Ketones are reduced to secondary alcohols. For example:

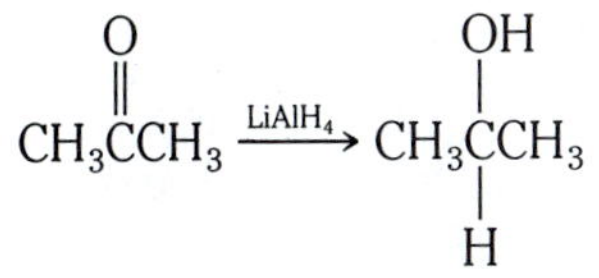

$$CH_3\overset{O}{\underset{\|}{C}}CH_3 \xrightarrow{LiAlH_4} CH_3\overset{OH}{\underset{H}{\overset{|}{C}}}CH_3$$

acetone   *2-propanol* (a secondary alcohol)
(propanone)

Notice that these are the reverse of the oxidation reactions that convert alcohols into aldehydes or ketones (Section 14.4).

Figure 15.10 summarizes the oxidation and reduction reactions involving aldehydes and ketones.

Reduction occurs by transfer of a hydride ion [H:]⁻ (see Chemistry Spotlight in Section 5.3). NaBH$_4$ transfers one of its hydrogens as a hydride ion to the partially positive carbonyl carbon; a proton from an acid then adds to the negative oxygen atom to complete the reaction:

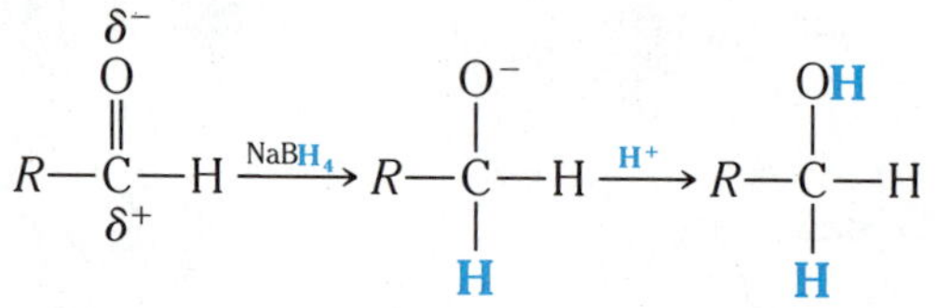

$$R-\underset{\delta^+}{\overset{\overset{\delta^-}{\overset{O}{\|}}}{C}}-H \xrightarrow{NaBH_4} R-\underset{H}{\overset{\overset{O^-}{|}}{C}}-H \xrightarrow{H^+} R-\underset{H}{\overset{\overset{OH}{|}}{C}}-H$$

Your body uses enzymes and hydride transfers to reduce carbonyl compounds to alcohols. For example, the enzyme lactate dehydrogenase (LDH) and a hydride ion donor called NADH reduce the keto group in pyruvate to an alcohol group in lactate (lactate is the salt of lactic acid; see Section 16.3 and the Chemistry Spotlight in Section 5.3).

EXAMPLE 15.5 Write structural formulas and names for the major organic product in the following reactions (write "no reaction" where appropriate):

(a) $CH_3CH_2CH_2CH_2CH_2\overset{\overset{O}{\|}}{C}CH_3 \xrightarrow{LiAlH_4}$

(b) $CH_3CH_2CH_2CH_2CHO + H_2 \xrightarrow{NaBH_4}$

SOLUTION (a) $CH_3(CH_2)_4\overset{\overset{OH}{|}}{C}HCH_3$, 2-heptanol;
(b) $CH_3CH_2CH_2CH_2CH_2OH$, 1-pentanol

---

2. Addition of H$_2$O (Hydration) Water adds to a carbonyl group to form a *diol* that has two alcohol groups on the same carbon. The product is usually called a **hydrate**. The general reaction is:

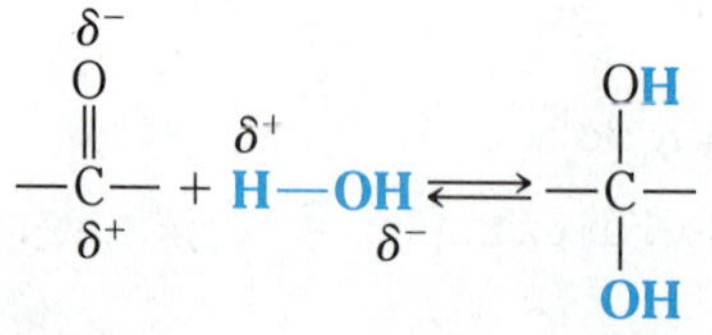

$$-\underset{\delta^+}{\overset{\overset{\delta^-}{\overset{O}{\|}}}{C}}- + \overset{\delta^+}{H}-\underset{\delta^-}{OH} \rightleftharpoons -\underset{OH}{\overset{\overset{OH}{|}}{C}}-$$

This reaction is between two polar groups because oxygen is much more electronegative than carbon and hydrogen (Section 4.6). Notice that the oppositely charged poles attract each other and add together.

Most aldehydes and ketones don't form stable hydrates. Two that do are formaldehyde and chloral. Formaldehyde forms a hydrate, HCH(OH)$_2$, that is the active ingredient in formalin solutions (Figure 15.15). Chloral hydrate forms as follows:

$$CCl_3\underset{O}{\overset{\parallel}{C}}-H + H-OH \rightleftharpoons CCl_3\underset{OH}{\overset{OH}{\underset{|}{\overset{|}{C}}}}-H$$

chloral   *chloral hydrate*

Figure 15.15 Formalin solution, used to preserve tissues, contains the hydrate of formaldehyde, $HCH(OH)_2$.

When taken internally, chloral hydrate and ethanol intensify each other's effects and rapidly put the user to sleep. In old movies and detective stories, this dangerous combination was called "knockout drops" or a "Mickey Finn." In high doses, it can prove fatal.

EXAMPLE 15.6 Write structural formulas for the hydrates of the following compounds: (a) $CH_3\overset{O}{\overset{\parallel}{C}}CH_3$ and (b) CH_3CH_2CHO

SOLUTION (a) $CH_3\underset{OH}{\overset{OH}{\underset{|}{\overset{|}{C}}}}CH_3$, (b) $CH_3CH_2\underset{OH}{\overset{OH}{\underset{|}{\overset{|}{C}}}}H$

---

3. Addition of Alcohols Alcohols ($R-O-H$) are the nearest organic relatives of water ($H-O-H$), and add to the carbonyl group the same way water does. But they form products that are more stable than hydrates. The general reactions are:

$$\underset{\underset{\delta^-}{O}}{\overset{\delta^+}{\underset{\parallel}{R-C-H}}} + \underset{\delta^+}{R'-\overset{\delta^-}{O}H} \rightleftharpoons R-\underset{OR'}{\overset{OH}{\underset{|}{\overset{|}{C}}}}-H \qquad \underset{\underset{\delta^-}{O}}{\overset{\delta^+}{\underset{\parallel}{R-C-R''}}} + \underset{\delta^+}{R'-\overset{\delta^-}{O}H} \rightleftharpoons R-\underset{OR'}{\overset{OH}{\underset{|}{\overset{|}{C}}}}-R''$$

an aldehyde   *a hemiacetal*   *a ketone*   *a hemiketal*

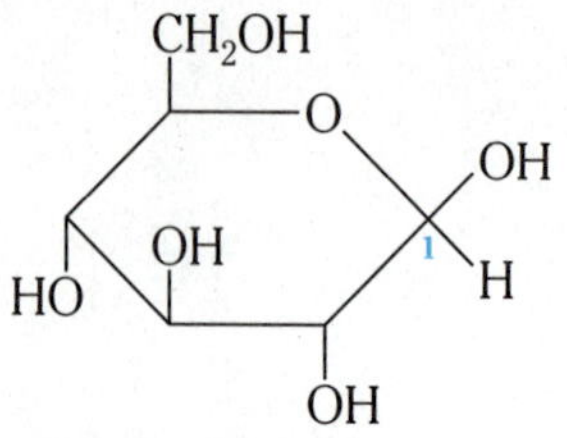

A compound with an alcohol (—OH), ether (—OR′), and H on the same carbon atom is called a **hemiacetal.** A compound with an —OH, —OR, and two carbon groups bonded to the same carbon is called a **hemiketal.**

Notice that the arrows go both ways in these reactions. Aldehydes and ketones are in equilibrium with their hemiacetal or hemiketal form, respectively. Most simple hemiacetals and hemiketals are unstable and cannot be isolated.

In your body, the main compounds that exist as cyclic hemiacetals or hemiketals are simple sugars (Chapter 19). Notice in Figure 15.16 that glucose forms a cyclic hemiacetal while fructose forms a cyclic hemiketal.

glucose

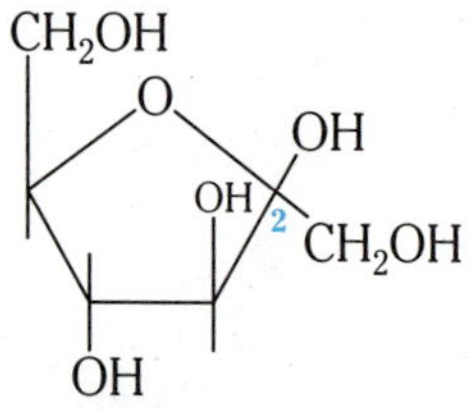

fructose

Figure 15.16 Cyclic structures of glucose and fructose.

> **EXAMPLE 15.7** From Figure 15.16, explain why the cyclic forms of glucose and fructose are a hemiacetal and hemiketal, respectively.
>
> **SOLUTION** The cyclic form of glucose has an —OH, —OR, C, and H bonded to carbon 1; this makes it a hemiacetal. The cyclic form of fructose has carbon 2 bonded to an —OH, —OR, and two other carbon atoms; this makes it a hemiketal.

The prefix *hemi-* means *half.* Just as a hemisphere is half a sphere, a hemiacetal (or hemiketal) is halfway to an acetal (or ketal). In the presence of an acid catalyst such as HCl, hemiacetals and hemiketals react with a second alcohol molecule to form *acetals* and *ketals*, respectively. The reactions are:

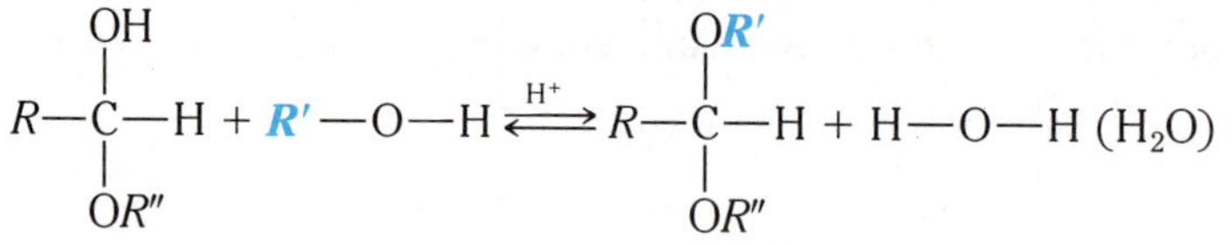

$$\underset{\text{a hemiacetal}}{R-\underset{\underset{OR''}{|}}{\overset{\overset{OH}{|}}{C}}-H} + R'-O-H \underset{}{\overset{H^+}{\rightleftharpoons}} \underset{\text{an acetal}}{R-\underset{\underset{OR''}{|}}{\overset{\overset{OR'}{|}}{C}}-H} + H-O-H \; (H_2O)$$

$$\underset{\text{a hemiketal}}{R-\underset{\underset{OR'}{|}}{\overset{\overset{OH}{|}}{C}}-R''} + R'-O-H \underset{}{\overset{H^+}{\rightleftharpoons}} \underset{\text{a ketal}}{R-\underset{\underset{OR'}{|}}{\overset{\overset{OR'}{|}}{C}}-R''} + H-O-H \; (H_2O)$$

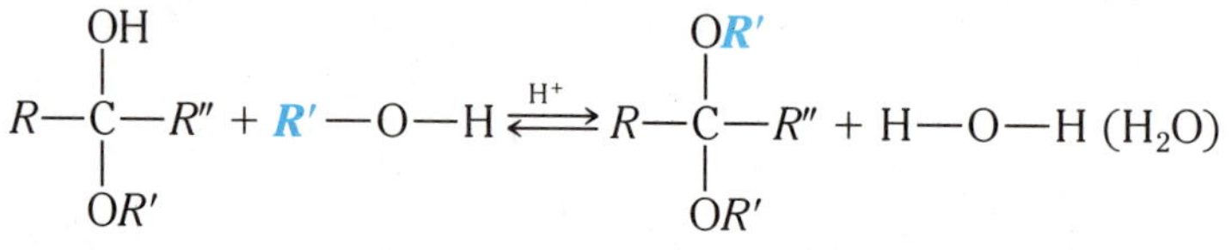

According to Le Châtelier's principle (Section 8.5), removing water drives the reactions to the right and favors the formation of acetals and ketals. Adding water has the opposite effect.

The alkyl group in the alcohol replaces H in the OH group of the hemiacetal or hemiketal. The products contain carbon bonded to two ether (—OR′) groups, plus either a hydrogen atom and a carbon group in the case of **acetals,** or two carbon groups in the case of **ketals.** Notice that these reactions are in equilibrium. Acetals and ketals are stable enough to be isolated in the absence of acid and water.

> **EXAMPLE 15.8** Write structural formulas for the (a) hemiacetal or hemiketal and (b) acetal or ketal formed by the reaction of ethanol with (1) butanone and (2) propanal.

SOLUTION (1) $\text{CH}_3\underset{\underset{\text{OCH}_2\text{CH}_3}{|}}{\overset{\overset{\text{OH}}{|}}{\text{C}}}\text{CH}_2\text{CH}_3$ (hemiketal) and $\text{CH}_3\underset{\underset{\text{OCH}_2\text{CH}_3}{|}}{\overset{\overset{\text{OCH}_2\text{CH}_3}{|}}{\text{C}}}\text{CH}_2\text{CH}_3$ (ketal)

(2) $\text{CH}_3\text{CH}_2\underset{\underset{\text{OCH}_2\text{CH}_3}{|}}{\overset{\overset{\text{OH}}{|}}{\text{CH}}}$ (hemiacetal) and $\text{CH}_3\text{CH}_2\underset{\underset{\text{OCH}_2\text{CH}_3}{|}}{\overset{\overset{\text{OCH}_2\text{CH}_3}{|}}{\text{CH}}}$ (acetal)

Notice that the general reactions above are reversible. The reverse reaction is the *hydrolysis* of an acetal or ketal to a hemiacetal or hemiketal, respectively. You will see examples of such hydrolysis reactions in Chapter 19 when we examine carbohydrates in more detail. You will also see how the hemiacetal or hemiketal form of a simple sugar reacts with the alcohol (—OH) group of another sugar molecule to form an acetal or ketal. But for now, you need to learn how to recognize these functional groups.

EXAMPLE 15.9 Identify each of the following as an acetal, hemiacetal, ketal, hemiketal, hydrate, or none of the above:

(a) $\text{CH}_3\text{CH}_2-\text{O}-\underset{\underset{}{\overset{\overset{\text{CH}_3}{|}}{}}}{\text{CH}}-\text{O}-\text{CH}_3$ (b) $\text{CH}_3\text{CH}_2\overset{\overset{\text{OH}}{|}}{\text{CH}}-\text{OH}$

(c) $(\text{CH}_3\text{O})_2\text{C}(\text{CH}_3)_2$ (d) $\text{CH}_3-\text{O}-\text{C}(\text{CH}_3)_3$

(e) $\text{CH}_3\text{CH}_2-\text{O}-\overset{\overset{\text{OH}}{|}}{\text{CH}}-\text{CH}_3$ (f) $(\text{CH}_3)_2\overset{\overset{\text{OH}}{|}}{\text{C}}-\text{O}-\text{CH}_2\text{CH}_3$

SOLUTION (a) acetal, (b) hydrate, (c) ketal, (d) none of the above (it is an ether), (e) hemiacetal, (f) hemiketal

4. Self Addition (Aldol Condensation) In the presence of a base such as dilute NaOH, an aldehyde or ketone may add to itself or to a different aldehyde or ketone. Carbonyl compounds that have H on a carbon next to the carbonyl carbon participate in this reaction, often called an **aldol condensation.** One example is:

$$\text{CH}_3-\overset{\overset{\text{O}}{\|}}{\text{C}}-\text{H} + \overset{\overset{\text{H}}{|}}{\text{CH}_2}-\overset{\overset{\text{O}}{\|}}{\text{C}}-\text{H} \xrightarrow{\text{OH}^-} \text{CH}_3\overset{\overset{\text{OH}}{|}}{\text{CH}}\text{CH}_2\overset{\overset{\text{O}}{\|}}{\text{C}}-\text{H}$$

3-hydroxybutanal

The base removes H (shown in color above); the C from which H was removed bonds to the carbonyl C in the other molecule; O in that carbonyl group then bonds to the H originally removed (in color).

Branched products often form. With acetone, for example, the reaction is:

$$CH_3\overset{O}{\overset{\|}{C}}CH_3 + CH_2\overset{O}{\overset{\|}{C}}CH_3 \xrightarrow{OH^-} CH_3\underset{CH_3}{\overset{OH}{\overset{|}{C}}}CH_2\overset{O}{\overset{\|}{C}}CH_3$$

(where the H in the second reactant and the OH in the product are highlighted)

4-hydroxy-4-methyl-2-pentanone

EXAMPLE 15.10 Using structural formulas, write the reaction for the aldol condensation of propionaldehyde. Name the product.

SOLUTION

$$CH_3CH_2CHO + CH_3\overset{H}{\overset{|}{C}}HCHO \xrightarrow{OH^-} CH_3CH_2\underset{CH_3}{\overset{OH}{\overset{|}{C}}}HCHCHO$$

3-hydroxy-2-methylpentanal

You will see in later chapters that your body uses aldol condensations to synthesize large molecules from smaller ones. This reaction produces, for example, a six-carbon sugar from two three-carbon compounds, and six-carbon citric acid by condensation of a four-carbon compound with a two-carbon carbonyl compound.

> Discovery consists of seeing what everybody has seen and thinking what nobody has thought.
>
> Albert Szent-Györgyi (1893–1986)

SUMMARY

Aldehydes and ketones contain a carbonyl ($\supset C=O$) group and are represented as $R-\overset{O}{\overset{\|}{C}}-H$ and $R-\overset{O}{\overset{\|}{C}}-R'$, respectively. The arrangement around a carbonyl carbon is flat and triangular, with 120° bond angles. IUPAC names for aldehydes and ketones use *-al* and *-one* suffixes, respectively. A number is used to locate the carbonyl carbon in a ketone. Aldehyde common names use special prefixes, corresponding to the number of carbons, plus *-aldehyde*. Ketone common names are the names of the groups bonded to the carbonyl carbon, plus *ketone*.

Aldehydes and ketones have a polar carbonyl group. They cannot form hydrogen bonds with like molecules, but can accept hydrogen bonds from H_2O. As a result, their solubility in water and melting and boiling points are higher than alkanes of similar molecular weights, but lower than alcohols.

Formaldehyde is widely used as a preservative. Acetone is a solvent and is exhaled by untreated diabetics and other people who have ketosis. Important aldehydes and ketones in the body include carbohydrates and certain hormones.

Aldehydes are produced by the oxidation of primary alcohols, and are themselves oxidized to carboxylic acids. Ketones are synthesized by oxidizing secondary alcohols but are not oxidized further. The Tollens, Benedict, and Fehling tests use mild oxidizing agents that give positive reactions with aldehydes but generally not with ketones.

Aldehydes and ketones participate in addition reactions with NaBH$_4$, H$_2$O, and alcohols; addition of one molecule of alcohol produces a hemiacetal and hemiketal, respectively, while reaction with a second alcohol molecule produces an acetal and ketal, respectively. Certain aldehydes and ketones react to form larger compounds by aldol condensation.

KEY TERMS

Acetal (15.4)
Aldehyde (15.1)
Aldol condensation (15.4)
Benedict test (15.4)

Fehling test (15.4)
Hemiacetal (15.4)
Hemiketal (15.4)
Hydrate (15.4)

Ketal (15.4)
Keto group (15.2)
Ketone (15.2)
Tollens test (15.4)

EXERCISES

Even-numbered exercises are answered at the back of this book.

Carbonyl Group

1. In formaldehyde, which type of orbitals are used by carbon to form covalent bonds with hydrogen atoms?

2. In a keto group, how many of each type of bond (sigma and pi) join carbon to oxygen?

3. In acetone, which atoms are not in the same plane as the carbon and oxygen in the carbonyl group?

Names and Structures of Aldehydes and Ketones

4. Write IUPAC names for the following: (a) CH$_3$CHO,

(b) CH$_3$CHICHCHO, with CH$_3$ branch, (c) Cl—⌬—CHO,

(d) CH$_3$CH$_2$CHCHBrCH$_3$ with CHO branch

5. Write IUPAC names for the following:
(a) (CH$_3$)$_2$CHCHO, (b) CH$_3$CH$_2$CHO,
(c) ⌬ with CHO and Cl substituents,

(d) CH$_3$CH$_2$CHCHCICHO with CH$_2$(CH$_2$)$_2$CH$_3$ branch

6. Write the common name for the compound in exercise 4a.

7. Write the common name for the compound in exercise 5b.

8. Write condensed structural formulas for the following: (a) butyraldehyde, (b) ethanal, (c) 3,4-dichlorobenzaldehyde, (d) 6-bromo-4-isopropyl-3-methylheptanal.

9. Write condensed structural formulas for the following: (a) 3-methylhexanal, (b) formaldehyde, (c) m-nitrobenzaldehyde, (d) 2,2-dichloropropanal.

10. Write IUPAC names for the following:

(a) CH$_3$(CH$_2$)$_2$ÖCCH$_2$CH$_3$, (b) (CH$_3$)$_2$CHÖCCH$_3$,

(c) ⌬—C(=O)—⌬,

(d) CH$_3$CHCICH$_2$ÖCCH(CH$_3$)$_2$.

11. Write IUPAC names for the following: (a) CH$_3$ÖCCH$_3$,

(b) (CH$_3$)$_2$CHCH$_2$ÖCCH$_3$, (c) ⌬—ÖCCH$_3$,

(d) CH$_3$CH$_2$ÖCCH$_2$CH$_3$

12. Write common names for the compounds in exercises 10b and c.
13. Write common names for the compounds in exercise 11.
14. Write condensed structural formulas for the following: (a) methyl phenyl ketone, (b) 2,4-dimethyl-3-pentanone, (c) cyclohexyl ethyl ketone, (d) cyclohexanone.
15. Write condensed structural formulas for the following: (a) ethyl benzophenone, (b) 3-hexanone, (c) diisopropyl ketone, (d) ethyl sec-butyl ketone.
16. Why is the "2-" in 2-butanone not really necessary in the name for this compound?
17. Can the compound 2-butanal exist? Explain.

Physical Properties and Uses of Aldehydes and Ketones

18. Which one of the following is the most soluble in water: (a) pentanal, (b) acetone, (c) diethyl ketone? Explain why.
19. Arrange the following in order of increasing boiling point (all have about the same molecular weight): (a) ethyl methyl ether, (b) propanal, (c) butane, (d) 1-propanol.
20. List the strongest type of attraction between molecules for each substance in exercise 19.
21. Arrange the compounds in exercise 19 in order of decreasing strength of their intermolecular attractions.
22. Which would be the most effective solvent for hexanal: (a) ethanol, (b) hexane, (c) water, or (d) acetone? Explain why.
23. What is the typical content of a formalin solution used to preserve biological specimens?
24. The sweetish odor on the breath of people who have ketosis comes primarily from the compound _____.
25. From the suffixes in the following compounds, identify the functional group they contain: (a) aldosterone (an adrenal hormone), (b) cholesterol, (c) furfural (produced by chemical reaction of corn cobs).

Chemical Reactions of Aldehydes and Ketones

26. Name the alcohol that is oxidized by $K_2Cr_2O_7$ to form butanone.
27. Identify which of the following can be oxidized to produce a ketone: (a) isopropyl alcohol, (b) ethanol, (c) acetaldehyde, (d) ethyl ether.
28. Write the reactions by which, in two steps, propene could be converted into acetone.
29. For exercise 28, explain why those two reactions wouldn't produce much propanal.
30. Write condensed structural formulas for the major organic products formed in each of the following reactions (write "no reaction" where appropriate):
    (a) $HCHO + H_2O \longrightarrow$
    (b) $CH_3CH_2\overset{\overset{O}{\|}}{C}CH_3 \xrightarrow{NaBH_4}$
    (c) $CH_3CH_2\overset{\overset{O}{\|}}{C}CH_3 \xrightarrow{(O)}$
31. Write condensed structural formulas for the major organic products formed in each of the following reactions (write "no reaction" where appropriate):
    (a) $CH_3(CH_2)_3CHO \xrightarrow{(O)}$
    (b) $CH_3CH_2CHO + 2\ CH_3OH \underset{}{\overset{H^+}{\rightleftharpoons}}$
    (c) $(CH_3)_2CH(CH_2)_3CHO \xrightarrow{NaBH_4}$
32. Which reactant in exercise 30 gives a positive Tollens test? Which gives a positive Fehling test?
33. Which reactants in exercise 31 give a positive Tollens test? Which give a positive Benedict test?
34. A patient's urine gives a negative Benedict test but her blood gives a positive Benedict test. Is this normal? Explain.
35. A substance with the formula C_4H_8O gives a negative Tollens test but is reduced by $LiAlH_4$ to form a product with a higher boiling point. What is the original substance?
36. A substance with the formula C_4H_8O gives a positive Tollens test and is reduced by $NaBH_4$ to form a product with a higher boiling point. What is the original substance?
37. Write the structural formula of the ketal that forms from 2-pentanone and ethanol in the presence of HCl.
38. Identify each of the following as a hydrate, hemiacetal, hemiketal, acetal, ketal, or none of the above:
    (a) $CH_3\underset{OH}{\overset{OCH_3}{\underset{|}{\overset{|}{C}}}}CH_2CH_3$, (b) $CH_3\underset{OH}{\overset{OH}{\underset{|}{\overset{|}{C}}}}CH_3$,
    (c) $(CH_3CH_2-O)_2CHCH_3$, (d) $CH_3OCH_2CH_2OH$
39. Identify each of the following as a hydrate, hemiacetal, hemiketal, acetal, ketal, or none of the above.
    (a) $(CH_3O)_2\overset{\overset{CH_3}{|}}{C}CH_2CH_3$, (b) $(CH_3CH_2)_2C(OH)_2$,
    (c) $CH_3CH_2\underset{OCH_3}{\overset{OH}{\underset{|}{\overset{|}{C}}}}H$, (d) $CH_3CH_2-O-\underset{CH_2CH_3}{\overset{OCH_3}{\underset{|}{\overset{|}{C}}}}CH_3$

40. Write a condensed structural formula for the (a) hemiacetal and (b) acetal that could form by reaction between formaldehyde and ethanol in embalming fluid.

41. Identify which combination could react to form a ketal: (a) propanone and propanal, (b) propanal and 1-propanol, (c) propanone and 1-propanol.

42. Identify which one of the following would *not* react in an aldol condensation with another like molecule:

(a) $CH_3\overset{\overset{O}{\|}}{C}CH_2CH_3$, (b) ⟨phenyl⟩—$CH_2CHO$

(c) ⟨phenyl⟩—CHO,

43. Write the structural formula for the product formed by an aldol condensation reaction involving compound (c) in exercise 42.

DISCUSSION EXERCISES

1. Review the three types of attractions between molecules—London forces, dipole–dipole forces, and hydrogen bonds (Section 6.1). Relate the type and strength of these intermolecular forces to (a) alkanes, (b) alcohols, (c) aldehydes, and (d) ketones of similar molecular weights. Explain how this affects a compound's melting point, boiling point, and solubility in water. Explain which type of intermolecular force increases in strength as molecular weight increases, and how this affects the physical properties above.

2. Compare the alkene functional group ($\supset C=C\subset$) with the carbonyl group ($\supset C=O$) in terms of (a) shape of the molecule, (b) polarity, and (c) types of reactions.

16

Carboxylic Acids and Esters

1. What are the names and structures of some carboxylic acids? Why are some carboxylic acids called fatty acids?
2. What are the physical properties and important uses of carboxylic acids?
3. Why are carboxylic acids classified as acids? Can they be neutralized (as inorganic acids are) by reaction with bases to form salts and water?
4. What are the main chemical reactions of carboxylic acids involving the formation and breakdown of esters?
5. What are the names and structures of some esters? What are their main physical properties and uses?
6. What are phosphate esters and anhydrides?

Kidney Stones and Gout

Like other acids, carboxylic acids (see Table 11.3) are neutralized by bases to form salts. Some of those salts are insoluble and deposit in the body as solids, causing health problems.

One example is oxalic acid, which is abundant in rhubarb leaves, strawberries, spinach, and unripe tomatoes. The neutralized, ionic form of this acid binds calcium ions in the body to form insoluble calcium oxalate. Sometimes this material deposits in the kidneys and urinary tract, creating a kidney stone (Figure 16.1). In addition to calcium oxalate, kidney stones may contain crystals of uric acid and calcium phosphate, $Ca_3(PO_4)_2$. These deposits damage nearby tissue, cause pain, can lead to infections, and sometimes block the flow of urine. Kidney stones are removed by surgery, or by suction through a tiny tube inserted into the kidney. They can also be treated with ultrasonic waves that shatter the stones.

Uric acid is produced naturally in the body and forms insoluble salts with sodium. People with gout have high concentrations of uric acid and sodium urate in their bodies. As a result, crystals of these substances deposit in the kidneys and joints, causing stiffness and soreness. The solubility of these compounds (like most solids) is lower at cooler temperatures (see Section 7.2). Since the limbs are slightly cooler than other parts of the body, deposits are most common in the feet, knees, elbows, and hands.

Figure 16.1 Kidney stones.

W hat do vitamin C, soap, ant stings, and some salad dressings have in common? They all contain *carboxylic acids* or closely related compounds (Figure 16.2). Carboxylic acids help make lemons and limes sour, give goats their special fragrance, make rhubarb leaves dangerous to eat, and account for the vinegary taste of wine exposed to air.

Carboxylic acids are related to another group of compounds called *esters*. Esters are found in perfumes, fats and oils, aspirin, and beeswax.

Figure 16.2 Oranges are a good source of vitamin C, a carboxylic acid.

16.1 CARBOXYLIC ACIDS: NAMES AND STRUCTURES

Carboxylic acids contain the
$$\overset{O}{\underset{\|}{-C}}-O-H$$
functional group (see Table 11.3), called the **carboxyl group**. Notice that the carboxyl group contains both a hydroxyl (—OH) and a carbonyl group ($>C=O$). Carbon in a carboxyl group has the same bond arrangement as in the carbonyl group of aldehydes and ketones. In each case, carbon uses its three sp^2 hybrid orbitals to form sigma bonds to three other atoms, and its p orbital to form a pi bond to the double-bonded oxygen. Carbon and its three bonded atoms lie in the same plane with 120° bond angles (Figure 16.3).

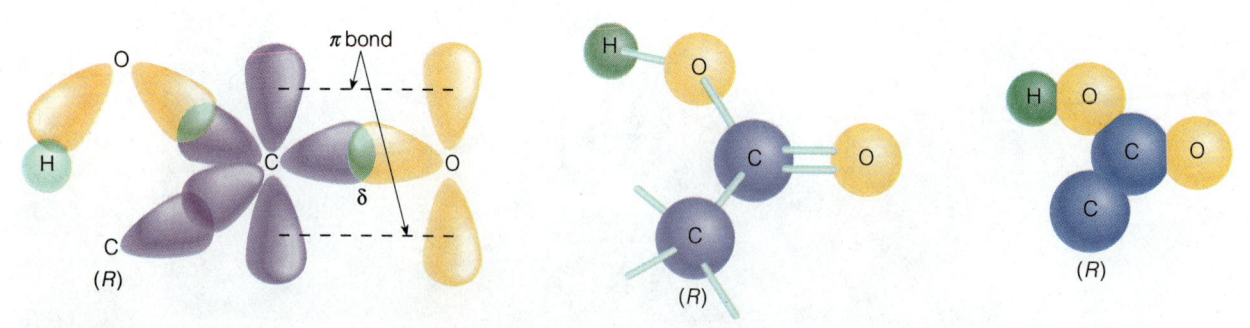

Figure 16.3 (Left) A carboxyl carbon forms three sigma bonds with its sp^2 orbitals, and one pi bond to double-bonded oxygen by side-to-side overlapping of p orbitals. The flat, triangular shape can be represented by a ball-and-stick model (center) or space-filling model (right). The same arrangement applies to esters except that H in the figures above is replaced by C from an R group.

Naming Carboxylic Acids. You can write IUPAC names for carboxylic acids by using three rules:

1. *Identify the longest continuous carbon chain containing the carboxylic acid group. The carboxyl carbon is number 1.*
2. *Change the final -e in the alkane name to -oic acid. Specify substituents in the usual way.* Two examples are:

$$\underset{\text{ethanoic acid}}{CH_3\overset{\overset{O}{\|}}{C}-OH} \qquad \underset{\text{hexanoic acid}}{CH_3(CH_2)_4\overset{\overset{O}{\|}}{C}-OH}$$

3. *Name derivatives of benzoic acid in the usual way for aromatic compounds, identifying other substituents on the benzene ring by number or by the o-, m-, p- system.* One example is:

m-nitrobenzoic acid

Now let's consider the following compound:

$$\underset{54321}{CH_3CHClCHClCH\overset{\overset{O}{\|}}{\underset{\underset{CH_2CH_3}{|}}{C}}-OH}$$

The longest continuous chain *containing the carboxyl group* has five carbons. Changing -e in pentane to -oic and adding *acid* gives the name pentanoic acid. Number from the carboxyl end and identify substituents to make the complete name: 3,4-dichloro-2-ethylpentanoic acid.

Table 16.1 gives names and structures of some carboxylic acids. Notice that condensed structural formulas represent the carboxyl group as —COOH (or —CO$_2$H). Long-chain acids (C$_{10}$ and longer) have considerable nonpolar, hydrocarbon material and are called *fatty acids*.

Many acids have common names (Table 16.1). Notice that they use the same prefixes you learned in Section 15.1 for the common names of aldehydes.

The common names for carboxylic acids use a special system to locate substituents. Instead of numbering carbon atoms as in the IUPAC system, Greek letters identify the carbons, beginning with the carbon *next to* the carboxyl carbon. In case your Greek is rusty, the sequence (which follows the Greek alphabet) is α (alpha), β (beta), γ (gamma), δ (delta), and ε (epsilon):

$$\underset{\epsilon\delta\gamma\beta\alpha}{CH_3-CH_2-CH_2-CH_2-CH_2-\overset{\overset{O}{\|}}{C}-OH}$$

The common name for CH$_2$BrCH$_2$COOH, then, is β-bromopropionic acid—since the common name for a three-carbon acid is propionic acid, and Br is bonded to the β carbon. The IUPAC name is 3-bromopropanoic acid. Be careful

Table 16.1 Some Carboxylic Acids

Name Common (IUPAC)	Origin of Common Name	Condensed Structural Formula	Melting Point (°C)	Boiling Point (°C)	State at 25°C
Formic acid (methanoic acid)	L. *formica* (ant)	HCOOH	8	101	liquid
Acetic acid (ethanoic acid)	L. *acetum* (vinegar)	CH$_3$COOH	17	118	liquid
Propionic acid (propanoic acid)	Gr. *proto pion* (first fat)	CH$_3$CH$_2$COOH	−21	141	liquid
Butyric acid (butanoic acid)	L. *butyrum* (butter)	CH$_3$(CH$_2$)$_2$COOH	−5	166	liquid
Valeric acid (pentanoic acid)	L. *valere* (strong)	CH$_3$(CH$_2$)$_3$COOH	−34	186	liquid
Caproic acid (hexanoic acid)	L. *caper* (goat)	CH$_3$(CH$_2$)$_4$COOH	−2	205	liquid
Lauric acid (dodecanoic acid)	laurel	CH$_3$(CH$_2$)$_{10}$COOH	44	299	solid
Palmitic acid (hexadecanoic acid)	palm	CH$_3$(CH$_2$)$_{14}$COOH	63	350	solid
Stearic acid (octadecanoic acid)	Gr. *stear* (solid)	CH$_3$(CH$_2$)$_{16}$COOH	71	360	solid
Benzoic acid	Gum benzoin	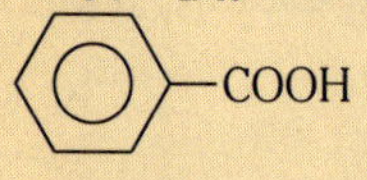 C$_6$H$_5$—COOH	122	249	solid
Salicylic acid	L. *salix* (willow)	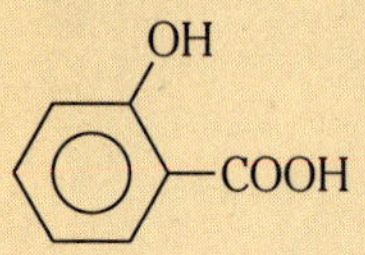 (2-hydroxy)C$_6$H$_4$—COOH	159	d*	solid

*decomposes

not to mix the two systems; 3-bromopropionic acid and β-bromopropanoic acid are *not* correct names.

EXAMPLE 16.1 Write IUPAC and common names for the following:

(a) (CH$_3$)$_2$CHCOOH   (b) CH$_3$CCl$_2$(CH$_2$)$_2$COOH

(c) 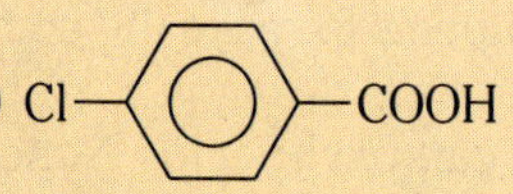 Cl—C$_6$H$_4$—COOH

SOLUTION (a) 2-methylpropanoic acid, α-methylpropionic acid; (b) 4,4-dichloropentanoic acid, γ,γ-dichlorovaleric acid; (c) 3-chlorobenzoic acid or *p*-chlorobenzoic acid

Dicarboxylic and **tricarboxylic acids** have two and three carboxyl groups, respectively. They are almost always known by common names. Figure 16.4 shows two examples. Oxalic acid, found in rhubarb and spinach leaves, is toxic and combines with calcium ion (Ca^{2+}) to form kidney stones (see Chapter Opener). Citric acid, a tricarboxylic acid, helps give citrus fruits such as lemons and limes their tart taste.

COOH
|
COOH

oxalic acid

CH$_2$—COOH
|
HO—C—COOH
|
CH$_2$COOH

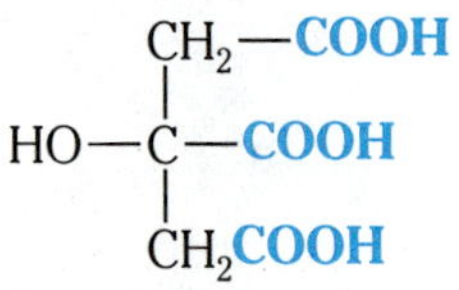

citric acid

Figure 16.4 Oxalic acid and citric acid are dicarboxylic and tricarboxylic acids, respectively.

16.2 ACIDIC PROPERTIES OF CARBOXYLIC ACIDS

Recall that acids in solution donate protons (H$^+$) to water and other proton acceptors (Section 9.1). A *strong acid* reacts almost completely with water molecules to form hydronium ions (H$_3$O$^+$). But only a small percentage of molecules of a *weak acid* donates protons to water molecules (Section 9.2). We can represent the reactions as:

Strong Acid $HNO_3 + H_2O \rightleftharpoons H_3O^+ + NO_3^-$

Weak Acid $CH_3\overset{O}{\overset{\|}{C}}-OH + H_2O \rightleftharpoons H_3O^+ + CH_3\overset{O}{\overset{\|}{C}}-O^-$

Carboxylic acids, such as acetic (ethanoic) acid, are weak acids.

Recall that bases react with acids to form salts plus water (Section 9.4). This *neutralization* occurs both with strong and weak acids. The reaction of sodium hydroxide (NaOH) with a carboxylic acid is:

$$R-\overset{O}{\overset{\|}{C}}-OH + Na^+OH^- \rightleftharpoons R-\overset{O}{\overset{\|}{C}}-O^-Na^+ + H_2O$$

carboxylic acid   base   salt   water

At alkaline or neutral pH, the concentration of OH^- ion is high enough to drive the reaction to the right, converting a carboxylic acid almost completely into its salt. Adding an acid such as HCl lowers the pH and neutralizes OH^-; removing OH^- ion then drives the reaction to the left, leaving a carboxylic acid almost completely in its acid form (Figure 16.5).

Recall that the pK_a for an acid is the pH at which there are equal concentrations of an acid and its conjugate base (Section 9.6). Acetic acid, a typical carboxylic acid, has a K_a value of 1.8×10^{-5} (see Table 9.5) and a pK_a of 4.7. Thus, at pH values less than 4.7, most acetic acid molecules are in the $CH_3\overset{O}{\overset{\|}{C}}-OH$ form; above pH 4.7, the ionized form $\left(CH_3\overset{O}{\overset{\|}{C}}-O^-\right)$ predominates; at pH 4.7, equal amounts of these two forms occur. In your body, then, where the pH is fairly neutral (except for your stomach), carboxylic acids exist mostly in their ionic, salt form.

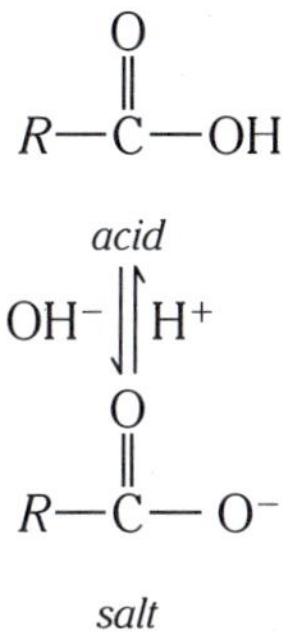

The anion (negatively charged ion) formed by neutralizing a carboxylic acid is called a **carboxylate ion.** To name these ions, delete *-ic acid* from the name of the acid and replace it with *-ate.* For example, the anion of acetic (or ethanoic) acid is called acetate (or ethanoate). As when naming other salts, simply name the positive ion and then the anion. Thus, $CH_3\overset{O}{\overset{\|}{C}}-O^-K^+$ is potassium acetate (or potassium ethanoate). Use common names when possible.

Strong inorganic acids include hydrochloric (HCl), hydroiodic (HI), hydrobromic (HBr), perchloric ($HClO_4$), nitric (HNO_3), and sulfuric (H_2SO_4) acid.

$R-\overset{O}{\overset{\|}{C}}-OH$

acid

$OH^- \updownarrow H^+$

$R-\overset{O}{\overset{\|}{C}}-O^-$

salt

Figure 16.5 Carboxylic acids are in the acid form at low pH and in the salt form at neutral and alkaline pH.

EXAMPLE 16.2 Write common names for compounds (a) and (b), and IUPAC names for all three:

(a) $CH_3(CH_2)_2COO^-Li^+$ (b) 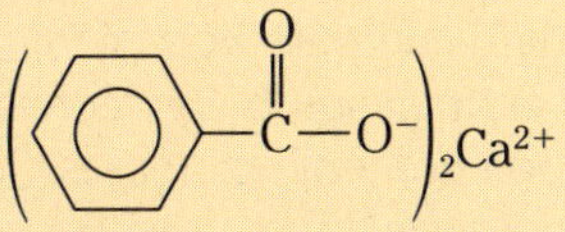 $\left(\langle O \rangle-\overset{O}{\overset{\|}{C}}-O^-\right)_2 Ca^{2+}$

(c) $(CH_3)_3CCH_2CHBrCOO^-Na^+$

SOLUTION (a) lithium butyrate, lithium butanoate (IUPAC); (b) calcium benzoate (common and IUPAC,); (c) sodium 2-bromo-4,4-dimethylpentanoate

16.3 PHYSICAL PROPERTIES AND USES OF CARBOXYLIC ACIDS AND THEIR SALTS

Notice the origins of the names of butyric and caproic acid in Table 16.1.

Physical Properties. Carboxylic acids with three to eight carbon atoms have foul odors. Butyric acid, for example, helps provide the stench of rancid butter, and caproic acid smells like goats.

The carboxyl group is polar because oxygen is much more electronegative than carbon and hydrogen. Carboxylic acids form two hydrogen bonds between like molecules (Figure 16.6). Because of these strong attractions, carboxylic acids are all liquids or solids at room temperature (see Table 16.1). In fact, carboxylic acids have the highest melting and boiling points of any organic compounds of similar molecular weights that we have examined so far, including alcohols. Figure 16.7 shows this comparison.

Figure 16.6 Two hydrogen bonds (dotted lines in color) form between carboxylic acid molecules.

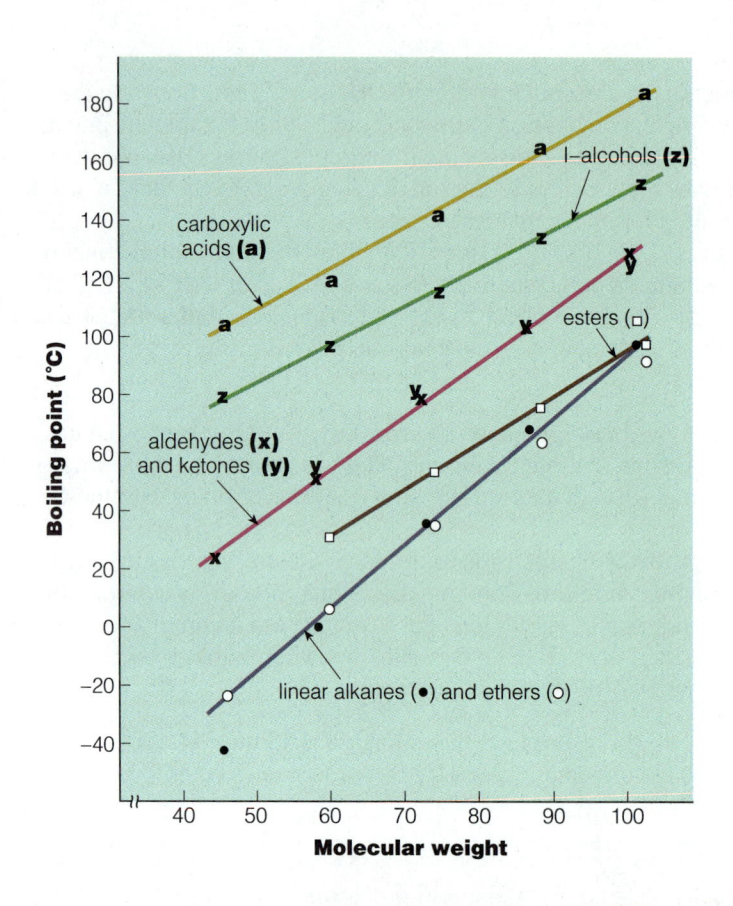

Figure 16.7 Carboxylic acids have higher boiling (and melting) points than other organic compounds of similar molecular weights. Esters (Section 16.5) have lower boiling (and melting) points than aldehydes and ketones of similar molecular weights.

Their ability to form hydrogen bonds also makes carboxylic acids more soluble in water than comparable organic compounds. Like other organic compounds, however, their solubility decreases as their molecular weight (and nonpolar hydrocarbon content) increases. So carboxylic acids with more than four carbons are not very soluble in water.

Salts of carboxylic acids have quite different properties. Since they are ionic compounds, with strong electrical attractions between ions, acid salts are solids at room temperature and have much higher melting and boiling points than the corresponding acids (Figure 16.8).

Figure 16.8 Butyric acid (left) is a liquid whereas its salt, sodium butyrate (right), is a solid.

Figure 16.9 At acidic pH (left) benzoic acid is in its acid form, which is not very soluble in water. At alkaline pH (right) benzoic acid is neutralized to its salt, sodium benzoate, which is more soluble in water.

Because their ions strongly attract polar water molecules, salts are much more soluble in water than their parent carboxylic acids. As a result, the acidity (pH) of a solution has a great influence on the amount of material that dissolves in water. Adding a base such as NaOH neutralizes a carboxylic acid and converts it into its salt, which is much more soluble in water than the parent acid (Figure 16.9).

EXAMPLE 16.3 Look at Figure 16.9. Predict what you would observe if sufficient HCl solution were added to the solution on the right. Explain why.

SOLUTION Adding HCl solution would lower pH, converting the salt back into the carboxylic acid, which is less soluble in water. Thus you would see solid, undissolved benzoic acid reappear (as in the photo on the left).

Uses. Your body produces many carboxylic acids. They exist mainly in their salt forms at the near-neutral pH values in your cells and tissues. One group of metabolic reactions, called the citric acid cycle (see Chapter 24), consists almost entirely of dicarboxylic and tricarboxylic acids, including citric acid (in its salt form). You also produce lactic acid (as lactate ion) when you exercise vigorously for a few minutes (see Chemistry Spotlight in Section 5.3).

Formic acid is one of the stinging substances produced by red ants, bees, and nettles. A 5% (w/v) solution of another acid, acetic acid, is vinegar.

Certain dicarboxylic acids and acids containing an alkene group are used to synthesize very large molecules called *polymers* (see Chapter 18). For example, $HOOC(CH_2)_4COOH$ is used to make nylon. $CH_2=CHCOOH$ (acrylic acid) and its derivatives are used to make textiles, paints, and other polymer products.

Salts of carboxylic acids have many uses. Several are used as food preservatives to inhibit the growth of molds and fungi (Figure 16.10). Salts of long-chain fatty acids are used as soaps (see Section 16.7).

The ability of negatively charged carboxylate ions to attract positive ions has several useful applications. Citrate and oxalate ions, which have three and

Notice the origins of the names formic and acetic acid in Table 16.1.

$CH_3CH_2\overset{\overset{O}{\|}}{C}-O^-Na^+$

sodium propionate

$CH_3CH=CHCH=CH\overset{\overset{O}{\|}}{C}-O^-K^+$

potassium sorbate

sodium benzoate

Figure 16.10 Three acid salts used to preserve food.

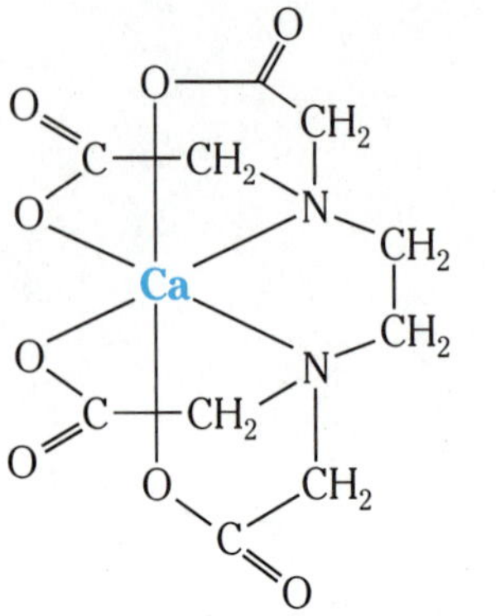

Figure 16.11 The calcium salt of EDTA.

two carboxylate groups, respectively, are added to blood samples to bind calcium ions (Ca^{2+}); this prevents blood from clotting during storage. Citrate and oxalate function as *chelating agents,* compounds with clawlike structures that bind metals effectively.

In medicine, chelating agents are used to bind toxic metals and prevent them from damaging the body. One example is EDTA (ethylenediaminetetraacetic acid), which binds metals to its nitrogen and carboxylate groups. EDTA is administered as the calcium salt (Figure 16.11) to treat poisoning by such metals as lead, beryllium, cadmium, or iron. Toxic metal ions replace calcium and form metal–EDTA complexes. The reaction with cadmium is:

$$Ca^{2+}\text{–EDTA} + Cd^{2+} \longrightarrow Cd^{2+}\text{–EDTA} + Ca^{2+}$$

16.4 CHEMICAL REACTIONS OF CARBOXYLIC ACIDS

You have already learned two of the three main reactions that involve carboxylic acids. First, recall that primary alcohols and aldehydes are oxidized to form carboxylic acids (Sections 14.4 and 15.4):

Oxidation $\quad R\text{—}CH_2OH \xrightarrow{(O)} R\text{—}CHO \xrightarrow{(O)} R\text{—}COOH$

Second, carboxylic acids are neutralized by reacting with bases such as NaOH or KOH to form water and a salt of the acid (Section 16.2):

Neutralization $\quad R\text{—}COOH + KOH \rightleftharpoons R\text{—}COO^-K^+ + H_2O$

The third reaction is the formation of **esters,** which have the functional group $R\text{—}\overset{\overset{\displaystyle O}{\|}}{C}\text{—}O\text{—}R'$. Esters form when a carboxylic acid reacts with an alcohol in the presence of an acid catalyst. The general reaction, called *esterification,* is:

$$\underset{\text{acid}}{R-\overset{\overset{\displaystyle O}{\|}}{C}-O-H} + \underset{\text{alcohol}}{H-O-R'} \underset{}{\overset{H^+}{\rightleftharpoons}} \underset{\text{ester}}{R-\overset{\overset{\displaystyle O}{\|}}{C}-O-R'} + H_2O$$

In effect, water forms from part of the acid group and part of the alcohol (shown in color), while the remainder of the reactant molecules join to form an ester. One example is:

$$\underset{\text{ethanoic acid}}{CH_3\overset{\overset{\displaystyle O}{\|}}{C}-OH} + \underset{\text{1-propanol}}{H-O-CH_2CH_2CH_3} \overset{H^+}{\rightleftharpoons} \underset{\text{an ester (propyl ethanoate)}}{CH_3\overset{\overset{\displaystyle O}{\|}}{C}-O-CH_2CH_2CH_3} + H_2O$$

Formation of an ester is a condensation reaction because H_2O is removed by reaction between two molecules (see Section 14.4).

EXAMPLE 16.4 Using structural formulas, write an equation for (a) the esterification reaction of butanoic acid with methanol, and (b) the breakdown (deesterification) of $CH_3CH_2\overset{\overset{\displaystyle O}{\|}}{C}-O(CH_2)_4CH_3$ into a carboxylic acid and an alcohol.

> **SOLUTION**
>
> (a) $CH_3(CH_2)_2\overset{O}{\overset{\|}{C}}OH + HOCH_3 \rightleftharpoons CH_3(CH_2)\overset{O}{\overset{\|}{C}}-OCH_3 + H_2O$
>
> (b) $CH_3CH_2\overset{O}{\overset{\|}{C}}-O(CH_2)_4CH_3 + H_2O \rightleftharpoons CH_3CH_2\overset{O}{\overset{\|}{C}}-OH + HO-CH_2(CH_2)_3CH_3$

Notice that the reaction goes in both directions. Since the ester keeps breaking down to form reactants, this reaction does not produce esters in high yields. One way to increase yield is to use an excess of one reactant, often the alcohol. According to Le Châtelier's principle (Section 8.5), this shifts the equilibrium to the right, producing more ester.

Another way to increase the yield of ester is to use a different, more reactive form of the acid. The two main possibilities are acyl halides and anhydrides.

Acyl Halides. An **acyl halide** has the structure $R-\overset{O}{\overset{\|}{C}}-X$, where X is a halogen, usually Cl or Br. Acyl halides are made from carboxylic acids, though we won't examine the details. $R-\overset{O}{\overset{\|}{C}}-$ is called an **acyl group,** and is named according to the acid from which it is made. Simply change the *-ic acid* in the acid name to *-yl* and then add the name of the halide. For example, $CH_3(CH_2)_2\overset{O}{\overset{\|}{C}}-Cl$ is made from butanoic (or butyric) acid, so its name is butanoyl (or butyryl) chloride.

Acid Anhydrides. The other form of the acid used to make esters is an **acid anhydride,** which has the structure $R-\overset{O}{\overset{\|}{C}}-O-\overset{O}{\overset{\|}{C}}-R'$. Think of an anhydride (which means "without water") as forming by condensation, the removal of water from between two molecules of acid:

$$R-\overset{O}{\overset{\|}{C}}-O-H + H-O-\overset{O}{\overset{\|}{C}}-R' \longrightarrow R-\overset{O}{\overset{\|}{C}}-O-\overset{O}{\overset{\|}{C}}-R + H_2O$$

The most common example, acetic anhydride, is made from acetic acid:

$$CH_3-\overset{O}{\overset{\|}{C}}-O-H + H-O-\overset{O}{\overset{\|}{C}}-CH_3 \longrightarrow CH_3-\overset{O}{\overset{\|}{C}}-O-\overset{O}{\overset{\|}{C}}-CH_3 + H_2O$$

acetic anhydride

Notice that both reactions have the arrow written in one direction to indicate that little of the ester changes back into reactants.

Neither acid anhydrides nor acyl halides occur in the body, but both are used in the laboratory to synthesize esters. The general reactions are:

$$\underset{\text{acyl chloride}}{R-\overset{\overset{O}{\|}}{C}-Cl} + \underset{\text{alcohol}}{H-O-R'} \longrightarrow \underset{\text{ester}}{R-\overset{\overset{O}{\|}}{C}-O-R'} + HCl$$

$$\underset{\text{acid anhydride}}{R-\overset{\overset{O}{\|}}{C}-O-\overset{\overset{O}{\|}}{C}-R} + \underset{\text{alcohol}}{H-O-R'} \longrightarrow \underset{\text{ester}}{R-\overset{\overset{O}{\|}}{C}-O-R'} + \underset{\text{acid}}{R-\overset{\overset{O}{\|}}{C}-O-H}$$

EXAMPLE 16.5 Using structural formulas, write equations for the reaction of (a) acetyl chloride and (b) acetic anhydride with ethanol to produce an ester.

SOLUTION

(a) $CH_3-\overset{\overset{O}{\|}}{C}-Cl + HO-CH_2CH_3 \longrightarrow CH_3-\overset{\overset{O}{\|}}{C}-O-CH_2CH_3 + HCl$

(b) $CH_3-\overset{\overset{O}{\|}}{C}-O-\overset{\overset{O}{\|}}{C}-CH_3 + HOCH_2CH_3$

$\longrightarrow CH_3-\overset{\overset{O}{\|}}{C}-O-CH_2CH_3 + HO-\overset{\overset{O}{\|}}{C}-CH_3$

EXAMPLE 16.6 Write IUPAC and common names for the following compounds.

(a) $(CH_3)_2CHCH_2COOH$, (b) $CH_3CH_2\overset{\overset{O}{\|}}{C}-Cl$,

(c) $CH_3CH_2\overset{\overset{O}{\|}}{C}-O-\overset{\overset{O}{\|}}{C}CH_2CH_3$

SOLUTION (a) 3-methylbutanoic acid (isovaleric acid or β-methylbutyric acid), (b) propanoyl chloride (propionyl chloride), (c) propanoic (or propionic) anhydride

EXAMPLE 16.7 Write a structural formula for the ester each of the compounds in Example 16.6 forms by reaction with methanol.

SOLUTION

(a) $(CH_3)_2CHCH_2\overset{\overset{O}{\|}}{C}-OCH_3$, (b) $CH_3CH_2\overset{\overset{O}{\|}}{C}-OCH_3$,

(c) $CH_3CH_2\overset{\overset{O}{\|}}{C}-OCH_3$

16.5 ESTERS: NAMES AND STRUCTURES

Esters have the functional group $R-\overset{\overset{O}{\|}}{C}-O-R'$ (Table 11.3). Their names come from the acids and alcohols from which they are made. The only difficulty in naming esters is to recognize which carbon group comes from the acid and which comes from the alcohol. Think of the oxygen in the chain $\left(R-\overset{\overset{O}{\|}}{C}-O-R'\right)$ as a bridge separating the original acid and alcohol. By studying the reactions above, you can see that the carbon group directly bonded to (and including) the carbonyl ($>C=O$) group came from the acid. The alcohol group is on the other side of the single-bonded oxygen. Figure 16.12 shows how this works.

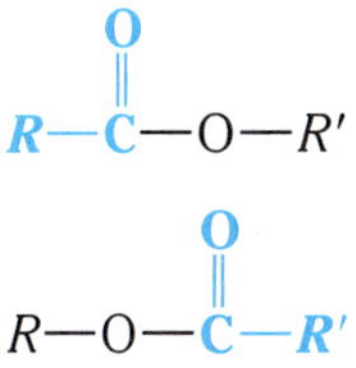

To name an ester, first name the part that came from the acid, in the same way you name acid salts, using the -ate suffix. Consider, for example, the following ester:

$$CH_3CH_2-O-\overset{\overset{O}{\|}}{C}-(CH_2)_2CH_3 \text{ (or } CH_3(CH_2)_2-\overset{\overset{O}{\|}}{C}-O-CH_2CH_3)$$

The group shown in color comes from a four-carbon acid, so the name of the salt is butanoate (IUPAC) or butyrate (common). To complete the name, identify the carbon group coming from the alcohol. The name of this ester, then, is ethyl butanoate (IUPAC) or ethyl butyrate (common).

$R-\overset{\overset{O}{\|}}{C}-O-R'$

$R-O-\overset{\overset{O}{\|}}{C}-R'$

Figure 16.12 Part of an ester molecule comes from an acid (shown in color). The remainder comes from an alcohol.

EXAMPLE 16.8 Write IUPAC and common names for the following esters:

(a) $CH_3\overset{\overset{O}{\|}}{C}-O-(CH_2)_3CH_3$ (b) $CH_3(CH_2)_2-O-\overset{\overset{O}{\|}}{C}H$

(c) 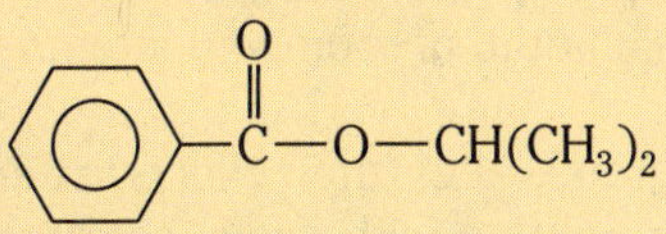 $\langle\text{C}_6\text{H}_5\rangle-\overset{\overset{O}{\|}}{C}-O-CH(CH_3)_2$

SOLUTION (a) butyl ethanoate (IUPAC), butyl acetate (common); (b) propyl methanoate (IUPAC), propyl formate (common); (c) isopropyl benzoate.

16.6 PHYSICAL PROPERTIES AND USES OF ESTERS

Physical Properties. Like aldehydes and ketones, esters are polar but don't form hydrogen bonds with each other because they lack a hydrogen atom bonded to a highly electronegative atom such as oxygen. Esters have melting and boiling points lower than aldehydes and ketones of similar molecular weights, and much lower than comparable carboxylic acids and alcohols (Figure 16.7). At room temperature most esters are liquids.

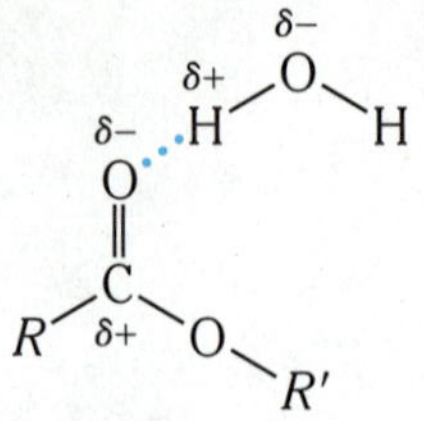

Figure 16.13 Esters accept hydrogen bonds (dotted line in color) with water.

Esters form hydrogen bonds with water (Figure 16.13), so compounds with up to four carbons are fairly (to barely) soluble in water. Larger esters have too much nonpolar, hydrocarbon material to dissolve well in water.

Uses. Many esters have sweet, fragrant aromas and occur naturally in fruits and flowers. Esters are used as flavoring agents in foods and as scents in perfumes and other personal products.

Esters with high molecular weights are solids that do not vaporize readily and thus have no aroma. Some are natural waxes. Substances with repeating ester groups are called *polyesters*. They have very large molecular weights. We discuss these synthetic fabrics, such as Dacron, in Chapter 18.

The most abundant esters in your body are *triglycerides*, which form by the esterification of glycerol (Section 14.3) with three long-chain fatty acid molecules (Figure 16.14). Triglycerides with saturated hydrocarbon chains, such as

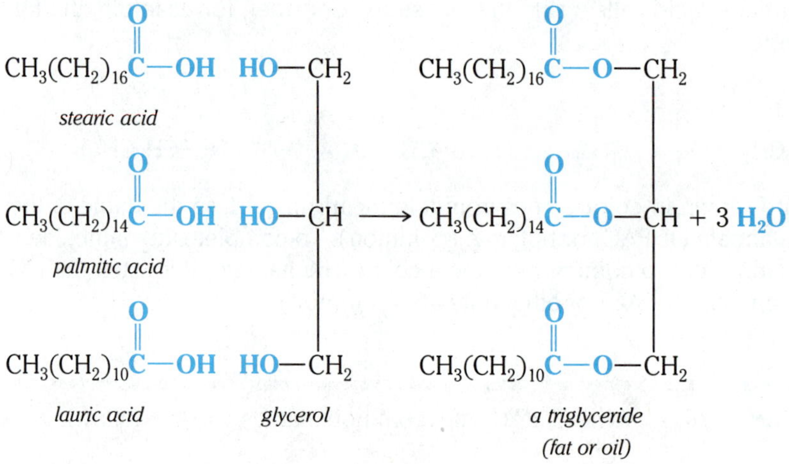

Figure 16.14 Esterification of fatty acids with glycerol produces a triglyceride.

those shown in Figure 16.14, are solid at room temperature and called *fats*. When hydrocarbon chains in triglycerides are highly unsaturated, as commonly occurs in plants, the material is likely to be a liquid at room temperature and is called an *oil* (see Figures 13.3 and 13.10).

Esterifying carboxylic acid or alcohol groups alters their effects in the body. For example, reacting alcohol groups in morphine with acetic acid produces the ester of morphine called heroin (Figure 16.15).

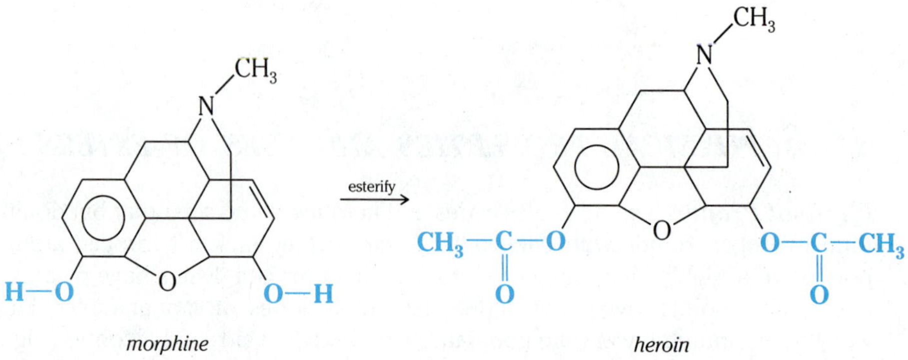

Figure 16.15 Esterification of morphine produces heroin.

CHEMISTRY SPOTLIGHT

Esters That Kill

To allow nerve impulses to pass from one cell to the next, your body relies on the action of chemicals called neurotransmitters. One neurotransmitter is acetylcholine, an ester (Figure 16.18). Once a nerve impulse has been transmitted, acetylcholine on the receiving cell is broken down by an enzyme to the parent acid and alcohol (choline) (see equation).

Breaking down and removing acetylcholine prevents the receiving cell from becoming overstimulated with additional nerve signals.

Certain types of nerve gas and pesticides (Figure 16.16) are esters that structurally resemble acetylcholine. By masquerading as acetylcholine in the cell, these compounds bind to and inhibit the enzyme that normally breaks down the neurotransmitter. As a result, acetylcholine builds up, disrupting the proper transmission of nerve impulses. The victims experience convulsions, tremors, paralysis, and then death—often by cardiac or respiratory failure.

$$CH_3\overset{O}{\overset{\|}{C}}-O-(CH_2)_2\overset{+}{N}(CH_3)_3 + H_2O \xrightarrow{enzyme} CH_3\overset{O}{\overset{\|}{C}}-OH + HO-(CH_2)_2\overset{+}{N}(CH_3)_3$$

acetylcholine → acetic acid + choline

acetylcholine (neurotransmitter)

mevinphos (insecticide)

carbaryl (Sevin) (pesticide)

GD (a nerve gas)

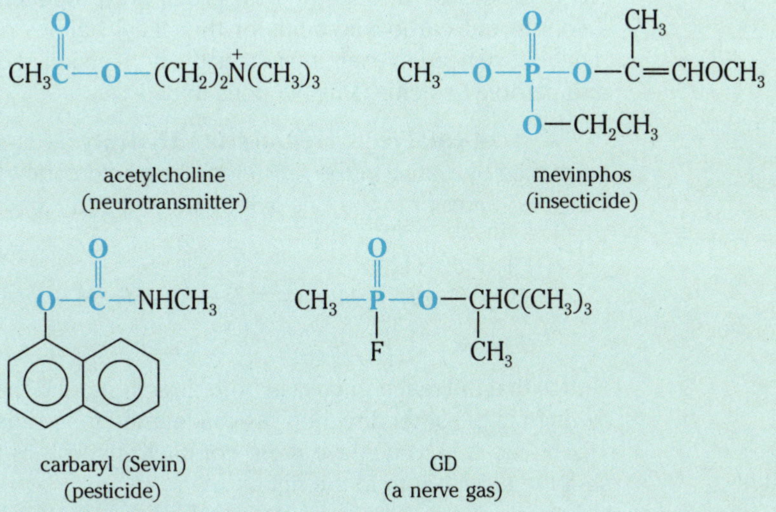

Figure 16.16 Carbon and phosphate ester compounds that paralyze victims by interfering with the normal action of the ester acetylcholine.

ibuprofen structure: CH₂CH(CH₃)₂ on benzene ring with CH₃CHCOOH

acetaminophen structure: OH on benzene ring with NH—C(=O)CH₃

Figure 16.18 Two widely used alternatives to aspirin.

Another example is aspirin. More than two centuries ago, people used salicylic acid from the bark of willow trees to relieve pain and fever. But salicylic acid is unpleasant to take orally, and it damages membranes lining the mouth, esophagus, and stomach. In 1899 Felix Hoffman, a young chemist at the Bayer Company in Germany, esterified the phenolic group in salicylic acid to produce acetylsalicylic acid, commonly called aspirin (Figure 16.17). Although

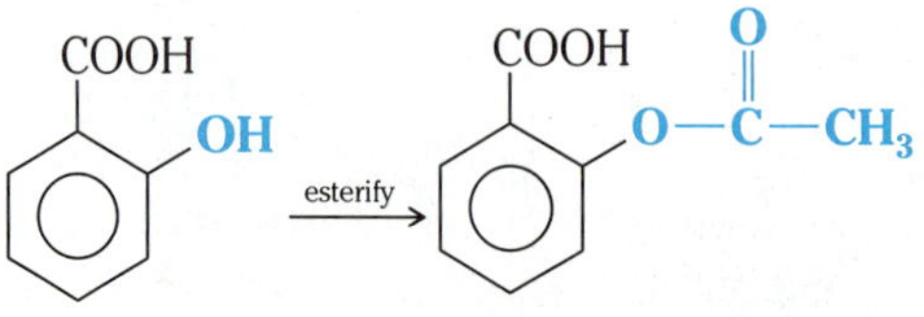

salicylic acid → (esterify) → acetylsalicylic acid (aspirin)

Figure 16.17 Esterification of salicylic acid produces aspirin.

still an acid, aspirin is less irritating than salicylic acid to membranes in the digestive tract. Ibuprofen, known by brand names such as Advil and Nuprin, is an even less irritating but more expensive acid (Figure 16.18). Another aspirin-substitute, acetaminophen (Tylenol or Datril) has phenol and amide (see Chapter 17) functional groups. Some people who are especially sensitive to aspirin use ibuprofen or acetaminophen instead.

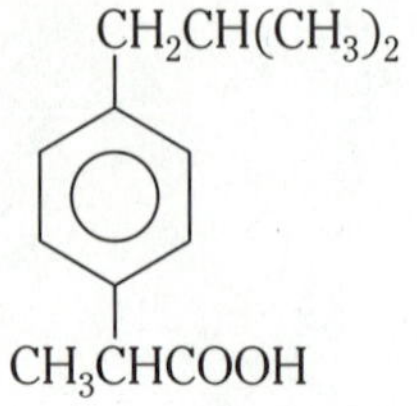

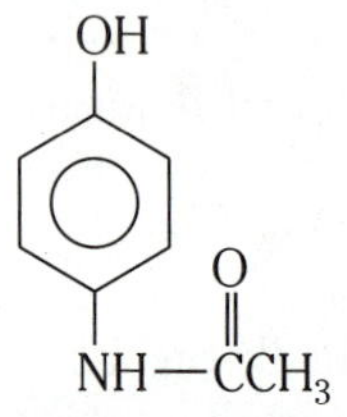

16.7 CHEMICAL REACTIONS OF ESTERS

As we discussed in Section 16.4, esters are produced by reactions between alcohols and carboxylic acids or their acyl halides or anhydrides. The only reaction we need to examine here is the conversion of esters back into alcohols and carboxylic acids. This happens in two ways.

1. Acid-catalyzed Hydrolysis Hydrolysis is the splitting apart of a substance by water. In the presence of an acid catalyst, water reacts with (hydrolyzes) an ester to form its parent alcohol and carboxylic acid:

$$R-\underset{\underset{ester}{}}{\overset{O}{\overset{\|}{C}}}-O-R' + H_2O \underset{}{\overset{H^+}{\rightleftharpoons}} R-\underset{\underset{acid}{}}{\overset{O}{\overset{\|}{C}}}-O-H + \underset{alcohol}{R'-O-H}$$

Notice that the reaction goes in both directions, and that it is the same reaction (written in the other direction) as you learned for synthesizing an ester (Section 16.4). One way to achieve more complete hydrolysis is to use excess H_2O to shift the equilibrium to the right.

EXAMPLE 16.9 Using structural formulas, write an equation for the acid-catalyzed hydrolysis of propyl acetate.

SOLUTION

$$CH_3CH_2CH_2-O-\overset{O}{\overset{\|}{C}}-CH_3 + H_2O \overset{H^+}{\longrightarrow} CH_3CH_2CH_2OH + HO-\overset{O}{\overset{\|}{C}}-CH_3$$

2. Saponification Esters also break down in the presence of a base such as NaOH or KOH. In this case, though, one of the products is not the carboxylic acid, but the neutralized salt of the acid. Think of the reaction as hydrolysis of an ester, followed by neutralization of the acid:

$$\text{Hydrolysis} \quad R-\underset{\underset{O}{\|}}{C}-O-R' + H_2O \longrightarrow R-\underset{\underset{O}{\|}}{C}-OH + R'-OH$$

$$\text{Neutralization} \quad R-\underset{\underset{O}{\|}}{C}-OH + NaOH \longrightarrow R-\underset{\underset{O}{\|}}{C}-O^-Na^+ + H_2O$$

$$\text{Net Reaction} \quad R-\underset{\underset{O}{\|}}{C}-O-R' + NaOH \longrightarrow R-\underset{\underset{O}{\|}}{C}-O^-Na^+ + R'-OH$$

$$\text{ester} \qquad\qquad\qquad \text{acid salt} \qquad \text{alcohol}$$

EXAMPLE 16.10 Using structural formulas, write an equation for the net reaction of propyl acetate ($CH_3CH_2CH_2-O-\underset{\underset{O}{\|}}{C}-CH_3$) with KOH solution.

SOLUTION $CH_3CH_2CH_2-O-\underset{\underset{O}{\|}}{C}-CH_3 + KOH$

$$\longrightarrow CH_3CH_2CH_2OH + K^+ \ {}^-O-\underset{\underset{O}{\|}}{C}-CH_3$$

This reaction is called **saponification,** which literally means "soap making." Soap is made from animal fats, which are mostly long-chain fatty acids esterified with glycerol to form triglycerides (Figure 16.14). Lye (NaOH) reacts with fat to produce the alcohol (glycerol) and sodium salts of the fatty acids:

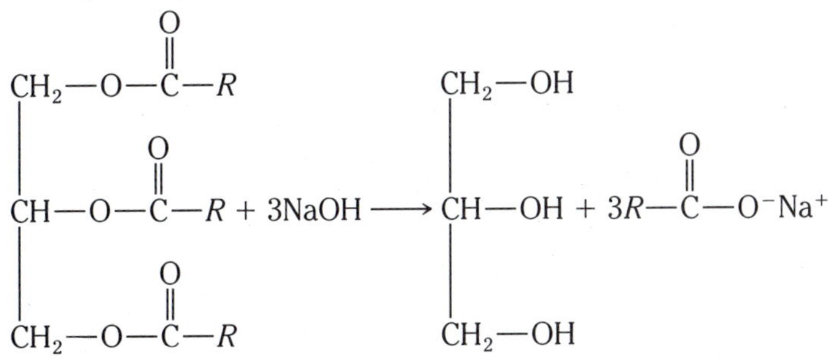

$$\begin{array}{l} CH_2-O-\underset{\underset{O}{\|}}{C}-R \\ | \\ CH-O-\underset{\underset{O}{\|}}{C}-R + 3NaOH \\ | \\ CH_2-O-\underset{\underset{O}{\|}}{C}-R \end{array} \longrightarrow \begin{array}{l} CH_2-OH \\ | \\ CH-OH \\ | \\ CH_2-OH \end{array} + 3R-\underset{\underset{O}{\|}}{C}-O^-Na^+$$

fat (triglyceride)   glycerol   fatty acid salts (soap)

Sodium salts of fatty acids such as lauric, palmitic, and stearic acid (Table 16.1) are soaps. Floating soap bars have air whipped into them to make them less dense than water. Potassium salts of fatty acids (made by saponification with KOH instead of NaOH) are used in shaving creams and liquid soap preparations.

By working as a *detergent* (see Chemistry Spotlight in Section 7.2), soaps help remove greasy and other nonpolar material that water alone cannot dissolve. Soaps do this because they have an ionic, water soluble end (the —CO$_2^-$Na$^+$ end) and a long hydrocarbon, nonpolar end (the *R* group). When nonpolar material such as grease mixes with water in the presence of soap, the nonpolar end of soap dissolves in the grease, while the ionic part of soap attracts polar water molecules. In this way, soap serves as a sort of "chemical diplomat" to produce a suspension of nonpolar material in water (Figure 16.19).

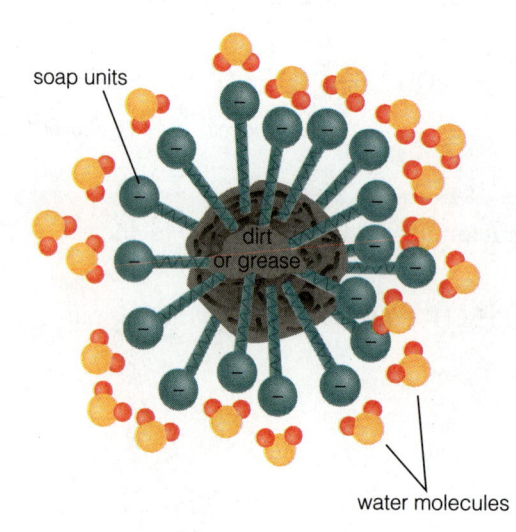

Figure 16.19 The ionic end of soap dissolves in polar water, while the nonpolar end dissolves in nonpolar dirt or grease. This breaks dirt and grease into smaller pieces that stay suspended in water until rinsed away.

16.8 PHOSPHATE ESTERS AND ANHYDRIDES

Beginning in Chapter 19, you will learn about two additional types of biochemicals: phosphate esters and phosphate anhydrides. They resemble the esters and acid anhydrides discussed in this chapter, but are based on phosphoric acid instead of a carboxylic acid.

Phosphoric acid, H$_3$PO$_4$, has three —OH groups. Each can be esterified with an alcohol to form a **phosphate ester:**

$$\text{HO}-\overset{\overset{\text{O}}{\|}}{\underset{\underset{\text{OH}}{|}}{\text{P}}}-\text{OH} + \text{H}-\text{OR} \longrightarrow \text{HO}-\overset{\overset{\text{O}}{\|}}{\underset{\underset{\text{OH}}{|}}{\text{P}}}-\text{OR} + \text{H}_2\text{O}$$

phosphoric acid   *phosphate ester*

Further reactions form phosphate di- and triesters:

$$\text{RO}-\overset{\overset{\text{O}}{\|}}{\underset{\underset{\text{OH}}{|}}{\text{P}}}-\text{OR} \qquad \text{RO}-\overset{\overset{\text{O}}{\|}}{\underset{\underset{\text{OR}}{|}}{\text{P}}}-\text{OR}$$

phosphate diester   *phosphate triester*

In later chapters you will see examples of carbohydrate and lipid phosphate esters. DNA and RNA are phosphate diesters you will study.

Phosphate anhydrides, like acid anhydrides, form by removal of water in a condensation reaction:

$$\text{HO}-\underset{\underset{\text{OH}}{|}}{\overset{\overset{\text{O}}{\|}}{\text{P}}}-\text{OH} + \text{HO}-\underset{\underset{\text{OH}}{|}}{\overset{\overset{\text{O}}{\|}}{\text{P}}}-\text{OH} \longrightarrow \text{HO}-\underset{\underset{\text{OH}}{|}}{\overset{\overset{\text{O}}{\|}}{\text{P}}}-\text{O}-\underset{\underset{\text{OH}}{|}}{\overset{\overset{\text{O}}{\|}}{\text{P}}}-\text{OH} + \text{H}_2\text{O}$$

phosphoric acid   *pyrophosphoric acid*

Notice the similarity of pyrophosphoric acid, a phosphate anhydride, to the structure of an acid anhydride.

Further reaction produces another anhydride linkage:

$$\text{HO}-\underset{\underset{\text{OH}}{|}}{\overset{\overset{\text{O}}{\|}}{\text{P}}}-\text{O}-\underset{\underset{\text{OH}}{|}}{\overset{\overset{\text{O}}{\|}}{\text{P}}}-\text{OH} + \text{HO}-\underset{\underset{\text{OH}}{|}}{\overset{\overset{\text{O}}{\|}}{\text{P}}}-\text{OH}$$

pyrophosphoric acid

$$\longrightarrow \text{HO}-\underset{\underset{\text{OH}}{|}}{\overset{\overset{\text{O}}{\|}}{\text{P}}}-\text{O}-\underset{\underset{\text{OH}}{|}}{\overset{\overset{\text{O}}{\|}}{\text{P}}}-\text{O}-\underset{\underset{\text{OH}}{|}}{\overset{\overset{\text{O}}{\|}}{\text{P}}}-\text{OH} + \text{H}_2\text{O}$$

triphosphoric acid

Phosphate anhydrides, like phosphoric acid, react with alcohols to form phosphate esters. For example:

$$\text{HO}-\underset{\underset{\text{OH}}{|}}{\overset{\overset{\text{O}}{\|}}{\text{P}}}-\text{O}-\underset{\underset{\text{OH}}{|}}{\overset{\overset{\text{O}}{\|}}{\text{P}}}-\text{OH} + \text{H}-\text{O}R \longrightarrow \text{HO}-\underset{\underset{\text{OH}}{|}}{\overset{\overset{\text{O}}{\|}}{\text{P}}}-\text{O}-\underset{\underset{\text{OH}}{|}}{\overset{\overset{\text{O}}{\|}}{\text{P}}}-\text{O}R + \text{H}_2\text{O}$$

In Section 23.1 you will learn about *nucleotides* (such as ATP), organic di- and triphosphate anhydrides that have very important roles in the body.

> Would you be calm and placid
> If you were full of formic acid?
>
>   *Ogden Nash (1902–1971)*

SUMMARY

Carboxylic acids have the general formula $R-\overset{\overset{\text{O}}{\|}}{\text{C}}-\text{O}-\text{H}$, which may be written $R-CO_2H$ or $R-COOH$. Carbon in the carboxyl group (—COOH) has a flat, triangular arrangement with its three bonded atoms. IUPAC names change -*e* in the alkane name to -*oic acid,* with the carboxyl carbon number 1. Common names use the same prefixes as for common names of aldehydes; Greek letters locate substituents.

Like all acids, carboxylic acids donate protons in water and are neutralized by bases to form salts. In solution, carboxylic acids are mostly in the acid form at acidic pH and in the salt form at neutral and alkaline pH. The carboxylate ion in a salt is named by replacing *-ic acid* in the acid name with *-ate*.

Because they form two hydrogen bonds per molecule, carboxylic acids are more soluble in water and have higher melting and boiling points than other organic compounds of similar molecular weights. Acid salts, having ionic attractions, have even higher melting and boiling points and solubilities in water.

Carboxylic acids are made by the oxidation of primary alcohols or aldehydes. They (or their more active acyl halide or anhydride forms) react with alcohols to form esters. Esters are converted back into alcohols and either carboxylic acids or acid salts by acid-catalyzed hydrolysis or saponification, respectively.

To name esters, name the portion that came from the acid as the acid salt; precede this with the name of the alkyl group from the alcohol. Esters accept hydrogen bonds but cannot form them with like molecules. Their melting and boiling points are lower than aldehydes and ketones of similar molecular weights, but higher than alkanes. Esters with four or fewer carbons are fairly soluble in water. Important esters include waxes, fats, and the neurotransmitter acetylcholine.

Phosphate esters and anhydrides are made from phosphoric acid, H_3PO_4.

KEY TERMS

Acid anhydride (16.4)
Acyl group (16.4)
Acyl halide (16.4)
Carboxylate ion (16.2)
Carboxylic acid (16.1)

Carboxyl group (16.1)
Dicarboxylic acid (16.1)
Ester (16.4)
Hydrolysis (16.7)
Phosphate anhydride (16.8)

Phosphate ester (16.8)
Saponification (16.7)
Tricarboxylic acid (16.1)

EXERCISES

Even-numbered exercises are answered at the back of this book.

Carboxyl Groups

1. Name three other functional groups that have the same geometry as occurs around a carboxyl carbon.

2. Four of the five atoms in a molecule of formic acid, HCOOH, lie in the same plane. Which atom does not have to lie in that plane?

3. Which atom in a carboxyl group is joined to carbon by a pi bond?

4. Which type of hybrid orbital does the carbon in a carboxyl group use for bonding?

Names and Structures of Carboxylic Acids

5. Write IUPAC names for the following:

   (a) $CH_3(CH_2)_3CO_2H$, (b) [benzene ring with COOH and Br substituents],

   (c) $CH_2ClCHClCCH_2COOH$ with two CH_3 groups on the central carbon.

6. Write IUPAC names for the following:

   (a) CH_3CHCH_3 with CO_2H substituent, (b) [benzene ring with OH and COOH substituents],

   (c) $CH_3(CH_2)_{13}CHClCOOH$.

7. Write common names for compounds (a) and (c) in exercise 5.

8. Write a common name for each compound in exercise 6.

9. Write condensed structural formulas for the following: (a) *p*-nitrobenzoic acid, (b) β-methylcaproic acid, (c) trichloroacetic acid, (d) lauric acid.

10. Write condensed structural formulas for the following: (a) α-chlorobutyric acid, (b) 4,5-dimethyl-3-phenyloctanoic acid, (c) 2,4-dichlorobenzoic acid, (d) oxalic acid.

11. Write IUPAC names for compounds (b) and (c) in exercise 9.
12. Write the IUPAC name for compound (a) in exercise 10.
13. Is propionic acid a strong or a weak acid?

Salts of Carboxylic Acids

14. Is propionic acid mostly in its salt form at (a) pH 2 or (b) pH 8?
15. Write a structural formula for the salt that forms when acetic acid is neutralized with NaOH solution.
16. Write a structural formula for the salt that forms when 2-methylpropanoic acid is neutralized with KOH solution.
17. Write condensed structural formulas for the following: (a) sodium acetate, (b) lithium formate, (c) potassium stearate.
18. Write condensed structural formulas for the following: (a) sodium benzoate, (b) calcium hexanoate, (c) potassium 4,4-diisopropyl-7-iododecanoate.
19. Write IUPAC names for the following:

(a) $CH_3CH_2CH_2\overset{O}{\underset{\|}{C}}-O^-Na^+$, (b) $CH_3\overset{CH_3}{\underset{|}{CH}}\overset{O}{\underset{\|}{C}}-O^-K^+$,

(c) $(CH_3CH_2CH_2CH_2\overset{O}{\underset{\|}{C}}-O^-)_2Mg^{2+}$.

20. Write IUPAC names for the following:

(a) $CH_3CHBr\overset{O}{\underset{\|}{C}}-O^-K^+$, (b) $H\overset{O}{\underset{\|}{C}}-O^-Na^+$,

(c) $(CH_3\overset{O}{\underset{\|}{C}}-O^-)_2Ca^{2+}$.

21. Write common names for the compounds in exercise 19.
22. Write common names for the compounds in exercise 20.

Physical Properties of Carboxylic Acids and Their Salts

23. Arrange the following in order of increasing solubility in water: (a) butyric acid, (b) potassium butanoate, (c) hexanal, (d) hexanoic acid.
24. Arrange the following in order of increasing boiling point (all have similar molecular or formula weights): (a) 2-hexanone, (b) pentanoic acid, (c) heptane, (d) sodium butyrate, (e) 1-hexanol.
25. Would lauric acid dissolve better in water or in hexane? Why?
26. If you added a concentrated solution of HCl to a saturated solution of sodium benzoate in water until the pH dropped to 3, what would you expect to observe? Why?
27. Suppose you have been trying unsuccessfully to dissolve a small amount of palmitic acid in water. Adding which one of the following would most likely produce a solution: (a) CH_3COOH, (b) KCl, (c) HCl, (d) NaOH, (e) oxalic acid?
28. Write the structural formula and name for the organic substance that forms from the reaction of palmitic acid with NaOH solution.
29. Carboxylate ions that function as chelating agents do so by binding to (a) soaps, (b) dicarboxylic acids, (c) fats, (d) metal ions, (e) esters.

Acyl Halides and Acid Anhydrides

30. Write common names for the following:

(a) $CH_3-\overset{O}{\underset{\|}{C}}-O-\overset{O}{\underset{\|}{C}}-CH_3$ (b) $CH_3(CH_2)_3\overset{O}{\underset{\|}{C}}-Br$

(c) $H\overset{O}{\underset{\|}{C}}-Cl$.

31. Write structural formulas for the following: (a) butyric anhydride, (b) acetyl chloride, (c) propanoyl bromide.
32. An alcohol reacts with a carboxylic acid in the presence of an acid catalyst to produce (a) another acid, (b) another alcohol, (c) an acid anhydride, (d) an acyl halide, (e) an ester.

Reactions of Carboxylic Acids

33. Which compound in exercise 30, and which alcohol, would react together to produce the ester ethyl pentanoate (ethyl valerate)?
34. Write condensed structural formulas for the esters that would be produced by reaction of 1-propanol with each substance in exercise 31.
35. Write condensed structural formulas for the esters that would be produced by reaction of ethanol with each substance in exercise 30.
36. Write condensed structural formulas for the major organic products formed in each of the following reactions (write "no reaction" where appropriate):

(a) $C_6H_5-\overset{O}{\underset{\|}{C}}-Cl + (CH_3)_2CHCH_2OH \longrightarrow$

(b) $CH_3CH_2\underset{\underset{CH_2CH_3}{|}}{CH}CH_2COOH + KOH \longrightarrow$

(c) $CH_3COO^-Na^+ + NaOH \longrightarrow$

37. Write condensed structural formulas for the major organic products formed in each of the following reactions (write "no reaction" where appropriate):

(a) $CH_3CH_2COOH + \langle\bigcirc\rangle-CH_2OH \xrightleftharpoons{H^+}$

(b) $CH_3-\overset{O}{\underset{\|}{C}}-O-\overset{O}{\underset{\|}{C}}-CH_3 + \langle\bigcirc\rangle\overset{OH}{-COOH} \longrightarrow$

(c) $(CH_3)_2CHCOOH + NaOH \longrightarrow$

38. Write IUPAC and common names for each ester in exercise 34.

39. Write IUPAC and common names for each ester in exercise 35.

40. Write IUPAC and common names for the following:

(a) $CH_3(CH_2)_7-O-\overset{O}{\underset{\|}{C}}-H$

(b) $(CH_3)_2CH-O-\overset{O}{\underset{\|}{C}}-(CH_2)_{10}CH_3$

(c) $CH_3-\overset{O}{\underset{\|}{C}}-O-CH_3$.

41. Write IUPAC and common names for the following:

(a) $CH_3CH_2-O-\overset{O}{\underset{\|}{C}}-CH_3$

(b) $\langle\bigcirc\rangle-\overset{O}{\underset{\|}{C}}-O-(CH_2)_3CH_3$

(c) $CH_3(CH_2)_3-\overset{O}{\underset{\|}{C}}-O-\langle\bigcirc\rangle$.

42. Write the name of the acid and alcohol used to make each compound in exercise 40.

43. Write the name of the acid and alcohol used to make each compound in exercise 41.

44. Write structural formulas for the following: (a) ethyl benzoate, (b) *t*-butyl hexanoate, (c) isopropyl formate.

45. Write structural formulas for the following: (a) methyl octanoate, (b) ethyl valerate, (c) isobutyl acetate, (d) butyl laurate.

46. Arrange the following in order of increasing solubility in hexane: (a) ethyl propanoate, (b) valeric acid, (c) potassium pentanoate, (d) heptane.

47. Write the structural formula for a triglyceride made from glycerol and one molecule each of lauric, palmitic, and stearic acid.

Reactions of Esters

48. Using structural formulas, write reactions for the acid-catalyzed hydrolysis of each compound in exercise 41.

49. Using structural formulas, write reactions for the acid-catalyzed hydrolysis of each compound in exercise 40.

50. Using structural formulas, write a saponification reaction with NaOH for the compound in exercise 47.

51. Using structural formulas, write saponification reactions with NaOH for (a) *t*-butyl hexanoate and (b) butyl laurate.

52. Write structural formulas for the organic products of the following reactions:

(a) $CH_3CH_2CH_2-O-\overset{O}{\underset{\|}{C}}CH_2CH(CH_3)_2 + H_2O \xrightleftharpoons{H^+}$

(b) $CH_2ClCH_2\overset{O}{\underset{\|}{C}}-O-C(CH_3)_3 + KOH \longrightarrow$

Phosphate Esters and Anhydrides

53. Identify each of the following as a phosphate ester, phosphate anhydride, both, or neither:

(a) $CH_3-O-\overset{O}{\underset{\underset{OH}{|}}{\overset{\|}{P}}}-OH$, (b) $^-O-\overset{O}{\underset{\underset{OH}{|}}{\overset{\|}{P}}}-O^-$,

(c) $CH_3-O-\overset{O}{\underset{\underset{OH}{|}}{\overset{\|}{P}}}-O-\overset{O}{\underset{\underset{OH}{|}}{\overset{\|}{P}}}-OH$.

54. Using structural formulas, write reactions for the synthesis of compounds (a) and (c) in exercise 53.

DISCUSSION EXERCISES

1. Write an equation for the reaction of acetic acid with water. Then write an equation for the acid ionization constant, K_a, for acetic acid. [Review Section 9.2 if necessary.] K_a for acetic acid is 1.8×10^{-5} (see Table 9.5). Is formic acid, which has a K_a of 1.8×10^{-4}, a stronger or weaker acid than acetic acid?

2. Predict whether a monocarboxylic acid or a dicarboxylic acid of similar molecular weight has a higher **(a)** melting point, **(b)** boiling point, and **(c)** solubility in water. Explain why.

3. Suppose you have a solution consisting of decanoic acid, hexane, and hexanal. All are nonpolar and insoluble in water. What chemical treatment could you use to convert decanoic acid into a substance you could remove from the mixture by dissolving it in water? What chemical treatment(s) could you then use to remove hexanal from hexane in a similar way?

17

Nitrogen-Containing Organic Compounds

1. What are the names and structures of some amines? How are they related to ammonia (NH_3)?
2. What are the physical properties and uses, especially in the body, of amines?
3. What are the main chemical reactions of amines?
4. What are the names and structures of amides?
5. What are the main physical properties, uses, and chemical reactions of amides?
6. What are nitrile and nitro compounds, and what are a few practical examples?

The Invention of Dynamite

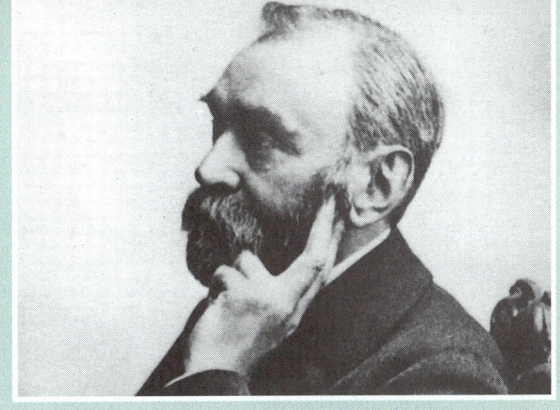

Figure 17.1 Alfred Nobel (1833–1896).

In 1847, Italian chemist Ascanio Sobrero discovered nitroglycerin, a powerful explosive containing the nitro ($-NO_2$) functional group joined to glycerin (glycerol). The trouble with nitroglycerin was its unreliability. Sometimes it didn't explode on demand, and too often it exploded without warning. Factories and people handling this thick, pale-yellow liquid had a tendency to blow up unexpectedly.

In 1866, a young Swedish inventor—whose brother had died in just such an explosion—discovered that nitroglycerin was safer to handle when soaked into an absorbent, claylike material called kieselguhr. This material was made into sticks and called dynamite.

The new product became widely used for military purposes and for railroad construction and mining. The Swedish inventor made a fortune. Though not a pacifist, he thought dynamite was so powerful that it would make wars too horrible to fight. He declared, "On the day when two armies will be able to annihilate each other in one second all civilized nations will recoil from war in horror and disband their forces."

He soon learned how wrong he was. As a humanitarian, however, he used some of his considerable wealth to establish prizes, for outstanding contributions to humanity in many areas, including peace. Those prizes have been awarded annually since 1901. The inventor's name was Alfred Nobel (Figure 17.1).

Figure 17.2 Earth's atmosphere is mostly N_2, a form of nitrogen our bodies cannot use directly.

Nitrogen is essential to life. Proteins, nucleic acids (DNA and RNA), and many vitamins contain nitrogen. Yet your body does not use nitrogen gas (N_2), which makes up about 80% of our atmosphere (Figure 17.2). Neither do you use nitrogen in the form of ammonia gas (NH_3) or ammonium (NH_4^+) salts, since ammonium ion is toxic to the central nervous system. Instead, other living things supply you with nitrogen in usable organic forms.

In this chapter, we examine several types of organic nitrogen compounds. The most common functional groups are *amines* and *amides,* which occur in amino acids and proteins. Sometimes, as in DNA and RNA, nitrogen is a member of a ring. We also briefly discuss compounds containing *nitrile* (cyano) or *nitro* functional groups.

17.1 AMINES: NAMES AND STRUCTURES

Nitrogen has five valence electrons ($\cdot \ddot{N} \cdot$), so it forms three covalent bonds to become stable. The simplest example is ammonia (NH_3). VSEPR theory (Section 4.5) predicts that the three hydrogen atoms in NH_3 and the pair of valence electrons not involved in a covalent bond are arranged around nitrogen with tetrahedral bond angles of about 109° (Figure 17.3). Measurements show the actual bond angles are 107°.

Amines are organic derivatives of ammonia (NH_3) in which one or more of the hydrogens is replaced with carbon. Amines, then, have the general structure

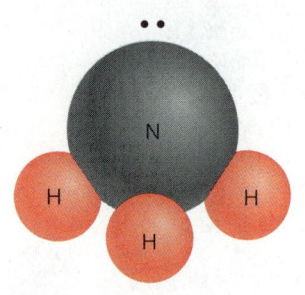

$$R-\underset{\underset{H \text{ (or } R'')}{|}}{N}-H \text{ (or } R')$$

Figure 17.3 The three hydrogen (or other) atoms bonded to N form approximately 109° bond angles.

where nitrogen bonds directly to at least one carbon (in an *R* group). Like ammonia, amines have a tetrahedral bond angle (about 109°) around N.

Amines are classified according to the number of carbons bonded directly to nitrogen:

primary amines—nitrogen bonded to one carbon atom
secondary amines—nitrogen bonded to two carbon atoms
tertiary amines—nitrogen bonded to three carbon atoms

$$CH_3-NH_2 \qquad CH_3-NH-CH_3 \qquad CH_3-\underset{\underset{CH_3}{|}}{\overset{\overset{CH_3}{|}}{N}}-CH_3$$

primary amine   *secondary amine*   *tertiary amine*

EXAMPLE 17.1 Identify the following as primary, secondary, or tertiary amines:

(a) $CH_3CH_2NH_2$ (b) ⬡—$NHCH_3$ (c) $(CH_3CH_2CH_2)_3N$

(d) ⬡—NH_2

SOLUTION (a) primary, (b) secondary, (c) tertiary, (d) primary

Naming Amines. Common names for amines consist of the organic group(s) bonded to nitrogen, followed by *-amine*. For example, $CH_3CH_2NH_2$ is ethylamine, $CH_3CH_2NHCH_3$ is ethylmethylamine, and $(CH_3)_2N(CH_2)_3CH_3$ is butyldimethylamine.

The IUPAC system considers —NH_2 as a substituent group, called an **amino group.** Consider, for example, the compound:

$CH_3(CH_2)_2NH_2$

Its common name is propylamine. If you treat —NH_2 as a substituent, however, you get the IUPAC name, 1-aminopropane. Another example is:

$CH_3CHCH_2CH_3$ *2-aminobutane* (IUPAC name)
|
NH_2 or *sec-butylamine* (common name)

Aromatic amines with an amino group bonded directly to benzene are named as a derivative of aniline (Figure 13.16). Two examples are:

aniline   m-*nitroaniline*

When N in aniline has substituent groups, we use the prefix *N-* to specify that they are bonded directly to N. Three examples are:

N-*methyl aniline*   N,N-*dimethylaniline*   N-*ethyl-N-methylaniline*

EXAMPLE 17.2 Write common names for the following amines and IUPAC names for compounds (b) and (e):

(a) CH_3CH_2NH (b) $(CH_3)_2CHNH_2$ (c) $Cl-\langle\bigcirc\rangle-NH_2$
|
CH_3

(d) [phenyl]—N—$(CH_2)_4CH_3$ / $C(CH_3)_3$ (e) $(CH_3)_2CHCH_2NH_2$

SOLUTION (a) ethylmethylamine, (b) isopropylamine, 2-aminopropane (IUPAC), (c) *p*-chloroaniline (3-chloroaniline), (d) *N-t*-butyl-*N*-pentylaniline, (e) isobutylamine (common), 1-amino-2-methylpropane (IUPAC)

Some amines contain N as part of a ring. Rings containing one or more noncarbon atoms are called *heterocyclic*, and rings containing the N atom of an amine are called **heterocyclic amines.** Structures of three biologically important heterocyclic amines are shown in Figure 17.4.

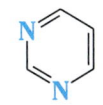

pyrimidine
(in DNA and RNA)

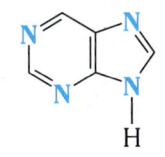

purine
(in DNA and RNA)

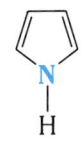

pyrrole
(in hemoglobin)

Figure 17.4 Three heterocyclic amines.

17.2 PHYSICAL PROPERTIES AND USES OF AMINES

Physical Properties. Nitrogen is highly electronegative (Section 4.6), so amines containing at least one N—H bond form hydrogen bonds. Figure 17.5 shows how primary and secondary (but not tertiary) amines hydrogen bond to like molecules.

Figure 17.5 Primary (left) and secondary (center) amines form hydrogen bonds (dotted lines in color). Tertiary amines (right) do not.

Their hydrogen bonds give amines higher melting and boiling points than many organic compounds of similar molecular weights (Figure 17.6). But their melting and boiling points are lower than alcohols and carboxylic acids, whose

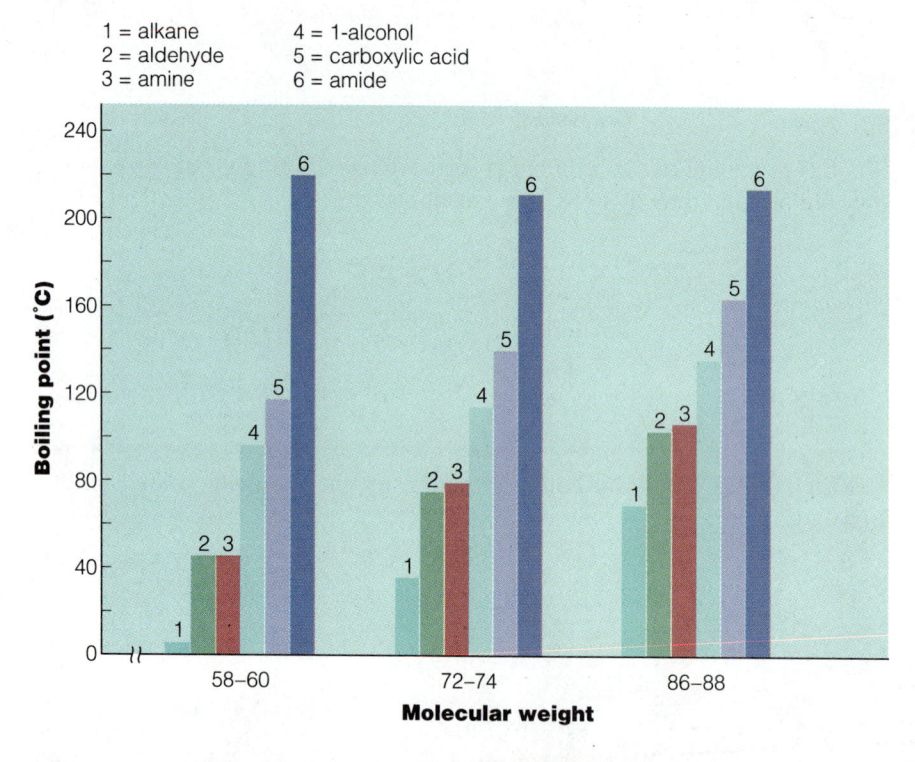

Figure 17.6 Comparison of the boiling points of organic compounds of similar molecular weights.

hydrogen bonding is stronger because O is more electronegative than N. Hydrogen bonding with water molecules makes small amines (up to six carbon atoms) soluble in water (Figure 17.7).

Ethyl- and methylamines smell a bit like ammonia, but some larger amines have an unpleasant, "fishy" smell. Figure 17.8 shows structural formulas of two amines whose common names attest to their disagreeable aromas.

Uses. Amines have many roles in the body. Two amine hormones are epinephrine (adrenalin) and thyroxine (thyroid hormone). DNA, RNA, and heme in hemoglobin all are heterocyclic amines (Figure 17.4). Many vitamins are

Nitrogen-Containing Organic Compounds   Chapter 17   375

amines; in fact, the word *vitamin* comes from an early (mistaken) belief that all such compounds were amines.

Many nitrogen-containing compounds affect the nervous system. Amines such as acetylcholine function as neurotransmitters (see Chemistry Spotlight in Section 16.6). Amines are used as drugs (legal and illegal) to relieve pain, treat mental disorders, induce sleep, produce hallucinations, and stimulate physical and mental activity.

Alkaloids are basic compounds in plants that affect the body. Because of the effects they produce, many cultures have used alkaloids for centuries. As you can see in Figure 17.9, alkaloids such as nicotine, morphine, and caffeine are amines. You might have expected this because their names all end in *-ine*, a common suffix for amines.

The first local anesthetic used successfully in modern medicine was the alkaloid cocaine (see the Chemistry Spotlight in Section 17.3). Many of the modern local anesthetics, such as Novocaine, are closely related to this drug. Pharmaceutical companies often chemically alter known drugs to try to discover more effective compounds. Several examples are shown in Figure 17.10, which displays the structures of cocaine and some synthetic amines or amides.

Certain amines stimulate activity of the brain and nervous system. Another group of amines, the hallucinogens, are mind-affecting chemicals that cause

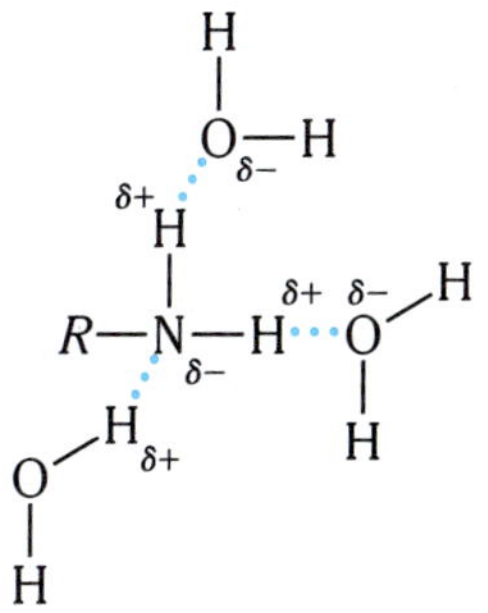

Figure 17.7 Amines form hydrogen bonds (dotted lines in color) with water.

$NH_2-(CH_2)_4-NH_2$

putrescine

$NH_2-(CH_2)_5-NH_2$

cadaverine

Figure 17.8 Two smelly amines in rotting flesh.

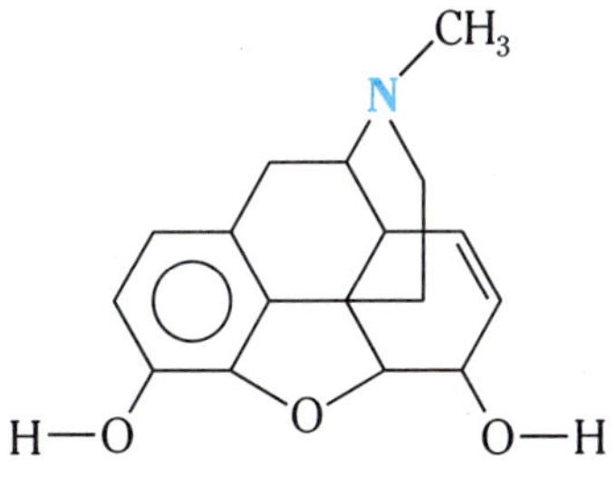

morphine 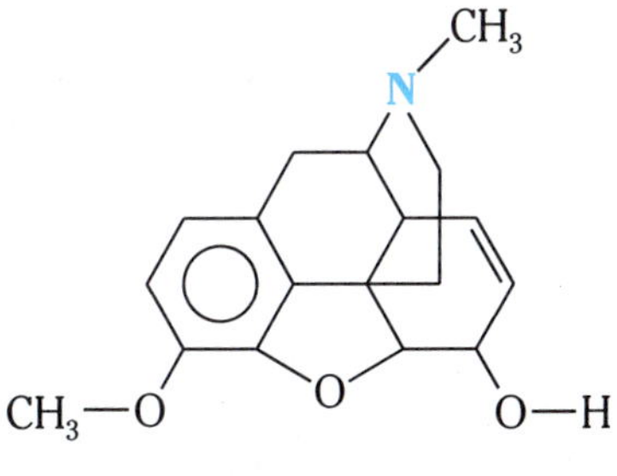   *codeine* 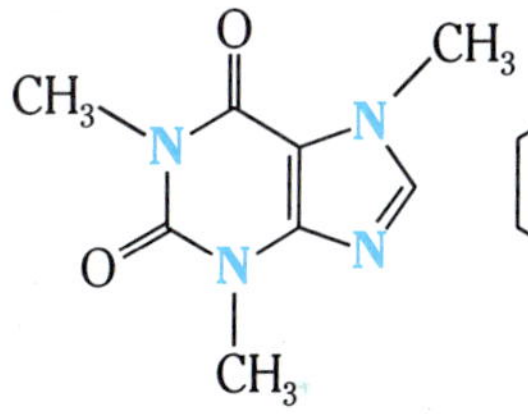   *caffeine* 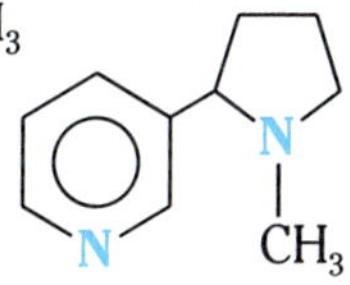   *nicotine*

Figure 17.9 Four alkaloids. All are heterocyclic amines.

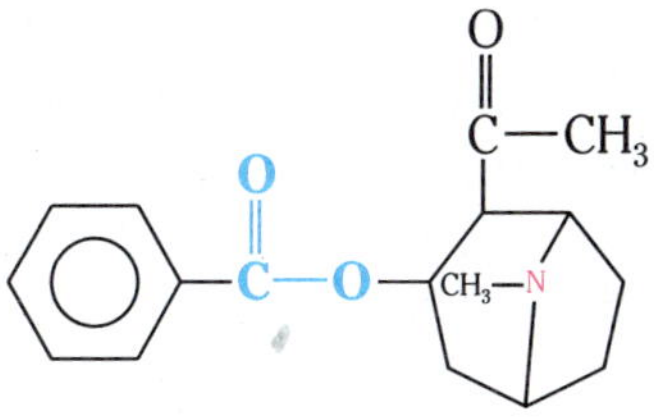

cocaine 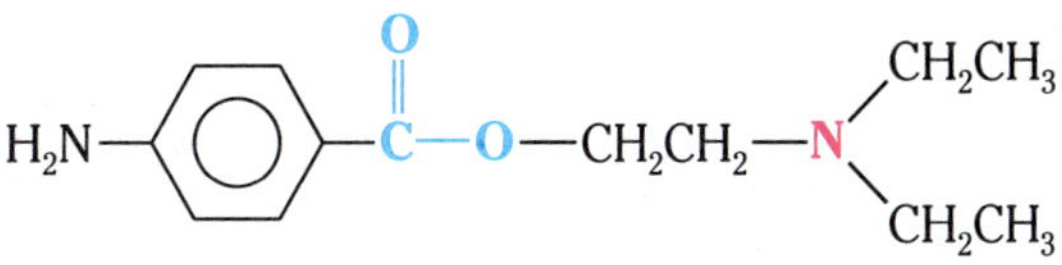   *procaine (Novocaine)*

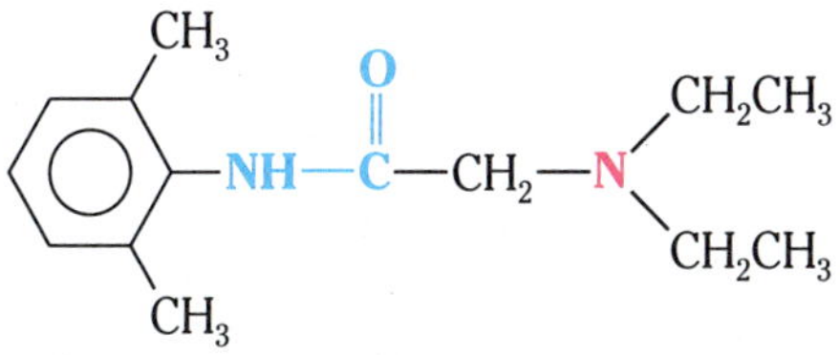

lidocaine (Xylocaine) 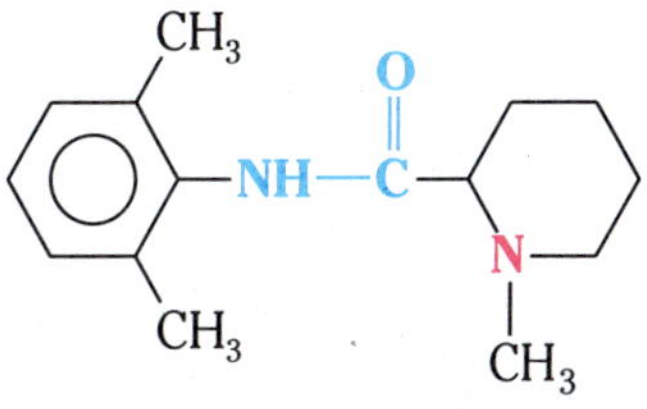   *mepivacaine (Carbocaine)*

Figure 17.10 Some local anesthetics. All have an aromatic ring joined to an ester or amide (blue) and a tertiary amine group (red).

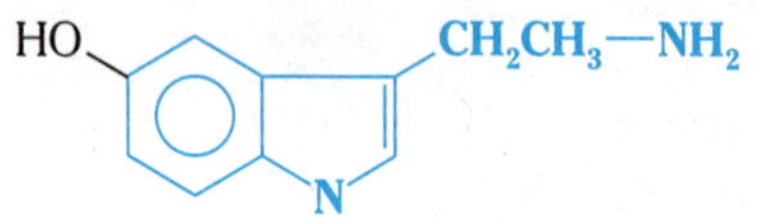

[Structures shown:
- serotonin (a neurotransmitter)
- psilocybin (a hallucinogen) 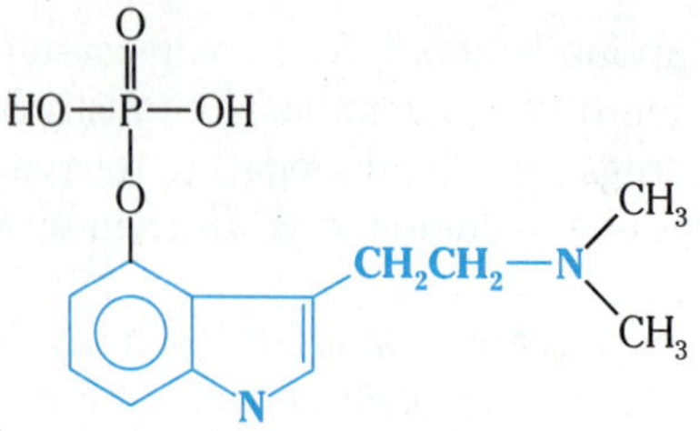
- lysergic acid diethylamide (LSD) (a hallucinogen) 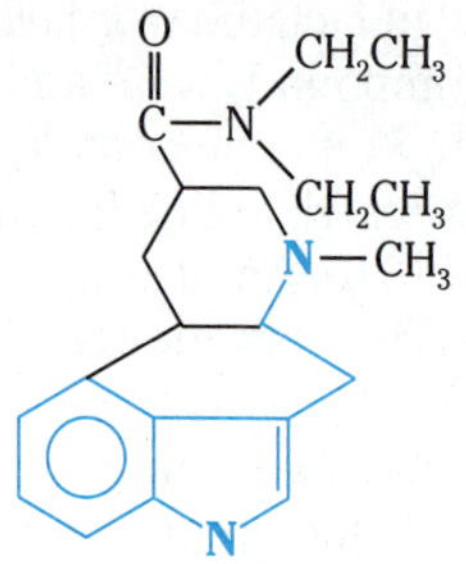
- norepinephrine (a neurotransmitter) 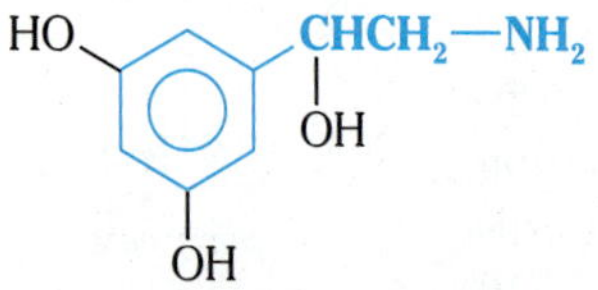
- amphetamine (a stimulant) 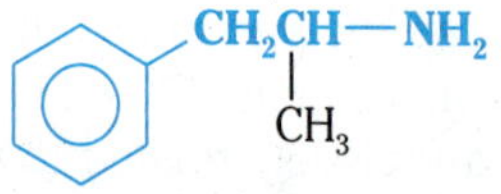
- mescaline (a hallucinogen)] 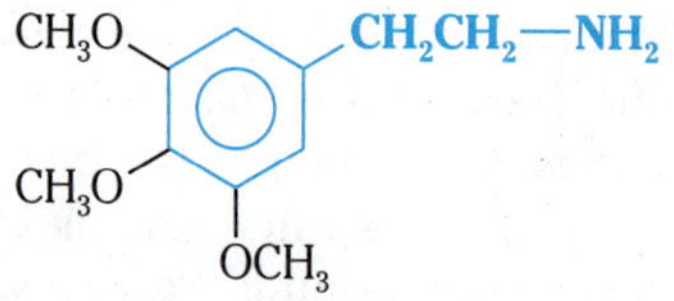

Figure 17.11 Some hallucinogens structurally resemble the neurotransmitter serotonin (top row; common structure in color). Amphetamines and some hallucinogens structurally resemble the neurotransmitter norepinephrine (bottom row; common structure in color).

Recall that pure water contains a small but equal concentration (about 1×10^{-7} M) of H^+ (as H_3O^+) and OH^- ions (Section 9.2).

vivid illusions, fantasies, and hallucinations. Notice in Figure 17.11 that the structures of several stimulants and hallucinogens resemble those of the neurotransmitters serotonin and norepinephrine.

17.3 CHEMICAL REACTIONS OF AMINES

Amines as Bases. If you dissolved propylamine in water and measured the pH of the solution, you would discover that the pH was alkaline. Apparently the amine reacts with H_2O molecules to bond hydrogen ions (H^+), leaving an excess of hydroxide ions (OH^-) in the process:

$$CH_3CH_2CH_2-\overset{..}{\underset{H}{N}}-H + H_2O \longrightarrow CH_3CH_2CH_2-\overset{H\,+}{\underset{H}{\overset{..}{N}}}-H + OH^-$$

This happens because N in an amine (or ammonia) has a pair of valence electrons not used for bonding. Those electrons are available to form a covalent bond with any available H^+ (which seeks a pair of electrons to become more stable). An amine (or NH_3) reacts with water to bind H^+, leaving an excess of OH^- and making the solution alkaline. Since amines (like NH_3) are proton acceptors, they are classified as bases (Section 9.1).

Like other bases, amines neutralize acids to form salts. For example:

$$CH_3CH_2-\overset{..}{\underset{H}{N}}-H + HCl \longrightarrow CH_3CH_2-\overset{H\,+}{\underset{H}{\overset{..}{N}}}-H\ Cl^-$$

ethylamine   ethylammonium chloride

The product is named as an organic derivative of the salt ammonium chloride ($NH_4^+Cl^-$). Secondary and tertiary amines have similar reactions because they

CHEMISTRY SPOTLIGHT

Cocaine

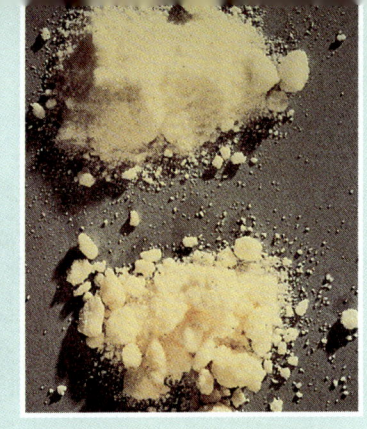

Figure 17.12 Leaves of the coca plant (left) contain as much as 2% extractable cocaine (right).

Cocaine comes from the leaves of the coca plant (Figure 17.12), which grows in the Andes Mountains of South America. It is processed into white crystals and typically used as the ammonium salt, cocaine hydrochloride. This salt is converted into the base form by a process called "free basing":

cocaine hydrochloride + NaOH →

cocaine (base form) + H_2O + NaCl

Since cocaine base is not ionic, it vaporizes faster than the salt form and acts faster in the body.

Besides being an anesthetic (Figure 17.13), cocaine is a stimulant that triggers the release of certain neurotransmitters (dopamine, serotonin, and norepinephrine) in the brain. Because its euphoric effects last an hour or less, some users snort or inject cocaine many times a day. Intense psychological dependence develops with regular use (physical dependence probably occurs too). An overdose tightens blood vessels, depresses the heart and respiration, and can be fatal.

Cocaine abuse used to be more common among wealthy individuals who could afford its high price. But the price has decreased for a smokable derivative called "crack." Crack cocaine increases the chances for dependence by providing a rapid and intense euphoria lasting about 30 minutes, followed by depression, irritation, and a craving for the drug. More people, including the very young, are paying the price.

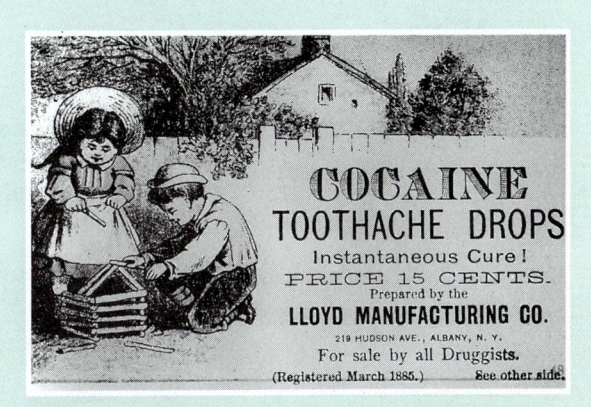

Figure 17.13 Cocaine once was used as a local anesthetic for toothaches.

also have an N atom with two valence electrons available to bond hydrogen ions.

Being ionic, amine salts are more soluble in water than their parent amines. Figure 17.14 shows the effect on solubility of neutralizing an amine with HCl to form its salt. Amine drugs usually are used as the salt because that form is more soluble in water.

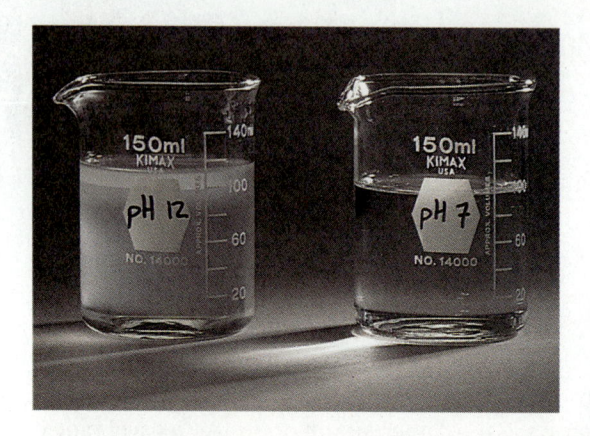

Figure 17.14 At alkaline pH (left) triethylamine amine is less soluble in water than its neutralized, salt form (right).

EXAMPLE 17.3 Write names and condensed structural formulas for the salts formed by neutralizing each of the following with HCl: (a) propylamine, (b) trimethylamine, (c) isobutylamine.

SOLUTION (a) $CH_3(CH_2)_2-\overset{\overset{H}{|}}{\underset{\underset{H}{|}}{N^+}}-H\ Cl^-$, propylammonium chloride

(b) $(CH_3)_3-N^+-H\ Cl^-$, trimethylammonium chloride

(c) $(CH_3)_2CHCH_2-\overset{\overset{H}{|}}{\underset{\underset{H}{|}}{N^+}}-H\ Cl^-$, isobutylammonium chloride

---

Synthesis of Amides. Ammonia and primary and secondary amines react with carboxylic acids—particularly their acyl halides, acid anhydrides, or esters—to form **amides**. The general reactions are:

$$R-\overset{\overset{O}{\|}}{C}-Cl + H-\underset{\underset{H\ (or\ R'')}{|}}{N}-H\ (or\ R') \longrightarrow R-\overset{\overset{O}{\|}}{C}-\underset{\underset{H\ (or\ R'')}{|}}{N}-H\ (or\ R') + HCl$$

acyl chloride　　　　amine　　　　　　　　amide

$$R-\overset{\overset{O}{\|}}{C}-O-\overset{\overset{O}{\|}}{C}-R + H-\underset{\underset{H\ (or\ R'')}{|}}{N}-H\ (or\ R') \longrightarrow$$

acid anhydride　　　amine

$$R-\overset{\overset{O}{\|}}{C}-\underset{\underset{H\ (or\ R'')}{|}}{N}-H\ (or\ R') + R-COOH$$

　　　　　　amide　　　　　　　　carboxylic acid

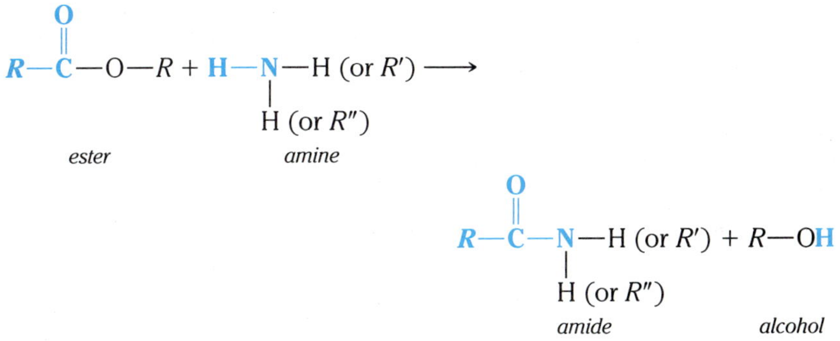

$$R-\overset{\overset{O}{\|}}{C}-O-R + H-\underset{\underset{H \text{ (or } R'')}{|}}{N}-H \text{ (or } R') \longrightarrow$$

ester   amine

$$R-\overset{\overset{O}{\|}}{C}-\underset{\underset{H \text{ (or } R'')}{|}}{N}-H \text{ (or } R') + R-OH$$

amide   alcohol

The first two reactions resemble those for synthesizing esters from carboxylic acids and alcohols. Think of an amine as substituting for an alcohol, with the product being an amide instead of an ester. In both cases, the reaction works best when the carboxylic acid is in a more reactive form—an acyl halide or an anhydride.

EXAMPLE 17.4 Write condensed structural formulas for the major organic products from the following reactions:

(a) 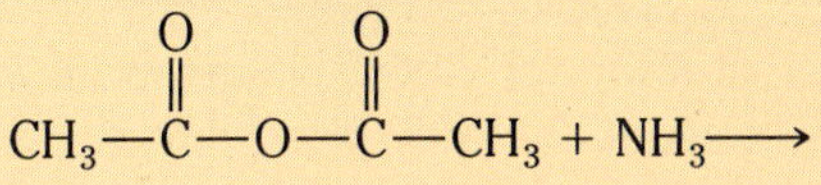 $CH_3-\overset{\overset{O}{\|}}{C}-O-\overset{\overset{O}{\|}}{C}-CH_3 + NH_3 \longrightarrow$

(b) 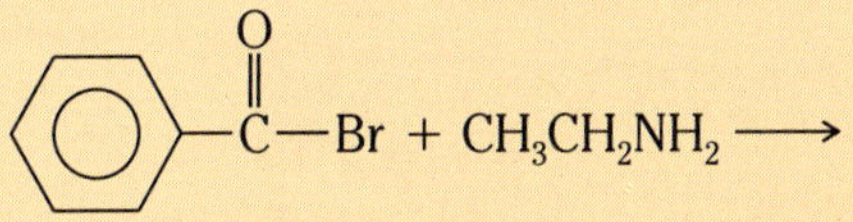 $\langle \bigcirc \rangle -\overset{\overset{O}{\|}}{C}-Br + CH_3CH_2NH_2 \longrightarrow$

SOLUTION (a) 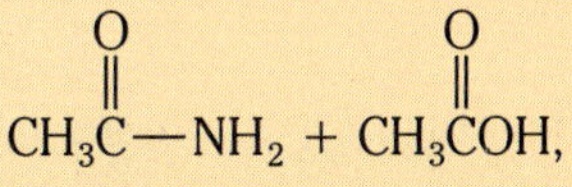 $CH_3\overset{\overset{O}{\|}}{C}-NH_2 + CH_3\overset{\overset{O}{\|}}{C}OH$,

(b) 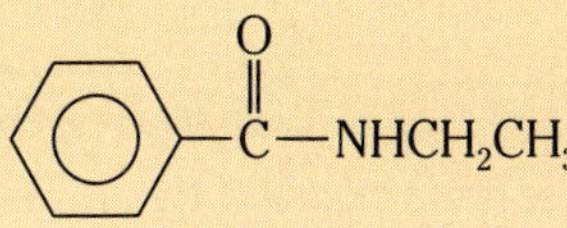 $\langle \bigcirc \rangle -\overset{\overset{O}{\|}}{C}-NHCH_2CH_3$

Amino acids contain both functional groups—amine and carboxylic acid —needed to form amides. Figure 17.15 shows the functional groups of an amino acid in their ionic (salt) forms, which occur at neutral pH. In your body, the amino group of one amino acid reacts (in the presence of an enzyme catalyst) with the carboxyl group of another amino acid to form an amide. Long chains of amino acids joined by amide groups are called *proteins* (see Chapter 21).

$$\underset{R}{\underset{|}{H_3\overset{+}{N}-\underset{|}{C}-H}}\overset{\overset{O}{\|}}{\overset{|}{C}-O^-}$$

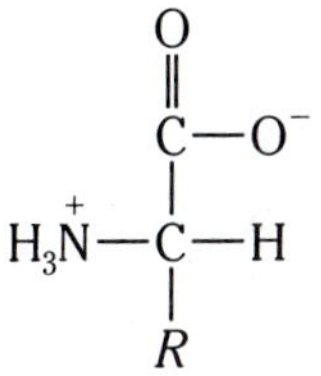

Figure 17.15 General structure of an amino acid at neutral pH.

17.4 AMIDES: NAMES AND STRUCTURES

Amides are made by reaction between amines and carboxylic acids (Section 17.3) and have the general structure

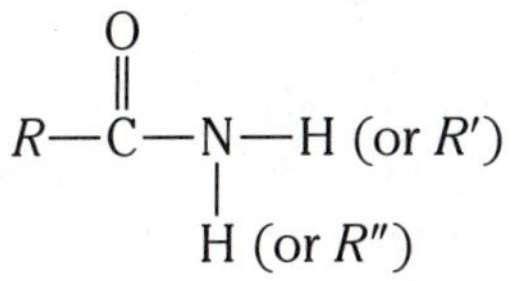

$$R-\overset{\overset{O}{\|}}{C}-\underset{\underset{H \text{ (or } R'')}{|}}{N}-H \text{ (or } R')$$

IUPAC and common names for amides with the structure

$$R-\overset{\overset{O}{\|}}{C}-NH_2$$

are easy to learn. For the IUPAC name, first determine the IUPAC name for the acid (including the carbonyl carbon) used to make the amide. Then replace *-oic acid* with *-amide*. Consider, for example:

$$CH_3CH_2CH_2\overset{\overset{O}{\|}}{C}-NH_2$$

The four-carbon acyl group comes from butanoic acid. Substituting *-amide* for *-oic acid* gives you the IUPAC name, butanamide.

Common names are based on the common name of the acid (see Table 16.1) supplying the acyl group. Then substitute *-amide* for *-ic acid* to name the amide. In the example above, the four-carbon acid has the common name butyric acid, so the common name of the amide is butyramide.

In addition to the acyl group, some amides have one or two other organic groups bonded directly to N. Their general structure is:

$$R-\overset{\overset{O}{\|}}{C}-\underset{\underset{H\ (or\ R'')}{|}}{N}-R'$$

Name these substituents (represented as R' and R'' above) in the usual way and precede each with the prefix *N-* to indicate that they bond directly to N. For example, the amide

$$CH_3CH_2\overset{\overset{O}{\|}}{C}-\underset{\underset{CH_2CH_3}{|}}{N}-CH_3$$

has the IUPAC name *N*-ethyl-*N*-methylpropanamide. Its common name is *N*-ethyl-*N*-methylpropionamide.

Recall that you also use the prefix N- to name derivatives of aniline containing organic groups bonded to N (Section 17.1).

> **EXAMPLE 17.5** Write IUPAC and common names for the following amides:
>
> (a) $HC-NH_2$ with $C=O$ (b) $CH_3\overset{\overset{O}{\|}}{C}-NHCH_3$ (c) $C_6H_5-\overset{\overset{O}{\|}}{C}-NHCH_2CH_3$
>
> (d) $CH_3(CH_2)_3\overset{\overset{O}{\|}}{C}-N(CH_2CH_3)_2$
>
> **SOLUTION** (a) methanamide, formamide (common); (b) *N*-methylethanamide, *N*-methylacetamide; (c) *N*-ethylbenzamide; (d) *N,N*-diethylpentanamide, *N,N*-diethylvaleramide

17.5 PHYSICAL PROPERTIES, USES, AND CHEMICAL REACTIONS OF AMIDES

Physical Properties. Like amines, amides with at least one N—H bond form hydrogen bonds (Figure 17.16). In amides, however, the carbonyl ($>C=O$) group next to the —NH$_2$ group makes the N—H bond more polar and in-

Figure 17.16 Amides having at least one N—H group form hydrogen bonds (dotted lines in color).

creases the strength of hydrogen bonding. In fact, amides exist partially in a second form, a dipolar (saltlike) form:

dipolar form

This partial double bond between C and N in amides restricts free rotation and fixes all six bonded atoms in the same plane. You will see in Chapter 21 how this affects the shapes of proteins, which are amides.

Bonds between amide molecules are so strong that only the smallest amide, formamide, and some of its N-substituted derivatives are liquids at room temperature; all other amides are solids. Indeed, amides have the highest melting and boiling points (relative to molecular weight) of all the organic functional groups we have examined (Figure 17.6). Strong hydrogen bonding also makes amides more soluble in water than other organic compounds of similar molecular weights.

Uses. We will examine in much greater detail the important role of polyamides as proteins (Chapter 21) and in synthetic materials such as nylon (Chapter 18). Urea, the amide made by Wöhler in his famous experiment (Chapter Opener in Chapter 11), is a waste material in urine and used as fertilizer. A common substitute for aspirin is the amide, acetaminophen (trade names include Tylenol and Datril; see Figure 16.19). Another amide, N,N-diethyl-m-toluamide (trade name meta-Delphene) is an insect repellent (Figure 17.17).

The best-known sedatives and sleep inducers (hypnotics) are the barbiturates, synthetic compounds made from barbituric acid, a cyclic amide. Table

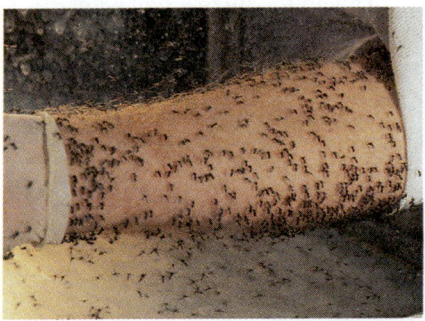

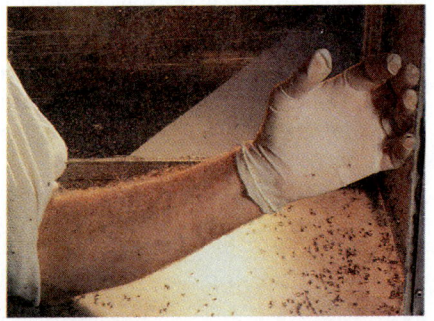

N,N-diethyl-m-toluamide

Figure 17.17 An arm exposed to mosquitoes (left) is protected (center) by a spray containing N,N-diethyl-m-toluamide (right).

Table 17.1 Four Widely Used Barbiturates

Name	Structure	Chief Uses	Effects
Pentobarbital (Nembutal)	(barbiturate ring with CH₂CH₃ and CHCH₂CH₂CH₃ / CH₃ substituents)	Hypnotic	Fast action and short duration
Secobarbital (Seconal)	(barbiturate ring with CH₂CH=CH₂ and CHCH₂CH₂CH₃ / CH₃ substituents)	Hypnotic	Fast action and short duration
Butabarbital (Butisol)	(barbiturate ring with CH₂CH=CH₂ / CH₃ and CH₂CH / CH₃ substituents)	Sedative, hypnotic	Intermediate action and duration
Phenobarbital (Luminal)	(barbiturate ring with CH₂CH₃ and phenyl substituents)	Sedative, hypnotic, anticonvulsant	Slow action and long duration

Adolph von Baeyer, who synthesized barbituric acid in 1864, named the compound after Barbara. Which Barbara is unclear. Evidence points to three possibilities: a waitress, his girlfriend, or Saint Barbara, the patron saint of artillerymen.

17.1 shows the structures and uses of common barbiturates. Some tranquilizers, which are used to treat anxiety, stress, and certain mental disorders, also are amines or amides (Table 17.2).

Chemical Reactions. In contrast with amines, amides are *not* bases. Amides have a polar carbonyl group (especially the very electronegative O atom) next to N that attracts the pair of valence electrons on N not used for bonding (Section 17.5). This makes the nonbonded electron pair on N unavailable to bond protons. As a result, amides are not bases (proton acceptors). They are neutral.

Amines react with carboxylic acids (particularly their acyl halide, acid anhydride, or ester derivatives) to form amides (Section 17.3). The other main reaction of amides is *hydrolysis*, the splitting of an amide by reaction with water.

Hydrolysis of Amides. Like esters, amides react with water in hydrolysis reactions to produce the amine (or ammonia) and carboxylic acid from which they were formed. The general reaction is:

$$R-\overset{O}{\underset{\|}{C}}-\underset{\underset{H\ (or\ R'')}{|}}{N}-H\ (or\ R') + H_2O \longrightarrow R-\overset{O}{\underset{\|}{C}}-OH + H-\underset{\underset{H\ (or\ R'')}{|}}{N}-H\ (or\ R')$$

Table 17.2 Three Tranquilizers and a Hypnotic

Name	Structure	Chief Uses	Effects
Chlorpromazine (Thorazine)	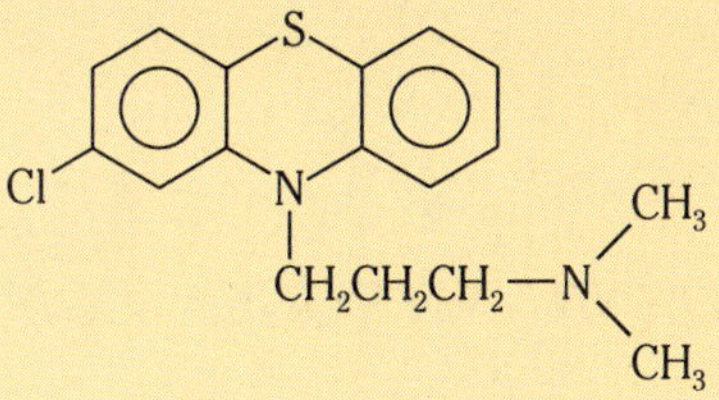 *phenothiazine structure with Cl and CH₂CH₂CH₂—N(CH₃)₂ side chain*	Treatment of psychotic disorders; has largely replaced electroconvulsive shock therapy and psychosurgery	Strong tranquilizer that affects the brain as well as the autonomic nervous system
Promazine (Compazine)	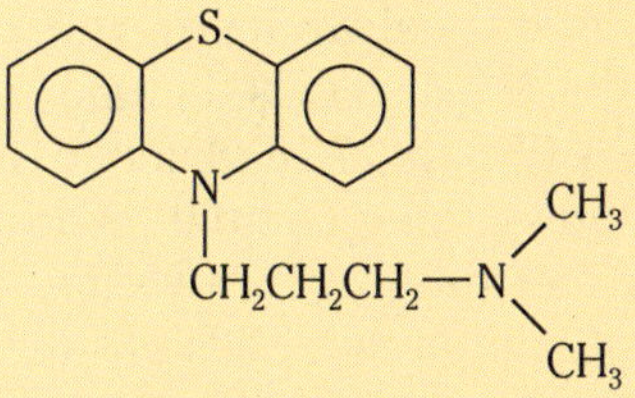 *phenothiazine structure with CH₂CH₂CH₂—N(CH₃)₂ side chain*	Similar to chlorpromazine	Most potent antipsychotic effects; also the most toxic
Diazepam (Valium)	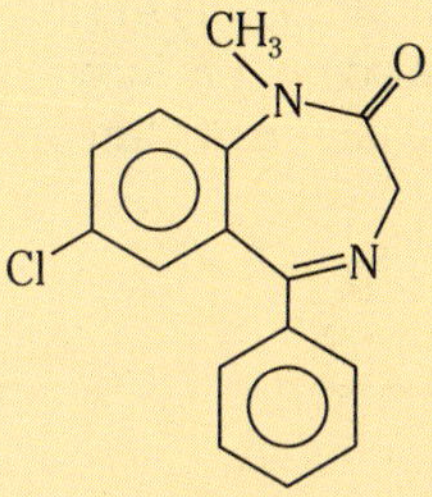 *benzodiazepine structure with CH₃, Cl, and phenyl substituents*	Relief from anxiety and tension; treatment of muscle tension and epilepsy	Mild tranquilizer; depresses central nervous system and relaxes skeletal muscles
Flurazepam (Dalmane)	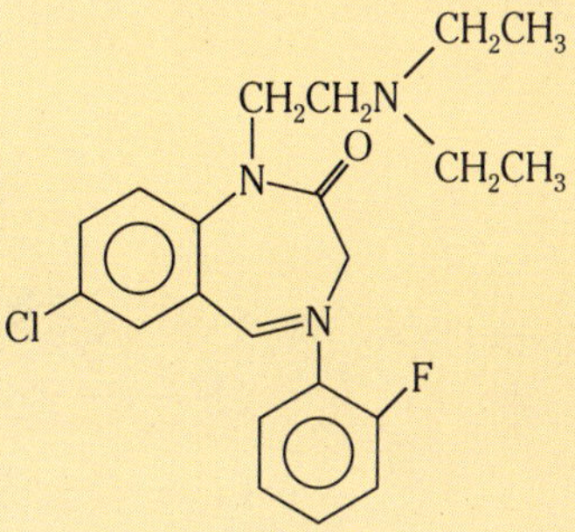 *benzodiazepine structure with CH₂CH₂N(CH₂CH₃)₂, Cl, and 2-fluorophenyl substituents*	Treatment of insomnia	Hypnotic; depresses central nervous system

This reaction is very slow unless a catalyst is present. Enzymes catalyze this reaction in your body. In the laboratory, either acids or bases accelerate the reaction, yielding slightly different products in each case. In the presence of acid, the amine product is neutralized to its salt form. In the presence of a base, the carboxylic acid product is neutralized to its salt form.

Recall that the carboxylic acid product is also neutralized in saponification reactions (Section 16.7).

EXAMPLE 17.6 Write condensed structural formulas for the major organic products formed from the following reactions:

(a) $(CH_3)_2CHC(=O)-NH_2 + H_2O \xrightarrow{OH^-}$

(b) 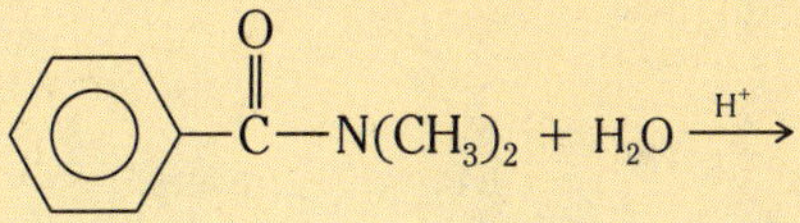 $C_6H_5-C(=O)-N(CH_3)_2 + H_2O \xrightarrow{H^+}$

384 Part II Organic Chemistry

SOLUTION (a) $(CH_3)_2CHCOO^-$ (+NH_3),

(b) 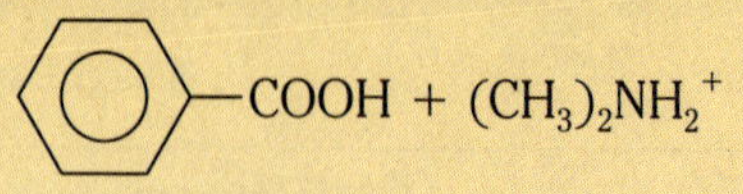 ⬡—COOH + $(CH_3)_2NH_2^+$

⬡—CH—CN
　　　|
　　　OH

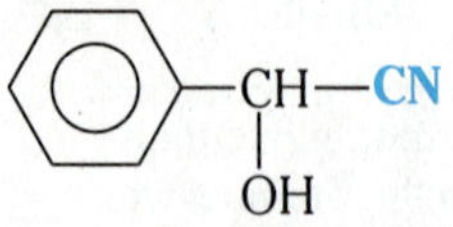

a component of amygdalin (Laetrile)

$NH_2CH_2CH_2$—CN

β-aminopropionitrile

Figure 17.18 Two naturally occurring nitriles.

17.6 OTHER ORGANIC NITROGEN COMPOUNDS

Two other nitrogen-containing functional groups are the **nitro group** (—NO_2) and the **nitriles** (R—CN), which are also called *cyano compounds*.

A few nitriles occur naturally (Figure 17.18). Laetrile (amygdalin) occurs in apricot pits. Although most studies show Laetrile to be ineffective, it is promoted and used as an anticancer drug by a few clinics outside the United States. β-Aminopropionitrile is found in the seeds of sweet peas; it blocks normal development of connective tissue and impairs the strength of rats, a condition called lathyrism. The nitrile group also occurs in synthetic polymers called acrylonitriles, which have trade names such as Orlon and Acrilan (see Chapter 18).

Nitro compounds contain the nitro group (—NO_2) directly bonded to carbon. They are used as solvents and as intermediates in the synthesis of dyes, drugs, and insecticides. Some compounds react with explosive force and are used for that purpose (Figure 17.19). The simplest compound is nitromethane, a fuel for drag racers and model engines. Another example is 2,4,6-trinitrotoluene, the explosive known as TNT.

CH_3—NO_2

2,4,6-trinitrotoluene structure with CH_3, three NO_2 groups

CH_2—O—NO_2
|
CH—O—NO_2
|
CH_2—O—NO_2

Figure 17.19 Three explosive nitro compounds.

nitromethane　　*2,4,6-trinitrotoluene (TNT)* 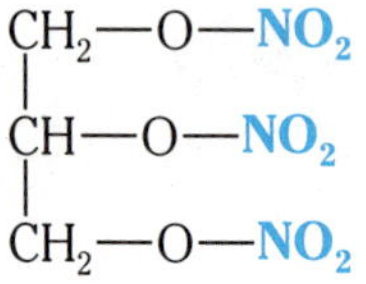　　*nitroglycerin (a nitrate ester)*

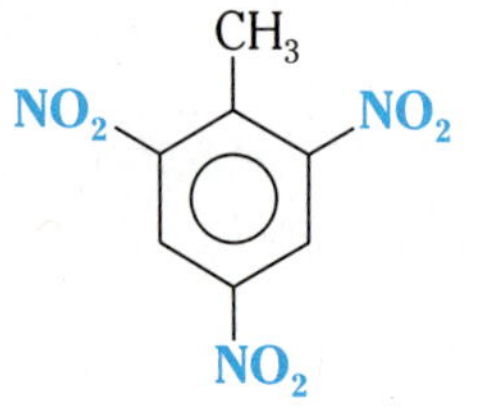

The nitro group also occurs in *nitrate esters* (R—O—NO_2), which are organic derivatives of nitric acid, HNO_3 (or HO—NO_2). Nitrocellulose, also known as guncotton, is one explosive example. Another is nitroglycerin (Figure 17.19), the active ingredient in dynamite (see Chapter Opener). Since it dilates arteries in the heart and relaxes smooth muscle, nitroglycerin also is used to treat angina pectoris, a heart disease that often produces severe chest pains due to tightening of cardiac blood vessels.

> Such I hold to be the genuine use of gunpowder; that it makes all men alike tall.
>
> *Thomas Carlyle (1795–1881)*

SUMMARY

Primary amines have the structure $R-NH_2$; secondary and tertiary amines have two and three R groups, respectively, bonded to N. Compounds containing N in a ring are heterocyclic amines. Common names consist of naming the R groups followed by -amine. Aromatic amines are named as derivatives of aniline, and the prefix *N*- designates additional organic groups bonded to N. Because they form hydrogen bonds, amines have higher melting and boiling points and greater solubility in water than organic compounds of similar molecular weights.

Amines have many practical uses and occur in many compounds found in the body. They include hormones, amino acids, vitamins, and DNA. Alkaloids are amines, and many are used as drugs or medicines because of their effects on the nervous system.

Amines are bases. They neutralize acids and form amine salts, which are more soluble in water than their parent amines. Amines react with carboxylic acids (especially their acyl halides, acid anhydrides, or esters) to form amides.

Amides have the structure
$$R-\overset{\overset{O}{\|}}{C}-NH_2;$$
one or both H atoms may be replaced by an R group. To name an amide, name the carboxylic acid from which it formed and replace *-ic acid* (common) or *-oic acid* (IUPAC) with *-amide*. The prefix *N*- designates additional organic groups bonded to N.

Because they form exceptionally strong bonds, amides have the highest melting and boiling points of all the organic compounds (of similar molecular weight) we have examined. Amides are relatively soluble in water. They occur in proteins, urea, nylon, and other products. Amides are hydrolyzed in the presence of acid or base catalyst.

Other nitrogen-containing functional groups are the nitro ($-NO_2$) and nitrile ($-CN$) groups.

KEY TERMS

Alkaloid (17.2)
Amide (17.3)
Amine (17.1)
Amino acid (17.3)

Amino group (17.1)
Heterocyclic amine (17.1)
Nitrile (17.6)
Nitro group (17.6)

Primary amine (17.1)
Secondary amine (17.1)
Tertiary amine (17.1)

EXERCISES

Even-numbered exercises are answered at the back of this book.

Names and Structures of Amines

1. In trimethylamine, $(CH_3)_3N$, the methyl groups are arranged around the N atom at approximately tetrahedral (109°) bond angles. Which type of orbital around N is most likely involved: **(a)** p, **(b)** s, **(c)** sp, **(d)** sp^2, **(e)** sp^3?

2. What inorganic compound most resembles amines in molecular shape and physical and chemical properties?

3. Identify each of the following as a primary, secondary, or tertiary amine:
   (a) $(CH_3)_3CNH_2$ **(b)** $(CH_3)_2CHNH_2$ **(c)** $(CH_3)_2NH$
   (d) ⌬—N(CH_3)_2.

4. Identify each of the following as a primary, secondary, or tertiary amine: **(a)** trimethylamine, **(b)** *p*-chloroaniline, **(c)** *sec*-butylamine, **(d)** ethylmethylamine.

5. Purine (Figure 17.4) is a heterocyclic amine. Is purine a primary, secondary, or tertiary amine?

6. Name the amines in exercise 3.

7. Name the following: **(a)** CH_3CHCH_3,
   |
   NH_2

   (b) $(CH_3)_2CHCH(CH_2)_5CH_3$,
   |
   NH_2

   (c) ⌬ with NH_2 substituent, **(d)** ⌬ with Br and $NHCH_3$ substituents.

8. Write condensed structural formulas for the following amines: **(a)** isobutylamine, **(b)** diisopropylamine, **(c)** *N,N*-diethylaniline, **(d)** *o*-chloroaniline.

9. Write condensed structural formulas for the following amines: **(a)** *p*-aminobenzoic acid, **(b)** 3-aminopentane, **(c)** diethylpropylamine, **(d)** *N*-propylaniline.

Physical Properties and Uses of Amines

10. Does 2-propanol or 2-aminopropane have stronger hydrogen bonds between like molecules? Explain why.
11. Which one of the following does *not* normally contain N: **(a)** DNA, **(b)** carbohydrates, **(c)** proteins, **(d)** vitamins.
12. Are the tranquilizers Thorazine and Compazine (see Table 17.2) primary, secondary, or tertiary amines?
13. Which local anesthetics shown in Figure 17.10 can be classified as heterocyclic amines?
14. Name the most widely abused local anesthetic.
15. An example of an alkaloid is **(a)** vitamin C, **(b)** amphetamine, **(c)** phenobarbital, **(d)** nicotine, **(e)** ethanol.
16. Compared with other compounds of similar molecular weights, amines typically have higher boiling points than **(a)** alcohols, **(b)** alkanes, **(c)** carboxylic acids, **(d)** amides.

Chemical Reactions of Amines

17. Write condensed structural formulas for the salts formed by the neutralization reaction of HCl with each compound in exercise 3.
18. Name each compound in exercise 17.
19. Suppose a substance that is either an amine, amide, or carboxylic acid partially dissolves in water, making the solution pH 9. Which type of substance is it?
20. What will happen to the solubility of the compound in exercise 19 if HCl solution is added until the pH is 3? Explain.
21. Write condensed structural formulas for **(a)** diethylammonium chloride and **(b)** hexylammonium bromide.
22. Write condensed structural formulas for **(a)** trimethylammonium bromide and **(b)** isobutylammonium chloride.
23. Write condensed structural formulas for the major organic products formed in each of the following reactions (write "no reaction" where appropriate):

    (a) $CH_3(CH_2)_4NH_2 + NaOH \longrightarrow$

    (b) $CH_3CH_2NH_2 + CH_3-\overset{\overset{O}{\|}}{C}-O-\overset{\overset{O}{\|}}{C}-CH_3 \longrightarrow$

24. Write condensed structural formulas for the major organic products formed in each of the following reactions (write "no reaction" where appropriate):

    (a) $(CH_3CH_2)_3N + HNO_3 \longrightarrow$

    (b) $(CH_3)_2NH + CH_3CH_2\overset{\overset{O}{\|}}{C}-Cl \longrightarrow$

Names and Structures of Amides

25. Write IUPAC and common names for the following:

    (a) $CH_3CH_2\overset{\overset{O}{\|}}{C}-NH_2$ **(b)** C₆H₅$-\overset{\overset{O}{\|}}{C}-NHCH_3$

    (c) $CH_3\overset{\overset{O}{\|}}{C}-N(C_6H_5)(CH_2CH_3)$

    (d) $CH_3(CH_2)_3\overset{\overset{O}{\|}}{C}-N(CH_3)_2$.

26. Write IUPAC and common names for the following:

    (a) $CH_3\overset{\overset{O}{\|}}{C}-NH_2$, **(b)** $CH_3\overset{\overset{O}{\|}}{C}-NH(CH_2)_3CH_3$,

    (c) 3-Cl-C₆H₄$-\overset{\overset{O}{\|}}{C}-NH_2$,

    (d) $CH_3CH_2\overset{\overset{O}{\|}}{C}-N(CH_3)_2$.

27. Write structural formulas for the following:
    (a) *N*-pentylpentanamide, **(b)** *N,N*-dipropylvaleramide, **(c)** benzamide, **(d)** *N*-phenylformamide.
28. Write structural formulas for the following:
    (a) *N*-methylacetamide, **(b)** butyramide,
    (c) *N*-isopropylethanamide, **(d)** *N,N*-dimethylbenzamide.
29. Write the common and IUPAC names and structural formula for the lowest molecular weight amide.

Physical Properties and Uses of Amides

30. Arrange the following in order of decreasing boiling point (all have similar molecular weights): **(a)** butanamide, **(b)** hexane, **(c)** pentylamine, **(d)** *N,N*-dimethylpropylamine.
31. Explain, in terms of chemical bonding, why ethanamide is a solid at room temperature while propyl-

mine, which has the same molecular weight, is a liquid.

32. For each compound in exercise 30, list the strongest type of attraction between like molecules.

33. Which local anesthetics shown in Figure 17.10 are amides?

34. Alkaloids are basic compounds in plants that have physiological effects. Use this definition to explain why alkaloids are amines but not amides.

35. If you dissolved *N*-methylacetamide in water, the pH of the solution would be closest to **(a)** 3, **(b)** 7, **(c)** 9.

Chemical Reactions of Amides

36. Write condensed structural formulas for the major organic products formed in each of the following reactions (write "no reaction" where appropriate);

(a) $\text{C}_6\text{H}_5-\overset{\overset{\text{O}}{\|}}{\text{C}}-\text{NHCH}_3 + \text{H}_2\text{O} \xrightarrow{\text{H}^+}$

(b) $(\text{CH}_3)_2\text{CHC}\overset{\overset{\text{O}}{\|}}{}-\text{NH}_2 + \text{H}_2\text{O} \xrightarrow{\text{OH}^-}$

37. Write structural formulas for the products formed by HCl-catalyzed hydrolysis of each compound in exercise 27.

38. Write structural formulas for the products formed by HCl-catalyzed hydrolysis of each compound in exercise 28.

39. Write structural formulas for the products formed by NaOH-catalyzed hydrolysis of each compound in exercise 25.

40. Write structural formulas for the products formed by NaOH-catalyzed hydrolysis of each compound in exercise 26.

41. In addition to carbon and nitrogen, nitro compounds contain the element **(a)** P, **(b)** S, **(c)** O, **(d)** Cl, **(e)** none of the above.

42. A substance commonly used to treat angina pectoris is **(a)** Laetrile, **(b)** mescaline, **(c)** codeine, **(d)** nitromethane, **(e)** none of the above.

43. Match each drug with its use:
    — phenobarbital   a. antipsychotic
    — diazepam (Valium)   b. mild analgesic
    — acetaminophen   c. diuretic
       (Tylenol)   d. alkaloid stimulant
    — caffeine   e. sleep inducer
    — lidocaine   (hypnotic)
       (Xylocaine)   f. hallucinogen
    — mescaline   g. mild tranquilizer
    — chlorpromazine   h. general anesthetic
       (Thorazine)   i. local anesthetic

DISCUSSION EXERCISES

1. Notice in Section 17.5 that amides exist partially in a dipolar (saltlike) form. Taking this into account, describe the types of bonds between like amide molecules, and between amide and water molecules.

2. Using the general structure of an amino acid (Figure 17.15), draw a structural formula for a substance (called a peptide) consisting of three amino acids joined together by amide bonds. (You will learn in Chapter 21 that this is the bonding arrangement in proteins.)

3. Describe tests you could do to determine whether an unknown compound is an amine or an amide.

4. What are the categories for which Nobel prizes are awarded?

18

Polymers

1. What are the main characteristics of polymers such as elastomers, fibers, plastics, copolymers, thermoplastic polymers, and thermosetting polymers?
2. How do addition polymers form? What are some specific examples of vinyl and other addition polymers? What are some of their important uses?
3. How do condensation polymers such as polyamides and polyesters form? What are some specific examples and their uses?
4. What are some types and uses of silicone polymers?

Synthesis of Nylon

In the 1920s, chemists at the Du Pont Company began trying to develop a synthetic material to replace silk or cotton. Their first materials—polyesters—were unstable in water and melted at warm temperatures (about 70°C). Finally, in 1935, they synthesized a polymer with repeating amide functional groups. This product, made from six-carbon dicarboxylic acid and six-carbon diamine molecules, was called nylon-6,6.

When World War II cut off U.S. supplies of silk and hemp (for rope) from the Far East, the demand for substitute materials increased. Nylon was rushed into production, and became an important material for parachutes, rope, and other military articles. Nylon also became a glamorous new fabric for clothing and consumer products. In 1940, nylon stockings went on sale nationwide. The manufacturers had an inventory of four million pairs. Within four days, nearly all of them had been sold. When the sale of stockings resumed after the war, some stores needed mounted police and patrolmen to control the crowds.

The Du Pont chemist most responsible for developing nylon, Wallace Carothers, never witnessed the sensational success of his discovery. In 1937, depressed by feelings of failure, he drank a mixture of lemon juice and potassium cyanide. His body was found in a Philadelphia hotel room. He was 41.

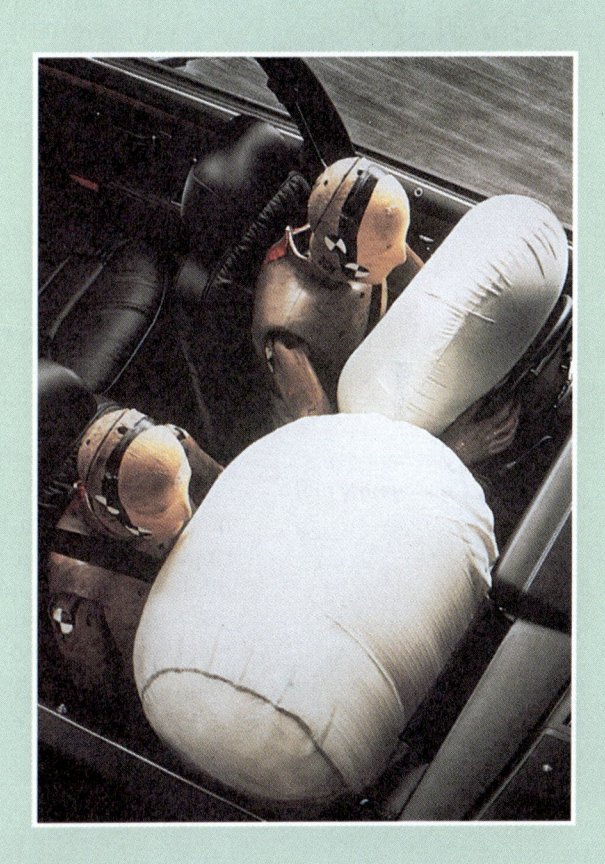

Figure 18.1 An egg won't stick to a Teflon frying pan.

The word polymer *comes from two Greek words that mean "many parts."*

What do palm trees, lizards, raincoats, and Silly Putty have in common? They all are constructed from very large organic molecules, called *macromolecules*. *Polymers* are macromolecules made from "building blocks" of smaller units, like the bricks that make up a wall.

Many of the molecules of life—proteins, DNA, RNA, starch, and cellulose—are polymers. So are the natural fabrics we use, such as cotton, leather, silk, and wool. In addition, chemists have discovered how to make a wide variety of synthetic polymers to serve our wants and needs. These products include Styrofoam ice chests, hair spray, Saran food wrap, Teflon frying pans, and nylon jackets (Figure 18.1).

In Chapter 19, we begin to examine the macromolecules of life—their structures and functions. But first, we look at different types of organic polymers (mostly synthetic ones), how they form, and some practical uses.

18.1 POLYMERS: BASIC TERMS

The organic compounds you have studied so far have fewer than 100 atoms, and most have fewer than 50. But the ability of carbon to form stable covalent bonds with other carbon atoms can also produce molecules containing hundreds or even thousands of atoms. These large molecules are called **macromolecules.**

Most macromolecules are **polymers,** substances made from small molecules that join together with covalent bonds. The small molecules that serve as building blocks for polymers are called **monomers.** Polymers made from a single monomer substance are called **homopolymers** (Figure 18.2). Polymers made from two or more different monomers are called **copolymers.**

Many natural substances are polymers, and some of the first synthetic polymers were designed to replace natural ones: expensive ivory for billiard balls, hemp for rope, or natural rubber for automobile tires.

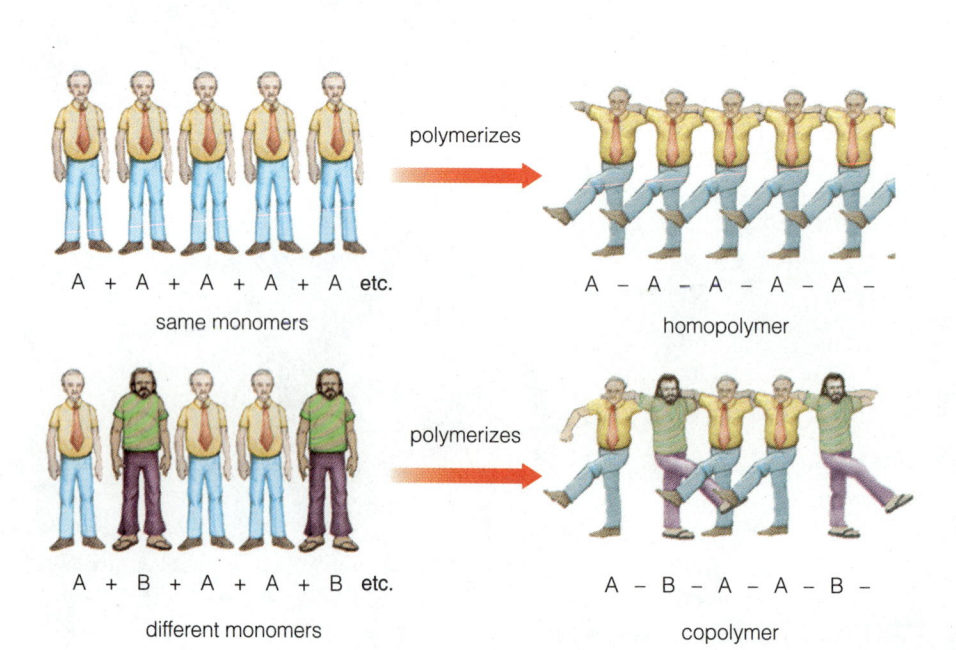

Figure 18.2 Formation of a homopolymer (top) and a copolymer from two different monomers (bottom). The copolymer shown is only one of many possible arrangements.

A + A + A + A + A etc.
same monomers

polymerizes

A – A – A – A – A –
homopolymer

A + B + A + A + B etc.
different monomers

polymerizes

A – B – A – A – B –
copolymer

During the second half of this century, the polymer industry has developed a remarkable array of entirely new materials and products. There are three important reasons for this development:

1. Monomer materials, mostly from petroleum, have been abundant and inexpensive.
2. Extensive research has revealed the molecular structures of polymers, their properties, and the reactions that produce them.
3. Improved technology has made it possible to tailor-make polymers with desirable properties for specific consumer products.

Before examining specific polymers, we need to know a few basic terms. Synthetic polymers include elastomers, fibers, and plastics. **Elastomers,** like rubber, are elastic; stretch them, and they will spring back to their original shape. **Fibers** are polymers that have high tensile strength; that is, they don't break easily or deform when you pull along the length of the material. Nylon is one example of a synthetic fiber (Figure 18.3). **Plastics,** such as polyethylene and Saran wrap, are polymers that can be molded or made into sheets or other shapes (Figure 18.4).

a linear chain polymer

a branched-chain polymer

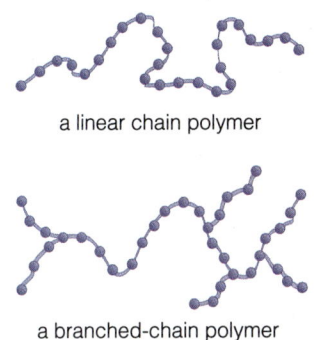

Figure 18.5 Different structural possibilities for polymer chains.

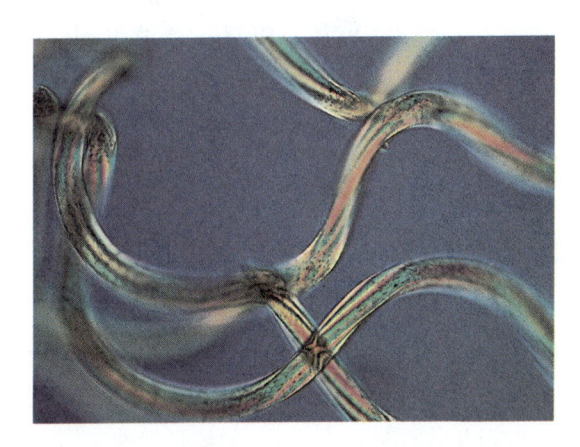

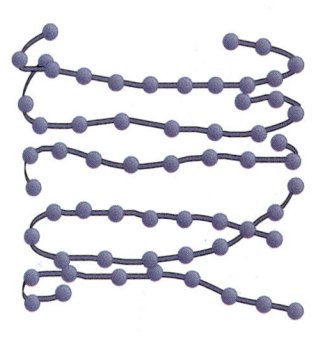

Figure 18.3 Magnified nylon fibers.

Figure 18.4 Producing sheets of Saran wrap.

linear polymer
(high melting point, fairly hard, rigid and tough)

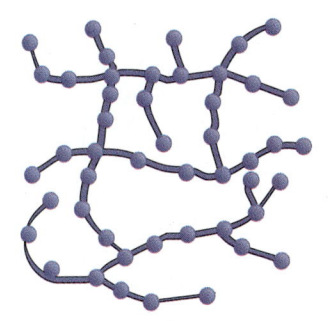

branched-chain polymer
(no definite melting point, soft and flexible)

Figure 18.6 Linear polymers (top) have different properties than branched-chain polymers (bottom).

Polymers can be linear or branched (Figure 18.5). In a linear polymer, monomer units join end-to-end to produce a very long, continuous chain. Branched polymers, in contrast, have chains extending out from other chains, like branches from a tree trunk. Individual molecules of linear polymers (like

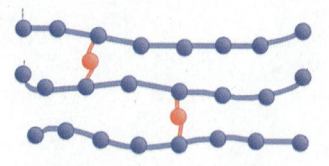

cross-linked polymer chains

Figure 18.7 Cross-links (in red) hold polymer molecules together and make the material stronger.

Table 18.1 Some Organic Polymers

Natural Addition Polymer
  rubber
Synthetic Addition Polymers
  polyethylene
  Saran
  Styrofoam
  synthetic rubber
Natural Condensation Polymers
  cotton
  DNA
  leather
  starch
Synthetic Condensation Polymers
  Dacron
  nylon
  Plexiglas
  Spandex

tree trunks with their limbs sawed off) pack closely together, making the material dense and rigid. Molecules of branched polymers (like tree trunks with their limbs intact) cannot pack as tightly and thus produce a softer, less dense material (Figure 18.6).

Individual polymer molecules, depending on their functional groups, attract each other by London forces, dipole–dipole forces, or hydrogen bonds (Section 6.1). But they are held even more firmly in place when they join together with covalent bonds, which are called **cross-links** (Figure 18.7). Controlling the extent of branching and cross-linking are two ways that polymer chemists can develop products with desired properties for specific uses.

Polymers are also classified according to what happens to them during heating. Some, such as polyethylene and nylon, soften and then regain their original texture when cool; these are **thermoplastic polymers.** Others, such as Bakelite (used in radios, buttons, and plastic jewelry) and Melmac (tableware), get harder and more rigid when heated and stay that way after cooling; these are **thermosetting polymers.** Hardening occurs because heat causes permanent cross-links to form and join polymer molecules into larger, more rigid networks.

Two general classes of polymers are *addition polymers* and *condensation polymers* (Table 18.1). As the name suggests, **addition polymers** form when monomer units join together (add) without losing any atoms. **Condensation polymers,** in contrast, form when monomer units bond and eliminate a small molecule, usually water or an alcohol.

18.2 ADDITION POLYMERS

Polymerization. The monomers of addition polymers typically are alkenes, often ethene (ethylene) or one of its derivatives. Pi electrons in the double bond provide a source of electrons for electron-seeking substances.

Polymerization proceeds by a chain reaction. First, ethylene (or a derivative) is mixed with an *initiator*. Some initiators are acids or bases. Others are peroxides or other substances that produce *free radicals,* molecular fragments with an odd, unpaired number of electrons (see Section 12.5). The mechanism of the chain reaction depends on the type of initiator. We will consider only the case of a free-radical reaction.

Figure 18.8 shows how this reaction works. The initiator (X·, where the dot represents the unpaired electron) attacks the pi electrons in the double bond of

Figure 18.8 (a) A free-radical initiator (X·) begins polymerization. The next steps (b and c) extend the polymer chain to some number (n) of repeating monomer units. (d) The chain stops growing when it becomes stable by reacting with an inhibitor substance or another free radical (Y·).

Initiation

a. $X\cdot + CH_2::CH_2 \longrightarrow X:CH_2:\dot{C}H_2 \quad (X-CH_2-\dot{C}H_2)$

Chain lengthening

b. $X-CH_2-\dot{C}H_2 + CH_2::CH_2 \longrightarrow X-CH_2-CH_2-CH_2-\dot{C}H_2$

c. (repeating step b) $\longrightarrow X-(CH_2-CH_2)_n-CH_2-\dot{C}H_2$

Termination

d. $X-(CH_2-CH_2)_n-CH_2-\dot{C}H_2 + \cdot Y \longrightarrow X-(CH_2-CH_2)_{n+1}-Y$

ethylene, attracting one pi electron to join the previously unpaired electron to form a covalent bond. This stabilizes the free radical and breaks the pi bond, leaving only a single (sigma) bond. As a result, one odd, unpaired electron is left on an end carbon atom. This carbon then becomes a free radical and attacks another ethylene molecule to become stable; in the process, the chain grows longer by two carbon atoms, again leaving an unpaired electron on the end carbon.

The cycle continues, building carbon chains of hundreds, or even thousands, of carbon atoms. The reaction stops when the free radical on the chain encounters an inhibitor substance or another free radical, either from another growing chain or from the initiator. Two free radicals stabilize each other by forming a covalent bond, with each contributing an unpaired electron to the bond.

The product of this chain reaction is a polymer. The repeating unit in the chain consists of a monomer molecule with a single bond instead of its original double bond:

$$n\ CH_2{=}CH_2 \xrightarrow{polymerize} {-}(CH_2{-}CH_2)_n{-}$$

Notice that every atom in the monomer remains in the repeating unit; this makes the product an addition polymer.

By changing reaction conditions such as temperature, pressure, and presence of catalysts, polymer chemists can control the average length and branching in a chain. Branching occurs when the free radical in the growing chain relocates to an internal carbon instead of an end carbon. The chain then begins growing from the new, internal point. By controlling the branching and average molecular weight of a polymer molecule, chemists can incorporate the desired density, rigidity, and certain other properties.

Common Addition Polymers. Many addition polymers form from derivatives of ethene (ethylene). Table 18.2 shows some examples. Notice that in all cases the repeating units consist of the monomer with a single bond instead of the original double bond. Most monomers contain the *vinyl* ($CH_2{=}CHZ$) or the *vinylidene* ($CH_2{=}CZ_2$) group (*Z* represents a substituent). One example is vinyl polymer, which has a methyl group as the substituent, polypropylene (Figure 18.9). Another is vinyl chloride, $CH_2{=}CHCl$, which forms polyvinyl chloride (PVC; Figure 18.10).

Table 18.2 lists some uses of addition polymers. Polyvinylpyrrolidene (PVP), for example, is used in hair sprays; once the solvent evaporates, it leaves a plastic film that holds your hair in place. Polymethyl methacrylate (Lucite, Plexiglas) is derived from acrylic acid ($CH_2{=}CHCOOH$), and used to make "unbreakable" windows and contact lenses. Copolymers of acrylonitrile and vinyl chloride (Acrilan, Orlon) are spun into fibers and used as fabrics in clothing.

Figure 18.9 The exceptional strength and flexibility of polypropylene allow it to be used in batteries.

Elastomers. Natural rubber is an elastomer made from isoprene:

$$n\ CH_2{=}\underset{\underset{CH_3}{|}}{C}{-}CH{=}CH_2 \xrightarrow{polymerize} {-}(CH_2{-}\underset{\underset{CH_3}{|}}{C}{=}CH{-}CH_2)_n{-}$$

isoprene → polyisoprene (natural rubber)

Table 18.2 Structures and Uses of Some Synthetic Addition Polymers

Monomer Name (Common)	Monomer Formula	Polymer Formula	Polymer Name (Trade)	Uses
ethene (ethylene)	$CH_2=CH_2$	$-(CH_2-CH_2)-$	polyethylene (Polythene)	bags, films, molded objects, containers, toys
propene (propylene)	$CH_2=CH-CH_3$	$-(CH_2-CH(CH_3))-$	polypropylene (Prolene, Herculon)	carpeting, rope, upholstery, films, bottles
chloroethene (vinyl chloride)	$CH_2=CHCl$	$-(CH_2-CHCl)-$	polyvinyl chloride, PVC (Koroseal)	house siding, pipes, garden hoses, upholstery
phenylethene (styrene)	$CH_2=CH-C_6H_5$	$-(CH_2-CH(C_6H_5))-$	polystyrene (Styrofoam, Lustrex)	insulation, ice chests, packaging material, cups
1,1-dichloroethene (vinylidene chloride)	$CH_2=CCl_2$	$-(CH_2-CCl_2)-$	polyvinylidene chloride (Saran)	food wrap, container liners
(vinyl pyrrolidene)	$CH_2=CH-N(\text{pyrrolidone})$	$-(CH_2-CH-N(\text{pyrrolidone}))-$	polyvinyl pyrrolidene	hair sprays, textiles, adhesives
tetrafluoroethene (tetrafluoroethylene)	$CF_2=CF_2$	$-(CF_2-CF_2)-$	polytetrafluoroethylene (Teflon)	pan coatings, bearings, tape, insulation
(acrylonitrile)	$CH_2=CH-CN$	$-(CH_2-CH(CN))-$	polyacrylonitrile (Acrilan, Orlon)	clothing, rugs, yarn, draperies
(methyl methacrylate)	$CH_2=C(CH_3)-COOCH_3$	$-(CH_2-C(CH_3)(COOCH_3))-$	polymethyl methacrylate (Lucite, Plexiglas)	windows, latex paints, molded items

Figure 18.10 Polyvinyl chloride (PVC) pipe is lightweight, resistant to chemicals, and easy to install.

CHEMISTRY SPOTLIGHT

The Accidental Discovery of Teflon

In 1938, some Du Pont chemists were trying to make new fluorine compounds for use as refrigerants. In one experiment, they stored tetrafluoroethylene gas in a cold cylinder. That cylinder happened to contain small amounts of O_2 gas as an impurity. When they opened it, they discovered a waxy, white, slippery solid that was heat-resistant and insoluble in almost all solvents. The material was a polymer of tetrafluoroethylene, which we now know as Teflon:

$$n\ CF_2\!=\!CF_2 \xrightarrow{\text{polymerize}} (\!CF_2\!-\!CF_2\!)_n$$

tetrafluoroethylene   *Teflon*

The first Teflon-coated cookware went on sale in 1960. But the material didn't adhere well to the pans' metal surfaces and peeled badly. That problem was solved in 1962. Now Teflon is used not only for nonstick cookware, but also for Gore-Tex jogging suits, chainsaw blades, specialized tubing, and even the underpinnings of the Statue of Liberty. In medicine, Teflon is a crucial component in the artificial blood vessels that have been implanted in thousands of patients.

The repeating unit contains one double bond instead of the two in isoprene. Polyisoprene can take either the *cis* or *trans* form:

cis-polyisoprene   and   *trans*-polyisoprene

Natural rubber is *cis*-polyisoprene. The *trans* isomer is harder; it is used for golf-ball covers and to seal root canals in teeth.

Certain plant cells use isoprene to make large molecules such as vitamin A. Your cells use a derivative of isoprene to make cholesterol; in fact, all 27 carbon atoms in cholesterol come from isoprene (see Chapter 26).

Modified isoprene molecules are used to make various types of synthetic rubber. Replacing the methyl group in isoprene with chlorine produces chloroprene, $CH_2\!=\!CCl\!-\!CH\!=\!CH_2$, the monomer unit for polychloroprene (trade names Neoprene and Duprene). Because it resists aging and doesn't dissolve in oils, gasoline, and other organic solvents, Neoprene is used for garden hoses, shoe soles, conveyor belts, seals, and gaskets.

Joseph Priestley (1733–1804) is best known for discovering oxygen. He also discovered that "caoutchouc," a material used by South American Indians, could rub out pencil marks. So in 1770 he named it rubber.

Replacing the methyl group in isoprene with a hydrogen atom makes 1,3-butadiene, $CH_2=CHCH=CH_2$. A copolymer of about 75% butadiene and 25% styrene (Table 18.2) is called SBR (styrene-butadiene rubber), the most common synthetic rubber. It is used in automobile tires and water-based latex paints.

Elastomers stretched to several times their normal length spring back to their original shape. This happens because their polymer molecules are arranged in a fairly random way, with enough freedom to move past each other when stretched, but also with a few cross-links between molecules to keep them from being stretched too far. In 1839, Charles Goodyear discovered that adding sulfur to rubber at 140°C gave it additional cross-links, making it stronger and more durable (Figure 18.11). This process, called *vulcanization*, is used to produce automobile tires and other rubber products.

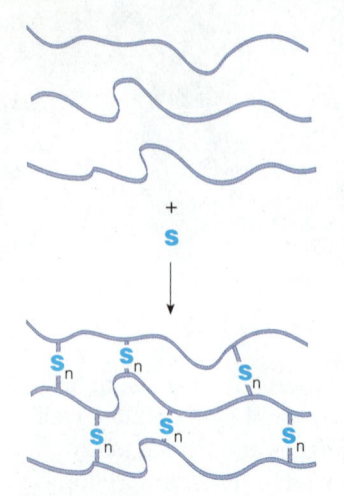

Figure 18.11 Vulcanization cross-links rubber molecules with sulfur. The subscript *n* indicates a variable number of S atoms.

18.3 CONDENSATION POLYMERS

Polyesters. A condensation polymer with regular, repeating ester linkages is a **polyester**. Recall that a carboxylic acid (especially its acyl halide or anhydride) reacts with an alcohol to form an ester and water (Section 16.4). Polyesters are copolymers made from a monomer with two carboxylic acid groups (a dicarboxylic acid), and a monomer with two alcohol groups (a diol). An alcohol group in one monomer reacts with an acid group in the other monomer, forming an ester and eliminating water:

$$HO-\overset{O}{\underset{\|}{C}}-R-\overset{O}{\underset{\|}{C}}-OH + HO-R'-OH$$
$$\longrightarrow HO-\overset{O}{\underset{\|}{C}}-R-\overset{O}{\underset{\|}{C}}-O-R'-OH + H_2O$$

Though his name and process are famous, Charles Goodyear reaped few benefits from his discovery. He spent years defending his patent, and ended up in debtor's prison more than once. He died a poor man in 1860.

The elimination of H_2O when their monomer units combine makes polyesters condensation polymers.

The product still has an alcohol group on one end and an acid group on the other; these groups can react with an appropriate (acid or alcohol) group of another monomer unit to extend the chain in both directions and produce a polyester.

A common polyester, Dacron (Terylene), is made by the reaction of ethylene glycol with terephthalic acid:

HOOC—⟨benzene⟩—COOH + HO—CH₂CH₂—OH $\xrightarrow{polymerize}$

terephthalic acid   *ethylene glycol*

···—O—C(=O)—⟨benzene⟩—C(=O)—O—CH₂CH₂—O—C(=O)—⟨benzene⟩—C(=O)—O—··· + H_2O

Dacron (a polyester)

Dacron fibers (Figure 18.12) are used widely in clothing and fabrics. In medicine, a Dacron coating is used on artificial heart valves and artery

Figure 18.12 Dacron fibers being spun.

segments. When this polyester is spun into a thin film, it is called Mylar, a common material for computer diskettes, recording tape, and packaging for frozen foods.

Polyesters that form cross-links are hard and tough. They include alkyd resins used in the lacquers and paints that give automobiles and major appliances a strong, tough coating (Figure 18.13).

Figure 18.13 Alkyd resins give this automobile a tough, shiny coating.

In Chapter 23, you will learn that DNA and RNA are natural polymers whose monomer units are joined together by phosphate ester groups.

Polyamides. Condensation polymers with repeating amide linkages are called **polyamides**. Recall that a carboxylic acid (especially as an acyl chloride, anhydride, or ester) reacts with an amine to form an amide and eliminate water (Section 17.4). Polyamides are copolymers made from a dicarboxylic acid monomer and a diamine monomer. The most important polyamide, nylon,

Removal of H₂O when its monomers combine makes nylon a condensation polymer.

forms from the following two monomers:

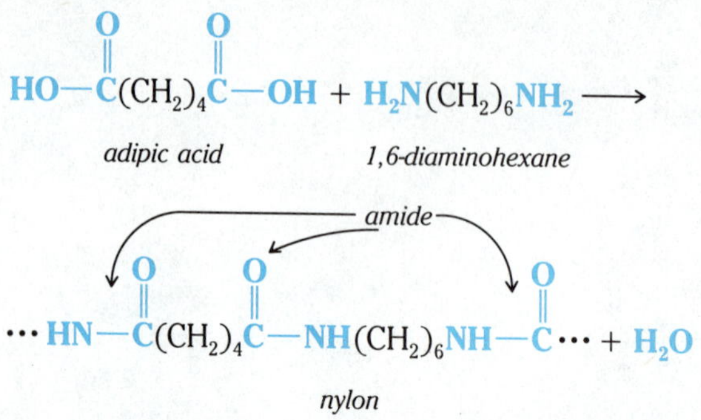

$$HO-\overset{O}{\underset{}{C}}(CH_2)_4\overset{O}{\underset{}{C}}-OH + H_2N(CH_2)_6NH_2 \longrightarrow$$
     adipic acid   1,6-diaminohexane

$$\cdots HN-\overset{O}{\underset{}{C}}(CH_2)_4\overset{O}{\underset{}{C}}-NH(CH_2)_6NH-\overset{O}{\underset{}{C}}\cdots + H_2O$$
       nylon

Nylon is very versatile. It can be woven into fabrics or molded into various shapes. Its strength comes from hydrogen bonds between amide groups of nearby molecules (Figure 18.14). Because its polar groups are tied up in hydrogen bonds, nylon doesn't readily form hydrogen bonds with water. This makes nylon garments easy to dry. (It also makes nylon a poor material for towels.)

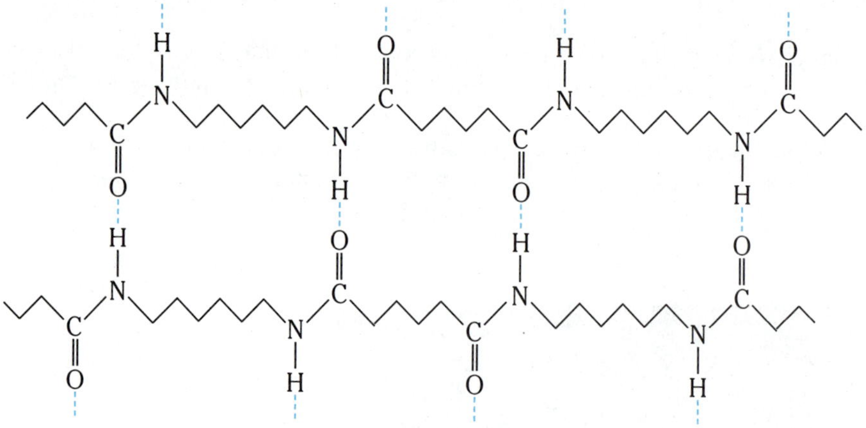

Figure 18.14 Hydrogen bonds (dotted lines in color) between polymer molecules in nylon. Each corner in the zigzag lines represents —CH₂—.

The structure shown above is for nylon-6,6 (6 carbons in both monomer units), but chemists vary the length of the carbon chain in the monomer units to produce nylons with different properties. Some types, for example, are very resistant to wear and hard enough to use as gears in machinery.

Nylon and other polymers such as Dacron and Orlon can be spun into fibers (Figure 18.3), and twisted into threads. The polymer molecules lie along the direction of the fiber, providing considerable strength during stretching. Extensive side-to-side bonding between molecules keeps them from slipping past each other. This makes nylon useful for such things as fishing line and tire cord.

In Chapter 21 you will learn about proteins, which are natural polyamides.

Other Condensation Polymers. The first thermosetting plastics were phenol-formaldehyde resins, developed in 1909 and sold under the trade name Bakelite. They form in a complex reaction (involving several intermediate chemicals) between phenol and formaldehyde, with water being split out and driven off by heat as the polymer sets:

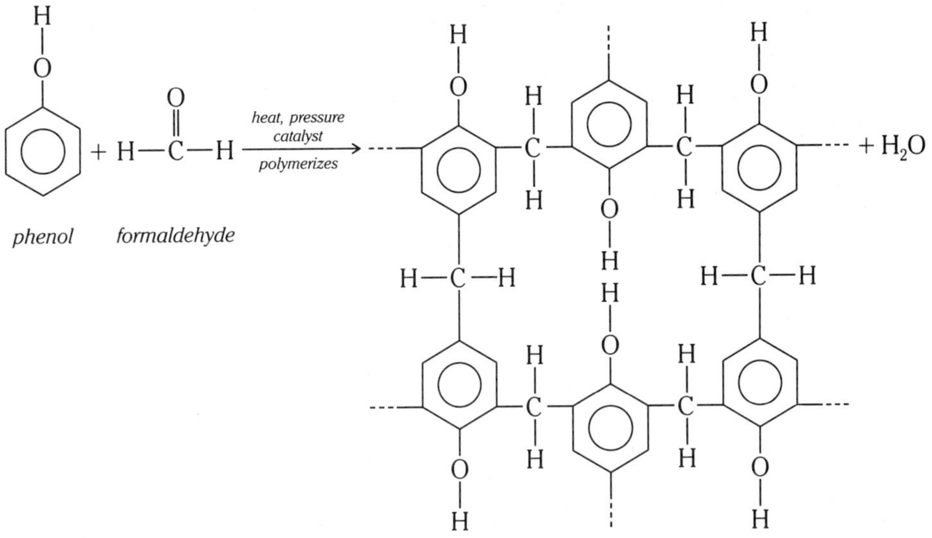

phenol-formaldehyde polymer
(Bakelite)

Its stiff, rigid chains and multiple cross-links make this polymer very hard, rigid, and heat-resistant. It is widely used for radios, television circuit boards, plywood adhesive, electrical insulation, buttons, cookware handles, and distributor caps.

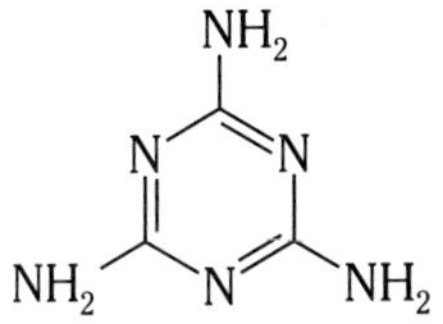

Using melamine (Figure 18.15) instead of phenol produces thermosetting melamine-formaldehyde resin (trade name Melmac). Its main use is in high-quality, everyday tableware that is dishwasher-safe and nearly unbreakable. It also is one of several polymers used to make permanent-press fabrics. When fabrics are impregnated with polymers such as melamine-formaldehyde resin, their fibers are cross-linked and fixed in their original, pressed shape. As a result, clothes remain unwrinkled even after repeated washings.

Figure 18.15 Melamine, a monomer unit for Melmac.

Polyurethanes are thermosetting condensation polymers made from monomers with two isocyanate (—N=C=O) groups and monomers with two or more alcohol (—OH) groups, such as ethylene glycol or glycerol:

$$O=C=N-R-N=C=O + HO-R-OH$$

a di-isocyanate   *a dihydroxy alcohol*

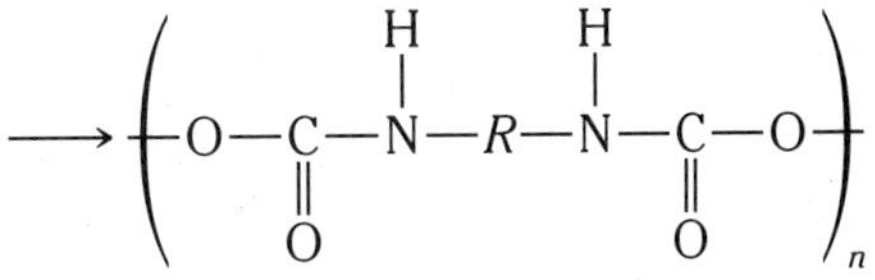

a polyurethane

When a foaming agent is added, tiny gas bubbles form in the plastic mass to make a rigid, quick-setting foam known as "foam rubber." Polyurethanes are also used in stretch fabrics such as Spandex and Lycra, which are both rigid and elastic.

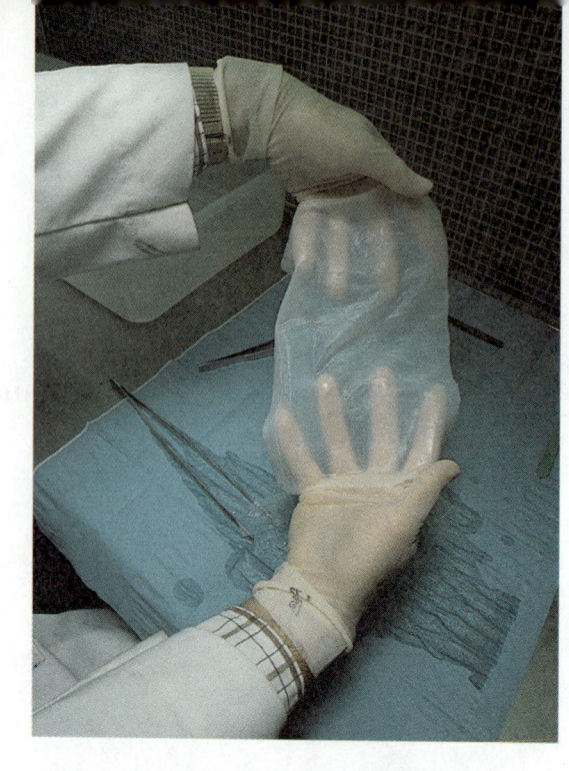

Figure 18.16 Membranes made of silicone polymer are used as a skin substitute for burn victims during their recovery.

18.4 OTHER POLYMERS

Although most polymers are based on carbon, a few are based on the element nearest carbon in the same group in the periodic table—silicon (Si). Unlike carbon, silicon doesn't form long chains of atoms bonded to each other. But it does form long chains of alternating silicon and oxygen atoms.

Dihydroxysilicon compounds are the monomer units for **silicone polymers** (siloxanes). A typical reaction is:

$$\text{HO}-\underset{\underset{\text{CH}_3}{|}}{\overset{\overset{\text{CH}_3}{|}}{\text{Si}}}-\text{OH} + \text{HO}-\underset{\underset{\text{CH}_3}{|}}{\overset{\overset{\text{CH}_3}{|}}{\text{Si}}}-\text{OH} \longrightarrow -\text{O}-\underset{\underset{\text{CH}_3}{|}}{\overset{\overset{\text{CH}_3}{|}}{\text{Si}}}-\text{O}-\underset{\underset{\text{CH}_3}{|}}{\overset{\overset{\text{CH}_3}{|}}{\text{Si}}}-\text{O}- + \text{H}_2\text{O}$$

Depending on the average chain length and degree of cross-linking, silicone polymers can be oils, greases, waxes, or solids. Because they aren't chemically reactive and don't decompose at high temperatures or thicken at low temperatures, silicone oils are effective engine lubricants. Other oils are used in cosmetics, car waxes, and waterproof coatings. One kind of silicone oil, mixed with chalk, is known as Silly Putty.

A variety of silicone rubbers can be made by cross-linking silicone chains. Silicone rubber stays flexible at freezing temperatures and is more stable than other rubbers when exposed to heat, oxidizing agents, and other reactive materials. Like silicone oils, silicone rubber is useful in airplanes and spacecraft, which experience a wide range of temperatures. Some types are also used for electrical insulation, gaskets, O-rings, or as skin substitutes for burn victims (Figure 18.16).

Plastic surgeons frequently implant silicone material to improve the appearance of the ears, chin, nose, breasts, or other body parts. U.S. companies, however, no longer manufacture silicone-gel breast implants because they may cause neurological and autoimmune disorders. One alternative is saline implants, which consist of salt solutions encased in silicone.

Besides carbon- and silicon-based polymers, chemists have made a few polymers containing boron, phosphorus, sulfur, and other elements. For example, boron–boron and boron–nitrogen polymers may have even greater heat resistance than silicone polymers. Most of these other polymers, however, are still in the experimental stages of development.

> My tie is made of terylene [Dacron];
> Eternally I wear it,
> For time can never wither, stale,
> Shred, shrink, fray, fade or tear it . . .
>
> *"In Praise of $(C_{10}H_9O_5)_x$," by John Updike (1932–   )*

SUMMARY

Very large molecules are macromolecules. Most macromolecules are polymers, which are made by joining large numbers of monomer units together. Polymers made from one monomer substance are homopolymers; those having two or more monomers are copolymers. Branched polymers are softer and less dense. Cross-linking makes them more rigid. Thermoplastic polymers soften when heated and regain their original texture when cooled. When thermosetting polymers are heated, however, they form permanent cross-links and harden.

Monomers that join without losing any atoms form addition polymers. Polymerization can be initiated by free radicals. Ethene (ethylene) and its derivatives are typical monomers. Some examples include polyethylene and vinyl polymers such as polyvinyl chloride (PVC). Elastomers are elastic addition polymers often made from isoprene. Cross-linking rubber with sulfur (vulcanization) makes rubber more durable.

When monomers join with the loss of a small molecule (usually H_2O), the product is a condensation polymer. Polyesters (such as Dacron) and polyamides (such as nylon) are the most common types.

Silicone polymers are based on silicon. A few other polymers are based on such elements as boron, phosphorus, and sulfur.

KEY TERMS

Addition polymer (18.1)
Condensation polymer (18.1)
Copolymer (18.1)
Cross-link (18.1)
Elastomer (18.1)
Fiber (18.1)

Homopolymer (18.1)
Macromolecule (18.1)
Monomer (18.1)
Plastic (18.1)
Polyamide (18.3)
Polyester (18.3)

Polymer (18.1)
Silicone polymer (18.4)
Thermoplastic polymer (18.1)
Thermosetting polymer (18.1)

EXERCISES

Even-numbered exercises are answered at the back of this book.

Types of Polymers

1. Are diamonds and graphite macromolecules? Are they polymers? Explain. [Review Section 11.4 if necessary for the structures of diamond and graphite.]

2. Identify which of the following are homopolymers: **(a)** Teflon, **(b)** SBR rubber, **(c)** polyvinylchloride, **(d)** Dacron.

3. Which of the polymers in exercise 2 are copolymers?

4. Identify which one of the following is a copolymer: **(a)** polyethylene, **(b)** nylon, **(c)** Teflon, **(d)** Styrofoam, **(e)** Saran.

5. Polyethylene can be made in high-density and low-density forms. The high-density form (a) is thermosetting, (b) contains oxygen, (c) uses a different monomer unit, (d) has less cross-linking, (e) none of the above.

6. Cross-links are (a) ionic bonds, (b) covalent bonds, (c) hydrogen bonds, (d) London forces, (e) dipole–dipole forces.

7. If plastic seat covers temporarily soften on a warm day, are they made of a thermoplastic or a thermosetting polymer?

Addition Polymers

8. Write electron dot structures for the following and determine which one can be classified as a free radical: (a) hydrogen molecule, (b) helium atom, (c) chlorine molecule, (d) chlorine atom.

9. Are organic peroxide compounds that produce free radicals more useful in catalyzing the synthesis of an addition or a condensation polymer?

10. During polymerization initiated by a free radical, what terminates the growth of a polymer chain?

11. Explain why alkenes are more likely than alkanes to react with a free radical and polymerize.

12. Why is a polymer such as polyethylene called an *addition* polymer?

13. The main monomer unit in butyl rubber is 2-methylpropene (isobutylene). Write the structural formula for this monomer and for the repeating unit in butyl rubber based on this monomer.

14. Write structural formulas for the monomers used to produce each of the following addition polymers:

    (a) ... —CH$_2$CHCH$_2$CHCH$_2$CH— ... ,
            | | |
            CN   CN   CN

    (b) ... —CH$_2$CCl$_2$CH$_2$CCl$_2$CH$_2$CCl$_2$— ...

15. Write structural formulas for the monomers used to produce each of the following addition polymers:

    (a) ... —CH$_2$—C—CH$_2$—C—CH$_2$—C—CH$_2$—...
              | | |
             CH$_3$ CH$_3$ CH$_3$
         (with COOH on each C)
         (polymethacrylic acid)

    (b) ... —CH$_2$—CH—CH$_2$—CH—CH$_2$—...
                   | |
                   O—CCH$_3$ O—CCH$_3$
                   ‖   ‖
                   O   O
         (polyvinyl acetate)

16. Write the structural formula for the repeating unit in the addition polymer that could form from 2-chloropropene.

17. Vinyl chloride, the monomer for polyvinyl chloride (PVC), has been found to cause a rare form of liver cancer. Write the structural formula for vinyl chloride.

18. Among other things, polyvinyl chloride (PVC) is used for raincoats and plastic pipes. Explain why PVC, in terms of its chemical structure, does not dissolve in water.

19. Which polymer in exercise 4 is aromatic? Which ones contain an element besides C and H?

20. Identify which of the following are elastomers: (a) polyvinyl chloride (PVC), (b) Neoprene, (c) natural rubber, (d) Plexiglas.

21. The process of cross-linking rubber with sulfur is called _____.

22. Which of the compounds in exercise 2 are addition polymers?

Condensation Polymers

23. Which of the compounds in exercise 2 are condensation polymers?

24. Would acetic acid work as a monomer unit to form a polyester? Explain why or why not.

25. Identify which of the following could work as a monomer to form a polyester: (a) HO—CH$_2$(CH$_2$)$_3$CH$_3$,
                                                                                                 O
                                                                                                 ‖
    (b) HOOC—CH$_2$CH$_2$—COOH, (c) CH$_3$C—OCH$_3$,
    (d) HO—CH$_2$CH$_2$CH$_2$—OH, (e) CH$_3$CH$_2$—NH$_2$.

26. Which of the compounds in exercise 25 could work as a monomer to form a polyamide?

27. What does the "6,6" in "nylon-6,6" represent?

28. Identify which is the most likely substance to use as a monomer in place of the dicarboxylic acid, HOOC(CH$_2$)$_4$COOH, in making nylon:
    (a) HO(CH$_2$)$_6$OH,   (b) CH$_3$(CH$_2$)$_4$COOH,
    (c) HOOCCH$_2$CH═CHCH$_2$COOH,
            O   O
            ‖   ‖
    (d) Cl—C(CH$_2$)$_4$C—Cl,
                O   O
                ‖   ‖
    (e) CH$_3$—O—C(CH$_2$)$_2$C—O—CH$_3$

29. Bakelite, a thermosetting polymer made from the monomer units formaldehyde and phenol, has the

following structure:

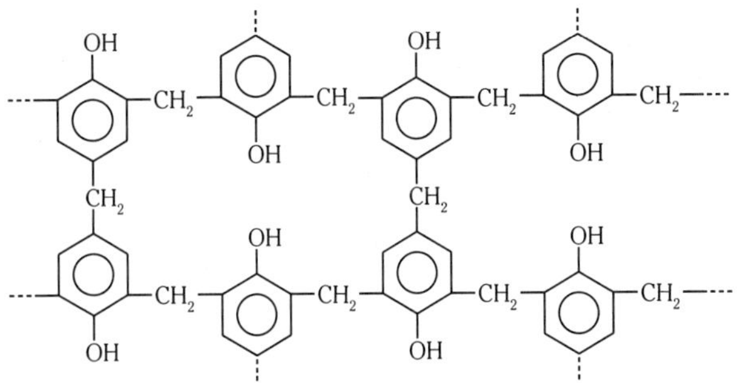

[structure of Bakelite showing phenol rings with OH groups connected by CH₂ bridges]

Is Bakelite most likely an addition or a condensation polymer?

30. What chemical feature accounts for the hard, rigid properties of Bakelite (see structure in exercise 29)?
31. Why do polymers, even highly polar ones, not dissolve readily in water?
32. Compare "foam rubber" with natural rubber in terms of their chemical structures.
33. Proteins have the general structure:

$$\cdots -\overset{O}{\underset{\|}{C}}-NH-CHR-\overset{O}{\underset{\|}{C}}-NH-CHR'-\overset{O}{\underset{\|}{C}}- \cdots$$

Identify which of the following terms apply to proteins: **(a)** polyester, **(b)** condensation polymer, **(c)** homopolymer, **(d)** polyamide, **(e)** addition polymer, **(f)** copolymer, **(g)** elastomer, **(h)** vinyl polymer.

34. The general structure of proteins (see exercise 33) indicates that protein molecules can attract each other with **(a)** dipole–dipole forces, **(b)** London forces, **(c)** hydrogen bonds, **(d)** covalent bonds.

35. During tanning, reagents such as formaldehyde produce cross-links in leather (which is mostly protein). What effect does this have on the properties of leather?
36. Many esters have a fragrant aroma. Why don't polyesters typically have an aroma?
37. Cotton is a polymer of glucose, which contains multiple —OH groups. In terms of chemical structure, is polypropylene or cotton more likely to get, and stay, wet? Why?

Silicones

38. Compare silicon with carbon in terms of **(a)** atomic size and **(b)** electronegativity.
39. In comparison with silicone oils, silicone rubber is likely to be **(a)** lower molecular weight, **(b)** more cross-linked, **(c)** a copolymer, **(d)** more polar, **(e)** more than one of the above.
40. Are silicones organic compounds?

DISCUSSION EXERCISES

1. Are all macromolecules polymers? Are all polymers macromolecules? Explain both answers.
2. Do you predict synthetic polymers will be more abundant or less abundant in consumer products 100 years from now? Explain why.
3. Are there chemical or other reasons to prefer natural rubber (*cis*-polyisoprene) to synthetic *cis*-polyisoprene (or vice versa)? Explain.

III

BIOCHEMISTRY

19

Carbohydrates

1. What are the main types and structures of carbohydrates?
2. What are the most common simple carbohydrates in your body? What are their major sources and uses?
3. What is optical isomerism? How does this apply to carbohydrates?
4. What are some typical reactions of simple carbohydrates?
5. What are the structures, sources, and functions of disaccharides and polysaccharides in your body?

Sweet Taste

While recovering from cancer, I spent many days lying in a hospital bed with chemicals streaming through a needle into my arm. Those chemicals saved my life. But they made me so nauseous I could hardly eat. Even putting a toothbrush in my mouth triggered stomach spasms.

One of the few foods I could tolerate was applesauce. Its light, sweet taste was good. And I liked the taste of a few other carbohydrate foods.

No one knows for sure why simple carbohydrates taste sweet, but it has to do with their ability to bind to protein receptors in taste buds. Nerves passing through the base of each taste bud send signals to the brain that enable us to detect four main flavors—salty, sour, bitter, and sweet.

Many different types of chemicals taste sweet, and scientists have discovered several chemical features they have in common. Each has a polar group that forms a hydrogen bond and another group nearby that accepts a hydrogen bond. One theory is that the receptor protein in a taste bud has complementary functional groups that form hydrogen bonds with two groups in the sweet substance (Figure 19.1).

It turns out, then, that carbohydrates aren't the only substances that taste sweet. On another night during my recovery, I had a sudden urge to drink a Diet Coke. Its artificial sweetener, though not a carbohydrate, also tasted good. I enjoyed my drink, but the caffeine—which I hadn't had in months—kept me awake all night.

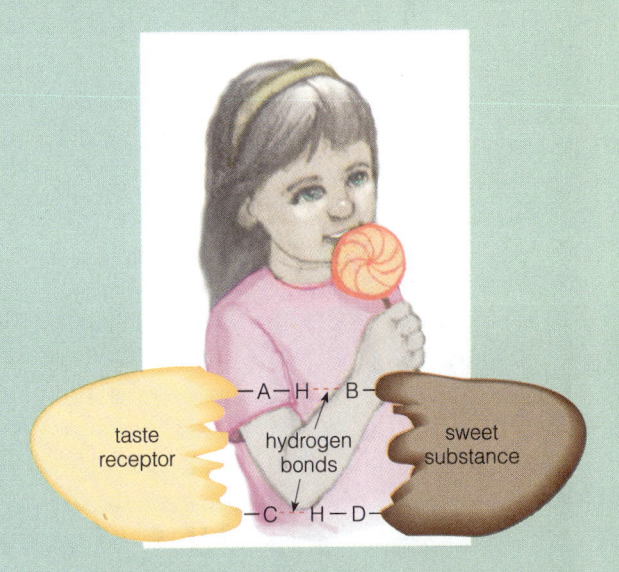

Figure 19.1 According to one theory, sweet substances form hydrogen bonds to specific receptors in taste buds.

What do you have in common, chemically, with monkeys, pine trees, and hummingbirds (Figure 19.2)? More than you might think. All living things—no matter how different they look and act—are made from the same basic ingredients.

The recipe for living organisms calls for generous portions of carbohydrates, lipids, proteins, nucleic acids, minerals, and water, plus a sprinkling of

Figure 19.2 This white uakari lives in the Amazon region of South America.

vitamins. Organisms differ in the types and amounts of these materials, to be sure, but they all use some of each. Of these ingredients, only two—minerals and water—are common to nonliving things. The others are organic compounds that rarely appear in nature except where there is life.

With this chapter, you begin the study of **biochemistry,** the chemistry of living organisms. You will see that the same laws of nature operate in organisms as in everything else, and you can understand much about living things by understanding their chemistry. Each of you is a walking, talking, thinking bundle of chemicals undergoing thousands of reactions to keep you alive and healthy. And when you aren't healthy, it's because some other chemical events are occurring. If you understand the chemistry of illness, you can often take appropriate measures to restore health.

19.1 INTRODUCTION TO CARBOHYDRATES

Bread, potatoes, noodles, corn, candy, and fruits all are rich in carbohydrates (Figure 19.3). Although you get some energy from fats and proteins, carbohydrates are your major source. In fact, they supply 50–60% of the kilojoules (or Calories) of a typical American diet and up to 80% of the energy intake for people in developing nations.

Carbohydrates are aldehydes or ketones containing multiple alcohol (—OH) groups. The name comes from an early belief that carbohydrates consisted of hydrated carbon, that is, carbon bound to water molecules. Indeed, the formulas of most carbohydrates can be written as $C_x(H_2O)_y$, where x and y may be the same or different numbers. For example, the formula for the most common simple sugar in our bodies, glucose, can be written as $C_6(H_2O)_6$ or $C_6H_{12}O_6$. Sucrose (table sugar) has the formula $C_{12}H_{22}O_{11}$ or $C_{12}(H_2O)_{11}$.

We classify carbohydrates according to the number of sugar units in their molecules. **Monosaccharides** such as glucose, fructose, and ribose are the simplest kind of carbohydrate and consist of one sugar unit. Two monosaccha-

Figure 19.3 Corn is a rich source of carbohydrates.

Carbohydrates *are not simple hydrates of carbon, but the name remains.*

ride units bonded together make a **disaccharide.** The most common examples are sucrose (table sugar), lactose (milk sugar), and maltose (malt sugar). **Polysaccharides** are polymers containing hundreds or even thousands of monomer sugar units. The most familiar examples are cellulose, starch, and glycogen.

19.2 MONOSACCHARIDES

Monosaccharides typically have three to seven carbon atoms. We classify them according to this number of carbon atoms, using the suffix -*ose*. So a sugar is a *triose, tetrose, pentose, hexose,* or *heptose,* depending on whether it has 3, 4, 5, 6, or 7 carbons, respectively.

Pentoses and hexoses are the most common. Two important pentoses, *ribose* and *2-deoxyribose* (Figure 19.4), are in RNA and DNA, respectively. When the aldehyde group in a pentose called xylose is reduced to an

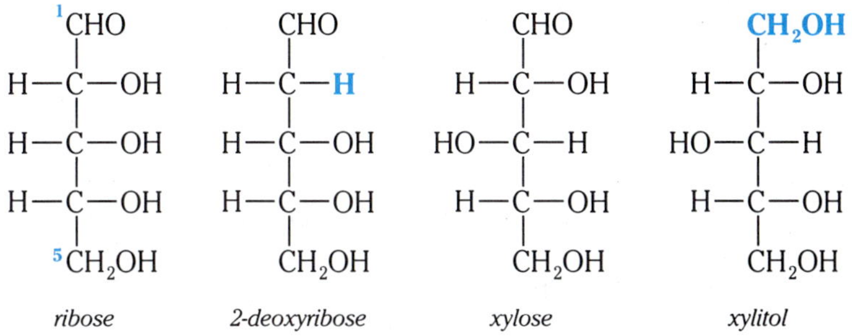

Figure 19.4 Some pentoses of practical importance. Differences from the parent pentose are in color.

alcohol, the product is xylitol (Figure 19.4). Xylitol occurs naturally in carrots, spinach, and plums and is used as a sweetening agent in "sugarless" chewing gum.

The three main hexoses in your body are *glucose, fructose,* and *galactose* (Figure 19.5). You don't get much galactose directly in your diet. But you digest the disaccharide lactose (milk sugar) in dairy products into its monosaccharide units—galactose and glucose (see Section 19.5). Glucose occurs naturally in most fruits, especially figs, dates, raisins, and grapes, while fructose is in various fruits, vegetables, and honey.

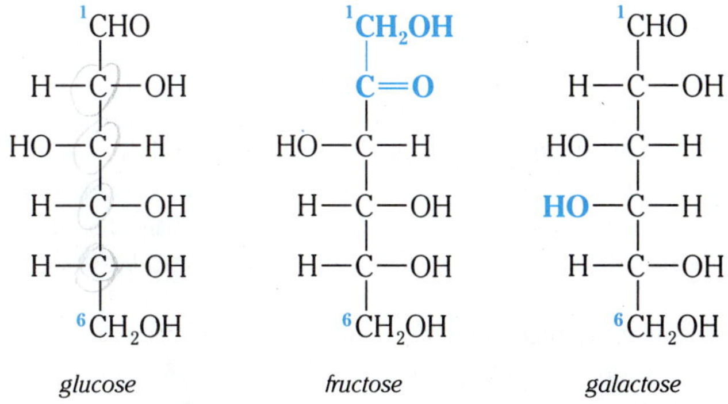

Figure 19.5 The three major hexoses in the body. Differences from glucose are in color.

Glucose and galactose are aldehydes; fructose is a ketone. As with other aldehydes and ketones, carbon atoms are numbered from the end nearer the carbonyl ($\!>\!\!C\!\!=\!\!O$) group. These forms are sometimes called open-chain structures.

In your body, however, monosaccharides are almost entirely in cyclic form. The ring forms when the polar carbonyl group in the open-chain form reacts with the polar alcohol group on carbon 5. The reaction with glucose is:

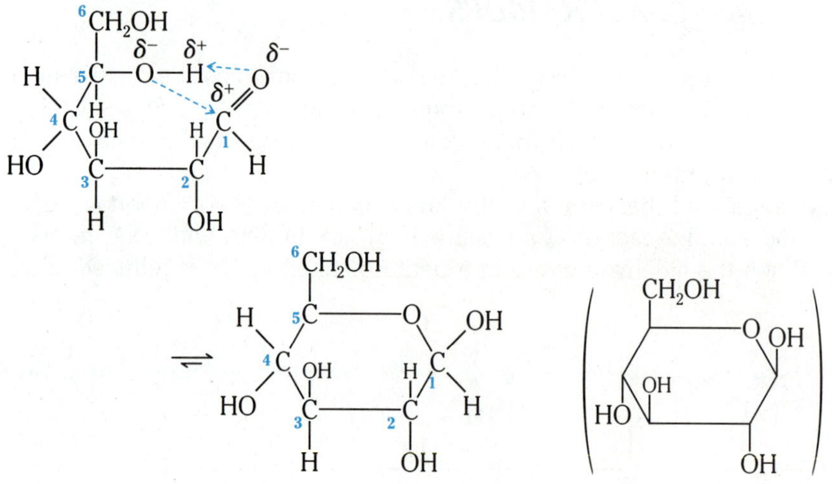

The —OH group at carbon 1 can be either up or down (see Sections 19.3 and 19.4).

You have seen this type of reaction before. Recall that an aldehyde reacts with an alcohol to produce a hemiacetal, and a ketone reacts with an alcohol to produce a hemiketal (Section 15.4). In the reaction above, notice that carbon 1 in the cyclic form is bonded to —H, —OH, —R, and —OR. This makes the ring form of glucose a cyclic hemiacetal.

EXAMPLE 19.1 Draw a structural formula for the cyclic form of galactose. Is galactose (like glucose) a cyclic hemiacetal?

SOLUTION Galactose is a cyclic hemiacetal (see carbon 1).

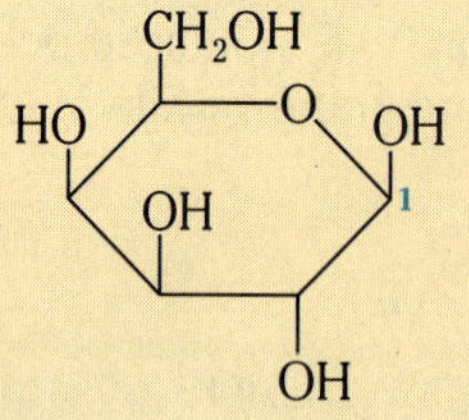

The most common way to represent cyclic sugars (as with other cyclic organic compounds) is simply to write a pentagon or hexagon that doesn't show ring carbons or their bonded hydrogens (Figure 19.6). Ring carbons that project toward the viewer may be indicated by shading (although this is usually omitted), and substituents to ring carbons are shown above or below the general plane of the ring.

A regular hexagon, however, has 120° bond angles. Recall that nonaromatic six-carbon rings exist in chair and boat forms and are not completely flat

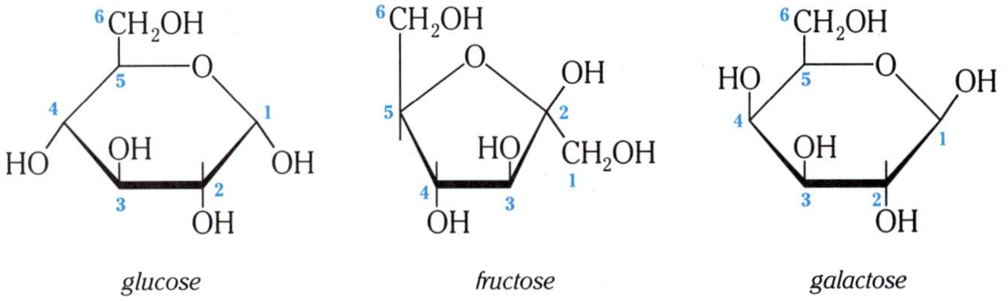

Figure 19.6 Haworth structures for three hexoses.

(Section 11.2). This is also the arrangement with six-member carbohydrate rings, which are made mostly of carbon. A five-atom ring, as in fructose, is flat and stable because the 108° bond angles in a regular pentagon nearly match the 109° tetrahedral bond angles around carbon.

The cyclic structures in Figure 19.6 are called **Haworth structures.** Converting a structural formula from the type in Figure 19.5 into a Haworth structure is fairly simple. The main rules to remember are:

Haworth structures were devised by the English chemist Walter Haworth, who received the Nobel prize in chemistry in 1937.

1. *Groups on the right in the structural formula project down in a Haworth structure; groups on the left project up.*
2. *Carbon 6, which is not in the ring, projects up from carbon 5.*

EXAMPLE 19.2 Write Haworth structures for (a) ribose (use the structure in Figure 19.4) and (b) mannose, which has the same structure as glucose except for the opposite orientation of —H and —OH on carbon 2.

SOLUTION

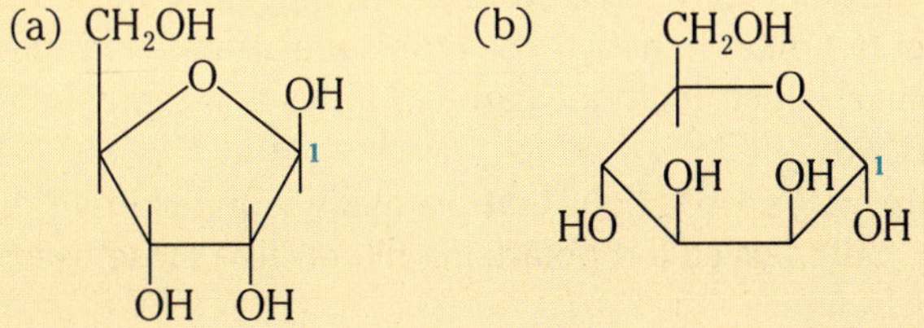

(The —OH group on carbon 1 can be either up or down; see Sections 19.3 and 19.4.)

Because of their polar alcohol groups, monosaccharides (and disaccharides such as sucrose) form hydrogen bonds and dissolve in water (Figure 19.7). Glucose, for example, dissolves in blood, enabling it to travel throughout the body and provide a constant supply of energy to organs that don't store their own fuel, such as the brain.

Although your body uses glucose directly, it converts other dietary carbohydrates into glucose derivatives before using them for energy. This is why glucose is the carbohydrate given in intravenous solutions to people in life-threatening situations. In such extreme stress, the body needs a direct source of energy.

Fructose is sweeter than glucose (Figure 19.8). Since glucose and fructose both have the same energy value (about 17 kilojoules or 4 Calories per gram), fructose provides more sweetening per kJ (or Cal) than glucose. Some food manufacturers take advantage of this difference by using fructose as a natural sweetener.

412   Part III   Biochemistry

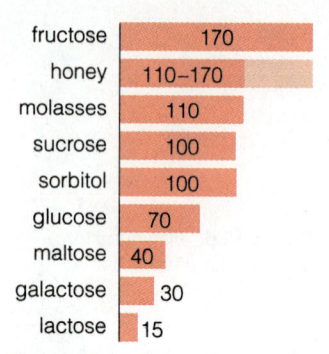

fructose	170
honey	110–170
molasses	110
sucrose	100
sorbitol	100
glucose	70
maltose	40
galactose	30
lactose	15

Figure 19.8 Approximate sweetness in comparison to sucrose, which is assigned a value of 100.

Figure 19.7 Simple sugars form hydrogen bonds (dotted lines in color) with water.

Notice that the structures of galactose and glucose (Figures 19.5 and 19.6) differ only in their orientation of —H and —OH at carbon 4. This makes glucose and galactose isomers—a new type of isomer that we discuss in the next section.

19.3 OPTICAL ISOMERS

Optically Active Compounds.   You have learned about two kinds of isomers so far: structural isomers (Section 12.2) and geometric isomers (Section 13.3), which exist in *cis* and *trans* forms.

Geometric isomers are **stereoisomers;** they have the same atoms bonded to each other but differ in how those atoms are arranged in space. Another type of stereoisomer, which includes carbohydrates and many other molecules are the *optical isomers.*

As Figure 19.9 shows, light that normally vibrates in all directions can be screened by materials called polarizing filters (like those used in Polaroid sun-

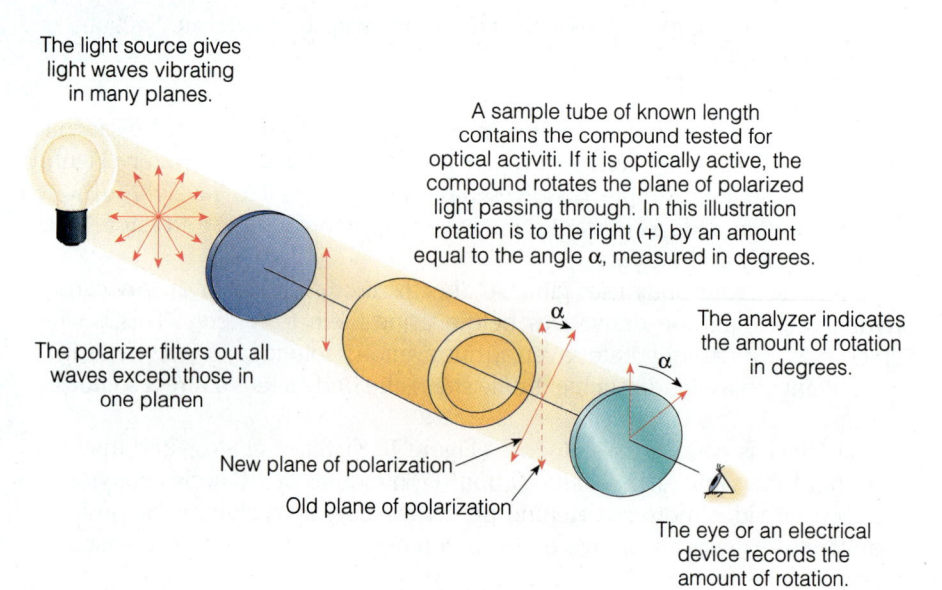

The light source gives light waves vibrating in many planes.

A sample tube of known length contains the compound tested for optical activiti. If it is optically active, the compound rotates the plane of polarized light passing through. In this illustration rotation is to the right (+) by an amount equal to the angle α, measured in degrees.

The polarizer filters out all waves except those in one planen

New plane of polarization
Old plane of polarization

The analyzer indicates the amount of rotation in degrees.

The eye or an electrical device records the amount of rotation.

Figure 19.9 How a polarimeter works.

glasses) so that the emerging light vibrates only in one plane. When this *plane-polarized light* travels through a tube, it emerges at the other end still vibrating in the same plane. However, when solutions of some compounds are placed in the path of plane-polarized light, they cause the plane of light to shift (rotate) in a clockwise or counterclockwise direction. Such compounds are said to have **optical activity. Optical isomers** are stereoisomers that are optically active.

A *polarimeter* measures the rotation of plane-polarized light as it passes through a sample. The amount of rotation (at a specified wavelength and temperature) depends on the compound, its concentration, and the length of the polarimeter tube. The standard measurement, called *specific rotation*, is the rotation measured for a 1 g/mL solution with a tube length of 1 dm (10 cm). The formula is:

$$\text{specific rotation} = \frac{\text{observed rotation}}{(\text{g/mL sample})(\text{dm of light path})}$$

A positive sign (+) indicates rotation to the right (clockwise); a negative sign (−) means rotation to the left (counterclockwise).

EXAMPLE 19.3 What is the specific rotation for a substance if a rotation of −64.6° is observed for a 3.20 g/mL sample using a light path of 20.0 cm?

SOLUTION $20.0 \text{ cm} \times \dfrac{1 \text{ m}}{100 \text{ cm}} \times \dfrac{10 \text{ dm}}{1 \text{ m}} = 2.00 \text{ dm}$

$$\text{specific rotation} = \frac{-64.6°}{(3.20)(2.00)} = -10.1°$$

What kinds of compounds have optical activity? Geometric isomers such as *cis*- or *trans*-2-butene are stereoisomers, but they are not optically active. Many organic compounds—such as 2-chlorobutane, CHFClBr, 2-pentanol, and monosaccharides—are optically active. It turns out that optically active compounds exist in forms that are nonidentical mirror images. The two forms are like your left and right hands (see Figure 19.13). Although they look alike, you know your hands are different because you cannot superimpose one hand on the other.

You can see this same pattern in CHFClBr (Figure 19.10). No matter how the molecules are rotated in space, this compound has two different structures

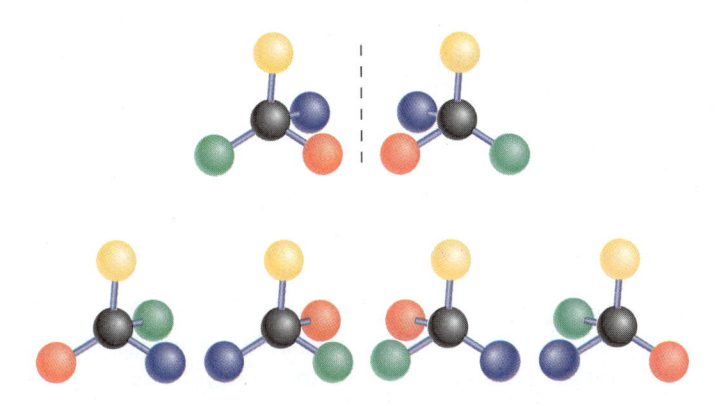

Figure 19.10 Ball-and-stick models of the two optical isomers of CHFClBr. Three arrangements are shown; each is a nonidentical mirror image of an arrangement of the other isomer. Color key: carbon = black; hydrogen = yellow; fluorine = violet; chlorine = green; bromine = red.

Another term for a chiral carbon is an asymmetric *carbon.*

The word chiral *comes from* cheir, *a Greek word that means* hand.

CH₃CH₂—C(OH)(H)—CH₃

2-butanol

CH₂Cl—C(OH)(H)—C(OH)(H)—CH₃

1-chloro-2,3-butanediol

Figure 19.11 Two compounds that have chiral carbons (in color).

that are mirror images of each other but not identical. Each of the two forms is an optical isomer and is optically active. In fact, these two isomers rotate plane-polarized light the same amount, but in opposite directions.

Organic compounds that occur as optical isomers typically have at least one carbon atom bonded to four different groups (Figure 19.10). Such a carbon atom is called a **chiral carbon.** Molecules with at least one chiral carbon usually are optically active.

A compound can exist as a maximum of 2^n different stereoisomers, where n is the number of chiral carbons. The number is a maximum because a plane of symmetry appears in some molecules; this makes some compounds identical to their mirror images and reduces the number of different stereoisomers and optical isomers.

One example is tartaric acid, which has two chiral carbons and thus can exist in four (2^2) different forms:

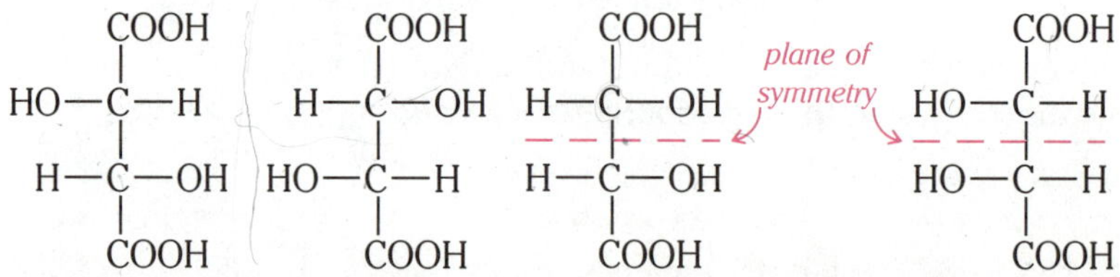

```
   COOH         COOH         COOH    plane of      COOH
    |            |            |      symmetry       |
HO—C—H       H—C—OH       H—C—OH                HO—C—H
    |            |            |     - - - -        |
 H—C—OH       HO—C—H       H—C—OH                HO—C—H
    |            |            |                    |
   COOH         COOH         COOH                 COOH
```

Two forms have an internal plane of symmetry in which the top half is the mirror image of the bottom half. This internal symmetry makes these two structures optically inactive and identical to each other. Tartaric acid, then, consists of three—not four—different stereoisomers.

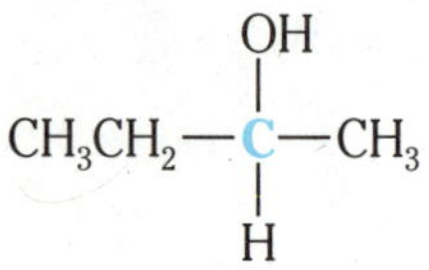

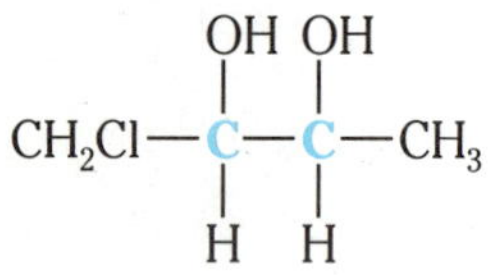

EXAMPLE 19.4 Identify which of the following have chiral carbons and exist as optical isomers: 2-propanol, 1-butanol, 2-butanol, and 1-chloro-2,3-butanediol. For those with isomers, how many different stereoisomers exist?

SOLUTION 2-Propanol and 1-butanol have no chiral carbons, so no optical isomers exist. 2-Butanol has one chiral carbon (carbon 2) so $2^1 = 2$ stereoisomers exist. 1-Chloro-2,3-butanediol has two chiral carbons (carbons 2 and 3) and no internal plane of symmetry, so $2^2 = 4$ stereoisomers exist. See Figure 19.11.

Carbohydrates and Optical Isomerism. Examine the open-chain structure of glucose (Figure 19.5). Carbons 2–5 are each bonded to four dissimilar groups, and are therefore chiral. Carbon 1, the carbonyl carbon, is only bonded to three groups, so it isn't chiral. Carbon 6 is bonded to two H atoms, so it is not chiral, either. Since it has four chiral carbons, glucose is one of 16 (2^4) different stereoisomers for open-chain hexoses that are aldehydes.

Figure 19.12 shows all 16 isomers of glucose in the condensed versions called **Fischer projection formulas.** Each intersection point represents a chiral carbon; groups on the left and right (—H and —OH) project toward the viewer, while vertical groups (—CHO and —CH₂OH) project back from the viewer.

The names of eight pairs of hexoses in Figure 19.12 differ only in the prefix L- or D-. If you examine the structures for each pair, you will see that their chiral carbons have opposite arrangements of their —H and —OH groups. The L

Carbohydrates Chapter 19

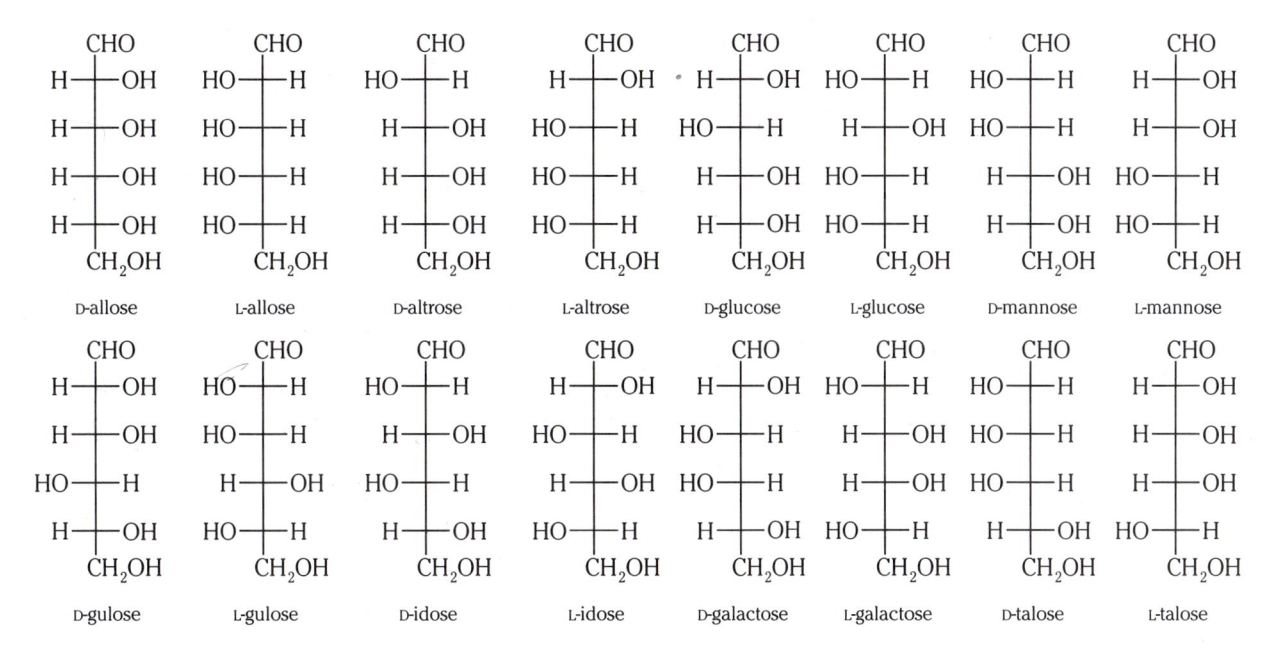

CHO	CHO	CHO	CHO	CHO	CHO	CHO	CHO
H—OH	HO—H	HO—H	H—OH	H—OH	HO—H	HO—H	H—OH
H—OH	HO—H	H—OH	HO—H	HO—H	H—OH	HO—H	H—OH
H—OH	HO—H	H—OH	HO—H	H—OH	HO—H	H—OH	HO—H
H—OH	HO—H	H—OH	HO—H	H—OH	HO—H	H—OH	HO—H
CH₂OH	CH₂OH	CH₂OH	CH₂OH	CH₂OH	CH₂OH	CH₂OH	CH₂OH
D-allose	L-allose	D-altrose	L-altrose	D-glucose	L-glucose	D-mannose	L-mannose

CHO	CHO	CHO	CHO	CHO	CHO	CHO	CHO
H—OH	HO—H	HO—H	H—OH	H—OH	HO—H	HO—H	H—OH
H—OH	HO—H	H—OH	HO—H	H—OH	HO—H	HO—H	H—OH
HO—H	H—OH	HO—H	H—OH	HO—H	H—OH	HO—H	H—OH
H—OH	HO—H	H—OH	HO—H	H—OH	HO—H	H—OH	HO—H
CH₂OH	CH₂OH	CH₂OH	CH₂OH	CH₂OH	CH₂OH	CH₂OH	CH₂OH
D-gulose	L-gulose	D-idose	L-idose	D-galactose	L-galactose	D-talose	L-talose

version looks like the D form reflected in a mirror. This relationship is like that of your right and left hand (Figure 19.13). Indeed, your right hand is a mirror image of your left hand, but your right and left hands do not have the same arrangement in space; you cannot fit them into the same glove, for example.

Figure 19.12 Fischer projection formulas for the 16 optical isomers of hexoses with an aldehyde group.

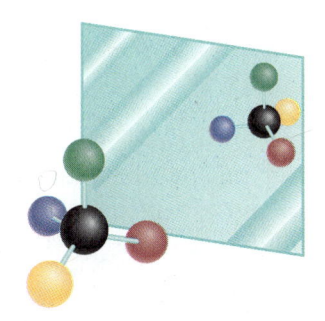

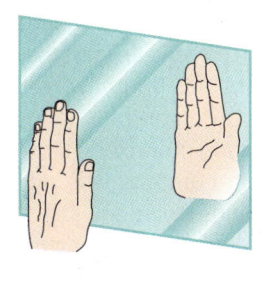

Figure 19.13 The enantiomer of a chiral compound is its mirror image (left) just as the right hand is the mirror image of the left hand (right). In both cases, the object and its mirror image do not have the identical arrangement in space and cannot be superimposed on each other.

A pair of optical isomers that are mirror images of each other but not identical in space, are called **enantiomers.** For example, the 16 isomers in Figure 19.12 consist of 8 pairs of enantiomers. Because a pair of enantiomers, such as D-mannose and L-mannose, have the same chemical and physical properties, they have the same name and are specified only by D or L.

Stereoisomers that are not enantiomers are called **diastereomers.** For example, D-glucose is a diastereomer of all the other compounds in Figure 19.12 except L-glucose, which is its enantiomer. Diastereomers have different properties and thus different names.

EXAMPLE 19.5
(a) Name (1) the enantiomer and (2) a diastereomer of L-mannose (Figure 19.12).
(b) Are the isomers separated by a dotted line in Figure 19.10 enantiomers or diastereomers?

SOLUTION
(a) (1) D-mannose, (2) any of the 14 substances in Figure 19.12 not having the name mannose; L-galactose, for example.
(b) enantiomers

D or L refers to two groups of compounds based on the orientation of —OH on the highest-numbered chiral carbon in the monosaccharide (carbon 5 for hexoses). If that —OH is on the left (as shown in standard open-chain formulas), the sugar is an L isomer. If that —OH is on the right, it is a D isomer. Examine Figure 19.12 to see how this pattern works.

EXAMPLE 19.6 Is the following pentose a D or L sugar?

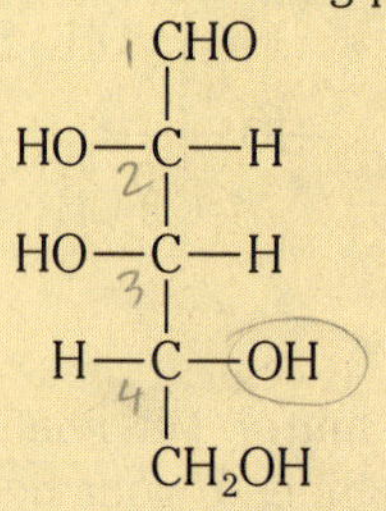

$$\begin{array}{c} \text{CHO} \\ | \\ \text{HO—C—H} \\ | \\ \text{HO—C—H} \\ | \\ \text{H—C—OH} \\ | \\ \text{CH}_2\text{OH} \end{array}$$

SOLUTION D (because —OH on carbon 4 is on the right)

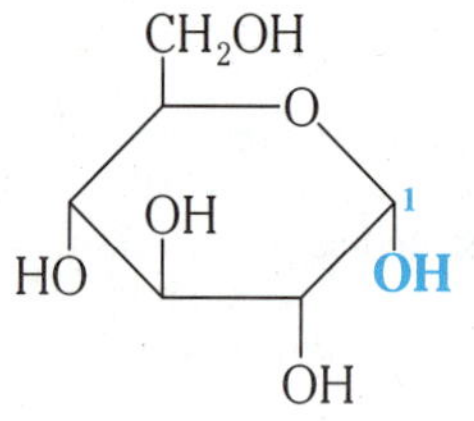

Since our bodies use only D sugars, we will not consider L sugars in the remainder of this book. When the letter isn't specified, a name such as galactose is understood to mean D-galactose.

Because of their mirror-image structures, the two members of a pair of enantiomers have an equal but opposite effect on plane-polarized light. One member causes plane-polarized light to rotate a certain amount to the right (clockwise); such a compound is said to be **dextrorotatory** (+). The other member rotates plane-polarized light the same amount to the left (counterclockwise); it is called **levorotatory** (−).

Because of the letters D and L, you might expect that all D sugars are dextrorotatory and all L sugars are levorotatory. This is *not* so. For example, D-fructose has the common name *levulose* because it is levorotatory; D-glucose, on the other hand, is sometimes called *dextrose* and is dextrorotatory. You cannot tell by its structure whether a D (or an L) sugar is dextrorotatory or levorotatory. The only way to tell is to measure the sugar's optical activity.

Carbohydrates have one more type of stereoisomerism. In their straight-chain structures (Figures 19.5 and 19.12), aldehyde hexoses have four chiral carbons (carbons 2–5). But cyclic monosaccharides have an additional chiral carbon. Compare the straight-chain structures of glucose, fructose, and galactose (Figure 19.5) with their cyclic forms (Figure 19.6). Notice that the nonchiral carbonyl carbon in the straight-chain form (carbon 1 for the aldehydes glucose and galactose; carbon 2 for the ketone fructose) is chiral in the cyclic form. This additional chiral carbon is called an *anomeric carbon,* and the two arrangements around it produce a pair of stereoisomers called **anomers.**

We use α and β to distinguish the two anomers. Figure 19.14 shows structures of α-D-glucose and β-D-glucose. For D sugars, the α isomer has —OH down

α-D-glucose

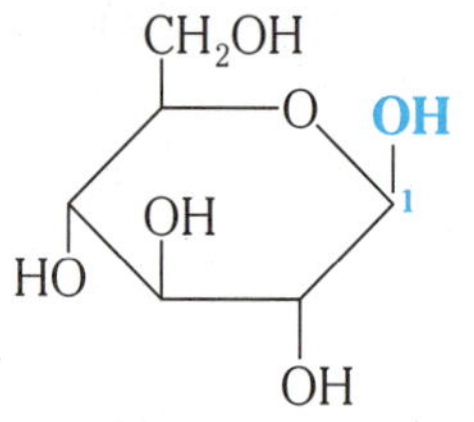

β-D-glucose

Figure 19.14 The two anomers of D-glucose differ in the orientation of —OH (in color) bonded to the anomeric carbon.

Because L sugars don't occur in the body, we won't be concerned with writing structural formulas for their anomers.

CHEMISTRY SPOTLIGHT

Optical Isomerism and Life

Living things specialize in optical isomers. George Wald (see Chemistry Spotlight in Section 13.3) wrote: "No other chemical characteristic is as distinctive of living organisms as is optical activity." Though D and L sugars have the same chemical properties, living organisms use the D version. You will see in Chapter 21 that amino acids also come in D and L varieties which have the same chemical properties. Yet here again, living things specialize; your body uses only L amino acids.

This specialization doesn't occur in a test tube. If you synthesize a compound in the laboratory that has one chiral carbon (and thus two different optical isomers), you are likely to make equal amounts of each isomer. Analyses of meteorites, for example, reveal the presence of amino acids, but the equal amounts of D and L forms suggest those compounds formed in space from conventional chemical reactions, not from living things.

When your body synthesizes a compound with a chiral carbon, you use special enzyme catalysts to make just one of the isomers. Why is that so? You will learn later that enzymes are chiral themselves. But how did your body come to have chiral catalysts, and to specialize in just D sugars and L amino acids when chemical reactions normally produce or consume equal amounts of both types of isomers?

No one really knows. A few theories point to special radiation or temperature conditions in which more of one isomer forms than the other. Once specialization began, we can imagine the development of catalysts to metabolize that type of isomer. Eventually, metabolism could become based only on the type of optical isomer first selected. But how it all got started remains a delightful mystery.

---

in the Haworth structures; β isomers have —OH projecting up. A pair of anomers are diastereomers, not enantiomers, so they do not have equal-but-opposite effects on plane-polarized light.

EXAMPLE 19.7

(a) Write Haworth structures for α-D-galactose, β-D-fructose, and β-D-ribose (see Figure 19.4).

(b) Is the structure shown for glucose in Figure 19.6 the α or β anomer?

418   Part III   Biochemistry

> **SOLUTION**
>
> (a)
>
> α-D-galactose   β-D-fructose   β-D-ribose
>
> (b) α anomer

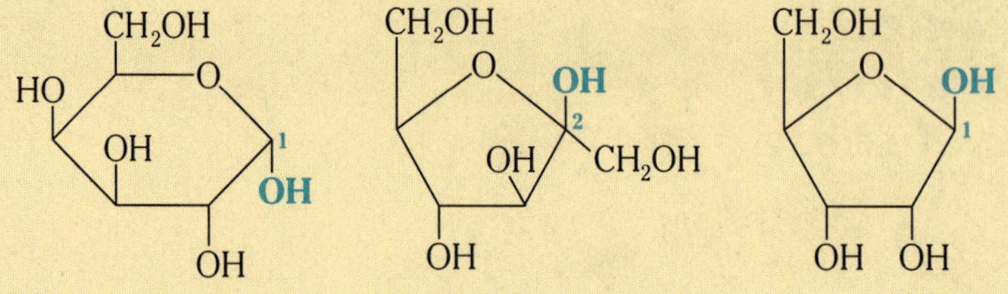

19.4 CHEMICAL REACTIONS OF CARBOHYDRATES

Oxidation. Although less than 1% of a monosaccharide in solution exists in the open-chain form, enough is there to give typical aldehyde or ketone reactions. Recall that aldehydes are oxidized by Tollens reagent (Section 15.4), reducing $Ag(NH_3)_2^+$ to metallic silver. Aldehydes are also oxidized in the Benedict and Fehling tests, reducing Cu (II) (as blue Cu^{2+}) to Cu (I) (as red Cu_2O; see Figure 15.12). Ketones with —OH bonded to the carbon next to the carbonyl carbon—such as fructose—also give positive Benedict and Fehling tests.

Sugars that give a positive Benedict or Fehling test are called **reducing sugars.** Reducing sugars must have at least a small amount of their open-chain form to provide an aldehyde (or ketone) group to react. Once the reaction begins, the equilibrium between the cyclic and open-chain forms (Figure 19.15) shifts to replenish the open-chain form that reacted, in accordance with

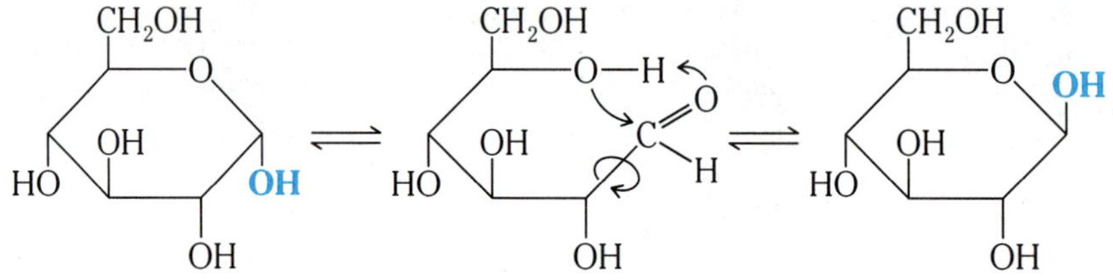

Figure 19.15 Both anomeric forms of glucose in solution are in equilibrium with a small amount (<1%) of the open-chain form, which reacts as a reducing sugar.

α-D-glucose ⇌ D-glucose (open-chain form) ⇌ β-D-glucose

Le Châtelier's principle (Section 8.5). The reaction continues until nearly all of the sugar is oxidized.

All monosaccharides, plus the disaccharides lactose and maltose, are reducing sugars. You will see in Section 19.5, however, that one common disaccharide (sucrose) and the polysaccharides are not reducing sugars because they don't provide a carbonyl group to react.

Mutarotation. A fresh solution of α-D-glucose has a specific rotation of +112°. But the specific rotation of the solution decreases steadily, finally stabilizing at +53°. A fresh solution of β-D-glucose, in contrast, has a specific rotation of +19°; the specific rotation slowly rises until it stabilizes at +53°, the same value that the α-D-glucose solution reached.

What is happening here? The cyclic hemiacetal forms of α- and β-glucose are in equilibrium with the open-chain aldehyde (Figure 19.15). Whether you start with a pure α or pure β solution, each with its own specific rotation, one anomer reacts to form the other until the system reaches equilibrium, which has about 63% in the β form, 37% in the α form, and less than 1% in the open-chain form. This mixture gives the final specific rotation of +53°.

The interconversion between anomers, measured by a change in optical activity, is called **mutarotation.** All monosaccharides exhibit mutarotation. So do the disaccharides lactose and maltose (Section 19.5), each of which has an anomeric carbon and exists in solution in α and β forms. The disaccharide sucrose and the polysaccharides lack anomeric carbons and thus do not mutarotate.

Formation of Acetals and Ketals. Recall that hemiacetals and hemiketals react with alcohols in the presence of an acid catalyst to form acetals and ketals, respectively (Section 15.4). For example, α-D-galactose reacts with methanol as follows:

α-D-galactose + CH$_3$OH $\stackrel{H^+}{\rightleftharpoons}$ methyl α-D-galactoside + H$_2$O

Sugar acetals are called **glycosides,** and are named with the suffix *-ide*. The bond from the anomeric carbon to the —OR group is called a **glycosidic bond.**

Notice that carbon 1 in methyl α-D-galactoside has the characteristic structure of acetals:

$$R-\underset{\underset{OR}{|}}{\overset{\overset{OR}{|}}{C}}-H$$

EXAMPLE 19.8 Write the Haworth structural formula for methyl β-D-glucoside.

SOLUTION

(β-glycosidic bond shown on Haworth structure)

When a hemiacetal or hemiketal sugar reacts with an alcohol group from another monosaccharide, the product is an acetal or ketal; it is also a disaccharide. Figure 19.16 shows the formation of three common disaccharides: maltose, lactose, and sucrose.

Once an anomeric carbon forms a glycoside, its ring no longer can open to provide a reactive aldehyde or keto group. As a result, those compounds aren't reducing sugars and do not mutarotate.

Examine the disaccharides in Figure 19.16. Notice that both anomeric carbons in sucrose (carbon 2 in fructose and carbon 1 in glucose) are joined in a

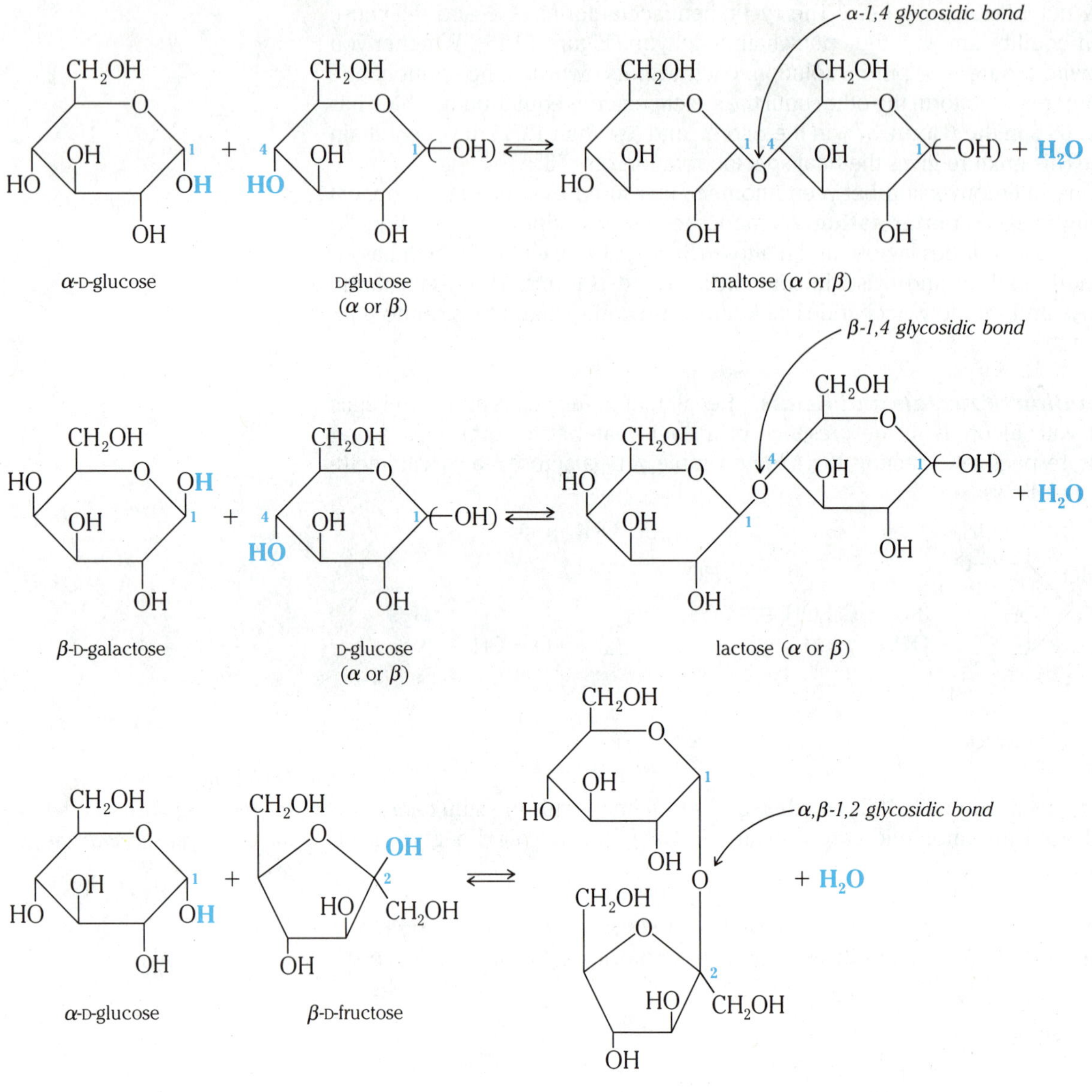

Figure 19.16 Formation of three common disaccharides.

glycosidic bond. Thus sucrose is not a reducing sugar and does not mutarotate. Maltose and lactose, in contrast, each have one anomeric carbon (carbon 1 in the glucose unit on the right) *not* joined in a glycosidic bond; therefore, that ring can open and these disaccharides form α and β isomers, undergo mutarotation, and are reducing sugars.

Notice also in Figure 19.16 that the reaction to form disaccharides is reversible. The reverse direction is a *hydrolysis* reaction (Section 15.4). Acids catalyze the hydrolysis of disaccharides (and polysaccharides) to produce their monosaccharide units. Your body, however, uses digestive enzymes to catalyze the hydrolysis of most complex carbohydrates that you eat (Chapter 25).

19.5 DISACCHARIDES AND POLYSACCHARIDES

Disaccharides. Table sugar is *sucrose*, a disaccharide made from glucose and fructose (Figure 19.16). Its glycosidic bond is oriented α with respect to glucose and β with respect to fructose, so we call it an α,β-1,2 glycosidic bond. Sucrose is abundant in sugar cane and sugar beets, and more widely produced than any other pure organic compound in the world. Its sweet taste is the standard for all sweeteners (Figure 19.8).

Lactose is made from joining carbon 1 of galactose to carbon 4 of glucose (Figure 19.16). The bond is β with respect to the anomeric carbon of galactose, so it is called a β-1,4 glycosidic bond. Lactose is milk sugar and makes up about 4–6% of cow's milk and 5–8% of human milk.

Maltose consists of two glucose units joined together in an α-1,4 glycosidic bond (Figure 19.16). It is produced in plants when seeds germinate, and occurs in malted milk and beer.

Polysaccharides. Polysaccharides are macromolecules made from monosaccharide units, usually glucose. The most common polysaccharides are starch, glycogen, and cellulose.

Glycogen and *starch* are made by joining hundreds, and usually thousands, of glucose units together by glycosidic bonds. Starch occurs in plants such as rice, potatoes, and wheat. Starch is a mixture of two polysaccharides. About 20–30% is amylose, a straight-chain polymer of dozens or hundreds of glucose units joined by α-1,4 glycosidic bonds (Figure 19.17). The remaining 70–80% of starch is amylopectin, a highly branched polymer. The branch points have α-1,6 glycosidic bonds, while linear glucose units join by α-1,4 bonds (Figure 19.17). Glycogen occurs in animals and is highly branched, with the same type of structure as amylopectin. It typically contains many thousands of glucose units per molecule.

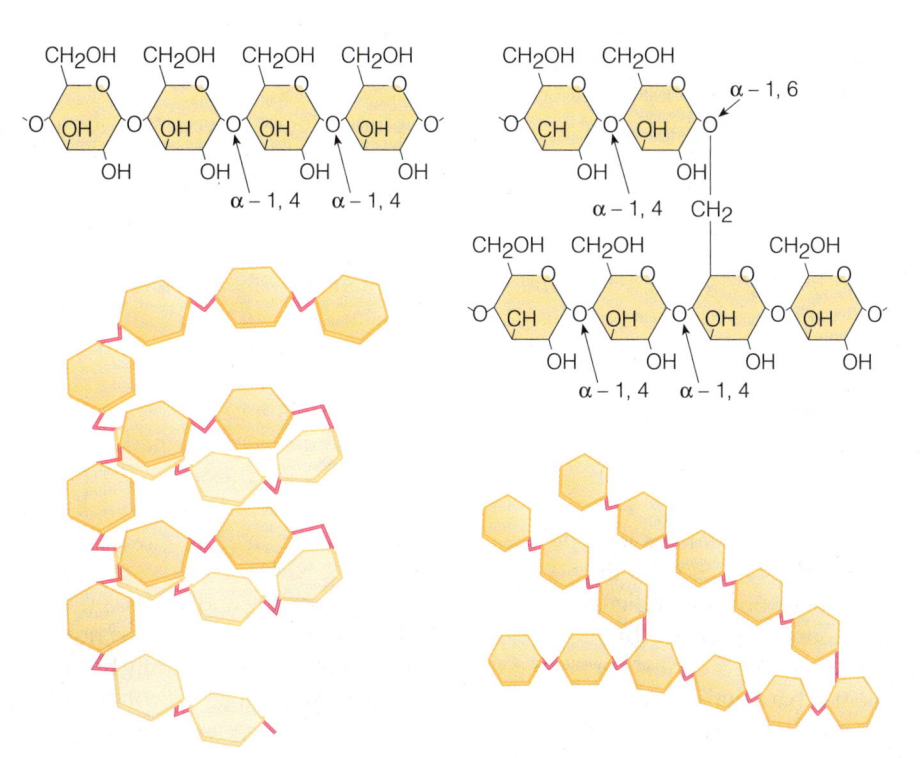

Figure 19.17 Amylose (left), a linear polymer of glucose containing α-1,4 glycosidic bonds, forms a coiled structure. Amylopectin (right) is a branched glucose polymer with branches at α-1,6 glycosidic bonds. Glycogen has a similar structure.

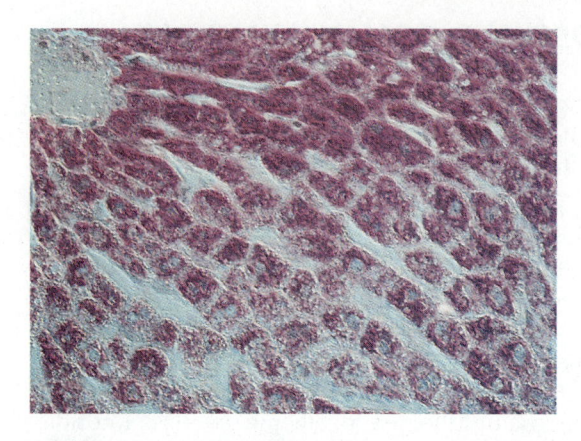

Figure 19.18 Glycogen granules are stored in human liver tissue.

Animals and plants store carbohydrate as dense, solid granules of glycogen and starch, respectively (Figure 19.18). You get much of your energy from eating starch and digesting it into glucose. Your body then converts extra glucose into glycogen and stores it, especially in your liver and muscle cells, for later use. At any given time, more than 90% of your glucose is in storage as glycogen.

When you need energy, your body hydrolyzes insoluble glycogen into soluble glucose. Muscle uses most of its glycogen for its own energy needs. The liver, however, uses its glycogen primarily to resupply the blood; this maintains blood glucose levels and provides organs such as the brain with a steady, constant energy supply. The liver is like a fuel storage tank for the body, with glycogen as the fuel.

Blood normally has 70–100 mg glucose per dL, which provides only about 170 kJ (40 Cal) of energy. You typically store 300–350 g of glycogen, which amounts to about 5500 kJ (1300 Cal). If you compare this with your daily energy needs (Section 2.4), you can see that you don't have enough carbohydrate in your body to meet the energy needs for even one full day. Most of your potential energy is stored as fat (Chapter 20).

Under conditions such as starvation or a low-carbohydrate diet, your body converts proteins into glucose to maintain blood glucose levels (see Section 25.2). But this causes acids and ketones to appear in blood and urine, an unpleasant condition called *ketoacidosis* (see Sections 15.3 and 26.2). To prevent this, you should include at least 100 g of carbohydrate per day in your diet.

Cellulose, the main structural material in plants, is the most abundant organic material in the world. Common sources are cotton, wood, and woody parts of plants. Cellulose is used for such things as paper, clothing (in linen, cotton, and rayon), furniture and buildings (wood), and even explosives (in the chemically modified forms guncotton and nitrocellulose).

Cellulose is a linear chain of 100–3000 glucose molecules joined by β-1,4 glycosidic bonds (Figure 19.19). The difference between these bonds and the α-1,4 bonds in glycogen and starch makes all the difference to someone trying to use polysaccharides for food. Your digestive enzymes hydrolyze α-1,4 bonds to produce glucose, but they don't break β-1-4 bonds.

Since you cannot digest cellulose, foods such as celery and lettuce provide necessary "roughage" but no kilojoules (or Calories). Animals such as horses, cows, and goats can use grass and other plant materials for food only because special bacteria in their digestive tracts make enzymes that hydrolyze β-1,4 glycosidic bonds. One idea, still in the experimental stage, is to isolate large

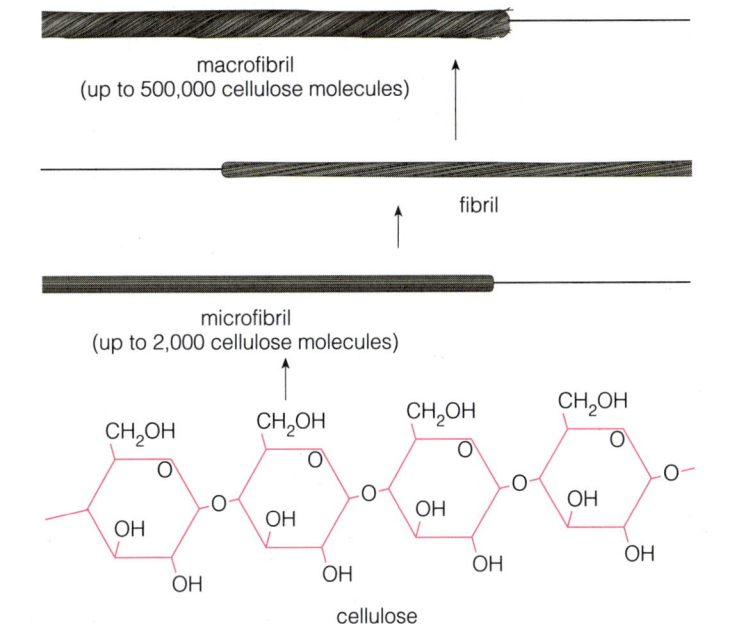

Figure 19.19 Cellulose molecules form hydrogen bonds to assemble into fine strands (microfibrils), which twist together into threadlike fibrils and macrofibrils. Macrofibrils are as strong as steel thread of the same thickness.

amounts of those bacterial enzymes and use them commercially to convert cellulose into edible glucose.

Except for a very few units at the ends of chains, glucose units in polysaccharides have their anomeric carbons tied up in glycosidic bonds. So polysaccharides are not reducing sugars and don't mutarotate.

Polysaccharides also don't dissolve well in water. One reason is that macromolecules are harder for water molecules to keep suspended. Another reason is that polar alcohol groups tend to form hydrogen bonds with each other within the molecule, making them unavailable to form hydrogen bonds with water. But it's probably for the best. Cellulose, the main structural material in wood, wouldn't do a tree much good if it dissolved in the rain.

> I never met a carbohydrate I didn't like.
>
> *David Lygre (with apologies to Will Rogers)*

CHEMISTRY SPOTLIGHT

John Dillinger and the Potato Pistol

Knowing a little chemistry can be useful to anyone, even criminals. For example starch can be detected chemically by the *iodine test,* in which iodine (I_2) solution reacts with starch to form a dark blue or violet complex of iodine trapped within starch molecules. (As starch is hydrolyzed into smaller units, it gradually loses its ability to react with iodine in this way.)

While serving a jail sentence in the 1930s, notorious American gangster John Dillinger allegedly put this chemical knowledge to dramatic use. First he asked for a potato. Then he deliberately cut his finger, and asked for iodine solution to use as a disinfectant on the cut. After carving the potato into the shape of a pistol, he painted it with iodine until it resembled blue steel. Then he called a guard and—using his potato pistol—made one of the most famous "chemical escapes" in history.

SUMMARY

Carbohydrates are aldehydes and ketones with multiple alcohol groups. The most important monosaccharides in the body are the pentoses ribose and 2-deoxyribose, and the hexoses glucose, fructose, and galactose. Monosaccharides occur as cyclic hemiacetals or hemiketals, which can be represented by Haworth structures.

Carbohydrates are optical isomers, a type of stereoisomer. They contain chiral carbon atoms and rotate plane-polarized light. Those that rotate plane-polarized light to the right (clockwise) are dextrorotatory (+); those that rotate it to the left (counterclockwise) are levorotatory (−). The maximum number of stereoisomers for a compound is 2^n, where *n* is the number of chiral carbons.

Carbohydrate isomers consist of pairs of enantiomers that are mirror images of each other but not identical (superimposable) in space. The letters D or L refer to a family of sugars based on the orientation of —OH on the highest-numbered chiral carbon. The chiral carbon formed when the open chain becomes a ring is the anomeric carbon; isomers that differ only in their configuration around this carbon are anomers. Mutarotation is the change in specific rotation as anomers interconvert until they reach equilibrium.

All monosaccharides are reducing sugars; they give positive Benedict and Fehling tests and are oxidized. They all mutarotate and form acetals or ketals by reaction with an alcohol group. When the alcohol group comes from another monosaccharide molecule, the product is a disaccharide.

The most common disaccharides are sucrose, maltose, and lactose. Sucrose (table sugar) is made from glucose joined to fructose by an α,β-1,2 glycosidic bond. It is not a reducing sugar and doesn't mutarotate. Lactose (galactose joined to glucose by a β-1,4 glycosidic bond) and maltose (two glucose units joined by an α-1,4 glycosidic bond) are reducing sugars and mutarotate.

The most common polysaccharides are glycogen, starch, and cellulose. All are polymers of glucose. Glycogen has α-1,4 and α-1,6 glycosidic bonds and is highly branched. It is the main storage form for carbohydrates in animals. Starch has the same glycosidic bonds as glycogen and is the storage form for carbohydrates in plants. Cellulose, the main structural material in plants, is a linear polymer of glucose joined by β-1,4 glycosidic bonds. Your body has enzymes to hydrolyze starch and glycogen—but not cellulose—to glucose.

KEY TERMS

Anomer (19.3)
Biochemistry (19.1)
Carbohydrate (19.1)
Chiral Carbon (19.3)
Dextrorotatory (19.3)
Diastereomer (19.3)
Disaccharide (19.1)
Enantiomer (19.3)
Fischer projection formula (19.3)
Glycoside (19.4)
Glycosidic bond (19.4)
Haworth structure (19.2)
Levorotatory (19.3)
Monosaccharide (19.1)
Mutarotation (19.4)
Optical activity (19.3)
Optical isomer (19.3)
Polysaccharide (19.1)
Reducing sugar (19.4)
Stereoisomer (19.3)

EXERCISES

Even-numbered exercises are answered at the back of this book.

Monosaccharides

1. Which of the following could be classified as a carbohydrate in terms of being an aldehyde or ketone with multiple alcohol groups and having a $C_x(H_2O)_y$ formula?

    glyceraldehyde   glycerol

    acetone   dihydroxyacetone

2. Are the carbohydrates in exercise 1 (a) pentoses, (b) trioses, (c) hexoses, or (d) tetroses?

3. Name the compounds in exercise 1 that have at least one chiral carbon.

4. In exercise 1, is glyceraldehyde written as a D or L compound?

5. Draw an open-chain structure for a tetrose aldehyde.

6. Why are five- and six-member rings the most common rings for carbohydrates?

7. Would glyceraldehyde (see exercise 1) be likely to occur as a cyclic hemiacetal? Why?

8. Name a hexose that forms a hemiketal.

9. Should the absence of an oxygen bonded to carbon 2 prevent 2-deoxyribose (Figure 19.4) from forming a hemiacetal ring structure?

10. From its open-chain structure (Figure 19.4), write the Haworth structure for the α anomer of D-ribose. [The ring forms by reaction of the carbonyl group with the —OH on carbon 4.]

11. From their structures shown in Figure 19.12, write Haworth structures for (a) α-D-altrose, (b) β-D-idose, and (c) β-D-gulose.

12. In Figure 19.6, which anomer is shown for (a) glucose, (b) fructose, and (c) galactose?

13. Identify the number of the anomeric carbon in each of the following: (a) galactose, (b) fructose, (c) ribose, and (d) glucose.

14. Look at the structure of fructose in Figure 19.6. Write a Haworth structure for its anomer.

15. Describe the actual three-dimensional shape of the glucose ring.

16. The sweetest common monosaccharide is _____.

17. The most abundant monosaccharide in your body is _____.

Optical Isomers

18. Are *cis*-2-pentene and *trans*-2-pentene optical isomers? Explain.
19. Does either α-chloropropionic acid or β-chloropropionic acid consist of optical isomers? Explain.
20. What is the specific rotation for a substance if a 1.74 g/mL solution rotates plane-polarized light (with a light path of 10.0 cm) 24.3° counterclockwise?
21. If a substance has a specific rotation of +34.6°, how much (and in which direction) will a 212 mg/mL solution rotate plane-polarized light with a path length of 1.50 dm?
22. The specific rotation of α-D-galactose is +190.7°. Is this sugar dextrorotatory or levorotatory? Is β-D-galactose dextrorotatory or levorotatory?
23. Is the xylose shown in Figure 19.4 D-xylose or L-xylose?
24. Draw a straight-chain structure for the enantiomer of xylose shown in Figure 19.4. Name the enantiomer.
25. From the structure of 2-deoxyribose in Figure 19.4, draw a Fischer projection formula for this compound.
26. Identify the following pairs of compounds as anomers, enantiomers, diastereomers, or none of these: (a) D-glucose and D-galactose, (b) α-D-glucose and β-D-galactose, (c) D-fructose and L-fructose, (d) D-ribose and 2-deoxy-D-ribose, (e) α-D-mannose and β-D-mannose
27. A pair of compounds that have the same specific rotation but in opposite directions are (a) optical isomers, (b) anomers, (c) diastereomers, (d) stereoisomers, (e) enantiomers.
28. Which of the compounds in exercise 1 would be expected to be optically active?
29. Does 2-chlorobutane consist of stereoisomers that are optically active?
30. Does β-D-fructose in its cyclic form have the same number of chiral carbons as β-D-glucose in its cyclic form?
31. D-glucose is dextrorotatory. Identify which of the following statements, then, are necessarily true:
    (a) D-fructose is dextrorotatory.
    (b) A solution of D-glucose rotates plane-polarized light in a counterclockwise direction.
    (c) L-glucose is levorotatory.
    (d) D sugars are dextrorotatory and L sugars are levorotatory.
32. D-glucose and D-mannose differ only in the orientation of —H and —OH on carbon 2. Therefore, these two compounds are: (a) enantiomers, (b) diastereomers, (c) anomers, (d) none of the above.

Disaccharides and Polysaccharides

33. A disaccharide that does not mutarotate is _____.
34. Look at the structure of lactose in Figure 19.16. Identify which of the following functional groups are present in lactose: (a) hemiacetal, (b) hemiketal, (c) acetal, (d) ketal, (e) glycoside.
35. Write the structural formula for α-lactose.
36. Does α-lactose contain an α-1,4 or a β-1,4 glycosidic bond?
37. Name the type of glycosidic bonds that occur at branch points in glycogen.
38. Which common disaccharide is not a reducing sugar? Explain why.
39. Do all carbohydrates taste sweet?

Reactions of Carbohydrates

40. Suppose an unknown carbohydrate gives a positive reducing sugar test. Would you expect it to mutarotate? Why?
41. Are reducing sugars reduced? Explain.
42. Name the products from the acid-catalyzed hydrolysis of (a) sucrose and (b) maltose.
43. If you performed a Benedict test on milk (which contains lactose), what would you expect to observe?
44. Saliva contains an enzyme that hydrolyzes some of the α-1,4 glycosidic bonds in starch. Before treatment with saliva, does starch give a positive iodine test? Is it a reducing sugar? After lengthy treatment with saliva, how does starch react in these tests?
45. After treatment with saliva (see exercise 44), does cellulose give a positive test as a reducing sugar? Explain.
46. A disaccharide consisting of two glucose units joined by an α-1,6 glycosidic bond is called isomaltose. Its structure is:

    (a) Does isomaltose exist in anomeric forms?
    (b) Is isomaltose a reducing sugar?
    (c) Does this structure occur within the structure of cellulose, amylose, or amylopectin?

DISCUSSION EXERCISES

1. Why is it necessary to study organic chemistry before studying biochemistry?
2. Calculate the number of different stereoisomers for pentoses with an aldehyde group. Describe how these isomers are related in terms of being **(a)** diastereomers, **(b)** enantiomers, and **(c)** anomers of each other.
3. Suppose you replaced 30 g of carbohydrate per day with a noncaloric, artificial sweetener. How much weight reduction (in kg) would this accomplish in one month? [Review Section 2.4 if necessary.]
4. Suppose a newborn baby is at risk for galactosemia, a genetic disease characterized by high levels of galactose in blood and urine. What tests, mentioned in this chapter, could you use to help diagnose galactosemia? How certain would the diagnosis be? Could you distinguish between high levels of galactose and high levels of glucose? Explain.

20

Lipids

1. What are the main lipids in the body?
2. What are the names, structures, and common reactions of fatty acids?
3. What are the structures, functions, and reactions of fats and oils?
4. What are phosphorus- and sphingosine-containing lipids, and what are their roles in the body?
5. What are steroids? What are some important steroids and their functions?
6. What are the structures and functions of other lipids such as lipoproteins, prostaglandins, fat–soluble vitamins and terpenes?
7. What is the role of lipids in cell membranes?

Anabolic Steroids

One reason men tend to be larger than women is that they have higher levels of testosterone. This hormone stimulates the body to synthesize protein, the main material in muscle. Testosterone is a *steroid*, a lipid material containing four carbon rings fused together (Figure 20.1).

Anabolic steroids are hormones that increase the size, strength, and stamina of muscles. Some weight lifters, bodybuilders, football players, sprinters, and other track and field athletes use these steroids illegally. One study found that nearly 7% of 12th-grade males in the United States use anabolic steroids. Of those students, about half use steroids to improve athletic performance; one-fourth use them just to look more muscular.

Testosterone is an anabolic steroid; so are many synthetic compounds (Figure 20.1). Because they are nonpolar, these hormones stay in the body's fatty tissues for months. The most nonpolar steroids, such as nandrolone decanoate (Deca-Durabolin; Figure 20.1), are taken by injection and can be detected in urine 6–12 months after use. Oxandrolone (Anavar) is slightly more polar and taken orally; it can be detected up to ten weeks after use. Testosterone injections also are detected by elevated levels in urine.

People who take anabolic steroids do increase muscle mass. The user may look better, but actually is less healthy. These drugs alter the normal hormone balance in the body. In men, they can cause aggressiveness, irritability, reduced fertility, undersized testicles, and enlarged breasts. In women, they can cause menstrual irregularities and masculinizing effects such as a deeper voice and male-pattern balding. These drugs also cause temporary acne and increase the risk of liver tumors as well as heart and kidney disease.

testosterone

nandrolone decanoate (Deca-Durabolin)

oxandrolone (Anavar)

Figure 20.1 Testosterone and two synthetic anabolic steroids. The long hydrocarbon chain (in color) makes Deca-Durabolin even more nonpolar.

Part III Biochemistry

you may think of lipids mostly as things to avoid in your diet, fatty compounds that cause obesity and heart disease. But you couldn't stay alive and healthy without lipids. Fat typically provides 10–30% of your body weight, depending on age, sex, weight, and physical activity. In addition to energy, lipids provide you with necessary vitamins and hormones and are critical components of cell membranes.

Lipids is a catchall term for natural organic substances that dissolve in nonpolar solvents. They include fats, oils, the orange and gold pigments in autumn leaves, whale blubber, sex hormones, beeswax, cholesterol, and vitamin D. Though their structures vary considerably, lipids have one thing in common: they dissolve in nonpolar solvents. Since like dissolves like (Section 7.2), you can conclude that lipids themselves are nonpolar. As you look at structures of various lipids in this chapter, notice that they abound in nonpolar, hydrocarbon material.

20.1 FATS AND OILS

Fatty Acids. Many lipids are made from fatty acids esterified with another substance. **Fatty acids** are carboxylic acids containing a long hydrocarbon chain (usually 12–20 carbons). They are just like other carboxylic acids, except that the long hydrocarbon chain makes them essentially nonpolar—despite their polar carboxylic acid group.

Table 20.1 lists the most common fatty acids. Notice that they all have an even number of carbon atoms, and that the double bonds in polyunsaturated fatty acids are spaced at three-carbon intervals.

Saturated fatty acids have higher melting points than unsaturated ones and are solids at room temperature (Table 20.1). Natural unsaturated fatty acids

Fatty acids have an even number of carbons because cells synthesize fatty acids in two-carbon segments (see Chapter 26).

Table 20.1 Common Fatty Acids

Name Common (IUPAC)	Carbon Atoms	Double Bonds (Position)*	Condensed Structural Formula	Melting Point (°C)
lauric (dodecanoic)	12	0	$CH_3(CH_2)_{10}COOH$	44
myristic (tetradecanoic)	14	0	$CH_3(CH_2)_{12}COOH$	58
palmitic (hexadecanoic)	16	0	$CH_3(CH_2)_{14}COOH$	63
palmitoleic (hexadecenoic)	16	1 (9)	$CH_3(CH_2)_5CH=CH(CH_2)_7COOH$	−1
stearic (octadecanoic)	18	0	$CH_3(CH_2)_{16}COOH$	71
oleic (octadecenoic)	18	1 (9)	$CH_3(CH_2)_7CH=CH(CH_2)_7COOH$	16
linoleic (octadecadienoic)	18	2 (9, 12)	$CH_3(CH_2)_4CH=CHCH_2CH=CH(CH_2)_7COOH$	−5
linolenic (octadecatrienoic)	18	3 (9, 12, 15)	$CH_3CH_2CH=CHCH_2CH=CHCH_2CH=CH(CH_2)_7COOH$	−11
arachidonic	20	4 (5, 8, 11, 14)	$CH_3(CH_2)_4CH=CHCH_2CH=CHCH_2CH=CHCH_2CH=CH(CH_2)_3COOH$	−50

*Number in parentheses is for the lower-numbered carbon in the double bond; for example, (9) means the double bond is between carbons 9 and 10, numbering from the carboxyl end.

have a *cis* arrangement at each double bond, which puts a bend in the hydrocarbon chain (Figure 20.2). Just as untrimmed trees cannot pack as closely together as trimmed ones, the bent, unsaturated fatty acids cannot pack as closely together as saturated ones. So they are in a less dense, liquid state at room temperature.

$CH_3(CH_2)_{16}COOH$
stearic acid

$$CH_3(CH_2)_7 \overset{H}{\underset{}{\diagdown}} C=C \overset{H}{\underset{(CH_2)_7COOH}{\diagup}}$$
oleic acid

Figure 20.2 The double bond *(cis)* is the only difference in structure between solid stearic acid (left) and liquid oleic acid (right).

You will learn in later chapters that the body synthesizes many lipids from materials such as carbohydrates and amino acids. The only lipids you must have in your diet are the fat-soluble vitamins (see Section 20.6) and two polyunsaturated fatty acids—linoleic acid and linolenic acid (Table 20.1). These are called the **essential fatty acids.** Your body cannot synthesize enough of these fatty acids, required for cell membranes and to synthesize hormones called prostaglandins (see Section 20.6).

Waxes. **Waxes** are esters of fatty acids with nonpolar alcohols. One component of beeswax, for example, is an ester of palmitic acid and a C_{30} alcohol. Its structure is:

$$CH_3(CH_2)_{14}-\overset{\overset{O}{\|}}{C}-O-(CH_2)_{29}CH_3$$

In lanolin, an ester from lamb's wool, the alcohol group is furnished by a type of lipid called a *steroid* (see Section 20.4).

Waxes coat the leaves of many plants, protecting them against dehydration and infectious microbes. A plug of wax helps shield your inner ear from outside damage. And many animals have a slick, protective coating on their feathers, fur, or skin. When the nonpolar coating on the feathers of ducks and other birds dissolves, as occurs in oil spills, they cannot keep themselves warm and often drown. We use natural waxes such as lanolin in soaps and skin lotions, and carnauba wax (from a Brazilian palm tree) for polishes.

Fats and Oils. Molecules containing glycerol esterified with three fatty acid molecules are called **triglycerides** or **triacylglycerols.** The general reaction is:

$$\begin{array}{c}CH_2-OH \\ CH-OH \\ CH_2OH\end{array} + \begin{array}{c}RCOOH \\ R'COOH \\ R''COOH\end{array} \underset{}{\overset{H^+}{\rightleftharpoons}} \begin{array}{c}CH_2-O-\overset{O}{\underset{\|}{C}}-R \\ CH-O-\overset{O}{\underset{\|}{C}}-R' \\ CH_2-O-\overset{O}{\underset{\|}{C}}-R''\end{array} + 3H_2O$$

glycerol   fatty acids   triglyceride

Here R, R', and R'' represent the same or different fatty acids. When just one or two of the alcohol groups in glycerol are esterified with fatty acids, the product is a monoglyceride (monoacylglycerol) or diglyceride (diacylglycerol), respectively.

Fat is triglyceride material from animals. Because of its high content of saturated fatty acids, fat is solid at room temperature. Vegetable **oil,** in contrast, is liquid triglyceride material from plants that is highly unsaturated or contains shorter-chain fatty acids. Fats and oils are mixtures of many triglyceride molecules, each with its own combination of fatty acids. Table 20.2 shows how the types of fatty acids vary in some animal fats and vegetable oils.

Fat insulates you, cushions you against mechanical injury, and provides your main storehouse of energy. When oxidized to CO_2 and H_2O, fat provides

You store enough energy as glycogen to last less than a day, but most people can survive six to eight weeks on the energy stored in their fat.

Table 20.2 Average Fatty Acid Composition and Iodine Number of Various Fats and Oils

	\multicolumn{8}{c	}{Composition of Fatty Acids (%)}	Iodine
	Lauric	Myristic	Palmitic	Stearic	Palmitoleic	Oleic	Linoleic	Linolenic	Other	Number
Animal Fats										
Butter	2	11	29	9	5	27	4		13	36
Beef tallow		6	27	14		50	2		1	50
Lard		1	28	12	3	47	6		3	59
Human		3	24	8	5	47	10		3	68
Marine Oils										
Whale		9	16	3	14	35			23	120
Herring		7	13		5			21	54	140
Cod liver		6	8	1	20	←— 29 —→			36	165
Vegetable Oils										
Coconut	45	18	10	2		8			17	10
Corn		1	10	3	2	50	34			123
Linseed		6	3			19	24	47	1	179
Olive		7	2			84	5		2	81
Peanut		8	3			56	26		7	195
Safflower	←——— 7 ———→					19	70	3	1	145
Soybean			10	2		29	51	7	1	130
Sunflower seed		6	2			25	66		1	126
Wheat germ	←——— 16 ———→					28	52	4		125

CHEMISTRY SPOTLIGHT

Fats and Isomers

Fatty acids in natural fats and oils have a *cis* arrangement at their double bonds (Figure 20.2). When you eat meat or vegetables, your body metabolizes these *cis* fatty acids.

Partial hydrogenation of polyunsaturated oils (Figure 13.11) produces triglycerides whose fatty acids are more saturated, and also converts some of the remaining double bonds from a *cis* to a *trans* arrangement. These synthetic products are important ingredients in shortening, margarine, and commercial frying fats. Some studies show that *trans* fatty acids (unlike *cis* ones) increase blood cholesterol levels—especially LDL (low-density lipoprotein) cholesterol, the type most associated with an increased risk of heart attacks. They also decrease HDL cholesterol levels. Additional studies are in progress to determine the significance of these findings.

---

about 38 kJ (9 Cal) per g while carbohydrates and proteins provide only about 17 kJ (4 Cal) per g. This is because fats contain a large amount of hydrocarbon material in a highly reduced state; carbohydrates, in contrast, have oxygen-containing alcohol and carbonyl groups and are already partially oxidized. Since your body has to oxidize fats more extensively than carbohydrates (and proteins) before forming CO_2 and H_2O, fats produce more energy.

20.2 CHEMICAL REACTIONS OF TRIGLYCERIDES AND FATTY ACIDS

Hydrolysis of Triglycerides. Esters react with water in the presence of an acid (or base) to produce their parent alcohol and acid (or acid salt) (see Section 16.7). The reaction for acid-catalyzed hydrolysis of a triglyceride is:

Notice that this hydrolysis reaction is the reverse of the reaction shown for the synthesis of a triglyceride.

$$\begin{array}{c} CH_2-O-\overset{O}{\underset{\|}{C}}-R \\ | \\ CH-O-\overset{O}{\underset{\|}{C}}-R' \\ | \\ CH_2-O-\overset{O}{\underset{\|}{C}}-R'' \end{array} + 3H_2O \underset{}{\overset{H^+}{\rightleftharpoons}} \begin{array}{c} CH_2-OH \\ | \\ CH-OH \\ | \\ CH_2-OH \end{array} + \begin{array}{c} RCOOH \\ R'COOH \\ R''COOH \end{array}$$

  triacylglycerol   glycerol   fatty acids

When your body must use fat for energy, enzyme catalysts hydrolyze the stored fat. The resulting fatty acids and glycerol enter your bloodstream and travel to muscles and other organs to serve as fuel.

Base-catalyzed hydrolysis of an ester is called *saponification* and produces the neutralized salts of the acids (see Section 16.7). The reaction with a triglyceride is:

$$\begin{array}{c} \text{CH}_2\text{—O—C(=O)—}R \\ | \\ \text{CH—O—C(=O)—}R' \\ | \\ \text{CH}_2\text{—O—C(=O)—}R'' \end{array} + 3\,\text{NaOH} \longrightarrow \begin{array}{c} \text{CH}_2\text{OH} \\ | \\ \text{CHOH} \\ | \\ \text{CH}_2\text{OH} \end{array} + \begin{array}{c} R\text{—COO}^-\text{Na}^+ \\ R'\text{—COO}^-\text{Na}^+ \\ R''\text{—COO}^-\text{Na}^+ \end{array}$$

triacylglycerol   *glycerol*   *fatty acid salts*

Fatty acid salts are effective detergents (see Figure 16.19), and saponifying animal fat with lye (NaOH) has long been a way to make soap (Figure 20.3).

Figure 20.3 Making soap by saponifying animal fat with lye (NaOH).

Catalytic Hydrogenation. Alkenes and alkynes undergo catalytic hydrogenation in the presence of a catalyst such as Pt, Pd, or Ni (see Section 13.4). This reaction saturates some double bonds in vegetable oils and also converts a few *cis* into *trans* arrangements (see Chemistry Spotlight). The reaction produces a soft solid used as shortening.

Iodine Number. Alkenes readily participate in addition reactions (Section 13.4). The reaction with iodine is:

$$R\text{—CH}=\text{CH—}R' + I_2 \longrightarrow R\text{—CHI—CHI—}R''$$

Since I_2 itself isn't very reactive, the actual reaction is done with ICl or IBr and then calculated for an equivalent amount of I_2.

A common way to measure the unsaturation in a fat or oil is to determine its **iodine number,** the number of grams of I_2 that react with 100.0 g of triglyceride. You can calculate, for example, the iodine number for a triglyceride esterified to one molecule each of palmitoleic acid, stearic acid, and oleic acid:

$$\begin{matrix} CH_2-O-\overset{O}{\underset{\|}{C}}(CH_2)_7CH=CH(CH_2)_5CH_3 \\ CH-O-\overset{O}{\underset{\|}{C}}(CH_2)_{16}CH_3 \qquad + 2I_2 \\ CH_2-O-\overset{O}{\underset{\|}{C}}(CH_2)_7CH=CH(CH_2)_7CH_3 \end{matrix}$$

$$\longrightarrow \begin{matrix} CH_2-O-\overset{O}{\underset{\|}{C}}(CH_2)_7CHICHI(CH_2)_5CH_3 \\ CH-O-\overset{O}{\underset{\|}{C}}(CH_2)_{15}CH_3 \\ CH_2-O-\overset{O}{\underset{\|}{C}}(CH_2)_7CHICHI(CH_2)_7CH_3 \end{matrix}$$

Molar masses: triglyceride ($C_{55}H_{102}O_6$)
$$= 55(12.01) + 102(1.01) + 6(16.00) = 859.57 \text{ g}$$
$$I_2 = 2(126.9) = 253.8 \text{ g}$$

$$100.0 \text{ g triglyc} \times \frac{1 \text{ mol triglyc}}{859.57 \text{ g triglyc}} \times \frac{2 \text{ mol } I_2}{1 \text{ mol triglyc}} \times \frac{253.8 \text{ g } I_2}{1 \text{ mol } I_2}$$
$$= 59.05 \text{ g } I_2 \text{ (iodine number)}$$

The iodine number is 0 for a saturated fat because no I_2 reacts. The iodine number increases as the unsaturation increases (Table 20.2).

EXAMPLE 20.1 A sample of triglyceride containing only oleic acid has an iodine number of 87. Predict the iodine number for triglycerides containing only (a) linoleic acid, (b) stearic acid, (c) linolenic acid, and (d) equal amounts of stearic, linoleic, and linolenic acid.

SOLUTION (a) 174 (twice as many double bonds as in oleic acid, so twice the iodine number), (b) 0, (c) 261, (d) 145

20.3 PHOSPHOLIPIDS AND SPHINGOLIPIDS

Phosphoglycerides. Lipids containing phosphate esters are called **phospholipids**. Although phosphate is ionic, it is part of many molecules that are lipids because the rest of the molecule is so nonpolar. In phospholipids, phosphate is esterified either to glycerol or to sphingosine. Those esterified to glycerol are called **phosphoglycerides**.

Phosphoglycerides are diglycerides containing phosphate esterified to an end carbon of glycerol (Figure 20.4). The phosphate group is also esterified to another group, such as choline, serine, or ethanolamine.

To name a phosphoglyceride, identify the group esterified to phosphate and precede its name with *phosphatidyl*. The choline compound, for example, is called phosphatidyl choline. It is also known as *lecithin*.

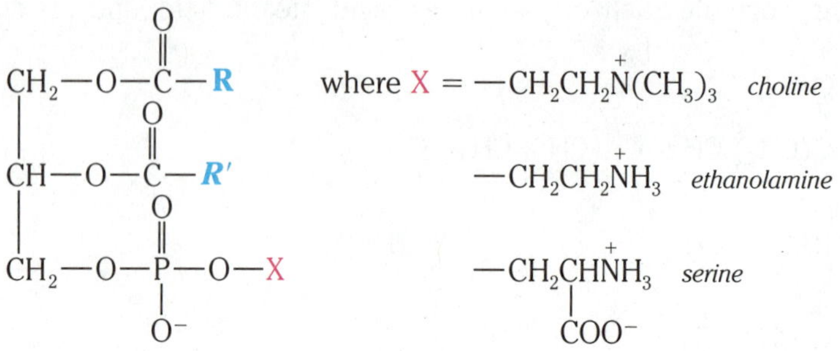

Figure 20.4 General structure of phosphoglycerides. Nonpolar regions are in blue; the remainder is polar or ionic.

$$\begin{array}{l} CH_2-O-\overset{\overset{O}{\|}}{C}-R \\ \overset{O}{\|} \\ CH-O-\overset{\|}{C}-R' \\ \overset{O}{\|} \\ CH_2-O-\overset{\|}{P}-O-X \\ \underset{O^-}{} \end{array}$$

where $X = -CH_2CH_2\overset{+}{N}(CH_3)_3$ choline

$-CH_2CH_2\overset{+}{N}H_3$ ethanolamine

$-CH_2\overset{}{C}H\overset{+}{N}H_3$
$|$
COO^- serine

You can see from their general structure (Figure 20.4) that phosphoglycerides have both nonpolar regions (long-chain hydrocarbon *R* groups) and polar or ionic groups (phosphate and choline, serine, or ethanolamine). These dual qualities enable phosphoglyceride detergents to fit at the boundaries between polar and nonpolar materials (Figure 16.19). In Section 20.7, you will see that phosphoglycerides are an important component of cell membranes, which provide boundaries to the watery interiors of cells.

Sphingolipids. The parent compound for another group of lipids, called **sphingolipids,** is sphingosine:

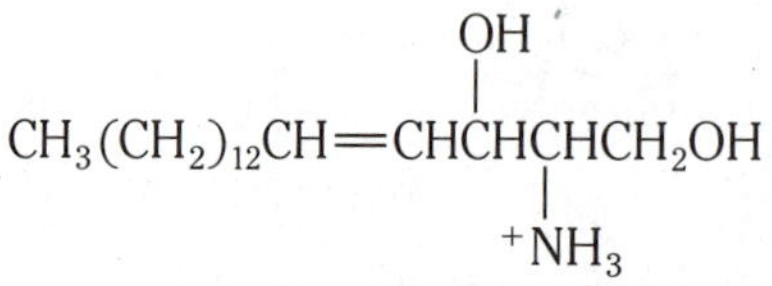

$$CH_3(CH_2)_{12}CH=CHCHCHCH_2OH$$
with OH on third carbon from right and $^+NH_3$ on second carbon from right.

In all sphingolipids, the amino group in sphingosine forms an amide bond with a fatty acid. But three different types of sphingolipids—sphingomyelins, cerebrosides, and gangliosides—differ in what bonds to the end alcohol group in sphingosine. Of these, only sphingomyelins contain phosphorus and can be classified as phospholipids.

Sphingomyelins have a phosphate and choline group esterified to the alcohol group in sphingosine (Figure 20.5). Sphingomyelins are the simplest and

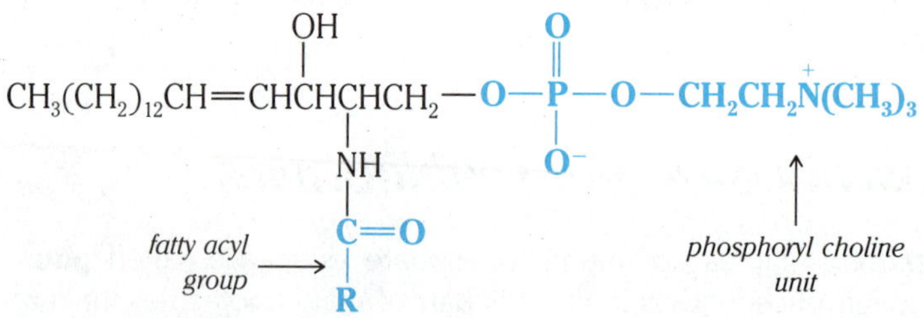

Figure 20.5 Structure of sphingomyelin.

$$CH_3(CH_2)_{12}CH=CHCHCHCH_2-O-\overset{\overset{O}{\|}}{\underset{O^-}{P}}-O-CH_2CH_2\overset{+}{N}(CH_3)_3$$

with OH on third carbon from right, NH on second carbon from right connected to C=O (fatty acyl group) with R below; phosphoryl choline unit on the right.

most abundant sphingolipids. They are found in most animal cell membranes and are especially abundant in the lipid sheath (called the myelin sheath) around certain nerve cells.

Cerebrosides have one or more monosaccharide units joined by a glycosidic bond to the end alcohol group in sphingosine (Figure 20.6). Those containing galactose typically occur in cell membranes in the brain; those containing glucose are in cell membranes in other tissues.

Gangliosides are like cerebrosides (Figure 20.6) but have a more complex chain of carbohydrate units joined by glycosidic bonds. Gangliosides are

CHEMISTRY SPOTLIGHT

Lipid Storage Diseases

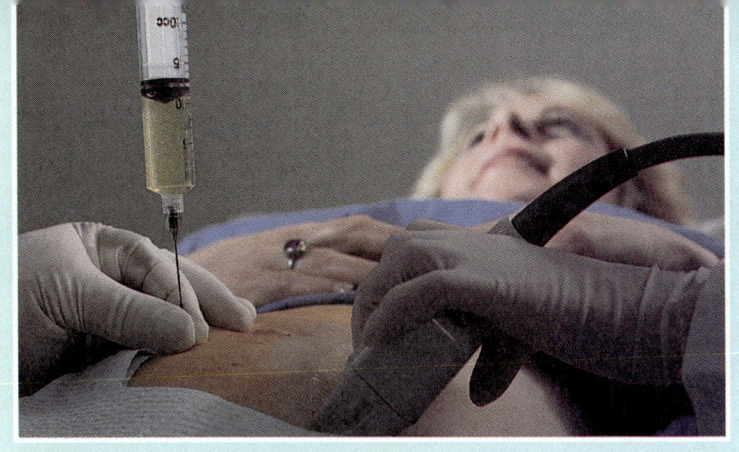

Figure 20.7 Withdrawing amniotic fluid by amniocentesis.

Sphingolipids in membranes are constantly synthesized and broken down by enzyme action. When an enzyme involved in the breaking-down process is absent due to a genetic disorder, large amounts of a sphingolipid (or one of its products) can accumulate. Ten different genetic disorders of this type, called **lipid storage diseases,** are known. We briefly discuss two.

Niemann-Pick disease occurs in people who lack an enzyme to hydrolyze and remove the phosphoryl choline portion of sphingomyelins (Figure 20.5). As a result, sphingomyelin accumulates in the brain, spleen, and liver, causing mental retardation and early death.

People with *Tay-Sachs disease* lack an enzyme that hydrolyzes a carbohydrate unit in gangliosides. The partially degraded ganglioside accumulates, causing the nervous system to deteriorate. Infants with this disease typically are blind and mentally retarded; they rarely live more than three years. No effective treatment is known.

Though very rare in the general population, Tay-Sachs disease occurs an average of once per 3600 births when both parents are Ashkenazic Jews. Prospective parents can be tested to determine if they carry the gene for this disease. In addition, fetal cells can be removed during pregnancy by a procedure called amniocentesis (Figure 20.7). Tests on those cells identify fetuses that lack the enzyme (and thus have the disease).

$$CH_3(CH_2)_{12}CH=CHCHCHCH_2-O-\text{glucose or galactose}$$

with OH on the third carbon, NH-C(=O)-R (fatty acyl group) on the second carbon.

Figure 20.6 Structure of a cerebroside. The structure of gangliosides is similar, except for a more complex chain of carbohydrate units.

abundant in the gray matter of the brain and at nerve endings. They also occur in the membranes of most other tissues. Cerebrosides and gangliosides can also be classified as **glycolipids** because they contain both carbohydrate and lipid material.

20.4 STEROIDS

Compounds with the basic four-ring structure shown in Figure 20.8 are **steroids.** Cholesterol is the most abundant steroid in our bodies.

Despite its one alcohol group, cholesterol (Figure 20.8), like other steroids, is essentially nonpolar and doesn't dissolve in water. Much of cholesterol in the body is esterified to fatty acids, further decreasing its solubility.

Figure 20.8 Steroids such as cholesterol (right) have a fused four-ring structure (left).

basic steroid structure

cholesterol

Because it is so insoluble in water, cholesterol forms deposits on blood vessel walls. When cholesterol and other lipids accumulate on the inner lining of arteries (Figure 20.9), the flow of blood is restricted and the heart must work harder to pump blood through the narrower vessels. This condition, called *atherosclerosis,* can cause a heart attack or stroke if a blood clot lodges near a clogged area and blocks the blood flow to the heart or brain.

High levels of cholesterol can also cause trouble in another way. Your gall bladder normally stores cholesterol and natural detergents (called bile salts) that help us digest lipids. But if there is too much cholesterol, some of it solidifies and forms stones (Figure 20.10). Gallstones cause abdominal pain, nausea, and a failure to digest lipids.

Despite these risks, you could not live long without cholesterol. It is an important component in cell membranes, and is used to make all the other steroids in the body, such as vitamin D, bile salts, and many hormones.

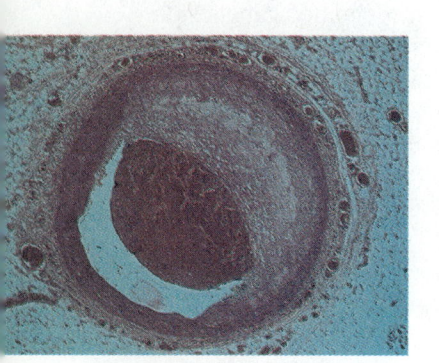

Figure 20.9 Cholesterol and triglyceride deposits in an artery cause atherosclerosis.

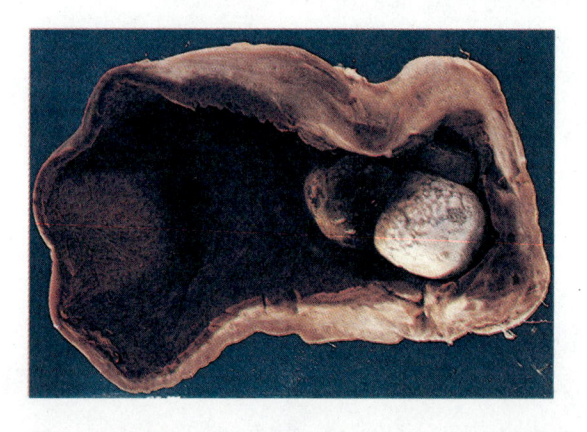

Figure 20.10 Gallstones are rich in cholesterol.

Figure 20.11 shows the structure of a bile salt. Notice the similarity to cholesterol; the main difference is in the group bonded to the five-carbon ring. Like other carboxylic acids (Section 16.2), bile acids such as cholic acid exist as the acid salt at neutral pH. This ionic group, in combination with the nonpolar steroid ring, makes bile salts detergents. Bile salts travel to the intestine through the bile duct and help emulsify dietary lipids so you can absorb them.

Steroid hormones include the male sex hormones (called androgens) and the female sex hormones (estrogens and progestogens; Figure 20.12). Synthetic estrogens and progestogens are used as birth control pills (Figure 20.12; opening of Chapter 13). The adrenal hormones cortisol, hydrocortisone, and aldosterone also are steroids.

Figure 20.11 A bile salt, the sodium salt of cholic acid.

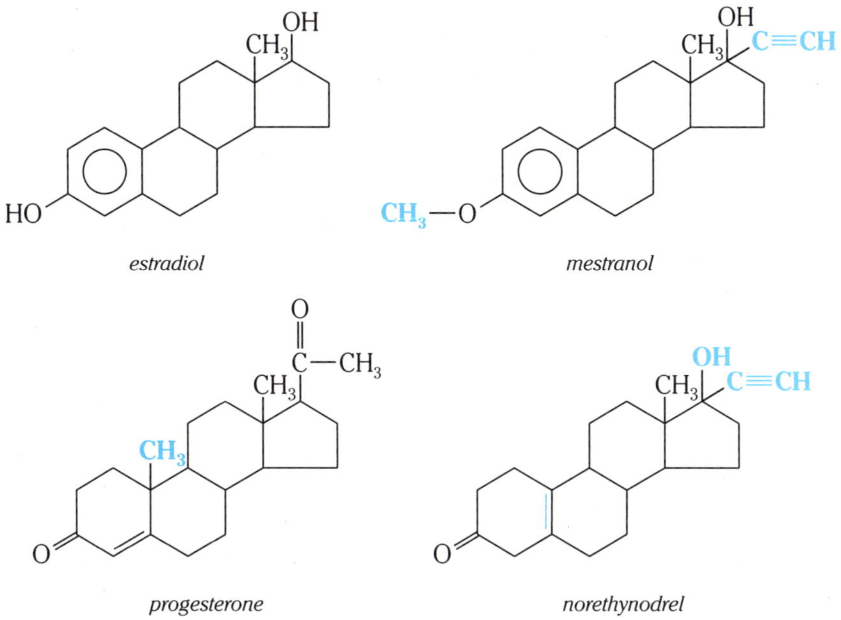

estradiol

mestranol

progesterone

norethynodrel

Figure 20.12 Two natural female sex hormones, estradiol and progesterone, and synthetic hormones used as oral contraceptives. Structural differences between the natural and synthetic compounds are in color.

20.5 LIPOPROTEINS

Because they are insoluble in water, lipids must combine with detergents to be transported in the bloodstream. Cholesterol and triglycerides in your blood are usually present as **lipoproteins,** a complex of nonpolar lipids, phospholipids, and protein.

Table 20.3 lists the four main types of lipoproteins and their compositions. Notice that their densities vary directly with protein content and inversely with triglyceride content. *Chylomicrons,* the least dense type, form in intestinal cells and transport digested lipids into the lymph and then into the blood (see Chapter 26).

Table 20.3 Approximate Composition of Lipoproteins

Lipoprotein	Density (g/mL)	Triglyceride (%)	Cholesterol (%)	Phospholipid (%)	Protein (%)
chylomicron	< 0.95	85	4	9	2
VLDL	0.95–1.01	52	20	18	10
LDL	1.01–1.06	10	46	21	23
HDL	1.06–1.21	4	17	24	55

Very low-density lipoproteins (VLDL) take up triglycerides from the liver and transport them to adipose tissue, where they are stored. The loss of triglycerides converts VLDL into *low-density lipoproteins (LDL),* the form with the highest cholesterol content (Table 20.3). LDL transports cholesterol to tissues other than the liver. Cells that have appropriate receptors on their membranes bind LDL and take in its cholesterol.

High-density lipoproteins (HDL) contain mostly protein and are synthesized in the liver. HDL receives cholesterol from other lipoproteins and delivers cholesterol to the liver for further metabolism.

Because of their effects on blood cholesterol levels, high levels of LDL increase the risk of heart disease, while high levels of HDL reduce the risk. Lower risk is associated with blood cholesterol levels below 200 mg/dL and a total cholesterol to HDL cholesterol ratio below 4.5. Normal HDL values are about 45 mg/dL for men and 55 mg/dL for women.

Effective methods to reduce the risk of heart attack include low-fat, low-cholesterol diets, exercise (which raises HDL), and prescribed drugs such as niacin (a B vitamin), cholestipol, and lovastatin (which reduces cholesterol synthesis in the body; see Section 26.4).

20.6 OTHER IMPORTANT LIPIDS

Because *lipids* have such a broad definition, many other natural substances are classified as lipids. Let's briefly examine a few of the most important ones.

Prostaglandins. Prostaglandins (Figure 20.13) are cyclic, 20-carbon fatty acids that your body synthesizes from arachidonic acid (Table 20.1). Although one of the first known sources was the prostate gland (for which these compounds are named), prostaglandins are produced in most of your tissues and affect almost every organ system. Some prostaglandins, for example, stimulate contraction of smooth muscles and have been used to induce labor in pregnant women. Others are being tested for treating hypertension, severe allergic reactions, asthma, and ulcers.

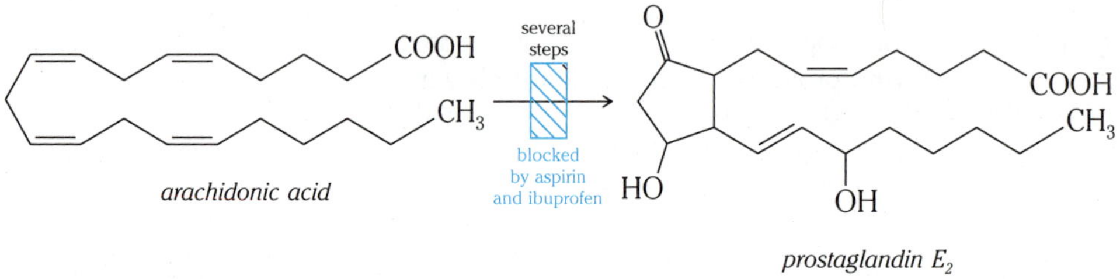

Figure 20.13 Many different prostaglandins are made from arachidonic acid. Aspirin and ibuprofen inhibit this synthesis. (In these condensed structures, each corner of the zigzag line represents a carbon atom with the appropriate number of bonded hydrogens.)

Although aspirin has been used since the turn of this century, no one knew how it worked until prostaglandins were discovered. Some prostaglandins tighten (constrict) blood vessels, which increases body temperature because heat cannot escape from tissues to the blood. Certain prostaglandins also increase the permeability of capillaries, allowing water to pass from capillaries into nearby tissues and cause swelling and pain. Aspirin inhibits the synthesis of these prostaglandins (Figure 20.13), thereby relieving pain, fever, and inflammation. Ibuprofen (two brand names are Advil and Nuprin) is similar to aspirin as a pain-reliever but less effective against inflammation, probably because it is a weaker inhibitor than aspirin of prostaglandin synthesis.

Fat-Soluble Vitamins. Vitamins are classified as water-soluble (polar) or fat-soluble (nonpolar). The latter group—vitamins A, D, E, and K—are lipids. Notice that their structures (Figure 20.14) are mostly hydrocarbon, and that vitamin D is like a steroid.

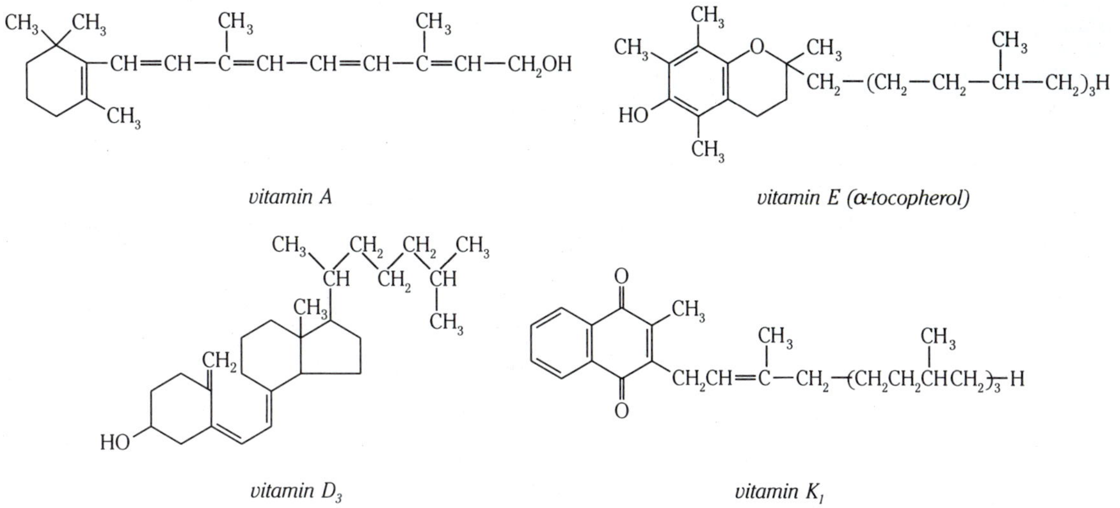

vitamin A

vitamin E (α-tocopherol)

vitamin D₃

vitamin K₁

Figure 20.14 The fat-soluble vitamins.

Because they are nonpolar, fat-soluble vitamins have certain characteristics in common. For one thing, they can accumulate to toxic levels in fatty tissues such as the liver. The main culprits are vitamins A and D. Excess vitamin A causes lethargy, headaches, and in extreme cases, death. Excess vitamin D causes high levels of calcium ion (Ca^{2+}) in blood, causing high blood pressure, and calcium deposits in soft tissues such as the kidneys.

Hypervitaminosis is a toxic condition due to too much of a vitamin.

Fat-soluble vitamins, like other lipids in the diet, need to be emulsified by detergents such as bile salts before they can be digested. If a blocked bile duct or other problem prevents this detergent action, the vitamins stay in the intestine until they are excreted, thus producing a deficiency of those vitamins.

Table 20.4 Fat-Soluble Vitamins Required in the Diet

Vitamin	Recommended Daily Allowances (RDA)* Male	Female	Sources	Deficiency Effects
A (retinol)	1000 μg	800 μg	Fish-liver oils, egg yolks, milk, yellow and green vegetables	Night blindness, scaly skin, acne
D (cholecalciferol)	10 μg	10 μg	Fish-liver oils, egg yolks, yeast, liver, fortified milk	Rickets
E (α-tocopherol)	10 mg	8 mg	Plant oils, wheat germ, liver, eggs, leafy vegetables	Not established
K	70 μg	60 μg	Leafy vegetables, beef liver	Hemorrhages

*Recommended by the Food and Nutrition Board of the National Academy of Sciences (Revised 1989).

Table 20.4 lists sources and deficiency diseases of the fat-soluble vitamins. A dietary deficiency of vitamin E has been demonstrated in laboratory animals, but not in humans. A deficiency of vitamin K is also rare because intestinal bacteria normally provide enough of this vitamin. A few people, however, cannot absorb vitamin K (and other lipids) due to a lack of bile salts in the intestine. Vitamin K has a critical role in causing blood to clot (see Section 28.4).

$$CH_3 \underset{|}{CH_3} \overset{CH_3}{\underset{|}{C}} \overset{CH_3}{\underset{|}{C}} \overset{O}{\underset{||}{C}}$$

CH=CHC=CHCH=CHC=CHC—OH

Figure 20.15 The structural difference between retinoic acid and vitamin A (Figure 20.14) is in color. Forty weeks of using a skin cream with retinoic acid caused dark liver spots on this hand (right) to fade.

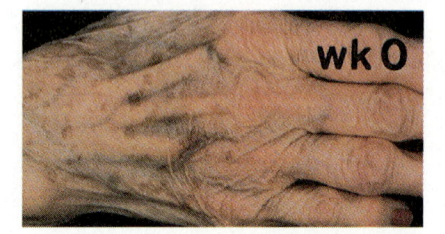

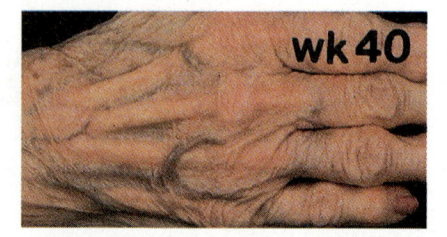

People who cannot see when they go from bright light to dim light have *night blindness.* This is caused by a deficiency of vitamin A, which plays a key role in vision (see Chemistry Spotlight in Section 13.3). A lack of this vitamin also causes scaly skin. Retinoic acid, a form of vitamin A, is used to treat acne. Sold under the trade name Retin-A, it also is used to reduce skin wrinkles and fade liver spots (Figure 20.15).

A lack of vitamin D causes *rickets,* a disease in which bone deformities develop in children (Figure 20.16). Because vitamin D normally increases calcium absorption from the intestine and calcium deposition in bones and teeth, bones weaken and bend when vitamin D is deficient. Rickets is less common in sunny regions because ultraviolet light from the sun synthesizes vitamin D from 7-dehydrocholesterol in the skin (Figure 20.17).

7-dehydrocholesterol

↓ UV light

vitamin D₃

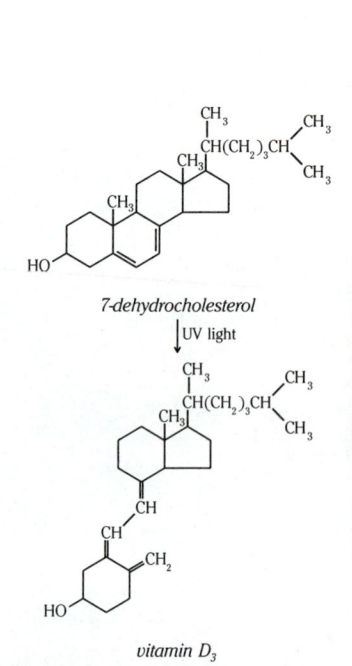

Figure 20.17 Solar ultraviolet light converts 7-dehydrocholesterol in skin into vitamin D₃.

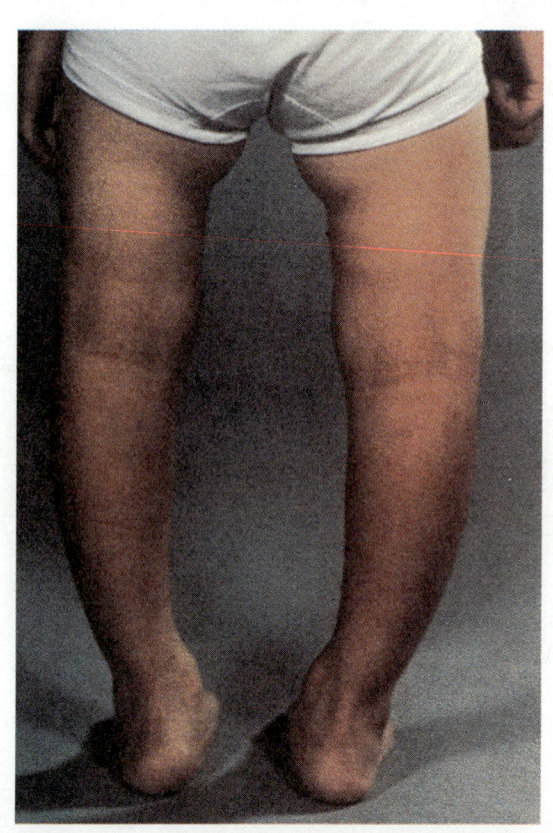

Figure 20.16 Child with bowlegs and knock-knees as a result of rickets.

Terpenes. Notice that the side chains of vitamins A, E, and K have repeating, five-carbon, branched structures (Figure 20.14). This is like the structure that occurs in natural rubber, which is made from the monomer isoprene (see Section 18.2). Isoprene is a monomer for steroids and for the nonpolar side chains in many other lipids, especially in plants.

Terpenes are lipids composed entirely of isoprene units. Figure 20.18 shows the structures of two terpenes, each composed of two isoprene units. Natural rubber (see Section 18.2) is also a terpene.

Chlorophyll and various plant pigments, called *carotenoids*, all contain isoprene units. The mixture of pigments accounts for the color in carrots and tomatoes. It is also responsible for the hues of some decorative plants, and the orange and yellow colors in autumn leaves.

myrcene (oil of bay)

limonene (oil of lemon)

Figure 20.18 Two terpenes.

20.7 CELL MEMBRANES

Lipids are important components of **membranes,** the boundary materials that surround cells and the compartments within cells. Membranes help control what goes in and out of a cell and its internal organelles. The membrane is also where substances such as hormones contact the cell with messages from other parts of an organism.

Membranes are mostly lipid and protein. The lipid portion is a double layer (bilayer) of phosphoglycerides, with nonpolar "tails" and polar "heads" (Figure 20.19). Phosphoglycerides align tail-to-tail inside the bilayer, leaving the heads to line both outer surfaces.

Depending on the particular membrane, other lipids also inhabit this bilayer. Cholesterol and sphingolipids, for example, fit in the bilayer with their

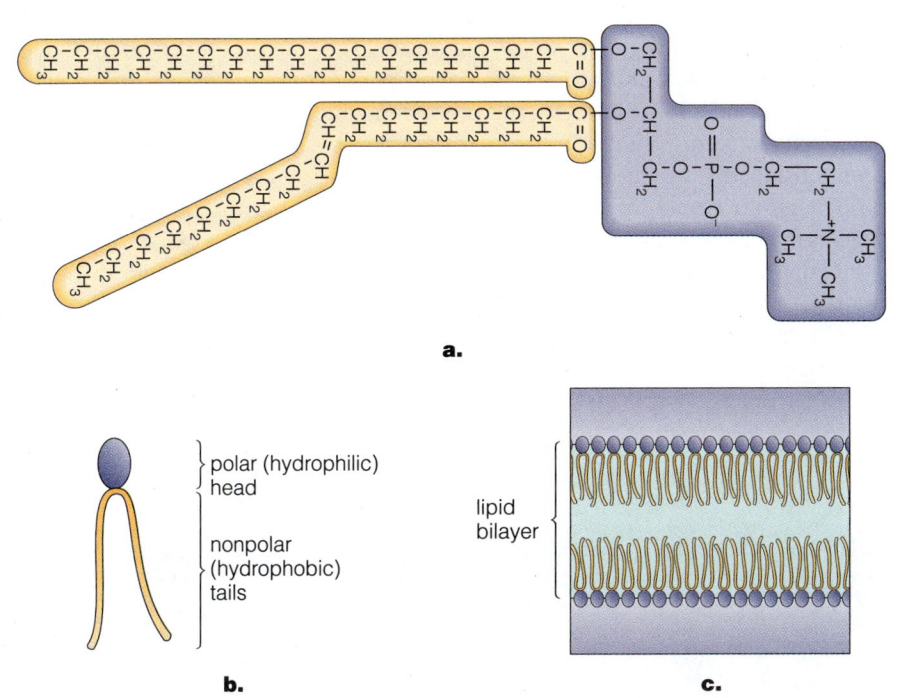

Figure 20.19 (a and b) Phosphoglycerides can be represented as having a polar (hydrophilic) head and a nonpolar (hydrophobic) tail. (c) The lipid bilayer in a membrane has an interior double layer of hydrophobic material.

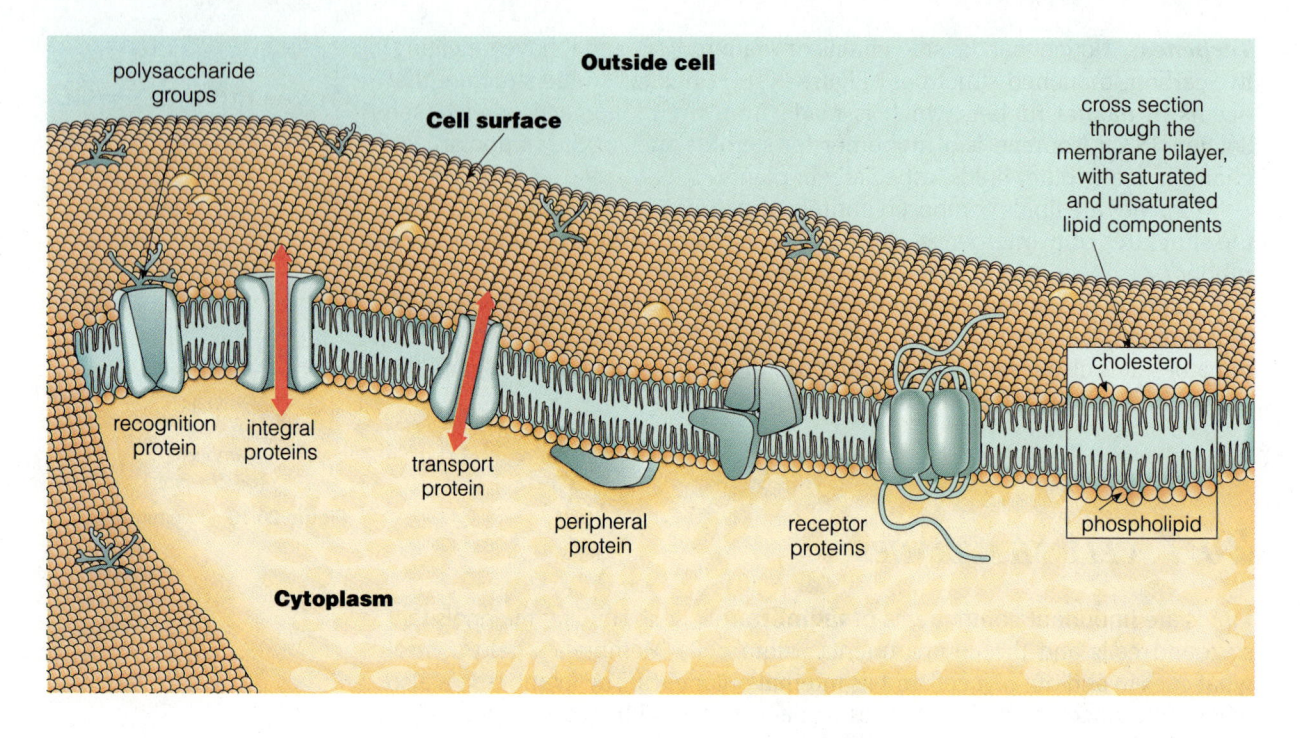

Figure 20.20 Liquid mosaic model of a membrane.

nonpolar material inside and any polar material (such as the carbohydrate portion of glycolipids and glycoproteins) outside.

Unsaturated fatty acids in the lipids make the membranes fluid. This helps nonpolar materials cross the membrane. Ionic and highly polar substances, however, do not readily dissolve in the nonpolar interior. They need a carrier substance to help them across the membrane. These carriers are proteins.

We discuss proteins in more detail in the next chapter. For now, it is enough to know that two general types of proteins are in membranes. According to the *liquid mosaic model* of membranes (Figure 20.20), some proteins coat only one surface of the membrane; these are called *peripheral proteins*. Others, called *integral proteins*, extend through the membrane and have access to both sides. Some have channels through which certain materials can pass. Proteins transport substances that cannot cross on their own. Proteins also serve as membrane receptors that bind to outside substances, such as hormones; this binding triggers specific changes within the cell.

Don't think of membranes as fixed structures. Within each layer of membrane, lipid and protein components are constantly in motion. In addition, membrane components are rapidly replaced by new ones. Many of the phosphoglyceride molecules now in the membranes of your red blood cells, for example, will no longer be there tomorrow.

> Hearken diligently unto me, and eat ye that which is good, and let your soul delight itself in fatness.
>
> *Isaiah 55:2*

SUMMARY

Lipids are natural organic compounds that dissolve in nonpolar solvents. Fatty acids are carboxylic acids with 12–20 carbon atoms. Three fatty acids esterified with glycerol make a triglyceride (triacylglycerol). Fats are relatively saturated triglycerides and are solids at room temperature. Oils are liquid triglycerides that are more unsaturated than fats. Waxes are esters with long hydrocarbon chains.

Acids and bases catalyze the hydrolysis of triglycerides to glycerol and fatty acids. Saponification with bases produces fatty acid salts, which are soaps. Unsaturated triglycerides undergo catalytic hydrogenation and participate in addition reactions with I_2.

Lipids containing phosphorus are phospholipids; those also containing glycerol are phosphoglycerides. The most common phosphoglycerides—phosphatidyl choline, ethanolamine, and serine—are detergents and important components in cell membranes.

Sphingolipids are derivatives of sphingosine. They include sphingomyelins, cerebrosides, and gangliosides. Genetic disorders of sphingolipid metabolism are known as lipid storage diseases.

Steroids are lipids with a fused four-ring structure. They include cholesterol, bile salts, and various hormones. Cholesterol, the most abundant steroid in the body, is transported in the blood bound to lipoproteins, especially LDL and HDL. Other important lipids include prostaglandins, terpenes, and vitamins A, D, E, and K.

Cell membranes have a lipid bilayer of phosphoglycerides, with the nonpolar material of each layer inside. Cholesterol and glycolipids are also present. Peripheral proteins are found on one surface, whereas integral proteins extend through a membrane. Membranes control the entry and exit of materials and respond to external substances such as hormones.

KEY TERMS

Essential fatty acid (20.1)
Fat (20.1)
Fatty acid (20.1)
Glycolipid (20.3)
Iodine number (20.2)
Lipid (20.1)

Lipid storage disease (20.3)
Lipoprotein (20.5)
Membrane (20.7)
Oil (20.1)
Phosphoglyceride (20.3)
Phospholipid (20.3)

Prostaglandin (20.5)
Sphingolipid (20.3)
Steroid (20.4)
Triacylglycerol (20.1)
Triglyceride (20.1)
Wax (20.1)

EXERCISES

Even-numbered exercises are answered at the back of this book.

Classification of Lipids

1. What structural feature do lipids typically have in common?
2. Identify which of the following would be the least effective solvent for lipids: **(a)** hexane, **(b)** methanol, **(c)** dibutyl ether, **(d)** carbon tetrachloride, **(e)** 1-hexanol.
3. Why is cholesterol, which has a polar alcohol group, classified as a lipid?

Fats, Oils, and Waxes

4. Do you expect the *cis* or the *trans* isomer of a fatty acid to have the higher melting point? Why?
5. Write structural formulas to distinguish between **(a)** *cis*-oleic acid and **(b)** *trans*-oleic acid.
6. Write a structural formula for a triglyceride made from three molecules of palmitoleic acid.
7. Write a structural formula for a wax made from stearic acid and $CH_3(CH_2)_{26}OH$.
8. Would a triglyceride made from one molecule each of arachidonic, oleic, and linolenic acid be a fat or an oil?

446   Part III   Biochemistry

9. Write a structural formula for a diglyceride containing palmitic and arachidonic acid.

10. According to Table 20.2, coconut oil has a very low iodine number. Why is coconut oil a liquid at room temperature instead of a solid?

11. Do you expect oils to have higher or lower iodine numbers than fats? Why?

12. A sample of triacylglycerol containing only oleic acid has an iodine number of 87. What is the iodine number for a triacylglycerol containing one residue each of stearic, linoleic, and linolenic acid?

13. What is the effect of catalytic hydrogenation of a vegetable oil on its iodine number?

14. Saponification literally means "soap making." What does soap have to do with saponification?

15. Write structural formulas for the products formed by the saponification with NaOH of the substance in exercise 6.

16. Write structural formulas for the products formed by the acid-catalyzed hydrolysis of the substance in exercise 6.

Phospholipids and Sphingolipids

17. Write a structural formula for phosphatidyl choline containing two oleic acid units.

18. In addition to carbon, hydrogen, and oxygen, phosphatidyl choline contains the element _____.

19. Some health food enthusiasts claim that eating large amounts of lecithin removes the fatty deposits in blood vessels that cause atherosclerosis. What property of lecithin is consistent with such a claim?

20. A phospholipid that doesn't contain glycerol is _____.

21. The most water-soluble part of phosphatidyl choline comes from (a) glycerol, (b) phosphate and choline, (c) two fatty acids, (d) sphingosine, (e) steroid ring.

22. Identify which of the following are glycolipids: (a) sphingomyelins, (b) steroids, (c) triglycerides, (d) gangliosides, and (e) phosphoglycerides.

23. Lipid storage diseases may be caused by (a) failure to synthesize gangliosides, (b) excess steroid synthesis, (c) excess breakdown of sphingomyelin, (d) failure to synthesize steroids, (e) failure to break down gangliosides.

24. Which type of lipid contains at least one glycosidic bond?

25. Could a sphingolipid participate in an addition reaction with I_2? Explain.

26. Why do babies born with Tay-Sachs disease have excess amounts of gangliosides in their neural tissues?

Steroids

27. How many ring carbons does a steroid have?

28. List the two key structural changes made in natural hormones that made birth control pills possible. [Review the opening to Chapter 13 if necessary.]

29. From the structure of its sodium salt (Figure 20.11), write the structural formula for cholic acid.

30. The sodium salt changes into cholic acid at pH (a) 2, (b) 7, (c) 10.

31. Anabolic steroids include the natural hormone _____.

32. Does lengthening the hydrocarbon chain of an anabolic steroid cause the compound to remain in the body a longer or shorter time?

Lipoproteins

33. What chemical property of proteins enables them to carry lipids such as cholesterol in blood?

34. If you have a (total cholesterol)/(HDL cholesterol) ratio of 4.0, what percentage of your cholesterol is bound to HDL? Is this a healthy ratio to have?

Other Lipids

35. Aspirin reduces fever by (a) stimulating steroid synthesis, (b) inhibiting prostaglandin synthesis, (c) inhibiting glycolipid synthesis, (d) inhibiting triacylglycerol breakdown, (e) stimulating prostaglandin synthesis.

36. Name the vitamins that must be emulsified by detergents in order to be absorbed from the intestine.

37. Predict which of the following natural substances are classified as lipids:

thiamine (vitamin B_1)

coenzyme Q

vitamin C

Cell Membranes

38. Are membranes symmetric or asymmetric? That is, are the two lipid layers and their associated proteins different or the same? Explain.
39. Are membranes in your cells predominantly in the (a) solid, (b) liquid, or (c) gas state?
40. Integral proteins, unlike peripheral proteins, occupy part of the interior of a membrane. Would you expect the protein material in the interior to be mostly polar or nonpolar? Why?
41. What are the main attractive forces between nonpolar molecules in the interior of a membrane?
42. What structural feature of membranes makes it difficult for Na^+ ions to pass freely through them?
43. Consider the part of a phosphatidyl choline molecule on the outside surface of a membrane. It could attract nearby molecules by which of the following: (a) hydrogen bonds, (b) dipole–dipole forces, (c) ionic bonds, (d) all of the above.

DISCUSSION EXERCISES

1. What specific solvents would you use to dissolve terpenes and other lipids to remove them from plant tissue? Explain.
2. Explain, in terms of geometry around the double bond, why *cis* arrangements in polyunsaturated fatty acids put more bends in the hydrocarbon chain than *trans* arrangements do. Would you expect *cis* or *trans* isomers to have higher melting and boiling points? Why?
3. Iodine number measures the amount of unsaturation in a fat by the amount of iodine that adds to its double bonds. What other substances (besides I_2) add quantitatively to double bonds? [Review Section 13.4 if necessary.]
4. Can sphingomyelins function as detergents? Can triglycerides? Explain both answers.

21

Proteins

1. What are the structures and types of amino acids that occur in proteins? Which isomeric and ionic forms occur in the body?
2. How do amino acids bond together to form proteins? What is the primary structure of proteins?
3. What complex types of structure are necessary for proteins to function in the body? What types of bonds are involved?
4. What are the main classes of proteins?
5. What are some specific proteins and their roles in the body?

Sickle-Cell Anemia

The sequence of amino acids in a large protein molecule determines the way the protein folds and twists into its elaborate three-dimensional shape. A change in the amino acid sequence, then, can change the shape of a protein.

A striking example occurs in *sickle-cell anemia.* People with this genetic disorder have hemoglobin that differs from normal hemoglobin by just one amino acid. The usual amino acid (glutamic acid), which is ionic and carries a negative electrical charge, is replaced in that position by a nonpolar amino acid (valine). The difference was discovered because sickle-cell hemoglobin moves differently in an electrical field than normal hemoglobin.

What effect does that have in the body? More than you might expect. Although only 1 of 146 amino acids in the protein chain in sickle-cell hemoglobin is different, this change is enough to give hemoglobin—and red blood cells—a different shape (Figure 21.1). The unusual, sickle-shaped cells are more fragile, less able to bind oxygen, and tend to clump together, forming networks that clog small blood vessels. This blocks blood circulation to some parts of the body and can be very painful. Their fragility causes sickled cells to decompose sooner, leaving the person with a lower concentration of red blood cells, a condition called anemia.

No one has found a truly satisfactory treatment for sickle-cell anemia, although several are in the experimental stage. In the Chemistry Spotlight in Section 28.2, you will read about one interesting new idea now being tested.

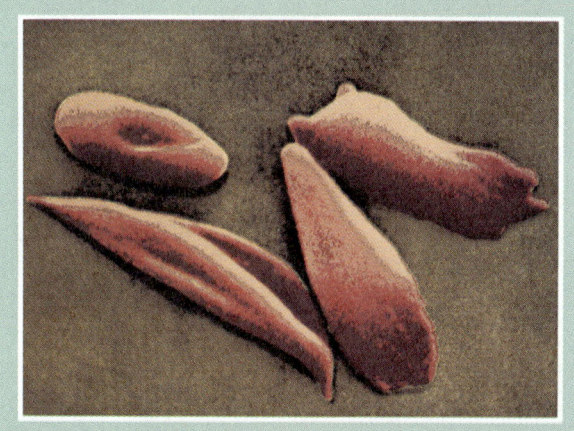

Figure 21.1 Comparison of normal red blood cells (left) and red blood cells from a person with sickle-cell anemia (right).

450 Part III Biochemistry

Protein comes from the Greek word proteios, *which means "of the first rank."*

I f we ranked chemicals according to their importance in the body, proteins would be at or near the top of the list. In fact, it's hard to find something important in living things that doesn't involve proteins (Figure 21.2). Your genes specify which proteins you have, and those proteins in turn provide all of your hereditary features. Your hair, nails, muscles, and skin are mostly protein. Proteins also serve as *enzymes* (see Chapter 22) to catalyze the reactions in your body. And proteins have critical roles in such diverse things as the immune system, muscle contraction, hormone action, and blood clotting.

Figure 21.2 A spider spins its silky web made of protein.

21.1 AMINO ACIDS

Proteins are polymers made from amino acids as the monomer units. Recall that **amino acids** (Section 17.3) contain an amino and a carboxylic acid group and have the general structure:

$$^+NH_3-CHR-COO^-$$

Compounds that have a positive charge on one atom and a negative charge on another atom are called *zwitterions* (from the German word *zwitter*, which means "hybrid"). Amino and carboxyl groups are in their ionic forms near neutral pH, so amino acids exist mainly as zwitterions in the body.

Table 21.1 shows structures of the 20 common amino acids found in proteins. All are α-amino acids because the amino group bonds to the α carbon.

Recall that in common names for carboxylic acids, the carbon next to the carboxyl carbon is the α carbon (Section 16.1).

EXAMPLE 21.1 Ignoring any amino groups in their side chains (R), classify (a) proline and (b) all other amino acids in Table 21.1 as primary, secondary, or tertiary amines. [Review Section 17.1 if necessary.]

SOLUTION (a) secondary amine, (b) primary amines

Amino acids differ in the side-chain (R) group bonded to carbon. These side-chain groups provide the basis for classifying amino acids (Table 21.1). Classes include nonpolar and aromatic amino acids, plus those based on general (acidic or basic) and specific (S- or OH-containing) functional groups.

Table 21.1 Structures of Amino Acids

Class	Side Chain (R) (at pH 7) for $^+NH_3-CHR-COO^-$	Name	Symbol	pI
Nonpolar	—H	glycine	gly	5.97
	—CH$_3$	alanine	ala	6.00
	—CH(CH$_3$)$_2$	valine	val	5.96
	—CH$_2$CH(CH$_3$)$_2$	leucine	leu	5.98
	—CH(CH$_3$)CH$_2$CH$_3$	isoleucine	ile	6.02
Aromatic	—CH$_2$—C$_6$H$_5$	phenylalanine	phe	5.48
	—CH$_2$—C$_6$H$_4$—OH	tyrosine	tyr	5.66
	—CH$_2$—(indole)	tryptophan	trp	5.89
Sulfur-containing	—CH$_2$SH	cysteine	cys	5.07
	—(CH$_2$)$_2$—S—CH$_3$	methionine	met	5.74
Hydroxyl-containing (besides tyrosine)	—CH$_2$OH	serine	ser	5.68
	—CH(OH)CH$_3$	threonine	thr	5.64
Acidic (or Amides)	—CH$_2$COO$^-$	aspartic acid	asp	2.77
	—(CH$_2$)$_2$COO$^-$	glutamic acid	glu	3.22
	—CH$_2$C(O)NH$_2$	asparagine	asn	5.41
	—(CH$_2$)$_2$C(O)NH$_2$	glutamine	gln	5.65
Basic	—CH$_2$—(imidazole)	histidine	his	7.59
	—(CH$_2$)$_4$NH$_3^+$	lysine	lys	9.74
	—(CH$_2$)$_3$NHC(=NH$_2^+$)NH$_2$	arginine	arg	10.76
Imino Acid (complete structure)	(pyrrolidine ring with $^+$NH$_2$ and COO$^-$)	proline	pro	6.30

Those containing only hydrogen and carbon are nonpolar. Look at the side-chain group for each amino acid. You should be able to see why each amino acid is classified as shown in Table 21.1.

> **EXAMPLE 21.2**
> (a) From its R group shown in Table 21.1, draw a structural formula for tyrosine.
> (b) Based on its R group, tyrosine could be listed in two different classes of amino acids. Name them.
>
> **SOLUTION**
>
> (a) $^+H_3N-CH(CH_2-C_6H_4-OH)-COO^-$ (b) aromatic, hydroxyl-containing

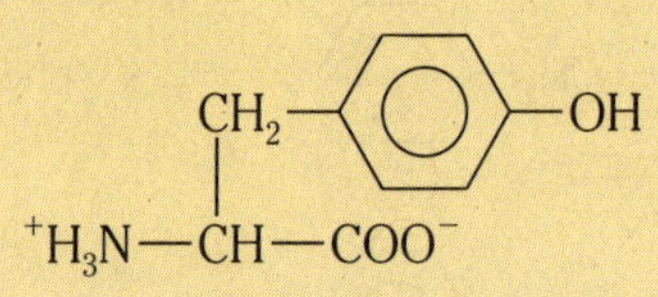

Isomers. You can see from the general structure of amino acids, $^+H_3N-CHR-COO^-$, that the α carbon is chiral when R is something other than $-H$, $-COO^-$, or $-NH_3^+$. Glycine, in which R is a hydrogen (H) is the only amino acid without a chiral carbon.

Recall that compounds with chiral carbons can exist as optical isomers (Section 19.3). Each amino acid (except glycine) occurs as a pair of enantiomers, which are designated D or L. Notice the comparison of the amino acid serine with the carbohydrate glyceraldehyde:

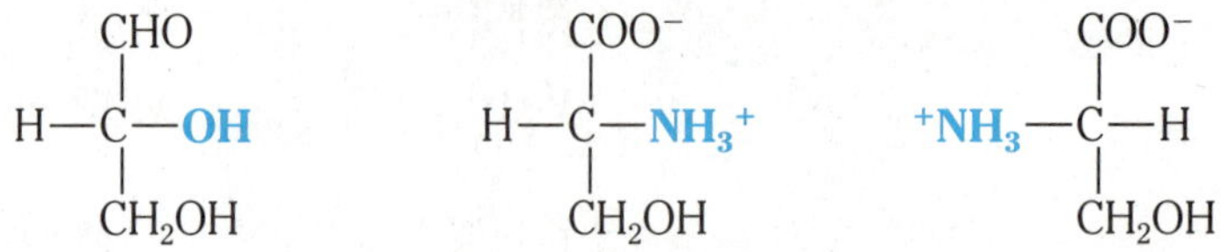

D-glyceraldehyde, D-serine, L-serine

Figure 21.3 shows another way to represent the two forms of an amino acid. Your body uses only L-amino acids.

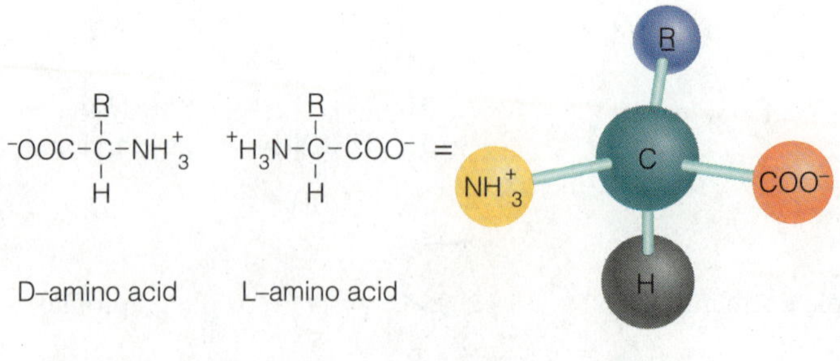

Figure 21.3 D- and L-amino acids are pairs of enantiomers. The structure of an L-amino acid is shown in two ways.

D–amino acid   L–amino acid

Acid–Base Properties. The ionic form of an amino acid varies with pH because the carboxyl group is an acid (H$^+$ donor) and the amino group is a base (H$^+$ acceptor). In addition, the amino acid side chain (R) may be acidic or basic.

Acidic and basic groups exist in protonated (carrying H⁺) and unprotonated (lacking H⁺) forms, depending on pH. Figure 21.4 shows how this works. At low pH (acidic) values, where a high concentration of H⁺ (as H_3O^+) occurs, all groups are protonated. At high pH (alkaline) values, where there is little H⁺, all groups are unprotonated. At intermediate pH values, acidic groups have donated their protons and thus are unprotonated; basic groups have accepted their protons and are protonated.

This is another example of Le Châtelier's principle (Section 8.5). Increasing the concentration of H⁺ (lowering pH) shifts the equilibrium to the right to favor the protonated form of the amino and carboxyl groups:

$$NH_2-CHR-COO^- + 2H^+ \rightleftharpoons {}^+NH_3-CHR-COOH$$

unprotonated   *protonated*

$^+NH_3-CHR-COOH$

acidic pH

⇅

$^+NH_3-CHR-COO^-$

intermediate pH (dipolar ion)

⇅

$NH_2-CHR-COO^-$

alkaline pH

Figure 21.4 Ionic forms of an amino acid at different pH values.

Reducing the concentration of H⁺ (raising pH) shifts the equilibrium to the left and favors formation of the unprotonated form.

The pK_a for an amino acid functional group tells you the pH at which equal amounts of the protonated and unprotonated forms exist. At pH values below pK_a, the protonated form predominates. At pH values above pK_a, the unprotonated form is more abundant.

The carboxylic acid and amino groups in valine, for example, have pK_a values of 2.3 and 9.7, respectively. From this, you can write the main ionic form at different pH values:

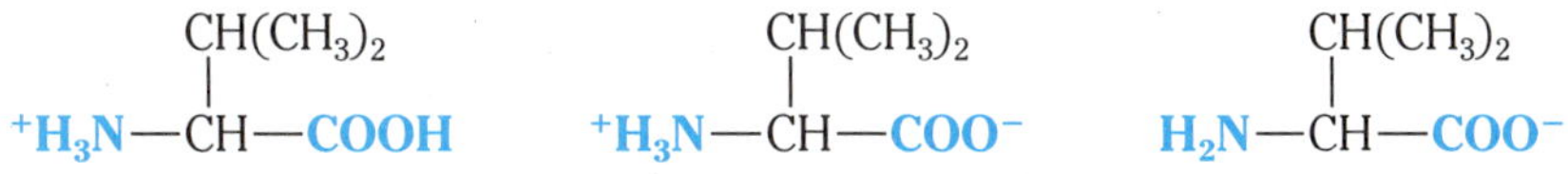

$^+H_3N-\underset{\underset{CH(CH_3)_2}{|}}{CH}-COOH$ $^+H_3N-\underset{\underset{CH(CH_3)_2}{|}}{CH}-COO^-$ $H_2N-\underset{\underset{CH(CH_3)_2}{|}}{CH}-COO^-$

below pH 2.3   between pH 2.3 and 9.7   above pH 9.7
net charge = +1   net charge = 0   net charge = −1

You can calculate the ratio of these ionic forms at any pH by using the Henderson–Hasselbalch equation (see Section 9.6).

The pH in your body is fairly neutral except in your stomach, which is highly acidic because it produces so much HCl. In the pH range 5–8, amino acids exist mostly in the form with an unprotonated carboxyl group and a protonated amino group (Figure 21.3). Substances like this, containing both a positive and a negative charge, are called **dipolar ions** (or *zwitterions*).

In their dipolar ion form, amino acids have no net charge, and thus aren't attracted to either the positive pole (anode) or negative pole (cathode) in an electrical field. Since the amount of each ionic form varies with pH, each amino acid has a specific pH at which it has a net charge of exactly zero; this is its **isoelectric point** (p*I*).

Look at the ionic forms of valine above. The isoelectric point is the pH at which most of valine is in its dipolar ion form, with equal (very small) amounts of +1 and −1 forms to give a net charge of zero. That point (p*I*) is midway between the two pK_a values. For valine, then,

$$pI = \frac{2.3 + 9.7}{2} = 6.0$$

Calculating pI values for amino acids with ionizable side-chain (R) groups is slightly more complex, and we won't consider it here.

Table 21.1 lists the isoelectric point for each amino acid. Most p*I* values are about 6, except for acidic and basic amino acids. At pH values below (more acidic than) their isoelectric point, amino acids have some amount of net positive charge and migrate toward the cathode (negative pole) in an electrical field. At pH values above (more alkaline than) p*I*, amino acids will have a negative charge and migrate toward the anode.

The method of separating amino acids (or other charged substances) by placing them in an electrical field is called **electrophoresis.** Because each amino acid has a different p*I* value, amino acids can be separated from each other by carefully selecting a pH value that discriminates between the amino acids in the mixture. In electrophoresis, each amino acid migrates a different distance toward the anode or cathode. Figure 21.5 shows an example of how this works.

Figure 21.5 Electrophoresis of aspartic acid (asp) and histidine (his) at pH 7.0. His (p*I* = 7.59) has a small positive charge at pH 7.0 and migrates toward the cathode, whereas asp (p*I* = 2.77) has a large negative charge at pH 7.0 and migrates toward the anode.

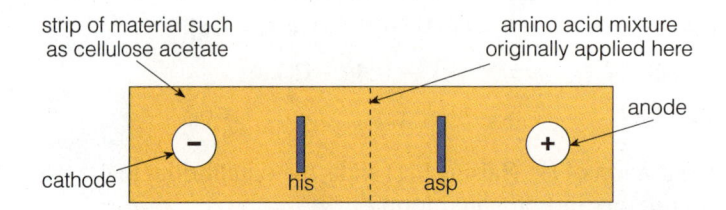

EXAMPLE 21.3 Using p*I* values in Table 21.1, predict whether cysteine migrates to the anode or cathode under each of the following: (a) pH 3.6, (b) pH 6.0, (c) pH 9.7.

SOLUTION (a) pH is below p*I*, so cysteine has a net positive charge and migrates toward the negative pole (cathode); (b) pH is above p*I*, so cysteine has a net negative charge and migrates toward the positive pole (anode); (c) anode

21.2 ESSENTIAL AMINO ACIDS

Amino acids in proteins are the most abundant nitrogen-containing materials in your body. One measure of health and physical development compares your intake of nitrogen with your excretion of nitrogen. If these two values are about the same, you are in *nitrogen balance.*

Healthy adults typically are in nitrogen balance. Growing children and people recovering from illness or injury are adding muscle and other protein mass, so they take in more nitrogen than they excrete. They have a positive nitrogen balance. People whose bodies are deteriorating because of starvation or disease have a negative nitrogen balance.

Protein provides 17 kJ (4 Cal) per gram and typically supplies 10–15% of the energy in the diet. You need to consume daily about 1 mg of protein per kg of body weight to stay in nitrogen balance. The amino acid content of that protein is also important. In order to synthesize its own proteins, your body requires 20 amino acids. Eight of those (ten in children) must be supplied in the diet because the body cannot synthesize them. These are called the **essential amino acids** (Table 21.2).

Animal proteins typically contain enough of all the essential amino acids; they are called *complete proteins.* Plant proteins, however, are often deficient in one or more essential amino acids; they are *incomplete proteins.* Notice in

Table 21.2 Essential Amino Acids and Their Content in Proteins

	Cow Milk %	Human Milk %	Egg (Whole) %	Meat %	Whole Wheat %	Soybean %	Corn (Zein) %
Arginine*	3.5	5.0	6.7	6.6	4.3	7.3	1.7
Histidine*	2.7	2.7	2.4	2.8	1.8	2.9	1.3
Isoleucine	6.5	5.2	6.9	4.7	4.4	6.0	7.3
Leucine	9.9	15.0	9.4	8.0	6.9	8.0	23.7
Lysine	8.0	7.2	6.9	8.5	2.5	6.8	0
Methionine	2.4	2.0	3.3	2.5	1.2	1.7	2.4
Phenylalanine	5.1	5.9	5.8	4.5	4.4	5.3	6.2
Threonine	4.7	4.6	5.0	4.6	3.9	3.9	3.5
Tryptophan	1.3	1.9	1.6	1.1	1.2	1.4	0.1
Valine	6.7	5.5	7.4	5.5	4.5	5.3	3.5

* Essential for infants only

Table 21.3 Plants and Their Amino Acid Deficiencies

Plant	Deficiency
corn	lys, trp
soybeans	met
rice	met
peas	met
beans	met, trp
peanuts	lys, met, thr
sunflower seeds	lys

Tables 21.2 and 21.3 that the deficient amino acids vary with different plants. People who eat proper combinations of plant proteins in the same meal—for example, soybeans and corn—receive adequate amounts of all the essential amino acids.

One of the world's major nutritional diseases is protein malnutrition. It occurs mainly in people who have a very limited protein intake, mostly from a single, incomplete plant source such as corn, beans, or rice.

Marasmus occurs when the diet is low in both energy and protein. An infant with marasmus typically has a bloated belly, thin body, shriveled skin, wide eyes, and an old-looking face (Figure 21.6, left). *Kwashiorkor* is a disease that

Marasmus *comes from a Greek word meaning "to waste away."* Kwashiorkor *is a Ghanian word meaning "displaced child."*

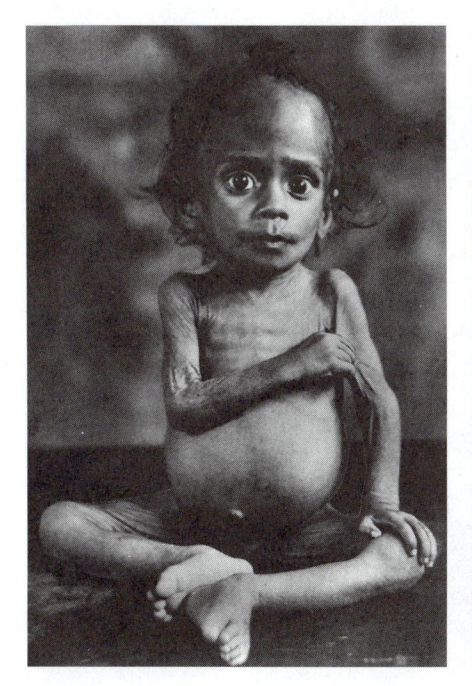

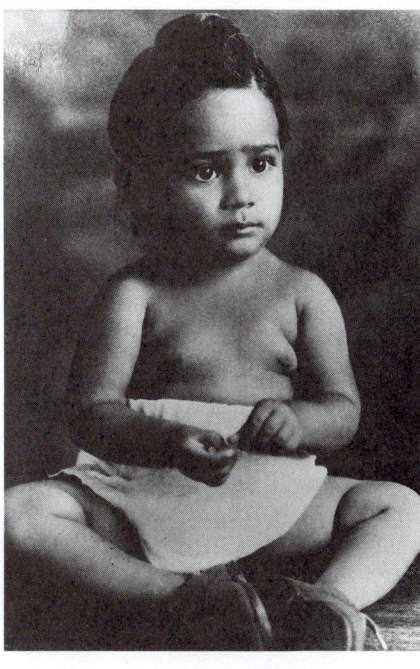

Figure 21.6 This two-year-old Venezuelan girl suffered from marasmus (left) but recovered after ten months of treatment and proper nutrition.

occurs in very young children who are displaced from mother's milk by younger siblings. When they change to a local diet that lacks one or more essential amino acids, these children develop skin sores, swollen stomachs, and liver degeneration. Many die because they are so vulnerable to infectious diseases. Both marasmus and kwashiorkor cause stunted growth, mental apathy, and possible mental retardation. But if the malnutrition isn't too severe, many symptoms can be reversed by an adequate diet (Figure 21.6, right).

21.3 PRIMARY PROTEIN STRUCTURE

The sequence of amino acids in a protein chain is called the **primary structure** of the protein. The chain then folds and bends into other types of structure that we discuss in Section 21.4.

In order to link two amino acids together, the carboxyl group in one reacts with the amino group of another as follows:

$$^+NH_3-CHR-\overset{O}{\underset{\|}{C}}-O^- + {^+NH_3}-CHR'-\overset{O}{\underset{\|}{C}}-O^- \longrightarrow$$

$$\text{amino acid} \qquad \text{amino acid}$$

$$^+NH_3-CHR-\overset{O}{\underset{\|}{C}}-\underset{\underset{H}{|}}{N}-CHR'-\overset{O}{\underset{\|}{C}}-O^- + H_2O$$

$$\text{a dipeptide}$$

Notice that the functional group joining the amino acids (in color) is an amide group (Section 17.2). Biochemists commonly call this linkage a **peptide bond**.

The product in the reaction above still has a free amino group (on the left) and a free carboxyl group (on the right) that are available to form peptide bonds with other amino acids. The chain grows in length as additional amino acids form peptide bonds.

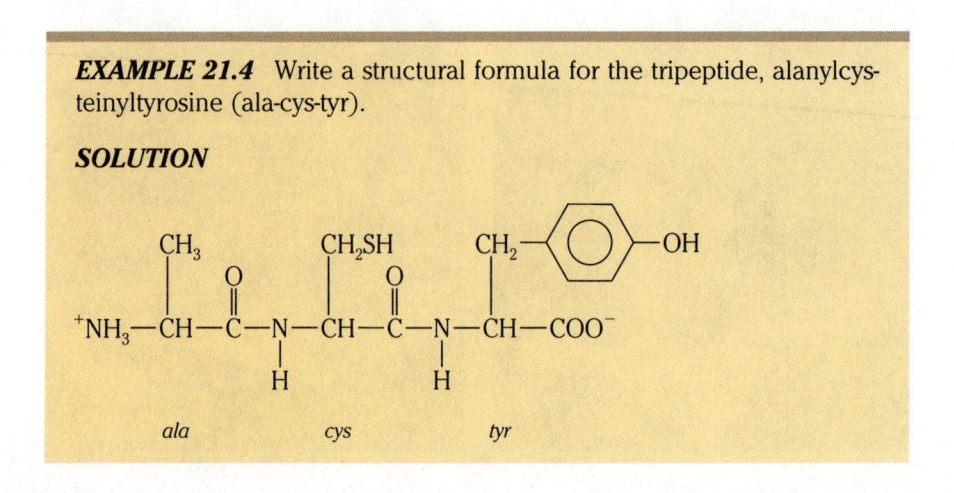

EXAMPLE 21.4 Write a structural formula for the tripeptide, alanylcysteinyltyrosine (ala-cys-tyr).

SOLUTION

$$^+NH_3-\underset{\underset{CH_3}{|}}{CH}-\overset{O}{\underset{\|}{C}}-\underset{\underset{H}{|}}{N}-\underset{\underset{CH_2SH}{|}}{CH}-\overset{O}{\underset{\|}{C}}-\underset{\underset{H}{|}}{N}-\underset{\underset{CH_2-\bigcirc-OH}{|}}{CH}-COO^-$$

$$\text{ala} \qquad\qquad \text{cys} \qquad\qquad \text{tyr}$$

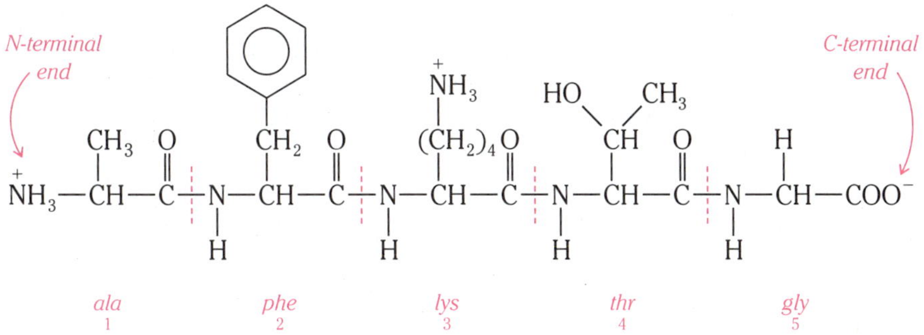

Figure 21.7 Structure, abbreviated structure, and numbering for a pentapeptide, alanylphenylalanyllysylthreonylglycine.

We use names to indicate the length of the chain. The product above is a *dipeptide* because it is made from two amino acid units. *Tripeptides* have three amino acids, *tetrapeptides* have four, and so on. **Peptides** typically have about 10 or fewer amino acids, **polypeptides** have more than 10, and proteins have more than 50 or so, though *protein* and *polypeptide* are sometimes used interchangeably.

Whatever its length, a protein chain has an amino acid with a free amino group on one end (the N-terminal end) and an amino acid with a free carboxyl group on the other (the C-terminal end). Standard practice names and numbers amino acid units starting at the N-terminal end. Figure 21.7 shows one example.

Notice in Example 21.4 that cysteine has a side-chain thiol (—SH) group. Biochemists often call this a **sulfhydryl group.** Recall that oxidizing two thiol (sulfhydryl) groups forms a disulfide bond (—S—S—) (Section 14.5). The general reaction is:

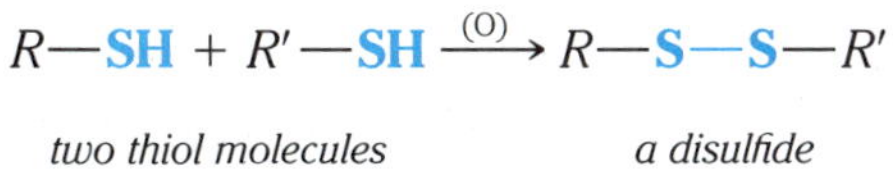

$$R-SH + R'-SH \xrightarrow{(O)} R-S-S-R'$$

two thiol molecules   *a disulfide*

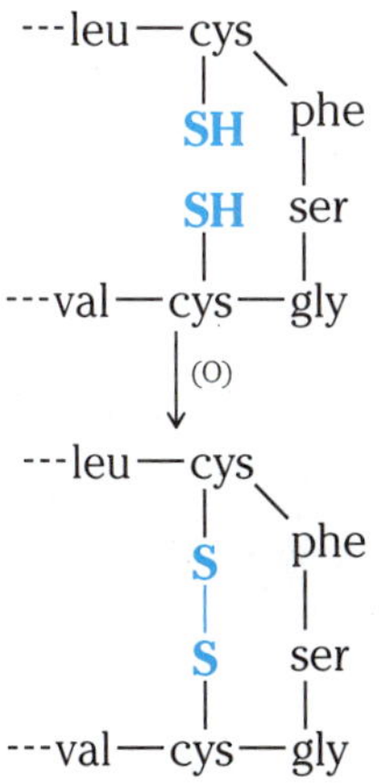

Side-chain sulfhydryl groups of two cysteine units in a protein can be oxidized to form a disulfide bond (Figure 21.8). Disulfide bonds are sometimes considered part of the primary structure of a protein. They also help determine higher levels of structure, as we discuss in Section 21.4.

Each unit in a protein chain can be any 1 of 20 different amino acids. It's like a chain of beads, with 20 different colors of beads. An incredibly large number of proteins with different primary structures is possible. For example, the number of possible sequences for a chain only 15 amino acids long is more than 30,000,000,000,000,000,000 (3.0×10^{19}). Imagine how many different combinations (primary structures) could occur in a chain 200 amino acid units long.

Figure 21.8 Oxidation of cysteine units in a protein forms a disulfide bond (color).

EXAMPLE 21.5 How many different dipeptides can form from the 20 amino acids listed in Table 21.1?

SOLUTION Each of the 20 amino acids can be on the free amino end, with any of the 20 others on the free carboxyl end. Altogether, this makes $20 \times 20 = 400$ different dipeptides.

Acid–Base Properties. Proteins, like individual amino acids, have functional groups that exist in different forms (protonated or unprotonated) at different pH values. Most are side chains of individual amino acid units; but proteins also have ionizable carboxyl and amino groups at their respective ends of the chain (Figure 21.7).

The net electrical charge on a protein at a particular pH depends on its amino acid composition. For example, the electrical charge on a protein containing only nonpolar amino acids (which have no ionizable side chains) comes entirely from the end carboxyl and amino groups. But, since most proteins have substantial numbers of amino acids with ionizable side chains, their net charges vary considerably with pH.

The pH at which there is no net change on a protein is its *isoelectric point* (p*I*). As with amino acids, a protein has a net positive charge when it is at a pH lower (more acidic than) its isoelectric point, and a negative charge at a pH higher (more alkaline than) its isoelectric point. Most proteins have p*I* values in the 4–6 range.

Because each protein has its own p*I* (which reflects its amino acid composition), a mixture of proteins can be separated by electrophoresis as long as the pH selected gives individual proteins different net charges (and thus a different attraction to the anode or cathode). One example, the separation of serum proteins, is shown in Figure 21.9.

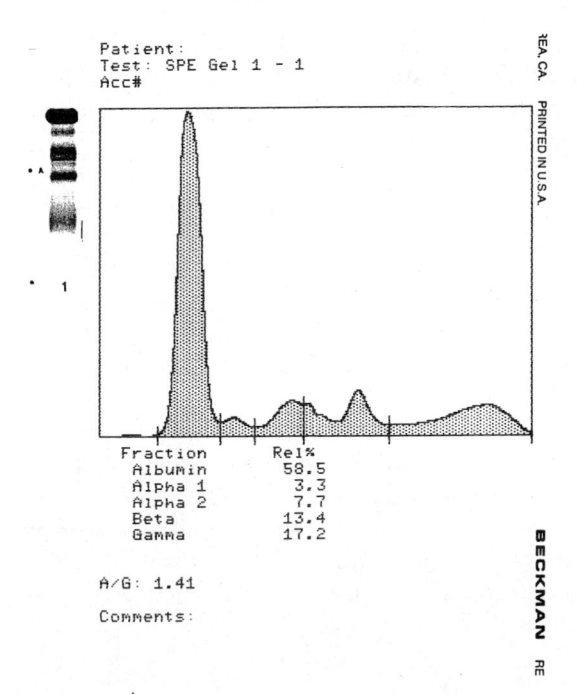

Figure 21.9 Separation of five serum protein fractions by electrophoresis. The area under each peak is the relative amount of each fraction.

Electrophoresis of proteins is a useful tool in clinical chemistry and biochemical research. Abnormal amounts of certain proteins are associated with specific disorders. With serum proteins, for example, excess gamma globulin can indicate an infection, while too little albumin may indicate a protein deficiency in the diet.

EXAMPLE 21.6 If electrophoresis were done at pH 6.0 on a mixture of protein A (pI = 4.7) and protein B (pI = 6.2), to which electrode (anode or cathode) would each protein migrate? Which protein would probably be attracted more strongly?

SOLUTION Protein A will have a negative charge and be attracted to the anode (positive pole). Protein B will have a slight positive charge and be weakly attracted to the cathode (negative pole).

The isoelectric point of a protein affects its solubility in water. Proteins are the least soluble at a pH equal to their pI. At this pH, protein molecules have no net charge, so they are less able to attract polar water molecules to carry them into solution. In addition, protein molecules are more likely to clump together and come out of solution because they lack a net charge that would cause them to repel each other.

One example is milk. Milk turns sour when bacteria produce acids, lowering the pH. The isoelectric point of casein, a major milk protein, is 4.7. When the pH of sour milk drops to that value, casein no longer remains suspended and separates out as curds.

Geometry of the Peptide Bond. Recall that a partial double bond occurs between amide (peptide) C and N atoms (Section 17.5). Because C and N cannot rotate freely around that bond, all six bonded atoms are fixed in the same plane:

The carbonyl carbon and amide nitrogen use sp^2 hybrid orbitals, which puts all bonded atoms in the same plane with 120° bond angles.

Figure 21.10 shows how a peptide chain forms a sequence of these plane structures. Bonded *R* groups are in a *trans* arrangement, which keeps them apart. By locking six atoms in a plane, a protein chain is limited in its three-dimensional shapes, as we discuss in the next section.

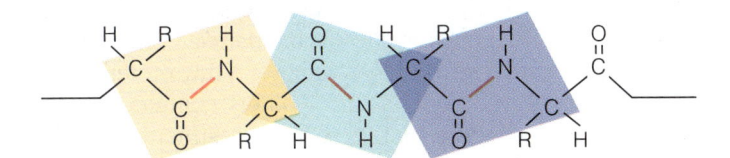

Figure 21.10 The lack of free rotation around the amide C—N bond (color) fixes the four amide atoms and the two adjacent C atoms in the same plane. *R* groups are *trans* to each other.

21.4 HIGHER LEVELS OF PROTEIN STRUCTURE

Each protein requires a particular three-dimensional shape (conformation) in order to function properly. The linear sequence of amino acids (primary structure) twists and bends into a wide array of conformations. These additional levels of protein structure are called *secondary, tertiary,* and *quaternary structure.*

Secondary Structure. **Secondary structure** is the twisting of the protein chain to accommodate maximum hydrogen bonding between atoms in nearby peptide bonds. The two main types of secondary structure are the *α-helix* and *β-pleated sheet.*

Figure 21.11 shows these structures. In an **α-helix,** the protein chain spirals like a corkscrew, usually in a right-handed direction. This enables all peptide groups to form hydrogen bonds with the peptide groups above or below them. The resulting hydrogen bonds are parallel to the helix axis. The most common arrangement has 3.6 amino acid units per turn, with side chain groups sticking out from the coil.

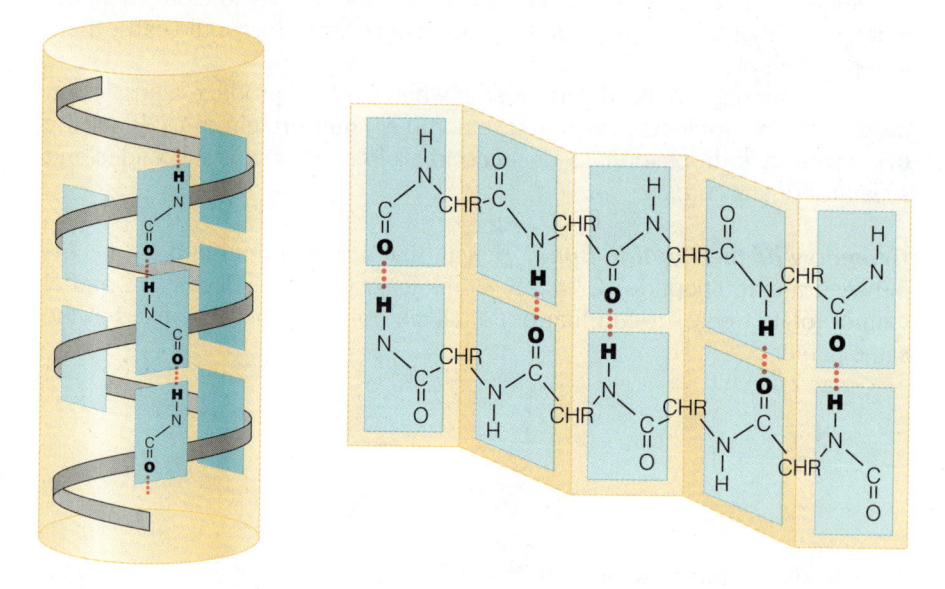

Figure 21.11 Secondary structure of a protein chain includes α-helix (left) and β-pleated sheet (right).

The **β-pleated sheet** (or *β*-sheet) structure aligns peptide-group atoms to form hydrogen bonds perpendicular to the chain (Figure 21.11). The zigzag sheet structures form from hydrogen bonding either between two regions of the same protein chain or between separate chains. When hydrogen-bonded peptide groups are arranged in the same direction (for example, their carbonyl carbons are both to the left of their nitrogens), the sheet is called *parallel.* When peptide groups run in opposite directions, the sheet is *antiparallel.*

Proteins vary considerably in their secondary structure. A few proteins are mostly α-helix or β-pleated sheet (see Section 21.5); others contain various amounts of one or both structures (Figure 21.12). Secondary structure that is neither α-helix or β-pleated sheet is said to be "random."

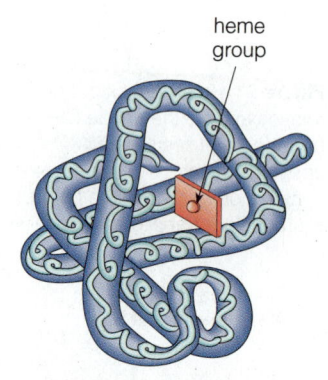

Figure 21.12 Model of a protein chain in hemoglobin that also contains nonprotein heme. Most of the protein is α-helix (shown as cylinders).

Tertiary Structure. The overall shape of a protein chain is its **tertiary structure.** Tertiary structure encompasses all the local regions of helix, sheet, and random arrangements, it also includes all other twists, turns, and linear sections that arise from other interactions between groups in the protein chain.

Figure 21.13 shows four important forces that contribute to tertiary structure. First, disulfide bonds between cysteine units hold together parts of the protein chain, even regions that are far apart in the primary structure. Hydrogen

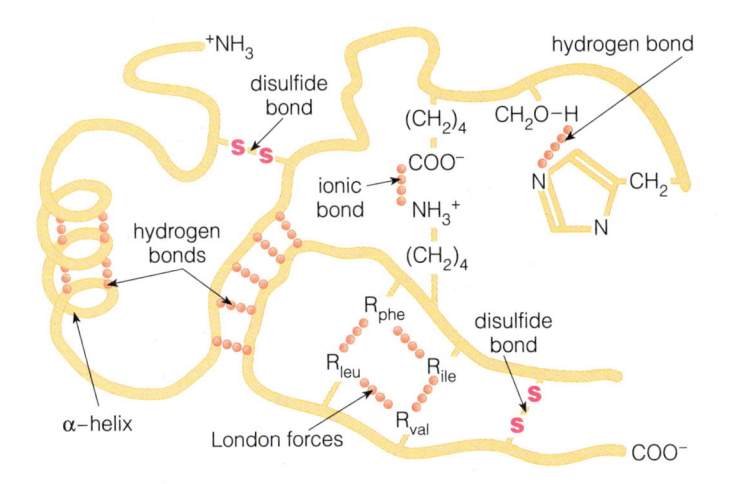

Figure 21.13 Hydrogen bonds, ionic bonds, disulfide bonds, and London forces all contribute to tertiary protein structure.

bonding between polar side-chain groups is a second factor. The third force is ionic bonding (sometimes called "salt bridges") between oppositely charged ionic side-chain groups. They occur, for example, between acidic amino acids (which have negatively charged side chains) and basic amino acids (which have positively charged side chains). The fourth interaction is between hydrophobic (nonpolar) side-chain groups. Although these groups attract each other with relatively weak London forces (Section 6.1), the large number of attractions makes this a major factor in many proteins.

The overall shape resulting from these interactions ranges from nearly spherical to very elongated and threadlike. In spherical proteins, nonpolar units tend to be buried inside, with polar and ionic side chains on the surface where they can form hydrogen bonds with water molecules.

Quaternary Structure. Some proteins are made up of multiple protein chains bonded together, usually by the same types of bonds that produce tertiary structure. The arrangement of individual protein chains into a functioning protein is called **quaternary structure.** Proteins that consist of a single protein chain do not have quaternary structure.

The quaternary structure of hemoglobin consists of four subunits (Figure 21.14). Each subunit has a protein chain and an iron-containing heme group (Figure 21.12). The subunits are of two types, called α and β, which differ in number and composition of amino acids. Hemoglobin consists of two α subunits and two β subunits joined together by ionic bonds, hydrogen bonds, and London forces. Since each subunit has a heme group that binds O_2, each hemoglobin molecule can bind four O_2 molecules.

Many types of quaternary structure exist. Protein chains within a protein may be the same or different (as in hemoglobin, which has two pairs of two different types). We indicate the number of protein subunits by a prefix and the suffix *-mer*. Hemoglobin, for example, is a *tetramer* because it has four protein chains. Myoglobin, an oxygen-binding protein in muscle that resembles one hemoglobin subunit (Figure 21.12), is a *monomer* and thus has no quaternary structure.

Denaturation. A protein must have the right shape in order to function. If an α subunit in hemoglobin unfolds, for example, or if the subunits separate, hemoglobin loses its ability to bind and release oxygen. Likewise, an enzyme

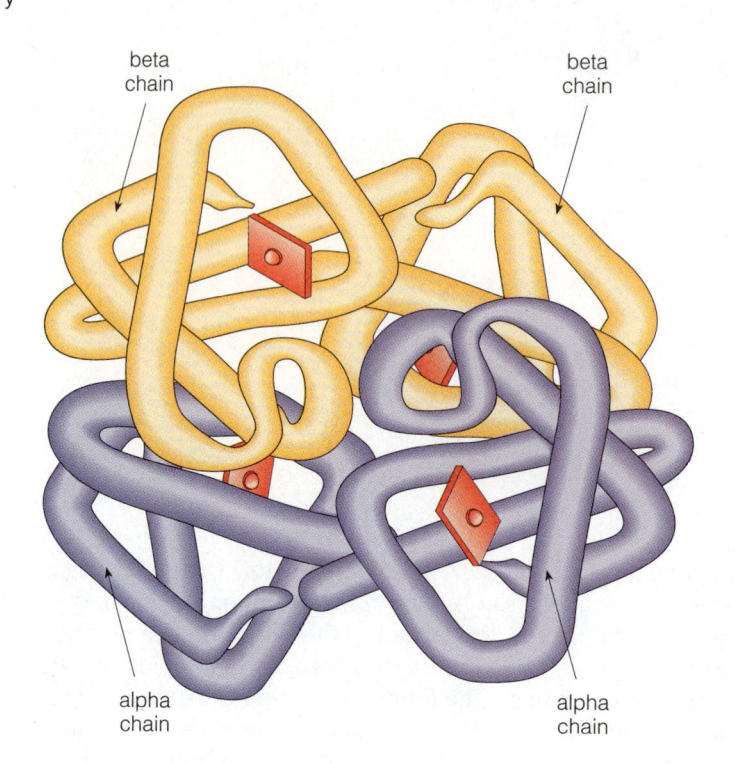

Figure 21.14 Model of quaternary structure of hemoglobin, a tetramer.

protein (see Chapter 22) loses its catalytic ability when its three-dimensional shape unravels.

When you digest food proteins, your body uses enzymes that hydrolyze peptide bonds. This action disconnects amino acids and destroys the primary level of protein structure—and all other levels, too. Those proteins not only cease to function; they cease to exist as proteins.

It's also possible to destroy protein function without completely destroying the proteins themselves. **Denaturation** is the disruption of secondary, tertiary, and quaternary (if any) protein structure, but not primary structure. A denatured protein still is a chain of amino acids, but it no longer twists and folds into the shape required to function.

Denaturing agents disrupt the forces responsible for higher levels of protein structure—ionic bonds, disulfide bonds, hydrogen bonds, and London forces—without breaking peptide bonds. Table 21.4 lists some denaturing agents.

Heat denatures proteins by disrupting relatively weak, noncovalent forces. Cooking food, for example, destroys protein function before your digestive tract destroys the proteins themselves. Strong acids or bases also denature proteins. Changing pH changes the ionic form of a protein's functional groups (Section 21.1), which changes its ability to form ionic and hydrogen bonds necessary for protein structure (Figure 21.15). Most proteins in your body function only at a fairly neutral pH. Heavy metals such as mercury and lead denature proteins by binding to cysteine sulfhydryl groups (see Chapter 14 opener).

Red blood cells give one dramatic example of how disulfide bonds affect protein structure. With their constant exposure to oxygen, sulfhydryl (—SH) groups in hemoglobin and other red blood cell proteins are vulnerable to oxidation to disulfide bonds; this would disrupt normal protein structure and destroy the cells. An enzyme system in red blood cells normally protects against this oxidation. In some people, however, this enzyme system may be genetically deficient, or inhibited by certain drugs or foods. As disruption increases, these

Table 21.4 Protein Denaturation Agents

Treatment	Method of Denaturation
Detergents	Suspend nonpolar amino acids in water, so they don't remain buried in the protein interior
Heat	Increases energy of the molecule, breaking weak bonds (such as hydrogen bonds) responsible for protein shape
Acids and bases	Alter the ionic form of amino acids by changing pH, thus disrupting hydrogen and ionic bonds
Mercaptoethanol (HOCH$_2$CH$_2$SH)	Reduces disulfide bonds that contribute to tertiary protein structure
Urea	Disrupts hydrogen bonds in proteins
Heavy metal ions (e.g., Pb^{2+}, Hg^{2+})	Bind free sulfhydryl (—SH) groups in proteins
Vigorous shaking or stirring	Disrupts weak bonds responsible for protein shape; may oxidize sulfhydryl groups

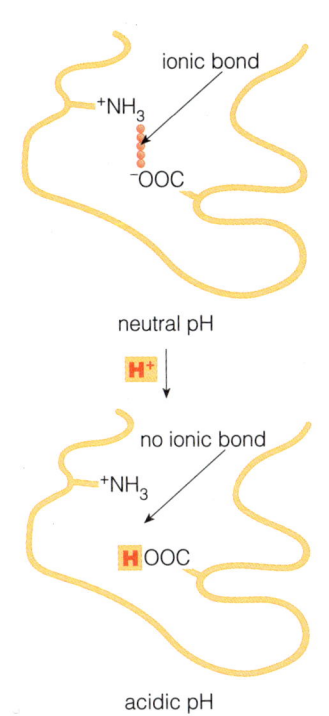

Figure 21.15 Changing pH changes the ionic form of side-chain groups, altering their ability to form ionic (and hydrogen) bonds.

people experience a shortage of red blood cells, a condition called *hemolytic anemia*.

Relationship Between Primary and Higher Levels of Protein Structure. You will learn in Chapter 23 how genes carry information for the proteins your cells make. But genes specify only the primary structure of proteins; they say nothing directly about secondary, tertiary, or quaternary structure.

So how do your proteins acquire the secondary, tertiary, and quaternary structures they need in order to function? It turns out that *primary structure determines secondary, tertiary, and quaternary structure.* In the cell, each protein chain naturally twists and turns as peptide groups and amino acid side-chains interact. Hydrogen, ionic, and disulfide bonds form, and nonpolar amino acids tuck inside to avoid water molecules near the protein surface. Eventually each protein finds the most stable shape for its amino acid sequence and its environment in the cell.

EXAMPLE 21.7 Identify which one of the following amino acid substitutions is the *least* likely to alter a protein's shape and ability to function (see Table 21.1): (a) ala for glu, (b) ser for lys, (c) phe for asp, (d) ile for val.

SOLUTION Substitution (d) because both are nonpolar amino acids of similar size.

21.5 CLASSIFICATION OF PROTEINS

Proteins are classified in several ways. We will use three categories—globular proteins, fibrous proteins, and conjugated proteins (Table 21.5).

Globular proteins, as the name implies, have somewhat spherical shapes and dissolve in water or salt solutions. Their shape enables them to bury nonpolar amino acids inside, away from water molecules. Polar and ionic amino acids are mostly on the surface, where they attract water molecules and help

Table 21.5 Classification of Proteins

Globular Proteins
Albumins—most abundant proteins in blood; bind and transport fatty acids and certain metal ions; soluble in water
Enzymes—trypsin, pepsin, ribonuclease, hundreds of others
Globulins—many different types; gamma globulins are antibodies; others are components of lipoproteins; soluble in salt solutions but not in water
Nutrient proteins—casein (milk), ovalbumin (egg), and gliadin (wheat)

Fibrous Proteins
Actin and myosin—proteins in muscle fibers
Collagen—major component in tendons, cartilage, bone, skin, and connective tissue
Elastins—in ligaments and lining of blood vessels
Keratins—major component of hair, nails, wool

Conjugated Proteins
Chromoproteins—contain colored heme or other groups; hemoglobin, myoglobin, cytochromes
Glycoproteins—contain carbohydrate; gamma globulins, interferon
Lipoproteins—contain lipid; VLDL, LDL, and HDL in blood
Metalloproteins—contain metal ions such as Mg^{2+}, Fe^{2+}, Mn^{2+}; hemoglobin, cytochromes, many enzymes
Nucleoproteins—contain nucleic acid; histones (proteins associated with DNA in chromosomes)

---

increase solubility. Globular proteins vary considerably in their amounts of α-helix and β-pleated sheet. They cannot have nearly 100% of either type and still fold into a spherical shape.

Most enzymes are globular proteins. So are most of the soluble proteins in blood. Two examples are *albumins* and *globulins* (Figure 21.9). Albumins, the most abundant proteins in blood, help transport metal ions and nonpolar fats. They also help maintain the osmotic pressure in blood (see Chemistry Spotlight in Section 7.5). One type of globulins—the gamma globulins—are produced by your immune system. Called *antibodies*, they protect against infectious agents (see Section 21.6).

Fibrous proteins, as their name suggests, have long, threadlike shapes. The shape often results from a rigid secondary structure, nearly 100% α-helix or β-pleated sheet. Most of your fibrous proteins are structural materials in the body. The most abundant is *collagen*, which occurs in connective tissue, skin, tendons, and bone. *Keratins* make up hair and nails, while *elastins* line blood vessels and are a major material in ligaments.

The long, rodlike shapes of fibrous proteins leave most of their amino acid units—including nonpolar ones—exposed on the surface. As a result, fibrous proteins don't dissolve in water. And you should be grateful. Imagine what would happen during a shower if your hair, skin, nails, muscles, and bones were made from water-soluble globular proteins.

Conjugated proteins contain nonprotein material. Hemoglobin, for example, contains the heme group. Because it contains iron (II) ion, hemoglobin is also one of many metalloproteins. Lipoproteins (such as low-density lipoproteins, LDL; Section 20.5) contain lipids. Glycoproteins (such as gamma globulins) contain carbohydrates, and nucleoproteins (chromosomes) contain nucleic acid.

21.6 A FEW IMPORTANT PEPTIDES AND PROTEINS

Collagen. About 30% of the protein in your body (and about 6% of your weight) is *collagen*, a fibrous protein that provides structure for your connective tissue, bones, tendons, teeth, cartilage, and skin.

Collagen consists of three protein chains wrapped around each other in a triple helix called *tropocollagen* (Figure 21.16). The primary structure of each

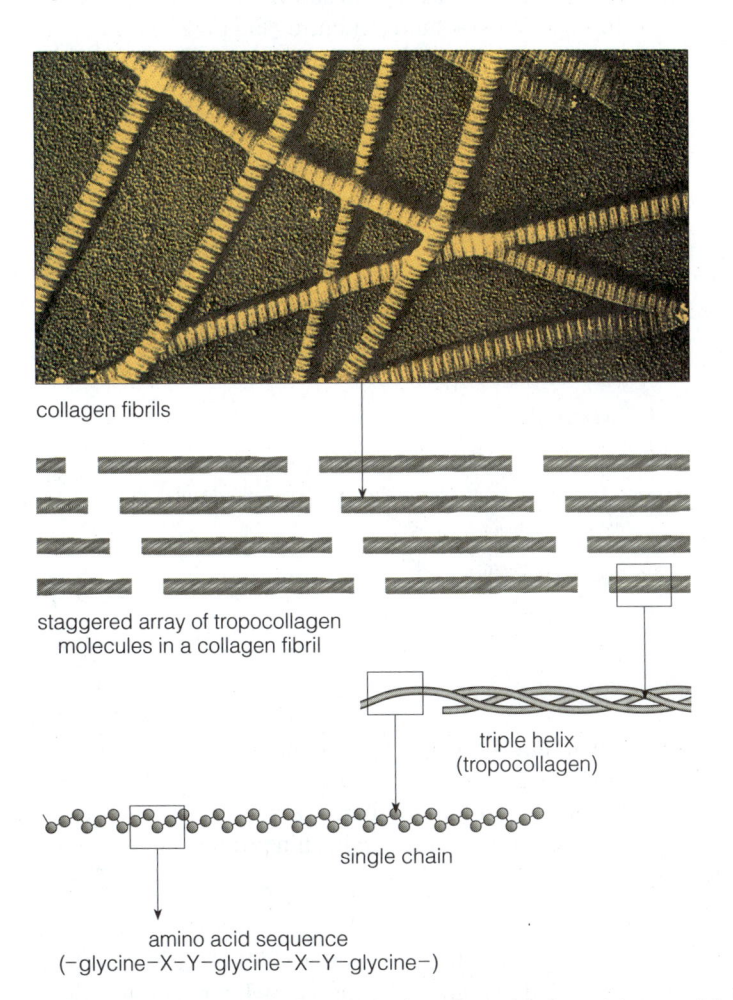

Figure 21.16 Structure of collagen, from collagen fibrils to the primary structure of a single chain. In the primary structure, X is any one of several amino acids; Y is usually proline or hydroxyproline.

chain typically has glycine units in every third position and large amounts of proline, lysine, and hydroxylated (—OH) derivatives of proline and lysine. Hydrogen and covalent bonds between side chains hold the three strands together. Tropocollagen molecules, in turn, bond with each other and pack into collagen fibers. Its ropelike structure makes collagen strong and light.

The triple helix isn't stable until lysine and proline units are oxidized to their hydroxylated (—OH) derivatives. The reaction with proline is:

proline unit →(O)/vitamin C→ 4-hydroxyproline unit

Notice that vitamin C (ascorbic acid) is needed in this reaction. People who lack vitamin C in their diets don't make stable collagen and suffer from deterioration of their collagen-rich materials. Symptoms of this disease, called *scurvy*, include spongy, bleeding gums and abnormal formation of bones and teeth.

Insulin. Insulin is a polypeptide hormone produced by the pancreas that helps regulate blood glucose levels and lipid metabolism. Pancreatic cells first synthesize a 110-amino acid chain called preproinsulin. Then they remove 59 amino acids, leaving insulin with a two-chain structure (Figure 21.17).

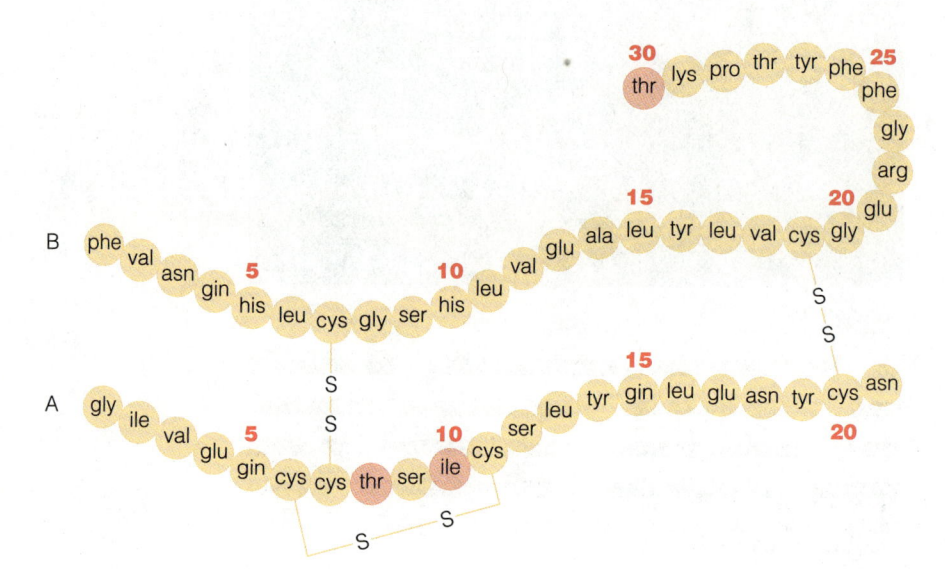

Figure 21.17 Human insulin contains two chains joined by two disulfide bonds. Pork insulin differs only in the amino acid at the end of the B chain (shaded). Beef insulin also has this difference, plus different amino acids at two other positions in the A chain (shaded).

People with insulin-dependent diabetes used to take injections of beef or pork insulin, which have slightly different structures from human insulin (Figure 21.17). In some patients, this caused mild allergic reactions. As we will see in Chapter 23, however, diabetics now use human insulin produced by bacteria as a result of genetic engineering.

Glutathione. Most (perhaps 90%) of the nonprotein thiol (—SH) groups in your cells are in a tripeptide called *glutathione*, which is γ-glutamylcysteinylglycine:

$$H_3^+N-CH-(CH_2)_2-\overset{O}{\underset{\|}{C}}+N-CH-\overset{O}{\underset{\|}{C}}+N-CH_2-COO^-$$
$$\underset{COO^-}{|} \quad \underset{H}{|} \underset{CH_2}{|} \quad \underset{H}{|}$$
$$\underset{SH}{|}$$

      r-glu   cys   gly

Glutathione readily reacts with oxidizing agents to form a disulfide product. If we represent glutathione as G—SH, the oxidation reaction is:

$$2\ G-SH \xrightarrow{(O)} G-S-S-G$$

Figure 21.18 Too little human growth hormone produces a dwarf, whereas too much produces a giant.

Because it is abundant and reactive, glutathione acts like a sponge in the cell to soak up (react with) oxidizing agents; this protects other, more critical materials (such as proteins, unsaturated fats, and DNA) from oxidative damage. For example, glutathione is used by the enzyme system that protects red blood cells against the oxidation and destruction that occurs in hemolytic anemia (Section 21.4).

Growth Hormone. Your pituitary gland (hypophysis) produces and secretes *growth hormone,* a small protein containing 191 amino acid units. Children who don't produce enough of this hormone are dwarfed, while overproduction results in very tall and large people (Figure 21.18). Growth hormones from other animals differ slightly from human growth hormone in their primary structures and are ineffective in humans.

Until recently, human growth hormone to treat dwarfed children was isolated from the pituitary glands of cadavers. This made it scarce and expensive—about $20,000 per year to treat one child. Now human growth hormone, like human insulin, is produced by genetically engineered bacteria (Section 23.6).

Immunoglobulins. The gamma globulin fraction in blood (Figure 21.9) contains several types of *antibodies,* also known as *immunoglobulins.* These antibodies attack and destroy foreign cells and substances, called *antigens.* Immunoglobulins are produced by your immune system to help defend you against infectious organisms and other harmful materials.

Immunoglobulins typically are Y-shaped glycoproteins. Figure 21.19 shows the structure of the most common type of antibody, called immunoglobulin G (IgG). Four protein chains are joined by disulfide bonds into a Y-shaped structure. IgG molecules bind and inactivate antigens at the ends of the "arms." The primary structure of IgG molecules varies greatly in these binding regions; this enables different antibodies to bind different antigens. The nonbinding region has the same amino acid sequence in all IgG molecules.

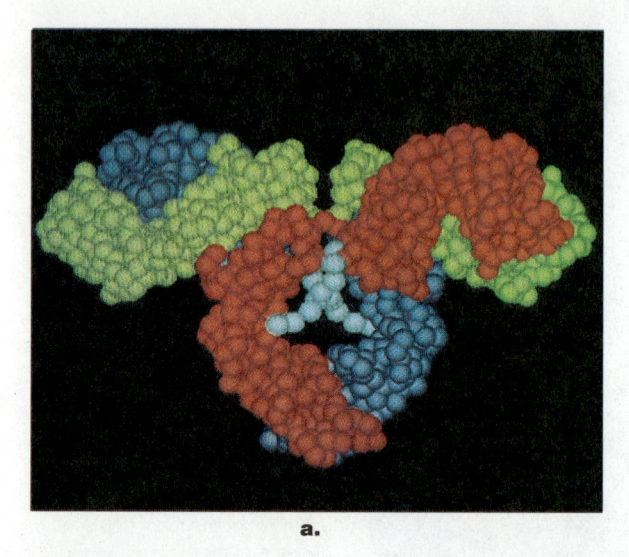

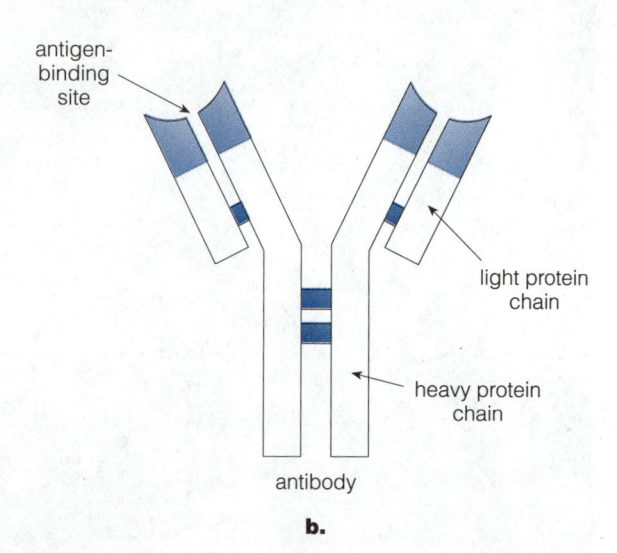

Figure 21.19 (a) Model of immunoglobulin G antibodies. Their Y-shaped structure (b) consists of four protein chains joined by disulfide bonds (blue). The antigen-binding regions (shaded) contain variable amino acid sequences.

Enkephalins and Endorphins. Some natural peptides bind to the same sites in the brain as opiates such as morphine. They are called *enkephalins*, which means "in the brain." Figure 21.20 shows the amino acid sequences of two enkephalins (both pentapeptides) that differ only in the amino acid (methionine or leucine) at the carboxyl end.

Longer peptides that bind to opiate sites in the brain are called *endorphins*. The structures of two endorphins, dynorphin and β-endorphin, are shown in Figure 21.20. Notice that they begin with the same pentapeptide sequences as enkephalins. Their tyrosine units at the amino end structurally resemble morphine (Figure 21.21), which may be why they bind to the same receptors in the brain.

tyr—gly—gly—phe—met^5

(met) enkephalin

tyr—gly—gly—phe—leu^5

(leu) enkephalin

tyr—gly—gly—phe—leu^5—arg—arg—ile—arg—pro^{10}—lys—leu—lys

dynorphin

tyr—gly—gly—phe—met^5—thr—ser—glu—lys—ser^{10}—gln—thr—pro—leu—val^{15}
                                                                                     |
                                                                                    thr
                                                                                     |
glu—gly^{30}—lys—lys—tyr—ala—asn^{25}—lys—ile—ile—ala—asn^{20}—lys—phe—leu

β-endorphin

Figure 21.20 Amino acid sequences of two enkephalins and two endorphins.

Like the opiates, these natural peptides are effective analgesics. Endorphins and enkephalins may help mediate the painkilling action of acupuncture and may mask fatigue during long periods of exercise. These compounds also may be involved in regulating mood and personality. Indeed, the story of these compounds is just beginning to be told.

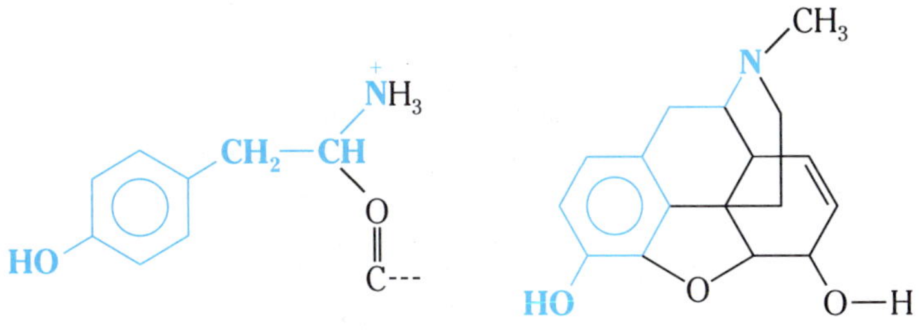

Figure 21.21 The tyrosine end of endorphins and enkephalins resembles morphine (similarities in color).

> Life is the mode of existence of proteins.
>
> *Friedrich Engels (1820–1895)*

SUMMARY

Proteins are polymers of amino acids joined by peptide (amide) bonds. Amino acids have the structure ^+H_3N—CHR—COO^- and are classified according to their side-chain (R) groups. Except for glycine, amino acids occur as pairs of enantiomers, D and L; only L-amino acids are used in the body.

Amino acids exist in different ionic forms depending on pH. At pH values more acidic than the pK_a value for a functional group, that group is protonated; at a pH more alkaline than pK_a, the functional group is unprotonated. At neutral pH most amino acids exist as dipolar ions (zwitterions), having charged carboxyl (—COO^-) and amino (—NH_3^+) groups. Amino acids with a net charge migrate toward the oppositely charged pole in electrophoresis. The pH at which an amino acid has no net charge is its isoelectric point, which for a simple amino acid (with no ionizable side chains) equals the average of its two pK_a values.

Primary protein structure is the sequence of amino acids; it sometimes includes disulfide bonds between cysteine sulfhydryl groups. Restricted rotation around the carbonyl C—N bond fixes six atoms in the same plane around a peptide bond. The net charge on a protein depends on pH. At its isoelectric point, a protein has no net charge, doesn't migrate in electrophoresis, and has minimum solubility in water.

Secondary protein structures are α-helix, β-pleated sheet, or random. Maximum hydrogen bonding between peptide groups produces the helix and sheet structures. Tertiary structure, the overall protein shape, results from disulfide bonds, ionic bonds, hydrogen bonds, and London forces between nonpolar amino acids in the protein interior. Some proteins have quaternary structure, an arrangement of separate protein subunits into a complex protein. Primary structure determines secondary, tertiary, and quaternary structure. Denaturation is the loss of higher levels of structure, but not primary structure.

Proteins are classified as globular or fibrous based on their shape and solubility in water. Conjugated proteins contain nonprotein material. A few important peptides and proteins are collagen, insulin, glutathione, growth hormone, immunoglobulins, and enkaphalins and endorphins.

KEY TERMS

Amino acid (21.1)
Conjugated protein (21.5)
Denaturation (21.4)
Dipolar ion (21.1)
Electrophoresis (21.1)
Essential amino acid (21.2)
Fibrous protein (21.5)

Globular protein (21.5)
α-helix (21.4)
Isoelectric point (21.1)
Peptide (21.3)
Peptide bond (21.3)
β-pleated sheet (21.4)
Polypeptide (21.3)

Primary structure (21.3)
Protein (21.1)
Quaternary structure (21.4)
Secondary structure (21.4)
Sulfhydryl group (21.3)
Tertiary structure (21.4)

EXERCISES

Even-numbered exercises are answered at the back of this book.

Amino Acids

1. Write the general structure for an α-amino acid.
2. The substance $^+H_3N-CH_2CH_2-COO^-$ is *not* an α-amino acid although it has both an amino and a carboxylic acid group. What kind of amino acid is it? [Review Section 16.1 if necessary.]
3. Do α-amino acids occur as optical isomers? Explain.
4. Which amino acid has no chiral carbon atom and thus does not have different optical isomer forms?
5. Consult Table 21.1 to determine which amino acids have more than one chiral carbon. How many stereoisomers exist for each of these amino acids?
6. Write structural formulas showing the ionic forms of leucine as it would exist at **(a)** pH 1, **(b)** pH 7, and **(c)** pH 12.
7. Aspartic acid has the following pK_a values: 2.0 for one carboxyl group, 4.0 for the side-chain carboxyl group, 10.0 for the amino group. Write structural formulas for the main ionic form that would occur at pH **(a)** 1, **(b)** 3, **(c)** 7, and **(d)** 10.
8. Write a structural formula for the form of aspartic acid (exercise 7) that would not migrate during electrophoresis.
9. From your answer to exercise 8, estimate the pH that would produce the maximum amount of the form of aspartic acid that has no net charge. Compare your answer with the isoelectric point (p*I*) for aspartic acid listed in Table 21.1.
10. Considering its p*I* value (Table 21.1), would aspartic acid migrate toward the anode or the cathode during electrophoresis at each of the following pH values: **(a)** 8.1, **(b)** 6.2, **(c)** 1.8.
11. Look at the p*I* values in Table 21.1. Select a pH for electrophoresis at which valine and lysine would migrate in opposite directions.
12. Look at the p*I* values in Table 21.1. Select a pH for electrophoresis at which cys and his would migrate in opposite directions.
13. Suppose a new amino acid had the side chain $-(CH_2)_2CHCH(CH_3)_2$. Classify it into one of the groups listed in Table 21.1.
14. The amino acid that contains S but not in the form of a thiol (sulfhydryl) group is _____.

Primary Protein Structure

15. Write a structural formula for phenylalanyllysine (phe-lys) as it would exist at pH 7.
16. Write a structural formula for lysylphenylalanine (lys-phe) as it would exist at pH 7.
17. After a protein is oxidized, are its cysteine S atoms mostly in the form of sulfhydryl groups or disulfide bonds?
18. In each of the following pairs, identify the larger molecule: **(a)** tetrapeptide or octapeptide, **(b)** protein or hexapeptide, **(c)** dipeptide or polypeptide.
19. Name the type of hybrid orbitals around the carbonyl C in a peptide bond.
20. There are *two* pairs of atoms in a peptide bond joined by bonds that don't have free rotation around them. Name both pairs of atoms.

Acid–Base Properties

The isoelectric points of some proteins are as follows: serum albumin (4.8), insulin (5.4), collagen (6.7), and hemoglobin (7.1). Use this information to complete exercises 21–24.

21. Which proteins migrate toward the anode (positive pole) during electrophoresis at pH 6.0?
22. Select a pH for electrophoresis at which all these proteins would migrate toward the cathode.
23. At what pH is insulin the least soluble in water?
24. Which of the four proteins has the largest number of basic amino acids and the fewest acidic ones in its primary structure?

Higher Levels of Protein Structure

25. What type of bonding is most responsible for secondary protein structure?
26. Is the β-pleated sheet structure shown in Figure 21.11 a parallel or antiparallel sheet?
27. In an α-helix or β-pleated sheet, hydrogen bonding is maximal between what functional groups?
28. Are hydrogen bonds parallel or perpendicular to the long axis of an α-helix?
29. Which of the following affect tertiary protein structure? **(a)** hydrogen bonds, **(b)** ionic bonds, **(c)** hydrophobic interactions, **(d)** disulfide bonds, **(e)** all of the above.
30. Fibrous and globular proteins are defined primarily in terms of the following level of structure: **(a)** primary, **(b)** secondary, **(c)** tertiary, **(d)** quaternary.
31. In a globular protein, polar and ionic amino acids are typically located on the _____ (outside or inside) of the protein, and nonpolar amino acids are mostly _____ (outside or inside).

32. The type of quaternary structure consisting of two protein subunits is called a _____.
33. How many different kinds of subunits are in hemoglobin?
34. Cooking food denatures its proteins. Why doesn't this make proteins nutritionally unsuitable?
35. Once a protein is denatured, can it ever regain its secondary, tertiary, and quaternary (if any) structure?
36. How do heavy metal ions such as Hg^{2+} and Pb^{2+} denature proteins?
37. Identify which of the following amino acid substitutions is the least likely to alter a protein's shape and ability to function: (a) tyr for lys, (b) arg for lys, (c) val for his, (d) trp for arg, (e) pro for glu.
38. In sickle-cell anemia, a valine unit substitutes for a glutamic acid unit on the surface of the β protein chain in hemoglobin.
    (a) What amino acid could substitute for glutamic acid and be the least likely to alter hemoglobin structure (see Table 21.1)?
    (b) Why are valine units usually *not* on the surface of proteins?

Classification of Proteins

39. An enzyme containing Zn^{2+} could be classified as: (a) fibrous protein, (b) globular protein, (c) conjugated protein, (d) more than one of the above, (e) none of the above.
40. Which categories of conjugated protein (Table 21.5) apply to hemoglobin?
41. Which type of protein is the principal component of hair: (a) globular, (b) fibrous, or (c) conjugated?
42. Why are fibrous proteins typically less soluble in water than globular proteins are?
43. What holds the three strands together in the collagen triple helix?

Important Peptides and Proteins

44. Is insulin (Figure 21.17) more commonly called a peptide or a polypeptide?
45. Which term (peptide or polypeptide) would you use to refer to (a) (met)enkephalin and (b) β-endorphin (Figure 21.20)?
46. What bonds join the two insulin chains together?
47. In the presence of a reducing agent, what would happen to the two chains in an insulin molecule?
48. Glutathione is γ-glutamylcysteinylglycine. What does the γ signify? [Hint: Review nomenclature for carboxylic acids in Section 16.1 if necessary.]
49. In the presence of a reducing agent, does glutathione exist mostly in its sulfhydryl or disulfide form?
50. List the following in order of increasing molecular weight: (a) growth hormone, (b) insulin, (c) glutathione, (d) immunoglobulin.
51. What region of primary structure in an immunoglobulin G molecule enables the antibody specifically to recognize an antigen?
52. The tyrosine unit at the amino end of enkephalins and endorphins structurally resembles what natural analgesic?

DISCUSSION EXERCISES

1. What important functions of the body do *not* involve proteins?
2. Do most proteins in the body have a net positive or negative charge? Explain. Where in the body would the answer be different? Explain.
3. Is a detergent likely to have a greater denaturing effect on globular or fibrous proteins? Explain.
4. Do amino acids and proteins function as buffers? [Review Section 9.6 if necessary.]

22

Enzymes

1. What are enzymes? What makes them such effective and specific catalysts?
2. What are the main nonprotein materials certain enzymes need to be effective catalysts? What is the role of water-soluble vitamins?
3. How are enzymes classified?
4. What factors change the rates of enzyme-catalyzed reactions?
5. How is enzyme activity regulated in the body? What are some practical examples of drugs that inhibit enzyme activity?
6. What are some clinical uses of enymes?

Cancer and Enzyme Inhibition

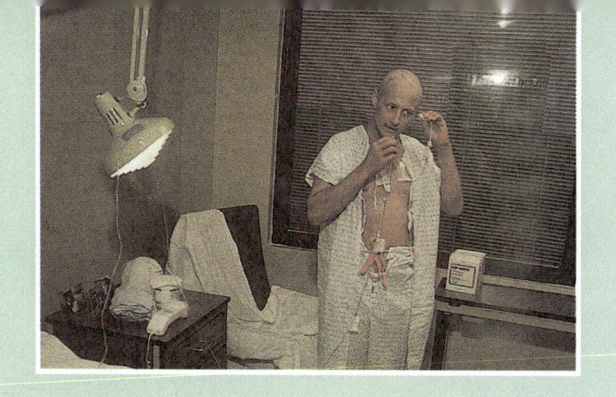

Cancer is the name for a group of diseases in which cells multiply uncontrollably, eventually invading the surrounding tissue and spreading to other parts of the body. One way to treat cancer is with chemicals, an approach called *chemotherapy*.

Anticancer drugs prevent cells from multiplying so rapidly. Some drugs work by inhibiting enzyme proteins that catalyze reactions necessary to make DNA. If a cell cannot make new DNA, it cannot multiply. Many anticancer drugs structurally resemble substances that the body normally makes into DNA. By masquerading as one of these substances, the drug binds to an enzyme and prevents it from catalyzing a reaction necessary to make DNA. Compounds that block metabolic processes in this way are called *antimetabolites*.

One example is 5-fluorouracil (5-FU). This antimetabolite resembles uracil (Figure 22.1), which an enzyme normally converts into thymine, an essential component of DNA. By inhibiting this enzyme, 5-FU blocks the synthesis of thymine and thus blocks DNA synthesis; this, in turn, prevents cell multiplication. One of the earliest anticancer drugs, 5-FU is used to treat colon, breast, ovarian, prostate, and stomach cancer.

Another example is 6-mercaptopurine (Figure 22.1), which resembles adenine, another normal component of DNA. Enzymes mistakenly incorporate this drug (instead of adenine) into newly synthesized DNA, thus preventing cell multiplication. 6-Mercaptopurine is used to treat acute lymphatic leukemia.

Because they prevent cell reproduction, anticancer drugs also lower the number of normal cells that multiply rapidly, such as intestinal cells and blood-forming bone marrow cells. As a result, cancer patients receiving chemotherapy often experience low blood-cell counts, infections, nausea, vomiting, and hair loss.

uracil   *thymine* (normal base in DNA)   *5-fluorouracil*

adenine (normal base in DNA)   *6-mercaptopurine*

Figure 22.1 Drugs used in cancer chemotherapy. Structural differences between the drugs and the normal compounds are in color.

If you want to speed up a chemical reaction in the laboratory, you can try several things. You can heat the reactants, increase their concentrations, or (in some cases) add an acid or base to help catalyze the reaction.

None of these techniques help much in the body. Your body is finely tuned to maintain a fairly constant temperature, concentration of each substance, and pH level. Even under these mild conditions, however, reactions need to go fast enough to sustain life.

Your body does this by producing *enzymes,* which catalyze (accelerate) the reactions that keep you alive. Enzymes also give your body a way to regulate its reactions, enabling you to adapt to different situations. When you don't eat for several hours, for example, enzymes accelerate energy-producing reactions to replenish your energy supply.

Enzymes are responsible for *metabolism,* the wide array of chemical reactions in the body (Figure 22.2). Although they are proteins, many enzymes need the assistance of nonprotein materials to work as catalysts. As you will see, those materials include metal ions and vitamins.

Figure 22.2 Enzymes convert carbohydrates into ethanol by fermentation.

22.1 ENZYMES: EFFECTIVE AND SPECIFIC CATALYSTS

Enzymes are proteins that catalyze chemical reactions. Reactants in enzyme-catalyzed reactions are called **substrates.** Virtually every reaction in your body has a particular enzyme to catalyze it. Like all other catalysts, enzymes are unchanged in the overall reaction and do not alter the equilibrium of a reaction, though they help the system reach equilibrium sooner. But enzymes tend to differ from other catalysts in two important ways—their effectiveness and their specificity.

Effectiveness. Consider the reaction:

$$CO_2 + H_2O \rightleftharpoons H_2CO_3$$

Recall that this reaction (Section 9.7) helps move CO_2 from tissues (where it is produced) into the blood, then releases CO_2 from the lungs into the air. An enzyme molecule converts 600,000 molecules of CO_2 to H_2CO_3 *per second.* Without this catalyst, the reaction in your body would proceed only 1/10,000,000 as fast.

Enzymes typically speed up reactions by at least a million times, and often more.

One way to explain how enzymes do this is to consider the activation energy of a reaction. Recall that the activation energy is the minimum amount of energy a reactant needs to react (Section 8.2); it is the energy barrier or "hill" a reactant has to overcome in order to react. Recall also that activation energy is related to the rate of a reaction; the lower the activation energy, the faster the reaction can occur.

An enzyme, like other catalysts (Figure 8.13 and Table 8.2), lowers activation energy, thus making a reaction go faster (Figure 22.3). An enzyme does this

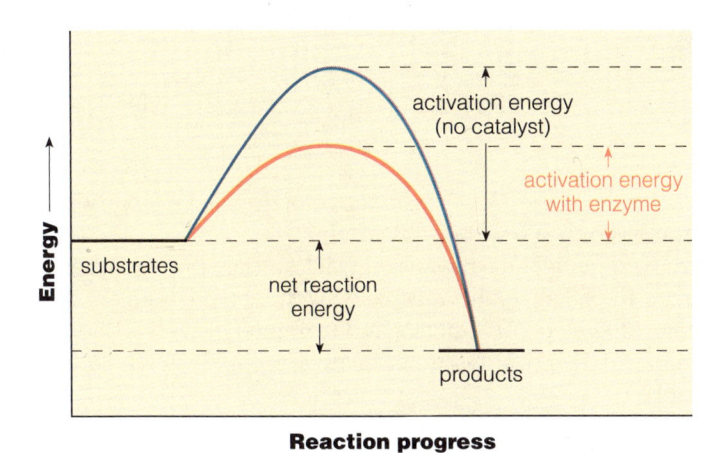

Figure 22.3 An enzyme accelerates a reaction by lowering its energy of activation.

by combining with the reactant (substrate) and aligning it in a way that favors the reaction. As a result, less energy is required than the reactant would need in the absence of a catalyst.

One example is the hydrolysis of sucrose (table sugar) to glucose and fructose. In the absence of a catalyst, sucrose in water hydrolyzes very slowly. An acid catalyst speeds things up, lowering the activation energy to about 110 kJ (26 kcal) per mol. But an enzyme called invertase is much more effective than acid, reducing the activation energy to 30–40 kJ (8–10 kcal) per mol.

Specificity. Enzymes are much more specific than most catalysts. Only a small number of substances typically work as substrates for a particular enzyme. An acid such as HCl catalyzes the hydrolysis of any disaccharide. But while the enzyme invertase is an excellent catalyst for hydrolyzing sucrose, it is not effective for hydrolyzing other disaccharides such as lactose and maltose. To hydrolyze those disaccharides, your body uses different enzymes (called lactase and maltase, respectively).

Why are enzymes so specific? According to one idea, the **lock-and-key model,** enzymes have a specific, rigid structure like a lock; to be a substrate, a substance must be like a key, with just the right shape to fit the lock (enzyme). Figure 22.4 illustrates this idea. The region of the enzyme where a substrate binds and is converted into product is called the **active site.**

The lock-and-key model explains enzyme specificity, but the notion of a rigid active site doesn't fit with evidence that enzymes often change shape a bit when substrates bind to them (Figure 22.5). Now biochemists prefer a revised model called the **induced-fit model.** According to this idea, the active site must have a specific orientation of catalytic groups to carry out a reaction; when it binds to the enzyme, a substrate induces (brings about) this proper

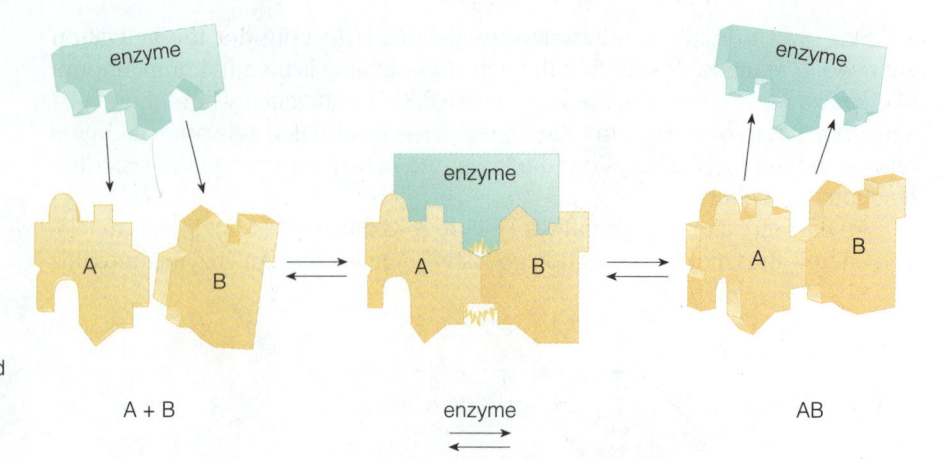

Figure 22.4 Substrates (A and B) fit the enzyme active site like keys fit a lock, according to the lock-and-key model.

A + B   enzyme   AB

orientation (Figure 22.5). Since very few substances can do this for a particular enzyme, each enzyme has only a few substrates.

Because they are so specific, enzymes usually act on only one member of a pair of enantiomers. Recall that enzymes selectively act on D-sugars and L-amino acids but have little effect on L-sugars and D-amino acids (see Chemistry Spotlight in Section 19.3). Enzymes themselves are chiral, made from L isomers of amino acids.

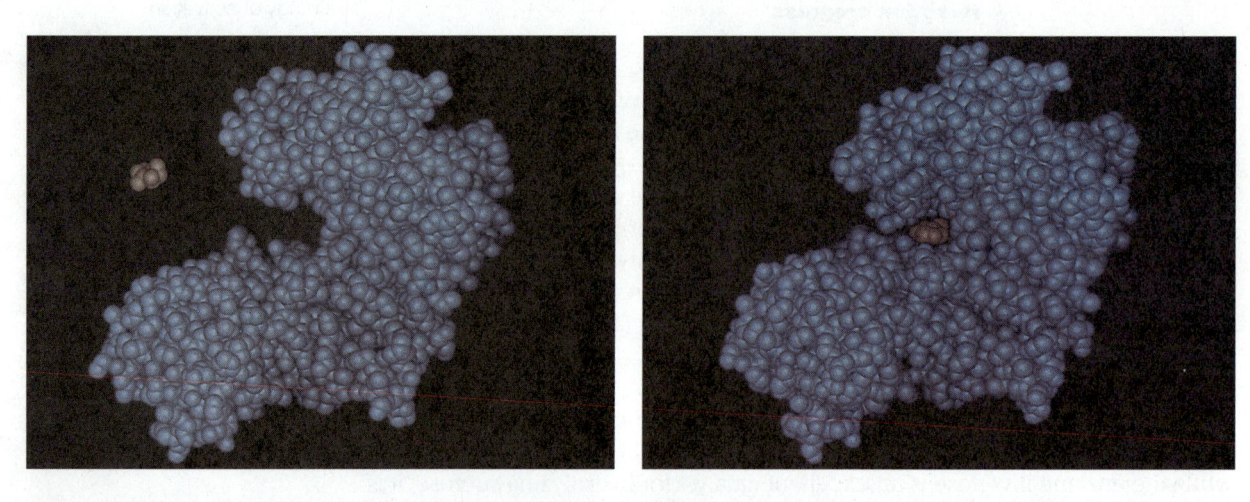

Figure 22.5 Model of how an enzyme (hexokinase) changes shape when it binds a substrate, glucose (red). Then the upper and lower parts of the enzyme close around the substrate, aligning catalytic groups in the active site to accelerate the reaction.

Enzymes also act on certain nonchiral substances to produce just one of two possible stereoisomer products. For example, an enzyme catalyzes the following reaction:

$$\begin{array}{c} CH_2-OH \\ | \\ HO-C-H \\ | \\ CH_2-OH \end{array} + R-OPO_3^{2-} \longrightarrow \begin{array}{c} CH_2-OH \\ | \\ HO-C-H \\ | \\ CH_2-OPO_3^{2-} \end{array} + R-OH$$

glycerol   glycerol-3-phosphate

Glycerol does not have a chiral carbon, but glycerol-3-phosphate has one chiral carbon (shown in color) and thus exists as a pair of enantiomers. An enzyme produces only one of the enantiomers of glycerol-3-phosphate (the L isomer, shown above).

How does an enzyme do this? One explanation is that a substrate binds to an enzyme at three (or more) points, one of which is where the substrate reacts. When this happens, only one of the two possible enantiomers can form (Figure 22.6).

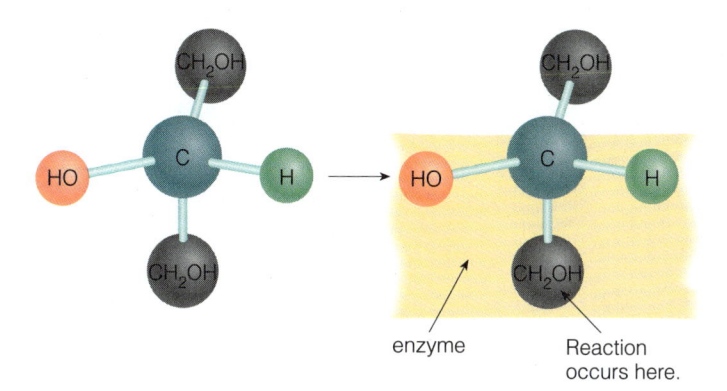

Figure 22.6 If a substrate such as glycerol binds to an enzyme at three points, the enzyme distinguishes between two identical groups (—CH$_2$OH) and produces just one member of a pair of enantiomers.

Enzymes are such effective catalysts that they keep your life-sustaining reactions going at appropriate rates even under the very mild reaction conditions in your body. And because they are so specific in the substrates they use, enzymes enable your body to maintain an orderly pattern of reactions, called metabolism, without hundreds of other side reactions.

22.2 COENZYMES AND VITAMINS

Coenzymes. In order to be effective catalysts, some enzymes require the assistance of nonprotein materials, called **cofactors.** Cofactors bind to an enzyme, usually in the vicinity of its active site, and help that site function more effectively. Many cofactors are metal ions, such as Zn^{2+}, Mn^{2+}, Co^{2+}, Fe^{2+}, Mg^{2+}, Cu^{2+}, and K^+.

One reason you need certain metals in your diet is to provide enzyme cofactors.

Other cofactors are organic substances called **coenzymes.** Most coenzymes are made from vitamins (Table 22.1), and this is the main reason you need those vitamins in your diet.

Table 22.1 Coenzymes

Coenzyme	Vitamin Source	Type of Reaction Catalyzed
NAD$^+$ (Nicotinamide adenine dinucleotide)	Niacin (B$_5$)	Oxidation–reduction
NADP$^+$ (Nicotinamide adenine dinucleotide phosphate)	Niacin (B$_5$)	Oxidation–reduction
FAD (Flavin adenine dinucleotide); FMN (Flavin mononucleotide)	Riboflavin (B$_2$)	Oxidation–reduction
CoA (Coenzyme A)	Pantothenic acid	Transfer acyl groups
Lipoic acid		Transfer acyl groups
Thiamine pyrophosphate	Thiamine (B$_1$)	Transfer acyl groups
Biotin	Biotin	CO$_2$ fixation
Pyridoxal phosphate	Pyridoxine (B$_6$)	Transfer amino groups
Tetrahydrofolic acid	Folic acid	Transfer various one-carbon groups
Cobalamin	B$_{12}$	Various specialized reactions
Ascorbic acid	C	Hydroxylation reactions

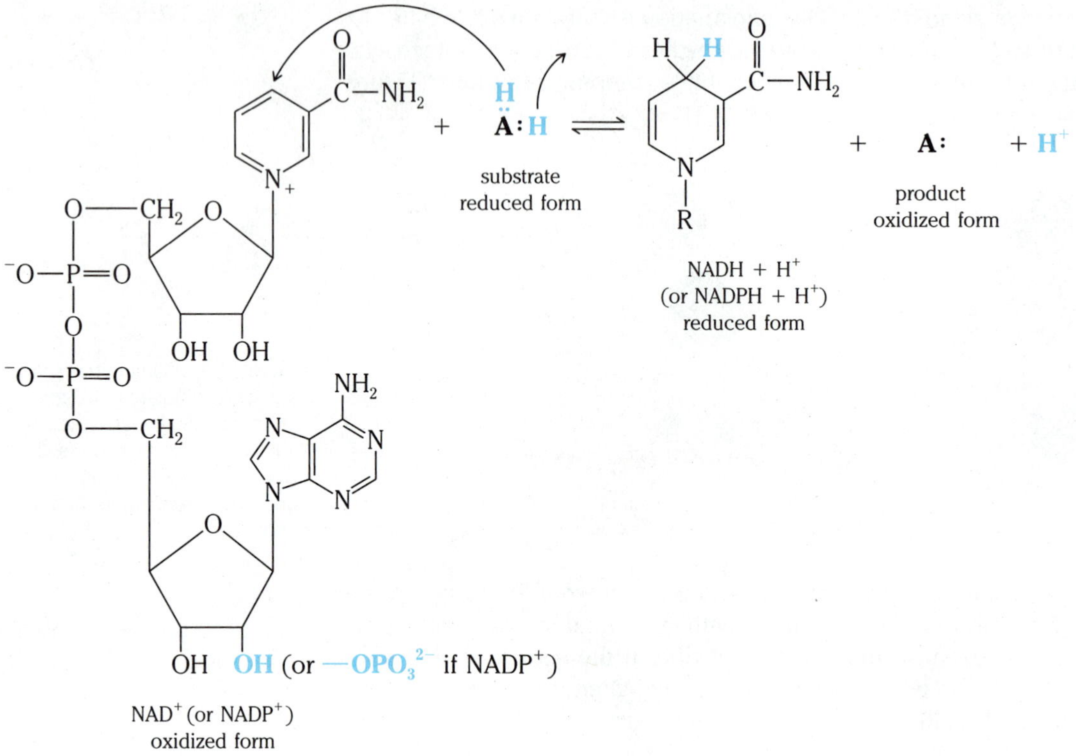

Figure 22.7 Oxidized and reduced forms of NAD⁺ and NADH, respectively. These structures are repeated for the coenzymes NADP⁺ and NADPH, except for the presence of a phosphate ester (—OPO₃²⁻) as shown.

A single molecule of NAD⁺ can participate in many enzyme-catalyzed reactions, continually shuttling back and forth between its oxidized and reduced forms.

Coenzymes vary in how tightly they bind to enzymes. Some remain tightly bound at all times. Others are like substrates in that they bind during the reaction, are converted into a different form, then leave the enzyme when the reaction is completed. Unlike substrates, however, coenzymes revert to their original form, sometimes in a subsequent reaction with another enzyme.

One example is the coenzyme NAD⁺ (Nicotinamide Adenine Dinucleotide), which is made from the vitamin niacin and participates in oxidation–reduction reactions (see Chemistry Spotlight in Section 5.3; Table 22.1). In the presence of a suitable enzyme, NAD⁺ accepts two electrons and one proton (H⁺) from the substrate being oxidized to become NADH, the reduced form of the coenzyme (Figure 22.7):

$$\text{AH}_2 + \text{NAD}^+ \xrightarrow{\text{enzyme 1}} \text{A} + \text{NADH} + \text{H}^+$$

substrate   coenzyme   product   coenzyme
(reduced)   (oxidized)   (oxidized)   (reduced)

NADH then leaves the enzyme and is available for use by the same or another enzyme that reduces a substrate, regenerating the original (oxidized) form of the coenzyme, NAD⁺:

$$\text{B} + \text{NADH} + \text{H}^+ \xrightarrow{\text{enzyme 2}} \text{BH}_2 + \text{NAD}^+$$

substrate   coenzyme   product   coenzyme
(oxidized)   (reduced)   (reduced)   (oxidized)

Although several different enzymes may use a particular coenzyme, each coenzyme has chemical properties that make it most suitable for one type of reaction. So each enzyme using that particular coenzyme is likely to catalyze that type of reaction. Table 22.1 lists the types of reactions associated with each coenzyme.

Vitamins. **Vitamins** are a group of about 20 organic compounds that you need in your diet for health and that your body cannot synthesize. The name is a contraction of "vital amine" and comes from an earlier (mistaken) belief that all these substances are amines.

Vitamins were discovered largely because diseases could be traced to deficiencies of these compounds. Although many such diseases have been known for centuries, only in the 20th century have scientists identified the actual substance that cures each disease.

Vitamins are classified as *water-soluble* (polar or ionic) or *fat-soluble* (nonpolar; see Section 20.6). The B vitamins and vitamin C are water soluble (Figure 22.8). Because these vitamins are rapidly excreted in urine you need to replenish them daily. As a result, these vitamins rarely accumulate to toxic levels.

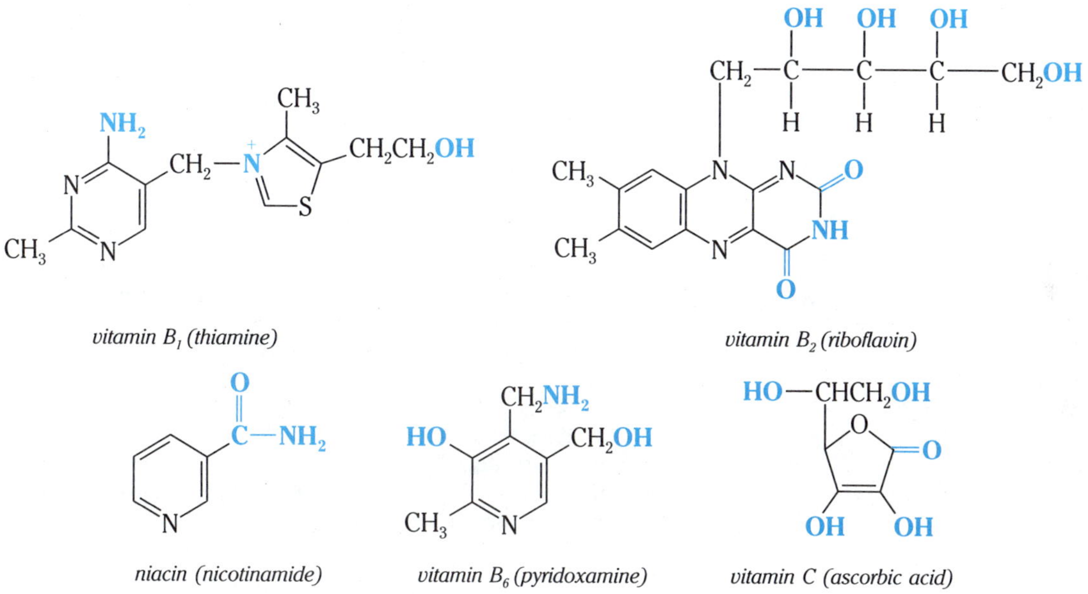

vitamin B$_1$ (thiamine)

vitamin B$_2$ (riboflavin)

niacin (nicotinamide)

vitamin B$_6$ (pyridoxamine)

vitamin C (ascorbic acid)

You may have heard that vitamins give you energy, but this is misleading. Many B vitamins (which often are modified chemically in the body) and vitamin C serve as coenzymes (Table 22.1). By themselves, vitamins aren't a significant source of energy, but as coenzymes they help metabolize carbohydrates, lipids, and proteins to produce energy.

Figure 22.8 Some water-soluble vitamins. Polar and ionic groups are in color.

Table 22.2 lists some sources and deficiency diseases of water-soluble vitamins. *Beriberi* is most common in parts of Asia where people eat large amounts of polished rice, which has the thiamine-containing outer hull removed. This deficiency of thiamine (vitamin B$_1$) produces stiff limbs, an enlarged heart, paralysis, pain, and eventual deterioration of the nervous system.

People who lack niacin (vitamin B$_5$) in their diets develop *pellagra*. Some symptoms of pellagra include scaly skin, diarrhea, an inflamed mouth, and an impaired central nervous system (dementia). Pellagra may occur in people

CHEMISTRY SPOTLIGHT

Albert Szent-Györgyi and Vitamin C

In the 1930s, Albert Szent-Györgyi, was trying to unravel the structure of vitamin C.

A Hungarian-born scientist who won the 1937 Nobel prize in physiology or medicine, Szent-Györgyi had discovered that the compound had carbohydratelike features (Figure 22.8), but he was ignorant of its complete structure. Since carbohydrates are commonly named with the suffix -*ose*, he submitted his results in a paper and suggested calling the vitamin *ignose*. Because of the proposed name, the paper—and all of its significant data—was rejected. So Szent-Györgyi resubmitted it with a new proposed name, *godnose*.

Today we know vitamin C as *ascorbic acid* because of its antiscorbutic (anti-scurvy) effects.

Table 22.2 Water-Soluble Vitamins Required in the Diet

Vitamin	RDA* Male	RDA* Female	Sources	Deficiency Effects
B_1 (thiamine)	1.5 mg	1.1 mg	Cereal grains, organ meats, vegetables	Beriberi
B_2 (riboflavin)	1.7 mg	1.3 mg	Liver, milk, eggs, yeast, vegetables	Dermatitis
B_5 (niacin)	19 mg	15 mg	Liver, vegetables, whole grains	Pellagra
B_6 (pyridoxine)	2.0 mg	1.6 mg	Eggs, liver, whole grains, fish	Dermatitis, nerve disorders
B_{12} (cobalamin)	2.0 µg	2.0 µg	Liver, seafood, eggs	Pernicious anemia
Folate	200 µg	180 µg	Whole-wheat products, green vegetables	Anemia
Pantothenic acid	**	**	Cereals, organ meats	Neural disorders
Biotin	**	**	Liver, eggs	Dermatitis
C (ascorbic acid)	60 mg	60 mg	Citrus fruits, raw vegetables	Scurvy

* Recommended by the Food and Nutrition Board of the National Academy of Sciences (Revised 1989).
** Not established.

who eat mostly corn. Corn lacks the essential amino acid tryptophan (see Tables 21.2 and 21.3), which is metabolized to niacin in the body.

Centuries ago, sailors who traveled for long periods suffered from bleeding and sore gums, loose teeth, weight loss, weakness, slow-healing wounds, and skin sores. This condition is known as *scurvy* (see Chemistry Spotlight). In the 19th century, British sailors were called "limeys" because they took daily rations of lime or lemon juice to prevent scurvy. Now we know that vitamin C is the substance in fresh citrus juices that prevents scurvy. Vitamin C is a coenzyme that enables the body to convert the amino acid proline into 4-hydroxyproline, which is necessary to form collagen for connective tissue (see Section 21.6).

22.3 NAMING AND CLASSIFYING ENZYMES

Enzyme names usually end in *-ase*. The main exceptions are digestive enzymes (such as pepsin and trypsin) that were named before the *-ase* suffix came into common use.

Enzymes are named for the reactions they catalyze. Common names for enzymes usually are based on the name of the substrate, the type of reaction, or a combination of the two, plus *-ase*. Figure 22.9 shows how three different enzymes, each having glucose-6-phosphate as a substrate, are named.

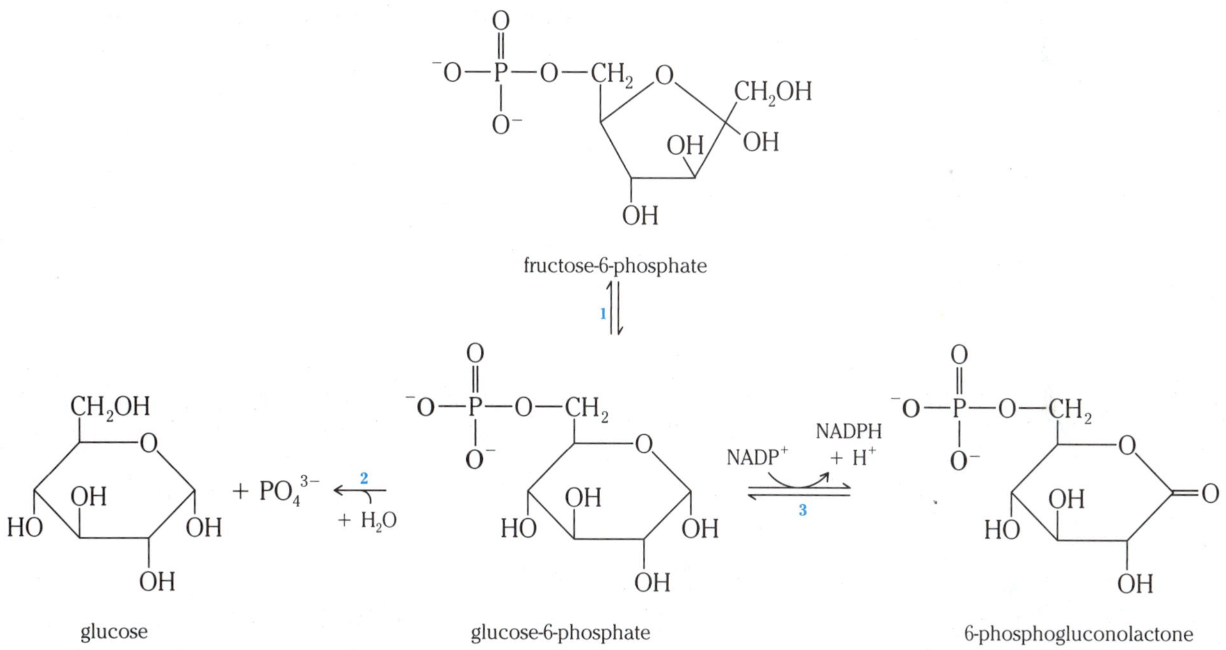

Biochemists gave common names to enzymes as they were discovered. But eventually the IUB (International Union of Biochemistry) adopted systematic names for enzymes—names that more clearly identified the reactions they catalyzed. The IUB established six general classes of enzymes:

1. *Oxidoreductases* catalyze oxidation–reduction reactions with the transfer of electrons and usually H or O atoms.
2. *Transferases* catalyze transfer of phosphate, methyl, amino, acyl and other groups.
3. *Hydrolases* catalyze hydrolysis—that is, addition of H_2O to a bond, thus breaking the bond.
4. *Lyases* catalyze removal of a group from a substrate without hydrolysis, usually leaving products containing a double bond.
5. *Isomerases* catalyze interconversion of isomers.
6. *Ligases* catalyze joining of two compounds into one product.

Figure 22.9 Three different enzymes, each having glucose-6-phosphate as substrate, are named (1) phosphoglucose isomerase, (2) glucose-6-phosphatase, and (3) glucose-6-phosphate dehydrogenase.

EXAMPLE 22.1 Identify the general class of each of the enzymes shown in Figure 22.9.

SOLUTION Phosphoglucose isomerase (1) is an isomerase; glucose-6-phosphatase (2) is a hydrolase; glucose-6-phosphate dehydrogenase (3) is an oxidoreductase.

22.4 RATES OF ENZYME-CATALYZED REACTIONS

It isn't enough just to know that a reaction occurs in your body. The reaction has to go fast enough to be useful. It doesn't do you much good, for example, to have an enzyme system that metabolizes and detoxifies ethanol if it operates so slowly that it takes a month to rid your body of the ethanol in one glass of wine. So let's examine five important ways to change the rates of enzyme-catalyzed reactions.

Substrate Concentration. Recall that reactions go faster in the presence of higher concentrations of reactants (Section 8.3). At higher concentrations, more reactant molecules collide with each other per second, increasing the number of molecules that actually react per second.

The same pattern occurs in enzyme-catalyzed reactions. Figure 22.10 shows how changes in substrate concentration typically affect the rate of an enzyme-catalyzed reaction. Notice that at low substrate concentrations, increases in substrate cause a proportional increase in rate. As the substrate concentration increases even more, this relationship changes. Finally, at high substrate concentrations, additional increases in substrate have virtually no effect on the rate.

How can we explain this? First, think of an enzyme (E) binding to its substrate (S) to form product (P). We represent the interaction as:

$$E + S \rightleftharpoons ES \rightleftharpoons EP \rightleftharpoons E + P$$

where ES and EP represent two (of many) intermediate forms. If the concentration of enzyme is constant and the concentration of substrate increases, at some point the enzyme becomes saturated with substrate; that is, it operates at full capacity in binding substrate molecules and converting them into product. Increasing substrate concentrations even more cannot increase the reaction rate; the enzyme is already working as rapidly as it can. So at very high levels of substrate, the reaction rate becomes virtually independent of substrate concentration.

One type of enzyme, however, doesn't show the usual effect of substrate concentration (Figure 22.11). **Allosteric enzymes** have other sites (besides the active site)—called *regulatory sites*—where substrates and other substances bind reversibly. Binding to a regulatory site changes the shape of the enzyme and alters its catalytic activity.

The S-shaped pattern in Figure 22.11 occurs because at low substrate levels, substrate molecules bind to a regulator site on an allosteric enzyme and make the active site more catalytically active. At higher substrate concentrations, the activated enzyme shows a pattern more like that in Figure 22.10.

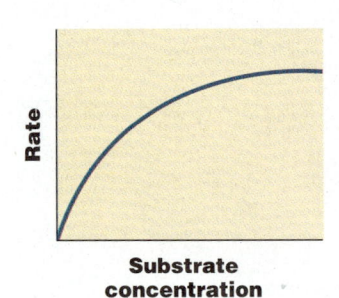

Figure 22.10 Effect of substrate concentration on reaction rate for most enzymes.

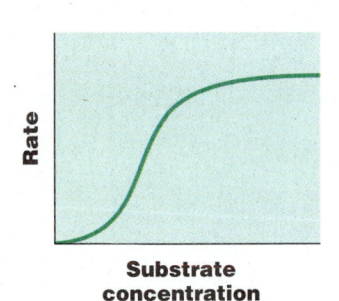

Figure 22.11 Effect of substrate concentration on reaction rate for allosteric enzymes.

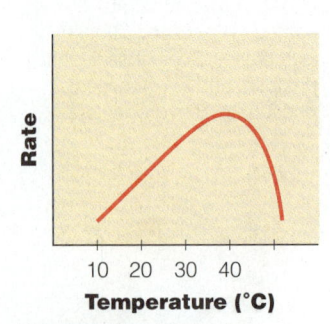

Figure 22.12 Effect of temperature on the rates of enzyme-catalyzed reactions.

Temperature. Like other reactions, enzyme-catalyzed reactions go faster at warmer temperatures as reacting substances collide more frequently and with greater energy (Section 8.3). But when the temperature exceeds about 30–40°C, reaction rates typically decrease (Figure 22.12). Enzymes, like all proteins, denature at high temperatures (Section 21.4); when this happens, they lose their catalytic activity.

You take advantage of these effects in several ways. Refrigerating food slows the enzyme action in organisms that cause spoilage. And cooking food destroys unwanted organisms by (among other things) denaturing their enzymes and other proteins. Your normal body temperature (37°C), however, is just about right for enzymes—warm enough for reactions to be fast but cool enough to avoid denaturation.

pH. Enzyme rates vary considerably with pH. Enzymes typically are most effective at fairly neutral pH values (Figure 22.13). This matches the conditions in your cells and tissues, most of which have pH values in the 6–8 range.

Why does enzyme action vary with pH? First, recall that proteins denature at extremes of pH (Section 21.4). In addition, ionizable groups in the substrate and in the enzyme active site have pK_a values (Section 9.6) and exist in different ionic forms at different pH values. Figure 22.14 shows one example. If a particular ionic form of the substrate or group in the active site is necessary for effective binding between substrate and enzyme, the enzyme is most effective in the pH range where that ionic form exists.

Because your cells and tissues keep fairly constant pH values, pH doesn't change enzyme activities in your body very much. One exception is an enzyme in saliva called amylase that hydrolyzes (digests) starch. Since it works most efficiently at neutral pH, amylase hydrolyzes starch while you are chewing your food. But once you swallow, amylase passes with the food into your stomach and is denatured there by strong acidity.

Activators and Inhibitors. Many substances bind to enzymes and increase or decrease catalytic activity. As we discussed in Section 22.2, many enzymes become active only after metal ions or vitamin derivatives bind near the active site. And substances can activate or inhibit allosteric enzymes by binding to their regulatory sites.

Competitive inhibition occurs when a substance binds to the active site of an enzyme and prevents the substrate from binding. Such inhibitors usually resemble the substrate; this is why they bind to the active site (Figure 22.15). One example is malonate, which structurally resembles succinate enough to be a competitive inhibitor of an enzyme that metabolizes succinate:

```
COO⁻            COO⁻
 |               |
CH₂            (CH₂)₂
 |               |
COO⁻            COO⁻

malonate        succinate
```

Many drugs are competitive inhibitors of enzymes. Recall that ethanol is an antidote to methanol and ethylene glycol poisoning (Chemistry Spotlight in Section 14.4). Ethanol structurally resembles those alcohols and is a competitive inhibitor of the enzyme that metabolizes them (Figure 22.16). This inhibi-

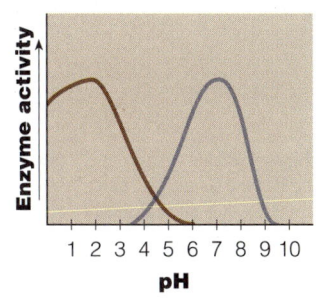

Figure 22.13 Effect of pH on enzyme activity. Most enzymes have optimal activity in the pH 6–8 range (violet line). A few enzymes have optimal activity at acidic pH values (brown line); one example is pepsin, which operates in the stomach at pH 1–3.

Figure 22.14 Glucose-6-phosphate, the substrate for reactions shown in Figure 22.9, exists in different ionic forms at different pH values.

Figure 22.15 An inhibitor (dark color) binds at the active site in competitive inhibition and at a different site in noncompetitive inhibition.

competitive inhibition   enzyme   noncompetitive inhibition

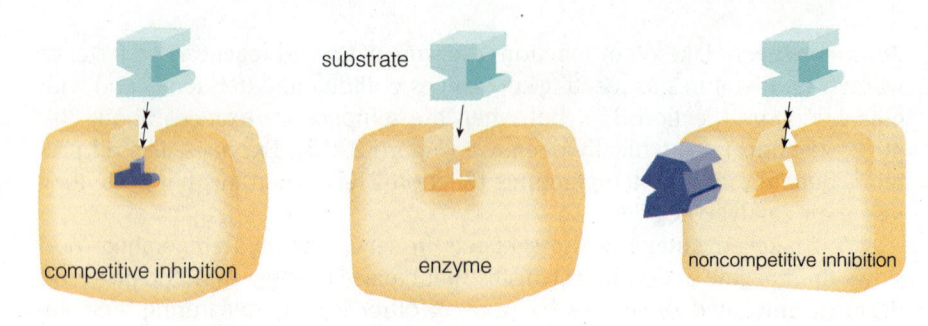

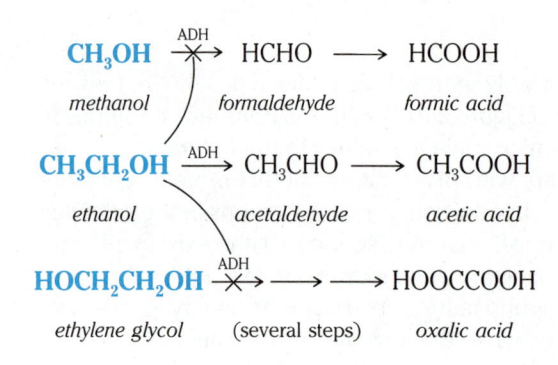

$$CH_3OH \xrightarrow{ADH} \text{HCHO} \longrightarrow \text{HCOOH}$$
methanol   formaldehyde   formic acid

$$CH_3CH_2OH \xrightarrow{ADH} CH_3CHO \longrightarrow CH_3COOH$$
ethanol   acetaldehyde   acetic acid

$$HOCH_2CH_2OH \xrightarrow{ADH} \longrightarrow \longrightarrow \text{HOOCCOOH}$$
ethylene glycol   (several steps)   oxalic acid

Figure 22.16 Alcohols are oxidized by a common enzyme system to their respective carboxylic acids. Because it resembles the other substrates and uses the same enzyme, alcohol dehydrogenase (ADH), ethanol is a competitive inhibitor of methanol and ethylene glycol oxidation.

H_2N—⬡—C(=O)—OH $\xrightarrow{\text{(many steps)}}$ folic acid

p-aminobenzoic acid

H_2N—⬡—SO_2—NH_2

sulfanilamide

Figure 22.17 Sulfanilamide competitively inhibits the conversion of p-aminobenzoic acid into folic acid by certain bacteria. Structural similarities are in color.

CH_3CH_2 \ / CH_2CH_3
N
|
C=S
|
S
|
S
|
C=S
|
N
CH_3CH_2 / \ CH_2CH_3

Figure 22.18 Disulfiram (Antabuse)

tion slows the production of toxic products, giving the body time to excrete the toxic alcohol.

Another example is sulfanilamide, a drug used to treat certain bacterial infections. Sulfanilamide resembles p-aminobenzoic acid, a compound certain bacteria use to make the vitamin folic acid (Figure 22.17). Sulfanilamide binds to the active site of the enzyme that normally metabolizes p-aminobenzoic acid; this kills the bacteria because they can no longer make folic acid. Since our cells don't make folic acid from p-aminobenzoic acid, sulfanilamide doesn't harm them.

Noncompetitive inhibition occurs when a substance binds reversibly to an enzyme somewhere other than at the active site and reduces its catalytic activity (Figure 22.15).

One example is disulfiram (Antabuse; Figure 22.18), a drug used to treat ethanol addiction. Your enzymes normally metabolize ethanol to acetaldehyde and then to acetic acid (Figure 22.16). As a noncompetitive inhibitor of the second reaction, disulfiram causes high levels of acetaldehyde to build up whenever ethanol is consumed; this causes pounding headaches, nausea, vomiting, palpitations, and other very unpleasant effects. So people who take disulfiram have a powerful incentive not to drink alcoholic beverages.

Irreversible inhibition occurs when an inhibitor forms a stable covalent bond with an enzyme, sometimes at the active site. Such inhibitors are usually toxic. Certain nerve gases and pesticides irreversibly inhibit an enzyme that breaks down acetylcholine, a neurotransmitter substance (see Chemistry Spotlight in Section 16.6). Toxic metals such as mercury, lead, and arsenic (a metalloid) react with sulfhydryl (—SH) groups in proteins (see opening to Chapter 14). This changes the shape of enzymes and inhibits their activity.

22.5 REGULATION OF ENZYME ACTIVITY

Enzymes are a bit like light switches; when they are *on* (catalytically active), certain reactions occur; when they are *off* (catalytically inactive), these reactions don't occur. In order to function effectively and adapt to their environment, living things have to be able to turn off reactions at certain times, and turn them on at other times. During starvation, for example, reactions that store fat are turned off and reactions that metabolize fat and generate energy are turned on.

Organisms regulate their reactions by regulating their enzymes. But your body needs a regulation system far more sophisticated than an on/off switch. You need something more like a dimmer switch, a way to increase or decrease enzyme activity in varying degrees, fine-tuning chemical reactions to fit the situation at the moment. To meet this need, your body regulates enzyme activity in several different ways.

Proenzymes. Recall that to make insulin (Section 21.6), your pancreatic cells first synthesize preproinsulin, which contains 110 amino acids. Then they remove peptide material to produce insulin, an active hormone with 51 amino acids.

Your body synthesizes certain enzymes, particularly digestive enzymes, in a similar way. First, cells make substances with extra peptide material that blocks enzymatic activity. Later, this extra material is removed, activating the enzymes. Such enzymes in their inactive form are called **proenzymes** or *zymogens*. You can recognize the names of these forms by the prefix *pro-* or the suffix *-ogen*.

One example is pepsinogen, the proenzyme for pepsin. Pepsin helps digest food proteins in the stomach by hydrolyzing certain peptide bonds. Pepsinogen is produced by cells in the stomach lining and passes from these cells into the interior cavity (lumen) of the stomach. There H^+ or existing pepsin molecules catalyze removal of peptide material, producing the active enzyme, pepsin:

$$\text{pepsinogen} \xrightarrow[\text{or } H^+]{\text{pepsin}} \text{pepsin} + \text{several peptides}$$

molecular wt = 40,000   *molecular wt = 32,700*

This mechanism provides a way to have an enzyme potentially available for action, but muzzled by extra peptide material; removing the muzzle (peptide material) unleases enzyme action. This is especially useful for digestive enzymes; if cells synthesized such enzymes in a catalytically active form, the cells would be destroyed by them. It's far better to synthesize proenzymes, export them to where the food is, then activate them there.

Isoenzymes. Some enzymes exist in multiple forms, differing only in the combination of subunits in their quaternary structure. These different forms are called **isoenzymes** or *isozymes*.

One example is the enzyme lactate dehydrogenase (LDH). LDH has two types of protein subunits that differ in their number and composition of amino acids. The H type is common in heart tissue; the M type is common in muscle. The five isoenzymes of LDH, each consisting of four subunits, contain different combinations of these two types. We can represent them as:

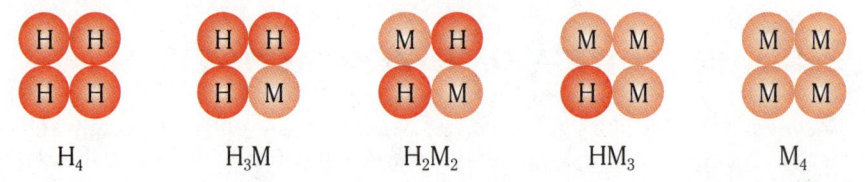

H_4 H_3M H_2M_2 HM_3 M_4

Isoenzymes enable different tissues to carry out distinct metabolic tasks. Each tissue has its own types of isoenzymes, and each isoenzyme has slightly different properties. For example, a shortage of oxygen is likely to occur in muscle tissue during intense exercise, and its M_4 type of LDH produces lactic acid under these conditions. The heart, in contrast, isn't likely to lack oxygen because it receives oxygenated blood directly from the lungs; its H_4 isoenzyme is different from the M_4 in muscle and doesn't readily produce lactic acid.

Isoenzymes also are useful in diagnosing diseases (see Section 22.6).

Covalent Modification of Enzymes. The catalytic activity of some enzymes changes when a group is covalently bonded to, or removed from, them. The most common group is phosphate, which can be esterified to any of the amino acids that contain a hydroxyl (—OH) side-chain group (serine, threonine, and tyrosine). The general reaction is:

$$\text{enzyme} + R\underset{\underset{\text{OH}}{|}}{-}O-PO_3^{2-} \longrightarrow \text{enzyme} + R\underset{\underset{O-PO_3^{2-}}{|}}{-}OH$$

The enzyme–phosphate form, in turn, is converted back into the enzyme—OH form by hydrolysis (reaction with water).

Some enzymes are more active when they have a bound phosphate group; others are less active. So attaching or removing phosphate groups is an effective way to regulate the activity of these enzymes. Often such reactions are themselves catalyzed by other enzymes, which are in turn regulated by hormones or other substances.

In Chapters 24–27, you will see other examples of allosteric enzymes regulated by feedback inhibition.

Allosteric Enzymes. Substances that bind to regulatory sites on allosteric enzymes activate or inhibit the enzymes to varying degrees. As a result, allosteric enzymes have a key role in regulating metabolism.

In **feedback inhibition,** the product of a sequence of reactions inhibits an allosteric enzyme that catalyzes an earlier reaction in the sequence. Figure 22.19 shows one example of this mechanism. When the product (isoleucine) reaches a sufficiently high concentration, it binds to a regulatory site on the allosteric enzyme catalyzing the first reaction; this inhibits the enzyme and prevents additional production of isoleucine. As the concentration of isoleu-

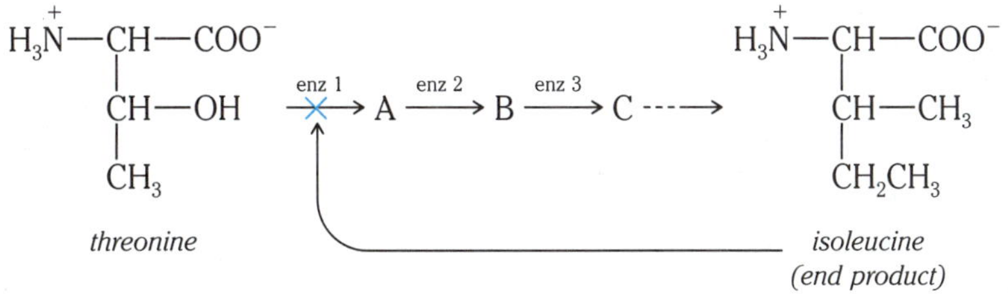

Figure 22.19 Feedback inhibition by isoleucine of an allosteric enzyme (enz 1).

cine subsequently decreases, the inhibition is lifted, enabling isoleucine to be produced again. This arrangement helps keep the concentration of isoleucine from getting too high (which might be toxic) or too low (which might produce a deficiency problem).

Enzyme Repression. Regulation usually occurs by activating or inhibiting enzyme activity; this provides an immediate change in enzyme action. Another strategy—and one which is slower to take effect—involves changing the amount of enzyme present.

Repression is a mechanism for blocking the synthesis of protein molecules, particularly enzymes. When enzyme molecules are no longer synthesized in the cell, their concentration gradually dwindles and catalytic activity decreases. Substances that block (repress) enzyme synthesis are often the end products of a sequence of reactions. In this respect, repression is like feedback inhibition. Chapter 23 discusses how repression works in more detail.

22.6 MEDICAL USES OF ENZYMES

Drugs. One of the most effective ways to alter chemical events in the body is to inhibit the enzymes responsible for those reactions. Sometimes this is hazardous. Certain nerve gases and insecticides, for example, work by irreversibly inhibiting an enzyme that breaks down the neurotransmitter acetylcholine (Section 22.4).

One type of reversible enzyme inhibitor is used after surgery or heart attacks to prevent blood clots. Since vitamin K is needed for an enzyme-catalyzed step in blood clotting (see Table 20.4), drugs that resemble it—called anticoagulants—inhibit that enzyme. Too much anticoagulant can cause internal bleeding (hemorrhaging), but an overdose usually can be corrected by giving the patient extra vitamin K.

Figure 22.20 shows two anticoagulants and their structural similarity to vitamin K. Dicoumarol was discovered because cows who ate spoiled sweet clover

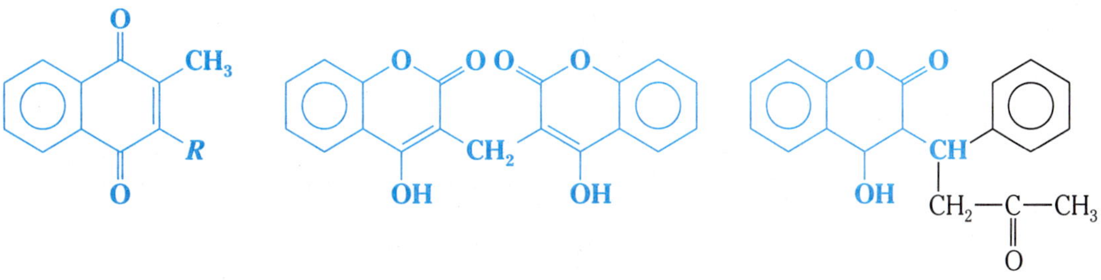

Figure 22.20 Vitamin K and two anticoagulants. Structural similarities are in color.

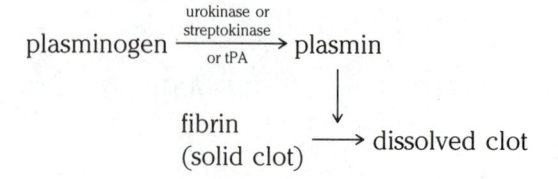

Figure 22.21 Urokinase, streptokinase, and tissue plasminogen activator (tPA) each convert plasminogen into plasmin, which causes blood clots to dissolve.

$$\text{plasminogen} \xrightarrow{\text{urokinase or streptokinase or tPA}} \text{plasmin}$$

$$\text{fibrin (solid clot)} \longrightarrow \text{dissolved clot}$$

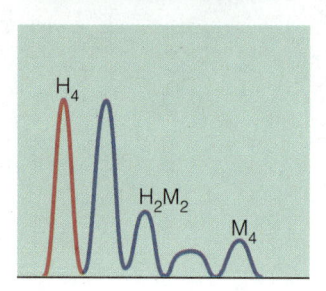

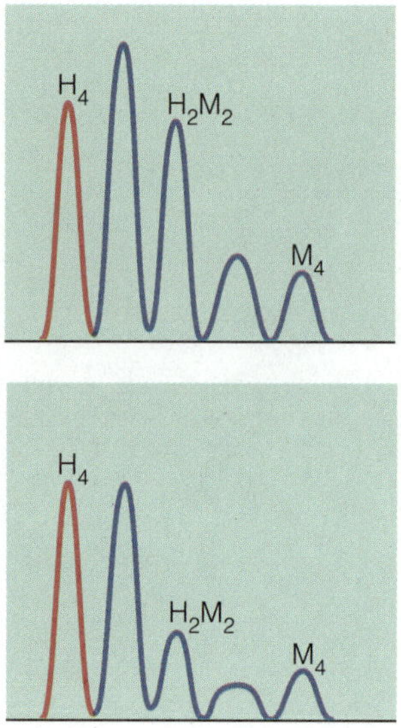

Figure 22.22 Pattern of lactate dehydrogenase isoenzymes in the blood of a normal person (top) and following a myocardial infarction (bottom). Notice the increased amount of H_4 isoenzyme on the bottom. The isoenzymes were separated by electrophoresis.

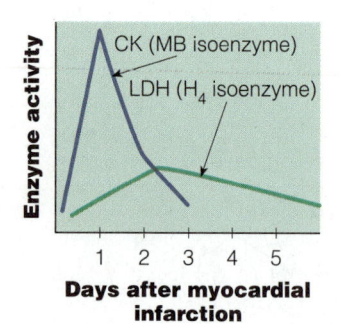

Figure 22.23 Increased amounts of H_4 isoenzyme of lactate dehydrogenase (LDH) and MB isoenzyme of creatine kinase (CK) appear in blood after a heart attack.

(which contains dicoumarol) suffered serious, and sometimes fatal, internal bleeding. Warfarin is used as a commercial rat poison that causes its victims to hemorrhage to death.

Blood clots are one of the few conditions for which enzymes are administered as part of the treatment. Another enzyme, plasmin, normally breaks down blood clots within a few days. Plasmin is made from the proenzyme plasminogen by the action of enzymes such as urokinase and streptokinase (Figure 22.21). To help dissolve clots that block blood vessels during heart attacks, these enzymes are sometimes injected directly into the bloodstream. Another, more expensive drug that has a similar effect is tissue plasminogen activator (tPA), a protein.

Enzymes and Clinical Chemistry. Since enzymes catalyze chemical reactions, they can help measure the amounts of substances in the body (usually in blood or urine) for medical diagnoses. Relatively pure samples of some enzymes are commercially available for these purposes.

One example is glucose oxidase, which clinical laboratories use to measure the amount of glucose in blood or urine. Recall that this enzyme oxidizes glucose, and that a second reaction produces a colored dye which shows how much glucose is present (Chemistry Spotlight, Section 15.4). Recall also that paper strips containing glucose oxidase and dye can be used to measure glucose levels. Thus glucose oxidase is useful in diagnosing diabetes or other diseases in which abnormal amounts of glucose appear in blood or urine.

Measuring the amount of certain enzymes in blood and tissue samples is another important tool for medical diagnosis. Enzyme levels in cells normally stay fairly constant, with only small amounts appearing in blood. But diseased or deteriorating tissues spill enzymes into blood and other fluids.

Some tissues contain enzymes, or isoenzymes, that are uncommon in other parts of the body. Those enzymes are especially useful as markers, since their sudden appearance in blood can be traced to a disease in a specific tissue.

Isoenzyme levels are used to diagnose myocardial infarction, a condition in which a clot or other material blocks a blood vessel and causes heart tissue to deteriorate. During a heart attack, increased amounts of the H_4 isoenzyme of lactate dehydrogenase (LDH) rapidly appear in blood and persist for several days (Figures 22.22 and 22.23).

Another useful diagnostic enzyme is creatine kinase (CK), which contains two protein subunits. The two types of subunit are called M (for muscle) and B (for brain). The three isoenzymes are MM, MB, and BB, with MB found mostly in heart tissue. Elevated blood levels of MB occur right after a heart attack and remain for 2–3 days (Figure 22.23).

Table 22.3 lists some enzymes whose elevated blood levels are associated with certain diseases. Clinical chemists are working to identify more "marker" enzymes whose blood concentrations are linked to specific diseases.

Table 22.3 Some Enzymes Used in Clinical Diagnosis

Enzyme	Disease
Acid phosphatase	Prostate cancer
Alanine aminotransferase	Hepatitis
Aldolase	Muscle diseases
Alkaline phosphatase	Bone or liver disease
Amylase	Pancreatic diseases
Aspartate aminotransferase	Heart attack, hepatitis
Creatine kinase (CK)	Heart attack
Elastase	Collagen diseases
Lactate dehydrogenase (LDH)	Heart attack
Plasmin	Blood-clotting disorders
Trypsin	Pancreatic diseases

An exchange at a U.S. Senate hearing on the National Science Foundation (NSF) budget:

Senator: I hear you biochemists want to adulterate bread by adding vitamin B to it. If the good Lord had wanted vitamin B in bread He would have put it there.

Philip Handler (then President of NSF): The good Lord did put vitamin B in bread. It was man who took it out in order to make white bread. The scientists who want to put it back are doing God's work.

SUMMARY

Enzymes are proteins that catalyze chemical reactions. Compared with other catalysts, enzymes are remarkably effective and specific. Their specificity is explained by the induced-fit model, a modified version of the lock-and-key model. Enzymes are chiral and typically act on only one member of a pair of enantiomers.

Many enzymes require the presence of nonprotein cofactors to be active. Many cofactors are metal ions. Organic cofactors, usually derivatives of water-soluble vitamins, are called coenzymes. Among the most important coenzymes are NAD^+ and $NADP^+$ (oxidized forms) and NADH and NADPH (reduced forms).

Enzyme names usually end in -*ase*. The IUB established six classes of enzymes based on the types of reactions they catalyze.

Enzyme-catalyzed reactions are faster at higher substrate concentrations but approach a constant rate at very high concentrations. In addition to their active sites, allosteric enzymes have regulatory sites to which substrates and other compounds bind, altering the reaction rate. Increasing temperature increases the rate of enzyme-catalyzed reactions until a temperature is reached that denatures the enzyme. Enzymes typically are most active near neutral pH, where denaturation is not a problem. The pH also may alter the ionic form of a substrate or group in the active site. Competitive inhibitors resemble the substrate and bind to the active site. Noncompetitive inhibitors bind elsewhere and decrease the activity of the active site. Irreversible inhibitors are usually toxic.

Enzyme activity is regulated by the existence of proenzymes and isoenzymes. Attaching or removing phosphate groups alters the activity of some enzymes. Allosteric enzymes are regulated by substances that bind to their regulatory sites. Feedback inhibition of allosteric enzymes maintains consistent concentrations of metabolic compounds.

Many drugs act as enzyme inhibitors. Enzymes also are used to measure concentrations of substances in the body. Levels of certain enzymes in blood are measured to diagnose specific diseases.

KEY TERMS

Active site (22.1)
Allosteric enzyme (22.4)
Coenzyme (22.2)
Cofactor (22.2)
Competitive inhibition (22.4)

Enzyme (22.1)
Feedback inhibition (22.5)
Induced-fit model (22.1)
Irreversible inhibition (22.4)
Isoenzyme (22.5)

Lock-and-key model (22.1)
Noncompetitive inhibition (22.4)
Proenzyme (22.5)
Substrate (22.1)
Vitamin (22.2)

EXERCISES

Even-numbered exercises are answered at the back of this book.

Enzyme Specificity and Effectiveness

1. How does an enzyme affect the equilibrium of a chemical reaction it catalyzes?

2. How does an enzyme alter the energy of activation for a reaction, and how does this affect the rate of the reaction?

3. According to the lock-and-key model of enzyme specificity, what substance is analogous to a lock and what substance is like a key?

4. What model has largely replaced the lock-and-key model?

5. The enzymes in your body act on ___ (D or L) sugars and ___ (D or L) amino acids.

6. Suppose you had your choice of using $KMnO_4$ or an enzyme to oxidize 2-methyl-1-butanol. Would the product (2-methylbutanoic acid) have the identical structure in either case? Why?

7. Would the answer to exercise 6 be different if the reactant was 1-butanol instead of 2-methyl-1-butanol? Why?

Coenzymes and Vitamins

8. How is a coenzyme different from a substrate?

9. How is a coenzyme different from a cofactor?

10. Under anaerobic conditions your muscles convert pyruvate into lactate:

$$CH_3-\overset{O}{\underset{\|}{C}}-COO^- \longrightarrow CH_3-\overset{OH}{\underset{|}{CH}}-COO^-$$

The coenzyme NAD^+ also is involved. Write a balanced equation for this reaction, including the oxidized and reduced forms of NAD^+ on appropriate sides of the equation.

11. Which substance in your answer to exercise 10 is oxidized in the reaction? Which substance is reduced?

12. Use Table 22.1 to predict a coenzyme that participates in each of the following reactions (complete reactions are not necessarily shown):

(a) $CH_3\overset{O}{\underset{\|}{C}}COO^- + CO_2 \longrightarrow {}^-OOCCH_2\overset{O}{\underset{\|}{C}}COO^-$

(b) $^-OOCCH_2CH_2COO^- \longrightarrow {}^-OOCCH=CHCOO^-$

(c) $CH_3\overset{O}{\underset{\|}{C}}COO^- + {}^-OOCCHRNH_3^+$

$\longrightarrow CH_3CHCOO^- + {}^-OOC\overset{O}{\underset{\|}{C}}R$
$\quad\quad\quad\quad |$
$\quad\quad\quad\quad NH_3^+$

13. Table 22.1 lists many vitamins that function as coenzymes. Are these vitamins classified as fat-soluble or water-soluble?

14. What vitamin is needed for the coenzymes (a) $NADP^+$ and (b) FAD?

Enzyme Classification

15. From the six classes of enzymes listed in Section 22.3, write the class of enzyme catalyzing each reaction in exercise 12.

16. Which class of enzymes catalyzes the reaction in exercise 10?

Rates of Enzyme-Catalyzed Reactions

17. What would be the effect on the curve in Figure 22.10 if the reaction rates were measured at the same substrate concentrations but in the presence of a greater amount of enzyme?

18. Does an increase in substrate concentration cause a greater increase in reaction rate at low or high substrate concentration?

19. The word *allosteric* means "other site." How does this apply to allosteric enzymes?

20. An allosteric enzyme can be identified by how its reaction rate changes with (a) substrate concentration, (b) enzyme concentration, (c) temperature, (d) pH, (e) none of the above.

21. Explain, on a molecular basis, why an increase in temperature normally accelerates a reaction.
22. Does your answer to exercise 21 apply equally to enzyme-catalyzed reactions? Explain.
23. Certain enzymes that help dissolve stains are used in laundry presoak products. Why should presoaking be done in cold water?
24. Are changes in body temperature and pH an effective way to regulate enzyme activity?
25. Why do most enzymes not have optimal catalytic activity at very acidic or alkaline pH?
26. Glucose-6-phosphate is the substrate for an enzyme that is inhibited by galactose-6-phosphate. Do you think galactose-6-phosphate is a competitive or noncompetitive inhibitor? Why?
27. If the concentration of competitive inhibitor stays constant while the substrate concentration increases, what will be the effect on the amount of inhibition? Why?
28. If the concentration of noncompetitive inhibitor stays constant while the substrate concentration increases, what will be the effect on the amount of enzyme inhibition? Why?
29. Do heavy metal ions such as Pb^{2+} need to structurally resemble the substrate in order to inhibit an enzyme? Why?

Regulation of Enzyme Activity

30. Since catalysis occurs at the active site of an enzyme, how can a substance that binds at a different site affect catalysis?
31. Does the molecular weight increase or decrease when a proenzyme is made into an enzyme?
32. From their names, identify which of the following are proenzymes: **(a)** aminopeptidase, **(b)** trypsinogen, **(c)** procarboxypeptidase, **(d)** chymotrypsin, **(e)** lactase.
33. Suppose an enzyme occurs as isoenzymes consisting of trimers (a total of three subunits) made from two types of subunits, A and B. How many different isoenzymes could exist?
34. In which of the following ways do isoenzymes differ from each other: **(a)** quaternary structure, **(b)** reaction they catalyze, **(c)** distribution in body tissues, **(d)** IUB name, **(e)** all of the above.
35. What is the difference between feedback inhibition and repression?
36. In feedback inhibition, does the inhibitor have to resemble the substrate in structure? Why?
37. What would be the effect of an inhibitor on the curve in Figure 22.11? Draw the curve as it might appear in the absence and in the presence of an inhibitor.

Medical Uses of Enzymes

38. Are dicoumarol and warfarin antimetabolites?
39. According to the caption for Figure 22.22, different isoenzymes of lactate dehydrogenase (LDH) separate during electrophoresis. Explain why.
40. What are the relative advantages and disadvantages of using LDH or CK for diagnosing a myocardial infarction?

DISCUSSION EXERCISES

1. What are some possible advantages and disadvantages of enzymes being proteins rather than carbohydrates or lipids?
2. What are the advantages of the body using enzymes instead of catalysts such as metals, inorganic acids, and bases?
3. Do all the catalytic groups in the active site need to be located near each other in the primary structure of the enzyme? Explain.
4. Compare the patterns you would expect to occur in a graph of reaction rate *vs.* substrate concentration for a normal enzyme in the presence of **(a)** no inhibitor, **(b)** a competitive inhibitor, and **(c)** a noncompetitive inhibitor. Explain.

23

Nucleic Acids

1. What are the chemical structures of DNA and RNA?
2. How do cells synthesize DNA when they divide?
3. How do cells make RNA from DNA? What are the different types of RNA?
4. How do cells make proteins from DNA? How do they regulate this process?
5. What are mutations? What are their causes and effects?
6. What are some common techniques for genetic engineering? What are some medical uses for this technology?

Genetic Engineering

In his best-selling book, *The Double Helix,* James Watson recalls: "I felt slightly queasy when at lunch Francis [his coworker, Francis Crick] winged in . . . to tell everyone within hearing distance that we had found the secret of life." That "secret of life" was the structure of DNA, the hereditary material in your chromosomes that specifies your genetic characteristics.

But genetic information also can specify a disorder, such as sickle-cell anemia, cystic fibrosis, hemophilia, or muscular dystrophy. Our growing knowledge of DNA, however, helps us understand, and sometimes even fix, the chemical problem in some genetic disorders.

adenosine → inosine

Figure 23.1 This child inherited an immune deficiency and lives in an isolated environment.

For example, several children (including newborns) have been given gene therapy for adenosine deaminase (ADA) deficiency. ADA catalyzes the reaction shown. The lack of this enzyme disables the immune system, making its victims unable to fight infectious diseases. As a result, they have to live in an isolated environment (Figure 23.1). You may have heard about a young boy with a similar disorder, who had to live in a huge, plastic bubble to protect him against infectious agents.

To treat ADA deficiency, scientists remove some of each patient's white blood cells (or, in the case of newborns, the cells that form white blood cells). Then they splice the gene coding for ADA into those cells. A supply of these new cells is grown outside the body and injected into the patient. The recipient then produces enough ADA to have a functioning immune system—and lead an essentially normal life.

493

494 Part III Biochemistry

Why are your eyes, hair, and skin the color that they are (Figure 23.2)? Why are you as tall, as bright, as healthy, as talented as you are? At least part of the answer is in your *genes*, the bits of chromosomal material in your cells that you inherited from your parents. Some traits, such as eye color, are determined entirely by your genes. Others, such as height, intelligence, and musical talent, depend both on your genes and on your environment.

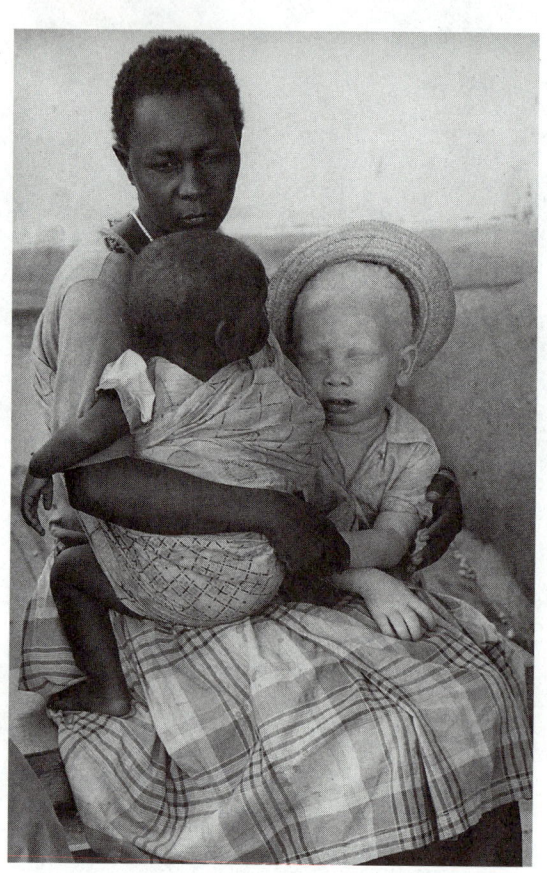

Figure 23.2 Although his mother has dark skin, the boy (right) lacks skin pigment because he inherited a gene for albinism from both parents.

How can genes do all this? You will see that genes specify what proteins your cells can make; proteins, in turn, cause your genetic features to appear. In this chapter, we examine the chemistry of substances called *nucleic acids*, and see how cells use genetic information to make proteins. We also examine ways to change the genetic makeup of organisms.

23.1 STRUCTURES OF DNA AND RNA

In 1868 Friedrich Miescher, a Swiss physician, isolated an acidic material from the nuclei of pus cells taken from bandages discarded by a hospital. He called the material *nuclein;* later, it was called **nucleic acid.** Now we recognize two types of nucleic acids—**deoxyribonucleic *a*cid (DNA),** which Miescher isolated, and **ribo*n*ucleic *a*cid (RNA).**

Almost all of a cell's DNA is in the nucleus. A small amount (less than 5%) occurs in membrane-enclosed units called mitochondria (see Appendix E). RNA is found in various locations—in the nucleus, in dense granules called ribosomes, and elsewhere in cytoplasm.

β-D-ribose

β-D-2-deoxyribose

Figure 23.3 Pentoses in RNA and DNA, respectively.

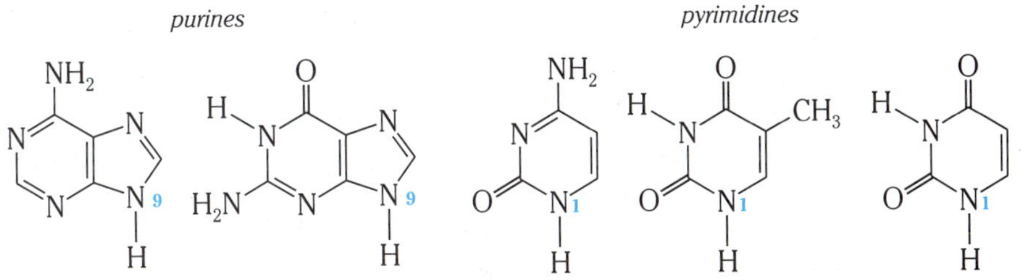

Figure 23.4 The five common nitrogen bases in DNA and RNA. Thymine (T) occurs only in DNA and uracil (U) only in RNA.

Components of DNA and RNA. DNA and RNA are made from three components—phosphate, a pentose, and nitrogen-containing bases. The pentose varies with the type of nucleic acid. As their names suggest, RNA contains ribose while DNA contains 2-deoxyribose (Figure 23.3).

Nitrogen-containing bases are of two types—purines and pyrimidines (Figure 23.4). DNA and RNA each contain four bases. The purines, adenine (A) and guanine (G), occur in both DNA and RNA. Of the pyrimidines, cytosine (C) is in both DNA and RNA, thymine (T) is mainly in DNA, and uracil (U) is mainly in RNA. Small amounts of a few other bases also occur in DNA and RNA, but we will not discuss them here.

Nucleosides and Nucleotides. In RNA and DNA the pentose bonds directly to a base. Recall that a bond from the anomeric carbon of a carbohydrate to another group is called a *glycosidic bond* (Section 19.4), and the orientation can be α or β. Carbon 1' of ribose and 2-deoxyribose is the anomeric carbon. So in nucleic acids, the bond joining carbon 1' of the pentose to a purine (at N-9) or pyrimidine (at N-1) is a β-glycosidic bond.

Compounds consisting of a pentose bonded to a base in this way are called **nucleosides** (Figure 23.5). Names of nucleosides are listed in Table 23.1.

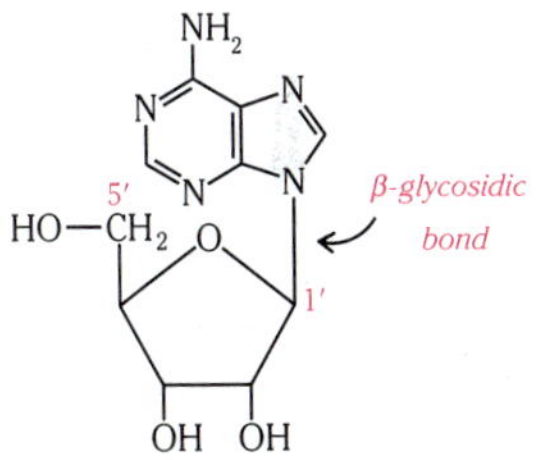

When the pentose in a nucleoside is also bonded to one or more phosphate groups, the compound is a **nucleotide.** In nucleotides, phosphate can bond to any carbon in the pentose that supplies an —OH group (3' or 5' in deoxyribose and 2', 3', or 5' in ribose) to form a phosphate ester (see Section 16.8). Most nucleotides have phosphate esterified to carbon 5' of the pentose. Such phosphates (but not those at 2' or 3' positions) are often joined to one or two

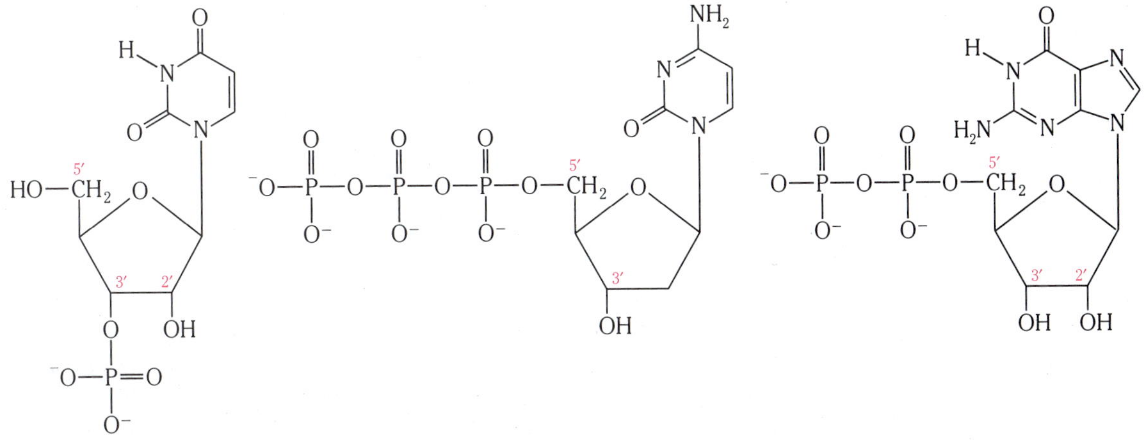

The number of a carbon atom in a pentose is designated with a prime (such as 1') when the pentose is bonded to a purine or pyrimidine (see Figure 23.5).

Figure 23.5 Adenosine, a nucleoside.

3'-UMP
(uridine 3'-monophosphate)

dCTP
(deoxycytidine triphosphate)

GDP
(guanosine diphosphate)

Figure 23.6 Three nucleotides.

Table 23.1 Names of Nucleosides Containing Ribose

Base	Name of Nucleoside
adenine	adenosine*
cytosine	cytidine
guanine	guanosine
thymine	thymidine
uracil	uridine

*Nucleosides containing deoxyribose have the prefix *deoxy-* (for example, deoxyadenosine).

additional phosphate groups (Figure 23.6). Those nucleotides are phosphate anhydrides (see Section 16.8).

We usually name nucleotides by using a simple, three-letter abbreviation. The first letter identifies the base—A, C, G, T, or U. The next two letters identify the number of phosphate groups (MP for monophosphate, DP for diphosphate, and TP for triphosphate). Preceding the three-letter abbreviation is a number (2' or 3') if the phosphate bonds to a pentose carbon other than the 5' carbon (no number is used to indicate the 5' position), and the letter *d* if the pentose is deoxyribose. Figure 23.6 shows three examples.

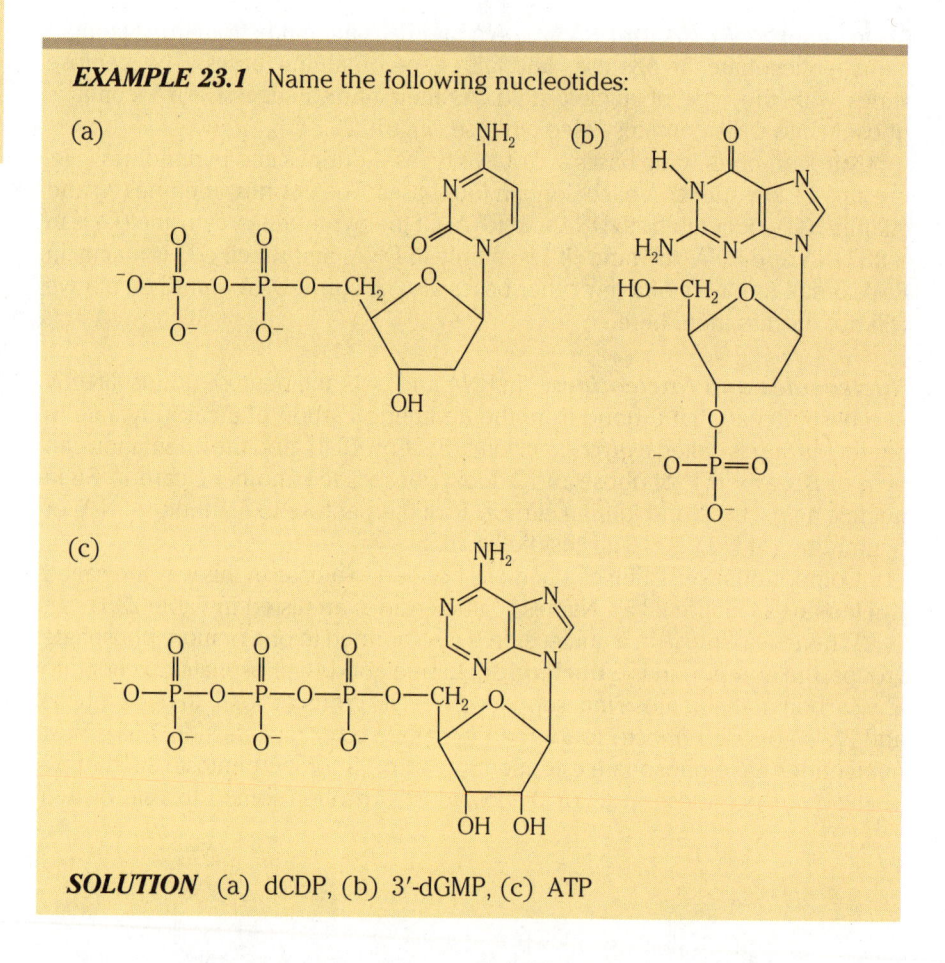

EXAMPLE 23.1 Name the following nucleotides:

(a)

(b)

(c)

SOLUTION (a) dCDP, (b) 3'-dGMP, (c) ATP

You will learn in the next chapter that ATP, a critical substance in providing for your energy needs, is one of the most important nucleotides in your body. In this chapter, however, our main interest in nucleotides is as building blocks for DNA and RNA.

Polynucleotide Chains. Because they are polymers of nucleotide units, DNA and RNA are *polynucleotides*. Their monomer units are 5' monophosphate nucleotides. In a polynucleotide chain, phosphate on the 5' carbon in one nucleotide bonds to the 3' carbon in a neighboring nucleotide (Figure 23.7). The resulting functional group is a phosphate diester (see Section 16.8), and the bond is called a *3',5'-phosphodiester bond*.

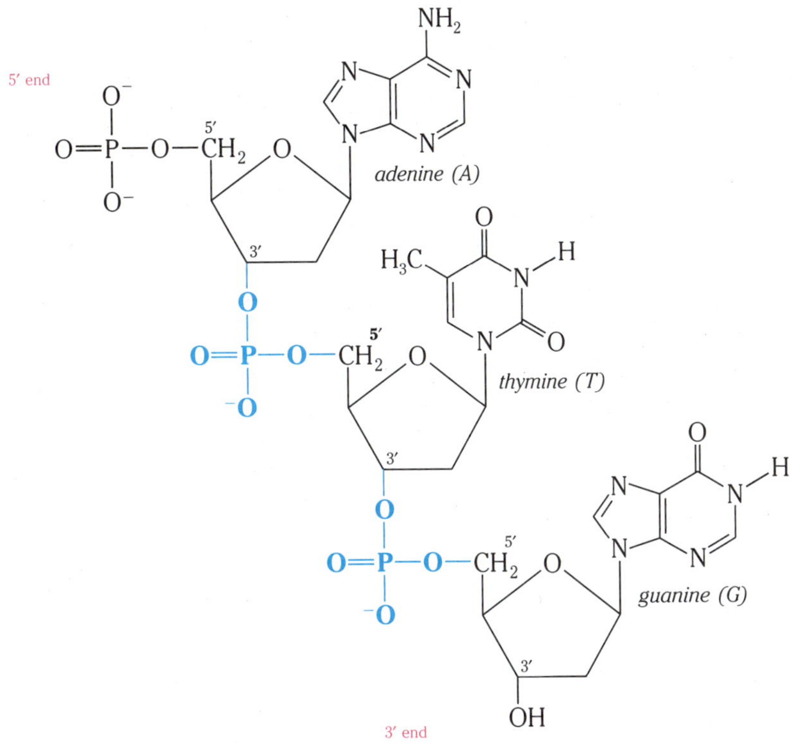

Figure 23.7 A chain of three nucleotide units joined by 3′,5′-phosphodiester bonds (in color). The chain could be extended by joining additional nucleotide units at the 3′ and/or 5′ ends.

Notice in Figure 23.7 that no matter how long the chain, one end has a 5′ phosphate group available to join another nucleotide unit; and the other end has a 3′ —OH group available to bond to a nucleotide that furnishes a 5′ phosphate. We call these the 5′ and 3′ ends, respectively.

The polynucleotide chains of DNA and RNA differ in two main ways: DNA contains deoxyribose while RNA contains ribose; and DNA contains thymine but not uracil, whereas RNA contains uracil but not thymine.

The other important difference is that RNA typically is a single polynucleotide chain (see Section 23.3). DNA, in contrast, is usually two polynucleotide chains wrapped around each other in a double helix.

*Because it contains deoxyribose, DNA is sometimes called a poly*deoxy*nucleotide chain.*

Structure of DNA. In 1953 James Watson and Francis Crick (using X-ray data provided by Maurice Wilkins and Rosalind Franklin) proposed a double helical structure for DNA. They (and Wilkins) received a 1962 Nobel prize for their work. According to their model, two DNA strands wrap around each other like a spiral staircase as shown in Figure 23.8.

The key features of the double helix are:

1. The two strands are antiparallel, which means they run in opposite directions. At each end of the helix is the 5′ end of one strand and the 3′ end of the other strand.

2. Polar pentose and phosphate components alternate on the outside of the helix, like handrails on a staircase. Phosphate groups carry a negative charge because at neutral pH the —OH group loses (donates) a proton (H⁺). This is why DNA is called a nucleic *acid*.

3. Bases are inside the helix. Each base on one strand pairs with a base on the other strand—adenine with thymine, cytosine with guanine. Each base pair

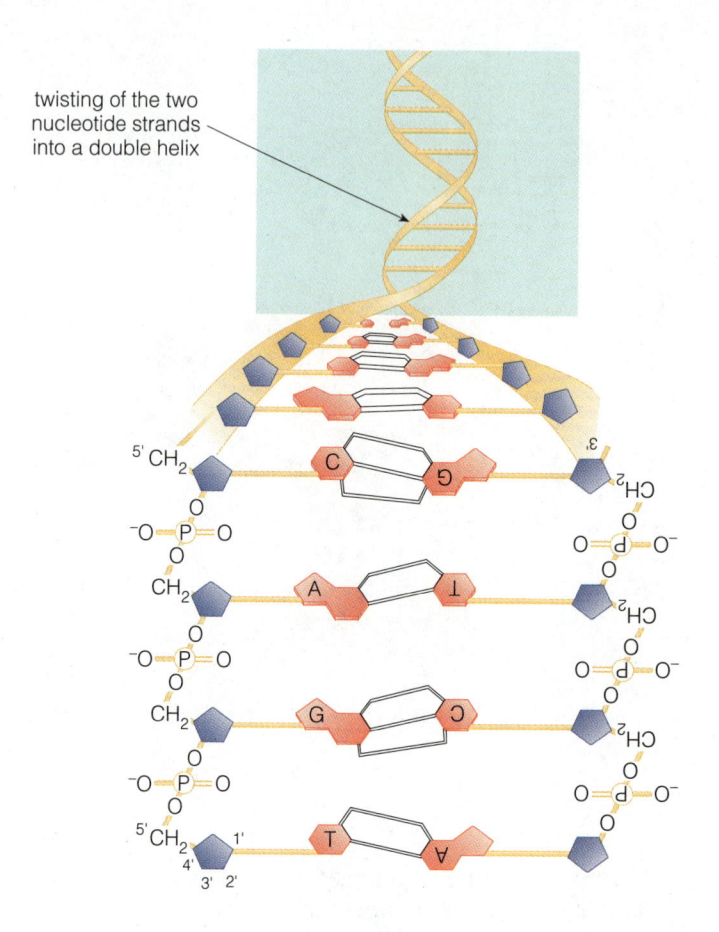

Figure 23.8 Representation of a DNA double helix. The two strands are joined by hydrogen bonds between complementary base pairs (see Figure 23.9).

is held together by hydrogen bonds (Figure 23.8). Base pairs are flat and stack above and below each other inside the helix. They also are parallel to each other and perpendicular to the axis of the helix, like steps in a spiral staircase. DNA may have as many as 30,000 base pairs, making it the largest of all natural organic molecules.

4. The helix is usually right-handed, with 10 base pairs per complete turn.

Other forms of DNA (including circular, single-stranded, left-handed, and supercoiled) also occur, but we will not be concerned with them here.

The Watson–Crick model fits with several observed characteristics of DNA. First, it explains why DNA from different organisms differs in its composition of bases A, C, G, and T, but *always has about equal amounts of A and T, and equal amounts of C and G* (Table 23.2). This is because A always pairs with T, and C always pairs with G.

Table 23.2 Base Composition in DNA from Various Organisms

Organism	%A	%C	%G	%T	A/T	C/G
Human	30.9	19.8	19.9	29.4	1.05	1.00
Sea urchin	32.8	17.3	17.7	32.1	1.02	1.02
E. coli	24.7	25.7	26.0	23.6	1.04	1.01
Wheat germ	27.3	22.8	22.7	27.1	1.01	1.00
Bacteriophage T7	26.0	24.0	24.0	26.0	1.00	1.00

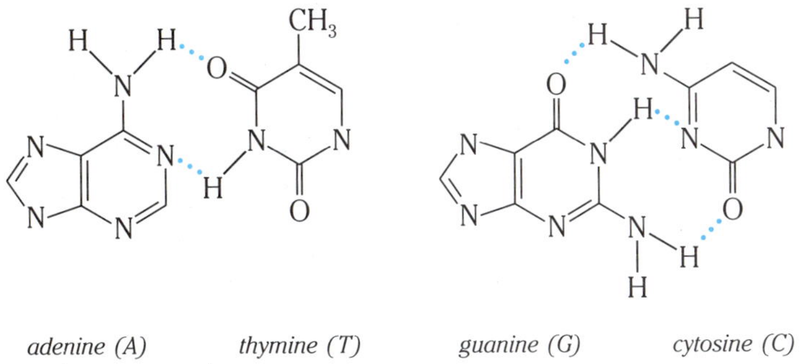

adenine (A)   thymine (T)   guanine (G)   cytosine (C)

Figure 23.9 In opposite strands of DNA, two hydrogen bonds (dotted lines in color) form between A and T, and three form between C and G.

The model also explains why DNA typically denatures at temperatures about 70°C. The "glue" holding the two strands together is hydrogen bonding between base pairs. Since hydrogen bonds are much weaker than ionic or covalent bonds (Section 6.1), heat breaks those bonds and allows the strands to separate. It takes more heat to denature DNA as its percentage of C and G increases. Because C and G form three hydrogen bonds while A and T form only two (Figure 23.9), higher content of those two bases means more hydrogen bonds to break.

But the greatest triumph of the double helix structure is that the base-pairing arrangement suggests how DNA works in the cell. The two most important reactions involving DNA are its copying during cell division and the making of RNA. Now we examine these two processes.

23.2 SYNTHESIS OF DNA

Replication. Before dividing, a cell synthesizes a copy of its DNA so that both of the resulting cells will have the same genetic material. This process is called **replication.** Replication occurs in the cell nucleus, and produces an identical copy of the original DNA.

How does replication occur? Is the new DNA made entirely of new nucleotide units? Or do the two DNA molecules each contain one strand of original DNA and one strand of new material? Or do bits and pieces of original and new material appear in both copies?

The answer came from experiments that identified both the original and new DNA (Figure 23.10). When bacterial cells are first grown in a medium where the only source of nitrogen is the isotope ^{15}N, all their DNA (with its nitrogen-containing bases) becomes labeled with ^{15}N. If the cells then divide in a medium containing only ^{14}N, any newly synthesized DNA will contain only ^{14}N. Since ^{15}N is heavier than ^{14}N, the two types of DNA separate when centrifuged at high speeds, with ^{15}N material traveling farther down in the centrifuge tube.

Figure 23.10 shows the results of such experiments. After one replication, all DNA was a hybrid of ^{15}N-DNA and ^{14}N-DNA. After two replications, half of the DNA was hybrid ^{15}N–^{14}N and half was all ^{14}N. These results show that when DNA replicates, the two resulting DNA molecules each have one strand of original DNA and one strand of newly synthesized material.

500   Part III   Biochemistry

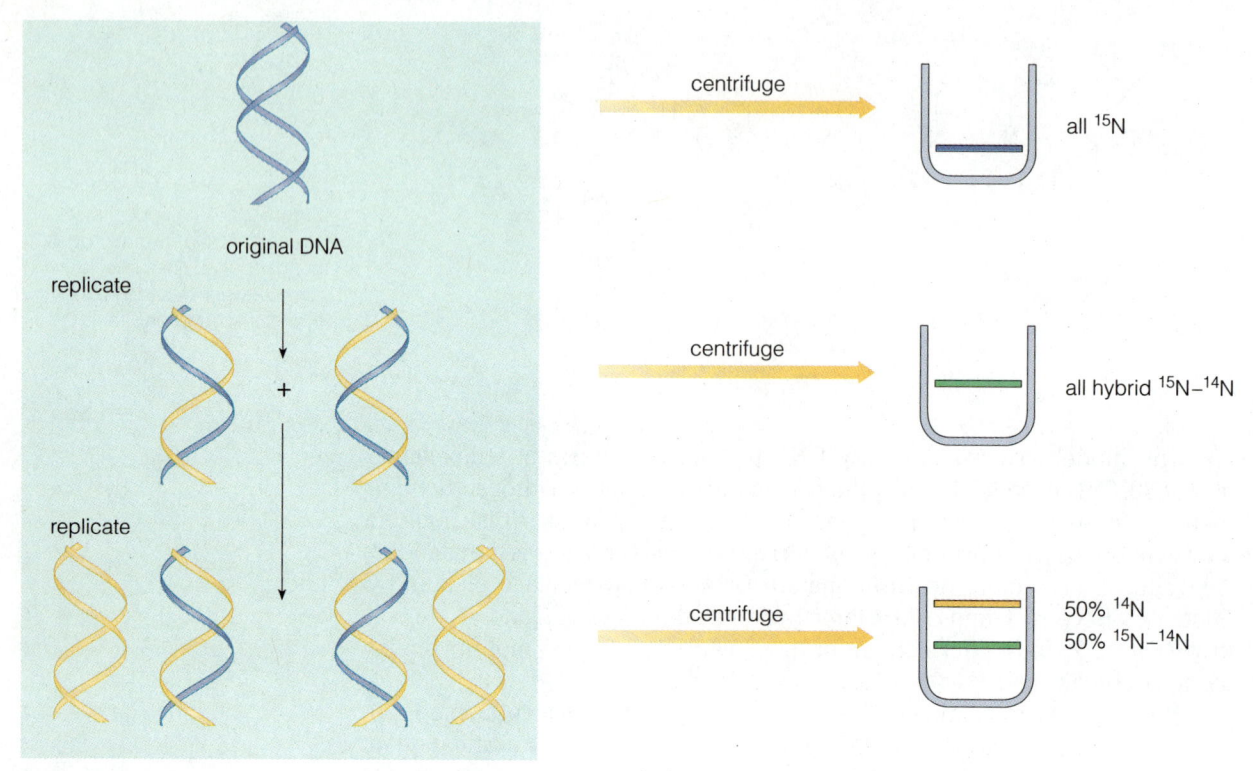

Figure 23.10   Replication of ^{15}N-DNA (blue) in ^{14}N media (yellow).

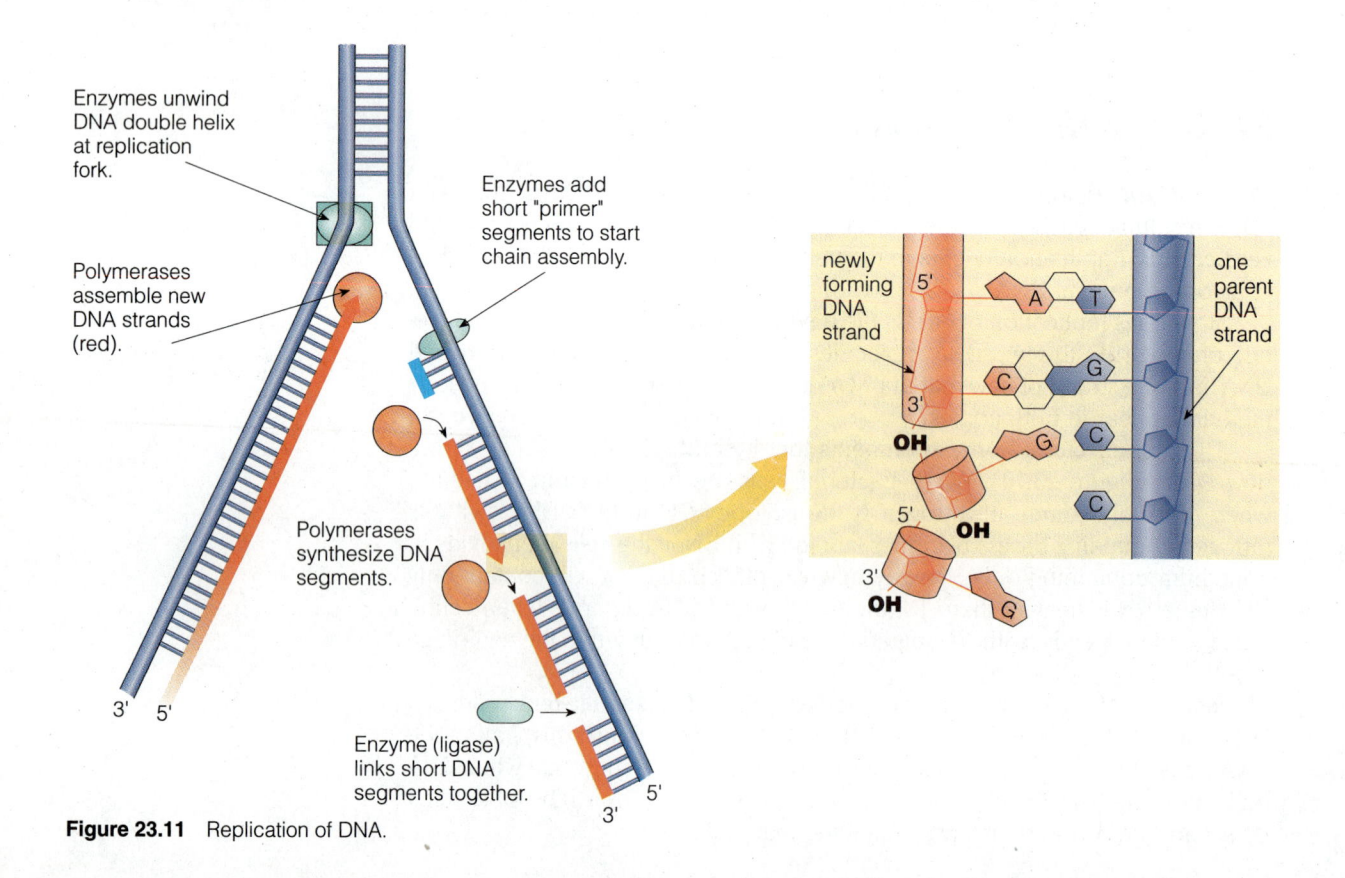

Figure 23.11   Replication of DNA.

EXAMPLE 23.2 If all ^{15}N-DNA replicates three times in ^{14}N medium, what is the percentage of DNA as (a) all ^{15}N, (b) hybrid ^{15}N–^{14}N, and (c) all ^{14}N?

SOLUTION If you extend Figure 23.10 to include one more replication in ^{14}N, the results are (a) 0% ^{15}N, (b) 25% hybrid ^{15}N–^{14}N, (c) 75% ^{14}N.

In your cells, replication begins at many distinct sites in DNA, called replication forks. Here enzymes unwind local regions of the double helix, separating the strands (Figure 23.11).

Deoxynucleotide units (as triphosphates) are attracted by hydrogen bonds to pair with appropriate bases on both strands. Then enzymes, called DNA polymerases, catalyze removal of the last two phosphate units and join

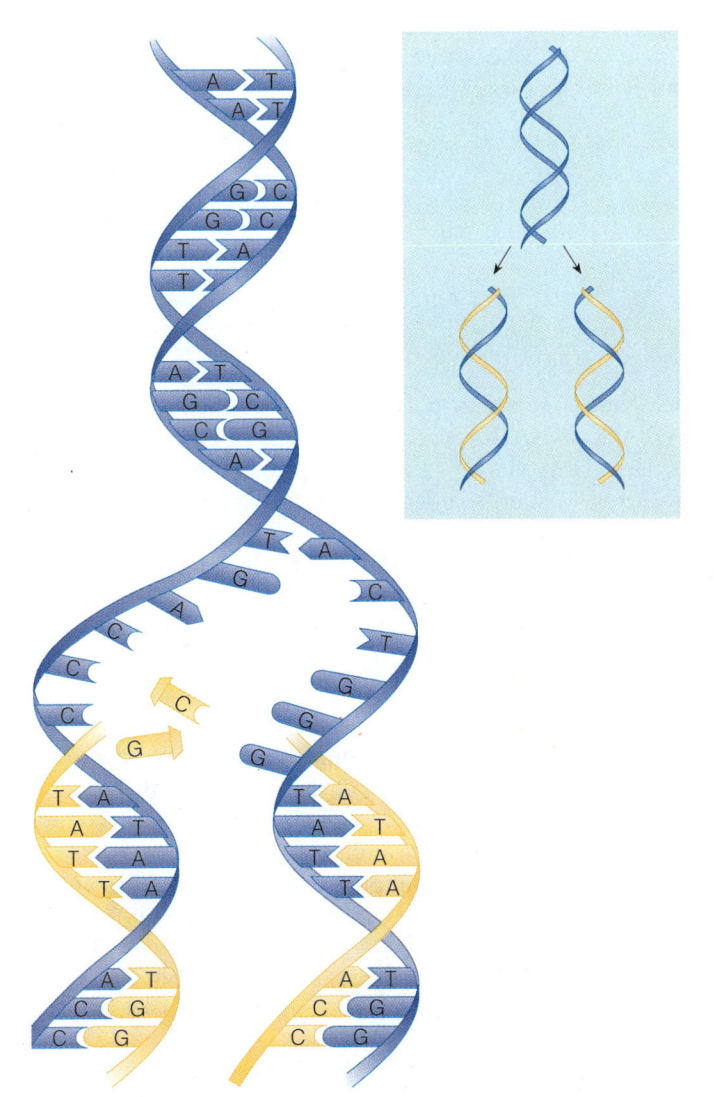

Figure 23.12 DNA strands separate during replication. The product DNA molecules each have one strand of the original DNA molecule.

CHEMISTRY SPOTLIGHT

Discovery of an Anticancer Drug

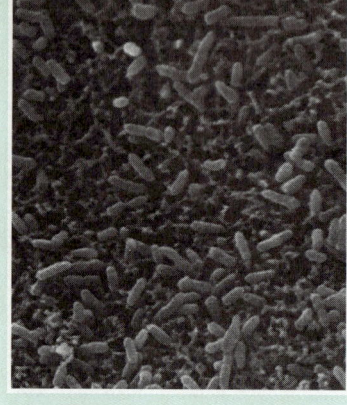

Figure 23.13 Electron micrograph of normal E. coli (left) and E. coli in the presence of cisplatin (right).

In 1961 Barnett Rosenberg, a biophysicist at Michigan State University, was studying the effects of electrical fields on cell division. He placed platinum electrodes in a colony of bacteria and noticed that the bacteria developed long, filament-like shapes (Figure 23.13). They were growing, but they weren't dividing.

What was blocking cell division? It turned out that the electrical field wasn't the culprit. Instead, it was a compound formed from the platinum electrodes, diaminedichloroplatinum, $Pt(NH_3)_2Cl_2$, or cisplatin.

Additional research showed that cisplatin blocks cell division by cross-linking DNA. When cisplatin was tested as an anticancer drug, it worked. Now it is one of the most effective drugs for treating ovarian, head and neck, bladder, testicular, prostate, and several other types of cancer.

Many people, including this author, are alive today because scientists followed a clue left by an experiment with platinum.

---

monophosphate nucleotides into new polynucleotide strands complementary to the original DNA strands. The fork moves along DNA, continuing to separate the strands. New DNA is synthesized continuously along one original strand, but synthesized in short segments along the other strand (Figure 23.11).

DNA is simultaneously synthesized at many forks. All the pieces of new DNA are joined together by the enzyme DNA ligase to produce the final new strand. The original strands separate completely, each wrapping around its newly synthesized, complementary strand to produce two double helix molecules (Figure 23.12).

Replication and Cancer. *Cancer* is characterized by uncontrolled cell growth and reproduction. Since cells must copy DNA before they divide, one way to treat cancer is with drugs that prevent replication.

Two major classes of anticancer drugs are antimetabolites and cross-linking agents. Recall that antimetabolites such as 5-fluorouracil and 6-mercaptopurine resemble normal bases in DNA and block DNA synthesis (replication) (see opening to Chapter 22).

Cross-linking agents (see the Chemistry Spotlight above) bind to both strands of double helical DNA and thus prevent replication. Strangely enough, the first known drugs of this type, called nitrogen mustards (Figure 23.14), are closely related to mustard gas, a chemical warfare agent used in World War I.

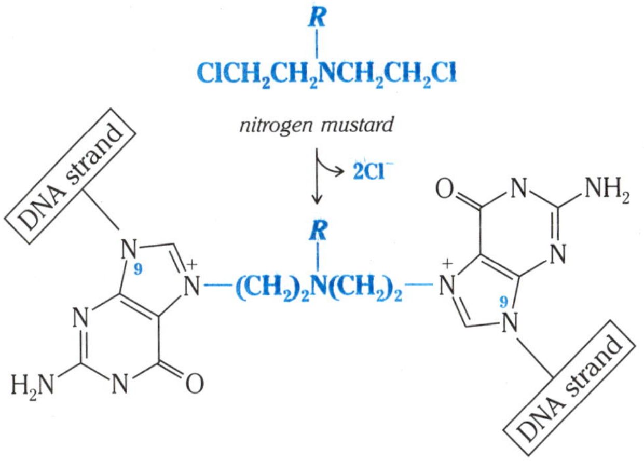

Figure 23.14 A nitrogen mustard binds to guanine (G) units on both DNA strands, cross-linking DNA so it cannot replicate.

23.3 SYNTHESIS OF RNA

Transcription. Your cells use DNA not only to make more DNA, but also to make RNA. Synthesizing RNA from DNA is called **transcription.** Transcription occurs in the cell nucleus and is a bit like replication except that the product, RNA, is a single polynucleotide strand containing the pentose ribose and the base uracil instead of thymine.

Transcription begins when the enzyme RNA polymerase binds to one of the DNA strands containing a *promoter,* a particular sequence of bases. The enzyme opens a local region of the double helix and synthesizes RNA complementary to the DNA strand containing the promoter (Figure 23.15).

As triphosphate nucleotides pair by hydrogen bonding with appropriate bases on DNA, an RNA strand forms. The base pairings are the same as in DNA except that adenine in DNA pairs with uracil (not thymine) in RNA (Figure 23.15). RNA polymerase removes the last two phosphate units from each nucleotide and joins the monophosphate nucleotides by 3',5'-phosphodiester bonds. The single RNA strand then separates from DNA.

RNA is chemically modified after leaving DNA, and passes out of the nucleus to function elsewhere in the cell. The chemical alterations and final structure depend on the type of RNA. We'll discuss each type separately.

Transfer RNA. The smallest type of RNA is **transfer RNA (tRNA),** which has about 75–90 nucleotide units and a molecular weight of about 25,000. As many as 60 different tRNA molecules occur in a cell.

Figure 23.16 shows structures of a single-stranded tRNA molecule. Base pairing within the strand produces a cloverleaf structure with three major loops. The 3' end has the sequence ACC. At the other end (loop II) is a sequence of three bases called the *anticodon.* As we discuss in Section 23.4, the 3' end binds amino acids for protein synthesis and the anticodon helps align amino acids in the proper sequence.

Many tRNAs are synthesized as part of a larger RNA molecule in the nucleus. After transcription, enzymes hydrolyze that RNA into smaller tRNA units and, when the sequence isn't already there, adds ACC to the 3' end. Other enzymes chemically modify bases in various ways (such as methylation or hydrogenation).

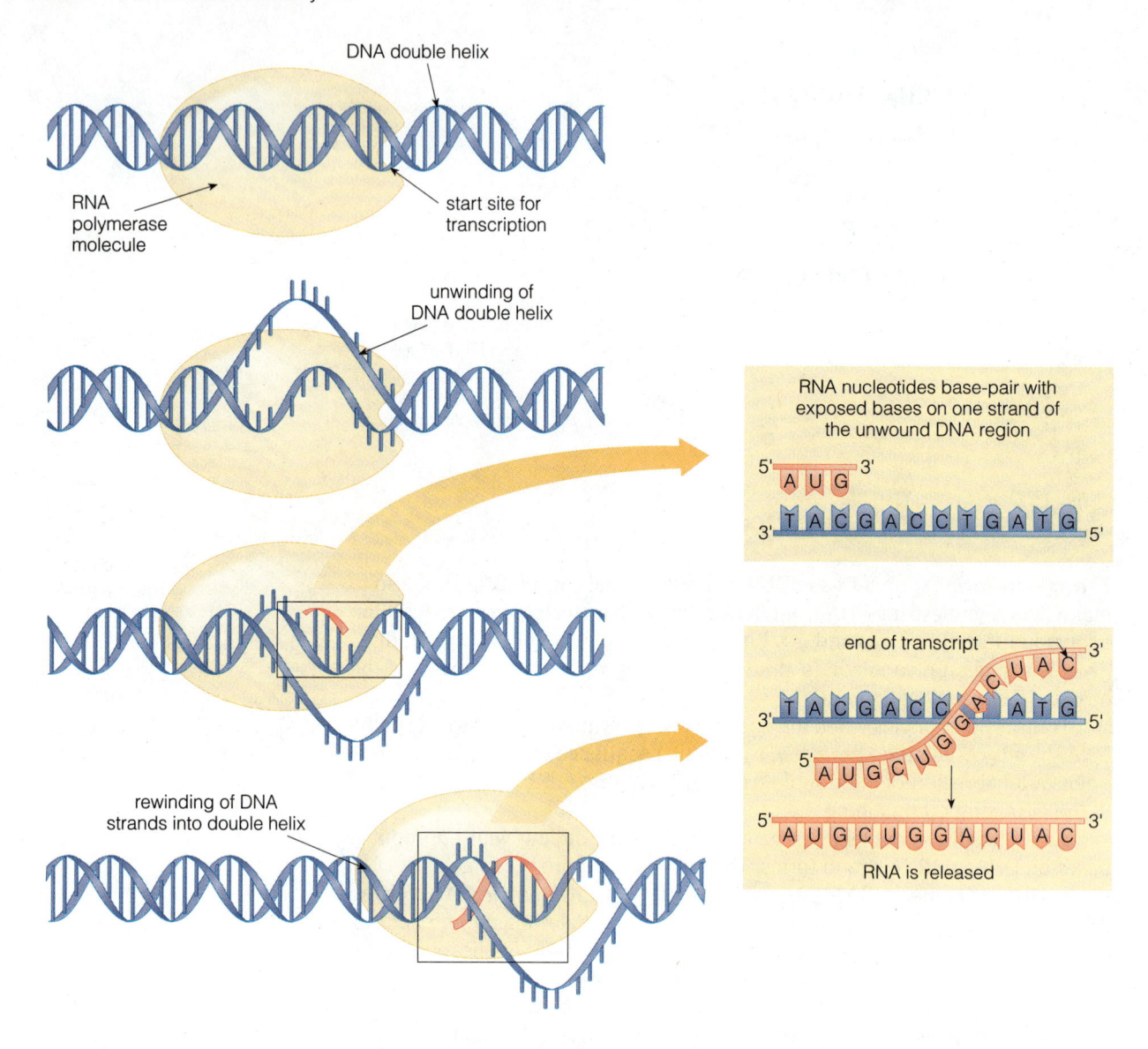

Figure 23.15 Transcription: synthesis of RNA from DNA. Following transcription, the unwound region of DNA winds up again into a double helix.

Ribosomal RNA. **Ribosomes** are small granules found either free in the cytoplasm or attached to the endoplasmic reticulum (see Appendix E). Ribosomes consist of two subunits (Figure 23.17). In animal cells, the larger subunit is called 60S[1] and the smaller 40S. In bacteria, they are called 50S and 30S, respectively.

Each subunit is made from proteins and a type of RNA called **ribosomal RNA (rRNA).** A ribosome is about 2/3 rRNA and 1/3 protein. We examine in Section 23.4 how ribosomes function in protein synthesis.

Messenger RNA. DNA in the nucleus contains the information for the proteins the cell can make. But in order for that information to be used, it has to travel from the nucleus to ribosomes, where proteins are assembled. This message-carrying function is done by **messenger RNA (mRNA).**

---

[1] *S* stands for a *S*vedberg unit, which is related to molecular size. It was named after the Swedish chemist who invented the ultracentrifuge.

Nucleic Acids   Chapter 23   505

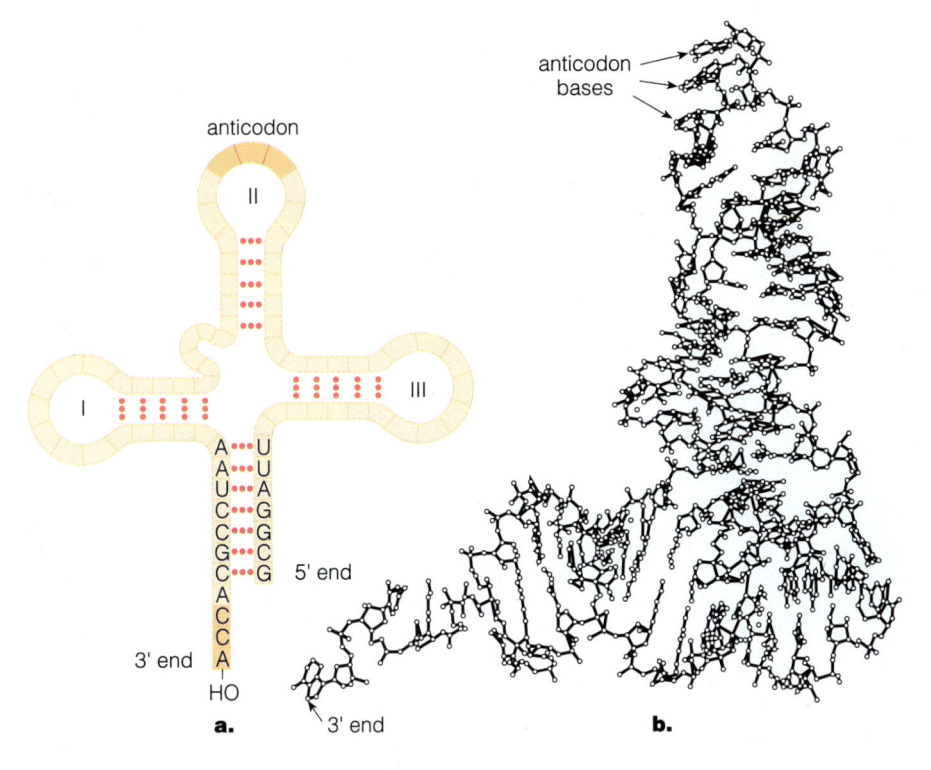

Figure 23.16 (a) Two-dimensional representation of a tRNA molecule and (b) a three-dimensional wire model of the same tRNA.

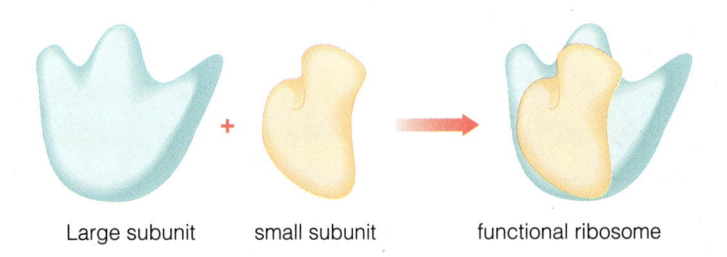

Figure 23.17 Model of a ribosome and its subunits.

Region of DNA template strand to be copied:

Region of Transcription
exon intron exon intron exon

Transcription

Modifications to the transcript:

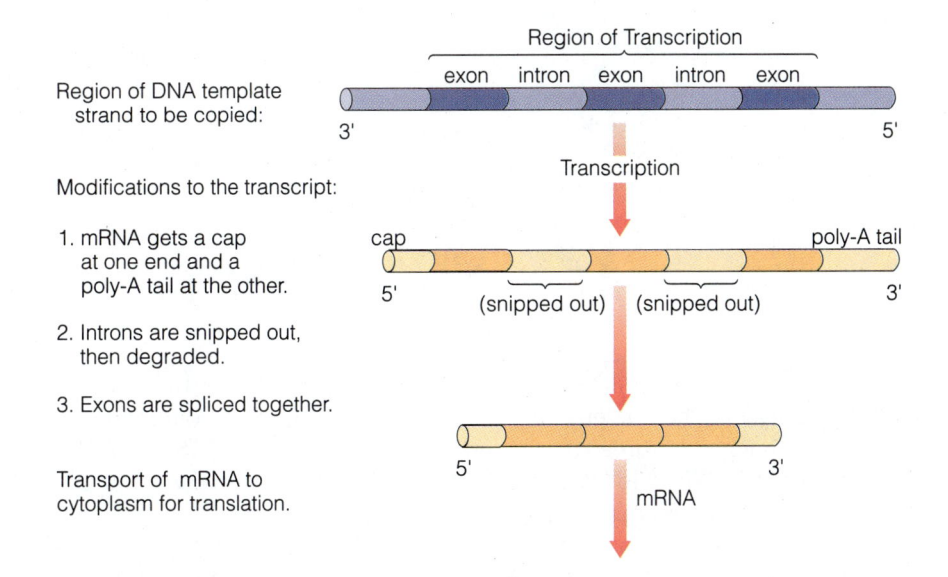

1. mRNA gets a cap at one end and a poly-A tail at the other.

cap — (snipped out) — (snipped out) — poly-A tail

2. Introns are snipped out, then degraded.

3. Exons are spliced together.

mRNA

Transport of mRNA to cytoplasm for translation.

Figure 23.18 Steps in processing mRNA.

CHEMISTRY SPOTLIGHT

Reverse Transcriptase

Your cells normally make RNA from DNA by transcription using the enzyme RNA polymerase. But in certain cases, this process is reversed: DNA is synthesized from RNA. An enzyme that catalyzes this process is called *reverse transcriptase*.

Reverse transcriptase is more than just a biological oddity. Certain viruses (called retroviruses) inject their RNA into infected cells and provide the cell with reverse transcriptase to make new DNA from viral RNA. This disrupts the cell's normal functioning and makes it produce more virus particles, thus spreading the infection.

One example is the virus that causes AIDS (acquired immune deficiency syndrome). Since reverse transcriptase is a key to the infectious action of the virus, scientists are developing drugs that inhibit this enzyme. Two such drugs are AZT (azidothymidine) and DDI (dideoxyinosine), which masquerade as normal components of DNA (Figure 23.19). Notice that both drugs lack an —OH group on the 3' position. This blocks formation of a 3', 5'-phosphodiester bond, a necessary step in DNA synthesis.

These drugs cause cells to make faulty DNA, and prevent the AIDS virus from multiplying. This prolongs the lives of many patients. Since they don't kill the virus, however, AZT and DDI don't cure AIDS.

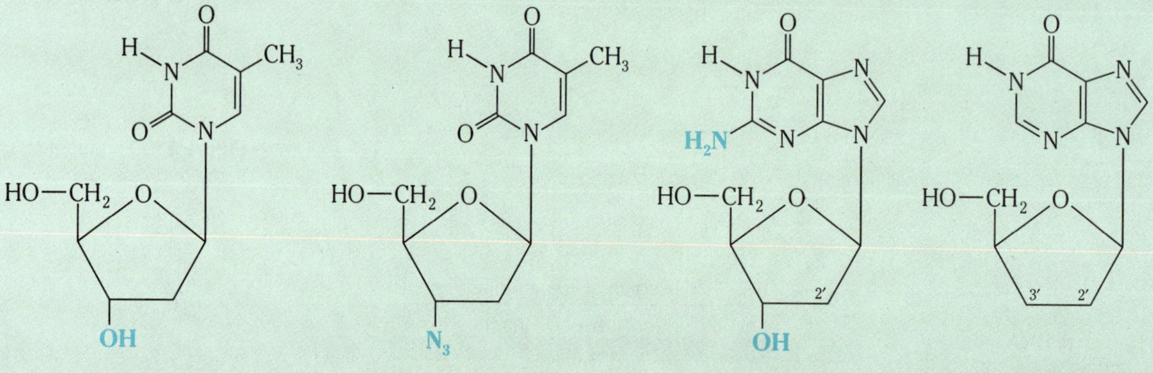

Figure 23.19 Two drugs used to treat AIDS patients resemble normal components in DNA. Structural differences are in color.

thymidine (normal component in DNA)

azidothymidine (AZT)

2'-deoxyguanosine (normal component in DNA)

2',3'-dideoxyinosine (DDI)

Like rRNA and tRNA, mRNA is made in the nucleus from DNA by transcription. Before leaving the nucleus, however, mRNA in animal and other eukaryotic cells is changed in three ways by enzyme action (Figure 23.18). At the 5' end a methylated GTP "cap" is attached. At the 3' end a long chain of 150–200 AMP units is added, perhaps to make mRNA more stable.

The third change removes nucleotide segments that don't code for proteins. mRNA transcribed from DNA contains one or more intervening segments, called *introns,* that are not used in protein synthesis. These segments are removed by hydrolase enzymes. The remaining segments, called *exons* because they are expressed during protein synthesis, are spliced together. The mRNA then leaves the nucleus, carrying information to the ribosomes for protein synthesis.

23.4 PROTEIN SYNTHESIS

A **gene** is a section of DNA that has a specific function in protein synthesis. Many genes code for the synthesis of mRNA molecules, from which specific proteins are made; other genes code for rRNA or tRNA molecules. Still other genes help regulate which proteins a cell actually makes.

Human cells normally have 46 chromosomes: 22 pairs plus an X and a Y chromosome for males (Figure 23.20) or two X chromosomes for females. Altogether these chromosomes contain about 3 billion base pairs and 50,000 to 100,000 genes. So one chromosome typically contains many millions of base pairs and more than a thousand genes.

The information in DNA for a specific protein first passes to mRNA by transcription in the nucleus. The sequence of nucleotides in mRNA reflects the nucleotide sequence in the DNA from which it was transcribed. Then mRNA carries this information out of the nucleus to a ribosome, where the sequence of nucleotides in mRNA is converted into a sequence of amino acids that makes up a specific protein. Because the information changes from a sequence of nucleotides into a sequence of amino acids, this process is called **translation.**

Genetic Code. Translating a sequence of nucleotides into a sequence of amino acids proceeds according to a *genetic code*. The entire code is known, and it is the same in nearly all organisms.

A sequence of three nucleotide bases, called a **codon,** carries the information for a specific amino acid. Table 23.3 lists codons as they appear in mRNA. By knowing the base pairings, you can figure out the corresponding three-base sequences in DNA. For example, UUU on mRNA codes for phenylalanine; this corresponds to an AAA sequence in the DNA from which that mRNA was transcribed.

Notice in Table 23.3 that many amino acids have several codons. Notice also that four codons are used as start (AUG) or stop (UAA, UAG, and UGA) signals. These signals specify the length of the amino acid chain. Codons lying between a start and stop signal are all translated to produce the sequence of amino acids in the protein.

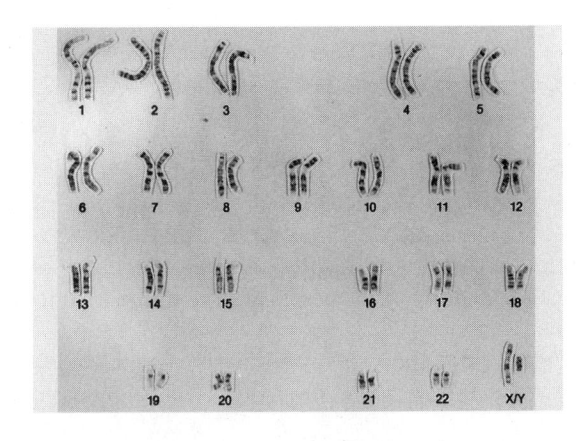

Figure 23.20 Normal chromosome pattern for a male.

Table 23.3 The Genetic Code (for mRNA)

First Base	Middle Base U	Middle Base C	Middle Base A	Middle Base G	Third Base
U	phe	ser	tyr	cys	U
	phe	ser	tyr	cys	C
	leu	ser	stop	stop	A
	leu	ser	stop	trp	G
C	leu	pro	his	arg	U
	leu	pro	his	arg	C
	leu	pro	gln	arg	A
	leu	pro	gln	arg	G
A	ile	thr	asn	ser	U
	ile	thr	asn	ser	C
	ile	thr	lys	arg	A
	start(met)	thr	lys	arg	G
G	val	ala	asp	gly	U
	val	ala	asp	gly	C
	val	ala	glu	gly	A
	val	ala	glu	gly	G

EXAMPLE 23.3

(a) List codons for the amino acid alanine.
(b) List amino acids specified by the following codons and determine the nucleotide sequence in DNA that corresponds to those codons in mRNA: (1) AUA, (2) CCC, (3) GAG.

SOLUTION

(a) GCU, GCG, GCA, GCC
(b) (1) ile (isoleucine), TAT in DNA; (2) pro (proline), GGG; (3) glu (glutamic acid), CTC

PP$_i$ is pyrophosphate, the neutralized salt of pyrophosphoric acid (see Section 16.8).

Translation. Translating a sequence of codons in mRNA into a sequence of amino acids in a protein chain is a complex process, and we will not examine all the details here. There are four steps:

1. *Joining amino acids to tRNA molecules.* Enzymes called aminoacyl-tRNA synthetases join an amino acid to a tRNA using the nucleotide ATP as an energy source. The reaction is:

$$\text{amino acid} + \text{tRNA} + \text{ATP} \xrightarrow{\text{aminoacyl-tRNA synthetase}} \text{aminoacyl-tRNA} + \text{AMP} + \text{PP}_i$$

These enzymes are highly specific; each one joins only one particular amino acid to one particular tRNA. In this reaction, ribose bonded to the 3' end of adenine in tRNA provides an —OH group that reacts with the carboxyl group of the amino acid to form an ester (Figure 23.21). The product, an aminoacyl-tRNA, has an amino acid bonded at the 3' end of tRNA and a three-base sequence, the **anticodon,** at the other end. tRNA carries its bound amino acid to the ribosome for assembly into a protein chain.

2. *Initiation of protein synthesis.* To begin protein synthesis, a small (40S) ribosome subunit binds mRNA near a "start" codon (AUG). The aminoacyl-

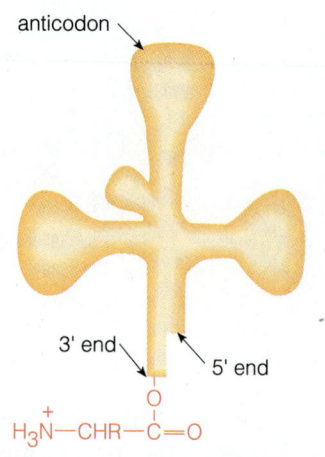

Figure 23.21 An aminoacyl-tRNA molecule.

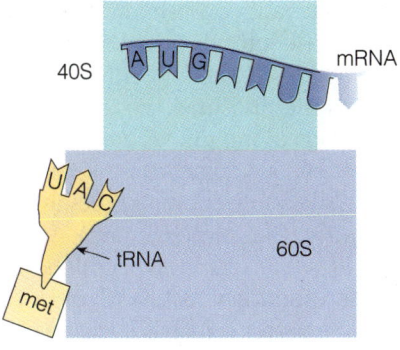

a. Translation begins with convergence of an initiator tRNA, an mRNA molecule, and the small and large ribosomal subunits.

b. Chain elongation begins when the second tRNA moves into the next site and base-pairs to the second mRNA codon. As it does, its attached amino acid aligns with the amino acid of the initiator, tRNA.

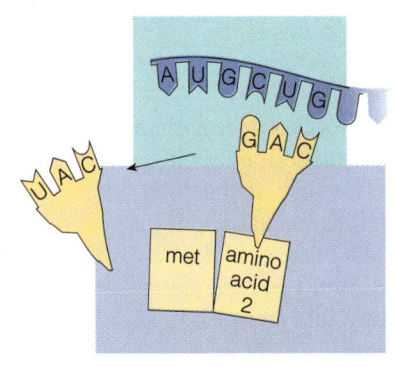

c. The bond between the first tRNA and its amino acid (met) is broken. A peptide bond forms between the two amino acids. The first tRNA leaves the ribosome.

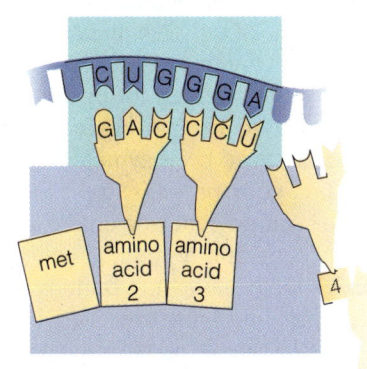

d. The third tRNA base-pairs with the third codon. A peptide bond forms between amino acids 2 and 3. Through repetitions of these steps, a polypeptide chain grows until a stop codon in the mRNA is reached. Then the chain is released from the ribosome.

Figure 23.22 Translation.

tRNA having the anticodon (UAC) that base pairs with AUG forms hydrogen bonds with that codon on mRNA. In animal cells, that tRNA carries the amino acid methionine. Finally, the large (60S) ribosome subunit binds to the complex (Figure 23.22a).

3. *Synthesizing the protein chain.* The next codon on mRNA is then read. An aminoacyl-tRNA with the appropriate anticodon bonds to the codon on mRNA next to AUG. A subunit (peptidyl transferase) of the 60S ribosome subunit then joins methionine by a peptide bond to the amino acid on this next tRNA (Figures 23.22b and c); in the process methionine is released from its tRNA, which then leaves the ribosome. The reaction is:

$$NH_2-CHR-\underset{\underset{tRNA_1}{O^-}}{\overset{\overset{O}{\|}}{C}}_{\delta^+} :NH_2-CHR-\underset{\underset{tRNA_2}{O^-}}{\overset{\overset{O}{\|}}{C}} \xrightarrow{\text{peptidyl transferase}} NH_2-CHR-\overset{\overset{O}{\|}}{C}-\underset{H}{N}-CHR-\underset{\underset{tRNA_2}{O^-}}{\overset{\overset{O}{\|}}{C}} + tRNA_1$$

met-tRNA   second AA-tRNA   *peptide bond*

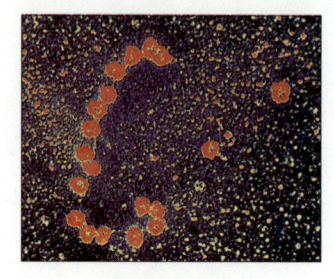

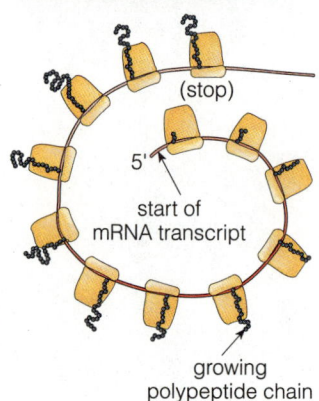

Figure 23.23 Electron micrograph (top) and drawing (bottom) of a group of ribosomes bound to a mRNA molecule (polyribosome).

Because of the way it works, induction *is sometimes called* derepression.

Next, another aminoacyl-tRNA with the appropriate anticodon hydrogen bonds with the next codon on mRNA. Peptidyl transferase joins that next amino acid to the two already in place, forming a tripeptide bonded to tRNA (Figure 23.22d). Now the second tRNA is released from the ribosome.

The pattern continues over and over. Codons along mRNA are "read" in order, and amino acids are joined in sequence. It's something like a tape (mRNA) passing through a tape recorder (the ribosome subunits), being converted (translated) into a beautiful piece of music (the protein chain).

4. *Termination and release of the protein chain.* As mRNA codons are "read," eventually the ribosome reaches a "stop" codon (UAA, UAG, or UGA). No aminoacyl-tRNA binds to such a codon, so the protein chain is released from the last tRNA, and that tRNA leaves mRNA. The ribosome leaves mRNA and separates into its subunits. The ribosome subunits and released tRNA molecules can be used again in protein synthesis from an appropriate mRNA.

A few other details are important. During protein synthesis, several ribosomes may be bound to different regions of a mRNA molecule (always starting at an AUG codon), simultaneously translating its codons into protein molecules. A strand of mRNA with several bound ribosomes is called a *polysome* or *polyribosome* (Figure 23.23). This enables cells to rapidly make multiple molecules of a protein coded for by a mRNA molecule.

The released protein chain is chemically modified. One or more amino acids are removed by enzyme action from the amino (methionine) end of the chain. Thus your proteins don't all have methionine at the N-terminal end. Sometimes amino acid side chains are modified. One example is the hydroxylation of proline and lysine units in collagen (see Section 21.6).

DNA specifies only the primary structure (amino acid sequence) of a protein. The protein chain, however, spontaneously folds into its most stable secondary, tertiary, and quaternary (if any) structure.

Regulation of Protein Synthesis. Virtually all your cells have the same DNA, yet they vary considerably in the proteins they make. Your pancreatic cells, for example, don't make hemoglobin and your blood cells don't make insulin. That must mean there is a way to regulate which genes a cell uses.

Figure 23.24 shows an example of gene regulation in prokaryotic cells. Control occurs at the level of transcription (mRNA synthesis); transcribed genes are expressed (that is, the products they code for are synthesized) while genes that aren't transcribed cannot be expressed. One or more *structural genes* (genes that code for a protein chain) comprise an *operon,* a set of genes that are all transcribed (or not transcribed) together. Near the operon is a *promoter,* the region of DNA where RNA polymerase binds to transcribe structural genes in that operon.

Protein synthesis depends on whether or not RNA polymerase transcribes the structural genes. A *regulator gene* codes for a protein, called a *repressor,* that blocks transcription by binding to an *operator,* which is located between the structural genes and the promoter. When a repressor binds to the operator, RNA polymerase cannot bind to the promoter and cannot transcribe the structural genes. This process of turning off transcription (which prevents protein synthesis) is called **repression.** You can think of a repressor binding to the operator gene as an "off" switch for a gene.

But some substances, called *inducers,* block repression. In other words, they inactivate the "off" switch and turn gene expression "on." This is called **induction.**

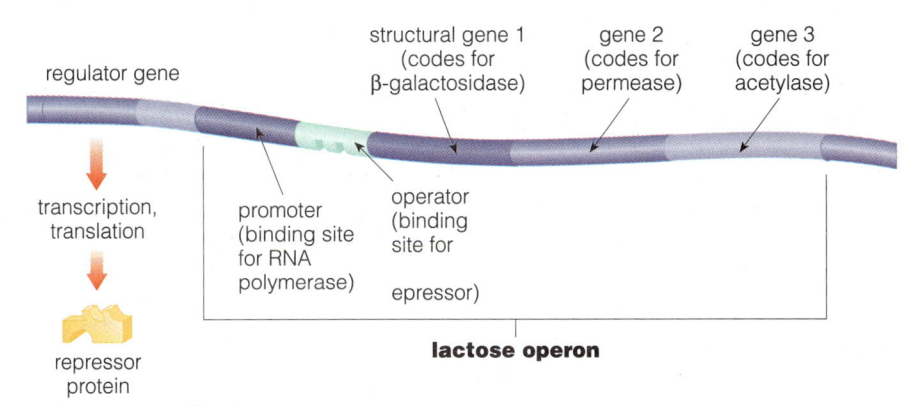

a. Components of the lactose operon.

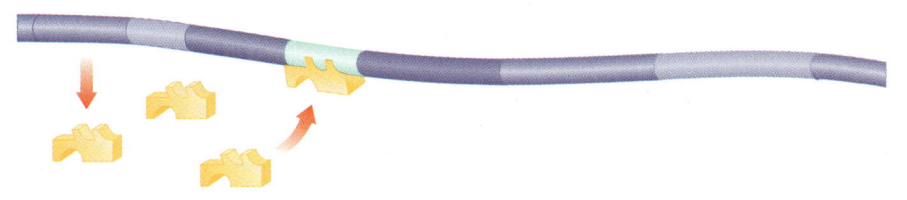

b. When lactose is absent, a repressor protein binds to the operator and so prevents RNA polymerase from binding to DNA and initiating transcription. Lactose-metabolizing enzymes (which are not needed) are not produced.

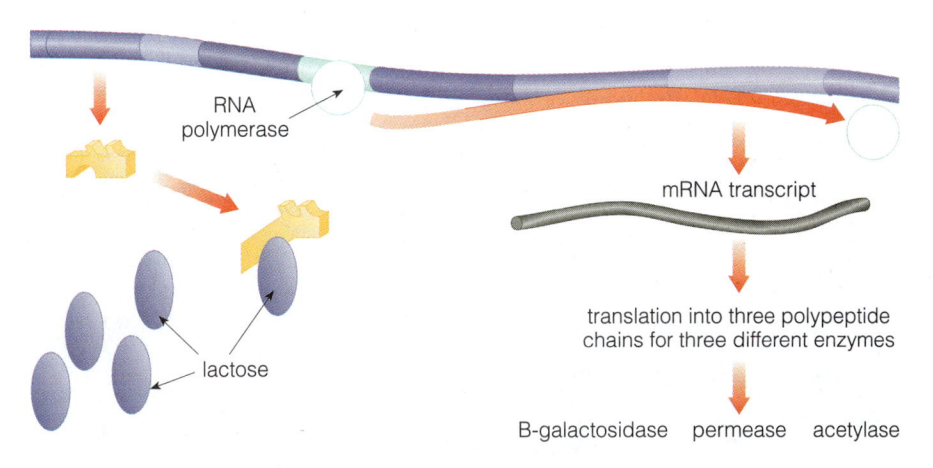

c. When lactose is present, it binds to the repressor and prevents it from binding to the operator. The promoter site is now exposed and RNA polymerase starts transcription. Thus lactose induces transcription of the lactose operon.

Figure 23.24 Induction and repression of the lactose operon in *E. coli*.

Figure 23.24 shows an example. Certain bacteria grown on glucose normally don't synthesize an enzyme (β-galactosidase) to hydrolyze lactose into glucose and galactose. But when glucose is absent and lactose is their only carbon source, the bacteria start synthesizing β-galactosidase and two other enzymes that help metabolize lactose.

Lactose itself is the inducer substance. Lactose binds to the repressor and prevents it from binding to the operator. When repression is stopped, the structural genes are expressed. Induction is an efficient way for a cell to produce proteins only when they are needed. In this example, bacteria produce lactose-metabolizing enzymes only when lactose is present.

23.5 MUTATIONS

A **mutation** is a change in the nucleotide (base) sequence in DNA. Mutations occur naturally from errors during replication. They also arise naturally during chromosome rearrangements when portions of DNA relocate on the same or another chromosome. Radiation and certain chemicals also cause mutations.

A mutated gene often codes for an abnormal protein. If the affected cells die and are replaced by normal cells, mutations produce little damage. But a tumor or other damage may occur when mutated DNA is copied into other cells during cell division. Mutations in an egg or sperm cell may pass on to the next generation in the form of genetic disorders. In sickle-cell anemia, for example, the gene for an abnormal hemoglobin protein passes from one generation to the next.

Chemicals that cause mutations are called **mutagens.** Mutagens work in several different ways. Some react directly with bases to chemically change them. Nitrous acid (HNO_2), for example, oxidizes amino groups in bases:

cytosine → uracil

Many chemicals may cause mutations in humans, but the evidence is often indirect. Potential mutagens are tested in other organisms; those that cause mutations are then prime suspects for human mutagens. The most widely used test, originated by Bruce Ames at the University of California at Berkeley, detects chemicals that cause mutations in the bacterium *Salmonella typhimurium* (Figure 23.25).

Chemicals that cause cancer are called **carcinogens.** A common estimate is that more than half of all cancers are triggered by environmental factors such as carcinogens. Though researchers know relatively little about how to identify

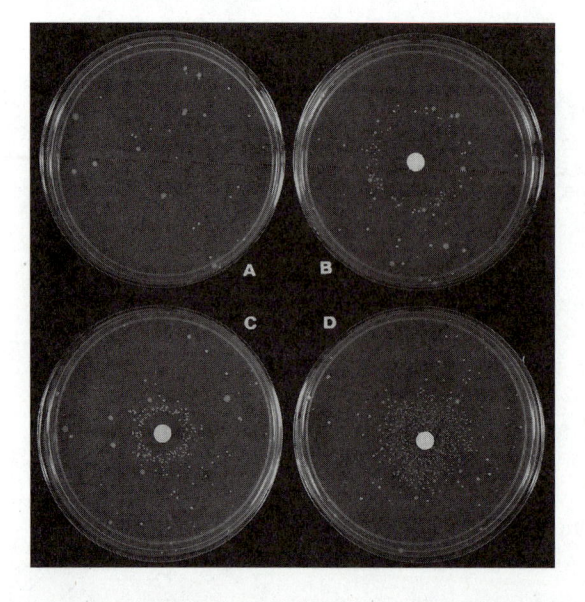

Figure 23.25 Plates containing bacteria in the absence (A) or presence (B-D) of a mutagen. In the Ames test, mutagens reverse a genetic error in the bacteria and thus enable them to grow around the disc (in the center of plates B-D) that contains a mutagen.

carcinogens in general, they have discovered a relationship between mutagens and carcinogens: about 80% of mutagens identified by the Ames test are carcinogens.

Scientists have also discovered dozens of cancer-causing genes, called *oncogenes*. Some oncogenes normally code for cell growth and development; when those genes become abnormal, cells grow in an uncontrolled way. Other oncogenes (for example, one involved in a rare type of eye cancer) cause cancer by their absence, or mutation. Such genes normally code for a growth-control substance; when the gene is absent or inactivated by a mutation, cells grow uncontrollably and cancer develops.

23.6 GENETIC ENGINEERING

Genetic engineering is a way to change the genes in cells so that cells make new proteins. Inserting genes into cells produces genetic combinations that don't occur in nature. Frogs and flies, for example, don't reproduce with each other, but scientists can splice together pieces of frog and fly DNA to produce cells having both kinds of genes. DNA containing material from two or more different sources is called **recombinant DNA.** Any combination is possible, including bacterial genes in plants, plant genes in animals, and animal (including human) genes in bacteria, plants, or other animals (Figure 23.26).

Figure 23.26 This tobacco plant, redesigned to contain a firefly gene, glows in the dark. (Courtesy Dr. Keith V. Wood)

Genetic engineering became possible in the early 1970s because of two developments. One was the discovery of suitable materials (or *vectors*) to carry new genes into cells. In animal cells, the vectors often are chemically modified viruses that insert DNA into cells they infect. The usual vectors for bacterial cells are **plasmids.** These are small, freely floating ringlets of DNA that are separate from chromosomal DNA (Figure 23.27). Chemists discovered simple chemical treatments to isolate plasmids from cells and insert them into other cells without harming them.

The second development was a way to splice new genes into vectors. Scientists discovered enzymes, called **restriction enzymes,** that cut double-stranded DNA at predictable places and leave it with ends containing a short segment of just one strand (Figure 23.28). These are sometimes called *sticky ends* because the single strand has a nucleotide sequence that spontaneously pairs with the bases of another such sticky end.

The idea, then, is to cut the plasmid (or another vector) with restriction enzyme and then add new genes containing the same kind of sticky ends. The DNA segments naturally join together because of hydrogen bonding between bases in the sticky ends (A with T; C with G). DNA ligase seals DNA pieces together to produce a hybrid plasmid containing new DNA (Figure 23.28). The hybrid plasmid then is put into bacteria, giving them some new genes.

But obtaining genes to splice into vectors can be a problem. One solution, if mRNA for a desired protein can be isolated, is to use reverse transcriptase (see Chemistry Spotlight, Section 23.3) to make the DNA (gene) from that mRNA. Another way is to chemically synthesize the gene from nucleotide units, since once the amino acid sequence of the desired protein is known, scientists can use the genetic code to devise a nucleotide sequence that codes for that protein.

In the past two decades, scientists have synthesized plant, animal (including human), and bacterial genes, spliced them into plasmids, inserted the hybrid plasmids into bacteria, and found that the bacteria synthesized the proteins

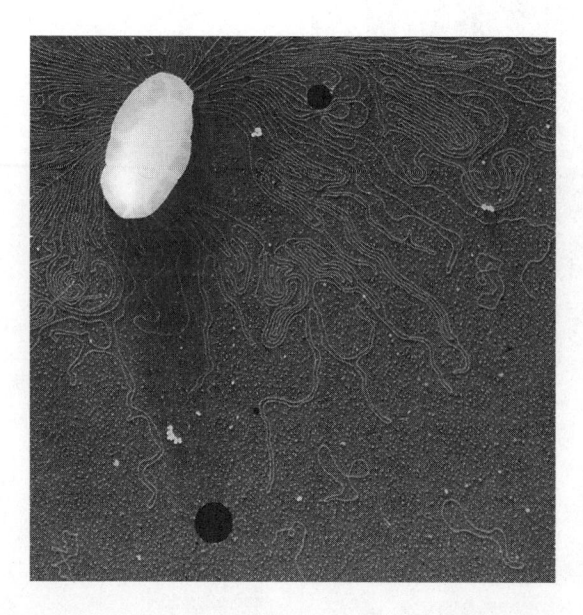

Figure 23.27 A ruptured *E. coli* cell releases its chromosomal DNA and several plasmids (arrows).

specified by the synthetic genes. If, for example, the bacteria receive a human gene, they produce a human protein.

Medical Uses. Genetic engineering is a way to make cells into miniature factories to produce needed proteins. The first genetically engineered product, human insulin, was made by bacteria that were given synthetic genes for insulin's two chains. Scientists then isolated the individual protein chains from the bacteria and joined them with disulfide bonds to produce human insulin (Figure 21.17).

Many other products are on the way. Genetically engineered bacteria now make human growth hormone (HGH) to treat children who are dwarfed because they produce too little HGH (Figure 21.18). Bovine growth hormone also is available; injections into dairy cattle increase their milk production by 10–15%. Recombinant DNA also is used to produce larger amounts of proteins such as interferon and interleukin-2, which are being tested for the treatment of several diseases, including cancer.

Another product, human tissue plasminogen activator (tPA), is used to treat patients during heart attacks (Section 22.6). Scientists have incorporated the gene for this protein into one-celled mouse embryos; those cells, implanted into a female mouse, develop into mice that produce tPA in their milk. If this can be done with cows or goats, milk from a herd of several dozen animals may be enough to provide the world's supply of tPA.

Vaccines are another important product. When a disease-causing bacterium or virus infects you, your immune system recognizes certain proteins on the surface of the infectious agent and produces antibodies to destroy it (Section 21.6). Traditional vaccines expose the immune system to a weakened form of the infectious agent. The antibodies your body produces help protect you against a real infection. But, given the right recombinant DNA, bacteria now can manufacture those surface proteins that your immune system recognizes. Using these proteins as the vaccine should be safer than the traditional method. Recombinant DNA now is used, for example, to produce protein vaccines for one type of hepatitis and for hoof-and-mouth disease, a viral disease that afflicts livestock.

Re-engineering whole animals is also possible. For example, scientists have given mouse embryos a rat gene for growth hormone. A few embryos, implanted in female mice, developed into baby mice that produced rat growth hormone. The extra hormone made the mice an average of 50% larger than normal (Figure 23.27).

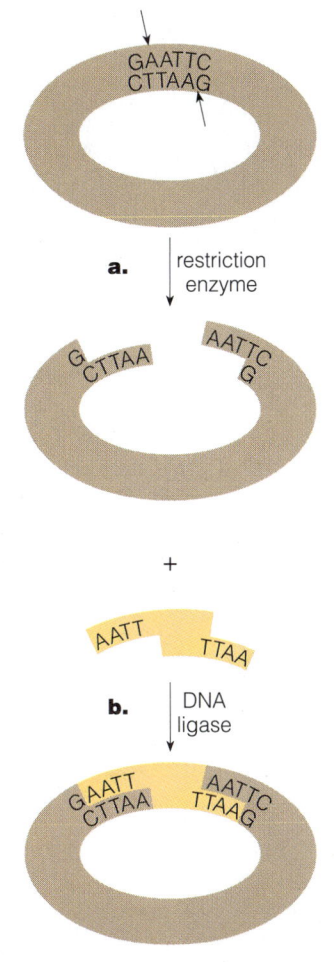

Figure 23.28 (a) Restriction enzyme cuts plasmid DNA, leaving "sticky ends." (b) The opened plasmid reacts with new DNA having similar "sticky ends" to form a hybrid plasmid of recombinant DNA.

Figure 23.29 The mouse on the left has a gene for rat growth hormone and is nearly twice as large as its normal-size littermate, who doesn't have the rat gene. Both mice are the same age (about 10 weeks). (Courtesy Dr. R. L. Brinster)

CHEMISTRY SPOTLIGHT

A New Way to Tell Whodunit

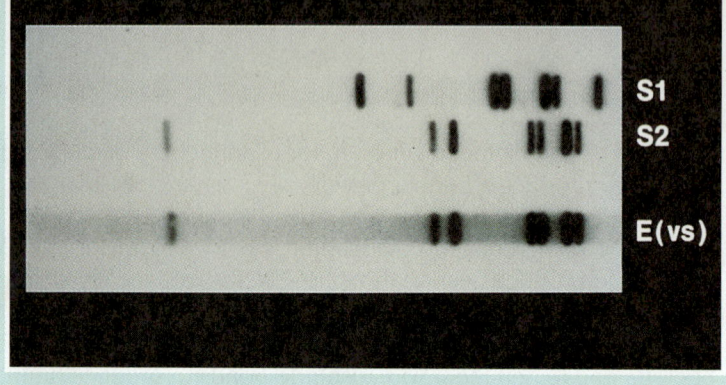

Figure 23.30 DNA fingerprints of two people (S1 and S2). Which one matches the blood sample (E)?

Restriction enzymes cut DNA only where specific nucleotide sequences occur. If several different restriction enzymes cut a person's DNA and the pieces are separated, the pattern of DNA fragments looks like the bar codes on supermarket products (Figure 23.30). The DNA fragment pattern is like a fingerprint; it is virtually unique for each person, except for identical twins.

This new method of chemical sleuthing works on DNA from such things as blood, hair, and semen. As little as a single hair is enough to do the analysis. DNA "fingerprinting" already has been used to convict (or acquit) people of murder and rape. The technique has also helped identify biological parents. It settled one dispute, for example, that enabled a boy to rejoin his true family in another country. Many more applications are on the way.

Table 23.4 Types of Genetic Disorders and Some Examples

Type of Disorder	Examples
Chromosome disorders (abnormal amounts of chromosome material)	Down's syndrome Klinefelter's syndrome Turner's syndrome
*Recessive disorders	Cystic fibrosis Albinism Sickle-cell anemia Phenylketonuria (PKU) Tay-Sachs disease Gaucher disease β-thalassemia Galactosemia Adenosine deaminase (ADA) deficiency
*Recessive disorders carried on the X chromosome	Hemophilia Muscular dystrophy Color blindness
Dominant disorders	Huntington's disease Marfan syndrome
Multigenic disorders	Club feet Cleft palate Spina bifida

* These disorders are more likely to respond to gene therapy.

Another application is to treat people with genetic disorders by supplying them with needed genes. The best candidates are people with disorders caused by the lack of one gene. Table 23.4 lists a few examples. Human gene therapy is just beginning.

> Perhaps when we've mutated the genes and integrated the neurons and refined the biochemistry, our descendants will come to see us as we see Pooh: frail and slow in logic, weak in memory and pale in abstraction, but usually warmhearted, generally compassionate, and on occasion possessed of innate common sense and uncommon perception.
>
> The Brain of Pooh *by Robert Sinsheimer (1920–)*

SUMMARY

DNA (deoxyribonucleic acid) contains 2-deoxyribose, phosphate, and the nitrogen bases A (adenine), C (cytosine), G (guanine), and T (thymine). RNA (ribonucleic acid) contains ribose, phosphate, and A, C, G, and U (uracil). A base bonded to a pentose by a β-glycosidic bond is a nucleoside. A nucleoside with 1–3 phosphate groups bonded to the pentose is a nucleotide. DNA and RNA are polynucleotides, with monophosphate nucleotide units joined by 3′,5′-phosphodiester bonds.

DNA commonly is a double helix of two polynucleotide strands held together by hydrogen bonds between complementary base pairs (C with G; A with T). Base pairs are inside the helix, parallel to each other, and perpendicular to the helix axis. When DNA replicates, the new double helix contains one strand of original DNA and one strand of newly synthesized DNA. DNA polymerases and DNA ligase catalyze replication. Some anticancer drugs block replication.

Single-stranded RNA is made by transcription of DNA using the enzyme RNA polymerase. Reverse transcriptase catalyzes the reverse—making DNA from RNA. There are three main types of RNA: transfer RNA (tRNA), messenger RNA (mRNA), and ribosomal RNA (rRNA). tRNA is the smallest; it carries amino acids to the ribosome for protein synthesis, and its anticodon binds to complementary codons on mRNA to align amino acids in the correct sequence. mRNA carries information from DNA specifying the length and primary structure (amino acid sequence) of the protein to be synthesized. Synthesis occurs at ribosomes, which consist of rRNA and proteins. The genetic code consists of three-base sequences called codons. Regulation of protein synthesis occurs at the level of transcription. Synthesis of some proteins is blocked by repressors and activated by inducers.

Mutations are changes in the nucleotide sequence in DNA. Substances that cause mutations are mutagens, and some are also carcinogens. Oncogenes play important roles in many types of cancer.

Genetic engineering is a way to provide new genes to cells as recombinant DNA. Genes are spliced into vectors (plasmids or viruses) using restriction enzymes. Uses include synthesizing needed proteins, DNA "fingerprinting," and gene therapy.

KEY TERMS

Anticodon (23.4)
Carcinogen (23.5)
Codon (23.4)
Deoxyribonucleic acid (DNA) (23.1)
Gene (23.4)
Genetic engineering (23.6)
Induction (23.4)
Messenger RNA (mRNA) (23.3)

Mutagen (23.5)
Mutation (23.5)
Nucleic acid (23.1)
Nucleoside (23.1)
Nucleotide (23.1)
Plasmid (23.6)
Recombinant DNA (23.6)
Replication (23.2)
Repression (23.4)

Restriction enzyme (23.6)
Reverse transcriptase (23.3)
Ribonucleic acid (RNA) (23.1)
Ribosomal RNA (rRNA) (23.3)
Ribosome (23.3)
Transcription (23.3)
Transfer RNA (tRNA) (23.3)
Translation (23.4)

EXERCISES

Even-numbered exercises are answered at the back of this book.

Structures of DNA and RNA

1. How do DNA and RNA differ in their pentose units?
2. How do DNA and RNA differ in their nitrogen bases?
3. Which type of nucleic acid occurs in (a) chromosomes and (b) ribosomes?
4. Draw structural formulas for (a) a nucleoside containing adenine and deoxyribose, and (b) 3'-CMP.
5. Draw structural formulas for (a) a nucleoside containing uracil and ribose, and (b) dADP.
6. Which base in exercises 4 and 5 is common in RNA but not in DNA?
7. If a partial sequence of bases on one strand of DNA is CATTGAC, what is the sequence on the other, complementary strand?
8. Base pairs in DNA are _____ (outside or inside) the double helix and _____ (parallel or perpendicular) to the helix axis.
9. What is the relationship among the amounts of the bases A, C, G, and T in DNA? Explain.
10. What is the relationship among the amounts of the bases A, C, G, and U in RNA? Explain.
11. Examine Table 23.2. At warm temperatures, would the double helix strands separate more readily in human or in *E. coli* DNA? Explain.
12. Why is DNA called an acid?

Synthesis of DNA

13. In what region of the cell does replication occur?
14. In the experiment in which bacteria containing ^{15}N-DNA replicate in a medium containing only ^{14}N (Section 23.2), why couldn't the type of DNA (all ^{15}N, ^{15}N–^{14}N hybrid, or all ^{14}N) be determined by radioactivity instead of by density differences?
15. During replication, nucleotides originally must have ____ phosphate group(s). Once incorporated into a DNA chain, each nucleotide unit has ____ phosphate group(s).
16. How do cross-linking agents work as anticancer drugs?

Synthesis of RNA

17. In what region of the cell does transcription occur?
18. What will be the base composition of RNA transcribed from a strand of DNA containing 21% A, 26% C, 30% G, and 23% T?
19. Suppose a section of DNA on one strand has the composition 18% A, 28% C, 31% G, and 23% T. The other (complementary) strand in the DNA double helix is transcribed. What is the base composition of the resulting RNA?
20. Identify which of the following are made by transcription: (a) mRNA, (b) rRNA, (c) tRNA, (d) all of the above, (e) none of the above.
21. Which type of RNA has the most internal hydrogen bonding between complementary base pairs?
22. Identify the region of tRNA that binds to amino acids.
23. List three chemical changes that occur in mRNA after transcription from DNA.
24. Genetic information normally flows from DNA to RNA to protein. What enzyme causes this information to flow in a different direction?
25. Look at Figure 23.19. How does AZT most likely inhibit reverse transcriptase: (a) competitive inhibition, (b) noncompetitive inhibition, (c) feedback inhibition? [Review Chapter 22 if necessary.]

Protein Synthesis

26. Distinguish between a gene and a chromosome.
27. Look at Figure 23.20. Do you think there are more genes on an X or Y chromosome? Explain.
28. Translation is the synthesis of _____ from _____.
29. There are a total of ____ codons, which code for ____ different amino acids. How does the genetic code handle the difference in these two numbers?
30. How many codons code for something besides an amino acid? What are their functions?
31. List the enzyme catalyzing each of the following: (a) formation of peptide bonds during translation, (b) replication, (c) transcription, (d) formation of aminoacyl-tRNA.
32. Which type of RNA contains codons? Which type contains anticodons?
33. Does the final amino acid incorporated into a protein chain during protein synthesis end up at the N-terminal or C-terminal end of the protein?
34. In prokaryotic cells, translation occurs on mRNA while mRNA is being synthesized from DNA. Could this happen in animal (eukaryotic) cells? Why? [See Appendix E if necessary.]
35. Use the genetic code (Table 23.3) to determine the amino acid sequence coded by the following nucleotide sequence in mRNA: CCGAUACUCGGAUCU.
36. Explain what is happening at each of the following during induction of the lactose operon: (a) structural gene for β-galactosidase, (b) regulator gene, (c) operator, (d) promoter.

37. Would you expect the product of a metabolic sequence of reactions to induce the synthesis of enzymes producing that product? Why?
38. Suppose a mutation resulted in a nonfunctional operator. How would this affect protein synthesis from the structural genes in that operon?
39. Distinguish between repression and feedback inhibition. [Review Section 22.5 if necessary.]

Mutations

40. If one base in a gene undergoes a mutation to a different base, what are the possible consequences in the primary structure of the protein coded by that gene?
41. Recall the amino acid substitution that occurs in sickle-cell hemoglobin (Chapter 21 opener). Look at the genetic code (Table 23.3) and determine if that particular amino acid change could have occurred by mutation of one base in an appropriate codon.
42. Distinguish between a mutagen and a carcinogen.

Genetic Engineering

43. What are the most common vectors for carrying new genes into cells during genetic engineering?
44. In what way are vaccines produced by recombinant DNA likely to be safer than conventional vaccines?
45. Why do the "sticky ends" from DNA pieces cut by the same restriction enzyme stick together?
46. What are some advantages of using genetically modified bacteria to produce insulin instead of isolating insulin from beef and pork pancreases?
47. When human growth hormone (HGH) was isolated from the pituitary glands of corpses, some HGH samples were contaminated with infectious viruses from those corpses. Would this be a hazard in HGH made by genetically engineered bacteria?
48. Most people who have Down's syndrome have an extra chromosome (#21) in their cells. Is gene therapy a promising way to treat this disorder? Why?

DISCUSSION EXERCISES

1. Which type of substance is more important to life—proteins or nucleic acids? Defend your answer.
2. One human cell's DNA contains about 3 billion base pairs. Estimate the formula weight of this DNA.
3. The Human Genome Project, now in progress, has a goal of identifying all the genes and the entire nucleotide sequence of human DNA. What are some uses for this information?
4. One current use of human growth hormone produced by genetic engineering is to enable short children to be taller. Do you favor this practice? What guidelines, if any, would you favor?

24

Metabolism and Energy

1. What are catabolism and anabolism, and how are they related?
2. What is the role of ATP in metabolism and energy balance? What nutrients does your body use to produce ATP?
3. What is the citric acid cycle? How does it generate energy?
4. What is electron transport? How does your body use it to make ATP?
5. What are some ways your body uses energy?

Cyanide Poisoning and How to Treat It

Figure 24.1 Cassava leaves contain cyanide.

Cyanide ion (CN$^-$) is very toxic. A tiny amount of KCN—about 70 mg (0.0025 oz.)—has a 50% chance of killing a 70-kg (154-lb) person.

CN$^-$ is lethal because it blocks the body's use of O$_2$ gas in a process called *electron transport*. When electron transport is blocked, cells don't make enough ATP, their main energy source. It's as if the body suffers a sudden power outage. The victim dies a painful death, usually within 15 minutes.

Patients suffering from cyanide poisoning are quickly injected with a solution of both sodium nitrite (NaNO$_2$) and sodium thiosulfate (Na$_2$S$_2$O$_3$). Nitrite ion (NO$_2^-$) oxidizes hemoglobin to another form, called methemoglobin, whose heme iron is in the iron (III) (Fe^{3+}) form. Because it has a strong attraction for CN$^-$, methemoglobin binds CN$^-$ and keeps it from blocking electron transport, allowing ATP synthesis to resume. Thiosulfate ion (S$_2$O$_3^{2-}$) helps remove CN$^-$ by changing it into thiocyanate ion (SCN$^-$), which the body rapidly excretes in urine:

$$CN^- + S_2O_3^{2-} \longrightarrow SCN^- + SO_3^{2-}$$

cyanide ion   *thiosulfate ion*   *thiocyanate ion*   *sulfite ion*

Treating cyanide poisoning is tricky because too much nitrite will oxidize too much hemoglobin to methemoglobin. When this happens, the patient can die from having too little hemoglobin to carry O$_2$ to cells. Sometimes, as in this case, the antidote to a poison is itself a poison.

You use gasoline to fuel your car, and food to fuel your body. To generate the energy that makes it run, your car burns gasoline to CO_2 and H_2O (and some CO). Similarly, your body generates the energy it needs by converting food mostly into CO_2 and H_2O. You use this energy to dance, keep warm, pump blood, grow hair, and do all the things that make you alive and human.

Figure 24.2 This cheetah runs mostly on carbohydrate fuel.

The chemical reactions in your body are collectively called **metabolism.** Metabolism includes digesting food, synthesizing DNA and proteins, and oxidizing food to CO_2 and H_2O. All these chemical reactions either consume or generate energy. As a result, your body has to operate on an energy budget; it cannot spend more energy than it produces.

24.1 ENERGY BALANCE

Catabolism and Anabolism. Chemical reactions either consume or give off energy (Section 5.6). Reactions release energy if reactants have more energy than products; if products have more energy than reactants, the reaction consumes energy (see Figure 5.17).

Metabolism is divided into two categories (Table 24.1). Reactions (and sequences of reactions) that give off energy are part of **catabolism.** These are typically oxidation reactions that convert organic compounds into simpler, lower-energy products such as H_2O and CO_2.

Table 24.1 Comparison of Catabolism and Anabolism

Characteristic	Catabolism	Anabolism
Type of products	simpler	more complex
Energy change	generates energy	consumes energy
Metabolism of ATP	produces ATP	consumes ATP
Type of process	oxidation	reduction
Effect on coenzymes	$NADP^+ \rightarrow NADPH$	*$NADPH \rightarrow NADP^+$
	$NAD^+ \rightarrow NADH$	
	$FAD \rightarrow FADH_2$	

*NADPH is the main reduced coenzyme used directly in anabolism.

The other category, **anabolism,** consists of metabolic reactions that require energy. These are typically reduction reactions that produce more complex, higher-energy products. One example we have discussed is protein synthesis (Section 23.4).

In living things, catabolism must be in balance with anabolism. The energy you produce during catabolism enables you to move, stay warm, and synthesize body materials. Synthesis reactions are part of anabolism. But you can only carry out as much anabolism as your energy supply allows. During starvation, for example, the only energy the victim can generate is by catabolism (breakdown) of the body itself; there isn't enough energy to synthesize new tissues.

Energy Sources. Three nutrients provide energy—carbohydrates (17 kJ or 4 Cal per g), lipids (38 kJ or 9 Cal per g), and proteins (17 kJ or 4 Cal per g). U.S. residents typically get 45–60% of their energy from carbohydrates, 30–45% from lipids, and 10–15% from proteins (Figure 24.3). To generate energy by catabolism, your body oxidizes these nutrients to CO_2 and H_2O. The nitrogen in proteins is converted mostly into urea ($NH_2-\overset{\overset{O}{\|}}{C}-NH_2$), which is excreted in urine (Section 27.3).

You also need vitamins, minerals, and water in your diet, but not directly for energy.

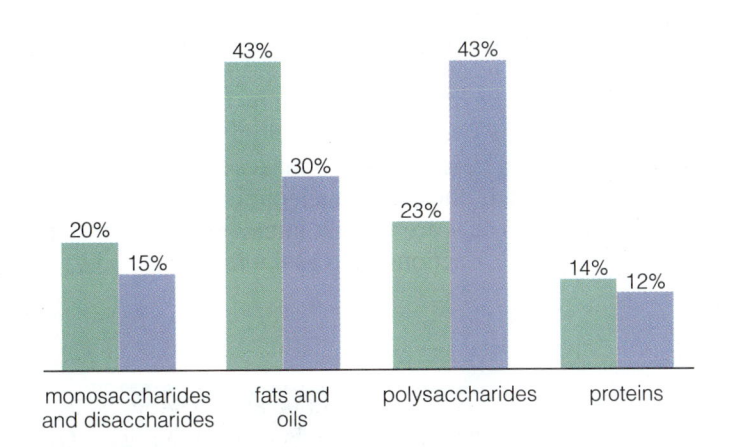

Figure 24.3 Distribution of kJ (or Cal) in a typical cheeseburger-shake-fries meal (green) compared to the distribution recommended by the U.S. Senate Select Committee on Nutrition and Human Needs (violet).

ATP. Your body uses energy from catabolism for many anabolic processes. For example, the energy produced by oxidizing glucose to CO_2 and H_2O can be used to snythesize fatty acids or proteins. You can do this because catabolism generates energy in a portable form. Although your body uses several chemicals for doing this, the most common one is the nucleotide adenosine triphosphate, ATP (Figure 24.4).

Your body uses some energy from catabolism to make ATP from ADP:

$$ADP + P_i + energy \longrightarrow ATP + H_2O$$

When ATP changes back into ADP or AMP, it releases energy to drive anabolic reactions:

$$ATP + H_2O \longrightarrow ADP + P_i + energy$$

$$ATP + H_2O \longrightarrow AMP + PP_i + energy$$

P_i is an abbreviation for phosphate (PO_4^{3-}) ion.
PP_i is an abbreviation for pyrophosphate ($P_2O_7^{4-}$) ion.

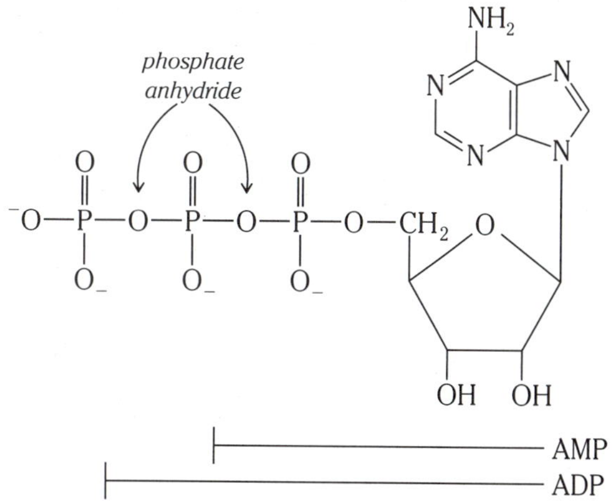

524 Part III Biochemistry

Figure 24.4 Structure of ATP. AMP and ADP have one and two phosphate groups, respectively, bonded to ribose.

Table 24.2 Energy from Hydrolysis of Phosphate Compounds under Standard Conditions	
Compound	Energy (kcal/mol)
phosphoenolpyruvate	14.8
1,3-bisphosphoglycerate	11.8
creatine phosphate	10.3
pyrophosphate (PP$_i$)	8.0
ATP (to AMP + PP$_i$)	7.7
ATP (to ADP + P$_i$)	7.3
glucose-1-phosphate	5.0
glucose-6-phosphate	3.3
glycerol-1-phosphate	2.2

Although ATP is sometimes called a "high-energy" compound, this name is misleading. The hydrolysis of ATP to ADP and phosphate is of intermediate energy compared to the hydrolysis of other phosphate compounds (Table 24.2). This enables other, higher-energy systems to synthesize ATP when needed. The energy from ATP hydrolysis is, however, enough to support many anabolic reactions.

Notice from its structure that ATP (and also ADP, but not AMP) is a phosphate anhydride (see Section 16.8). The four negatively charged groups on the phosphate chain of ATP repel each other, making the compound less stable (higher energy). For this and other reasons, substantial amounts of energy are released when the phosphoanhydride linkage is broken by hydrolysis.

Since the energy released by ATP hydrolysis comes from the phosphoanhydride structure, the nitrogen base bonded to ribose makes little difference in terms of energy. Since all the di- and triphosphate nucleotides release about the same energy as ATP and ADP do, they are readily interconverted with little change in energy. Enzymes catalyze reactions such as the following:

$$ADP + GTP \rightleftharpoons ATP + GDP$$

$$ADP + CTP \rightleftharpoons ATP + CDP$$

Thus all triphosphate nucleotides are energetically equivalent to ATP.

ATP plays an important role in regulating catabolism and anabolism. Low ATP levels (and thus high ADP and AMP levels) activate key allosteric enzymes in catabolism and inhibit anabolic enzymes. This replenishes ATP. At high ATP (and low ADP and AMP) levels the pattern is reversed: anabolic enzymes are activated and catabolic enzymes are inhibited, thus reducing the level of ATP. You will see specific examples of these effects in Chapters 24–27.

This type of feedback inhibition (Section 22.5) helps your body keep a fairly steady supply of ATP, not too much or too little.

24.2 METABOLISM: AN OVERVIEW

Carbohydrate, Lipid, and Protein Metabolism. Figure 24.5 summarizes the main metabolic pathways for carbohydrates, lipids, and proteins. Digestion consists of hydrolysis reactions that convert complex materials into simpler substances. Anabolic reactions then can reassemble these simpler building

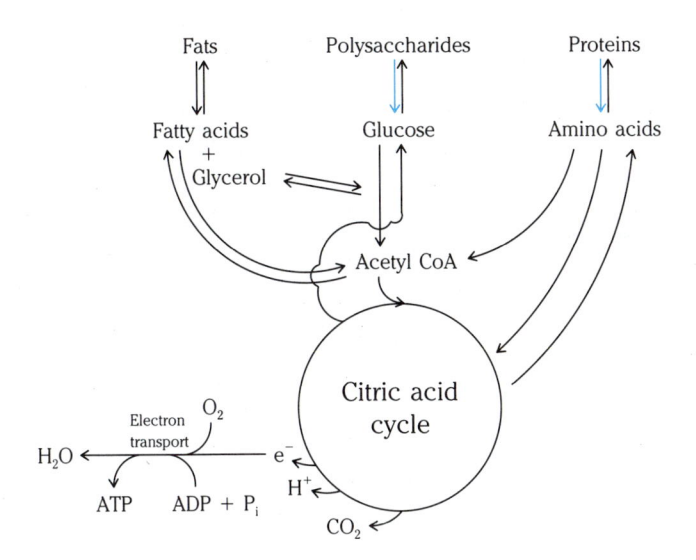

Figure 24.5 Summary of metabolism. Reaction sequences going down are catabolism; those going up are anabolism. Digestion is represented by arrows in color.

blocks into the fats, polysaccharides, and proteins your body needs. Or your body can oxidize these compounds to CO_2 and H_2O by catabolism.

The catabolism of fats, carbohydrates, and proteins ends in a sequence of reactions called the *citric acid cycle*. In this cycle, electrons (e^-) and protons (H^+) removed during oxidation reactions are passed to the coenzymes NAD^+ (Figure 22.7) and FAD (Figure 24.6), reducing them to NADH and $FADH_2$, respectively. We examine this cycle in Section 24.3.

What happens to these electrons and H^+ ions? Through a process called *electron transport*, NADH and $FADH_2$ release their electrons and H^+ ions to certain compounds. Through a series of reactions, those compounds then pass the electrons (and some H^+ ions) to O_2, making H_2O. This generates energy, some of which is used to make ATP. It also oxidizes NADH and $FADH_2$ back to NAD^+ and FAD, respectively, making them available to receive more electrons and H^+ ions from the citric acid cycle. In Section 24.4, we examine how this works.

Also notice in Figure 24.5 how your body interconverts carbohydrates, fats, and proteins. For example, if you eat too much carbohydrate, your body metabolizes glucose to acetyl CoA, then makes acetyl CoA into fat. So you store excess carbohydrate as fat, not carbohydrate.

But there are limits to these interconversions. For example, your body cannot make all 20 amino acids from other substances. Neither can it make all the polyunsaturated fatty acids you need. That is why you need to have essential

Figure 24.6 FAD is reduced to $FADH_2$ when it gains two protons (H^+) and two electrons (e^-). Only the reactive part of the coenzymes are shown.

526 Part III Biochemistry

amino acids (see Section 21.1) and essential fatty acids (see Section 20.1) in your diet.

Some conversions between simple and complex molecules occur in different parts of the cell. For example, fatty acids are broken down into acetyl CoA in mitochondria, while acetyl CoA is made into fatty acids in cytoplasm. Keeping these pathways in separate cell compartments lets them operate independently and be regulated separately.

As you examine specific metabolic pathways in Chapters 24–27, keep in mind their locations in the cell. Also focus on how each pathway is regulated (activated or inhibited), and how it relates with other pathways.

24.3 CITRIC ACID CYCLE

The Central Cycle. Notice in Figure 24.5 that catabolism of fats, carbohydrates, and proteins ends in the *citric acid cycle*. Here carbon and oxygen atoms from these compounds are made into CO_2, which you exhale. Hydrogen atoms, including their electrons, reduce the coenzymes NAD^+ and FAD to NADH and $FADH_2$, respectively. (Later, in electron transport, these coenzymes release hydrogen ions and pass electrons on to O_2, producing H_2O.)

The citric acid cycle has a central role in metabolism and also serves as a major link between catabolism and anabolism. Besides carrying out the final

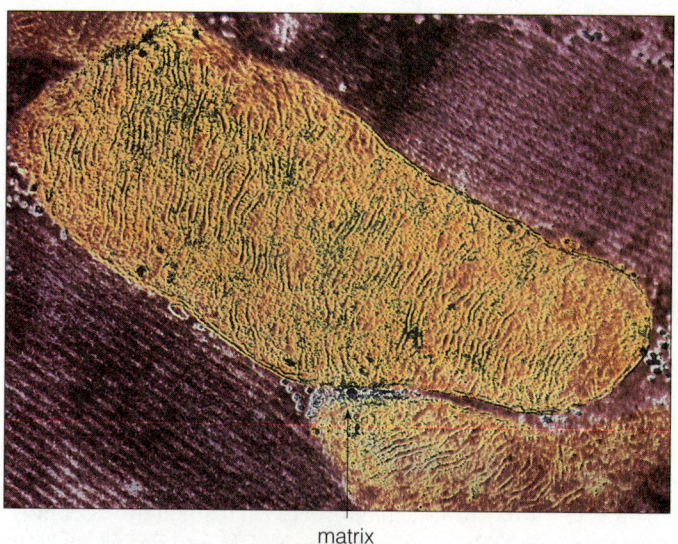

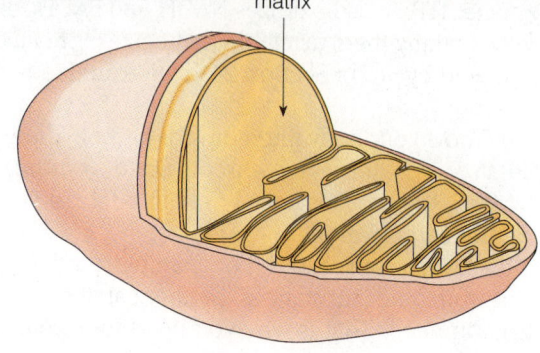

Figure 24.7 Electron micrograph (top) and a drawing (bottom) of a mitochondrion.

oxidation of fats, carbohydrates, and proteins, it provides materials for synthesizing these compounds.

The reactions take place inside cell mitochondria. Mitochondria have an outer membrane and an inner, highly folded membrane (Figure 24.7). Citric acid cycle reactions occur in the space inside the inner membrane, called the *matrix*.

Figure 24.8 shows the reactions of the **citric acid cycle,** named because the first reaction produces the salt of citric acid (citrate). Other names are the **Krebs cycle** (after Hans Krebs, who discovered the sequence of reactions; Figure 24.9) and the **tricarboxylic acid (TCA) cycle** (because citric acid has three carboxylic acid groups).

Figure 24.8 The citric acid cycle. Other reactants and products are shown only for the forward (clockwise) direction.

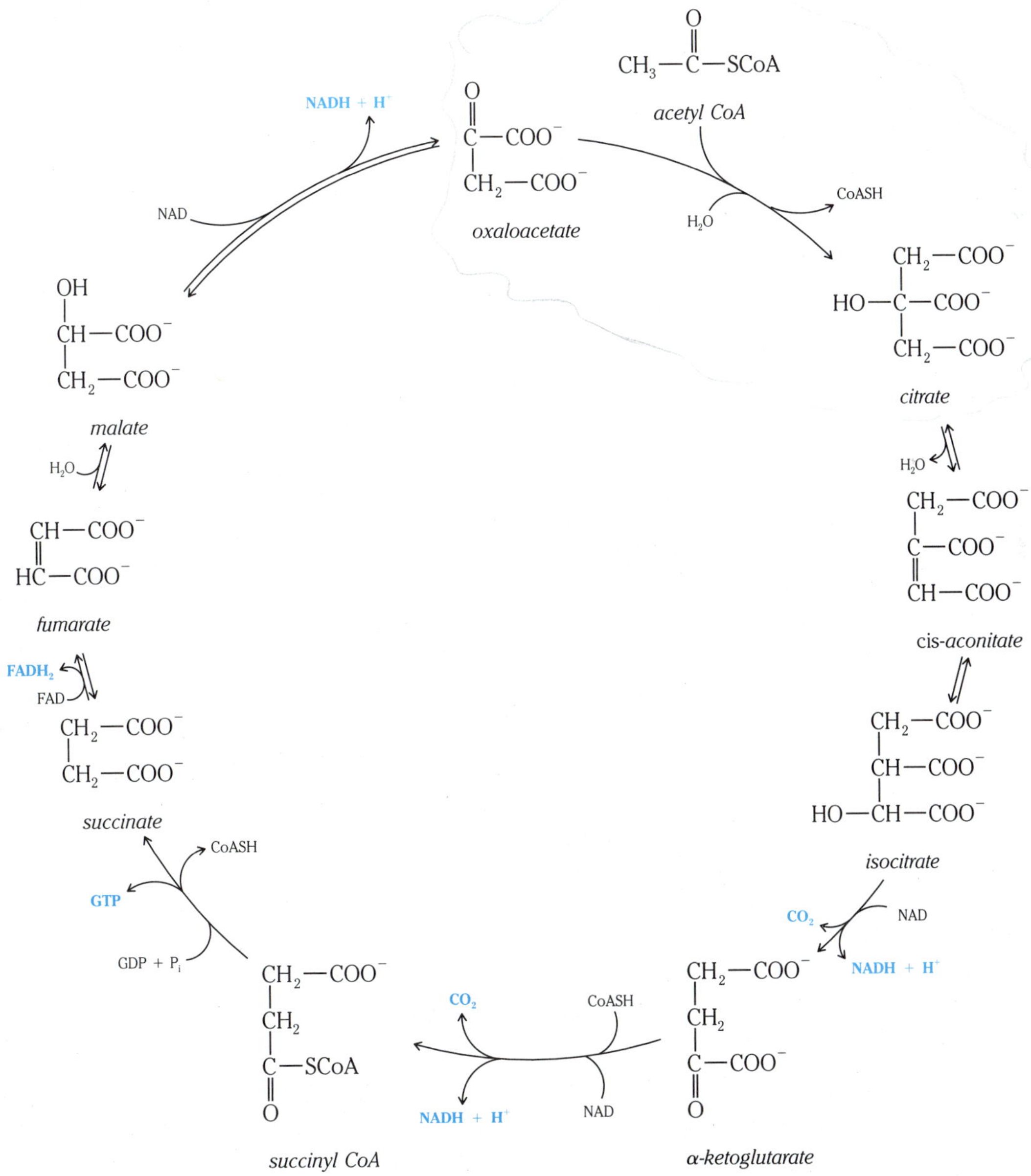

Figure 24.9 Hans Krebs (1900–1981) received the 1953 Nobel prize in medicine or physiology for research on the metabolic cycle that bears his name.

CoA—SH

coenzyme A

CoA—S—C(=O)—CH₃

acetyl CoA

Figure 24.10 Coenzyme A (top) is a complex molecule that has a reactive thiol (—SH) group. This thiol group bonds to acyl groups such as the acetyl group (bottom, in color).

Carbohydrates, fats, and some amino acids are oxidized to two-carbon acetyl (CH₃—C(=O)—) groups, which bond to coenzyme A (CoASH) to form acetyl CoA (Table 22.1; Figure 24.10). Oxidation of fatty acids and certain amino acids in the mitochondrial matrix produces acetyl CoA directly. Carbohydrates and other amino acids are oxidized in cytoplasm to produce pyruvate. Pyruvate then enters the mitochondrial matrix and is converted there into acetyl CoA (Section 25.2). Acetyl CoA itself cannot cross the inner mitochondrial membrane.

In the first reaction of the cycle, acetyl CoA reacts with four-carbon oxaloacetate to produce six-carbon citrate:

CH₃—C(=O)—SCoA + C(=O)(COO⁻)(CH₂—COO⁻) + H₂O ⟶ CH₂—COO⁻ | HO—C—COO⁻ | CH₂—COO⁻ + CoASH + H⁺

acetyl CoA   oxaloacetate   citrate

The reaction is an enzyme-catalyzed aldol condensation between two carbonyl compounds (see Section 15.4).

This is one of two key reactions in the cycle regulated by ATP. ATP inhibits the allosteric enzyme citrate synthase, which catalyzes this reaction. Since the cycle (coupled with electron transport) generates ATP, this feedback inhibition enables ATP to block its own production when levels are high. When ATP levels fall, inhibition is lifted and the citric acid cycle replenishes the supply.

NADH also inhibits citrate synthase, whereas NAD⁺ stimulates it. Since NADH production is linked to ATP production through electron transport (Section 24.4), this arrangement also ensures that the cycle operates when ATP is needed.

The next reaction, catalyzed by the enzyme aconitase, proceeds in two steps that change citrate into isocitrate:

CH₂—COO⁻ | HO—C—COO⁻ | CH₂—COO⁻   ⇌(−H₂O)   CH₂—COO⁻ | C—COO⁻ ‖ CH—COO⁻   ⇌(+H₂O)   CH₂—COO⁻ | CH—COO⁻ | HO—CH—COO⁻

citrate   cis-aconitate   isocitrate

Isocitrate then is oxidized to α-ketoglutarate, producing CO₂ and NADH. The net reaction is:

CH₂—COO⁻ | CH—COO⁻ | HO—CH—COO⁻ + NAD⁺ ⟶ CH₂—COO⁻ | CH₂ | C(=O)—COO⁻ + CO₂ + NADH + H⁺

isocitrate   α-ketoglutarate

The CO₂ produced in this and the next reaction represents much of the CO₂ that you exhale.

The allosteric enzyme catalyzing this reaction, isocitrate dehydrogenase, is the second key enzyme regulating the cycle. Like citrate synthase, this enzyme is inhibited by ATP and NADH, and stimulated by ADP, AMP, and NAD$^+$. Again, high levels of ATP inhibit its additional production, thus keeping a steady energy supply in the cell.

In the next reaction, catalyzed by a complex of three different enzymes known as α-ketoglutarate dehydrogenase, more CO_2 and NADH form:

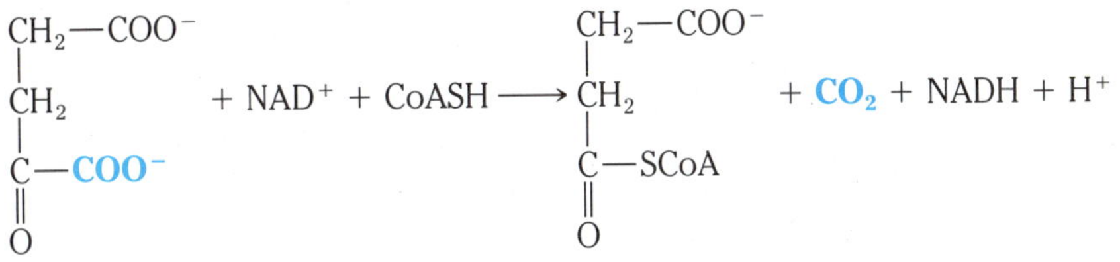

$$\begin{array}{c} CH_2-COO^- \\ | \\ CH_2 \\ | \\ C-COO^- \\ \| \\ O \end{array} + NAD^+ + CoASH \longrightarrow \begin{array}{c} CH_2-COO^- \\ | \\ CH_2 \\ | \\ C-SCoA \\ \| \\ O \end{array} + CO_2 + NADH + H^+$$

α-ketoglutarate   *succinyl CoA*

The net reaction shown above is actually a series of reactions involving five coenzymes: thiamine pyrophosphate, lipoic acid, FAD, NAD$^+$, and CoA (see Table 22.1).

Next, succinyl CoA synthetase releases the succinyl group from CoA and uses the energy to synthesize GTP from GDP and phosphate:

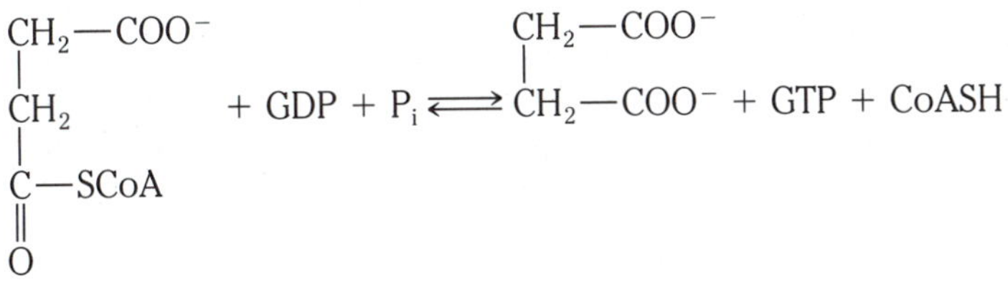

$$\begin{array}{c} CH_2-COO^- \\ | \\ CH_2 \\ | \\ C-SCoA \\ \| \\ O \end{array} + GDP + P_i \rightleftharpoons \begin{array}{c} CH_2-COO^- \\ | \\ CH_2-COO^- \end{array} + GTP + CoASH$$

succinyl CoA   *succinate*

Recall that GTP has the same energy as ATP and can be used to make ATP from ADP: GTP + ADP ⇌ GDP + ATP (Section 24.1).

Three more reactions convert succinate back into oxaloacetate, which was used to initiate the cycle. Succinate is oxidized to fumarate, catalyzed by succinate dehydrogenase:

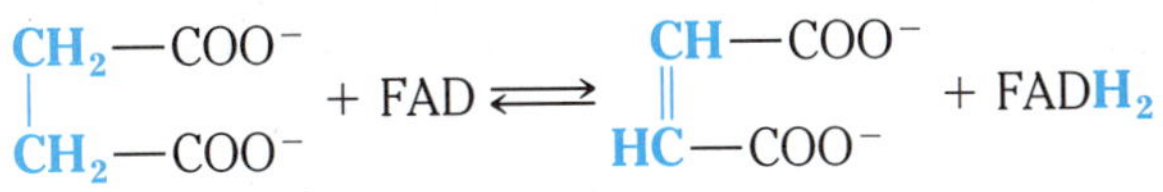

$$\begin{array}{c} CH_2-COO^- \\ | \\ CH_2-COO^- \end{array} + FAD \rightleftharpoons \begin{array}{c} CH-COO^- \\ \| \\ HC-COO^- \end{array} + FADH_2$$

succinate   *fumarate*

Fumarate reacts with water to form malate, catalyzed by fumarase:

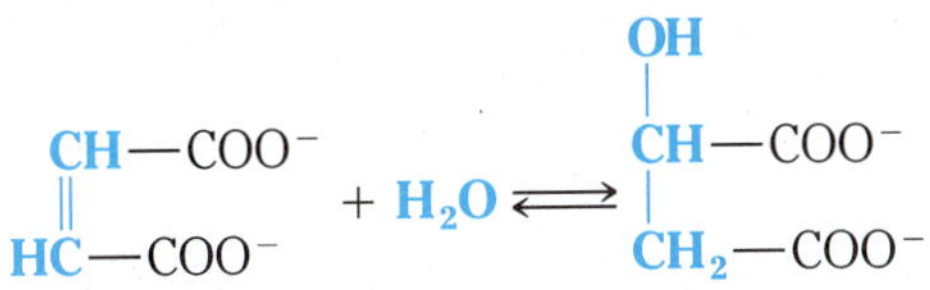

$$\begin{array}{c} CH-COO^- \\ \| \\ HC-COO^- \end{array} + H_2O \rightleftharpoons \begin{array}{c} OH \\ | \\ CH-COO^- \\ | \\ CH_2-COO^- \end{array}$$

fumarate   *malate*

Then malate dehydrogenase catalyzes the oxidation of malate to oxaloacetate, completing the cycle:

$$\begin{array}{c} \text{OH} \\ | \\ \text{CH}-\text{COO}^- \\ | \\ \text{CH}_2-\text{COO}^- \end{array} + \text{NAD}^+ \rightleftharpoons \begin{array}{c} \text{O} \\ \| \\ \text{C}-\text{COO}^- \\ | \\ \text{CH}_2-\text{COO}^- \end{array} + \text{NADH} + \text{H}^+$$

malate   oxaloacetate

Summary. An acetyl group (as acetyl CoA) reacts with oxaloacetate to begin the citric acid cycle. The cycle then regenerates oxaloacetate. The net effect is that an acetyl group is oxidized:

acetyl CoA + 3 NAD$^+$ + FAD + GDP + P$_i$ + 2 H$_2$O ⟶
2 CO$_2$ + 3 NADH + 3 H$^+$ + FADH$_2$ + GTP + CoA

You exhale the CO$_2$. GTP is an energy source equivalent to ATP. The electrons and protons removed during oxidation are carried as NADH (+H$^+$) and FADH$_2$. These reduced coenzymes ultimately pass electrons and protons to oxygen in electron transport, which generates much ATP (Section 24.4). The citric acid cycle, coupled to electron transport and ATP synthesis, is the most powerful metabolic system in your body for making ATP.

ATP regulates the citric acid cycle by inhibiting two key allosteric enzymes, citrate synthase and isocitrate dehydrogenase. In contrast, ADP and AMP activate these enzymes. Both enzymes are also inhibited by NADH and activated by NAD$^+$. All these actions keep the cycle operating to sustain ATP concentrations in the cell.

Because they produce so much ATP, mitochondria are often called the "powerhouse of the cell."

24.4 ELECTRON TRANSPORT AND OXIDATIVE PHOSPHORYLATION

Oxygen: The Ultimate Oxidizing Agent. Although you breathe in more O$_2$ than you exhale, if you looked at a chart of all the metabolic reactions in your body, you would find few that use O$_2$ directly. Most of your O$_2$ is consumed by something else: electron transport.

Figure 24.11 Sequence of carriers in electron transport. Regions in red indicate where protons (H$^+$) are pumped from the matrix of the mitochondrion into the intermembrane space.

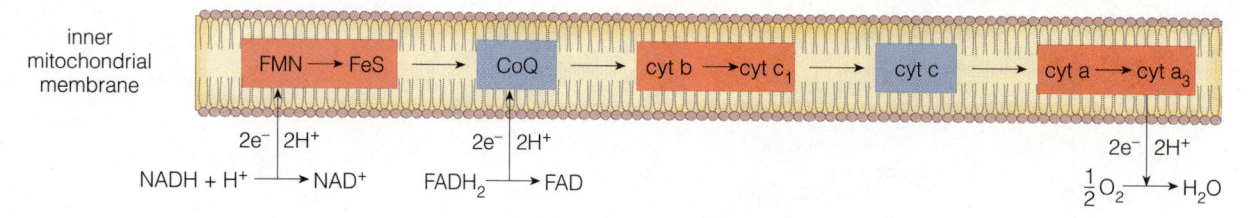

Electron transport is a series of reactions in which electrons and H^+ ions are passed to O_2 to produce H_2O. The electrons and H^+ ions come from two coenzymes generated in the citric acid cycle, NADH and $FADH_2$.

The synthesis of ATP, a process called **oxidative phosphorylation,** is connected to electron transport. Although ATP is made directly in a few metabolic reactions (called *substrate-level phosphorylation*), you make most of your ATP by oxidative phosphorylation.

Electron Transport. Electron transport occurs in the inner mitochondrial membrane. Figure 24.11 shows its steps. First NADH ($+H^+$), produced in the mitochondrial matrix by the citric acid cycle, passes two protons (H^+) and two electrons to a protein that contains the coenzyme FMN (Table 22.1). This oxidizes NADH to NAD^+ and reduces FMN to $FMNH_2$.

In the next step, $FMNH_2$ releases its protons and passes two electrons (one at a time) to a protein containing iron (III) ions (Fe^{3+}) and sulfur atoms. Then the electrons go to coenzyme Q (Figure 24.12), which also binds two protons from the mitochondrial matrix.

Notice in Figure 24.11 that $FADH_2$, produced in the citric acid cycle, passes two protons and two electrons directly to coenzyme Q. From this point on, electrons from NADH and $FADH_2$ pass through the same reactions in the electron transport chain.

Coenzyme Q passes two electrons, one at a time, to a series of proteins called *cytochromes*. These proteins contain a heme group, as in hemoglobin, and Fe ions that alternate between +2 and +3 forms after they receive or pass on electrons to the next cytochrome (Figure 24.13). The final cytochrome (a_3) passes electrons to O_2. O_2 also combines with H^+ ions inside the matrix to form H_2O. The net reaction is:

$$2e^- + 2H^+ + \tfrac{1}{2}O_2 \longrightarrow H_2O$$

Several electron transport carriers bind H^+ from the matrix but release H^+ in the space between the mitochondrial membranes. This happens in three stages of electron transport beginning with NADH, but at only two stages beginning with $FADH_2$ (Figure 24.11). The resulting increase in H^+ concentration in the intermembrane space plays a key role in ATP synthesis (see below).

Electron transport accomplishes two things. First, it oxidizes coenzymes back to NAD^+ and FAD so they can participate again in the citric acid cycle. Second, it produces ATP. Now we examine how this happens.

Oxidative Phosphorylation. Passing electrons from NADH or $FADH_2$ generates large amounts of energy. Experiments show that some of this energy is used to make ATP. Each NADH oxidized by electron transport generates three ATP, while each $FADH_2$ oxidized produces two ATP.

How is ATP synthesis coupled to electron transport? This question puzzled scientists for decades. They proposed numerous hypotheses to explain it, but were unable to find convincing evidence. Even now the details are not clear.

Peter Mitchell (Figure 24.14), an English biochemist, received the 1978 Nobel prize in chemistry for his description of the process. According to his **chemiosmotic hypothesis,** electron transport generates a concentration difference (gradient) of protons (H^+) across the inner mitochondrial membrane (Figure 24.11); this proton gradient then drives ATP synthesis catalyzed by the

The reactive part of FMN is the same as for FAD, shown in Figure 24.6.

Figure 24.12 Oxidized (top) and reduced (bottom) forms of coenzyme Q during electron transport.

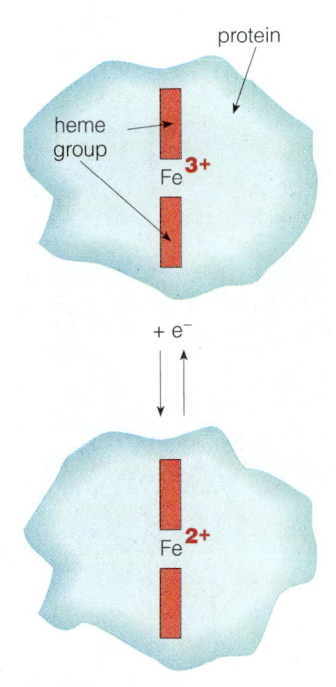

Figure 24.13 Oxidized (top) and reduced (bottom) forms of a cytochrome during electron transport.

CHEMISTRY SPOTLIGHT

Snake Venom and Oxidation Phosphorylation

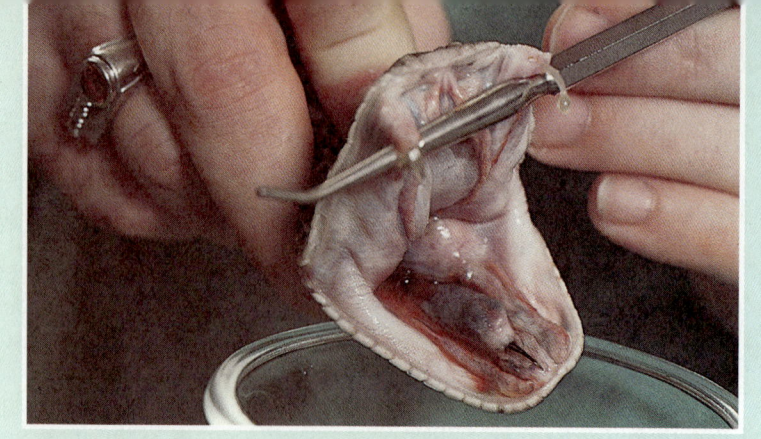

Figure 24.16 Obtaining snake venom.

Your body uses much of the energy from electron transport to make ATP. But this only happens when electron transport generates a concentration gradient of protons to drive ATP synthesis.

Snake venoms (Figure 24.16) are complex mixtures of materials. They contain many proteins, some of which are enzymes. When mitochondria are exposed to cobra venom, for example, they continue to carry out electron transport, but they don't make ATP. Substances that disconnect ATP synthesis from electron transport are called *uncouplers*.

How do uncouplers work? They let protons freely cross the inner mitochondrial membrane, destroying the proton gradient. As a result, electron transport cannot generate a higher concentration of protons outside the matrix than inside to drive ATP synthesis.

There are many known uncoupler substances including free fatty acids. Snake venoms may work by hydrolyzing phosphoglycerides (Section 20.3) to produce such fatty acids.

Figure 24.14 Peter Mitchell (1920–1992).

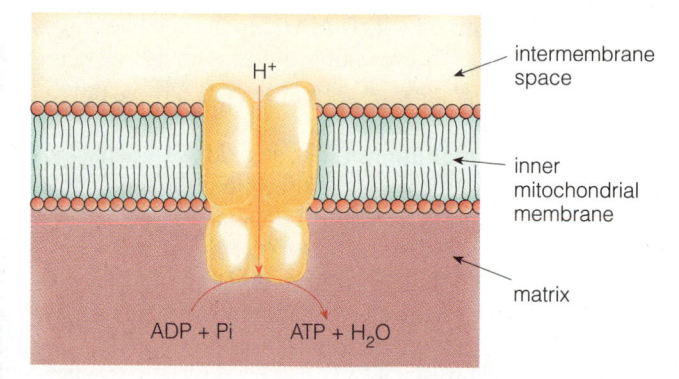

Figure 24.15 When protons (H$^+$) reenter the mitochondrial matrix, they activate ATPase (yellow) to synthesize ATP.

enzyme ATPase (Figure 24.15):

$$ADP + P_i \xrightleftharpoons{\text{ATPase}} ATP + H_2O$$

Although ATPase catalyzes the reaction in both directions, H$^+$ drives the reaction to the right to make ATP.

The basic idea is relatively simple. During electron transport, some carriers pick up H$^+$ ions in the matrix and release them on the outside of the inner membrane (Figure 24.11). In other words, electron transport causes protons (H⁺) to be pumped out of the matrix and into the space between the two membranes. Protons, like other materials, move spontaneously from areas of high concentration to areas of lower concentration. The higher concentration of protons outside the matrix, produced by electron transport, causes protons to reenter the matrix.

The enzyme ATPase is part of a complex assembly in the inner mitochondrial membrane (Figure 24.15). That assembly includes a channel for protons (H⁺) to enter the mitochondrial matrix. When they enter, protons stimulate the enzyme to make ATP.

More protons (H⁺) are pumped out of the matrix using NADH than $FADH_2$ because electrons from $FADH_2$ bypass the first carriers in the chain (Figure 24.11). This is why cells produce three ATP per NADH, but only two ATP per $FADH_2$.

24.5 ENERGY PRODUCTION AND USE IN THE BODY

Energy Produced by the Citric Acid Cycle. Using what you know about electron transport and oxidative phosphorylation, you can calculate how much ATP each round of the citric acid cycle produces. As Figure 24.8 shows, each acetyl group (as acetyl CoA) oxidized by the cycle generates 3 NADH, 1 $FADH_2$, and 1 GTP. Through electron transport and oxidative phosphorylation, each NADH produces 3 ATP, and each $FADH_2$ produces 2 ATP. GTP has the same energy value as ATP.

In each round of the cycle, then, one acetyl group is oxidized to CO_2 (in the cycle itself) and H_2O (through electron transport). In the process, a mitochondrion produces 12 ATP:

```
3 NADH   =  9 ATP
1 FADH₂  =  2 ATP
1 GTP    =  1 ATP
            12 ATP
```

Acetyl groups come from the catabolism of any of the major energy sources—carbohydrates, fats, or proteins (Figure 24.5).

In the following chapters, you will examine other catabolic pathways. But you will find none that produces so much ATP. Most of the energy that comes from carbohydrates, fats, and proteins is produced in mitochondria by the citric acid cycle coupled to electron transport and oxidative phosphorylation.

You might expect, then, that any substance that blocks the citric acid cycle or electron transport would be very toxic. One example is cyanide (CN^-) ion (see Chapter Opener). Another is fluoroacetate, which is metabolized to fluorocitrate (Figure 24.17). Because of its structural similarity to citrate (see Figure 24.8), fluorocitrate competitively inhibits (Section 22.4) the enzyme (aconitase) that metabolizes citrate in the citric acid cycle. As a result, fluorocitrate prevents the cell from generating normal amounts of ATP.

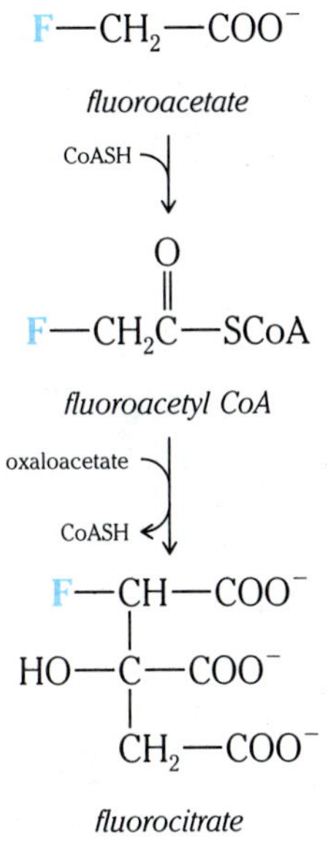

F—CH_2—COO⁻
fluoroacetate

CoASH ↓

F—CH_2C(=O)—SCoA
fluoroacetyl CoA

oxaloacetate ↓ CoASH

F—CH—COO⁻
|
HO—C—COO⁻
|
CH_2—COO⁻

fluorocitrate

Figure 24.17 Fluoroacetate is metabolized to fluorocitrate, which inhibits the citric acid cycle.

What difference does that make? As a commercial poison (called Compound 1080), fluoroacetate is used to kill rats and livestock predators. It causes seizures, coma, and death, usually from heart failure. Fluoroacetate is also toxic to humans.

Uses of Energy. Your body uses energy from catabolism for many things: to run, pump blood, blink your eyes, shiver, and breathe. You also use energy to transport substances in and out of cells. For example, nerve cells use ATP to remove Na^+ while keeping higher levels of K^+ inside the cell membrane than outside. This enables nerve impulses to travel from one nerve cell to another.

In the following chapters, you will see examples of how your body uses ATP for anabolism. ATP is a major go-between in metabolism. Catabolism produces ATP, while anabolism consumes it.

But ATP isn't the only useful substance produced by catabolism. In this chapter, you have seen that the citric acid cycle produces GTP, which has the same energy value as ATP. GTP is used in protein synthesis. In Chapters 25 and 26 you will see how NADPH, made by carbohydrate catabolism, is used for lipid synthesis (anabolism).

Not all of the energy produced during catabolism is used to make ATP. Your body, wonderful as it is, is not 100% efficient. Close to half of this energy is released as heat. However, since you use heat to keep warm, this is hardly a waste. In cold temperatures, your metabolic rate increases as your body tries to generate more heat. Animals that hibernate, however, lower both their body temperature and their metabolic rate (Figure 24.18). This is how they stay alive for long periods without eating or drinking.

Figure 24.18 A gray squirrel lowers its body temperature while hibernating.

"Life is very strange," said Jeremy.
"Compared with what?" replied the spider.

Men Who Play God *by Norman Moss (1928–)*

SUMMARY

Metabolism is the total set of chemical reactions in the body. Catabolism consists of oxidation reactions that give off energy and convert compounds into simpler products. Anabolism requires energy and consists of reduction reactions that produce more complex substances. Carbohydrates, fats, and proteins are the energy sources in the body. Much of the energy in the body is used as ATP.

Metabolism is summarized in Figure 24.5. Carbohydrates, fats, and proteins are ultimately oxidized in the citric acid cycle. The resulting electrons and protons are passed to O_2 through electron transport, which is coupled to ATP synthesis (oxidative phosphorylation). These reactions produce much energy and oxidize carbohydrates, fats, and proteins to CO_2 and H_2O.

The citric acid (also Krebs or tricarboxylic acid) cycle begins with the reaction of acetyl CoA with oxaloacetate to form citrate. Subsequent reactions regenerate oxaloacetate and produce two CO_2, three NADH, one $FADH_2$, and one GTP. All reactions take place in the mitochondrion matrix. The cycle is regulated by activation or inhibition of the allosteric enzymes, citrate synthase and isocitrate dehydrogenase. ATP itself, as a feedback inhibitor of these enzymes, helps maintain steady ATP levels in cells.

In electron transport, electrons and protons from NADH and FADH$_2$ pass through a series of carriers to O$_2$, producing H$_2$O and generating energy. The carriers are in the inner mitochondrial membrane. Some energy from electron transport is used to make ATP—three ATP per NADH and two ATP per FADH$_2$. According to the chemiosmotic hypothesis, electron transport generates a proton (H$^+$) concentration gradient across the inner mitochondrial membrane; protons then enter the matrix and activate the enzyme ATPase to synthesize ATP. Uncoupler compounds prevent this gradient from forming and thus block ATP synthesis.

The citric acid cycle coupled to electron transport and oxidative phosphorylation is the major way the body produces ATP. Twelve ATP are made per acetyl group oxidized to CO$_2$ and H$_2$O. Most of the energy not used to make ATP is released as heat, which helps maintain body temperature.

KEY TERMS

Anabolism (24.1)
Catabolism (24.1)
Chemiosmotic hypothesis (24.4)
Citric acid cycle (24.3)

Electron transport (24.4)
Krebs cycle (24.3)
Metabolism (24.1)

Oxidative phosphorylation (24.4)
Tricarboxylic acid (TCA) cycle (24.3)

EXERCISES

Even-numbered exercises are answered at the back of this book.

Energy Balance

1. Listed below are some metabolic processes you have studied or will learn about in Chapters 25–27. Classify each as anabolism or catabolism:
   (a) Digesting triglycerides into glycerol and fatty acids
   (b) Protein synthesis
   (c) Oxidizing glucose to CO$_2$ and H$_2$O
   (d) Making glycogen from glucose
   (e) Digesting carbohydrates
   (f) Producing cholesterol from acetyl CoA

2. Classify each of the following as anabolism or catabolism:
   (a) Producing glucose from oxaloacetate
   (b) Producing acetyl CoA from fatty acids
   (c) Producing fatty acids from acetyl CoA
   (d) Digesting proteins
   (e) Producing urea (CON$_2$H$_4$) from CO$_2$ and NH$_4^+$

3. Does starvation cause an increase or decrease in (a) anabolism and (b) catabolism?

4. What nutrients are the major sources of energy?

5. Write a balanced equation for the complete oxidation of glucose (C$_6$H$_{12}$O$_6$) with O$_2$ to produce CO$_2$ and H$_2$O.

6. Write a balanced equation for the complete oxidation of stearic acid (Table 20.1) with O$_2$ to produce CO$_2$ and H$_2$O.

7. Can a balanced equation be written for the complete oxidation of the amino acid valine (Table 21.1) with O$_2$ to produce CO$_2$ and H$_2$O? Explain.

8. Do carbohydrates actually contain kilojoules or Calories? What does it mean to say that a soft drink "contains 670 kJ (160 Cal)"?

9. Write structural formulas for ADP and ATP.

10. Do high levels of ATP stimulate or inhibit (a) anabolism and (b) catabolism?

Overview of Metabolism

11. Is digestion a part of anabolism or catabolism? Why?

12. Examine Figure 24.5. In what way is the citric acid cycle part of anabolism?

13. In what way is the citric acid cycle part of catabolism?

14. Using Figure 24.5, trace the metabolic route for converting polysaccharides into fatty acids. Does your experience show that your body really can change carbohydrates into fat?

15. Lipids are oxidized in reactions using coenzymes. Identify which of the following are forms of coen-

zymes that could result from such reactions: **(a)** NAD⁺, **(b)** NADH, **(c)** FAD, **(d)** FADH$_2$.

16. Identify which forms of the following coenzymes could be used for anabolic reactions: **(a)** NAD⁺, **(b)** NADH, **(c)** FAD, **(d)** FADH$_2$, **(e)** NADP⁺, **(f)** NADPH.

Citric Acid Cycle

17. In what region of the cell does the citric acid cycle occur?
18. Name the tricarboxylic acids that occur in the tricarboxylic acid cycle.
19. Name the enzymes that catalyze reactions in the Krebs cycle that produce CO_2.
20. Name the compounds in the citric acid cycle that contain chiral carbons.
21. Is succinate oxidized or reduced when it is converted into fumarate? How can you tell?
22. When ATP concentrations in a cell are high, are concentrations of ADP and AMP high or low?
23. Describe the effect of each of the following on the activity of the citric acid cycle: **(a)** ADP, **(b)** NADH, and **(c)** ATP.
24. Explain why regulation of the Krebs cycle by ATP is a type of feedback inhibition. [Review Section 22.5 if necessary.]
25. Name the enzymes whose activity is regulated by the compounds in exercise 23.
26. The structures of acids in the citric acid cycle are written as negative ions in Figure 24.8. In Chapters 25 and 26 you will see structures for other acids also written in this way. Explain why this is the form that exists in your body.
27. How many NADH, FADH$_2$, and GTP does each round of the citric acid cycle produce?
28. Does the citric acid cycle produce ATP in the absence of O_2?

Electron Transport and Oxidative Phosphorylation

29. In what part of the cell does electron transport occur?
30. Each component in the electron transport chain carries electrons released from NADH or FADH$_2$. Does each component also carry protons released from these coenzymes? Explain.
31. According to the chemiosmotic hypothesis, how does electron transport drive ATP synthesis?
32. What does an uncoupler do? According to the chemiosmotic hypothesis, how does an uncoupler work?
33. Rationalize how uncouplers such as snake venom can cause fever (see Chemistry Spotlight, Section 24.4).
34. Mechanical disruption of mitochondria can cause uncoupling. Rationalize why.
35. 2,4-Dinitrophenol (DNP) is a substance that binds protons (H⁺) and freely carries them across the inner mitochondrial membrane. Predict the effect of DNP on **(a)** electron transport and **(b)** oxidative phosphorylation.
36. During electron transport, are the coenzymes NADH and FADH$_2$ oxidized or reduced?
37. During electron transport, is O_2 oxidized or reduced?
38. An iron ion in a cytochrome has an electrical charge of ___ after it receives an electron, and a charge of ___ after it passes the electron on to the next carrier in the chain.
39. During cyanide poisoning (see the Chapter Opener), what is the electrical charge on iron in cytochrome a$_3$?
40. Each NADH produced in the citric acid cycle generates ___ ATP by electron transport. Each FADH$_2$ generates ___ ATP.
41. To what class of proteins (Table 21.5) do the cytochromes belong?
42. ATPase catalyzes the reaction ADP + P$_i$ ⇌ ATP + H$_2$O in both directions. How does an enzyme affect the equilibrium of a reaction? [Review Section 22.1 if necessary.]

Energy Production and Use in the Body

43. The equivalent of how many ATP are produced for each acetyl CoA molecule oxidized in the citric acid cycle?
44. Is electron transport a series of endothermic or exothermic reactions? [Review Section 5.6 if necessary.]
45. Do cold temperatures increase or decrease the metabolic rate of **(a)** warm-blooded animals such as humans, **(b)** cold-blooded animals such as flies, and **(c)** warm-blooded hibernating animals such as bears?
46. During intense exercise your muscles may develop a shortage of O_2. Would a shortage of O_2 cause high concentrations to develop of **(a)** NAD⁺ or **(b)** NADH? How would this affect the tricarboxylic acid cycle?
47. How would a shortage of O_2 affect ATP production?

DISCUSSION EXERCISES

1. Catabolism generates reduced coenzymes: NADH, NADPH, and FADH$_2$ (Table 24.1). How do each of these reduced coenzymes support anabolism?

2. If you went on a low-carbohydrate diet, would you still be able to generate enough ATP? Use Figure 24.5 to justify your answer.

3. Rationalize why the synthesis of ATP from electron transport is often called "oxidative phosphorylation."

4. Using the six classes of enzymes (see Section 22.3), classify each enzyme in the Krebs cycle. Could any be classified in more than one way?

25

Carbohydrate Metabolism

1. How do you digest carbohydrates?
2. What are glycolysis and gluconeogenesis? What do they accomplish and how are they regulated?
3. How does your body store and break down glycogen? How do hormones regulate this?
4. What is the pentose phosphate pathway? What does it accomplish?
5. How does your body regulate the levels of glucose in blood? What causes abnormal glucose levels?

Glycogen Loading

Your body uses two main types of fuel—carbohydrates and fats. Carbohydrate, in the form of glycogen, is stored directly in muscle and liver, whereas fat is deposited in many other places in the body. During exercise, muscles use mostly carbohydrates, which are immediately available. As exercise continues, muscles also use fats, which are removed from storage and carried in blood to muscles.

Since athletes with larger supplies of glycogen can exercise longer, those preparing for endurance competition often engage in a process called *carbohydrate* or *glycogen loading*.

Studies show that the body stores twice the normal amount of glycogen if glycogen is depleted before loading begins. So one week before competition, an athlete might deplete glycogen by three days of extensive exercise, or a low-carbohydrate diet, or a combination of both. The next three or four days might include little exercise and a high-carbohydrate diet to produce extra high levels of glycogen in muscle and liver.

Glycogen loading is typically used for events requiring 90 minutes or more of continuous exertion. For example, it is common to glycogen load before running a marathon (26.2 miles) or other long-distance race (Figure 25.1). But it isn't done for shorter events because the extra glycogen is not needed and adds unwanted weight.

Plants capture sunlight streaming to earth and use it to make glucose and life-giving (for us) oxygen gas (Figure 25.2). In this process, called **photosynthesis,** plants carry out the reaction:

$$6CO_2 + 6H_2O + \text{energy (sunlight)} \longrightarrow \underset{\text{glucose}}{C_6H_{12}O_6} + 6O_2$$

Your body does just the opposite. You use O_2 gas to oxidize glucose to CO_2 and H_2O, producing energy:

$$C_6H_{12}O_6 + 6O_2 \longrightarrow 6CO_2 + 6H_2O + \text{energy}$$

Glucose is a simple carbohydrate, and carbohydrates are the largest source of energy in most diets. Your blood carries glucose to be used as fuel throughout your body. When blood glucose levels fall too low, for example, you feel faint because your brain lacks the energy it needs.

Figure 25.2 These plants produce glucose and O_2 gas by photosynthesis.

25.1 CARBOHYDRATE DIGESTION

Monosaccharides and Disaccharides. Monosaccharides in your diet go through your digestive tract unchanged and enter your blood. Disaccharides —primarily sucrose, maltose, and lactose (Figure 19.16)—remain unchanged until they reach your small intestine. Intestinal enzymes (sucrase, maltase, and lactase) then catalyze the hydrolysis of disaccharides into their monosaccharide units (Figure 25.3; reverse reactions in Figure 19.16), which enter the blood.

$$\text{maltose} + H_2O \xrightarrow{\text{maltase}} 2\text{ glucose}$$

$$\text{sucrose} + H_2O \xrightarrow{\text{sucrase}} \text{glucose} + \text{fructose}$$

$$\text{lactose} + H_2O \xrightarrow{\text{lactase}} \text{glucose} + \text{galactose}$$

Figure 25.3 Intestinal enzymes hydrolyze disaccharides into their monosaccharide units during digestion.

Some people, particularly adults of certain ethnic groups—such as Thais, Chinese, and blacks—have difficulty digesting milk products because they have low levels of lactase. Undigested lactose attracts water into the small intestine. It then passes into the colon (large intestine), where bacteria ferment it into CO_2 and other products. The extra intestinal water and gas cause nausea, cramps, pain, and diarrhea.

Polysaccharides. As soon as food enters your mouth, an enzyme in saliva called amylase begins hydrolyzing glycogen and starch to produce glucose and maltose (Figure 25.4). If you chew starchy food such as bread for a long time, you can gradually detect a sweet taste as salivary amylase produces more glucose and maltose.

When food enters your stomach, amylase activity stops because stomach HCl denatures the enzyme. The pH increases in your small intestine because pancreatic secretions neutralize stomach acid. Those secretions also contain another amylase that completes the action begun by salivary amylase. An intestinal enzyme (maltase) hydrolyzes the remaining maltose into glucose, which enters your blood.

starch and glycogen
↓ amylases
glucose + maltose
↓ maltase
glucose

Figure 25.4 Digestion of starch and glycogen.

25.2 GLYCOLYSIS

Glycolysis. **Glycolysis** is a series of catabolic reactions that converts glucose into pyruvate. The reactions occur in the cytoplasm of cells and produce ATP.

Figure 25.5 shows the reactions of glycolysis. Two different enzymes, hexokinase and glucokinase, catalyze the transfer of phosphate from ATP to glucose in the first reaction:

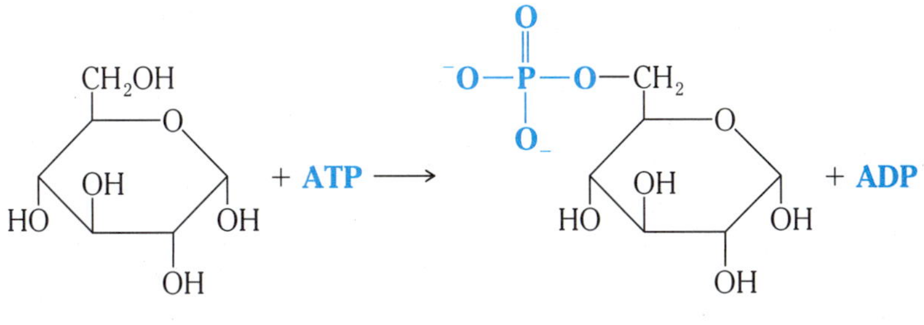

Glycolysis is also called the Embden–Meyerhof pathway in recognition of Gustav Embden and Otto Meyerhof, who discovered it in the 1930s.

glucose + ATP ⟶ glucose-6-P + ADP

Here ATP, a high-energy compound, is used to synthesize a lower-energy compound, glucose-6-P (see Table 24.2). Hexokinase catalyzes this reaction at all times. Glucokinase functions mainly when glucose concentrations in the cell are high.

Kinase is a common name for enzymes that transfer phosphate from ATP to another compound.

In the next reaction, glucose-6-P is converted into a structural isomer, fructose-6-P:

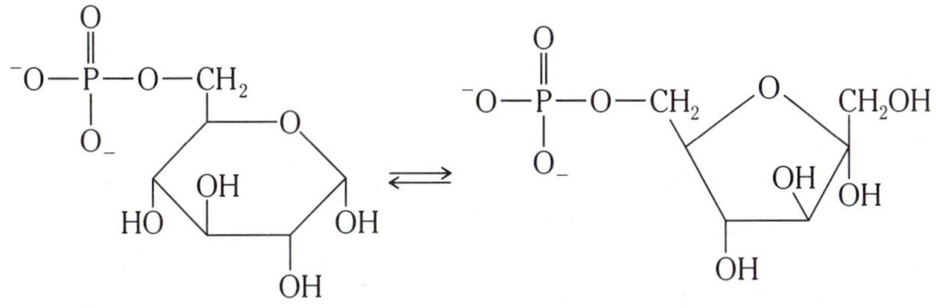

glucose-6-P ⇌ fructose-6-P

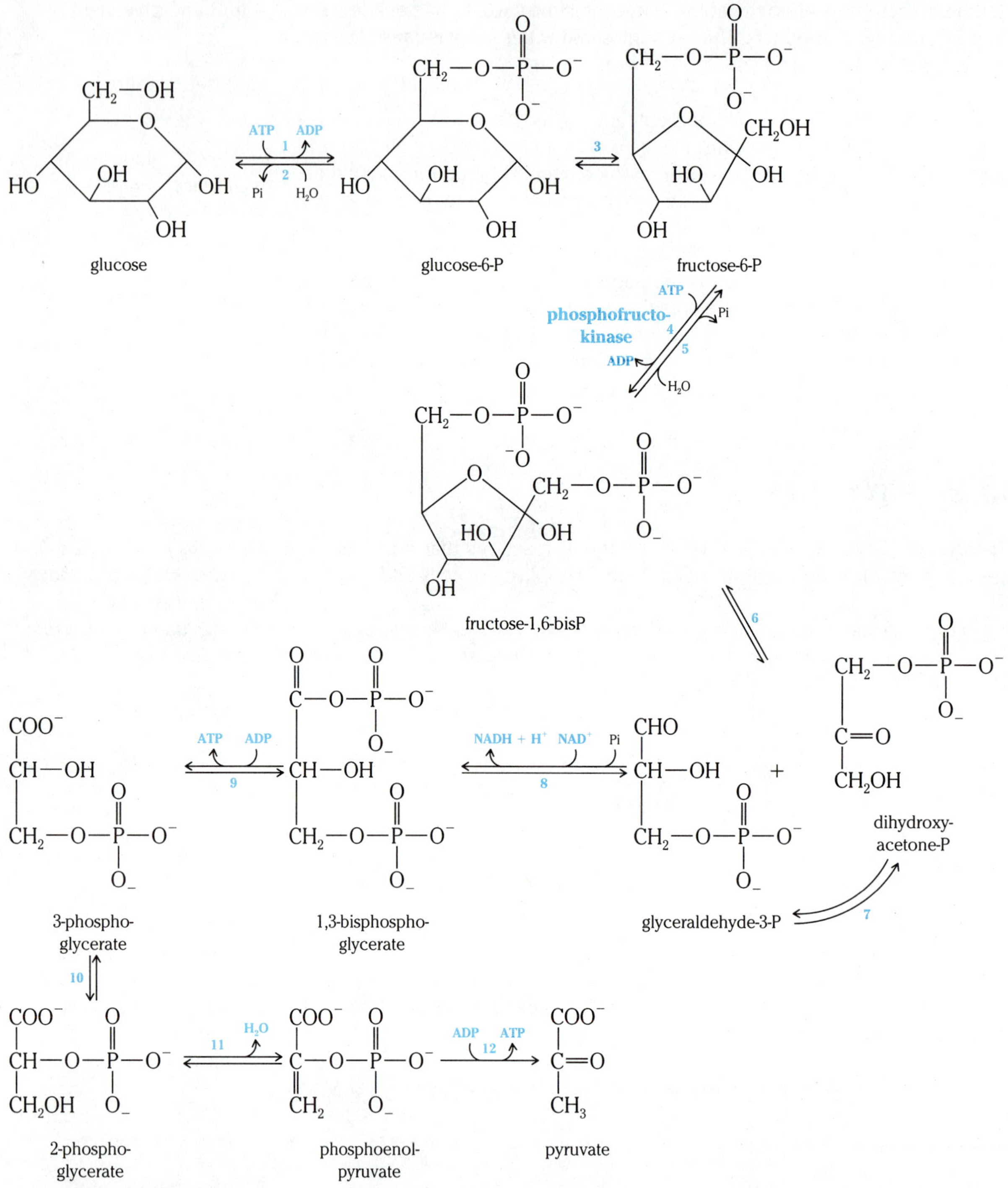

Figure 25.5 Glycolysis. Enzymes catalyzing these reactions are listed by number.

1. hexokinase, glucokinase
2. glucose 6-phosphatase
3. phosphoglucose isomerase
4. phosphofructokinase
5. fructose 1,6-bisphosphatase
6. aldolase
7. triosephosphate isomerase
8. glyceraldehyde-3-P dehydrogenase
9. phosphoglycerate kinase
10. phosphoglyceromutase
11. enolase
12. pyruvate kinase

Then phosphofructokinase catalyzes the transfer of another phosphate group from ATP:

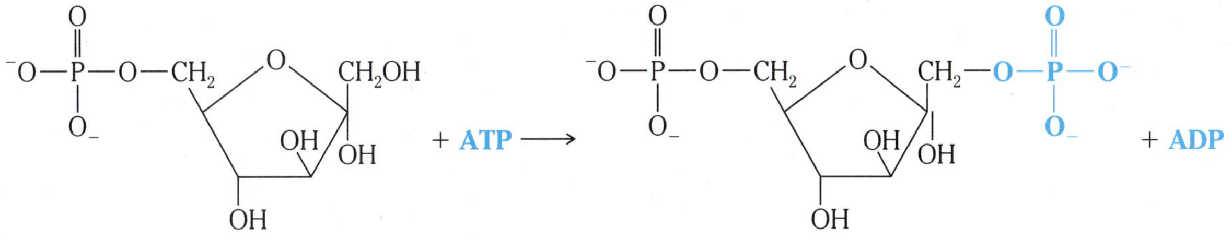

fructose-6-P + ATP ⟶ fructose-1,6-bisphosphate + ADP

Phosphofructokinase is the key regulatory enzyme in glycolysis. ATP inhibits this enzyme, while ADP and AMP activate it. This feedback inhibition by ATP ensures that glycolysis (which produces ATP) occurs only when ATP levels are low and need to be replenished.

The prefix bis- *is used instead of* di- *when the two phosphate groups are joined to different carbon atoms.*

Next, fructose-1,6-bisphosphate is split into two triose phosphates, and an isomerase converts dihydroxyacetone phosphate into glyceraldehyde-3-P:

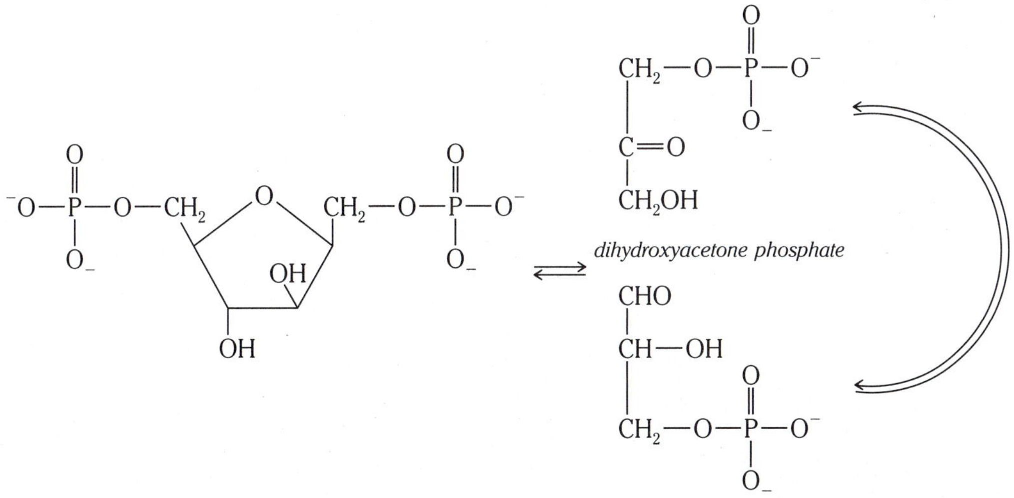

fructose-1,6-bisphosphate ⇌ dihydroxyacetone phosphate + glyceraldehyde-3-P

Thus two molecules of glyceraldehyde-3-P are produced from each molecule of fructose-1,6-bisphosphate.

Both glyceraldehyde-3-P molecules are oxidized and phosphorylated in the next reaction:

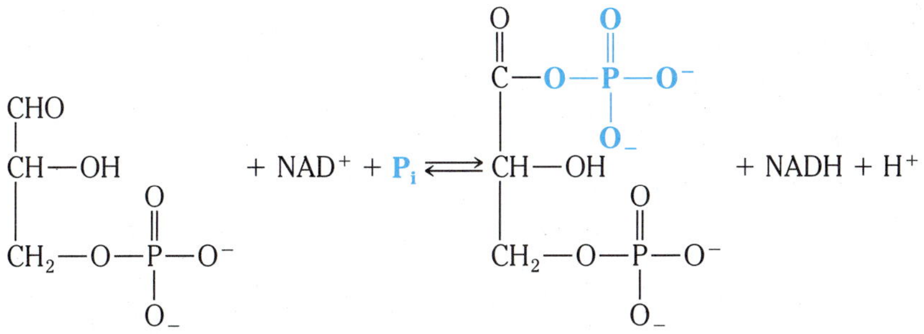

glyceraldehyde-3-P + NAD$^+$ + P$_i$ ⇌ 1,3-bisphosphoglycerate + NADH + H$^+$

1,3-Bisphosphoglycerate is a higher-energy compound than ATP (see Table 24.2). In the next reaction, phosphate is transferred from 1,3-bisphosphoglycerate to ADP to make ATP:

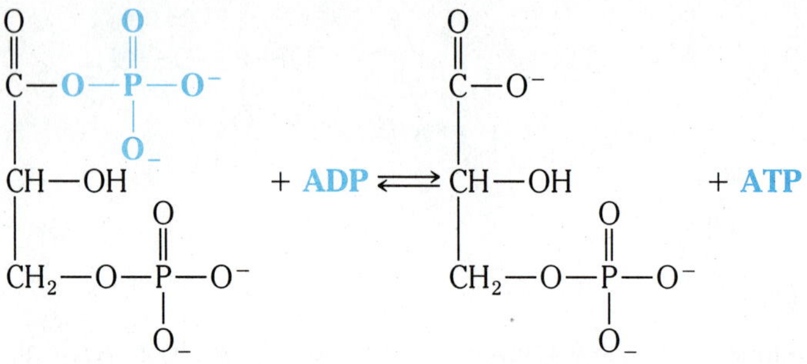

$$\begin{array}{c}\text{O} \quad \text{O} \\ \| \quad \| \\ \text{C}-\text{O}-\text{P}-\text{O}^- \\ | \quad | \\ \text{CH}-\text{OH} \quad \text{O}_- \\ | \\ \text{CH}_2-\text{O}-\text{P}-\text{O}^- \\ \| \\ \text{O}_-\end{array} + \text{ADP} \rightleftharpoons \begin{array}{c}\text{O} \\ \| \\ \text{C}-\text{O}^- \\ | \\ \text{CH}-\text{OH} \\ | \quad \text{O} \\ \text{CH}_2-\text{O}-\text{P}-\text{O}^- \\ \| \\ \text{O}_-\end{array} + \text{ATP}$$

1,3-bisphosphoglycerate   *3-phosphoglycerate*

Making ATP directly in a reaction is called *substrate-level phosphorylation*. This way to make ATP is different from using the electron transport chain, which is called *oxidative phosphorylation* (see Section 24.4).

Transferring phosphate from carbon 3 to carbon 2 is next:

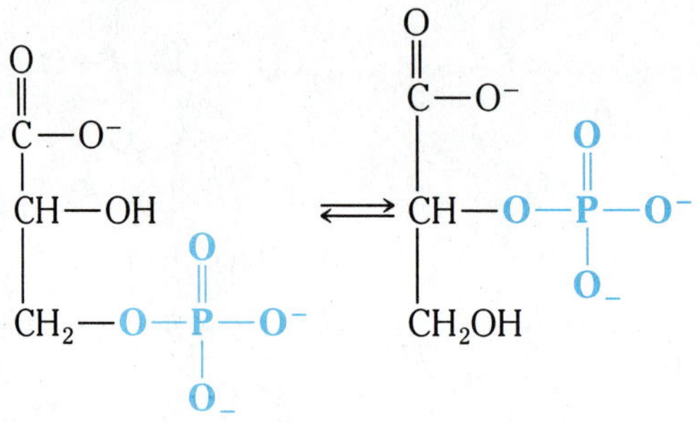

$$\begin{array}{c}\text{O} \\ \| \\ \text{C}-\text{O}^- \\ | \\ \text{CH}-\text{OH} \\ | \quad \text{O} \\ \text{CH}_2-\text{O}-\text{P}-\text{O}^- \\ \| \\ \text{O}_-\end{array} \rightleftharpoons \begin{array}{c}\text{O} \\ \| \\ \text{C}-\text{O}^- \\ | \quad \text{O} \\ \text{CH}-\text{O}-\text{P}-\text{O}^- \\ | \quad \| \\ \text{CH}_2\text{OH} \quad \text{O}_-\end{array}$$

3-phosphoglycerate   *2-phosphoglycerate*

Removal of H_2O from 2-phosphoglycerate produces phosphoenolpyruvate (PEP):

$$\begin{array}{c}\text{COO}^- \\ | \quad \text{O} \\ \text{CH}-\text{O}-\text{P}-\text{O}^- \\ | \quad \| \\ \text{CH}_2\text{OH} \quad \text{O}_-\end{array} \rightleftharpoons \begin{array}{c}\text{COO}^- \\ | \quad \text{O} \\ \text{C}-\text{O}-\text{P}-\text{O}^- \\ \| \quad \| \\ \text{CH}_2 \quad \text{O}_-\end{array} + \text{H}_2\text{O}$$

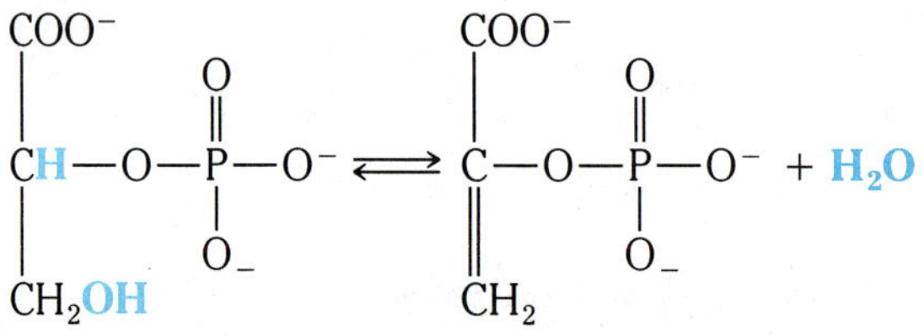

2-phosphoglycerate   *phosphoenolpyruvate*

PEP is a very high-energy compound (see Table 24.2), and this energy is used to transfer phosphate to ADP, making ATP:

$$\begin{array}{c}\text{COO}^- \\ | \quad \text{O} \\ \text{C}-\text{O}-\text{P}-\text{O}^- \\ \| \quad \| \\ \text{CH}_2 \quad \text{O}_-\end{array} + \text{ADP} \longrightarrow \begin{array}{c}\text{COO}^- \\ | \\ \text{C}=\text{O} \\ | \\ \text{CH}_3\end{array} + \text{ATP}$$

phosphoenolpyruvate   *pyruvate*

This is another example of substrate-level phosphorylation.

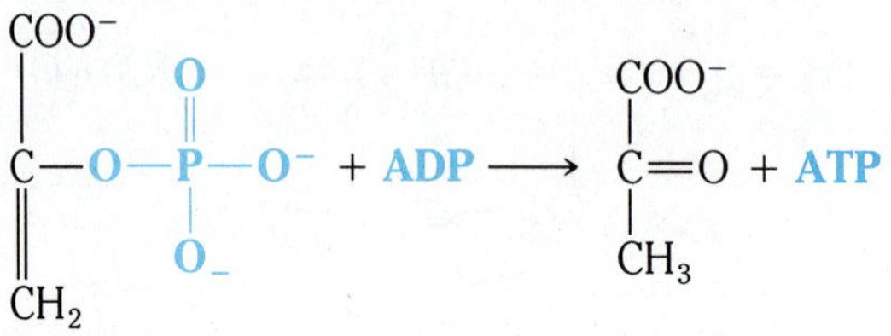

The net reaction for glycolysis is:

glucose + 2ADP + 2P$_i$ + 2NAD$^+$ ⟶ 2 pyruvate + 2ATP + 2NADH + 2H$^+$

Fructose is metabolized much the same way as glucose. When you eat foods that contain either fructose or sucrose (which you digest to fructose and glucose), each fructose molecule is metabolized to two glyceraldehyde-3-P molecules at the expense of two ATP (Figure 25.6). Glyceraldehyde-3-P then is oxidized to pyruvate as in glycolysis, producing the same net gain of two ATP and two NADH as with glucose.

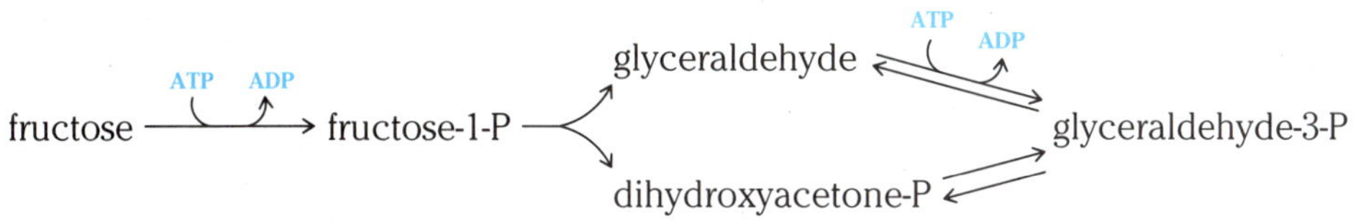

Two reactions in glycolysis consume 1 ATP each; two reactions produce 2 ATP each. The net gain, then, is 2 ATP.

fructose —ATP→ADP→ fructose-1-P → glyceraldehyde ⇌(ATP/ADP) glyceraldehyde-3-P
                                  dihydroxyacetone-P ⇌

Metabolism of Pyruvate. Pyruvate produced by glycolysis can be metabolized in several ways. One way uses the citric acid cycle (Section 24.3). Pyruvate enters the matrix of the mitochondrion and there is converted into acetyl CoA in an essentially irreversible exothermic reaction:

$$CH_3-\underset{\underset{O}{\|}}{C}-COO^- + CoA + NAD^+ \longrightarrow CH_3-\underset{\underset{O}{\|}}{C}-CoA + CO_2 + NADH + H^+$$

pyruvate — *acetyl CoA*

Figure 25.6 Metabolism of fructose to glyceraldehyde-3-P, which is metabolized further by glycolysis (Figure 25.5).

Acetyl CoA, in turn, is oxidized by the citric acid cycle.

When you exercise vigorously, however, your muscles don't always have enough O$_2$ to oxidize pyruvate in the citric acid cycle. Under these *anaerobic* conditions (lack of oxygen), your muscles convert pyruvate into lactate (see Chemistry Spotlight in Section 5.3):

$$CH_3-\underset{\underset{O}{\|}}{C}-COO^- + NADH + H^+ \rightleftharpoons CH_3-\underset{\underset{OH}{|}}{CH}-COO^- + NAD^+$$

pyruvate — *lactate*

Lactate then is absorbed into your blood, taken up by other tissues, and eventually converted back into pyruvate for further metabolism. Athletic trainers sometimes monitor the appearance of lactate in blood as athletes exercise; the amount of work that can be done before lactate appears is a measure of an athlete's aerobic capacity.

Why do oxygen-deprived muscle cells change pyruvate into lactate instead of just leaving it as pyruvate? Recall that glycolysis produces NADH from NAD$^+$ while making pyruvate. In order for glycolysis to continue, cells have to regenerate NAD$^+$ from NADH. Under anaerobic conditions, cells cannot use electron transport to do this. But by making pyruvate into lactate, cells regenerate NAD$^+$ (Figure 25.7). As a result, muscles keep carrying out glycolysis and producing a small amount of energy (two ATP per glucose).

Figure 25.7 Converting pyruvate into lactate regenerates NAD⁺ so glycolysis can continue to produce small amounts of ATP under anaerobic conditions.

$$\text{glucose} + 2\ ADP + 2\ P_i \xrightarrow{\text{glycolysis}} 2\ \text{pyruvate} + 2\ ATP$$

2 NAD⁺ → 2 NADH + 2H⁺

2 NAD⁺ ← 2 NADH + 2H⁺

2 lactate ⇌ 2 pyruvate

Yeast and certain other microorganisms have a different way to regenerate NAD⁺ under anaerobic conditions: They reduce pyruvate to ethanol:

$$\underset{\text{pyruvate}}{CH_3-\overset{O}{\underset{\|}{C}}-COO^-} \xrightarrow[CO_2]{H^+} \underset{\text{acetaldehyde}}{CH_3-\overset{O}{\underset{\|}{C}}-H} \xrightarrow[NAD^+]{NADH + H^+} \underset{\text{ethanol}}{CH_3-CH_2-OH}$$

Fermentation is one of the oldest described chemical processes. It was known in Egyptian and other ancient cultures.

Metabolizing glucose to ethanol under anaerobic conditions is called **fermentation**, a common way to make alcoholic beverages.

Energy from Oxidizing Glucose. You can calculate how much ATP your body makes from oxidizing glucose (or fructose) to CO_2 and H_2O. Notice in the net reaction for glycolysis above that one glucose produces two ATP plus two NADH. Recall that each NADH in the mitochondrion generates three ATP by electron transport (Section 24.4). But NADH produced by glycolysis is in cytoplasm, not mitochondria, and NADH cannot freely enter mitochondria. Instead, the protons and electrons carried by cytoplasmic NADH are transferred into mitochondria by shuttle mechanisms.

Figure 25.8 shows how the shuttle system works. In cytoplasm, NADH transfers electrons and protons to a carrier (A); in the process, NADH is oxi-

cytoplasm: NADH + H⁺ + A ⟶ AH₂ + NAD⁺

mitochondrion: NADH + H⁺ + A ⟵ AH₂ + NAD⁺
electron transport ↓
3 ATP

cytoplasm: NADH + H⁺ + A ⟶ AH₂ + NAD⁺

mitochondrion: FADH₂ + A ⟵ AH₂ + FAD
electron transport ↓
2 ATP

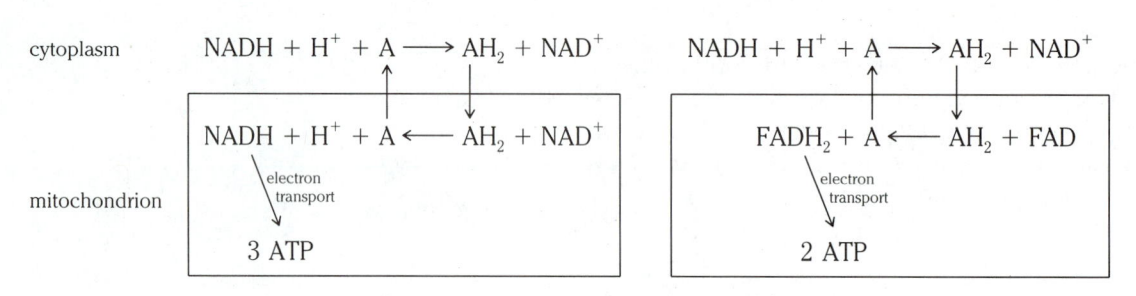

Figure 25.8 Shuttle mechanisms to generate ATP from NADH + H⁺ in cytoplasm. (Left) System generates 3 ATP/NADH in liver, kidney, and heart (A = oxaloacetate). (Right) System generates 2 ATP/NADH in muscle and brain (A = dihydroxyacetone-P).

dized back to NAD⁺. The reduced carrier (AH₂) then enters a mitochondrion and transfers its protons and electrons to NAD⁺ or FAD. The carrier A returns to the cytoplasm while the coenzyme in the mitochondrion carrying its protons and electrons (NADH or FADH₂) is metabolized by electron transport to produce ATP. It turns out, then, that each cytoplasmic NADH generates two or three ATP, depending on the tissue and the carrier used for the shuttle.

Now let's calculate the ATPs produced per glucose in muscle, where each cytoplasmic NADH generates 2 ATP (Figure 25.8). Glycolysis produces 2 ATP directly and 2 cytoplasmic NADH (each worth 2 ATP) for a total of 2 + (2 × 2) = 6 ATP.

Metabolizing pyruvate to acetyl CoA (see the reaction above) produces NADH in mitochondria, which each generate 3 ATP by electron transport. Recall that each acetyl CoA oxidized in the citric acid cycle (Section 24.5) generates an additional 12 ATP. So each pyruvate generates a total of 15 (3 + 12) ATP, and 2 pyruvates formed per glucose produce 30 ATP. Adding the 6 ATP from glycolysis to these 30 gives a total of 36 ATP per molecule of glucose oxidized to CO_2 and H_2O.

How efficient is this process? The total energy available from oxidizing glucose to CO_2 and H_2O is 686 kcal/mol. Each ATP has an energy value of 7.3 kcal/mol under standard conditions (see Table 24.2), so the total amount of energy generated as ATP is:

36 mol ATP × 7.3 kcal/mol = 270 kcal

$$\text{efficiency} = \frac{270 \text{ kcal}}{686 \text{ kcal}} \times 100\% = 39\%$$

Energy not saved as ATP provides heat to help maintain body temperature.

Because they use a different shuttle system (Figure 25.8, left), tissues such as liver, kidney, and heart generate 38 ATP per glucose oxidized to CO_2 and H_2O.

25.3 GLUCONEOGENESIS

Tissues such as muscle and brain primarily use glucose for fuel; they oxidize glucose to CO_2 and H_2O by glycolysis and the citric acid cycle to produce ATP. But liver synthesizes glucose by reactions that are largely the reverse of glycolysis. The synthesis of glucose from noncarbohydrate materials is called **gluconeogenesis.**

By looking at Figure 25.5, you can identify enzymes that convert phosphoenolpyruvate (PEP) into glucose. But notice that cells cannot make PEP directly from pyruvate. The conversion of PEP into pyruvate in glycolysis is irreversible; no enzyme catalyzes the reverse reaction.

In gluconeogenesis, the problem is producing PEP. This is done by an enzyme (called PEP carboxykinase) that converts oxaloacetate into PEP (Figure 25.9):

Gluconeogenesis in liver is an important way for your body to maintain its blood glucose levels (see Section 25.6).

$$\underset{\text{oxaloacetate}}{\begin{array}{c} O \\ \| \\ C-COO^- \\ | \\ CH_2-COO^- \end{array}} + \textbf{GTP} \longrightarrow \underset{\text{phosphoenolpyruvate (PEP)}}{\begin{array}{c} COO^- \\ | \quad\quad O \\ C-O-\overset{\|}{P}-O^- \\ \| \quad\quad | \\ CH_2 \quad\quad O_- \end{array}} + CO_2 + \textbf{GDP}$$

Recall that oxaloacetate is a component of the citric acid cycle (Figure 24.8). Thus, any substance that is metabolized to a citric acid cycle intermediate can be made into oxaloacetate, then into PEP, and then into glucose.

What materials can your body use for gluconeogenesis? As Figures 24.5 and 25.9 show, most amino acids can be made into citric acid cycle intermediates. Since they increase the amount of oxaloacetate, amino acids are one important source for gluconeogenesis.

Pyruvate (and lactate, which is made back into pyruvate) also provides material for gluconeogenesis. The enzyme pyruvate carboxylase converts pyruvate directly into oxaloacetate (Figure 25.9):

$$\underset{\text{pyruvate}}{\begin{array}{c} COO^- \\ | \\ C=O \\ | \\ CH_3 \end{array}} + \textbf{HCO}_3^- + ATP \longrightarrow \underset{\text{oxaloacetate}}{\begin{array}{c} O \\ \| \\ C-COO^- \\ | \\ CH_2-COO^- \end{array}} + ADP + P_i$$

This additional oxaloacetate then can be used to make glucose.

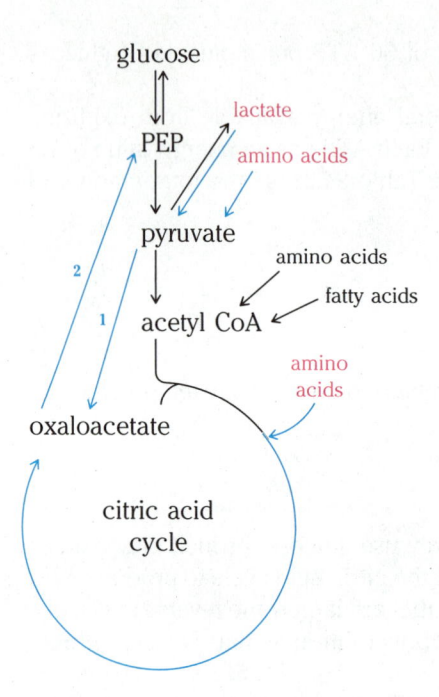

Figure 25.9 Gluconeogenesis (arrows in color). Pyruvate carboxylase (1) and PEP carboxykinase (2) are key enzymes. Sources for the net synthesis of glucose are in red.

If, however, pyruvate is metabolized to acetyl CoA and sent into the citric acid cycle, it cannot be used for gluconeogenesis. Recall that the citric acid cycle begins with a molecule of oxaloacetate reacting with acetyl CoA (Section 24.3). Subsequent reactions regenerate a molecule of oxaloacetate, but produce no additional oxaloacetate. Thus acetyl CoA *cannot* be used to synthesize glucose. Since fatty acids and certain amino acids are metabolized directly to acetyl CoA (see Figure 24.5), they are not potential sources for gluconeogenesis (Figure 25.9).

25.4 GLYCOGEN METABOLISM

Recall that glycogen (Section 19.5) is a polymer of glucose units joined by α-1,4 glycosidic bonds, and by α-1,6 bonds at branch points. Your body stores excess glucose as glycogen, mostly in liver and muscle.

Figure 25.10 shows how you synthesize and break down glycogen. Synthesis of glycogen is called **glycogenesis**; its breakdown is called **glycogenolysis.**

Glycogen has a different function in liver than in muscle. Liver breaks down glycogen to glucose-6-P; glucose 6-phosphatase (see Figure 25.5) then hydrolyzes glucose-6-P to glucose to maintain blood glucose levels.

Muscle lacks the enzyme glucose 6-phosphatase, so it cannot make glucose from glycogen. Instead, muscle tissue uses its glycogen directly for fuel. When glycogen breaks down to glucose-6-P, muscle metabolizes glucose-6-P by glycolysis and the citric acid cycle, producing ATP for muscle action.

Two hormones affect glycogen metabolism: epinephrine (adrenalin), produced by the adrenal glands; and glucagon, produced by the pancreas. Both

Recall that adding or removing phosphate is an important way to regulate an enzyme's activity (Section 22.5).

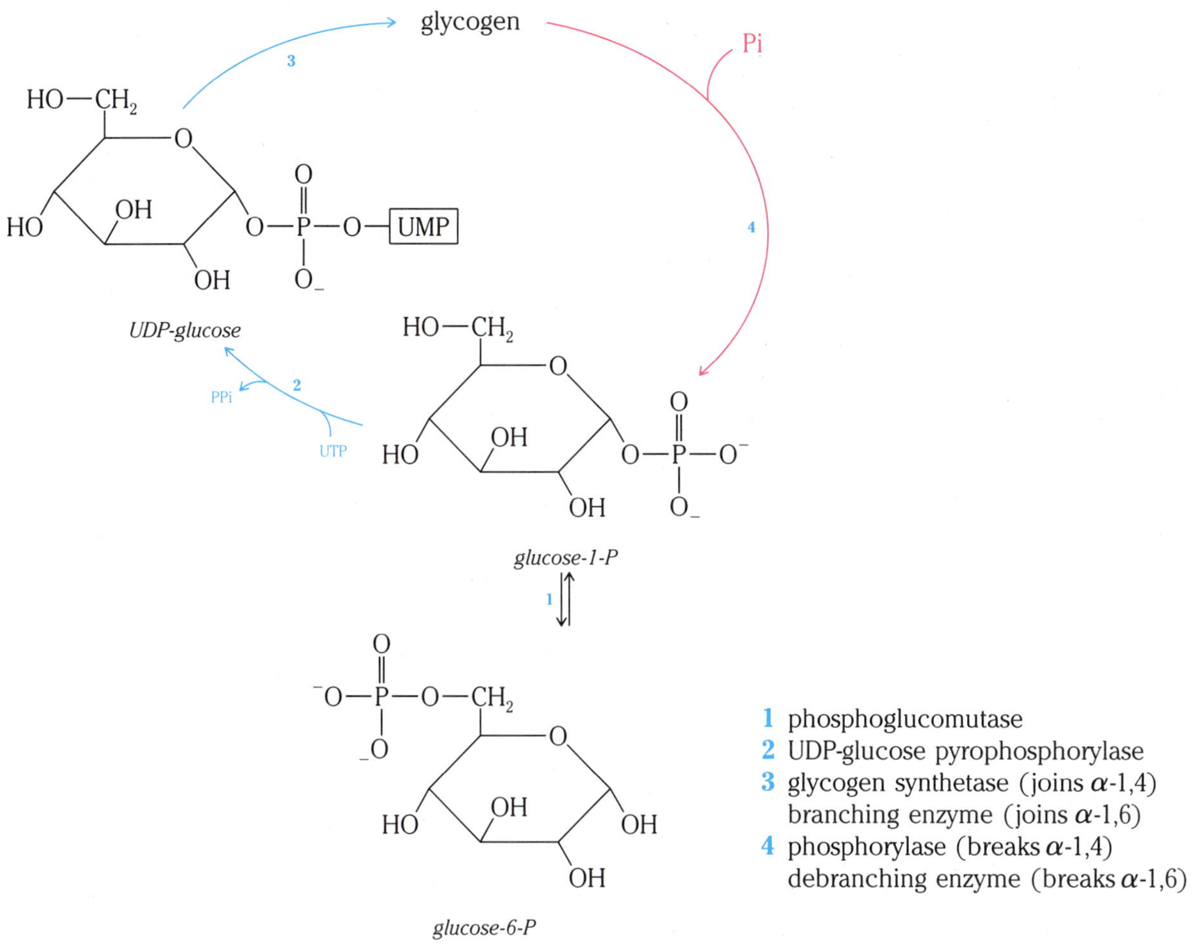

Figure 25.10 Glycogenesis (blue arrows) and glycogenolysis (red arrow). Enzymes catalyzing these reactions are listed by number.

1. phosphoglucomutase
2. UDP-glucose pyrophosphorylase
3. glycogen synthetase (joins α-1,4)
   branching enzyme (joins α-1,6)
4. phosphorylase (breaks α-1,4)
   debranching enzyme (breaks α-1,6)

hormones stimulate glycogenolysis and inhibit glycogenesis. They do this by binding to the cell membrane and activating adenylate cyclase, an enzyme that synthesizes cyclic AMP (Figure 25.11) from ATP (Figure 25.12). Cyclic AMP, in turn, triggers a series of reactions that attach phosphate groups to enzymes; this action activates a key enzyme (phosphorylase) in glycogenolysis and inhibits a key enzyme (glycogen synthetase) in glycogenesis.

Epinephrine mostly affects muscle tissue. When you get excited or frightened, the epinephrine you produce stimulates the breakdown of glycogen in your muscles. This makes fuel immediately available for your muscles to act. Glucagon, on the other hand, stimulates glycogenolysis primarily in the liver. So glucagon has the effect of increasing your blood glucose levels.

The only carbohydrate digestion product we have not discussed is galactose. Galactose is metabolized to glycogen by first converting it into UDP-glucose (Figure 25.13). People who lack one of the enzymes in this series of reactions develop high levels of galactose and galactose-1-P in their blood, which can impair mental and physical development. It's important to diagnose this genetic condition, called *galactosemia*, early in life. A diet that avoids lactose and galactose is an effective treatment.

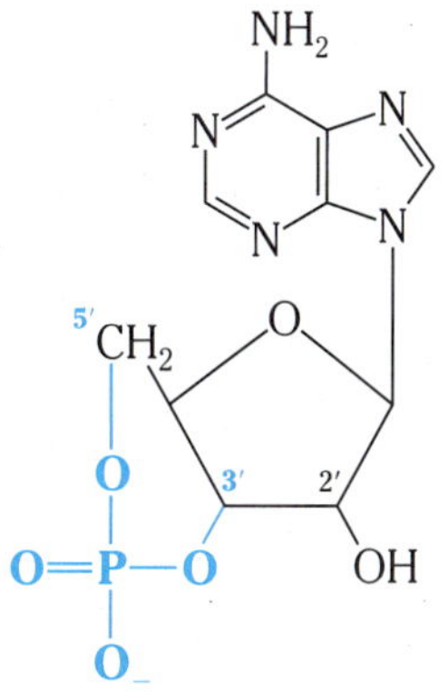

Figure 25.11 Cyclic AMP has phosphate esterified to both the 3' and 5' carbons of ribose.

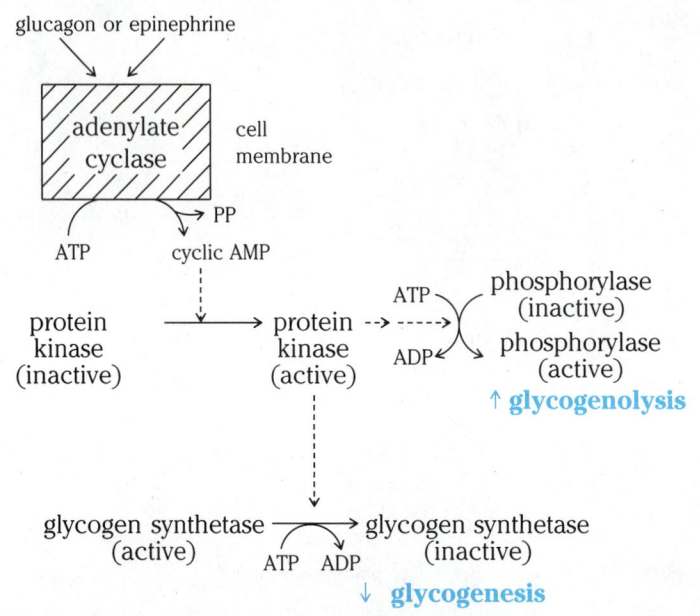

Figure 25.12 Action of glucagon and epinephrine on glycogen metabolism. All these actions of cyclic AMP are done by bonding phosphate to the affected enzymes.

Figure 25.13 Metabolism of galactose. The colored arrow represents the reaction lacking in galactosemia.

NADPH—unlike NADH and FADH$_2$—does not pass its protons and electrons through electron transport to generate ATP. Instead, NADPH is used directly for anabolism.

25.5 PENTOSE PHOSPHATE PATHWAY

Figure 25.14 shows a series of reactions known as the **pentose phosphate pathway.** These reactions occur in cytoplasm and have two main functions: (a) to produce ribose (and by related reactions, 2-deoxyribose) for nucleotides, DNA, and RNA; and (b) to produce NADPH (Figure 22.7).

Ribose-5-P (Figure 25.14) that is not incorporated into nucleotides undergoes additional reactions to regenerate glucose-6-P. For every six molecules of glucose-6-P metabolized in this way, the following overall reaction can be written:

$$6 \text{ glucose-6-P} + 12 \text{ NADP}^+ \longrightarrow 5 \text{ glucose-6-P} + 6\text{CO}_2 + \text{P}_i + 12 \text{ NADPH} + 12\text{H}^+$$

In effect, then, this catabolic pathway oxidizes glucose to produce CO$_2$ and NADPH. NADP$^+$ receives the electrons and hydrogen removed during oxidation in the form of a hydride transfer (see Chemistry Spotlight, Section 5.3).

You will see in Chapter 26 that NADPH is needed for lipid synthesis. So tissues that synthesize fats or steroids—especially the liver, adipose tissue, adrenal glands, and gonads—use this pathway to generate NADPH.

25.6 BLOOD GLUCOSE LEVELS

Effects of Eating and Fasting. The normal concentration of glucose in blood a few hours after eating is 70–100 mg/dL. When concentrations are lower, the condition is called **hypoglycemia.** An abnormally high blood glucose level is called **hyperglycemia.**

Figure 25.14 Key reactions of the pentose phosphate pathway.

Glucose in blood is a ready supply of fuel for other parts of the body. The brain, which doesn't store glycogen, relies on blood to supply glucose for its moment-to-moment actions. If blood glucose levels are too low, the brain cannot function normally.

Two pancreatic hormones, insulin and glucagon, help regulate your blood glucose levels. Glucagon, secreted by alpha cells of the pancreas, stimulates liver glycogenolysis and thus increases your blood glucose levels (Section 25.5).

Insulin, secreted by beta cells of the pancreas, has the opposite effect. This hormone binds to receptors in cell membranes and enables glucose to enter cells from the blood; this lowers blood glucose levels. Insulin also counteracts the stimulation of liver glycogenolysis by glucagon.

About 30 minutes after you consume carbohydrates, digested glucose begins to enter your blood. The resulting rise in blood glucose levels triggers the secretion of insulin, which enables glucose to enter cells and be metabolized by glycolysis, glycogenesis, or the pentose phosphate pathway. After enough glucose is oxidized to produce plenty of ATP, glycolysis is inhibited as ATP inhibits the key regulatory enzyme phosphofructokinase (Section 25.2).

During fasting, the body experiences a shortage of carbohydrate fuel. Muscle metabolizes its stored carbohydrate by glycogenolysis, glycolysis, and the citric acid cycle. The liver carries out glycogenolysis and gluconeogenesis to replenish blood glucose and prevent hypoglycemia.

Diabetes. When insulin is not produced in adequate amounts, glucose stays in the blood and results in hyperglycemia. This condition is called **diabetes.** In diabetes, blood glucose levels also rise because the liver produces more glucose by glycogenolysis and gluconeogenesis. The pattern is similar to what happens during fasting.

About 10% of diabetics have type I diabetes, which usually occurs by age 20 and is caused by the immune system attacking pancreatic beta cells. Since they cannot secrete enough insulin themselves, these diabetics must maintain their supply by injection. Since insulin is a protein (Figure 21.18), the digestive tract would destroy it if taken orally.

A diabetic who doesn't receive enough insulin can fall into a coma, often accompanied by the sweetish odor of acetone on the breath (see Sections 15.3 and 26.2). But a diabetic who injects too much insulin may go into shock due to acute hypoglycemia. The sweetish odor on the breath often can help you detect

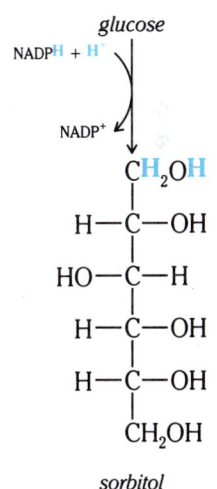

Figure 25.15 Synthesis of sorbitol.

Diabetes has been called "starvation in the midst of plenty." Glucose in blood surrounds the cells, but can't get in.

CHEMISTRY SPOTLIGHT

The Glucose Tolerance Test

Diabetics often have hyperglycemia; if blood glucose levels are high enough, some glucose appears in urine. Excessive urination and dehydration are other signs of the disease. But diagnosing diabetes can be difficult.

One diagnostic tool is the glucose tolerance test. The patient fasts overnight before taking the test. Blood glucose levels then are measured before the patient drinks a solution containing 75–100 g of glucose and at timed intervals afterward. Within an hour after the drink, most of the glucose enters the blood. This temporary hyperglycemia triggers the pancreas to secrete insulin to lower blood glucose levels.

Figure 25.16 shows the normal pattern as well as a typical pattern for a diabetic. Notice that a diabetic may have a higher initial blood glucose level. In addition, elevated blood glucose levels do not return as quickly to the initial one. This indicates a below-normal insulin response and is a sign of diabetes. The high levels also may exceed the kidney's ability to retain glucose in blood, so glucose appears in the urine.

In a shorter version of the glucose tolerance test, the patient consumes 75 g of glucose. Blood glucose levels are tested once or twice in the following two hours, then again at exactly two hours. Test results of 200 mg/dL or higher indicate diabetes (see Figure 25.16). Another, simpler test measures blood glucose levels after overnight fasting. Diabetes is indicated by readings of 140 mg/dL or higher two or more times.

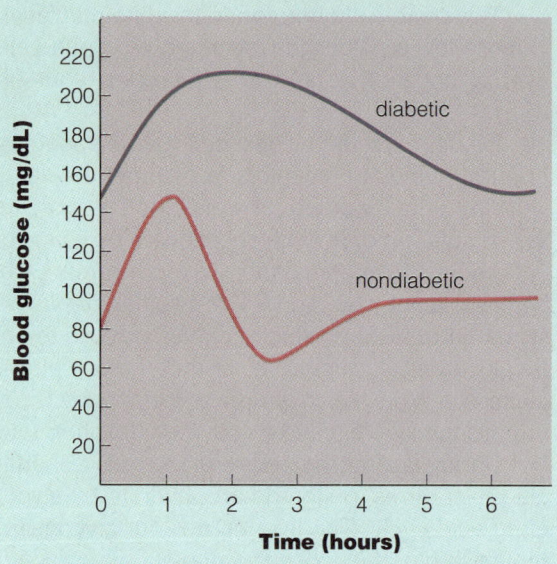

Figure 25.16 Typical patterns for a glucose tolerance test. Time is the period after ingesting glucose.

the difference between a diabetic coma, which requires insulin, and insulin shock, which requires a quickly digestible carbohydrate such as candy, fruit juice, or sugar.

Type II diabetes is more common. It usually occurs after age 30 or 40 and in people who are obese. Here the problem may be a combination of two factors: decreased insulin production, and cells with fewer receptors on their membranes to bind insulin and allow the hormone to work. Symptoms develop gradually and often can be controlled by diet and exercise. Some diabetics also take oral *hypoglycemic drugs* to lower their blood glucose levels. These drugs stimulate the pancreas to secrete more insulin.

One complication of diabetes is the formation of cataracts and other eye disorders. Excess glucose in the lens of the eye is converted into sorbitol (Figure 25.15). Sorbitol accumulates in the lens, attracting extra water, causing swelling, and initiating the formation of cataracts. Aspirin and ibuprofen (trade names Advil and Nuprin) may relieve this problem by inhibiting the enzyme that synthesizes sorbitol from glucose.

Hypoglycemia. After a carbohydrate-rich meal, blood glucose levels normally rise as digested glucose enters the blood. Your pancreas then secretes insulin to bring blood glucose levels back down. But a person with hypoglycemia may secrete too much insulin; this lowers blood glucose levels too much, making the person feel dizzy or even faint. To minimize an overreactive insulin response, such people often eat smaller, more frequent meals.

> The whole is more than the sum of its parts.
>
> *Aristotle (384–322 B.C.)*

SUMMARY

Carbohydrates are digested by hydrolysis to monosaccharides, mostly glucose. Glycolysis is a series of reactions in cytoplasm that converts glucose into pyruvate. Feedback inhibition of phosphofructokinase by ATP regulates glycolysis. Pyruvate may enter mitochondria, be converted into acetyl CoA, and then be oxidized in the citric acid cycle. When this occurs, muscle cells produce 36 ATP per glucose. Under anaerobic conditions, pyruvate is converted into lactate in muscle or into ethanol in yeasts and certain other microorganisms.

Gluconeogenesis is the synthesis of glucose from noncarbohydrate materials. Amino acids are a major source, but fatty acids are not. Gluconeogenesis occurs mostly in liver and is used to maintain blood glucose levels.

Glycogen is stored in muscle and liver. Glycogen synthesis (glycogenesis) and breakdown (glycogenolysis) are regulated by the hormones epinephrine (adrenalin) and glucagon. Epinephrine stimulates glycogenolysis and inhibits glycogenesis in muscle; glucagon has the same effect in liver. These actions are carried out by cyclic AMP, which regulates enzymes by phosphorylating them.

The pentose phosphate pathway occurs in cytoplasm. Its functions are to produce ribose for nucleotide and nucleic acid synthesis, and NADPH for anabolism, especially lipid synthesis.

Insulin lowers blood glucose levels largely by enabling glucose to enter cells from the blood. Glucagon increases blood glucose levels by stimulating liver glycogenolysis. These two pancreatic hormones are secreted as necessary to maintain blood glucose levels in the 70–100 mg/dL range. Diabetes occurs when insulin is not produced in adequate amounts. Oversecretion of insulin can cause hypoglycemia.

KEY TERMS

Diabetes (25.6)
Fermentation (25.2)
Gluconeogenesis (25.3)
Glycogenesis (25.4)

Glycogenolysis (25.4)
Glycolysis (25.2)
Hyperglycemia (25.6)

Hypoglycemia (25.6)
Pentose phosphate pathway (25.4)
Photosynthesis (25.1)

EXERCISES

Even-numbered exercises are answered at the back of this book.

Introduction

1. Write balanced chemical reactions for **(a)** photosynthesis, and **(b)** the oxidation of glucose by O_2 to form CO_2 and H_2O.
2. Which reaction(s) in exercise 1 require energy? Which release energy?

Carbohydrate Digestion

3. When you drink milk you consume lactose. What are the products of lactose digestion?
4. Cellulose is a polysaccharide in wood and paper. What would happen to the cellulose if you chewed and swallowed a toothpick? (Review Section 19.5 if necessary to answer this question.)
5. A clinical test for certain types of pancreatic disease is the presence of elevated levels of an enzyme in the blood. Do you think that enzyme is more likely to be sucrase or amylase? Why?
6. The enzyme maltase is needed to digest **(a)** glucose, **(b)** sucrose, **(c)** cellulose, **(d)** fructose, **(e)** starch.

Glycolysis

7. Look at Figure 25.5. List by number the reactions that consume ATP.
8. Look at Figure 25.5. List by number the reactions that produce ATP.
9. How many ATP are produced per molecule of glucose metabolized to pyruvate by glycolysis?
10. Is glyceraldehyde-3-P oxidized or reduced when it is converted into 1,3-bisphosphoglycerate? How can you tell?
11. If glycolysis begins with one glucose, what is the first reaction in which a coefficient of 2 can be written for each reactant and product?
12. In red blood cells, 2,3-bisphosphoglycerate is made from 1,3-bisphosphoglycerate produced in glycolysis. Write the structural formula for 2,3-bisphosphoglycerate.
13. What class of enzyme catalyzes the conversion of 1,3-bisphosphoglycerate into 2,3-bisphosphoglycerate? [Review Section 22.3 if necessary.]
14. Identify the class of enzyme that catalyzes the following reactions in glycolysis (numbers refer to reactions in Figure 25.5): **(a)** 1, **(b)** 2, **(c)** 3, **(d)** 6, **(e)** 8. [Review Section 22.3 if necessary.]
15. List the enzymes catalyzing reactions in glycolysis that are essentially irreversible.
16. An hour after a big meal, is glycolysis occurring in your muscles more rapidly or less rapidly than normal? How is the rate of glycolysis regulated in this situation?
17. During intense exercise, why is it advantageous for your muscles to produce lactic acid (lactate)?
18. How is lactate metabolized?
19. Under anaerobic conditions, why is it advantageous for yeast to convert pyruvate into ethanol?
20. How many ATP are made in muscle by metabolizing one glucose molecule to **(a)** lactate under anaerobic conditions, and **(b)** CO_2 and H_2O under aerobic conditions?
21. When you exercise intensely and your muscle cells become anaerobic, how does this affect **(a)** electron transport, and **(b)** the citric acid cycle? [Review Sections 24.3 and 24.4 if necessary.]

Gluconeogenesis

22. Classify each of the following as part of anabolism or catabolism: **(a)** glycolysis, **(b)** photosynthesis, **(c)** pentose phosphate pathway, **(d)** glycogenesis, **(e)** glycogenolysis, and **(f)** gluconeogenesis.
23. Does the production of one glucose molecule by gluconeogenesis from two oxaloacetate molecules give off or consume ATP? How many ATP?
24. Gluconeogenesis occurs mostly in which organ? Where does most of the glucose produced by gluconeogenesis go?
25. What enzyme produces fructose-6-phosphate in gluconeogenesis?

Glycogen Metabolism

26. Look at Figure 25.10. Some people genetically lack the enzyme phosphorylase. What effect would this have on their liver glycogen levels?
27. How would the genetic absence of the enzyme phosphorylase affect blood glucose levels? Explain.
28. Would the *removal* of phosphate from the enzyme increase or decrease the activity of **(a)** phosphorylase and **(b)** glycogen synthetase?
29. What is glycogen loading? Why is it used?
30. Caffeine has several effects on the body. One effect is to inhibit the breakdown of cyclic AMP. Use this information to rationalize why some athletes drink a cup of coffee an hour before a marathon.

31. People who have von Gierke's disease genetically lack the enzyme glucose-6-phosphatase (see Figure 25.5). What effect would this have on gluconeogenesis in muscle tissue? Why?
32. Von Gierke's disease (see exercise 31) causes very large amounts of glycogen to accumulate in the liver. Rationalize why.

Pentose Phosphate Pathway

33. CO_2 given off in the pentose phosphate pathway comes entirely from which carbon atom in glucose-6-P?
34. The pentose phosphate pathway is most active in tissues that have high levels of lipid synthesis (such as liver, adipose tissue, adrenal glands, and the gonads). Why do you think this is so?
35. Identify the part of the cell in which each of the following occur: **(a)** gluconeogenesis, **(b)** Kreb's cycle, **(c)** pentose phosphate pathway, **(d)** electron transport, **(e)** glycolysis.

Blood Glucose Levels

36. Name the two pancreatic hormones that have opposing effects on blood glucose levels.
37. Is the concentration of glucose in muscle and adipose tissue cells higher or lower than normal in a diabetic person? Why?
38. In diabetics, is gluconeogenesis in the liver increased or decreased?
39. List the two general types of diabetes. How is each treated?
40. Six hours after eating, is your pancreas more likely to be secreting extra insulin or extra glucagon? Why?
41. Why are people who have hypoglycemia more likely to feel dizzy or faint?
42. Under conditions of starvation, how does your body maintain a normal level of glucose in the blood? From what material is much of that glucose synthesized?

DISCUSSION EXERCISES

1. If carbon 1 in glucose is labeled with radioactive ^{14}C, and this glucose is metabolized by glycolysis, in which carbon atom(s) of pyruvate will ^{14}C appear?
2. How many molecules of ATP (or its equivalents) are needed to synthesize one molecule of glucose from pyruvate by gluconeogenesis?
3. During glycogen loading (see Chapter Opener) glycogen is stored both in liver and muscle. Can glycogen in both tissues be used for prolonged muscle activity? Explain.
4. Predict the pattern of a glucose tolerance test (see Figure 25.16) for a person who has hypoglycemia. Would you recommend doing this test on such a person? Why?

26

Lipid Metabolism

1. How are lipids digested?
2. How are triglycerides and fatty acids metabolized for energy? What is ketosis, and how does it develop?
3. How are fatty acids and triglycerides synthesized and stored? How is this regulated?
4. What happens to lipid metabolism under conditions such as fasting, eating, and diabetes?
5. How are cholesterol and other steroids synthesized? How are blood cholesterol levels controlled?

Some Unusual Polyunsaturated Fatty Acids

The Inuit peoples of Alaska and Canada eat a lot of fat but have a low incidence of heart disease. Why? One reason is that they eat large amounts of marine fish (Figure 26.1), which contain polyunsaturated fatty acids that aren't common in most diets.

Fish oils are rich in fatty acids having a double bond three carbon atoms from the methyl end; these are called n-3 fatty acids. One of the most abundant is eicosapentaenoic acid (EPA):

$$CH_3CH_2(CH=CHCH_2)_5 CH_2CH_2COOH$$

If you compare this structure to those given in Table 20.1, you will see that linolenic acid is the only common unsaturated fatty acid that is an n-3 compound; all the others are n-6 or n-9.

Furthermore, your enzymes cannot insert a double bond near the methyl end to produce n-3 compounds. The only way your body can get n-3 fatty acids is through diet.

What do EPA and other marine n-3 polyunsaturated fatty acids have to do with heart disease? Studies show that these compounds lower blood triglyceride levels and protect against blood clotting by altering the metabolism of arachidonic acid to form certain prostaglandins (see Section 20.6) that inhibit clotting. Fewer clots means less risk of heart attacks. Scientists are also examining the use of n-3 polyunsaturated fatty acids to treat arthritis, psoriasis, and inflammatory and immune diseases.

Some fats, then, may be good for your health.

Figure 26.1 This coho salmon is a good source of n-3 fatty acids.

Part III   Biochemistry

The metabolism of carbohydrates such as sugar and starch produces energy. If you eat and drink too many carbohydrates however (Figure 26.2), you will get fat, not sweet. Obviously, you can convert carbohydrates to fat.

You also can metabolize fat to produce energy. If you fasted or went on a low-carbohydrate diet, you would use mostly fat for energy. People typically have enough fat and protein stored in their bodies to survive about ten weeks of fasting. In contrast, their glycogen supply lasts only about one day.

Figure 26.2 The body metabolizes excess carbohydrates into fat.

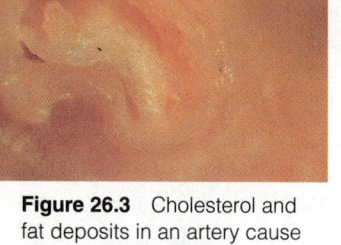

Figure 26.3 Cholesterol and fat deposits in an artery cause atherosclerosis. These deposits (yellow) can be seen by injecting dye into the blood, then taking X rays of the arteries.

Although lipids are a valuable fuel, too much fat is unhealthy. Overweight people have increased risks of various diseases, such as diabetes and heart disease due to atherosclerosis (Figure 26.3). In this chapter you will see how you metabolize lipids, and how this affects your health.

26.1 LIPID DIGESTION

Figure 26.4 summarizes what happens when you digest fat. When you eat fatty foods, a lipase enzyme in your stomach begins to hydrolyze triglycerides:

$$\begin{array}{c}CH_2-O-\overset{O}{\overset{\|}{C}}-R\\ |\\ CH-O-\overset{O}{\overset{\|}{C}}-R'\\ |\\ CH_2-O-\overset{O}{\overset{\|}{C}}-R''\end{array} + H_2O \xrightarrow{\text{lipase}} \begin{array}{c}CH_2-OH\\ |\\ CH-O-\overset{O}{\overset{\|}{C}}-R'\\ |\\ CH_2-O-\overset{O}{\overset{\|}{C}}-R''\end{array} + R-\overset{O}{\overset{\|}{C}}-OH$$

triglyceride   *diglyceride*   *fatty acid*

In your intestine, a lipase secreted by your pancreas continues this hydrolysis, producing a mixture of diglycerides, monoglycerides, glycerol, and fatty acids.

Detergents (see Chemistry Spotlight, Section 7.3) bring nonpolar lipids and polar H_2O together so that hydrolysis can occur. Your body uses ionic steroids called bile salts (Figure 20.11) as detergents. Bile salts enter the intestine through the bile duct. People who have a blocked bile duct cannot suspend

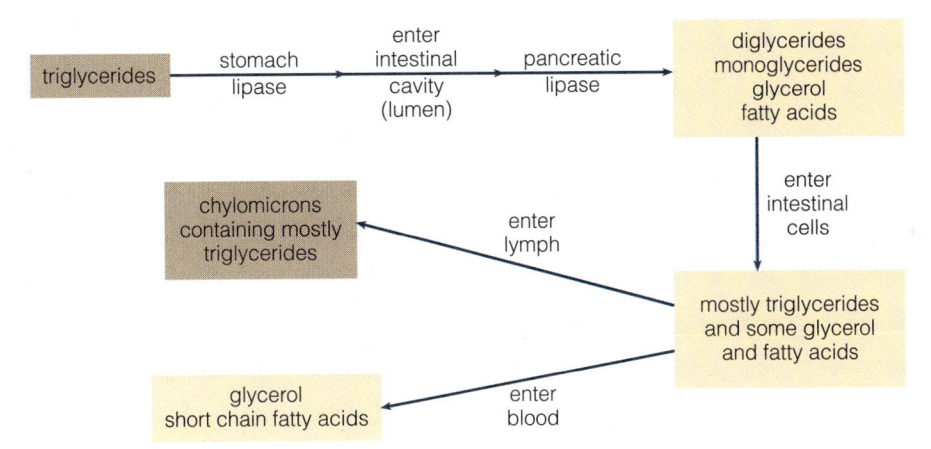

Figure 26.4 Summary of the digestion of triglycerides (fat).

dietary lipids in water to digest them. As a result, they excrete fat in their stools.

Hydrolyzed lipid material is absorbed into cells lining the intestine. There much of the lipid is converted back into triglycerides. The polar materials that remain, mostly glycerol and short-chain fatty acids, leave intestinal cells and enter the blood.

As triglycerides leave the intestinal cells, they are packaged into tiny spheres, about one micron in diameter, called **chylomicrons** (see Section 20.5). Surrounded by a polar coat of phospholipids and protein (see Table 20.3), chylomicrons enter the lymphatic circulation and then the blood. Adipose tissue takes up and stores most of these triglycerides.

When stored triglyceride is needed for energy, it is hydrolyzed in adipose tissue (Section 26.2). The resulting nonpolar fatty acids bind to albumin, an abundant protein in blood. Albumin then transports fatty acids to other tissues for metabolism.

26.2 CATABOLISM OF TRIGLYCERIDES

Triglycerides. Fat is a mixture of triglycerides. It functions as (a) a reservoir of energy (38 kJ or 9 Cal per g); (b) a cushion to protect internal body parts; and (c) an insulator to help maintain body temperature.

In order to use fat as a fuel, your body needs to remove it from storage in adipose tissue and transport it for metabolism. A lipase in adipose tissue hydrolyzes triglyceride to glycerol and free fatty acids:

triglyceride + 3H$_2$O $\xrightarrow{\text{lipase}}$ glycerol + fatty acids

Since it is polar, glycerol dissolves in blood and is taken up by tissues. It is converted into dihydroxyacetone-P (Figure 26.5), which is metabolized by

Figure 26.5 Metabolism of glycerol to dihydroxyacetone-P, an intermediate in glycolysis.

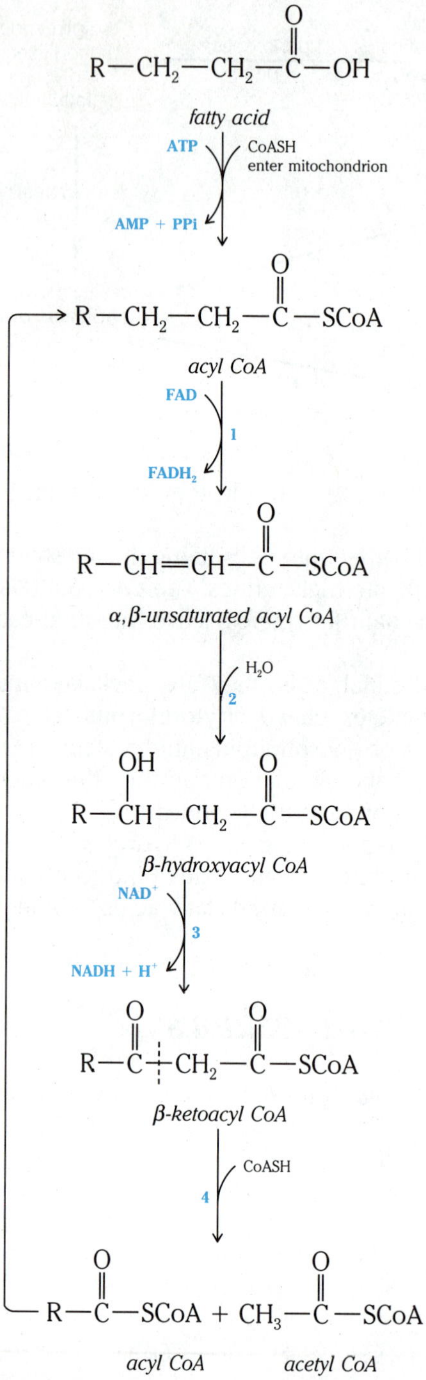

Figure 26.6 Fatty acid oxidation (β-oxidation). Enzymes catalyzing these reactions are listed by number.

1 acyl CoA dehydrogenase
2 enoyl-CoA hydratase
3 β-hydroxyacyl CoA dehydrogenase
4 thiolase

glycolysis (see Figure 25.5) to generate ATP, or by gluconeogenesis to produce glucose.

Fatty Acid Oxidation. Figure 26.6 shows the oxidation of fatty acids to acetyl CoA. First, ATP is used to convert fatty acids into their CoA derivatives and transport them into mitochondria. There fatty acyl CoA molecules go through a series of reactions that oxidizes the β-carbon. This is why **fatty acid oxidation** is sometimes called **β-oxidation**.

Recall that the β carbon is the second carbon from the carboxyl end (Section 16.1).

The first three reactions beginning with acyl CoA oxidize the β carbon atom (Figure 26.6). The final reaction splits off a two-carbon fragment as acetyl CoA. The remaining acyl CoA chain (two carbons shorter than the original acyl CoA) goes through the cycle again, with another round of oxidation and another acetyl CoA removed. The cycle continues until the original fatty acid chain is completely converted into acetyl CoA.

You now can calculate how much ATP your body produces when it completely oxidizes a fatty acid to CO_2 and H_2O (Figure 26.7). Let's consider palmi-

$CH_3(CH_2)_{14}COOH$ equivalents of ATP
palmitic acid

ATP, CoASH ...... −2 ATP
AMP + PPi

$CH_3(CH_2)_{14}\overset{O}{\underset{\|}{C}}-SCoA$

7 cycles ...... 7 × 5 = **35 ATP**

8 acetyl CoA

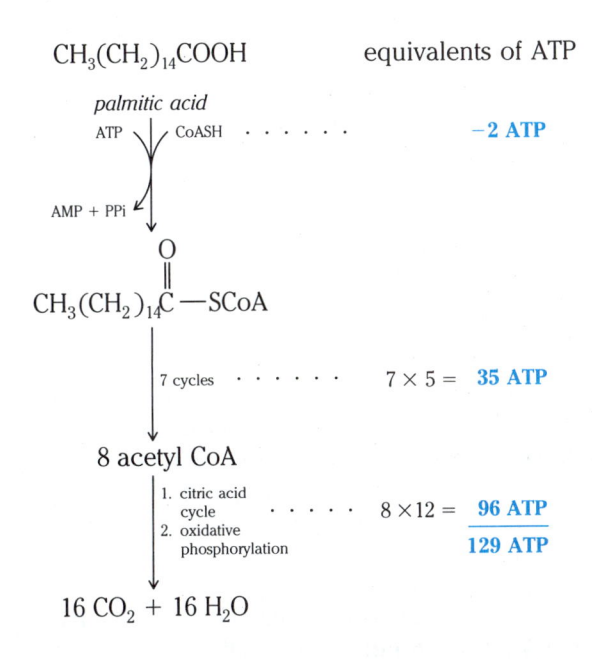

1. citric acid cycle
2. oxidative phosphorylation ...... 8 × 12 = **96 ATP**

129 ATP

$16\ CO_2 + 16\ H_2O$

Figure 26.7 Complete oxidation of palmitic acid to CO_2 and H_2O generates 129 ATP.

tic acid, a saturated, 16-carbon fatty acid (Table 20.1). Seven cycles of β-oxidation produce 8 acetyl CoA (2 acetyl CoA are produced in the final cycle). Each cycle produces 1 $FADH_2$ (worth 2 ATP by electron transport; see Section 24.4) and 1 NADH (worth 3 ATP by electron transport), for a total of 5 ATP per cycle. In addition, each acetyl CoA oxidized to CO_2 and H_2O by the citric acid cycle and electron transport generates 12 ATP (see Section 24.5). The cost of transporting palmitic acid into the mitochondrion and forming an acyl CoA is the equivalent of 2 ATP (since ATP is broken down to AMP instead of ADP). Altogether, the complete oxidation of palmitic acid to CO_2 and H_2O produces 129 ATP.

This calculation shows that fats are a rich source of energy in the body.

Ketosis. Sometimes acetyl CoA from fatty acid oxidation cannot be completely oxidized to CO_2 and H_2O. This happens in untreated diabetes, starvation, or low-carbohydrate diets. In these situations, cells lack glucose and rely more on fatty acid oxidation for energy. This generates large amounts of acetyl CoA.

In order to be oxidized to CO_2 and H_2O, acetyl CoA must react with oxaloacetate in the first reaction of the citric acid cycle. But under conditions of diabetes, starvation, or a low-carbohydrate diet, much of the oxaloacetate is metabolized to glucose by gluconeogenesis (Section 25.3). As a result, less oxaloacetate is available to react with the large amounts of acetyl CoA being produced.

Under these conditions, some acetyl CoA is metabolized to acetoacetic acid, β-hydroxybutyric acid, and acetone (Figure 26.8), which are known as

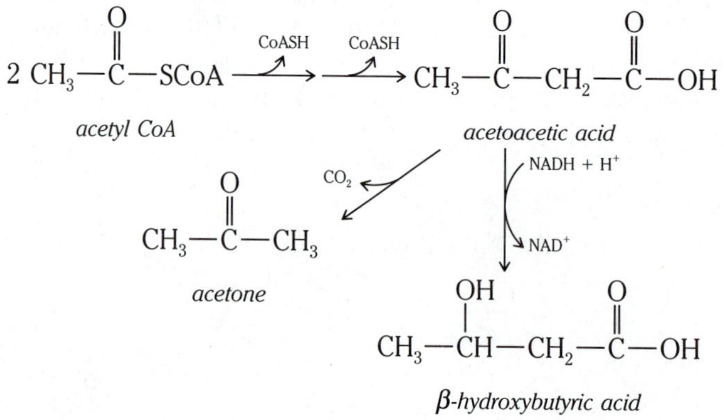

Figure 26.8 Metabolism of acetyl CoA to the three ketone bodies.

ketone bodies. People who have acetone on their breath, and ketone bodies in their blood *(ketonemia)* and in their urine *(ketonuria),* are said to have **ketosis** (see introduction to Chapter 15).

Notice that two ketone bodies are acids. Thus, people who have ketosis typically have excess acid in their blood and urine, a condition called *acidosis*. Acidosis resulting from ketosis is called **ketoacidosis.** The large volumes of urine excreted to remove the ketone bodies also can produce dehydration.

26.3 FATTY ACID SYNTHESIS

Making Acetyl CoA into Fatty Acids. Fatty acid synthesis is much like fatty acid oxidation in reverse. Just as your body breaks down fatty acids to acetyl CoA (Section 26.2), you synthesize fatty acids from acetyl CoA. But there are important differences. One is that fatty acid synthesis occurs in cytoplasm, whereas β-oxidation occurs in mitochondria. Another is that fatty acid synthesis uses NADPH in oxidation–reduction reactions, whereas β-oxidation uses the coenzymes NAD^+ and FAD for that purpose.

Figure 26.9 shows that **fatty acid synthesis** also uses different steps than fatty acid oxidation. The cycle of reactions builds fatty acid chains two carbons at a time, typically producing palmitic acid, a saturated, 16-carbon fatty acid.

In the first round of the cycle, the key regulatory enzyme acetyl CoA carboxylase converts acetyl CoA into malonyl CoA; then the malonyl group is transferred to a CoA-like carrier called acyl carrier protein (ACP). A second acetyl CoA is made into acetyl ACP. Then malonyl ACP reacts with acetyl ACP,

Because they are made in two-carbon segments, fatty acids almost always have an even number of carbon atoms.

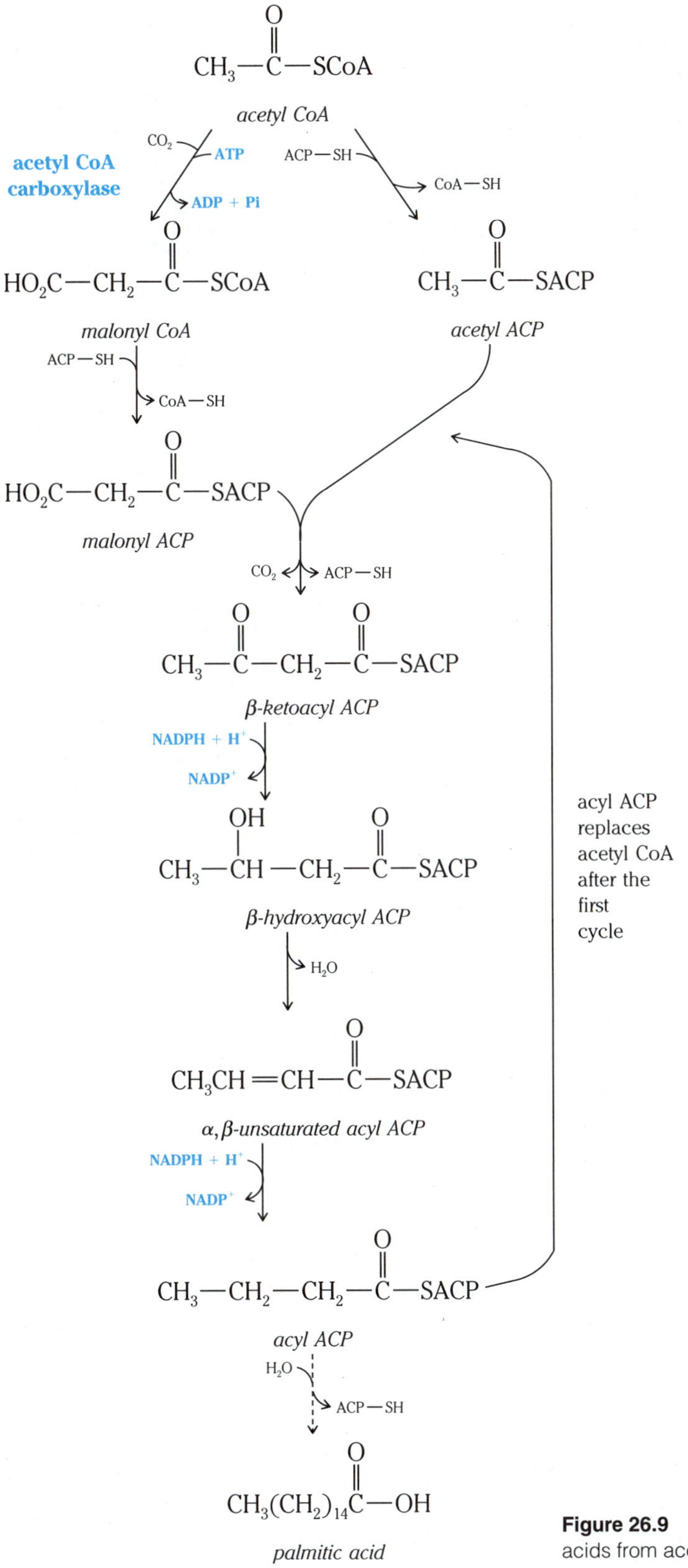

Figure 26.9 Synthesis of fatty acids from acetyl CoA.

eliminating one carbon as CO_2, to produce a four-carbon β-ketoacyl ACP. The next three reactions reduce the β carbon from a keto ($\supset$C=O) group to —CH_2— and produce butyryl ACP, completing the first cycle.

In the next round, butyryl ACP reacts with another malonyl ACP to produce a six-carbon β-ketoacyl ACP. After being reduced, the six-carbon ACP is recycled to react with yet another malonyl CoA, making an eight-carbon ACP. The cycle continues until a saturated, 16-carbon ACP is made. The final step is to hydrolyze the acyl group from ACP to produce palmitic acid.

Notice that two steps in each cycle require NADPH. Recall that the pentose phosphate pathway (Section 25.5) makes NADPH, largely for lipid synthesis. Because its reactions occur in cytoplasm, the pentose phosphate pathway produces NADPH in just the right part of the cell for fatty acid synthesis.

Your cells synthesize other fatty acids besides palmitic acid. Enzyme systems in your mitochondria and endoplasmic reticulum lengthen fatty acid chains (in two-carbon increments) and introduce double bonds in limited regions of the hydrocarbon chain. But your enzymes cannot synthesize enough of certain polyunsaturated fatty acids; they are the essential fatty acids (linoleic acid and linolenic acid) required in your diet (see Section 20.1).

Converting Glucose into Fat. Now you can see how your body makes excess carbohydrate into fat. You metabolize excess glucose to pyruvate by glycolysis, convert pyruvate into acetyl CoA, and make the acetyl CoA into fatty acids. Then you incorporate fatty acids into triglycerides and store the fat in adipose tissue until needed for energy.

There's just one problem: acetyl CoA is made from pyruvate in mitochondria, but fatty acid synthesis occurs in cytoplasm. Acetyl CoA cannot freely pass in and out of mitochondria, so it has to get out by a shuttle mechanism (Figure 26.10).

Figure 26.10 Citrate carries acetyl groups from the mitochondrion into cytoplasm.

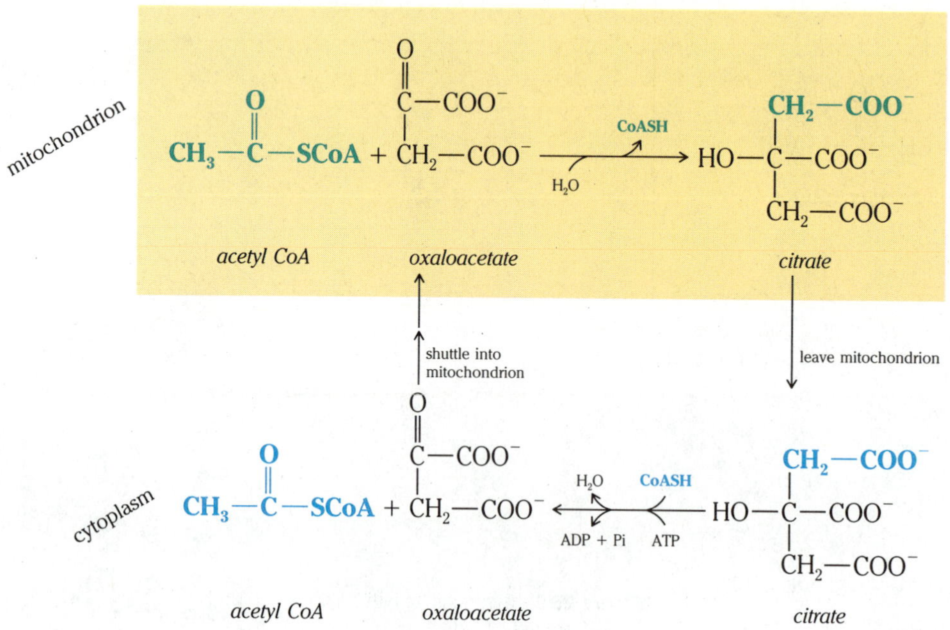

In the first reaction of the citric acid cycle (Figure 24.8), acetyl CoA reacts with oxaloacetate to form citrate. Citrate leaves the mitochondrion and is converted back into oxaloacetate and acetyl CoA, furnishing cytoplasmic acetyl CoA for fatty acid synthesis. Oxaloacetate in cytoplasm can be metabolized to PEP for gluconeogenesis (see Section 25.3), or transported back into mitochondria by another shuttle mechanism.

Citrate in cytoplasm also activates the key regulatory enzyme in fatty acid synthesis, acetyl CoA carboxylase (Figure 26.9). In addition, reactions of the citric acid cycle convert some citrate in mitochondria into isocitrate. Isocitrate (like citrate) passes out of the mitochondrion and activates acetyl CoA carboxylase.

The citrate shuttle arrangement not only furnishes acetyl CoA for fatty acid synthesis, but also activates the key enzyme for that synthesis.

26.4 CHOLESTEROL SYNTHESIS

Cholesterol Metabolism. The most abundant steroid in your body is cholesterol (Figure 20.18). Your body uses cholesterol to make steroids such as cortisol, bile salts, and male and female sex hormones. Most cholesterol is synthesized in the liver, where about 75% is further metabolized to bile salts.

Cholesterol has 27 carbon atoms, and each one comes from acetyl CoA. Figure 26.11 shows the metabolism of acetyl CoA to produce cholesterol. Notice that NADPH is used in two reactions. Much of this NADPH comes from the pentose phosphate pathway. Cholesterol synthesis is regulated by cholesterol, which exerts feedback inhibition on the enzyme HMG CoA reductase (Figure 26.11).

Isopentenyl PP (Figure 26.11) is a form of isoprene. Recall that isoprene units combine to form rubber (see Section 18.2), terpenes (see Section 20.6), and other natural products.

Lowering Blood Cholesterol Levels. High levels of cholesterol are associated with atherosclerosis, a condition in which cholesterol-rich lipid material builds up inside blood vessels, especially arteries (Figure 26.3). These fatty plaques partially block the flow of blood, increasing blood pressure and making the heart work harder. Sudden blockages cause heart attacks, in which heart tissue dies because it doesn't receive oxygen from blood. The form of cholesterol also is important. LDL cholesterol is more likely to produce atherosclerosis than cholesterol carried by high-density lipoproteins (HDL) (see Section 20.5).

Cholesterol levels less than 200 mg/dL of blood are associated with a lower risk of heart attacks. Low-fat diets can help, although feedback inhibition by cholesterol limits how much you can change your blood cholesterol levels by diet alone. If you eat large amounts of cholesterol, your body will synthesize less of it. On a low-cholesterol diet, however, you will synthesize more due to reduced feedback inhibition. Nevertheless, you can affect your blood cholesterol levels significantly through changes in eating habits and regular exercise.

Doctors sometimes prescribe drugs for patients with dangerously high cholesterol levels. One drug is niacin (see Figure 22.8), a B vitamin that in large doses reduces blood cholesterol levels. Another is cholestipol, which binds to bile salts in the intestine and removes them from the body; to replenish bile salts, cholesterol is then removed from the blood and metabolized in the liver.

Another drug, called mevinolin (trade name Lovastatin), blocks cholesterol synthesis by inhibiting the key enzyme HMG CoA reductase (Figure 26.11). Produced by a fungus, mevinolin is used to treat people who are genetically inclined to have high blood cholesterol levels.

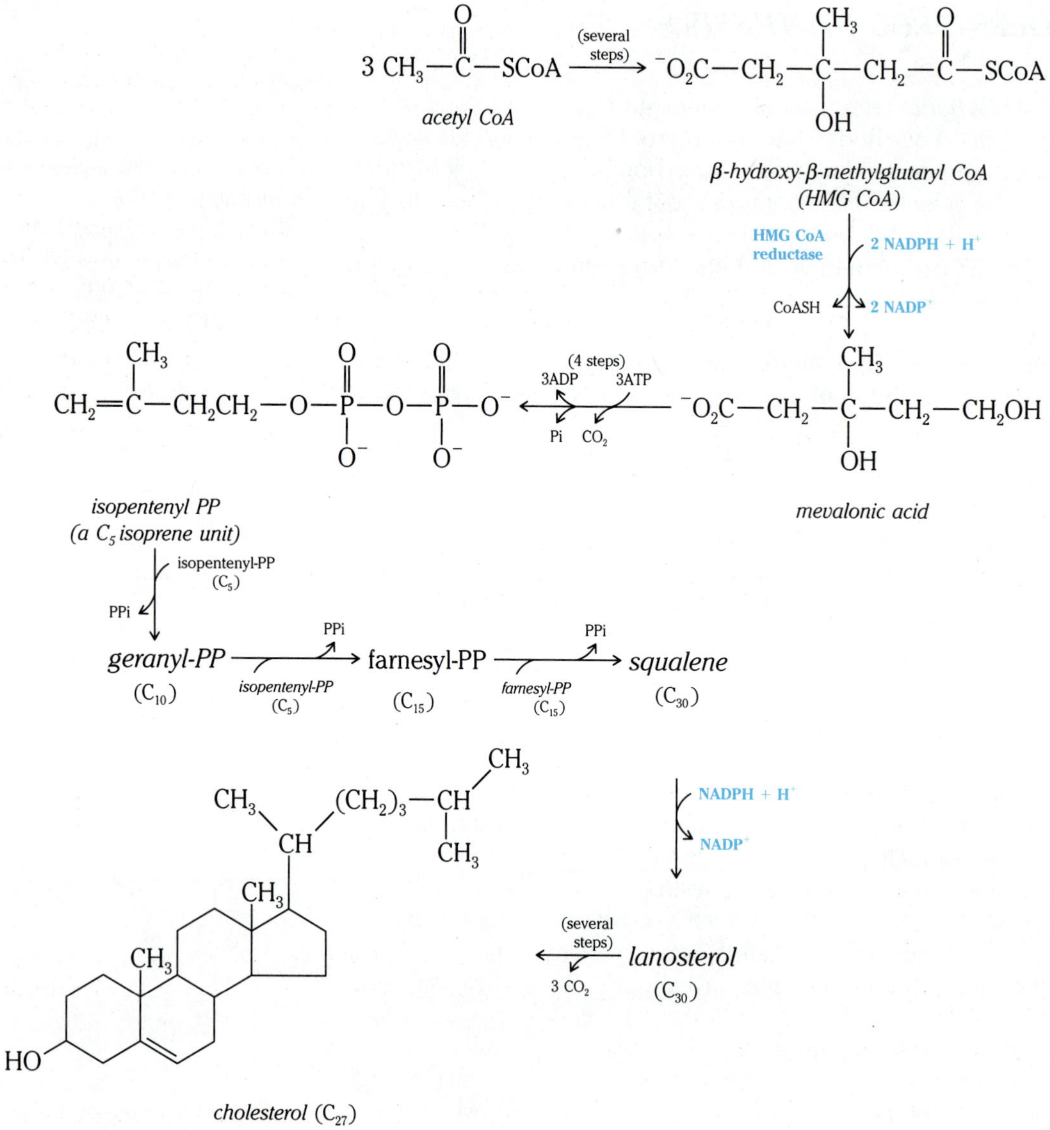

Figure 26.11 Synthesis of cholesterol from acetyl CoA.

> Our body is a machine for living.
>
> *War and Peace* by Leo Tolstoy (1828–1910)

SUMMARY

During digestion, triglycerides are hydrolyzed by lipases and then re-esterified. They leave the intestine as chylomicrons and eventually are stored in adipose tissue. When needed for energy, triglycerides are hydrolyzed by adipose tissue lipase. Glycerol is oxidized by glycolysis or made into glucose. Fatty acids are oxidized in mitochondria to acetyl CoA, which generates large amounts of ATP. Excess acetyl CoA forms the ketone bodies acetoacetic acid, β-hydroxybutyric acid, and acetone. Ketosis and ketoacidosis result from elevated levels of ketone bodies in blood and urine.

Excess glucose is stored as fat. Fatty acids, especially palmitic acid, are synthesized from acetyl CoA in cytoplasm. The reactions require ATP and NADPH. Acetyl groups are shuttled from mitochondria as citrate; citrate and isocitrate from mitochondria activate acetyl CoA carboxylase, a key enzyme. Other systems lengthen fatty acid chains and introduce double bonds in limited regions of the hydrocarbon chain.

Cholesterol, the most abundant steroid, is made from acetyl CoA. Feedback inhibition by cholesterol regulates the enzyme HMGCoA reductase. Blood cholesterol levels can be reduced by low-fat diets, exercise, and certain drugs.

KEY TERMS

Chylomicron (26.1)
Fatty acid oxidation (26.2)
Fatty acid synthesis (26.3)
Ketoacidosis (26.2)
Ketone body (26.2)
Ketosis (26.2)
β-oxidation (26.2)

EXERCISES

Even-numbered exercises are answered at the back of this book.

Lipids and Their Digestion

1. Define "lipid." [Review Chapter 20 if necessary.]

2. Why are triglycerides, fatty acids, and cholesterol classified as lipids? Is glycerol a lipid?

3. Write an equation to represent the complete hydrolysis of a triglyceride by pancreatic lipase.

4. Write an equation to represent the complete re-esterification of glycerol and three fatty acid molecules in an intestinal cell.

5. What is the function of bile salts?

6. Caffeine activates the lipase activity in adipose tissue. Use this information to rationalize why a cyclist might

drink caffeinated beverages before and during a long-distance race.

7. Arrange the following digestive stages of a triglyceride in chronological order: **(a)** chylomicron, **(b)** hydrolyzed triglyceride, **(c)** storage in adipose tissue, **(d)** re-esterification.

Triglyceride Catabolism

8. Can glycerol be metabolized to form acetyl CoA? Explain.

9. Does the number of ATP needed to convert fatty acids into their CoA derivatives and transport them into mitochondria vary with the fatty acid involved?

10. Calculate the number of ATP produced by the complete oxidation of stearic acid (a saturated, 18-carbon fatty acid) to CO_2 and H_2O.

11. Each double bond in a fatty acid reduces by 2 the number of ATP generated during β-oxidation. How many ATP, then, are produced by the complete oxidation of linolenic acid (Table 20.1) to CO_2 and H_2O?

12. When an odd-numbered fatty acid is oxidized, acetyl CoA and one molecule of propionyl CoA is produced. Using your knowledge of carboxylic acid nomenclature (Section 16.1), write a structural formula for propionyl CoA.

13. In what part of the cell does β-oxidation occur? How does this affect ATP production?

14. During starvation, fatty acids are oxidized to acetyl CoA, but acetyl CoA cannot be used for gluconeogenesis. Rationalize why. [Review Section 25.3 if necessary.]

Ketosis

15. Identify the kind of diet most likely to cause ketosis: **(a)** low-fat, **(b)** low-carbohydrate, **(c)** low-protein. Why is this true?

16. Rationalize why people going through the glycogen depletion phase of glycogen loading may show signs of ketoacidosis. [Review the Chapter 25 Opener if necessary.]

17. Write the names and structural formulas of the three ketone bodies. Are they all ketones?

18. Write structural formulas for **(a)** β-hydroxybutyric acid and **(b)** acetoacetic acid as they would occur in the body at neutral pH.

19. Write the IUPAC name for β-hydroxybutyric acid. [Review Section 16.1 if necessary and name —OH as a hydroxy substituent group.]

20. Why is ketonuria a common symptom of diabetes?

21. What are the three signs of ketosis?

Fatty Acid Synthesis

22. In what part of the cell are fatty acids synthesized?

23. Does your body convert more glucose into fatty acids, or more fatty acids into glucose?

24. Identify the conditions under which you would most likely have an increased rate of fatty acid synthesis: **(a)** diabetes, **(b)** during fasting, **(c)** after eating. Why is this so?

25. The pentose phosphate pathway is especially active in tissues (such as liver, adrenal glands, and gonads) that synthesize fatty acids and steroids. Rationalize why.

26. Write a structural formula for the α,β-unsaturated acyl ACP (see Figure 26.9) that forms during fatty acid synthesis after four molecules of acetyl CoA have been used.

27. Write a structural formula for a β,γ-unsaturated acyl ACP that has the same number of carbon atoms as the compound in exercise 26.

28. How many ATP are needed directly to synthesize palmitic acid from acetyl CoA?

29. Is arachidonic acid (Table 20.1) classified as an n-3, n-6, or n-9 fatty acid? [See the Chapter Opener if necessary.]

30. List the key regulatory enzyme in fatty acid synthesis and the substances that activate that enzyme.

31. Under conditions of fatty acid synthesis, what keeps isocitrate in mitochondria from being metabolized further in the citric acid cycle? [Review Section 24.3 if necessary.]

32. What coenzyme enables the enzyme acetyl CoA carboxylase to function as a catalyst? [Review Table 22.1 if necessary.]

Cholesterol Synthesis

33. Do you know your blood cholesterol, HDL, and LDL levels? Are they healthy?

34. List the key regulatory enzyme in cholesterol synthesis and the product that inhibits that enzyme.

35. Some researchers contend that extremely low blood cholesterol levels are unhealthy. Why could a lack of cholesterol cause problems?

36. Examine Figure 26.11. Rationalize why people with diabetes may have an increased rate of cholesterol synthesis.

DISCUSSION EXERCISES

1. If glycerol in a triglyceride has its carbon 2 labeled as the isotope ^{14}C, which carbon atom(s) in glucose will be ^{14}C when glycerol is made into glucose?

2. According to exercise 11, each double bond in a fatty acid reduces by 2 the number of ATP generated by β-oxidation. Use Figure 26.6 to rationalize why.

27

Metabolism of Nitrogen Compounds

1. How are proteins digested?
2. What are the main metabolic uses of amino acids?
3. How are nucleotides and heme compounds metabolized?
4. How are nitrogen compounds removed from the body?
5. What are some genetic disorders involving the metabolism of nitrogen compounds?

Parkinson's Disease

Parkinson's disease is a neural disorder. More than 1 million people in the United States, most of them 60 years or older, are afflicted with this disease. Patients have abnormally low concentrations of the neurotransmitter dopamine in certain regions of the brain, which causes them to lose some control over their muscles. They typically have uncontrolled tremors, especially in the limbs, stiffness, and difficulty in walking. Injections of dopamine don't help because this chemical cannot enter the brain from the blood.

Many doctors prescribe a drug called L-dopa (L-*d*ihydr*o*xy*p*henyl*a*lanine), which can enter the brain and is converted there into dopamine:

$$HO-C_6H_3(OH)-CH_2-CH(NH_3^+)-COO^-$$
L-dopa
$$\downarrow$$
$$HO-C_6H_3(OH)-CH_2-CH_2-NH_3^+ + CO_2$$
dopamine

L-Dopa is taken orally, often in combination with another substance that blocks conversion into dopamine until the drug reaches the brain.

This treatment has produced dramatic improvement in many patients. After several years of use, however, most progressively fail to respond to L-dopa and suffer from hallucinations and depression. Physicians now are experimenting with surgical implantation of dopamine-producing cells (from adrenal glands or fetal brain tissue) into the brains of patients with Parkinson's disease.

You can easily write balanced chemical equations for the oxidation of carbohydrates and fats. In your body, for example:

$$C_6H_{12}O_6 \text{ (glucose)} + 6O_2 \longrightarrow 6CO_2 + 6H_2O$$

$$C_{55}H_{100}O_6 \text{ (a triglyceride)} + 77O_2 \longrightarrow 55CO_2 + 50H_2O$$

Because proteins contain nitrogen, you might have more trouble writing an equation for the oxidation of a protein (Figure 27.1).

Figure 27.1 These are rich sources of protein in the diet.

In this chapter you will learn what happens to nitrogen when you oxidize proteins for energy. You digest dietary proteins to obtain their amino acids. Then you metabolize amino acids to synthesize glucose, ketone bodies, your own proteins, and other nitrogen compounds such as nucleotides and heme.

27.1 PROTEIN DIGESTION

The proteins you eat have different amino acid sequences (primary structures) than proteins in your body. In order to convert dietary proteins into the proteins your body needs, or to metabolize them to other compounds, you must first hydrolyze dietary proteins to their amino acids (see Figure 24.5).

Several enzymes in your digestive tract hydrolyze proteins (Table 27.1). Each enzyme acts on specific sites in the peptide chain. Notice that many of these enzymes are synthesized as proenzymes (see Section 22.5) and then are activated in the digestive tract.

Table 27.1 Protein-Digesting Enzymes

Enzyme	Organ of Origin	Site of Action	Region in Protein Where Peptide Bonds Hydrolyzed
* Pepsin	Stomach	Stomach	Many different sites
* Trypsin	Pancreas	Intestine	Arg, lys residues
* Chymotrypsin	Pancreas	Intestine	Phe, tyr, trp
* Carboxypeptidase	Pancreas	Intestine	C-terminal residues
Aminopeptidase	Intestine	Intestine	N-terminal residues
Dipeptidases	Intestine	Intestine	Various dipeptides

* Synthesized as a proenzyme

Amino acids from hydrolyzed proteins are water-soluble enough to leave the intestine and enter the blood, where they are taken up and metabolized by various tissues. A few small proteins and peptides, however, enter blood without being hydrolyzed. For a few days after birth, for example, babies absorb milk proteins into their blood without breaking them down. Some of these proteins are antibodies that give newborns temporary immunity against diseases. Later in life, you also absorb trace amounts of some dietary proteins without hydrolyzing them. The amount is trivial in terms of nutrition, but those proteins occasionally cause allergic reactions.

27.2 AMINO ACID METABOLISM

Major Uses. Amino acids are the most abundant source of nitrogen in your body. To keep you in nitrogen balance (see Section 21.2)—and to maintain an adequate pool of amino acids for metabolism—your intake of nitrogen needs to match your excretion of nitrogen.

Your body metabolizes amino acids in five ways: (1) protein synthesis (see Section 23.4); (2) conversion into other amino acids; (3) conversion into other nitrogen compounds; (4) oxidation to produce energy, and (5) conversion into glucose.

Conversion into Other Amino Acids. Recall that your proteins contain 20 different amino acids (see Section 21.1). All 20 must be present if protein synthesis is to occur. In addition, each of the 20 has its own metabolic pathways and products that your body needs.

How does your body get all 20 amino acids? Your main source is dietary proteins. In addition, some amino acids are converted into others as necessary. Your body cannot synthesize eight amino acids; these essential amino acids must be supplied in the diet (see Section 21.2).

Conversion into Other Nitrogen Compounds. You don't need nucleic acids, nucleotides, or heme compounds (such as hemoglobin and cytochromes) in your diet because your body synthesizes them. The nitrogen atoms in these compounds come from amino acids (Figure 27.2). The metabolic pathways are complex, and we won't examine them here.

Your body also uses amino acids to produce important nitrogen compounds that have hormonal or neural activity. Figure 27.3 shows several examples.

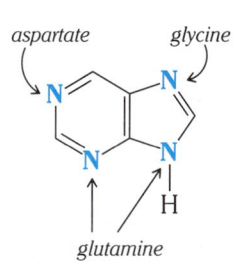

Figure 27.2 Origin of each N atom in a purine ring.

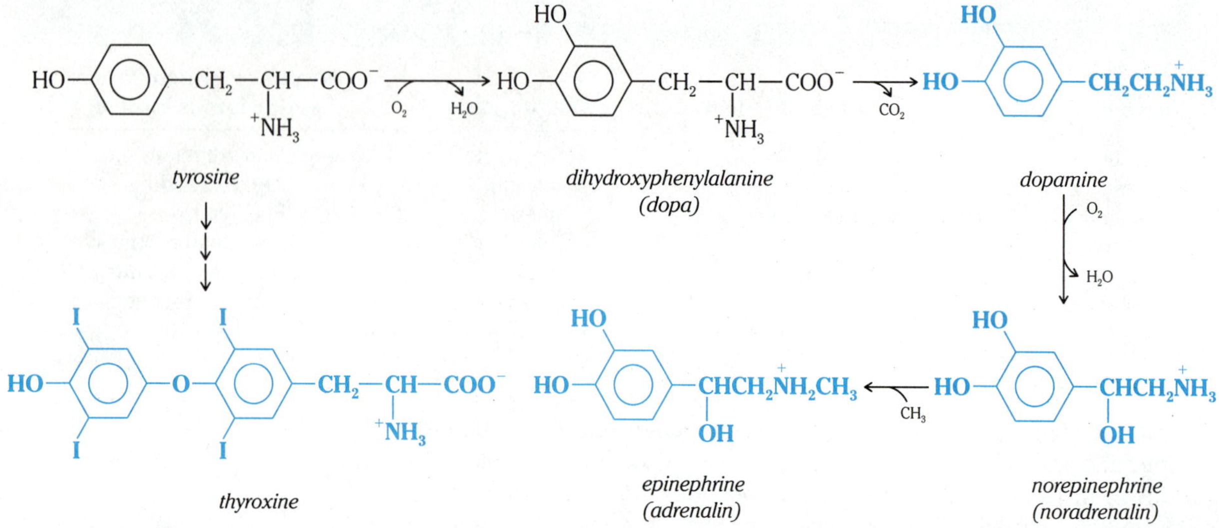

Figure 27.3 Tyrosine is metabolized to produce substances with hormonal or neural activity (in color).

The coenzyme in transamination reactions is pyridoxal phosphate, a derivative of vitamin B_6 (see Table 22.1).

Amino Acid Oxidation. Amino acids, like carbohydrates and fats, are oxidized in the citric acid cycle to produce energy (see Figure 24.5). But before it can enter the citric acid cycle, an amino acid must lose its nitrogen.

A common way to accomplish this is to transfer the amine group to an α-keto acid by a type of reaction called **transamination.** The reaction converts the amino acid into an α-keto acid, and changes the α-keto acid into an amino acid:

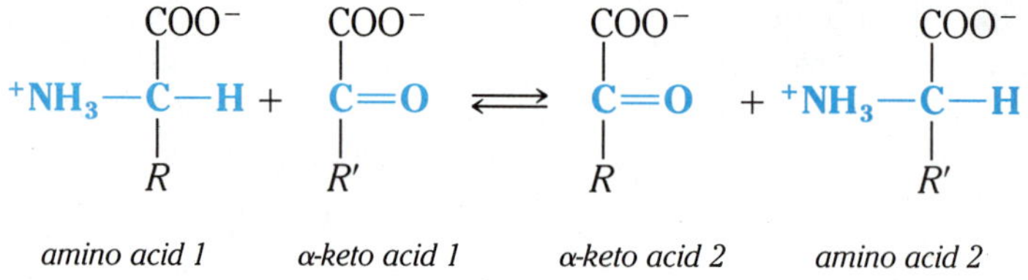

amino acid 1 + α-keto acid 1 ⇌ α-keto acid 2 + amino acid 2

At first glance it looks like nothing is accomplished. The reactants and products both consist of an amino acid and an α-keto acid, though different ones. The original amino acid, however, becomes an α-keto acid that can be metabolized to pyruvate or a compound directly in the citric acid cycle. For example, alanine is metabolized to form pyruvate:

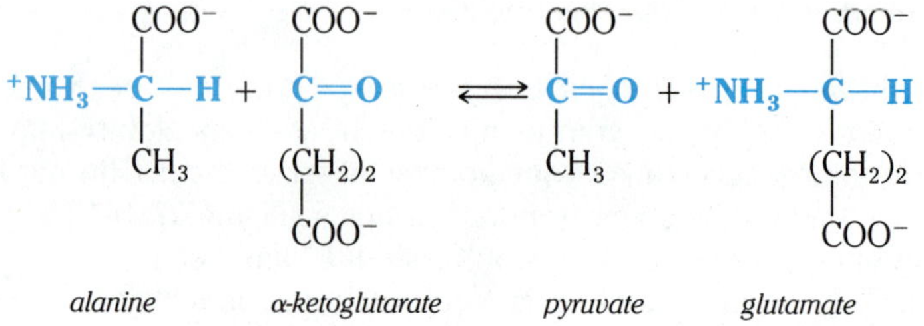

alanine + α-ketoglutarate ⇌ pyruvate + glutamate

Figure 27.4 shows how various amino acids are converted in this way.

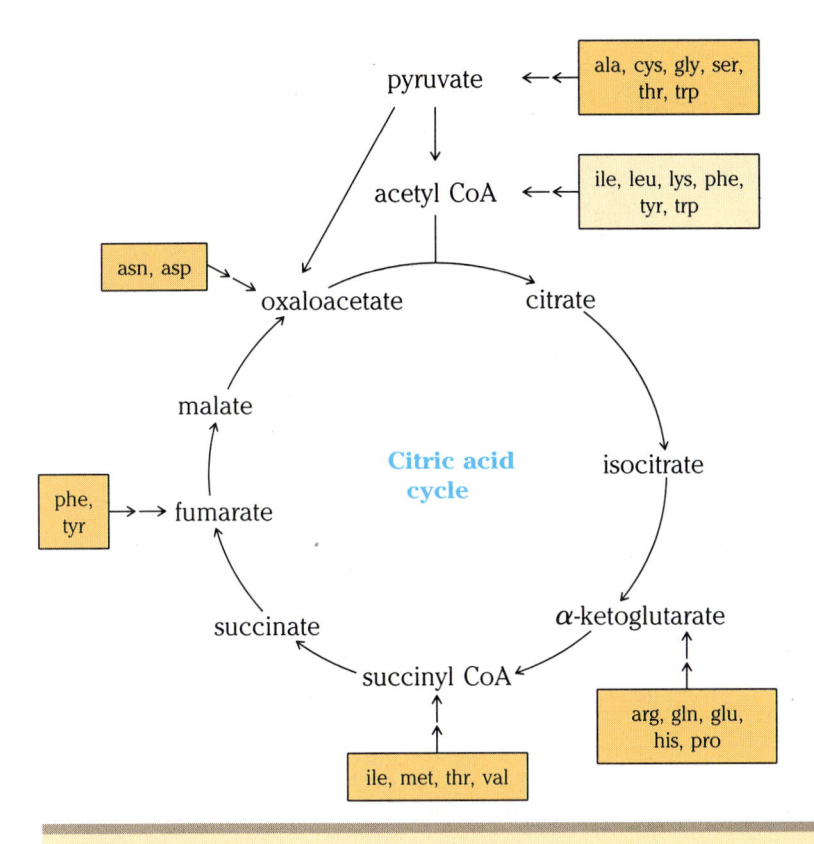

Metabolism of Nitrogen Compounds   Chapter 27   575

Figure 27.4 All amino acids can be metabolized to pyruvate, acetyl CoA, or a compound in the citric acid cycle.

EXAMPLE 27-1 From the structure of aspartate (aspartic acid) in Table 21.1, draw the structural formula and identify the product when aspartate is transaminated.

SOLUTION

$$\begin{array}{c} COO^- \\ | \\ {}^+NH_3-C-H \\ | \\ CH_2-COO^- \end{array} \longrightarrow \begin{array}{c} O \\ \| \\ C-COO^- \\ | \\ CH_2-COO^- \end{array}$$

aspartate   *oxaloacetate*

In transamination, the α-keto acid that receives the amine group typically is α-ketoglutarate, which is made in the citric acid cycle. After receiving the amino group, α-ketoglutarate becomes glutamate, an amino acid. As you will see later (Section 27.3), glutamate then unloads its amino group to make urea, which you excrete in urine. This recycles glutamate back to α-ketoglutarate, which can be used for another round of transamination (Figure 27.5).

Once amino acids are transaminated and enter the citric acid cycle, they are oxidized to CO_2 and H_2O, producing large amounts of ATP. But they are used less readily for energy than carbohydrates and lipids are, because amino acids are needed more for synthesizing proteins and other nitrogen compounds. This is why starving people deplete their glycogen before they use much of their fat and muscle (protein) reserves for energy.

Conversion into Glucose. Amino acids are classified as **glucogenic** (can be made into glucose by gluconeogenesis) or **ketogenic** (can be made into ke-

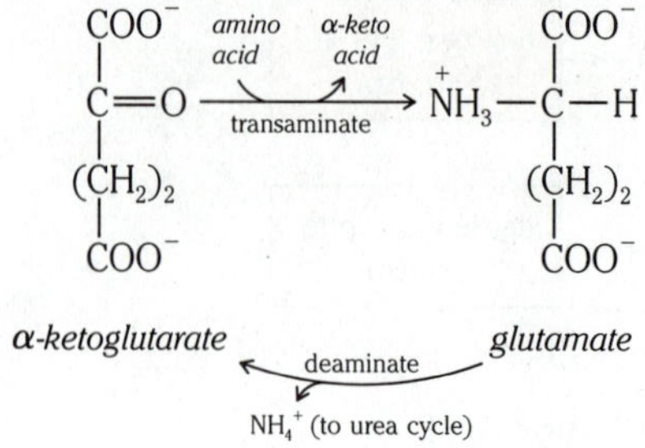

Figure 27.5 α-Ketoglutarate is used to transaminate amino acids. Glutamate then is deaminated to regenerate α-ketoglutarate (see Section 27.3).

tone bodies). Recall that any compound that increases the amount of oxaloacetate can be used to synthesize glucose (Section 25.3). This includes compounds metabolized to pyruvate or any citric acid cycle intermediate, but excludes compounds made only into acetyl CoA. These can be used to generate ATP and make ketone bodies (Section 26.2), but not glucose.

Notice in Figure 27.4 that all amino acids except leucine and lysine can be metabolized to pyruvate or a citric acid cycle intermediate. Thus, all amino acids except leucine and lysine are glucogenic. Isoleucine, phenylalanine, tryptophan, and tyrosine can be made into acetyl CoA as well as glucogenic products; these amino acids, then, are both glucogenic and ketogenic.

Since fatty acids are metabolized to acetyl CoA (which cannot be made into glucose), amino acids are a major source of glucose under conditions such as untreated diabetes, a low-carbohydrate diet, or starvation. In these situations, the liver has an increased rate of gluconeogenesis and sends glucose into the blood.

Genetic Disorders. Many genetic disorders of amino acid metabolism are known. The usual problem is a mutated gene that fails to code for a necessary enzyme. If the enzyme is absent, a needed reaction cannot occur. This can cause complications in two ways: a needed product may not be made, and a reactant may accumulate to toxic levels.

Let's consider the metabolism of just one amino acid, phenylalanine. Figure 27.6 shows some of its normal reactions. An average of about 1 baby born per 13,000 lacks the enzyme that converts phenylalanine into tyrosine. This disorder is called **phenylketonuria (PKU).**

Figure 27.6 Genetic disorders of phenylalanine metabolism.

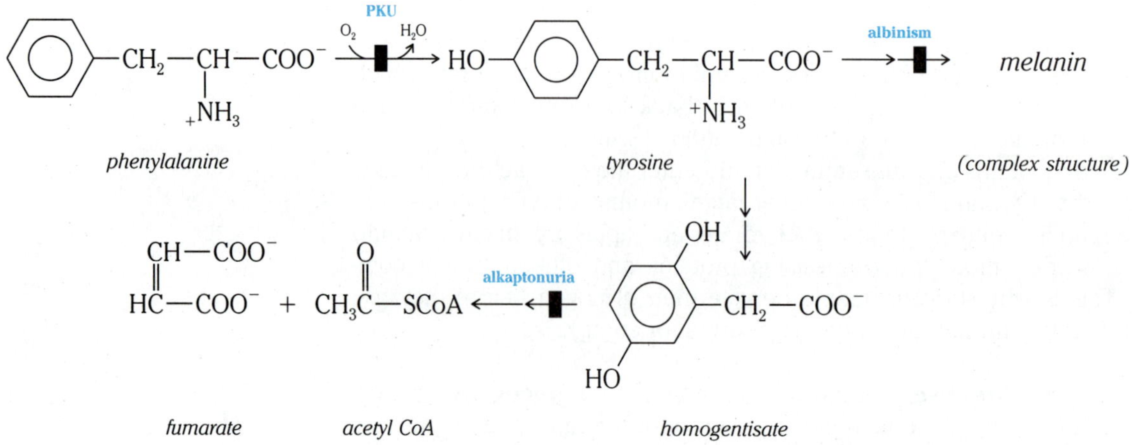

Babies born with PKU cannot metabolize phenylalanine in their diet, so phenylalanine accumulates to high levels. Excess phenylalanine is transaminated to its α-keto acid, phenylpyruvate (Figure 27.7). This phenylketone and several of its metabolic byproducts appear in blood and urine (hence the name *phenylketonuria*). These toxic compounds retard physical and mental development.

Babies are screened at birth for PKU. Their urine is tested for phenylketones and their blood for excess phenylalanine. Newborns with PKU are placed on a low-phenylalanine diet. They stay on the diet for about 10 years, until brain development is essentially complete; then they can change to less restrictive diets.

Some people cannot metabolize tyrosine to produce melanin, a normal pigment (Figure 27.6). People with this disorder, called **albinism** (see Figure 23.2), have white hair, light skin, and red or pink eyes (from underlying blood vessels).

Another genetic disorder occurs in people who lack the enzyme to metabolize homogentisate, a normal breakdown product of phenylalanine (Figure 27.6). These people excrete homogentisate in urine. As homogentisate is oxidized by O_2 in air, the urine turns black. This disorder, called *alkaptonuria*, is relatively harmless.

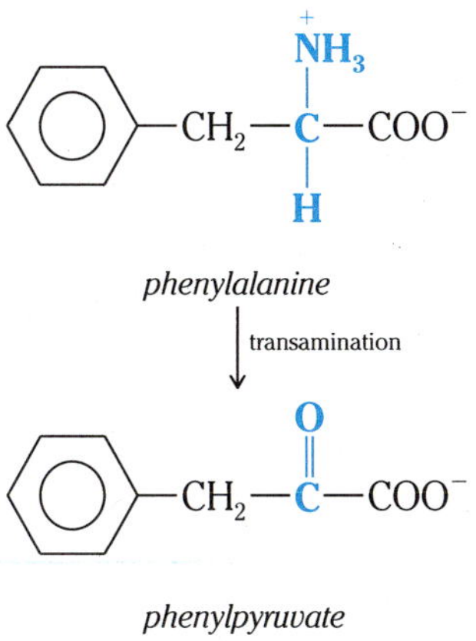

Figure 27.7 Metabolism of phenylalanine to a phenylketone.

27.3 UREA CYCLE

When you metabolize amino acids to produce energy, your body has to dispose of nitrogen. The nitrogen that remains from oxidizing amino acids to CO_2 and H_2O is like the ashes left over from burning wood.

You excrete small amounts of nitrogen as ammonium ion (NH_4^+) in your urine. But NH_4^+ is toxic to the central nervous system, so your body converts much of this nitrogen into urea by a sequence of reactions called the **urea cycle** (Figure 27.8).

The urea cycle takes place in the liver. Recall that many amino acids transfer their amino groups to α-ketoglutarate by transamination (Section 27.2). In the first reaction of the urea cycle, the glutamate that forms then unloads its amino group as ammonium ion:

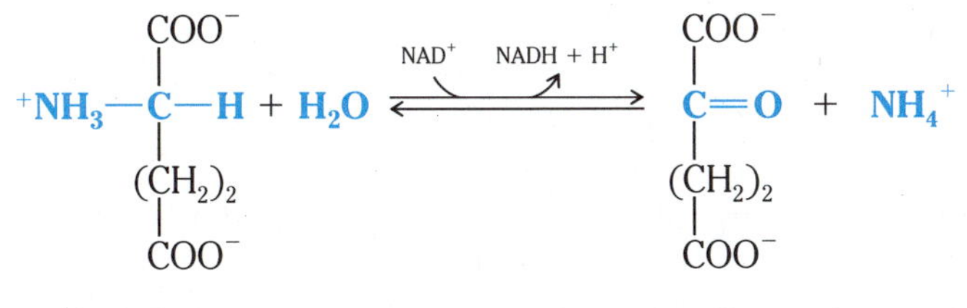

This oxidative *deamination* reaction recycles α-ketoglutarate for further transamination reactions (Figure 27.5).

Later in the cycle ammonium ion produced by deamination of glutamate is metabolized to produce urea, which is excreted in urine. Synthesizing one urea molecule costs four ATP (Figure 27.8), if we count the ATP converted to AMP as the equivalent of two ATP. This is the price of preventing the buildup of toxic NH_4^+.

People who genetically lack any enzyme in the urea cycle have increased amounts of ammonium ion in their blood and urine. This condition, called

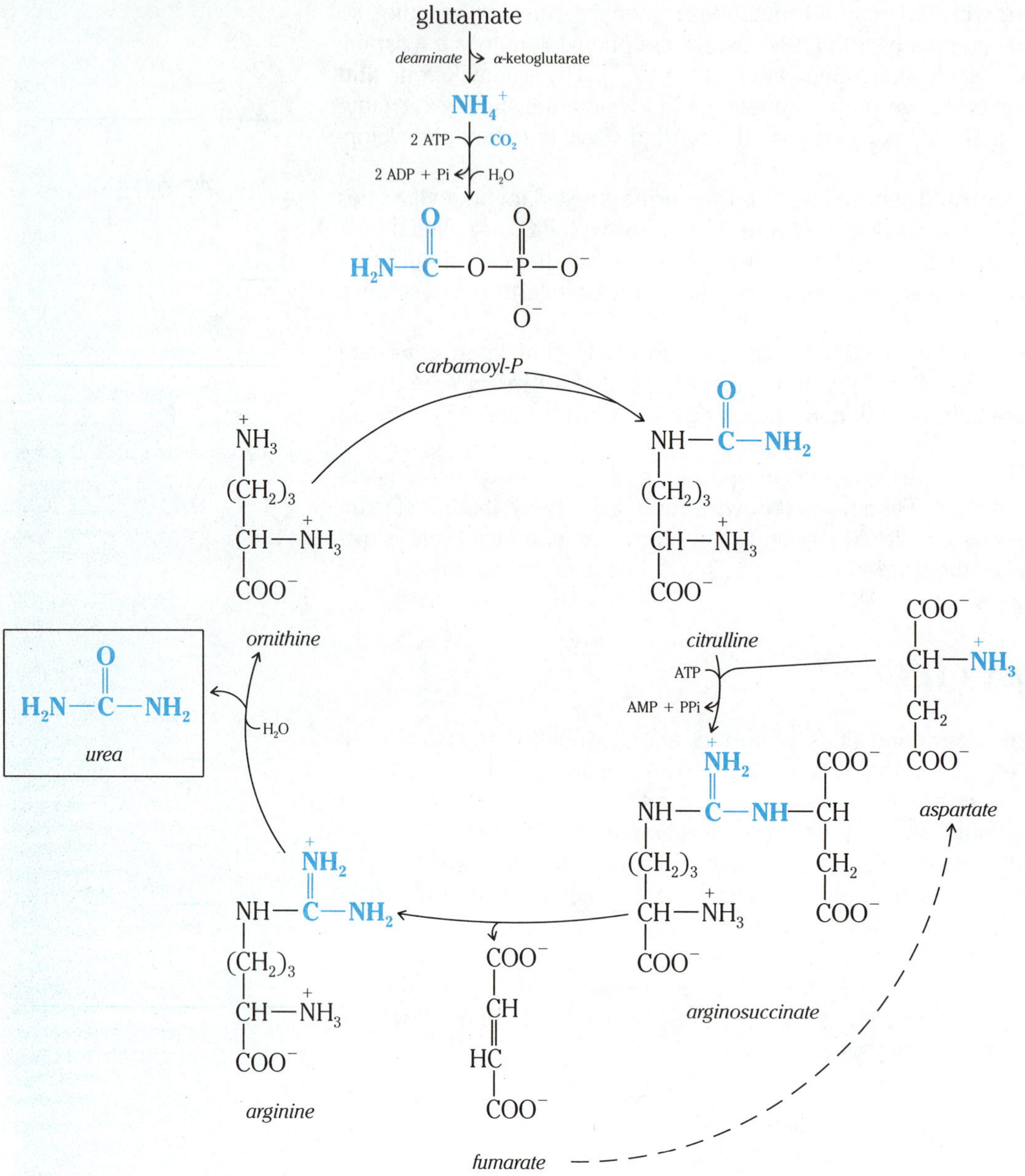

Figure 27.8 The urea cycle. Ammonium ion (NH_4^+) can come from other sources besides glutamate. Fumarate is converted back into aspartate by reactions of the citric acid cycle plus transamination.

hyperammonemia, produces lethargy, vomiting, mental retardation, and in the most severe cases, coma and death.

Other animals differ in the forms they use to excrete NH_4^+. Most terrestrial vertebrates, including humans, convert NH_4^+ into urea. But many aquatic mammals excrete NH_4^+ directly. Birds and reptiles convert NH_4^+ into uric acid (see Figure 27.9), which they excrete.

27.4 NUCLEOTIDE AND HEME METABOLISM

Nucleotides. Your body synthesizes purine and pyrimidine rings through a complex series of reactions. Nitrogen for those rings comes from amino acids (Figure 27.2). Additional reactions attach ribose (or deoxyribose) and phosphate groups to produce the nucleotides you use for energy or for synthesizing DNA and RNA.

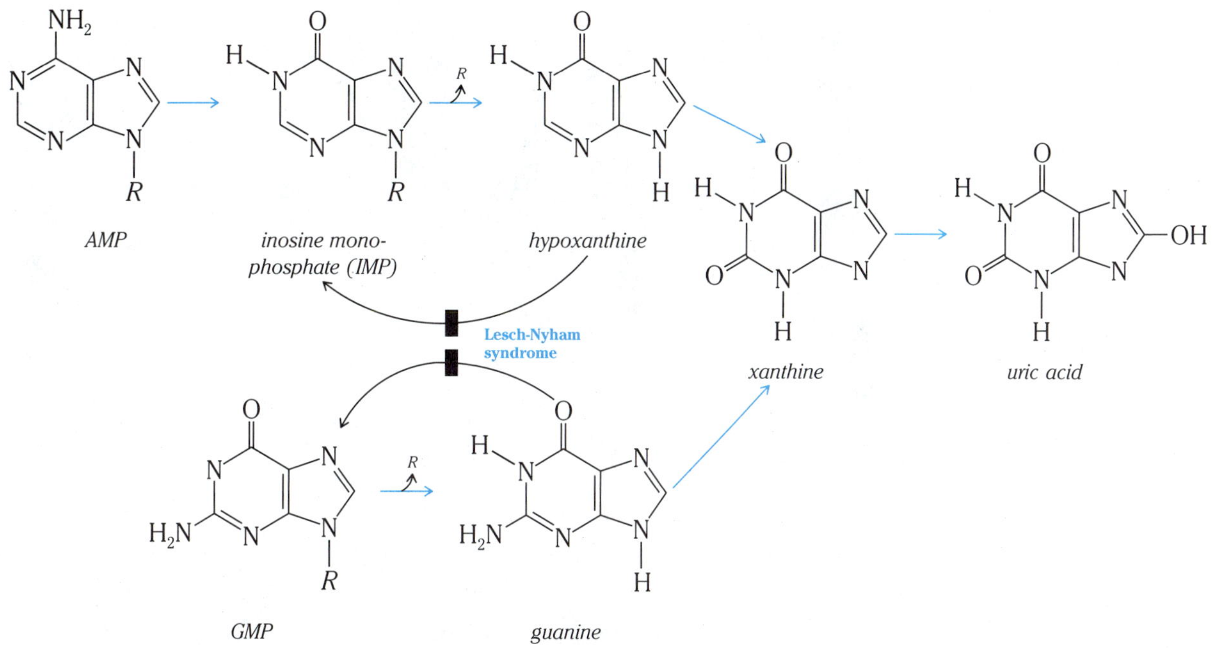

Pyrimidines are catabolized to products that are excreted or metabolized to other compounds:

thymine $\longrightarrow \longrightarrow CO_2 + NH_4^+ +$ succinyl CoA

cytosine $\longrightarrow$ uracil $\longrightarrow \longrightarrow CO_2 + NH_4^+ + \beta$-alanine

You exhale CO_2 and metabolize succinyl CoA in the citric acid cycle. The urea cycle metabolizes ammonium ion (NH_4^+) to urea, which is excreted. β-Alanine is incorporated into CoA.

Purine catabolism is a bit more complicated (Figure 27.9). You excrete the end product, uric acid. But you have an enzyme that recycles much of the hypoxanthine and guanine back to adenine and guanine nucleotides instead of converting it all into uric acid. People with *Lesch-Nyhan syndrome*, however, genetically lack this recycling enzyme and thus produce excess uric acid. Young children with this disorder bite their fingers and lips, are mentally deficient, and exhibit aggressive behavior.

Gout is a milder disorder due to increased production of uric acid. Uric acid in the blood is neutralized to form the salt, sodium urate. The slightly cooler temperatures in limb joints decrease the solubility of this salt just enough for it to solidify into sodium urate crystals. These solid deposits cause great discomfort (Figure 27.10). Allopurinol, which structurally resembles hypoxanthine (Figure 27.11), relieves symptoms of gout by inhibiting the enzyme that makes uric acid from hypoxanthine and xanthine.

Heme Compounds. Heme is an important component of hemoglobin and of cytochromes necessary for electron transport (Section 24.4). Since your body synthesizes heme (Figure 27.12) from amino acids (which supply nitrogen) and other compounds, you don't need heme compounds in your diet.

Figure 27.9 Catabolism of purines (colored arrows). An enzyme that recycles hypoxanthine and guanine is genetically lacking in Lesch-Nyhan syndrome.

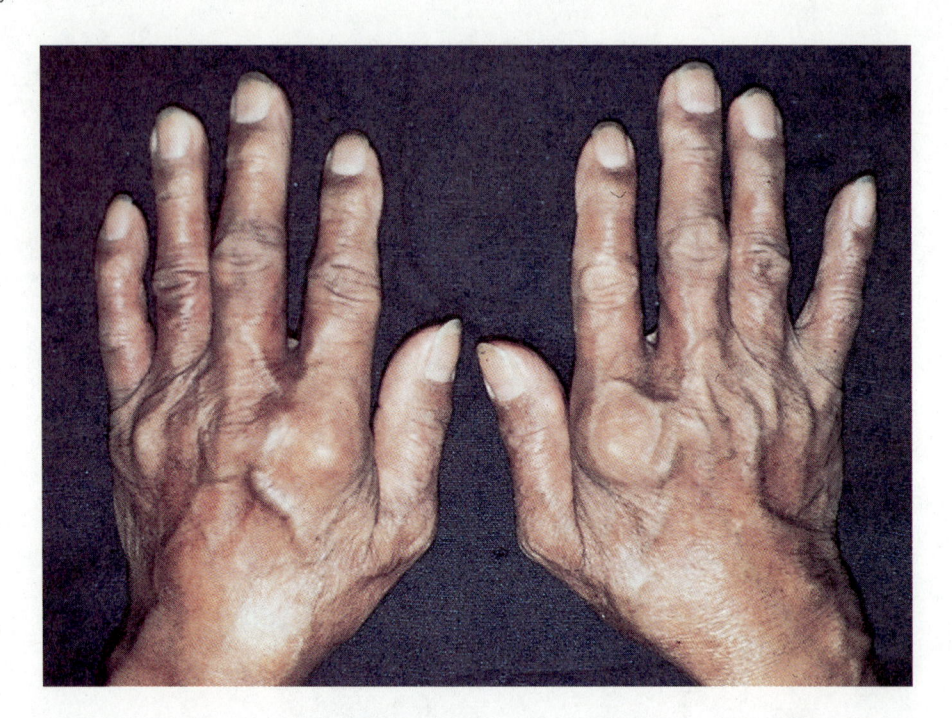

Figure 27.10 Because of gout, sodium urate crystals deposit in these hands.

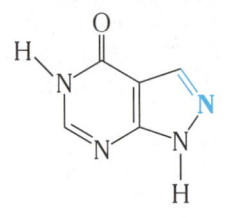

allopurinol

Figure 27.11 The structural difference between allopurinol, a drug used to treat gout, and hypoxanthine (Figure 27.9) is in color.

Heme is broken down, mostly by liver enzymes, to various colored products, such as biliverdin (green) and bilirubin (yellow-orange). These materials are metabolized to other pigmented products that give urine and stools much of their color.

When excessive, toxic amounts of bilirubin occur in blood, the skin and whites of the eyes take on a yellowish color; this condition is called **jaundice.** One cause is an excessive breakdown of red blood cells (called *hemolytic anemia*), which puts increased amounts of hemoglobin products in blood.

Jaundice also is caused by the liver failing to break down bilirubin and other heme by-products. This can occur in adults with liver disease, or in newborns with underdeveloped livers. One treatment for newborns is to place them under fluorescent light, which converts bilirubin into less toxic products that they excrete.

Figure 27.12 Heme.

The third cause of jaundice is an obstructed bile duct. This prevents the liver from sending heme by-products to the intestine for excretion. The accumulation of heme materials in the liver slows the metabolism of bilirubin in blood.

> Life is the garment we continually alter . . .
>
> *Whereas to Mr. Franklin* by David McCord
> (1897–   )

SUMMARY

Proteins are hydrolyzed during digestion to produce amino acids. Amino acids are used for protein synthesis, conversion into other nitrogen compounds (nucleotides, nucleic acids, and heme), energy production, and gluconeogenesis. Twelve of the 20 amino acids the body needs can be synthesized from other amino acids and are not required in the diet.

Amino acids are transaminated to form α-keto acids, which are oxidized in the citric acid cycle or made into glucose. Leucine and lysine are ketogenic amino acids; ile, phe, trp, and tyr are both ketogenic and glucogenic; the remaining amino acids are glucogenic. Genetic disorders of phenylalanine metabolism include phenylketonuria (PKU), albinism, and alkaptonuria.

Nitrogen from amino acid metabolism is incorporated into urea and excreted in urine. Ammonium ion (NH_4^+) is toxic to the central nervous system. A lack of any enzyme in the urea cycle results in hyperammonemia.

Pyrimidines are catabolized to CO_2, NH_4^+, and other products. Purines are catabolized to uric acid, which is excreted in urine. Elevated levels of uric acid occur in Lesch-Nyhan syndrome and in gout. Heme is broken down to pigments such as bilirubin. Elevated levels of bilirubin in blood result in jaundice.

KEY TERMS

Albinism (27.2)
Glucogenic amino acid (27.2)
Gout (27.4)
Hyperammonemia (27.3)
Jaundice (27.4)
Ketogenic amino acid (27.2)
Phenylketonuria (PKU) (27.2)
Transamination (27.2)
Urea cycle (27.3)

EXERCISES

Even-numbered exercises are answered at the back of this book.

Protein Digestion

1. Recall that if you chew your food many times, you will increase the breakdown of carbohydrates in your mouth (Section 25.1). Will this also increase the breakdown of proteins in your mouth? Why? [Review Table 27.1 if necessary.]

2. Using the information in Table 27.1, identify the region(s) in the peptide leu-trp-gly-phe-ile-arg-met that would be hydrolyzed by **(a)** trypsin, and **(b)** chymotrypsin.

3. The amino acid sequence of insulin is shown in Figure 21.17. At how many places in the two insulin chains would hydrolysis occur when insulin is exposed to trypsin (Table 27.1)?

4. What is the advantage of synthesizing digestive enzymes such as pepsin as proenzymes? [Review Section 22.5 if necessary.]

Amino Acid Metabolism

5. How many amino acids are not directly required in your diet because they can be synthesized in adequate amounts from other substances?

6. Of the two amino acids, tyrosine and phenylalanine, one is required in the diet and one is not. Use Figure 27.6 to determine which one is required.

7. Figure 27.3 shows the production of epinephrine (adrenalin) from tyrosine. What is the main metabolic effect of this hormone? [Review Section 25.4 if necessary.]

8. What nitrogen compound is present in abnormally low amounts in the brains of people with Parkinson's disease? From what amino acid is this substance synthesized?

9. Write an equation for a transamination reaction between glycine, $^+H_3N-CH_2-COO^-$, and α-ketoglutarate.

10. The amino acid alanine has a methyl group ($-CH_3$) as its R group. Write the structural formula for the α-keto acid formed from alanine by transamination. Use Figure 25.5 to identify this α-keto acid.

11. The amino acid aspartate has $-CH_2-COO^-$ as its R group. Write the structural formula for the α-keto acid formed from aspartate by transamination. Use Figure 24.8 to identify this α-keto acid.

12. Why is a compound with the structure
$$R-\overset{\overset{O}{\|}}{C}-COOH$$
called an "α-keto" acid? [Review Section 16.1 if necessary.]

13. What α-keto acid is the usual recipient of amino groups in transamination reactions? What amino acid does it become?

14. What coenzyme typically participates in transamination reactions? [Review Table 22.1 if necessary.]

15. Which amino acids are not glucogenic?

16. Use Figure 27.4 to explain why isoleucine is classified both as a ketogenic and a glucogenic amino acid.

17. How do starving people maintain their blood glucose levels?

18. Before they are metabolized to glucose, what element in amino acids is removed metabolically?

19. People with phenylketonuria (PKU) have toxic amounts of phenylpyruvate (Figure 27.7) in their blood and urine. Phenylpyruvate is metabolized further to produce phenyllactate and phenylacetate. Write structural formulas for these two compounds.

20. Rationalize why people who have phenylketonuria (PKU) sometimes have abnormally light skin. [See Figure 27.6.]

21. In the United States, nearly all babies are tested at birth for PKU. Why is it important to diagnose this disease right away?

22. The artificial sweetener aspartame (trade name Nutrasweet) is a methyl ester of the dipeptide made from aspartic acid and phenylalanine. Should a person with PKU avoid this sweetener?

23. Should people who have PKU be placed on a diet that has no phenylalanine at all? Why?

24. People with PKU can change to less restrictive diets after about age 10. Does a woman with PKU need to return to a more restrictive diet if she becomes pregnant? Why?

Urea Cycle

25. Hans Krebs is one of the people who discovered the urea cycle. What other cycle did he discover?

26. Is the synthesis of urea from ammonium ion (NH_4^+) and CO_2 part of anabolism or catabolism? Does it consume or generate energy?

27. As part of the urea cycle, fumarate is recycled to form aspartate (Figure 27.8). What compound in the citric acid cycle is formed from fumarate that is made into aspartic acid by transamination? [Hint: See exercise 11.]

28. What does the term *hyperammonemia* mean?

Nucleotide and Heme Metabolism

29. What are the normal breakdown products of (a) pyrimidines, and (b) purines?

30. Caffeine has the structure:

Rationalize why people who have gout might be advised not to consume large amounts of caffeinated products.

31. Do you expect allopurinol (Figure 27.11) to be a competitive or a noncompetitive inhibitor of the enzyme that metabolizes xanthine? Why? [Review Section 22.4 if necessary.]

32. Elevated levels of bilirubin in blood most likely indicate an impaired function of which tissue: (a) heart, (b) brain, (c) liver, or (d) muscle?

33. Recall that glutathione is a tripeptide that protects red blood cells against oxidation and destruction (Sec-

tion 21.6). A person whose red blood cells lack glutathione will develop jaundice. Explain why.

34. Rationalize why people with an obstructed bile duct have chalky, gray feces that lack normal pigmentation?

DISCUSSION EXERCISES

1. Rationalize why people who go on high-protein diets typically have increased urination.

2. Predict the effect (increase, decrease, or no effect) on each metabolic process below under conditions of diabetes, adequate nutrition, and starvation: **(a)** protein digestion, **(b)** amino acid oxidation, **(c)** gluconeogenesis from amino acids, **(d)** urea cycle, **(e)** purine and pyrimidine catabolism. Explain your predictions.

3. Babies are typically tested at birth for phenylketonuria (PKU). Should they be tested for all possible genetic diseases? If not, what should be the criteria for deciding which tests to do, and on whom?

28

Body Fluids

1. What are the main components of blood?
2. How does blood transport O_2 to tissues, maintain a consistent pH, and regulate body temperature?
3. How does the immune system work?
4. How does blood clot?
5. What are some important hormones, and how do they work?
6. How do kidneys regulate the pH, composition, and pressure of blood?
7. What are the main ingredients in urine?

Blood Typing

Table 28.1 Blood Group Compatibility

Blood Type	Can Donate To	Can Receive From
O	O, A, B, AB	O
A	A, AB	O, A
B	B, AB	O, B
AB	AB	O, A, B, AB

Your blood type depends on substances called antigens found on the surface of red blood cells. These antigens are sphingolipids that have the carbohydrate portion of the molecule on the outside cell surface.

All these sphingolipid antigens have a pentasaccharide (five-monosaccharide) chain ending with galactose. When no additional carbohydrate is present, the chain is called antigen O and the blood type is O (Figure 28.1). An additional carbohydrate unit is present in antigens A and B, which occur in type A and type B blood, respectively. A person who has both antigens A and B has type AB blood.

When antigens A and B are perceived by the body as "foreign" substances, the immune system produces antibodies to attack them. This arrangement limits what type of blood a person can receive. If you have type O blood, for example, you cannot receive types A, B, or AB; all would trigger an immune response against the "foreign" antigens on the new red blood cells. Antigen O, however, does not provoke this response. Type O blood can be received by a person with any blood type. Table 28.1 lists the safe combinations of blood transfusions.

Whenever you donate or receive blood, your blood is typed. In addition to O, A, B, or AB, blood is designated as positive (+) or negative (−), depending on whether or not it contains another antigen called Rh (first identified in rhesus monkeys). People with Rh⁻ blood should not receive Rh⁺ blood because their immune system would produce antibodies to attack and destroy the new red blood cells. Do you know your blood type?

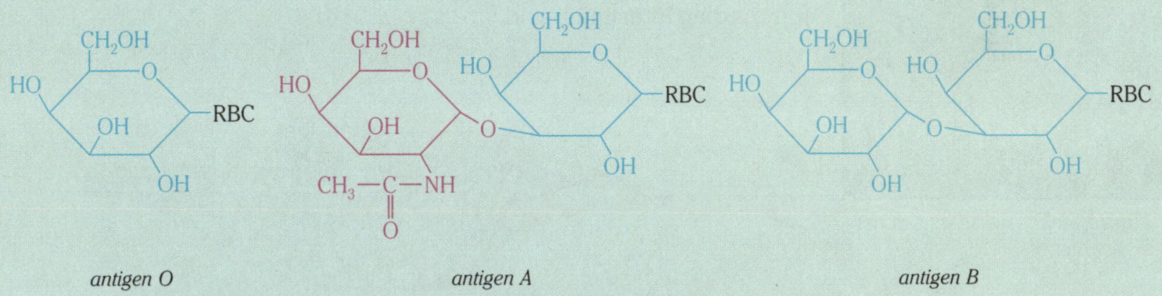

antigen O   *antigen A*   *antigen B*

Figure 28.1 Antigen O on the outer surface of a red blood cell (RBC) has an end galactose unit (blue). Antigens A and B have an additional *N*-acetylgalactosamine (red) or galactose (blue) unit, respectively.

586   Part III   Biochemistry

Table 28.2 pH of Body Fluids

Fluid	pH
Stomach contents	1.0–3.0
Duodenum contents	4.8–8.2
Feces	4.8–8.2
Urine	4.8–7.5
Sweat	6.1–6.7
Milk	6.6–7.6
Bile	6.8–7.0
Blood	7.3–7.5
Spinal fluid	7.3–7.5

About 70% of the earth's surface is water. Most of your body weight (60–65%) is also water. Like earth, your body is solid material interspersed in seas of water. Water coursing through your body carries nutrients, hormone messages, antibodies, and waste materials to their proper destinations, making life possible.

The water inside cells, called *intracellular fluid*, provides 70% of a cell's weight. All other water in your body is **extracellular fluid.** This includes **interstitial fluid,** the water surrounding and between cells, and **plasma,** the liquid portion of blood. Other body fluids include urine, lymph, cerebrospinal fluid, and synovial fluid.

These life-sustaining fluids bathe cells in substances they need and wash away waste and toxic materials. They also help buffer the body and remove excess acidity or alkalinity (Table 28.2).

28.1 COMPOSITION OF BLOOD

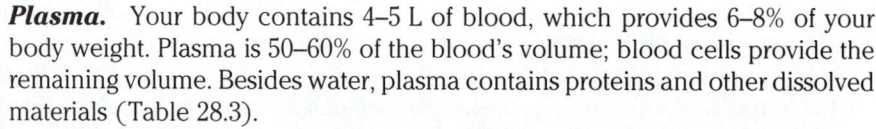

Plasma. Your body contains 4–5 L of blood, which provides 6–8% of your body weight. Plasma is 50–60% of the blood's volume; blood cells provide the remaining volume. Besides water, plasma contains proteins and other dissolved materials (Table 28.3).

Sodium (Na^+), chloride (Cl^-), and bicarbonate (HCO_3^-) ions are by far the most abundant ions (see Tables 4.4 and 7.6). The most abundant protein is albumin. Plasma proteins also include antibodies called gamma globulins that combat infections (see Sections 21.6 and 28.3). Another important plasma protein is fibrinogen, which helps blood clot (see Section 28.4).

Since plasma proteins are large molecules, they tend to remain inside blood vessels and increase the osmotic pressure of blood (see Chemistry Spotlight, Section 7.5). Albumin contributes the most to this effect. Because of its dissolved proteins, blood has an osmotic pressure about 18 torr higher than surrounding interstitial fluid.

Figure 28.2 A jellyfish is about 95% water.

Table 28.3 Components of Plasma

Component	Plasma Volume
Water	91–92%
Proteins	7–8%
Other solutes	1–2%*

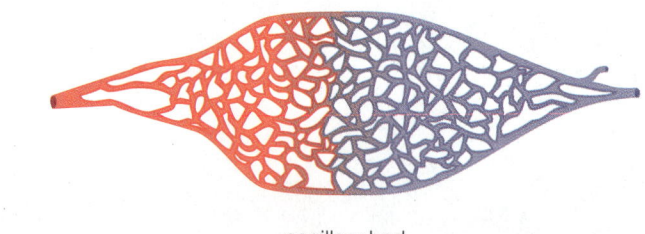

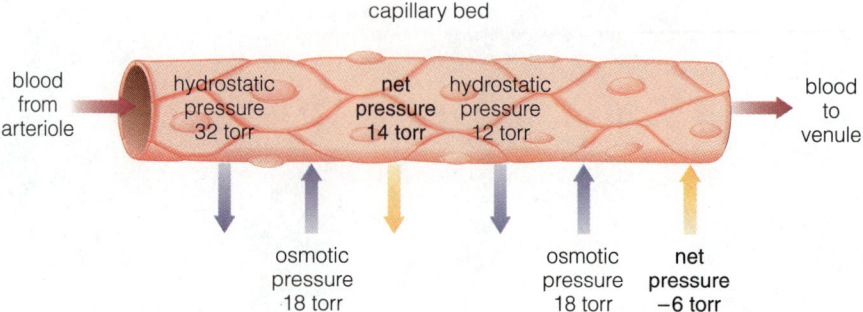

* See Table 7.6.

Figure 28.3 The difference between hydrostatic pressure and osmotic pressure in a capillary bed causes water to leave blood at the arterial end and enter blood at the venous end.

Now compare this osmotic pressure with the normal hydrostatic pressure from the pumping of the heart (Figure 28.3). The higher pressure in arteries (about 32 torr) forces fluids out of blood vessels and provides water and dissolved material to nearby cells. In venous blood, however, the pressure is only about 12 torr. Osmotic forces prevail, drawing water and dissolved material (including wastes) from interstitial fluid into blood. Wastes are eventually excreted in urine (see Sections 28.6 and 28.7).

Blood Cells. **Red blood cells,** or *erythrocytes,* are shaped like doughnuts without the hole (Figure 28.4). They are produced in bone marrow and have an average lifetime of about 120 days. Their main function is to carry O_2 from lungs to other tissues. They also help carry CO_2 from tissues to the lungs, where the gas is exhaled. Your blood has about 5 million red blood cells per microliter (μL).

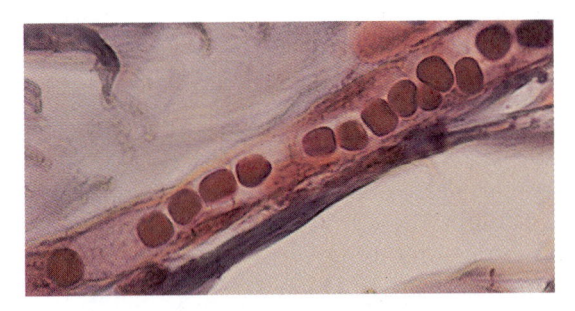

Figure 28.4 Red blood cells.

White blood cells, or *leukocytes,* also are made in bone marrow. There are five types: lymphocytes, monocytes, neutrophils, eosinophils, and basophils (Figure 28.5). White blood cells decompose cell debris and attack foreign agents such as bacteria and viruses. Lymphocytes play a key role in the immune system (see Section 28.3). These cells typically live a few days or less and are rapidly replaced. The most abundant types (neutrophils and lymphocytes) number a few thousand per μL of blood.

eosinophils

neutrophils

basophils

B lymphocytes (mature in bone marrow)

T lymphocytes (mature in thymus)

monocytes (in tissues)

platelets

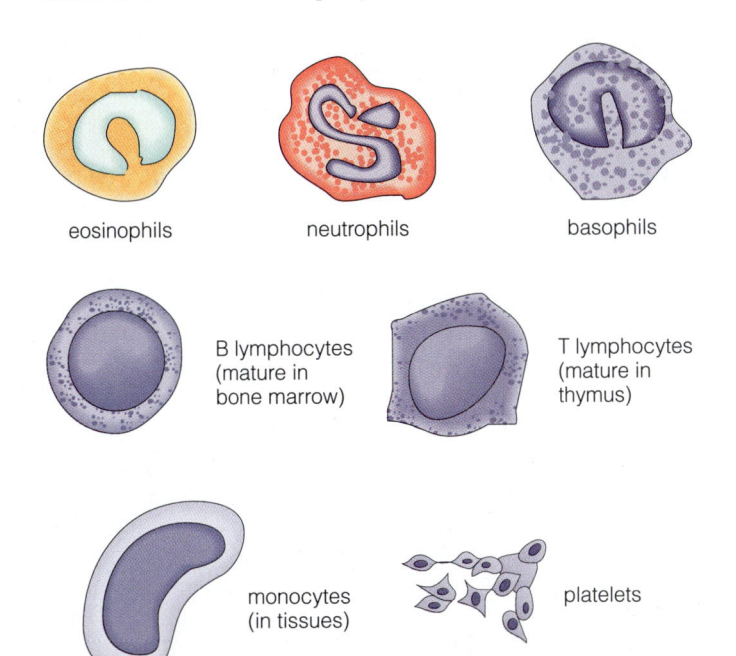

Figure 28.5 White blood cells and platelets.

Platelets (Figure 28.5), the third type of blood cells, play an important role in clotting. They are disc-shaped, produced in bone marrow, and live about one week. Blood has 250,000–300,000 platelets per μL.

28.2 FUNCTIONS OF BLOOD

Oxygen Transport. Red blood cells bind O_2 when they circulate through lung tissue. Recall that high O_2 concentrations in inhaled air push the equilibrium in the reaction below to the right as O_2 binds to hemoglobin (Hb) (see Section 9.7 and Figure 9.18):

$$HHb + O_2 \rightleftharpoons HbO_2 + H^+$$

The proton (H^+) released then drives two other equilibriums to the right:

$$H^+ + HbCO_2^- \rightleftharpoons HHb + CO_2$$
$$H^+ + HCO_3^- \rightleftharpoons H_2CO_3 \rightleftharpoons H_2O + CO_2$$

Both reactions release carbon dioxide (CO_2), which is exhaled from the lungs, ridding the body of this metabolic waste product.

Oxygenated blood goes from the lungs to other tissues, where it encounters acidity and CO_2 produced by metabolism. These products shift the equilibriums for all three reactions above to the left, releasing O_2 from hemoglobin and sending CO_2 into blood as bicarbonate ion (HCO_3^-) and bound directly to hemoglobin ($HbCO_2^-$). When blood reaches the lungs, the reactions go in the opposite direction again and the cycle repeats (see Chemistry Spotlight, Section 8.5).

Oxygenated blood is bright red. Blood partially depleted of O_2 is dark red but looks blue when seen through blood-vessel walls. This is why veins near the skin's surface look blue.

Blood pH. Blood is slightly alkaline, pH 7.40. Its pH is regulated by three buffers: protein, bicarbonate, and phosphate.

Hemoglobin is an important buffer protein. Notice in the two reactions below that Hb releases protons (H^+) when it binds O_2 or CO_2, and that it accepts protons in the reverse reactions:

$$HHb + O_2 \rightleftharpoons HbO_2^- + H^+$$
$$HHb + CO_2 \rightleftharpoons HbCO_2^- + H^+$$

Recall that a buffer is most effective in the pH range near its pK_a value. (Section 9.6).

A histidine group in Hb releases or binds a proton when Hb binds or releases O_2, respectively. The pK_a value of that histidine group is in the 7–8 range, which makes it an effective buffer at blood pH of 7.4.

Notice also that bicarbonate ion (HCO_3^-) neutralizes acidity in blood:

$$HCO_3^- + H^+ \rightleftharpoons H_2CO_3 \rightleftharpoons H_2O + CO_2$$

Chloride ion (Cl^-) is the most abundant negative ion in blood. But because it is the conjugate base of a strong acid (HCl), it is not an effective buffer (see Section 9.6).

Because its pK_a value is 6.4, bicarbonate is at the edge of its effectiveness as a buffer at blood pH and is more effective neutralizing excess acidity than excess alkalinity.

Though present at a lower concentration than bicarbonate ion (see Tables 4.4 and 7.6), hydrogen phosphate ion (HPO_4^{2-}) has a pK_a value of 7.2 and is an important buffer in blood:

$$HPO_4^{2-} + H^+ \rightleftharpoons H_2PO_4^-$$

CHEMISTRY SPOTLIGHT

A New Treatment for Hemoglobin Disorders?

Hemoglobin is a protein whose quaternary structure consists of two alpha (α) and two beta (β) chains (see Section 21.14). A fetus, however, synthesizes hemoglobin consisting of two α chains and two gamma (γ) chains. Fetal hemoglobin ($\alpha_2\gamma_2$) has a stronger attraction to O_2 than $\alpha_2\beta_2$ hemoglobin does (Figure 28.6). This enables a fetus to obtain O_2 bound to Hb in the mother's blood. Once a baby is born, the gene coding for the γ chain shuts off within six months as the baby begins to synthesize adult hemoglobin ($\alpha_2\beta_2$).

The β chain in hemoglobin is the site of several genetic disorders. An amino acid substitution of valine for glutamic acid in the β chain causes sickle-cell anemia (see opening to Chapter 21). Another genetic disorder, β-thalassemia, results from failure to synthesize a functioning β protein chain. Both of these disorders result in low red blood cell counts and a reduced supply of O_2 to cells. Patients lack energy because—without enough O_2—they cannot generate sufficient ATP by oxidative phosphorylation.

One idea to help such patients is to find a way to switch the gene that makes fetal hemoglobin back on; this might provide functioning hemoglobin that doesn't contain the faulty β chain. In 1993, a group of scientists reported that the gene for the γ chain is activated by butyrate, the neutralized salt of butyric (butanoic) acid, $CH_3CH_2CH_2COO^-$. Work is now in progress to determine if butyrate can help treat people who have sickle-cell anemia or β-thalassemia.

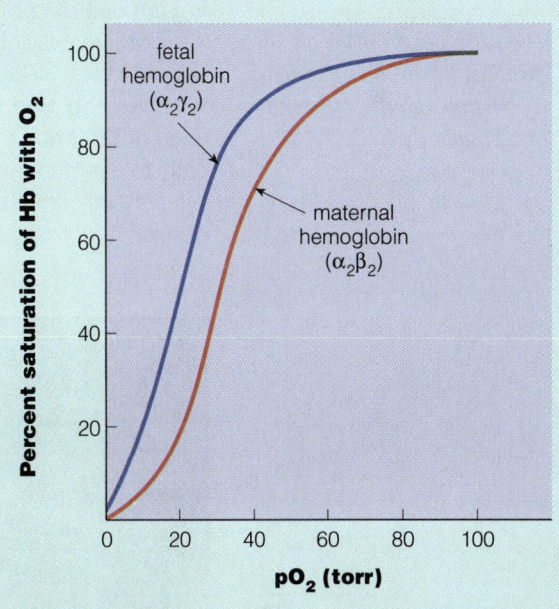

Figure 28.6 Fetal hemoglobin has a greater attraction to O_2 than adult hemoglobin does.

When blood pH is below 7.30 or above 7.50, the conditions are called *acidosis* and *alkalosis,* respectively. Causes, effects, and treatments of these pH imbalances are discussed in the Chemistry Spotlight in Section 9.7. Under these conditions, blood is less able to deliver O_2 to tissues and release CO_2 to lungs. In addition, the kidneys have to adjust what they excrete in urine (see Section 28.6).

Temperature Regulation. Due to the extra heat energy needed to break strong hydrogen bonds between water molecules, water has an unusually high specific heat and heat of vaporization (see Section 6.2). As a result, water is especially resistant to temperature changes, and it doesn't vaporize easily.

These properties of water in blood and in all your body fluids keep you from overheating on hot days and losing too much water by evaporation. Water helps keep your body temperature a steady 37° (see Chemistry Spotlight, Section 6.2).

28.3 IMMUNE SYSTEM

Your **immune system** protects you against foreign materials such as bacteria, viruses, and fungi. It consists of three types of white blood cells: macrophages (made from monocytes) and two kinds of lymphocytes, B cells and T cells (see Figure 28.5).

B and T cells are so named because they develop in bone marrow and the thymus gland, respectively.

Macrophages decompose the foreign material they encounter (Figure 28.7) and stimulate T cells to attack foreign materials that have entered cells. T cells bind to the surface of foreign cells and secrete proteins that hydrolyze the foreign cell membrane, making holes in it and destroying the cell.

The AIDS virus (HIV) disables the immune system by destroying macrophages and T cells.

Foreign cells circulating in blood are attacked by antibodies produced by B lymphocytes. Recall that antibodies are Y-shaped immunoglobulin glycoproteins (see Section 21.6). Antibodies differ in the amino acid sequence at the ends of the arms in the molecule (see Figure 21.19). Those ends bind specifically to a particular *antigen,* a protein or carbohydrate group on the surface of a foreign cell. Once an antibody binds to a foreign cell, macrophages attack and destroy it.

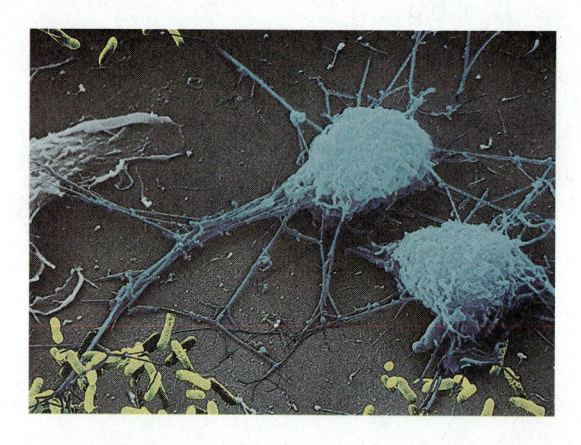

Figure 28.7 Scanning electron micrograph of a macrophage using cytoplasmic extensions to engulf and destroy bacteria.

Your immune system has two important qualities. First, it is highly specific; it distinguishes normal body cells from foreign cells and destroys only the latter type. Second, it has a "memory"; once it has produced antibodies against a particular antigen, it responds rapidly when that antigen is encountered again.

This second quality is the basis for vaccines. A vaccine contains a weakened form of an infectious agent that triggers an antibody response but doesn't transmit the disease. A vaccinated person's immune system then responds more rapidly and effectively if it encounters the infectious agent again.

28.4 BLOOD CLOTTING

Blood clotting is a delicate balancing act. If blood doesn't clot properly, a person may bleed to death from fairly minor injuries. But if blood clots too readily, the circulation in blood vessels can be blocked, causing a heart attack or stroke.

Blood clotting also is a spectacular example of proenzyme (zymogen) action (see Section 22.5). A long sequence of reactions produces the clot. In the final step, the proteolytic enzyme thrombin catalyzes the removal of peptide segments from fibrinogen, a blood protein, to produce fibrin:

$$\text{fibrinogen} \xrightarrow{\text{thrombin}} \text{fibrin} + \text{peptides (18–20 amino acids long)}$$

Attracted to each other by ionic and other bonds, fibrin molecules form covalent cross-links between amino acid side-chain groups. The resulting insoluble network of fibrous protein holds the clot together (Figure 28.8).

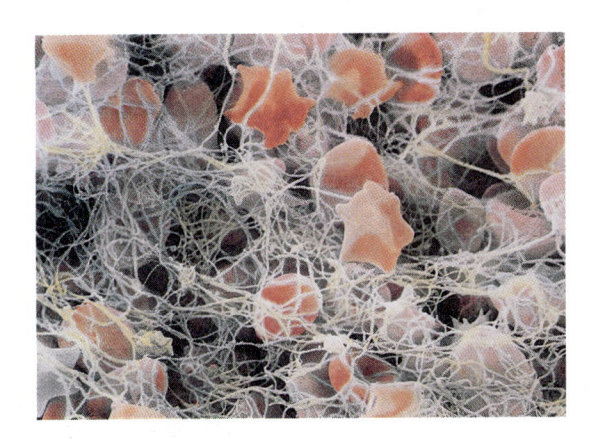

Figure 28.8 Fibrin threads entangle blood cells in a blood clot.

Figure 28.9 shows the cascade of reactions that leads to a blood clot. Each reaction involves activation of a proenzyme, which activates another proenzyme, and so on. The intrinsic pathway begins when tissue is damaged directly. The extrinsic pathway is triggered by trauma. In both pathways, tissues release factors to initiate the sequence of reactions, and the pathways converge at the reaction that activates factor X.

The activation of prothrombin to thrombin requires vitamin K, Ca^{2+} ion, and platelets. Vitamin K is a coenzyme for a reaction that chemically modifies prothrombin so that it binds Ca^{2+} ion. Ca^{2+} then helps bind prothrombin to platelets, which contain the activated factors V and X that catalyze the formation of thrombin (Figure 28.9).

If any one of these reactions fails, blood cannot clot. A deficiency of vitamin K, for example, causes excessive bleeding. Blood stored for later use is intentionally kept from clotting by adding chelating agents such as citrate or EDTA, which bind Ca^{2+} and make it unavailable to help activate prothrombin.

People who have classic *hemophilia* genetically lack factor VIII. As a result, tissue damage doesn't trigger blood clotting through the intrinsic pathway (Figure 28.9). Even gentle bumps can cause large, blue swellings from internal bleeding. Treatment consists of frequent infusions of a blood fraction containing concentrated factor VIII. Now that factor VIII can be synthesized by genetically engineered bacteria, patients should have access to a safer source of this clotting factor.

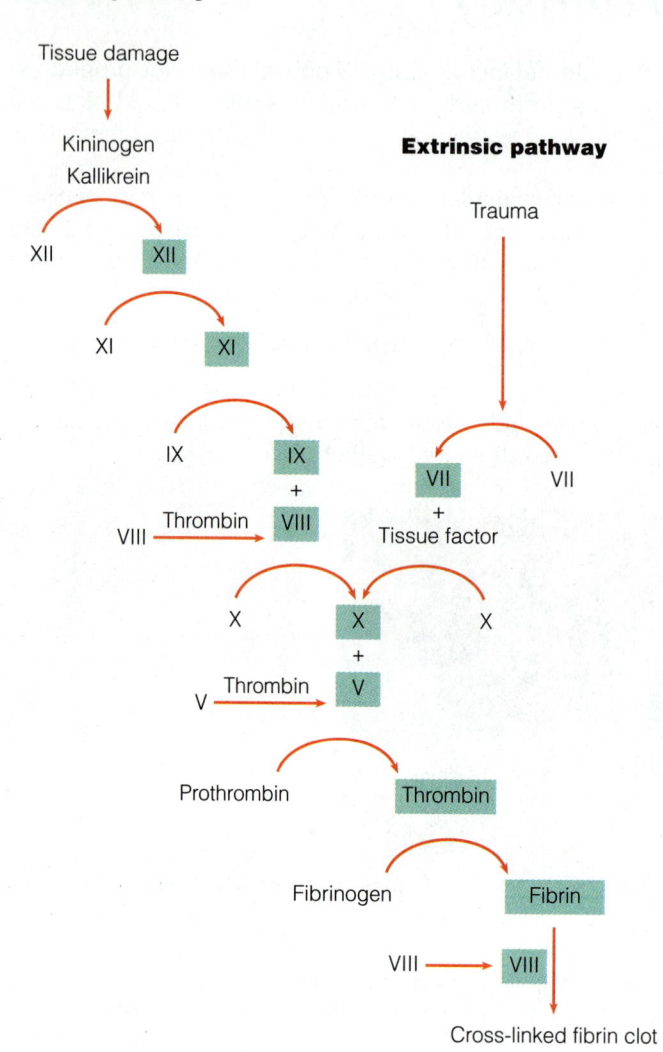

Figure 28.9 A cascade of reactions produces activated proteins (blue) to form blood clots.

28.5 HORMONES

Hormone comes from the Greek word horman, *which means to activate or excite.*

Certain glands in the body secrete **hormones,** chemical messengers that affect other parts of the body. Once it enters the blood, a hormone travels the watery highways until it reaches a type of cell with the specific receptor that binds it and responds to its message. Table 28.4 lists some important hormones and their effects.

Peptide hormones and epinephrine typically work by binding to the outer surface of cell membranes and initiating reactions that alter metabolism inside the cell. Recall that epinephrine and glucagon stimulate glycogenolysis in this way (see Section 25.3).

Insulin consists of a total of 51 amino acids arranged on two peptide chains joined by disulfide bonds (see Figure 21.17). Secreted by β cells in the pancreas in response to high blood glucose levels, insulin stimulates the entry of glucose into cells and promotes glucose metabolism by glycogenesis and glycolysis. It also inhibits lipase activity in adipose tissue.

Table 28.4 Some Important Hormones

Hormone	Secreted By	Major Functions
Peptides		
Glucagon	Pancreas (α cells)	Increase blood glucose and fatty acid levels
Insulin	Pancreas (β cells)	Decrease blood glucose and fatty acid levels
Vasopressin (ADH)	Posterior pituitary	Water retention, increase blood pressure
Steroids		
Aldosterone	Adrenal cortex	Sodium (Na^+) ion retention, increase blood pressure
Cortisol	Adrenal cortex	Decrease glucose catabolism, increase blood glucose levels
Estradiol	Ovary	Female sexual development
Progesterone	Ovary (Corpus luteum)	Regulate menstrual cycle and maintain pregnancy
Testosterone	Testis	Male sexual development
Amines		
Epinephrine (Adrenalin)	Adrenal medulla	Response to stress, increase carbohydrate and lipid catabolism
Thyroxine	Thyroid gland	Increase metabolic rate

Glucagon (Figure 28.10) is secreted by α cells of the pancreas when blood glucose levels are low, and has the opposite effects of insulin on carbohydrate and lipid metabolism. Glucagon increases blood glucose levels by stimulating glycogenolysis and gluconeogenesis in liver. In addition, it stimulates adipose tissue lipase; this releases fatty acids into blood to be taken up and used as fuel by other tissues. Epinephrine (see Figure 27.3) has the same effect on fatty acid metabolism, and also stimulates glycogenolysis in muscle.

$$\overset{+}{NH_3}-\underset{1}{His}-Ser-Gln-Gly-\underset{5}{Thr}-Phe-Thr-Ser-Asp-\underset{10}{Tyr}-Ser-Lys-Tyr-Leu-\underset{15}{Asp}$$
$$COO^--\underset{29}{Thr}-Asn-Met-Leu-\underset{25}{Trp}-Gln-Val-Phe-Asp-\underset{20}{Gln}-Ala-Arg-Arg-Ser$$

Steroid hormones and thyroxin cross cell membranes and bind to specific receptor proteins in the nucleus. There they work by altering gene expression. Because they are so nonpolar, steroid hormones in blood are bound to specific proteins to increase their solubility and transport them to target tissues.

Figure 28.10 Primary structure of glucagon, a peptide hormone.

28.6 KIDNEY FUNCTIONS

Structure of Kidneys. The kidneys (Figure 28.11) are a pair of organs that constantly filter water, ions, and organic materials from blood. About 180 L (45 gal) of blood go through your kidneys every 24 hours. The kidneys return most of the water and solute in incoming (arterial) blood to outgoing (venous) blood (Table 28.5). Material that isn't returned is excreted in urine. In this way, kidneys regulate the volume and solute concentration of blood, which in turn affects the volume and solute concentration of other extracellular fluids.

When blood enters a kidney, it goes through a renal capsule, where blood pressure forces out water and small solutes. This fluid continues through tubule regions surrounded by capillaries that reabsorb most of the tubule contents. The remaining waste goes through a collecting duct into the urinary bladder, where it is stored and eventually excreted as urine.

Table 28.5 Retention of Substances in Blood During Passage Through Kidneys

Substance	Retained in Blood
Glucose	100%
Sodium ion	99.5%
Urea	44%
Water	99%

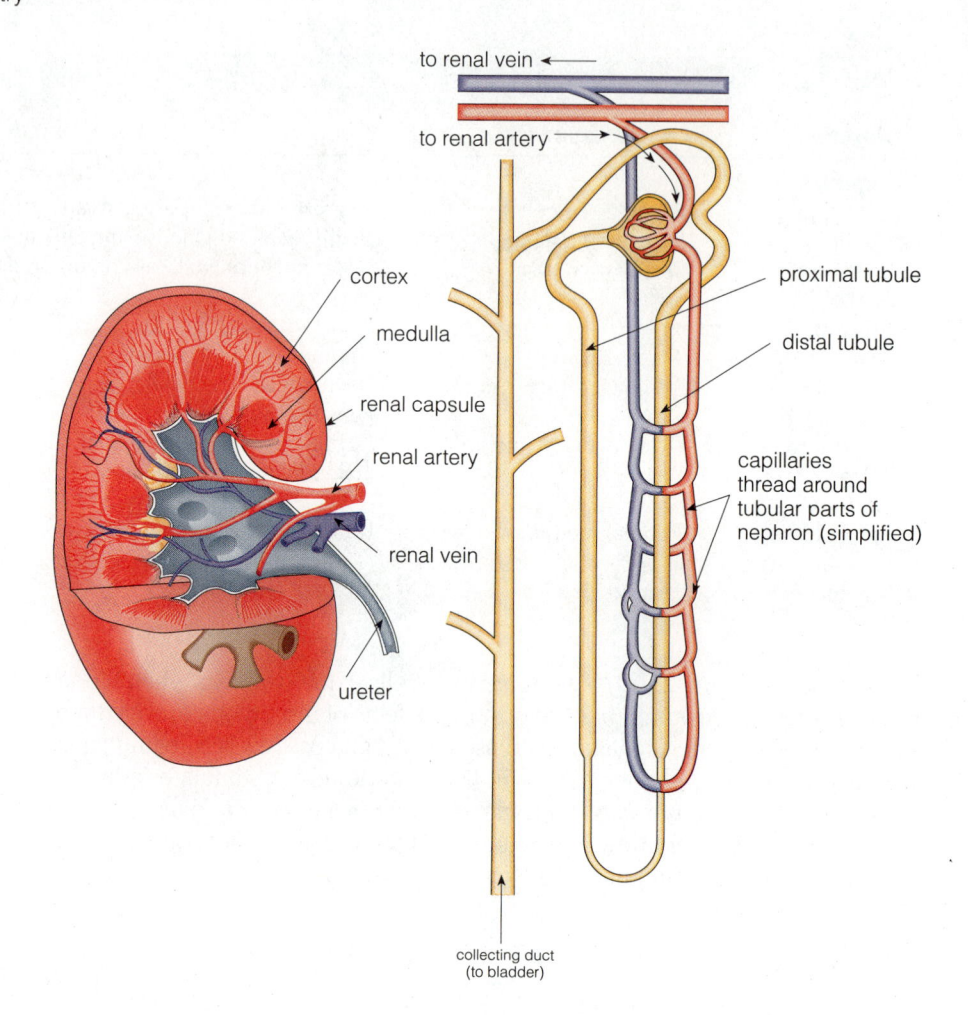

Figure 28.11 Structure of a kidney (left) and a functioning unit (called a nephron) within a kidney (right).

```
⁺NH₃
 |
₁Cys ─┐
 |    |
 Tyr  |
 |    |
 Phe  S
 |    |
 Gln  S
 |    |
₅Asn  |
 |    |
 Cys ─┘
 |
 Pro
 |
 Arg
 |
₉Gly
 |
 C=O
 |
 NH₂
```

Figure 28.12 Primary structure of vasopressin (ADH), a peptide hormone.

Salt and Water Balance. Two hormones help ensure that enough water and sodium are reabsorbed and retained in blood. The first is vasopressin, also called antidiuretic hormone (ADH; see Table 28.4). ADH is a nonapeptide (nine amino acid units; Figure 28.12) that stimulates H_2O reabsorption into blood from the distal tubules and collecting ducts of the kidney. As a result, the body produces less urine.

The secretion of ADH is a two-step process. A gland at the base of the brain called the hypothalamus senses when the concentration of solute in blood is high. The hypothalamus sends a peptide hormone to the posterior pituitary gland. The posterior pituitary then secretes ADH, which causes the kidneys to reabsorb water into the blood, thereby reducing the concentration of solute.

The action of ADH is inhibited by ethanol. As a result, drinking alcoholic beverages increases urine production in two ways. First, the intake of water increases the need to excrete water. Second, the kidneys reabsorb less water due to the inhibited action of ADH.

Aldosterone (Figure 28.13), a steroid hormone secreted by the adrenal glands (see Table 28.4), increases the reabsorption of Na^+ ion from the distal tubules and collecting ducts of the kidneys. Aldosterone is secreted when Na^+ concentrations are low in extracellular fluids, and during dehydration. When aldosterone causes blood to retain more Na^+, Cl^- is also retained to maintain electrical neutrality. Water, in turn, is retained to maintain normal osmotic pressure.

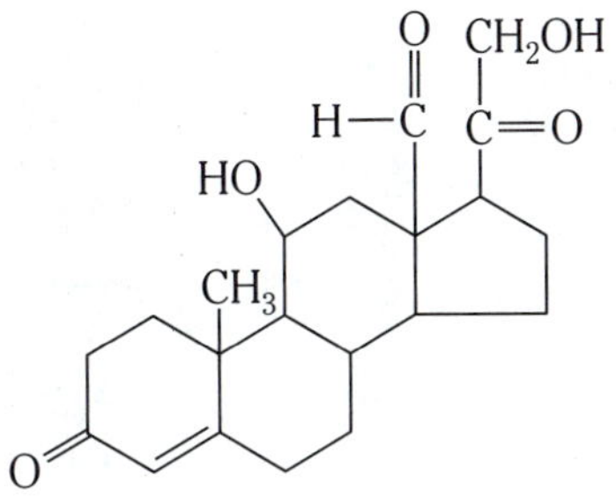

Figure 28.13 Structure of aldosterone, a steroid hormone.

Blood Pressure. Blood pressure depends on the pumping action of the heart, on resistance by muscles surrounding blood vessels, and on blood volume. ADH and aldosterone both increase blood volume, and thus increase blood pressure.

When blood pressure drops too low, the kidneys secrete renin, a proteolytic enzyme. Renin catalyzes the conversion of angiotensinogen in blood into angiotensin I, a decapeptide (10 amino acid units; Figure 28.14). Further hydrolysis produces the active hormone angiotensin II, an octapeptide.

angiotensinogen
↓ renin

asp—arg—val—tyr—ile—his—pro—phe—**his**—**leu**

angiotensin I
↓

asp—arg—val—tyr—ile—his—pro—phe + **his**—**leu**

angiotensin II

Figure 28.14 Activation of angiotensinogen by renin.

Angiotensin II increases blood pressure in two ways. First, it causes the walls of blood vessels to tighten (constrict). Second, it stimulates the secretion of aldosterone, which increases blood pressure by increasing the volume of blood.

A common way to treat high blood pressure (hypertension) is to use *diuretics,* drugs that increase urination and thus reduce blood volume. Diuretics work in a variety of ways, including:

- reducing Na^+ reabsorption from kidneys
- reducing Cl^- (and thus Na^+) reabsorption from kidneys
- inhibiting the production of angiotensin II

Maintaining Blood pH. Notice in Table 28.1 that the pH of urine varies considerably. Under conditions of acidosis or alkalosis, the kidneys respond by removing extra acid or base, respectively, from blood and excreting it in urine.

You excrete excess acidity in several forms. Some protons (H^+) are excreted as hydronium (H_3O^+) ion. Protons also react with ammonia (NH_3) and hydrogen phosphate (HPO_4^{2-}) ion to form ammonium (NH_4^+) and dihydrogen phosphate ($H_2PO_4^-$) ions, respectively, which are excreted:

$$NH_3 + H^+ \longrightarrow NH_4^+$$

$$HPO_4^{2-} + H^+ \longrightarrow H_2PO_4^-$$

CHEMISTRY SPOTLIGHT

Kidney Dialysis Methods

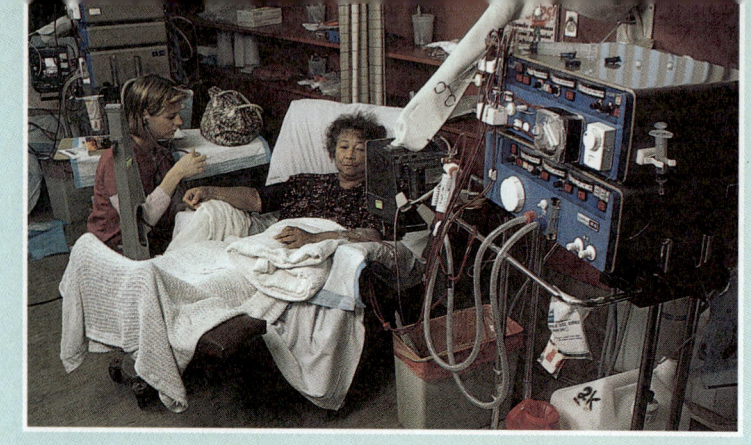

Figure 28.15 This patient's blood is being cleansed by dialysis.

About 13 million people in the United States have damaged or diseased kidneys. In these patients, waste and toxic materials accumulate in blood and aren't excreted. This upsets the pH and volume of blood and other extracellular fluids, and results in fatigue and nausea.

The best permanent solution for most patients is a kidney transplant. They may also be treated by *dialysis*, which removes certain small molecules and ions from blood (see Section 7.5).

In a kidney dialysis machine (Figure 28.15), blood is pumped through cellophanelike membranes bathed in a solution of glucose and various salts. Toxic and waste materials pass from blood through the membranes and into the surrounding solution. Blood remains inside the membranes and circulates back into the patient. Patients typically use dialysis machines at a medical facility for about four hours, three times a week.

Another option is peritoneal dialysis. Here the patient has a valve surgically installed in the abdomen. A fluid of proper composition is poured through the valve into the abdomen, where it remains for about 30 minutes. During this time, waste and toxic materials pass through the peritoneal membrane in the abdomen and enter the dialyzing fluid. The fluid then is drained from the body, removing those wastes. Patients typically do peritoneal dialysis at home at least three times a day.

Dialysis can be used to treat a patient until a kidney becomes available for transplantation. Even when it is the permanent treatment, dialysis enables many patients to enjoy years of fairly normal activity.

Table 28.6 Water Balance in the Body

Source	mL/24 hr
Water gain	
Solid intake	850
Liquid intake	1400
Produced by metabolism	350
Net gain	2600
Water loss	
Urine	1500
Feces	200
Evaporation	900
Net loss	2600

Kidneys also combat excess acidity by reabsorbing bicarbonate (HCO_3^-) ion, a buffer that neutralizes hydrogen ion:

$$HCO_3^- + H^+ \longrightarrow H_2O + CO_2$$

When blood is too alkaline, however, kidneys retain less bicarbonate ion and excrete more.

28.7 URINE

About 96% of urine is water. In moderate climates, urine is the body's largest source of water loss, often about 1.5 L per day (Table 28.6). In warmer climates, the volume of urine decreases as more water is lost by evaporation.

The remaining 4% of urine is dissolved material. High concentrations of dissolved material cause urine to have a high specific gravity, which can be measured by a specialized hydrometer called a urinometer (see Section 2.3). Urine typically has a specific gravity in the 1.004–1.030 range.

Body Fluids Chapter 28 **597**

Table 28.7
Components in Urine

Substance	Approx. Conc. (g/L)
Nitrogen compounds	
Urea	20
Creatinine	0.8
Uric acid	0.4
Positive ions	
Sodium	3.5
Potassium	1.5
Ammonium	0.3
Calcium	0.2
Negative ions	
Chloride	5.0
Phosphate	1.5
Sulfate	1.5

Table 28.8
Disorders Having Unusual Substances in Urine

Disorder	Substance
Albinism	Tyrosine
Alkaptonuria	Homogentisate
Diabetes	Glucose
Galactosemia	Galactose
Kidney disease (various)	Protein
Phenylketonuria (PKU)	Phenylpyruvate

Table 28.7 lists typical concentrations of solutes in urine. Urea, produced in the urea cycle (see Section 27.3), is the major solute. Creatinine (Figure 28.16) is a breakdown product of creatine, which is stored in muscle. People with greater muscle mass excrete more creatinine.

Abnormal compounds in urine are signs of disease. Table 28.8 lists a few examples.

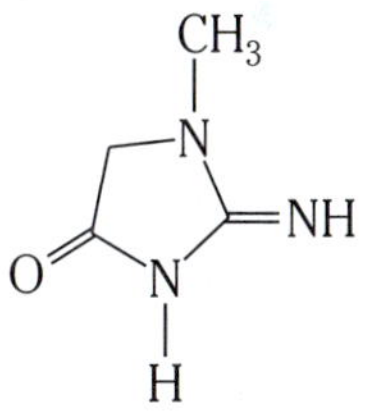

Figure 28.16 Creatinine.

> Almost all the chemical processes which occur in nature, whether in animal or vegetable organisms, or in the non-living surface of the earth . . . take place between substances in solution.
>
> *Wilhelm Ostwald (1853–1932)*

SUMMARY

Blood consists of plasma and blood cells. Plasma contains proteins, such as albumin and globulins, and other solutes. Red blood cells carry O_2 bound to hemoglobin. White blood cells defend the body against foreign substances. Platelets help blood clot.

Hemoglobin binds O_2 in inhaled air and releases O_2 to tissues for metabolism. Hemoglobin, bicarbonate (HCO_3^-), and hydrogen phosphate (HPO_4^{2-}) buffer blood and maintain its pH close to 7.40.

The immune system uses three types of white blood cells. Macrophages decompose foreign material in blood. T lymphocytes attack and destroy foreign material inside cells. B lymphocytes produce antibodies that attack foreign substances in blood. Antibodies selectively bind to specific antigens on the surface of the foreign substance.

Blood clotting is a series of reactions catalyzed by activated proenzymes. The final reaction, catalyzed by thrombin, converts fibrinogen into fibrin, which forms a clot. Hemophiliacs genetically lack factor VIII, so their blood doesn't clot.

Glands secrete hormones, which travel in blood to other parts of the body. Hormones bind to specific receptors and alter cell metabolism. Peptide hormones and epinephrine act by binding to cell membranes, while steroid hormones and thyroxin enter cells and alter gene expression in the nucleus.

Kidneys filter blood and remove waste materials to be excreted in urine. Retention of water and Na⁺ in blood are regulated by the hormones vasopressin (ADH) and aldosterone, respectively. These hormones and angiotensin II help regulate blood pressure.

Urine contains mostly water, plus nitrogen compounds and various ions. Abnormal compounds in urine are signs of disease.

KEY TERMS

Extracellular fluid (28.1)
Hormone (28.5)
Immune system (28.3)
Interstitial fluid (28.1)
Plasma (28.1)
Platelet (28.1)
Red blood cell (erythrocyte) (28.1)
White blood cell (leukocyte) (28.1)

EXERCISES

Even-numbered exercises are answered at the back of this book.

Introduction

1. Which body fluids can be the most alkaline?
2. Which body fluid is the most acidic?

Composition of Blood

3. People with anemia have a shortage of red blood cells. Does anemia increase or decrease the percentage of blood volume that is plasma?
4. Blood serum is the fluid remaining after blood has clotted. Name a protein present in plasma that is largely absent in serum.
5. What is the most abundant type of blood cell?
6. What effect does dehydration, which reduces blood volume, have on the osmotic pressure of blood?
7. How does the loss of protein from blood affect the flow of water between interstitial fluid and **(a)** arterial blood and **(b)** venous blood?

Functions of Blood

8. What chemical factors cause CO_2 to pass from blood into the lungs to be exhaled?
9. What chemical factors in tissues stimulate the release of bound O_2 from hemoglobin?
10. What substance in blood is most responsible for the color of blood?
11. What protein chain is present in both adult and fetal hemoglobin? Which chain is in adult Hb but not in fetal Hb? [See the Chemistry Spotlight in Section 28.2.]
12. Suppose it becomes possible to enable adults to produce fetal hemoglobin instead of their normal hemoglobin. Would this be a way to help people who have a genetic disorder in which they produce an abnormal α chain for hemoglobin? Explain. [See the Chemistry Spotlight in Section 28.2.]
13. Hydrogen sulfate ion (HSO_4^-) in plasma has a pK_a value of 1.9 for the reaction: $HSO_4^- \rightarrow SO_4^{2-} + H^+$. Is sulfate a useful buffer to help maintain blood pH at 7.4? Explain. [Review Sections 9.6 and 9.7 if necessary.]
14. Dihydrogen phosphate ion ($H_2PO_4^-$) has a pK_a value of 7.2 for the reaction: $H_2PO_4^- \rightarrow HPO_4^{2-} + H^+$. In blood at pH 7.4, is there more phosphate in the form HPO_4^{2-} or $H_2PO_4^-$? Explain. [Review Sections 9.6 and 9.7 if necessary.]
15. List the three major buffers in blood.
16. Assume that 62% of your body weight is water. Estimate the volume (in L) of all the water in your body.
17. Explain why water's high specific heat and heat of vaporization are advantages in maintaining normal body temperature.

Immune System

18. What type of blood cell produces antibodies?
19. A few vaccines now contain only the antigen of an infectious agent instead of a weakened form of the agent itself. What might be an advantage of such a vaccine?
20. Antibodies differ in the amino acid sequence in one region of an immunoglobulin molecule. How does this region of an immunoglobulin function?
21. If your blood type is A⁻, you could donate blood to people with which of the following blood types: **(a)** A⁻, **(b)** A⁺, **(c)** O⁻, **(d)** AB⁺, **(e)** B⁺, **(f)** AB⁻, **(g)** O⁺. [See the Chapter Opener.]

22. If your blood type is A⁻, you could receive blood from people with which of the following blood types: **(a)** A⁻, **(b)** A⁺, **(c)** O⁻, **(d)** AB⁺, **(e)** B⁺, **(f)** AB⁻, **(g)** O⁺. [See the Chapter Opener.]

Blood Clotting

23. What kinds of chemical bonds hold together fibrin molecules in a blood clot?
24. What reaction in blood clotting cannot occur if vitamin K is absent?
25. The cascade of reactions in blood clotting involves several **(a)** isoenzymes, **(b)** allosteric enzymes, **(c)** proenzymes.
26. Citrate prevents stored blood from clotting by binding to Ca^{2+} ion. Examine the structure of citrate in Figure 24.8 and predict how it binds to Ca^{2+}.

Hormones

27. What metabolic processes do epinephrine and glucagon alter when they bind to the membrane of a cell? [Review Section 25.4 if necessary.]
28. List the effect (stimulation or inhibition) of each of the following hormones on the release of fatty acids from storage as triglyceride in adipose tissue: **(a)** glucagon, **(b)** insulin, **(c)** epinephrine.
29. Which types of hormones enter their target cells?
30. Which hormone listed in Table 28.4 has the highest formula weight?
31. Which hormone listed in Table 28.4 has the lowest formula weight?

Kidney Functions

32. Explain why vasopressin is also called antidiuretic hormone (ADH).
33. How does the secretion of aldosterone affect the volume of urine produced?
34. Caffeine is a diuretic. Does it increase or decrease the volume of urine produced?
35. Which hormone—vasopressin or aldosterone—is synthesized in the body from cholesterol?
36. Ethanol inhibits the action of ADH. What is the effect of ethanol on blood pressure?
37. What would be the effect on blood pressure of a drug that inhibits renin?
38. List three chemical forms in which excess acidity is excreted in urine.
39. What happens to plasma proteins when blood is dialyzed? [See the Chemistry Spotlight in Section 28.6.]
40. In a kidney dialysis machine, why does the fluid bathing the membranes contain glucose? [See the Chemistry Spotlight in Section 28.6.]

Urine

41. How does dehydration affect the specific gravity of urine?
42. Why does tyrosine appear in the urine of people who have albinism? [Hint: see Figure 27.6.]

DISCUSSION EXERCISES

1. How do **(a)** acidosis and **(b)** alkalosis affect the ability of blood to take up O_2 from lungs and release CO_2 into lungs? Use chemical reactions to explain your answer.
2. Glucose can be synthesized in liver by gluconeogenesis, go into blood and travel to another tissue such as brain or muscle, and be metabolized to generate ATP. Can glucose be classified as a hormone? Why?
3. Athletes such as football players are disqualified from amateur competition if their urine is found to contain diuretics. Rationalize why.
4. What criteria, if any, should be used to determine which people will have access to kidney dialysis machines or peritoneal dialysis? Explain.

29

Medical Drugs

1. What are antibiotics? How do they kill infectious organisms?
2. What are antimetabolites? How do they treat disease?
3. How do nerve cells transmit nerve impulses? How do drugs affect the action of neurotransmitters?
4. How are drugs used to treat thyroid disorders and diabetes, and to control pregnancy?
5. How are drugs used to enhance athletic performance? Which compounds are illegal for this purpose?

The Discovery of Penicillin

In 1928, English bacteriologist Alexander Fleming (Figure 29.1) was studying the growth of bacteria that cause boils and other infections. A stray, airborne spore of a blue-green mold fell onto a culture plate where the bacteria were growing. Fleming noticed that bacteria near the mold were destroyed. Later work showed that the *Penicillium* mold contained an antibacterial agent, which he named *penicillin*. But Fleming couldn't purify the active substance and gave up on the project.

A decade after Fleming's discovery, two Oxford scientists, Howard Florey and Ernst Chain, took up the work and isolated a brownish product, penicillin G, from the mold. It cured certain bacterial infections in mice and was first tried on a human in 1941. The patient, a 43-year-old London policeman, had a serious case of blood poisoning from a shaving cut. He improved immediately and after four days had almost recovered. But then the tiny supply of penicillin ran out. The man's infection worsened, and he died.

Penicillin was available for clinical use by 1943. Two years later, there was enough for

Figure 29.1 Alexander Fleming (1881–1955).

worldwide use. A *Penicillium* strain discovered on a moldy cantaloupe in a Peoria, Illinois, market proved to be a particularly good source of the new antibiotic. By the end of World War II, penicillin had saved many lives threatened by pneumonia, bone infections, gonorrhea, gangrene, and other infectious diseases.

It was the beginning of a new era in using chemistry to treat disease. For their work, Fleming, Florey, and Chain received the Nobel prize for medicine or physiology in 1945.

Figure 29.3 A few medicinal plants from which drugs have been isolated.

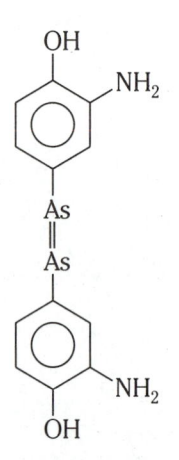

- foxglove (digitalis)
- willow bark (salicylic acid, also used to make aspirin)
- periwinkle (anti-cancer compounds)

arsphenamine (Salvarsan)

Figure 29.4 Arsphenamine, the first chemotherapeutic drug.

Although the term has become associated with cancer treatment, **chemotherapy** is the use of chemicals to treat *any* illness, not just cancer. People have used chemicals to treat aches, pains, and diseases since the dawn of recorded history. Many early drugs were natural animal and plant materials, and some seem a bit strange today. One remedy for blindness, for example, was to mix pigs' eyes, antimony, and honey and pour the concoction in the patient's ear.

Some ancient recipes worked well. Many contained drugs we now know are beneficial. An early Egyptian remedy for crying children, for example, contained poppy seeds, a source of the pain reliever morphine. And a cure for night blindness was "liver of ox, roasted and crushed," a good source of vitamin A.

Until well into the 19th century, physicians knew almost nothing about the chemistry of diseases. They also didn't know exactly what was in their potions, which ingredient was active, or why it worked. The drug era began in 1806 when the German pharmacist Friedrich Serturner isolated morphine (see Figure 16.15) from the opium poppy (Figure 29.2). Soon other drugs were isolated in

Figure 29.2 A field of opium poppies.

pure form from plants and animals (Figure 29.3). At last, physicians could identify the substance and dose they were giving their patients, and could study its effects on the body.

Our growing knowledge of biochemistry equips us to understand how many drugs work at the molecular level. In this chapter, we consider a **drug** to be any chemical used as medicine to treat illness.

29.1 ANTIBIOTICS

In 1865, the French scientist Louis Pasteur proposed that bacteria cause infectious diseases. During the next three decades, scientists confirmed that microorganisms cause many diseases. In 1867, Joseph Lister discovered that phenol kills microorganisms and could be used to prevent infections resulting from surgery. Although still used to disinfect tables, floors, and other surfaces in hospitals, phenol isn't used directly on skin because it causes blistering.

After Lister's discovery, scientists began searching for chemicals that kill bacteria inside the body without harming the body's own cells. In 1907 Paul Ehrlich, a German physician and chemist, synthesized the arsenic-containing compound arsphenamine (Salvarsan; Figure 29.4) and found it was toxic to the parasite that causes syphilis. Although it wasn't very effective in treating syphilis, arsphenamine was the first drug to kill harmful organisms without seriously damaging the human body. For this Ehrlich has been called the "father of chemotherapy."

Sulfa Drugs. In 1932, German bacteriologist Gerhard Domagk discovered that a red dye called prontosil killed certain infectious bacteria in mice, but didn't kill those bacteria in a test tube. It turned out that mice (but not test tubes) metabolize prontosil into sulfanilamide (Figure 29.5), the actual chemical that killed the bacteria.

For his discovery of sulfa drugs, Domagk received the Nobel prize in medicine or physiology in 1939.

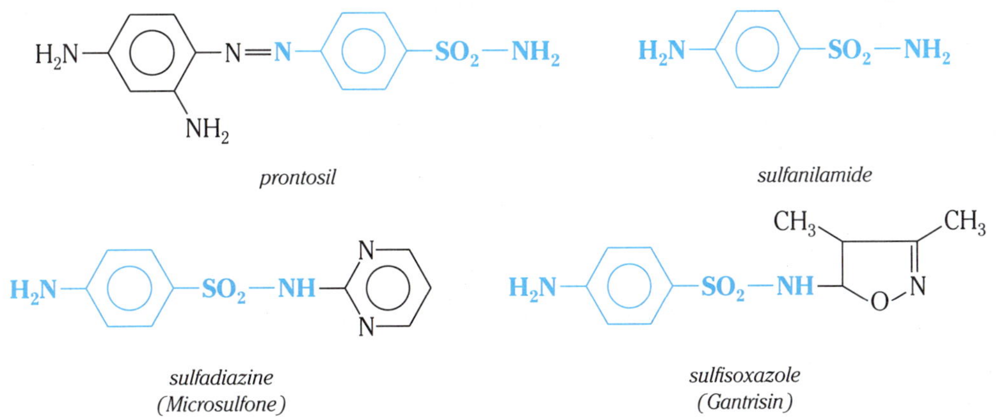

$H_2N-\langle\bigcirc\rangle(NH_2)-N=N-\langle\bigcirc\rangle-SO_2-NH_2$ $H_2N-\langle\bigcirc\rangle-SO_2-NH_2$

prontosil   *sulfanilamide*

$H_2N-\langle\bigcirc\rangle-SO_2-NH-\langle\text{pyrimidine}\rangle$ $H_2N-\langle\bigcirc\rangle-SO_2-NH-\langle\text{isoxazole with 2 CH}_3\rangle$

sulfadiazine (Microsulfone)   *sulfisoxazole (Gantrisin)*

Sulfanilamide kills streptococcal, staphylococcal, and certain other bacteria that cause diseases such as pneumonia, diphtheria, and gonorrhea. Because sulfanilamide has unwanted side effects, however, chemists used a common tactic in the drug industry: they synthesized new compounds by making structural changes in the original drug, then tested the new compounds to find ones that were safer or more effective. From more than 5400 such compounds tested, about a dozen proved to be useful drugs. Since they all contain an —SO$_2$— group, they are known as **sulfa drugs** (Figure 29.5).

Sulfa drugs kill bacteria by interfering with their metabolism. Cells need the vitamin folic acid (see Table 22.1 and Figure 29.6) to synthesize purine and

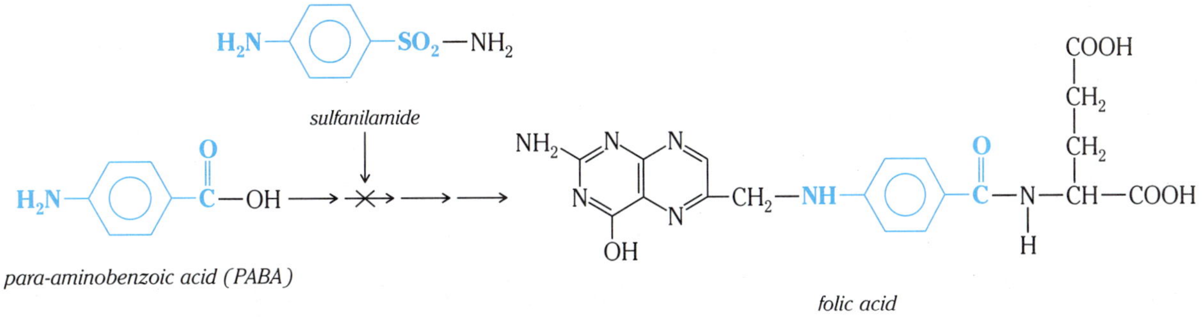

Figure 29.5 Three sulfa drugs and prontosil. Structural similarities are in color.

$H_2N-\langle\bigcirc\rangle-SO_2-NH_2$
sulfanilamide

$H_2N-\langle\bigcirc\rangle-\overset{O}{\underset{}{C}}-OH \longrightarrow \times \longrightarrow \longrightarrow$ [folic acid structure]

para-aminobenzoic acid (PABA)   *folic acid*

Figure 29.6 Because of its structural similarity to PABA (in color), sulfanilamide prevents bacteria from synthesizing folic acid from PABA.

pyrimidine rings for nucleotides; without nucleotides, cells can't make DNA and RNA. Certain bacteria (unlike human cells) cannot transport folic acid into their cells, so they must synthesize their own.

Bacteria synthesize folic acid from *p*-aminobenzoic acid (PABA). But sulfanilamide structurally resembles PABA and thus competitively inhibits an en-

Although they kill bacteria, sulfa drugs aren't usually considered antibiotics because they are synthetic compounds not produced by organisms.

zyme in PABA metabolism (Figure 29.6). When bacteria can't make folic acid, they can't multiply and soon die. Human cells aren't harmed by sulfanilamide because they don't synthesize folic acid; they receive it from the diet.

Sulfa drugs became the miracle drugs of the 1930s and 1940s. They lowered the death rate for streptococcal meningitis, for example, from 95% to less than 10%. Although now largely replaced by antibiotics, sulfa drugs are used to treat wounds, burns, and urinary tract infections.

Penicillin. **Antibiotics** are chemicals produced by organisms—usually bacteria, molds, or fungi—that kill or stop the growth of infectious bacteria. In order to do this, antibiotics attack a crucial chemical process that is different in bacteria than in human cells.

The first antibiotic was penicillin (see Chapter Opener), which is produced by a *Penicillium* mold (Figure 29.7). Penicillin kills bacteria by inhibiting an enzyme necessary to make cell walls. The wall consists of carbohydrate and protein. An enzyme normally catalyzes formation of a peptide bond that cross-links two peptide units in the wall. Because it has a ring amide group (Figure 29.8) that structurally resembles the substrate, penicillin binds to the enzyme and blocks its action. As a result bacteria make defective cell walls that cause their cells to rupture and die (Figure 29.9). Human cells don't have a cell wall like this, so penicillin doesn't harm them.

Figure 29.7 *Penicillium chrysogenum* is used to produce penicillin.

penicillin G
(original penicillin)

penicillin V
(acid resistant)

ampicillin
(acid resistant)

cloxacillin
(acid- and penicillinase resistant)

Figure 29.8 Structures of four types of penicillin. All have a ring amide group (blue). Differences from penicillin G are in red.

There are some drawbacks, however, to penicillin's original form (penicillin G). Because stomach acid destroys its activity, it must be injected. And some bacterial strains have become resistant to the drug by producing an enzyme (penicillinase) that hydrolyzes and destroys the antibiotic. In recent years, scientists have synthesized derivatives of penicillin that can be taken orally and are resistant to penicillinase (Figure 29.8).

Other Antibiotics. Penicillins, like sulfa drugs, are effective against only certain types of bacteria. So scientists looked for antibiotics that were effective against a wider variety of disease-causing bacteria. They discovered wide-spectrum antibiotics such as the tetracyclines Aureomycin and Terramycin (Figure 29.10). All tetracyclines have a fused four-ring (tetracyclic) structure; their differences are in the functional groups bonded to the rings.

Medical Drugs   Chapter 29   **605**

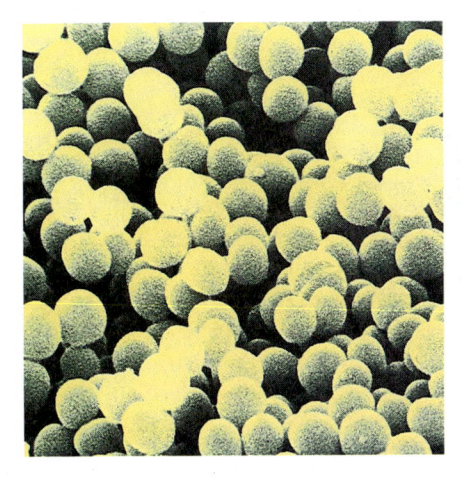

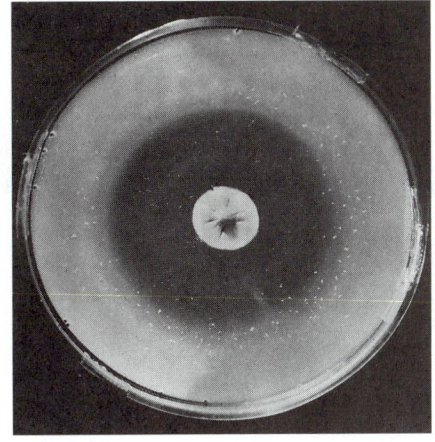

Figure 29.9 Electron micrographs of *Staphylococcus* bacteria in the growing phase (left) and (right) in a dish after exposure to penicillin.

Terramycin

Aureomycin

Figure 29.10 Two tetracyclines. Notice their four-ring or tetracyclic structures. The structural difference is in color.

Many antibiotics other than penicillins act by interfering with bacterial protein synthesis (Table 29.1). They inhibit either transcription (RNA synthesis; see Section 23.3) or translation (protein synthesis from DNA and RNA; see Section 23.4).

Puromycin, for example, inhibits translation because it structurally resembles a tRNA-amino acid complex (Figure 29.11). During translation, puromycin

**Table 29.1
Antibiotics That Inhibit Protein Synthesis**

Inhibit Transcription
actinomycin D
rifamycin
Inhibit Translation
chloramphenicol
erythromycin
puromycin
tetracycline

t-RNA-amino acid

puromycin

Figure 29.11 (Left) Structure of an amino acid bound to adenosine at the 3' end of tRNA. (Right) Puromycin has a similar structure. Differences are in color.

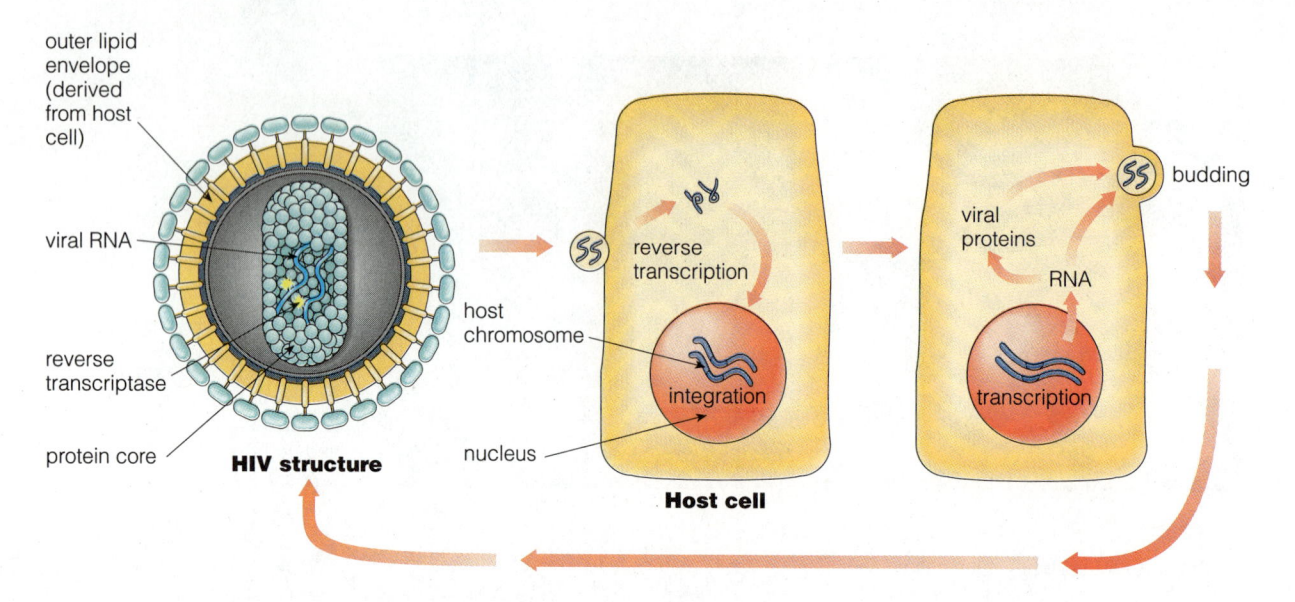

Figure 29.12 Life cycle of HIV.

guanosine
(normal component in DNA)

acyclovir

Figure 29.13 Acyclovir resembles guanosine. Differences are in color.

binds to mRNA and thus halts protein synthesis. Rifamycin, in contrast, binds to bacterial RNA polymerase and inhibits its action; without mRNA synthesis (transcription), bacteria cannot synthesize proteins and soon die.

Antiviral Drugs. Antibiotics don't work against viruses. **Viruses** typically consist of DNA or RNA enclosed in protein and sometimes other material.

For example, the AIDS virus (human immunodeficiency virus or HIV) contains RNA and a complex coat of protein and lipid (Fig. 29.12). Viral RNA enters a cell and causes complementary DNA to be synthesized by reverse transcriptase (see Chemistry Spotlight, Section 23.3). Once incorporated into the infected cell's own DNA, this new DNA directs the cell to synthesize more HIV viruses. Those viruses eventually leave the cell, spreading the infection. Recall that the AIDS drugs AZT and DDI (see Figure 23.19) chemically resemble normal nucleosides and thus block reverse transcriptase.

Another antiviral drug is acyclovir, which structurally resembles guanosine, a normal component of DNA and RNA (Figure 29.13). Acyclovir is used as an ointment to treat genital herpes. Although it doesn't cure herpes, acyclovir makes the periodic flare-up of sores less frequent and of shorter duration. Though now overshadowed by AIDS, genital herpes afflicts more than 20 million people in the United States.

29.2 ANTIMETABOLIC DRUGS

Many drugs work because they structurally resemble a natural substance and competitively inhibit an enzyme metabolizing that substance. Drugs that block metabolism in this way are **antimetabolites.**

We have already discussed many examples of antimetabolites. Sulfa drugs block the metabolism of *p*-aminobenzoic acid (PABA) to folic acid (Section 28.1). The antiviral drugs AZT, DDI, and acyclovir also are antimetabolites. Table 29.2 lists other antimetabolic drugs you learned about in previous chapters.

Table 29.2 Examples of Antimetabolic Drugs

Antimetabolite	Blocks the Metabolism of	Medical Use
Ethanol (Section 14.4)	Methanol, ethylene glycol	Treat poisoning by methanol, ethylene glycol, and other alcohols
5-Fluorouracil (Section 22.4)	Uracil	Several types of cancer
6-Mercaptopurine (Section 22.4)	Adenine	Acute lymphatic leukemia
Dicoumarol, warfarin (Section 22.6)	Vitamin K	Anticoagulant
Allopurinol (Section 27.4)	Hypoxanthine	Gout

Methotrexate (Figure 29.14) is an antimetabolite that resembles folic acid and inhibits its reduction to tetrahydrofolic acid (see Table 22.1), a coenzyme necessary for DNA synthesis. As a result, methotrexate prevents replication and cell division. It is used to treat acute leukemia and rheumatoid arthritis.

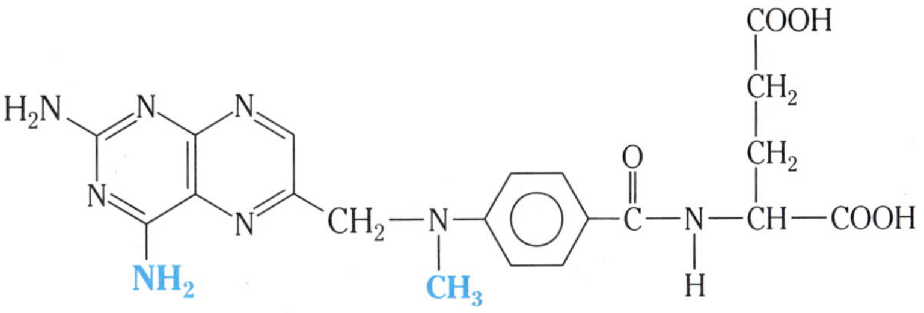

methotrexate

Figure 29.14 Structural differences between the antimetabolite methotrexate and folic acid (see Figure 29.6) are in color.

One experimental use of methotrexate brings cancer patients to the brink of death and then rescues them. These patients receive very high doses of the drug—doses that would ordinarily be fatal—and after several hours are rescued by large doses of a form of folic acid that counteracts methotrexate. This limited exposure to toxic levels of methotrexate takes a heavy toll on cancer (and other) cells that normally would multiply during that time.

Another group of antimetabolites are *antihistamines*, drugs that structurally resemble histamine and block its action (Figure 29.15). An allergic response or

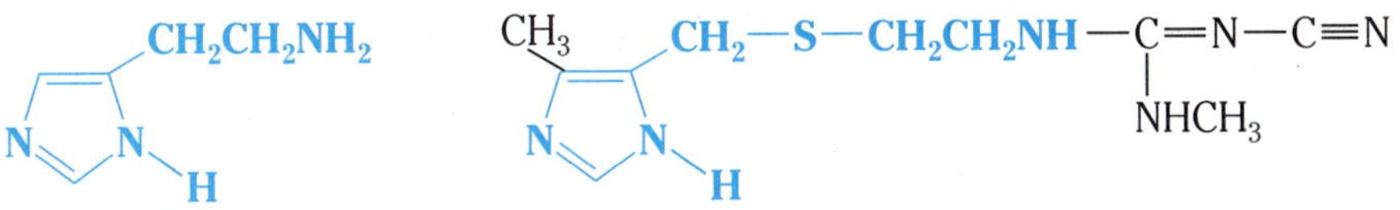

histamine

cimetidine (Tagamet)

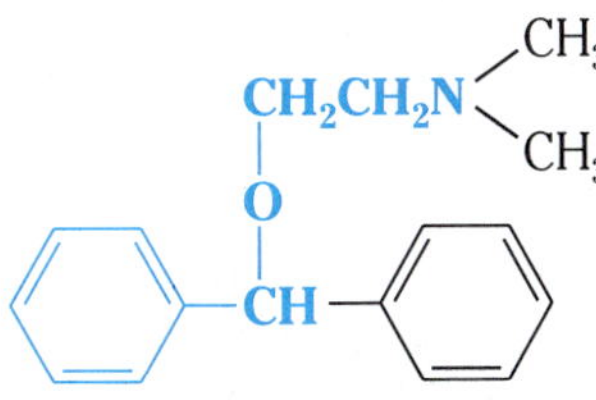

diphenhydramine (Benadryl)

Figure 29.15 Histamine and two antihistamines. Structural similarities are in color.

trauma causes cells to secrete histamine, which dilates capillaries and increases the flow of plasma materials into interstitial fluid. This produces swelling, inflammation, and low blood pressure. Allergy symptoms are treated with antihistamines such as diphenhydramine (trade name Benadryl; Figure 29.15), which binds to histamine receptors in the respiratory tract and blocks the effects of histamine.

Histamine also binds to receptors in stomach cells and stimulates them to secrete hydrochloric acid (HCl). Excess secretion of HCl produces stomach and duodenal ulcers. One drug used to treat ulcers is cimetidine (trade name Tagamet; Figure 29.15). This antihistamine binds to histamine receptors in stomach cells, preventing the secretion of excess HCl.

29.3 NERVOUS SYSTEM DRUGS

Nerve Impulses. Before you can understand how drugs affect the nervous system, you need to know what nerve cells are, how impulses are generated in the cell, and how impulses travel from one nerve cell, or *neuron,* to another on their way to target cells such as a gland, cardiac muscle, or skeletal muscle.

A typical neuron has a long fiber—an *axon*—at one end, and many branching, antennalike fibers—*dendrites*—at the other end (Figure 29.16). A nerve impulse travels from the axon of one neuron to a dendrite of a neighboring neuron. Since axons don't touch the dendrites of receiving neurons, nerve impulses must cross a tiny gap—a *synapse*—between the axon of one cell and a dendrite of another.

A neuron at rest is like a tiny battery waiting to discharge its energy. For a nerve signal to cross a synapse, tiny knobs at the end of an axon have to release chemicals known as **neurotransmitters.** A neurotransmitter crosses the synapse and binds to the dendrite of a receiving cell, altering its membrane so that Na^+ ions flow into the cell and K^+ ions flow out. The change in the membrane's electrical properties allows it to receive a nerve impulse across the synapse. Once a signal is received, the neurotransmitter is immediately removed from the dendrite so the neuron can recharge to receive another impulse.

Different neurons use different neurotransmitters. Common neurotransmitters include acetylcholine, dopamine, epinephrine, norepinephrine (see Figure 27.3), serotonin (Figure 29.17), histamine (Figure 29.15), and

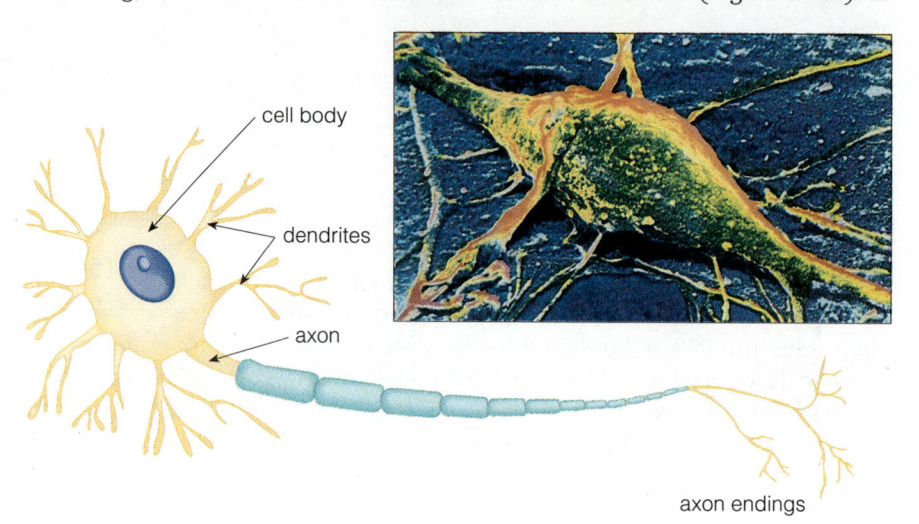

Figure 29.16 (Left) Structure of a neuron. (Right) Scanning electron micrograph of a neuron clearly shows the cell body and dendrites.

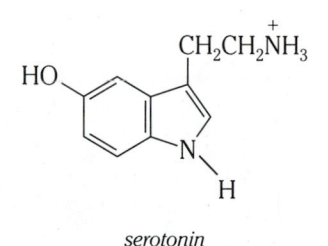

Figure 29.17 Two neurotransmitters.

gamma-aminobutyrate (GABA; Figure 29.17). Some (such as acetylcholine) excite a receiving cell to send the impulse on to another neuron, whereas others (such as GABA) inhibit the impulse from traveling farther.

Neurotransmitter Inhibitors. One of the most studied neurotransmitters is acetylcholine (ACH). As soon as they receive a nerve signal, neurons use the enzyme acetylcholinesterase (ACHase) to hydrolyze the ester ACH:

$$CH_3\overset{O}{\underset{\|}{C}}-O-CH_2CH_2\overset{+}{N}(CH_3)_3 + H_2O \underset{\text{choline acetylase}}{\overset{\text{ACHase}}{\rightleftharpoons}} CH_3COOH + HO-CH_2CH_2\overset{+}{N}(CH_3)_3$$

acetylcholine　　　　　　　　　　　　　acetic acid　　　choline

Choline acetylase catalyzes the reverse reaction to resynthesize acetylcholine so that it can be used again.

Many drugs and toxins inhibit acetylcholine action. Certain phosphate esters resemble acetylcholine enough to bind to acetylcholinesterase and inhibit the hydrolysis of acetylcholine (Figure 29.18). When this happens, ACH stays on the receiving neuron and overstimulates muscles, cells, glands, and nerves in an uncontrolled way, causing choking, convulsions, paralysis, and death. Two examples are sarin, a nerve gas, and parathion, an insecticide that kills its victims by disrupting their nervous systems.

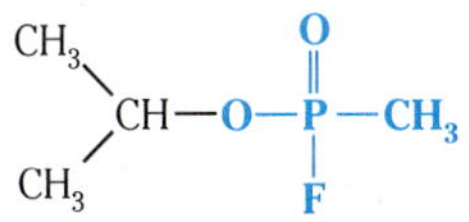

Some compounds inhibit choline acetylase. The toxins in dinoflagellates, microscopic marine organisms that cause "red tides," block ACH synthesis in this way. *Clostridium botulinum*, the bacterium which causes botulism, also produces a toxin that inhibits choline acetylase. By preventing normal transmission of nerve impulses with ACH, such neurotoxins paralyze key organs and can cause death by respiratory failure.

Other compounds block ACH from binding to dendrites of receiving neurons. This prevents transmission of nerve signals. Cobra venom works in this way, as do the local anesthetics nupercaine, procaine, cocaine, and tetracaine. Curare, a plant extract used by Amazon Indians to make poisoned arrows that paralyze their prey, is a muscle relaxant when administered in small doses. Atropine, another poison, is used at low concentrations to dilate the pupil of the eye during examination.

Atropine is also an antidote for poisoning by nerve gases or insecticides such as parathion. By blocking the binding of ACH to receptor neurons, atropine prevents the overstimulation of neurons caused by acetylcholinesterase inhibitors.

Neural and Mental Disorders. Recall that Parkinson's disease (see opening to Chapter 27) results from a shortage of the neurotransmitter dopamine; a common treatment is the drug L-dopa, which is metabolized to dopamine in the brain.

In schizophrenia, neurons that use dopamine transmit too many nerve signals. Chlorpromazine (trade name Thorazine) blocks the binding of dopamine to receptors in the brain. As a result, chlorpromazine lowers metabolic rate, reduces blood pressure, and gives the patient a calm, tranquil feeling. Chlorpromazine and similar drugs are used to treat schizophrenia and manic depression.

Some compounds that inhibit the breakdown of neurotransmitters also are useful antidepressant drugs. For example, some antidepressants inhibit monoamine oxidase (MAO), an enzyme that normally oxidizes neurotransmitters

sarin
(nerve gas)

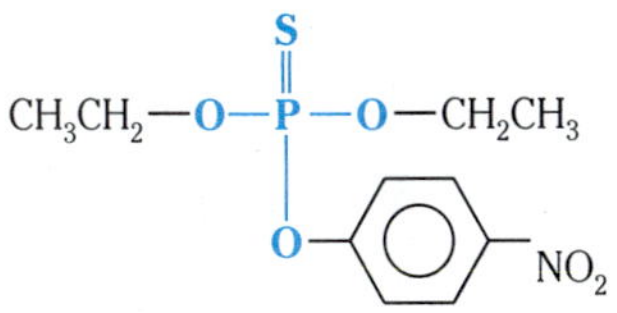

parathion
(insecticide)

Figure 29.18 Two phosphate esters (color) that inhibit acetylcholinesterase.

Botulin toxin is one of the most potent poisons known; a dose of 10^{-7} g can kill a person.

CHEMISTRY SPOTLIGHT

Alzheimer's Disease

Alzheimer's disease (AD) was first described in 1906 by the German physician Alois Alzheimer. This disease occurs primarily in people more than 50 years old. Patients typically suffer memory loss (especially of recent events), become easily confused, and gradually lose their ability to read, write, calculate, and speak.

Patients with AD have much fewer neurons and lower acetylcholine levels in a region of the brain called the basal nucleus. In addition, they have many plaques of a protein called β amyloid, and those plaques are surrounded by tangles of neurons.

One treatment strategy is to give AD patients drugs that increase acetylcholine levels in the brain. This is similar to the approach used with Parkinson's disease. Such drugs include choline and lecithin, both of which can be made into acetylcholine. Another compound, physostigmine (Figure 29.19), increases acetylcholine levels by inhibiting acetylcholinesterase. Tacrine (trade name Cognex), approved in 1993 for use in AD patients, acts in a similar way. The results with all of these compounds have been modest at best.

Scientists recently discovered a gene associated with AD. The gene codes for a protein (apolipoprotein E) that is part of lipoproteins in blood that transport cholesterol and other lipids. How—or whether—it causes AD is unknown. But the discovery may lead to an understanding of what causes Alzheimer's and to possible ways to treat the disease.

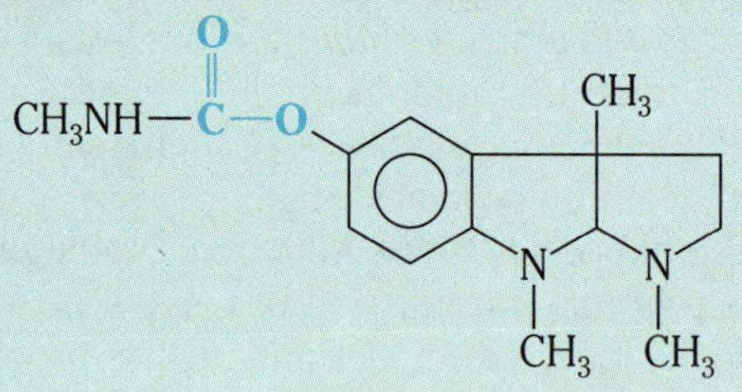

physostigmine

Figure 29.19 Physostigmine.

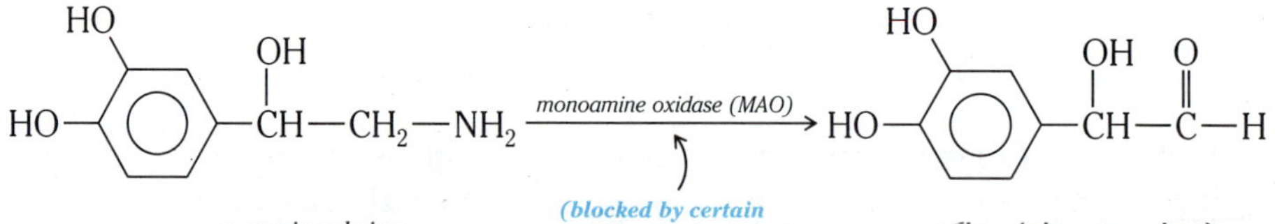

Figure 29.20 By inhibiting MAO, certain antidepressants increase the level of norepinephrine and other neurotransmitters in the brain.

norepinephrine → monoamine oxidase (MAO) → (breakdown product)
(blocked by certain antidepressant drugs)

such as serotonin and norepinephrine (Figure 29.20). By inhibiting MAO, drugs such as isocarboxazid (Marplan) and tranylcypromine (Parnate) increase the concentration of neurotransmitters in the brain, stimulating certain nerve pathways and relieving depression.

A variety of other compounds also affect nerve transmission. The hallucinogen LSD probably works by altering the action of serotonin. Both mescaline, a hallucinogen in peyote cactus, and amphetamine, a stimulant, resemble dopamine and alter its action (Figure 29.21). And γ-aminobutyrate (Figure 29.17) is sometimes used as a drug to treat epilepsy, a disease characterized by low levels of that neurotransmitter.

29.4 TREATING HORMONE DISORDERS

Thyroid Disorders. Your thyroid gland synthesizes thyroxin, a hormone that stimulates protein synthesis and catabolic pathways in muscle and liver. People with overactive thyroid glands, a condition called *hyperthyroidism*, tend to be slender and very active. Those with low thyroid activity, a condition known as *hypothyroidism*, tend to be obese and sluggish.

One reason for thyroxin deficiency, particularly in poorer countries, is a lack of iodide ion (I^-) in the diet. Your thyroid gland needs iodide ion to synthesize thyroxin from the amino acid tyrosine (see Figure 27.3). To try to compensate for a shortage of dietary iodide thyroid glands may enlarge; this condition is called *goiter* (see Figure 3.18).

The cure for goiter is simply to add iodide to the diet. A common source is *iodized salt*, which contains NaI or KI in addition to NaCl. Seafoods are also a rich source of iodide ion. People who have other types of hypothyroidism may be treated directly with thyroid hormones.

Hyperthyroidism may be treated with extra high doses of iodide (I^-) ion, which causes thyroid glands to secrete less thyroxin. Thiocyanate (SCN^-) and perchlorate (ClO_4^-) ions reduce thyroxin synthesis by competing with I^- for transport into thyroid cells. In addition, thiourea and thiouracil (Figure 29.22) inhibit oxidation reactions that incorporate iodide into thyroxin.

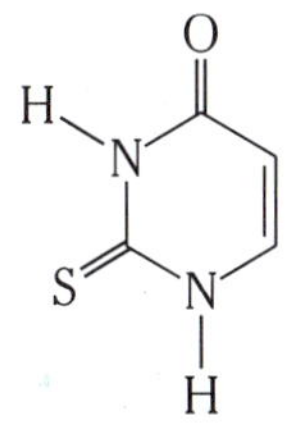

Graves' disease, a hyperthyroid condition, is sometimes treated by administering radioactive iodine-131, a β emitter that has a half-life of 8 days (see Section 10.3). Iodine-131 concentrates in the thyroid gland, where its emissions destroy active cells. This treatment, a chemical equivalent of surgery, leaves fewer thyroid cells to produce thyroxin.

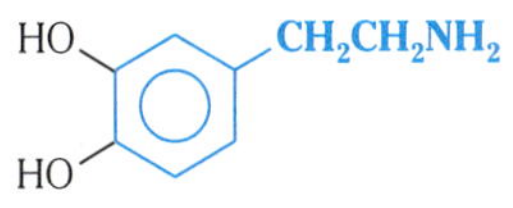

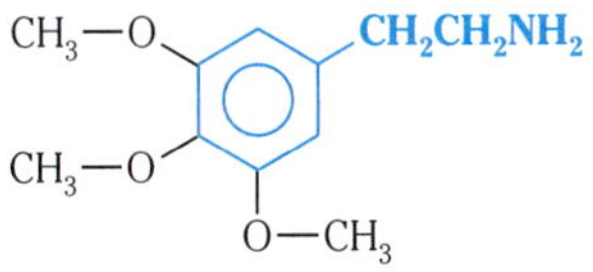

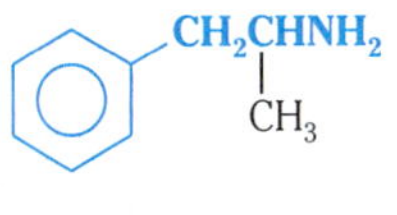

Diabetes. Diabetics don't secrete enough insulin. Those with type I diabetes typically get the disease by age 20 and must take regular injections of insulin (see Section 25.6). Type II diabetes typically occurs later in life than type I. Symptoms such as hyperglycemia and ketoacidosis develop gradually and often can be controlled by diet and exercise. But some diabetics also need to take oral *hypoglycemic drugs*, which lower blood glucose levels by stimulating the pancreas to secrete more insulin.

Pregnancy Control. *Progestins* are female sex hormones. The most important progestin is progesterone (see Figure 20.12), which is secreted by ovaries (see Table 28.4). Progesterone alters the lining of the uterus so that a fertilized egg can implant there. Once pregnancy begins, secretion of progesterone prevents ovulation.

Birth control pills typically contain a synthetic progestin, sometimes in combination with a synthetic estrogen. The drug mimics the natural action of progesterone in blocking ovulation. When ovulation doesn't happen, neither does pregnancy. Hormones used as oral contraceptives are chemically modified to prevent their breakdown in the digestive tract (see opening to Chapter 13).

RU-486 (Figure 29.23) is a drug that structurally resembles progesterone and binds to the same receptor in the uterus. As a result, RU-486 blocks the action of progesterone and prevents implantation of a fertilized egg. This drug is used to terminate pregnancy by releasing an implanted embryo from the uterus.

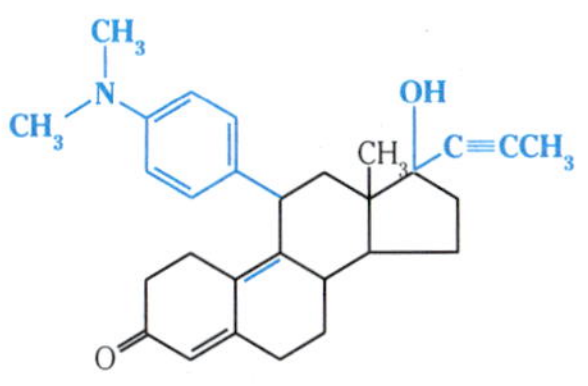

Figure 29.21 Structural similarities between dopamine, mescaline, and amphetamine are in color.

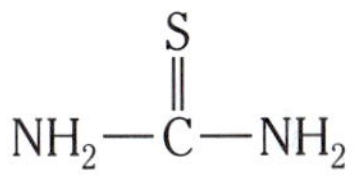

Figure 29.22 Two drugs that treat hyperthyroidism.

Figure 29.23 RU-486 resembles progesterone (see Figure 20.12). Structural differences are in color.

Figure 29.24 Exercise at high altitudes increases the concentration of red blood cells.

29.5 DRUGS AND ATHLETIC PERFORMANCE

Athletes use a wide variety of chemicals to improve performance. Recall that glycogen loading (see opening to Chapter 25) increases available carbohydrate fuel. Athletes also drink electrolyte fluids during training and competition (Table 29.3); this provides H_2O, salts such as Na^+ and K^+, and more carbohydrates.

Table 29.3 Contents of Sports Drinks per 227 g (8 oz)

Drink	Recommended Concentration (%)	Carbohydrate	Major Ions	Calories
Carbo Plus	16	Maltodextrin	Na^+ (5 mg) K^+ (100 mg) Mg^{2+} (100 mg)	170
E. R. G.	5.7	Glucose	Na^+ (70 mg) K^+ (100 mg)	45
Exceed	7.2	Glucose polymer	Na^+ (66 mg) K^+ (56 mg) Mg^{2+} (6 mg)	68
Gatorade	6	Glucose, sucrose	Na^+ (110 mg) K^+ (25 mg)	50
Max	7.5	Glucose polymer	Na^+ (15 mg)	70

Another approach is to increase the supply of O_2 to oxidize carbohydrates and fats and generate ATP. Regular exercise increases heart and lung capacity so that more O_2 binds to hemoglobin and reaches tissues. A greater number of red blood cells also increases the supply of O_2 to tissues. Training at high altitude accomplishes this naturally (Figure 29.24); the body responds to less O_2 at altitude by synthesizing more red blood cells.

But some athletes increase their red blood cell counts by a technique called *blood doping*. Although the details vary, they typically have a liter of their own blood removed and the red blood cells stored for 8–12 weeks while their bodies replenish their own supply. Then, about a day before competition, they are injected with their stored red blood cells. The extra cells instantly boost the amount of O_2 their blood carries—an advantage in endurance events such as the marathon or long bicycle races.

A natural kidney hormone, erythropoietin (EPO), now provides a new way to achieve this same effect. This hormone stimulates red blood cell production and increases the cell count by as much as 10%. EPO produced by recombinant DNA technology is used to treat anemia and to help some cancer patients replenish red blood cells. The users of this drug, however, also include some athletes.

Although illegal, blood doping is difficult to detect because no foreign substance is in the body.

Illegal Drugs. Five classes of drugs are illegal in athletic competition: narcotics and analgesics, stimulants, beta blockers, anabolic steroids, and diuretics.

Narcotics and analgesics relieve pain. Compounds such as morphine bind to receptor sites in the brain and reduce the perception of pain. Recall that enkephalins and endorphins are natural compounds that work in the same way (see Section 21.6). Analgesics such as aspirin and ibuprofen work by inhibiting the metabolism of certain prostaglandins (see Section 20.6). Also banned are compounds such as antihistamines (Section 29.2).

Figure 29.25 Beta blockers slow the pulse and increase shooting accuracy.

Amphetamines and other stimulants are used to reduce fatigue and to increase aggressiveness, attentiveness, and the perception of energy. Many enhance the action of neurotransmitters (Figure 29.20). Caffeine increases glycogenolysis in muscle and the removal of fatty acids from storage to be used as fuel. For this reason, coffee is a popular drink an hour or so before a marathon begins. The allowable limit for caffeine in athletic competition is the equivalent of six cups of coffee. The allowable limit for amphetamines is zero.

Beta blockers such as propranolol (Inderal) are used to treat high blood pressure. They bind to beta receptors in heart tissue and block the normal stimulatory action of epinephrine (which binds to the same receptors). As a result, the heart pumps slower and with less force. In sports such as archery and shooting, a slower pulse is an advantage because the competitor can more easily shoot between heartbeats; this provides a steadier aim and greater accuracy (Figure 29.25).

The most abused drugs in athletic competition are the *anabolic steroids*, which include testosterone and many synthetic derivatives (see opening to Chapter 20). These compounds stimulate protein synthesis and thus increase muscle mass. Testosterone injections can be detected by elevated levels of the hormone in urine compared to its metabolic product epitestosterone (Figure 29.26).

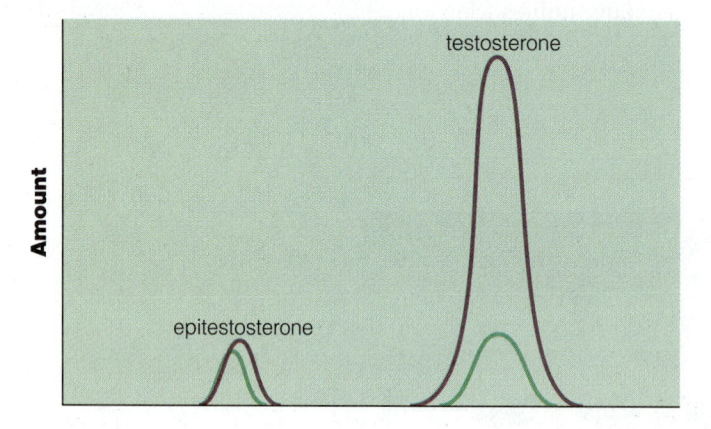

Figure 29.26 Graph of testosterone levels relative to epitestosterone, a metabolic product, in a normal subject (green line) and in a person taking testosterone injections (violet line). A testosterone–epitestosterone ration of 6.0 or more indicates the use of testosterone.

Diuretics (see Section 28.6) could be misused by athletes such as boxers and wrestlers, who have to reach a weight limit to compete. But the main reason diuretics are banned is that they increase the volume of urine enough to make the concentration of other illegal substances too low to detect.

> A drug is any substance which, when injected into a laboratory rat, produces a scientific paper.
>
> *Anonymous*

SUMMARY

Sulfa drugs are synthetic compounds derived from sulfanilamide that kill bacteria by inhibiting their synthesis of folic acid. Antibiotics are natural compounds, or their derivatives, that kill bacteria in various ways. Penicillins inhibit cell wall synthesis. Other antibiotics inhibit bacterial transcription or translation. Some antiviral drugs resemble nucleosides and inhibit reverse transcription in infected cells.

Antimetabolites block a normal metabolic process, usually by competitive inhibition of a key enzyme. Examples include the sulfa drugs, antiviral drugs, anticoagulants, certain anticancer drugs, and antihistamines.

Neurotransmitters enable nerve impulses to cross a synapse between the axon of one neuron and a dendrite of another neuron. Many drugs and toxins work by interfering with the action of a neurotransmitter; they may inhibit the synthesis or breakdown of a neurotransmitter, or block its binding to a receptor. Such drugs include local anesthetics, tranquilizers, antidepressants, stimulants, and hallucinogens.

Hypothyroidism may be treated by adding iodide ion (I^-) to the diet or administering the hormone thyroxin, depending on the cause of the disorder. Hyperthyroidism is treated by compounds that reduce thyroxin synthesis or destroy thyroid tissue. Diabetics are treated with insulin or with oral hypoglycemic agents, depending on the type of diabetes. Oral contraceptives work by creating a false pregnancy that prevents ovulation.

Athletes maximize their intake and storage of fuel, especially carbohydrates. They increase their ability to supply O_2 to cells by exercise, training at high altitude, and artificial methods such as blood doping and using erythropoietin. The types of drugs banned for use in competition are: narcotics and analgesics, diuretics, stimulants, beta blockers, and anabolic steroids.

KEY TERMS

Antibiotic (29.1)
Antimetabolite (29.2)
Chemotherapy (29.1)

Drug (29.1)
Neurotransmitter (29.3)

Sulfa drug (29.1)
Virus (29.1)

EXERCISES

Even-numbered exercises are answered at the back of this book.

Chemotherapy

1. Which of the following are examples of chemotherapy:
   (a) taking methotrexate to treat cancer
   (b) taking vitamin supplements for general health
   (c) training at high altitude
   (d) taking vitamin B_1 (thiamine) for beriberi
   (e) taking morphine to relieve acute pain
   (f) blood doping

2. Which of the following are examples of chemotherapy:
   (a) taking amphetamine to stay awake
   (b) taking amphetamine to lose weight
   (c) taking vitamin C to treat a cold
   (d) taking penicillin for an infection
   (e) taking electrolytes during athletic competition
   (f) taking chlorpromazine for schizophrenia

Antibiotics and Antiviral Drugs

3. Distinguish between a disinfectant and an antibiotic.
4. Why don't antibiotics kill normal body cells?
5. Why do some drugs kill microbes in the body but not outside the body?
6. How do sulfa drugs kill bacteria?
7. Bacteria normally synthesize folic acid from p-aminobenzoic acid (PABA). Write the structural formula for PABA.
8. Are sulfa drugs antibiotics? Explain.

616 Part III Biochemistry

9. How do penicillins kill bacteria?
10. Penicillin inhibits **(a)** an enzyme that synthesizes a glycosidic bond, **(b)** transcription, **(c)** translation, **(d)** binding to a receptor, **(e)** an enzyme that synthesizes a peptide bond.
11. Chemical modification of the original penicillin produced compounds with what other qualities of pharmaceutical value?
12. Which of the options in exercise 10 describes how puromycin works as an antibiotic?
13. Which of the options in exercise 10 describes how tetracyclines work as antibiotics?
14. Name the enzyme inhibited by AZT and DDI, two drugs used to treat AIDS.
15. Identify which of the following could be inhibited by acyclovir, an antiviral drug: **(a)** electron transport, **(b)** reverse transcription, **(c)** urea cycle, **(d)** replication, **(e)** digestion.

Antimetabolites

16. Identify which of the following are antimetabolites: **(a)** sulfanilamide, **(b)** morphine, **(c)** AZT, **(d)** methotrexate, **(e)** parathion.
17. Distinguish between an antimetabolite and an antibiotic.
18. What substance is used to "rescue" patients receiving a toxic dose of methotrexate?
19. Does methotrexate selectively kill cancer cells and not normal body cells? Explain.
20. How does histamine reduce blood pressure?
21. Which of the options in exercise 10 describes how antihistamines work?
22. In antihistamine preparations, diphenhydramine (Figure 29.15) is present as its HCl salt. Write the structural formula for diphenhydramine hydrochloride. [Review Section 17.4 if necessary.]
23. Antimetabolites typically function by **(a)** breaking down metabolic substances, **(b)** inhibiting enzymes, **(c)** denaturing proteins, **(d)** stimulating the synthesis of metabolic compounds.

Nervous System Drugs

24. Name the following structural features of neurons: **(a)** the extension that receives a nerve impulse, **(b)** the gap between nearby neurons, **(c)** the extension that sends a nerve impulse to the next neuron.
25. The concentrations of which ions in a neuron are altered when a neurotransmitter binds to a neuron?
26. What happens if a neurotransmitter is not removed from a receiving neuron?
27. Acetylcholine can be classified as **(a)** a primary amine, **(b)** a secondary amine, **(c)** a tertiary amine, **(d)** a quaternary amine.
28. Write the IUPAC name for the neurotransmitter gamma-aminobutyrate (Figure 29.17). [Review Section 16.1 if necessary.]
29. List the enzyme inhibited by **(a)** nerve gases and **(b)** botulin toxin.
30. Why is the concentration of atropine used in eye drops very important?
31. What does chlorpromazine (Thorazine) do that makes it useful as a treatment for schizophrenia?
32. Substances that inhibit monoamine oxidase (MAO) **(a)** kill bacteria, **(b)** block the action of morphine, **(c)** lower the levels of neurotransmitters, **(d)** increase the levels of neurotransmitters, **(e)** block cell division.

Treating Hormone Disorders

33. Name the main thyroid hormone. Which element in its structure is not common in other biochemicals?
34. Is an enlarged thyroid gland associated with hypothyroidism or hyperthyroidism?
35. Why is it important that iodine-131, a radioisotope used to treat Graves' disease, have a fairly short half-life? [Review Sections 10.3 and 10.5 if necessary.]
36. Why are drugs that stimulate the pancreas to secrete insulin called *hypoglycemic* drugs?
37. Oral contraceptives contain a substance that structurally resembles what natural hormone?
38. How does RU-486 terminate pregnancy?

Drugs and Athletic Performance

39. For which one of the following track and field events might training at high altitude be the most helpful: **(a)** throwing a javelin, **(b)** 100-meter dash, **(c)** 1500-meter run, **(d)** long jump, **(e)** pole vault.
40. What are the medical uses of erythropoietin?
41. Why are diuretics illegal for athletic competition?
42. Why are beta blockers illegal for athletic competition?
43. Identify which of the following are the most likely activities for which athletes would be tempted to take anabolic steroids: **(a)** soccer, **(b)** marathon, **(c)** weight lifting, **(d)** football, **(e)** diving.

DISCUSSION EXERCISES

1. Why does the extensive use of antibiotics increase the problem of antibiotic-resistant strains of bacteria?
2. Why do cancer patients who receive antimetabolite drugs typically experience side effects such as nausea, low blood counts, and hair loss?
3. Suppose a drug that relieves pain has been shown to cause cancer in rats. Would you favor banning the drug, restricting its use, or allowing free usage but with a warning label? Defend your answer.
4. Since no foreign substance is used, should "blood doping" be illegal for athletic competition? Defend your answer.

APPENDIX A
Significant Figures

When you make a measurement, the number of figures you report depends on the fineness, or detail, of the measurement. Suppose, for example, you measured a sidewalk to be 21.3 m long. This means your measuring device allows you to measure to the nearest tenth of a meter. This measurement has three significant figures: 2, 1, and 3.

You can convert this distance into other units, such as km or mm, by the following:

$$21.3 \text{ m} \times \frac{1 \text{ km}}{1{,}000 \text{ m}} = 0.0213 \text{ km}$$

or,

$$21.3 \text{ m} \times \frac{1000 \text{ mm}}{1 \text{ m}} = 21{,}300 \text{ mm}$$

But changing units cannot change the number of significant figures, for it doesn't change the fineness of the original measurement. In both calculations above, the number of significant figures remains 3, the same as in the original measurement.

This example raises two important questions. First, why aren't the answers limited to one significant figure since the conversion factors have terms (1 km or 1 m) with only one number? The answer is that *exact numbers have an indefinite number of significant figures.* Thus, they don't limit the number of significant figures in the calculated answer.

Conversion factors containing 1, 10, or multiples of 10 that are often used in metric conversions are exact and don't limit the number of significant figures in calculated answers.

The second question is more complex: How can you tell whether or not a number in a measurement is significant? The rules are:

- all nonzero numbers are significant
- zeros are significant *except* when they
  (a) precede the first nonzero number, or
  (b) follow the last nonzero number and no decimal point is present.

In the example above, the zeros in 0.0213 are not significant because of rule (a). The zeros in 21,300 are not significant because of rule (b). Both 0.0213 and 21,300 have 3 significant figures, just like the original, measured number: 21.3.

EXAMPLE A.1 Identify the number of significant figures in each of the following: (a) 21.04, (b) 00402, (c) 300, (d) 10,306, (e) 0.002, (f) 14.300, (g) 0.019840, (h) 300.7, (i) 21.0000.

SOLUTION (a) 4, (b) 3 (00**402**), (c) 1 (**3**00), (d) 5, (e) 1 (0.00**2**), (f) 5, (g) 5 (0.0**19840**), (h) 4, (i) 6.

When making calculations, you must be careful about writing figures that are not significant in your answer. This can easily happen, especially when using a calculator that displays eight or more numbers, whether they are significant or not. Three rules cover the appropriate number of significant figures in calculated answers.

Rule 1. When you add or subtract, identify the number which has the fewest significant figures past the decimal point. Limit your answer to that same number of significant figures past the decimal point. Two examples are shown below:

$$\begin{array}{r} 268.414 \\ +638.28 \\ \hline 906.694 = 906.69 \end{array} \qquad \begin{array}{r} 638.287 \\ -268.41 \\ \hline 369.877 = 369.88 \end{array}$$

Rule 2. When you multiply or divide, identify the number with the fewest significant figures. The number of significant figures that it has is the number your answer must have. Two examples follow.

(a) What is the length in cm of 16.03 m?

$$16.03 \text{ m} \times \frac{100 \text{ cm}}{1 \text{ m}} = 1603 \text{ cm}$$

(Metric conversion factors such as $\frac{100 \text{ cm}}{1 \text{ m}}$ are exact, and thus do not limit significant figures in the calculated value. So the significant figures allowed in the answer (4) are limited by the 4 significant figures in the measured value: 16.03 m.)

(b) The density of mercury is 13.6 g/mL. What is the volume in liters (L) of 56 g of mercury?

$$56 \text{ g mercury} \times \frac{1 \text{ mL mercury}}{13.6 \text{ g mercury}} \times \frac{1 \text{ L mercury}}{1{,}000 \text{ mL mercury}}$$

$$= 0.0041 \text{ L mercury}$$

(You are limited to 2 significant figures because 56 g mercury has only 2 significant figures.)

Rule 3. When rounding numbers, round the number preceding a final 5 (or a 5 followed only by one or more zeros) up or down to leave a final number that is even. For example:

$$\begin{array}{r} 447.24 \\ -\ 83.665 \\ \hline 363.575 = 363.58 \end{array} \qquad \begin{array}{r} 447.24 \\ -\ 83.675 \\ \hline 363.565 = 363.56 \end{array}$$

$$\frac{7435.8}{612} = 12.150 = 12.2$$

(limited to 3 significant figures)

EXAMPLE A.2 Solve the following, writing your answer in the correct number of significant figures:

(a) 612.5 − 36.57   (b) 16.4 × 1.7   (c) 4.8 / 82.33   (d) 25 / 2.0   (e) 47.18 + 8.135

SOLUTION (a) 575.9, (b) 28, (c) 0.058, (d) 12, (e) 55.32

Chapter 2, Appendixes B and C, and other sections in this book show more examples of calculations. Notice that each one follows rules 1–3 to express answers with the appropriate number of significant figures.

APPENDIX B
Exponents and Scientific Notation

Exponential Numbers. You can write any decimal number as the product of a number multiplied by 10 to some power or exponent. For example, 356 can be written as 3.56×10^2, 0.0036 as 3.6×10^{-3}, and 47,365,104 as 4.7365104×10^7. Numbers written in this way are called *exponential numbers*.

A positive power, or exponent, of 10 indicates how many times a number is multiplied by 10. For example, 3.56×10^2 means $3.56 \times 10 \times 10$, which equals 356. The number 10^4 (which you also can write as 1×10^4) means $10 \times 10 \times 10 \times 10$, which equals 10,000. Each time you multiply by 10, the number increases by moving the decimal point one place to the right. So a positive exponent of 10 in an exponential number tells you how many places to move the decimal point to the right. In the first example, 3.56×10^2, notice that you can convert the number into its full, nonexponential form by moving the decimal point two places to the right (because the exponent is +2): 3.56×10^2 becomes 356.

EXAMPLE B.1 Write the following exponential numbers in their full, nonexponential form: (a) 6.12×10^1, (b) 1.944×10^6, (c) 8.37×10^{11}.

SOLUTION (a) 61.2 (6.12×10), (b) 1,944,000, (c) 837,000,000,000

A negative power or exponent of 10 indicates how many times a number is divided by 10. For example,

$$1 \times 10^{-3} = \frac{1}{10 \times 10 \times 10} = \frac{1}{1000} = 0.001$$

$$6.53 \times 10^{-5} = \frac{6.53}{10 \times 10 \times 10 \times 10 \times 10} = \frac{6.53}{100,000} = 0.0000653$$

Each time you divide by 10, the number decreases by moving the decimal point one place to the left. So a negative exponent of 10 tells you how many places to move the decimal point to the left. Notice that you can convert 6.53×10^{-5} into its full, nonexponential form by moving the decimal point five places to the left (because the exponent is −5):

0 0 0 0 0 6.53 × 10⁻⁵ becomes 0.0000653.

EXAMPLE B.2 Write the following exponential numbers in their full, nonexponential form: (a) 8.29×10^{-6}, (b) 4.51×10^{-1}, (c) 1×10^{-8}.

SOLUTION (a) 0.00000829, (b) 0.451, (c) 0.00000001 or $\frac{1}{100,000,000}$

Notice in Example B.2 that a negative exponent could be written as a positive exponent in the denominator. You could write 1×10^{-8} for example, as:

$$\frac{1}{10^8}$$

A general rule is that you change the sign of the exponent (from positive to negative or *vice versa*) whenever you change an exponential number from numerator to denominator, or from denominator to numerator. We discuss this rule again in the section on multiplying and dividing exponential numbers.

Writing Numbers in Scientific Notation. Scientific measurements and calculations often use very small or very large numbers. For example, 12.0 g of carbon contains 602,000,000,000,000,000,000,000 atoms of carbon. One atom of hydrogen has a mass of about 0.00000000000000000000000167 g. It's inconvenient to write such numbers in this form, and easy to make an error in the number of zeros. To avoid these problems, scientists write such numbers in *scientific notation*.

To write in scientific notation, use exponential numbers, making sure that the decimal number is at least 1 but less than 10. In other words, the decimal point must come after the first nonzero digit.

The two very long numbers in the first paragraph of this section are written in scientific notation as 6.02×10^{23} atoms of carbon and 1.67×10^{-24} g. Notice in both cases that the decimal number is at least 1 but less than 10. The exponent or power of ten specifies how many places to move the decimal point to express the number in its original, nonexponential form.

Here is a simple procedure, using the first example above, to write numbers in scientific notation:

Step 1: Starting with the original number, count the number of places required to move the decimal point to the right of the first nonzero number. (If the original number has no decimal point, put a decimal point at the end of the number.)

6 0 2 0 0 0 0 0 0 0 0 0 0 0 0 0 0 0 0 0 0 0 0 0.

move the decimal point 23 places to the left

Step 2: Write the number in scientific notation except for the exponent of 10.

$6.02 \times 10^?$

Step 3: The exponent of 10 is the number of decimal places moved in Step 1. If the decimal point moved to the left, the exponent is positive. If the decimal point moved to the right, the exponent is negative.

6.02×10^{23} (decimal point moved 23 places to the left)

Now let's write 0.00000000000000000000000167 in scientific notation:

Step 1:

0.0 0 0 0 0 0 0 0 0 0 0 0 0 0 0 0 0 0 0 0 0 0 0 1 67

move the decimal point 24 places to the right

Step 2: $1.67 \times 10^?$

Step 3: 1.67×10^{-24} (decimal point moved 24 places to the right)

EXAMPLE B.3 Write the following numbers in scientific notation: (a) 6,210,000 J, (b) 0.00003002 kg, (c) 0.65 m.

SOLUTION (a) 6.21×10^6 J, (b) 3.002×10^{-5} kg, (c) 6.5×10^{-1} m

Multiplying and Dividing Exponential Numbers. When making calculations, terms with exponents will frequently occur in the numerator or denominator. A few simple rules will help you do these calculations:

1. Convert all exponential numbers into scientific notation (see the preceding section).

2. Move any powers of 10 in the denominator into the numerator, *changing the sign of the exponent.* For example:

$$\frac{4.2 \times 10^6}{1.2 \times 10^3} \text{ becomes } \frac{4.2 \times 10^6 \times 10^{-3}}{1.2}$$

3. Combine powers of 10 into a single term by adding the exponents. In the example above, $10^{6-3} = 10^3$, so the term becomes

$$\frac{4.2 \times 10^3}{1.2}$$

4. Multiply and/or divide the decimal numbers in the usual way and combine that answer with the combined exponent term from Step 3. In this example, $4.2/1.2 = 3.5$, so the answer becomes 3.5×10^3.

5. If necessary, adjust the answer to scientific notation. In this example, no adjustment is needed. Two more examples follow.

Example 1.
$$(0.0000035) \times (1.2 \times 10^2) = (3.5 \times 10^{-6}) \times (1.2 \times 10^2)$$
$$= (3.5 \times 1.2) \times (10^{-6+2}) = (3.5 \times 1.2) \times 10^{-4}$$
$$= 4.2 \times 10^{-4}$$

Example 2.
$$\frac{(258) \times (1.41 \times 10^{12})}{943,000,000} = \frac{(2.58 \times 10^2) \times (1.41 \times 10^{12})}{9.43 \times 10^8}$$
$$= \frac{(2.58 \times 10^2) \times (1.41 \times 10^{12}) \times 10^{-8}}{9.43}$$
$$= \frac{(2.58 \times 1.41) \times 10^{2+12-8}}{9.43} = \frac{(2.58 \times 1.41) \times 10^6}{9.43}$$
$$= 0.386 \times 10^6$$
$$= 3.86 \times 10^5 \text{ (scientific notation)}$$

EXERCISES

Answers to these exercises are at the back of this book.

1. Write out the following numbers in full, nonexponential form: (a) 6.4×10^{-1}, (b) 51.6×10^6, (c) 8.51×10^{-7}, (d) 9×10^9, (e) 10^{-3}

2. Write the following in scientific notation:
(a) 5,320,000, (b) 88, (c) 0.00000000000116, (d) 0.10030, (e) 633,100,000,000,000,000,000, (f) 100, (g) 0.000001

3. Multiply the following exponential numbers and write the answer in scientific notation: (a) $(3.5 \times 10^4)(6.0 \times 10^{-8})$, (b) $(4.04 \times 10^7)(5.50 \times 10^{12})$, (c) $(9.0 \times 10^{-5})(4.0 \times 10^{-2})$, (d) $(10^{-3})(2.15 \times 10^2)$, (e) $(5.4 \times 10^8)(6.5 \times 10^4)$

4. Multiply the following and write the answer in scientific notation: **(a)** $(0.00000035)(0.40)$, **(b)** $(5,000,000,000,000)(0.0000000006)$, **(c)** $(250)(70,000,000,000,000,000)$, **(d)** $(0.0000065)(0.0060)$, **(e)** $(1250)^2$

5. Divide the following exponential numbers and write the answer in scientific notation: **(a)** $(2.56 \times 10^5)/(1.60 \times 10^2)$, **(b)** $(1.98 \times 10^{-7})/(3.00 \times 10^4)$, **(c)** $(10^{-5})/(2.5 \times 10^{-7})$, **(d)** $(1.2 \times 10^{13})/(7.5 \times 10^{-10})$, **(e)** $(5.46 \times 10^{-3})/10^3$

6. Divide the following and write the answer in scientific notation: **(a)** $(255,000)/(0.00015)$, **(b)** $(0.0000000000100)/(0.0000333)$, **(c)** $(2400)/(960,000)$, **(d)** $(154)/(0.011)$, **(e)** $10,000/125$

APPENDIX C
Converting from One Unit to Another

Metric and SI Units of Measurement. When you measure the length, volume, mass, or temperature of an object, the resulting number must be accompanied by the appropriate unit (Chapter 2). For example, if your height is measured with a meter stick, then your height is reported in meters (m).

Most scientists use an updated version of the metric system called SI (Section 2.1) that uses fundamental units of measurement for quantities such as length, volume, mass, temperature, and energy. Chemists commonly use the meter (m) for length, the liter (L) for volume, the gram (g) for mass, the joule (J) for energy, and degrees Celsius (°C) or kelvins (K) for temperature (Section 2.2).

Other units of measurement are related to these fundamental units by positive or negative powers of 10 (such as 10^3 or 10^{-3}). A prefix before the unit corresponds to the power of 10. Table 2.1 lists the most commonly used prefixes, their abbreviations, and their values. The word *kilometer*, for example, literally means "1000 meters." The relationship, then, is:

$$1 \text{ km} = 1000 \text{ m} = 10^3 \text{ m}$$

The word *milligram* literally means "one-thousandth gram" or "0.001 gram." Therefore,

$$1 \text{ mg} = 0.001 \text{ g} = 10^{-3} \text{ g}$$

Conversion Factors. You can express the relationship between two different units as a conversion factor. You then can use this factor to convert from one unit to the other. We have just seen, for example, that 1 km = 1000 m, or 10^3 m. To change this relationship into a conversion factor, divide both sides of the equation by 1 km:

$$\frac{1 \text{ km}}{1 \text{ km}} = 1 = \frac{1000 \text{ m}}{1 \text{ km}} \text{ or } \frac{10^3 \text{ m}}{1 \text{ km}}$$

You can read this as 1000 or 10^3 meters per kilometer, where *per* means "divided by." In a conversion factor, the numerator must equal the denominator. That makes the ratio 1, which lets you multiply it by the given quantity and unit without changing the quantity; only the unit changes. Because this ratio equals 1, you can also write this or any other conversion factor in an inverted form that also equals 1:

$$1 = \frac{1 \text{ km}}{1000 \text{ m}} \text{ or } \frac{1 \text{ km}}{10^3 \text{ m}}$$

Table C1 lists some relationships between metric and English units. From these, you should be able to figure out the conversion factors, which are also listed in the table.

Table C1 Some Common Unit Conversion Factors

Property		Unit Conversion Factors
Length		
	Metric–Metric	
1 km = 1000 m, or 10^3 m		1000 m/km or 1 km/1000 m
1 cm = 0.01 m, or 10^{-2} m		0.01 m/cm or 1 cm/0.01 m
1 mm = 0.001 m, or 10^{-3} m		0.001 m/mm or 1 mm/0.001 m
	Metric–English	
1.61 km = 1 mi		1.61 km/mi or 1 mi/1.61 km
1 m = 39.4 in		1 m/39.4 in or 39.4 in/m
2.54 cm = 1 in		2.54 cm/in or 1 in/2.54 cm
Mass		
	Metric–Metric	
1 kg = 1000 g, or 10^3 g		1000 g/kg or 1 kg/1000 g
1 mg = 0.001 g, or 10^{-3} g		0.001 g/mg or 1 mg/0.001 g
	Metric–English	
1 kg = 2.20 lb		1 kg/2.20 lb or 2.2 lb/kg
454 g = 1 lb		454 g/lb or 1 lb/454 g
28.4 g = 1 oz		28.4 g/oz or 1 oz/28.4 g
Volume		
	Metric–Metric	
1 L = 1000 mL, or 10^3 mL		1000 mL/L or 1 L/1000 mL
	Metric–English	
1 L = 1.06 qt		1 L/1.06 qt or 1.06 qt/L
1 L = 0.265 gal		1 L/0.265 gal or 0.265 gal/L
Energy		
	Metric–Metric	
1 cal = 4.18 J		1 cal/4.18 J or 4.18 J/cal
1 kJ = 1000 J		1 kJ/1000 J or 1000 J/kJ
1 kcal = 1000 cal		1 kcal/1000 cal or 1000 cal/kcal

The Factor-Label Method. You can convert from one unit to another using the *factor-label method* (also called dimensional analysis). To do this, you multiply a given number and its unit by one or more conversion factors (see Table C1) to change to the desired unit.

given quantity and unit × conversion factor(s)
= same quantity with desired unit

First, read the problem and identify both the given (initial) unit and the desired unit. Then identify the conversion factors that relate the given unit to the desired unit. If no factor directly relates the two, try to find a conversion factor to change the given unit into a new one that you can use with a second conversion factor to change into the desired unit. In other words, figure out a sequence of conversion factors that will get you from the given unit to the desired one. Write out the sequence of units in the conversion. Your "road map" using one or two conversion factors, respectively, might look like this:

one conversion factor: given unit ⟶ desired unit

two conversion factors: given unit ⟶ new unit ⟶ desired unit

Then write the given value and unit and multiply it by the appropriate conversion factors. Set up your equations so that all the units you don't want cancel—put one in the numerator and one in the denominator—leaving only the desired unit. For example:

one conversion factor: $\text{given unit} \times \dfrac{\text{desired unit}}{\text{given unit}} = \text{desired unit}$

two conversion factors: $\text{given unit} \times \dfrac{\text{new unit}}{\text{given unit}} \times \dfrac{\text{desired unit}}{\text{new unit}}$
$= \text{desired unit}$

This method works for converting within the same system of measurements or between systems, as illustrated in the following examples.

Conversions within the Metric System. The examples below illustrate conversions from one metric unit to another.

EXAMPLE C.1 A steak has a mass of 255 g. What is its mass in kg?

SOLUTION given unit: g; desired unit: kg

conversion factors (Table C1): $\dfrac{1000 \text{ g}}{1 \text{ kg}}$ or $\dfrac{1 \text{ kg}}{1000 \text{ g}}$

Choose the conversion factor that has the given unit (g) in the denominator.

road map: g ⟶ kg

general approach: $\text{given unit} \times \dfrac{\text{desired unit}}{\text{given unit}} = \text{desired unit}$

calculation: $255 \text{ g} \times \dfrac{1 \text{ kg}}{1000 \text{ g}} = 0.255 \text{ kg}$

EXAMPLE C.2 A marathon is 42.2 km long. What is the distance in cm?

SOLUTION given unit: km; desired unit: cm

conversion factors (Table C1): $\dfrac{1 \text{ km}}{1000 \text{ m}}$ or $\dfrac{1000 \text{ m}}{1 \text{ km}}$;

$\dfrac{0.01 \text{ m}}{1 \text{ cm}}$ or $\dfrac{1 \text{ cm}}{0.01 \text{ m}}$

road map: km ⟶ m ⟶ cm

general approach: $\text{given unit} \times \dfrac{\text{new unit}}{\text{given unit}} \times \dfrac{\text{desired unit}}{\text{new unit}}$
$= \text{desired unit}$

calculation: $42.2 \text{ km} \times \dfrac{1000 \text{ m}}{1 \text{ km}} \times \dfrac{1 \text{ cm}}{0.01 \text{ m}} = \dfrac{42.2 \times 1000 \times 1 \text{ cm}}{1 \times 0.01}$
$= 4{,}220{,}000 \text{ cm (or } 4.22 \times 10^6 \text{ cm)}$

EXAMPLE C.3 The density of chloroform is 1.48 g/mL. What is the mass in g of 2.17 L of chloroform?

SOLUTION given unit: L; desired unit: g

conversion factors: $\dfrac{1.48 \text{ g}}{1 \text{ mL}}$ or $\dfrac{1 \text{ mL}}{1.48 \text{ g}}$; $\dfrac{1000 \text{ mL}}{1 \text{ L}}$ or $\dfrac{1 \text{ L}}{1000 \text{ mL}}$

road map: first convert liters into mL, then use density to convert mL into g; L ⟶ mL ⟶ g

general approach: $\text{given unit} \times \dfrac{\text{new unit}}{\text{given unit}} \times \dfrac{\text{desired unit}}{\text{new unit}} = \text{desired unit}$

calculation: $2.17 \text{ L} \times \dfrac{1000 \text{ mL}}{1 \text{ L}} \times \dfrac{1.48 \text{ g}}{1 \text{ mL}} = \dfrac{2.17 \times 1000 \times 1.48 \text{ g}}{1 \times 1}$

$= 3210 \text{ g}$

Conversions Between Systems. The examples below illustrate conversions between the English and metric systems.

EXAMPLE C.4 A young girl has a mass of 35 kg. What is her mass in pounds?

SOLUTION given unit: kg; desired unit: pounds (lb)

conversion factors (Table C1): $\dfrac{2.20 \text{ lb}}{1 \text{ kg}}$ or $\dfrac{1 \text{ kg}}{2.20 \text{ lb}}$

road map: kg ⟶ lb

general approach: $\text{given unit} \times \dfrac{\text{desired unit}}{\text{given unit}} = \text{desired unit}$

calculation: $35 \text{ kg} \times \dfrac{2.20 \text{ lb}}{1 \text{ kg}} = \dfrac{35 \times 2.20 \text{ lb}}{1} = 77 \text{ lb}$

EXAMPLE C.5 What is the speed in miles/hr of an automobile going 105,000 m/hr?

SOLUTION given unit: m/hr; desired unit: miles/hr

conversion factors (Table C1): $\dfrac{1000 \text{ m}}{1 \text{ km}}$ or $\dfrac{1 \text{ km}}{1000 \text{ m}}$;

$\dfrac{1.61 \text{ km}}{1 \text{ mi}}$ or $\dfrac{1 \text{ mi}}{1.61 \text{ km}}$

road map: m/hr ⟶ km/hr ⟶ mi/hr

general approach: given unit × $\dfrac{\text{new unit}}{\text{given unit}}$ × $\dfrac{\text{desired unit}}{\text{new unit}}$

= desired unit

calculation: $\dfrac{105{,}000 \text{ m}}{1 \text{ hr}} \times \dfrac{1 \text{ km}}{1000 \text{ m}} \times \dfrac{1 \text{ mile}}{1.61 \text{ km}} = \dfrac{105{,}000 \times 1 \times 1 \text{ mi}}{1 \times 1000 \times 1.61 \text{ hr}}$

= 65.2 mi/hr

EXERCISES

Answers to these exercises are at the back of this book.

1. Use the factor-label method and the information in Table 2.1 to make the following unit conversions:

   (a) 49 g = ____ mg
   (b) 14.8 mL = ____ L
   (c) 358 m = ____ cm
   (d) 68 mg = ____ µg
   (e) 0.023 mL = ____ nL
   (f) 0.000065 kg = ____ g
   (g) 864 mm = ____ cm
   (h) 1.25 L = ____ dL
   (i) 435 J = ____ cal
   (j) 277 kcal = ____ kJ

2. Use the factor-label method and the information in Table C1 to make the following unit conversions:

   (a) 10.0 cm = ____ inches (in)
   (b) 6.00 feet (ft) = ____ m
   (c) 16.0 gallons (gal) = ____ L
   (d) 875 g = ____ ounces (oz)
   (e) 107 miles (mi) = ____ km
   (f) 125 pounds (lb) = ____ kg
   (g) 459 ounces (oz) = ____ kg
   (h) 1.00 quart (qt) = ____ mL

3. How much would it cost to purchase 2.48 kg of a chemical that costs $1.25/g?

4. What is the volume in L of 2.65 kg of a liquid that has a density of 0.834 g/mL?

5. Suppose a person on a diet is limited to 1250 Cal per day. What is this amount of energy in joules (J) and in kilojoules (kJ)?

Additional exercises using the factor-label method appear at the end of Chapter 2.

APPENDIX D
The Elements

Listed below are the name and symbol of each element; the person, country, and date of discovery of the element; the origin of the name; and the appearance of the element at room temperature.

Actinium (Ac)   André Debierne in France (1899) and Friedrich Giesel in Germany (1902); Greek *aktinos,* ray; silvery solid

Aluminum (Al)   Hans Christian Oersted in Denmark (1825) and Friedrich Wöhler in Germany (1827); Latin *alumen,* alum (a mineral); silvery solid

Americium (Am)   Glenn Seaborg, Ralph James, Leon Morgan, Albert Ghiorso in U.S. (1944); America; silvery solid

Antimony (Sb)   known since antiquity; Greek *anti + monos,* not alone; Latin *stibium,* mark; silvery solid

Argon (Ar)   John William Strutt (Lord Rayleigh) and William Ramsay in U.K. (1894); Greek *a + ergon,* inactive; colorless gas

Arsenic (As)   isolated by Albertus Magnus in Germany (about 1250); Greek *arsenikos,* brave, male; gray or yellow solid (two forms)

Astatine (At)   Dale Corson, Kenneth MacKenzie, Emilio Segré in U.S. (1940); Greek *a + statos,* unstable; only trace amounts obtained

Barium (Ba)   Humphry Davy in U.K. (1808); Greek *barys,* heavy; silvery solid

Berkelium (Bk)   Stanley Thompson, Albert Ghiorso, Glenn Seaborg in U.S. (1949); Berkeley; silvery solid

Beryllium (Be)   Louis Nicolas Vauquelin in France (1797) and isolated by Friedrich Wöhler in Germany and Antoine-Alexandre Bussy in France (1828); Greek *beryllos,* beryl (a mineral); gray solid

Bismuth (Bi)   known in the 15th century, discoverer uncertain; German *weisse masse,* white mass; later German *wismut,* Latinized to *bisemutum;* silvery-pink solid

Boron (B)   Joseph Louis Gay-Lussac and Louis Jacques Thénard in France and Humphry Davy in U.K. (1808); Arabic *bauraq,* borax (a mineral); brown solid

Bromine (Br)   Antoine Jérôme Balard in France and Carl Löwig in Germany (1826); Greek *bromos,* stench; red-brown liquid

Cadmium (Cd)   Friedrich Strohmeyer in Germany (1817); Latin *cadmia,* calamine (a mineral); silvery-blue solid

Calcium (Ca)   Humphry Davy (1808) in U.K.; Latin *calx,* lime (a mineral); silvery solid

Californium (Cf)   Stanley Thompson, Kenneth Street, Jr., Albert Ghiorso, and Glenn Seaborg in U.S. (1950); California; only trace amounts obtained

Carbon (C)   known since antiquity; Latin *carbonis,* charcoal; black solid (graphite) or colorless solid (diamond)

Cerium (Ce) Jöns Jakob Berzelius and Wilhelm Hisinger in Sweden and Martin Heinrich Klaproth in Germany (1803); asteroid Ceres (discovered in 1801 and named after Ceres, Roman goddess of corn and harvest); gray solid

Cesium (Cs) Robert Bunsen and Gustav Kirchhoff in Germany (1860); Latin *caesius*, blue-gray; silvery solid

Chlorine (Cl) Karl Wilhelm Scheele in Sweden (1774); Greek *chloros*, greenish yellow; green-yellow gas

Chromium (Cr) Louis Nicolas Vaquelin in France (1797); Greek *chroma*, color; gray solid

Cobalt (Co) Georg Brandt in Sweden (1735); German *kobold*, evil spirit; Greek *cobalos*, mine; silvery solid

Copper (Cu) known since antiquity; Latin *cuprum*, Cyprus; reddish solid

Curium (Cu) Glenn Seaborg, Ralph James, Albert Ghiorso in U.S. (1944); Marie and Pierre Curie; only trace amounts obtained

Dysprosium (Dy) Paul-Émile Lecoq de Boisbaudran in France (1886); Greek *dysprositos*, difficult to obtain; silvery solid

Einsteinium (Es) Gregory Choppin, Stanley Thompson, Albert Ghiorso, Bernard Harvey in debris of a U.S. thermonuclear explosion in the Pacific (1952); Albert Einstein; only trace amounts obtained

Erbium (Er) Carl Gustav Mosander in Sweden (1842); Ytterby, Sweden; silvery solid

Europium (Eu) Eugène-Anatole Demarçay in France (1901); Europe; silvery solid

Fermium (Fm) Gregory Choppin, Stanley Thompson, Albert Ghiorso, Bernard Harvey in debris of a U.S. thermonuclear explosion in the Pacific (1953); Enrico Fermi; only trace amounts obtained

Fluorine (F) Henri Moissan in France (1886); Latin *fluere*, to flow; pale yellow gas

Francium (Fr) Marguerite Perey in France (1939); France; only trace amounts obtained

Gadolinium (Gd) Jean-Charles Galissard de Marignac in Switzerland (1880) and isolated by Paul-Émile Lecoq de Boisbaudran in France (1885); gadolinite (a mineral named after Johan Gadolin, a Finnish chemist); silvery solid

Gallium (Ga) Paul-Émile Lecoq de Boisbaudran in France (1875); Latin *Gallia*, France; silvery solid

Germanium (Ge) Clemens Alexander Winkler in Germany (1886); Latin *Germania*, Germany; gray solid

Gold (Au) known since antiquity; Anglo-Saxon *geolo*; Latin *aurum*, shining dawn; yellow solid

Hafnium (Hf) Dirk Coster and Georg von Hevesy in Denmark (1922); Latin *Hafnia*, Copenhagen; silvery solid

Helium (He) Pierre Janssen in France and Norman Lockyer and Edward Frankland in U.K. discover helium in the sun (1868); William Ramsay in U.K. discovers helium on earth (1895); Greek *helios*, sun; colorless gas

Holmium (Ho) Jacques-Louis Soret in Switzerland (1878) and Per Teodor Cleve in Sweden (1879); Latin *Holmia*, Stockholm; silvery solid

Hydrogen (H) Henry Cavendish in U.K. (1766); Greek *hydro* + *genes*, water forming; colorless gas

Indium (In) Ferdinand Reich and Hieronymous Richter in Germany (1863);

Latin *indicum,* indigo (named for the indigo line in its spectrum); silvery solid

Iodine (I)   Bernard Courtois in France (1811); Greek *iodes,* violet; violet-black solid

Iridium (Ir)   Smithson Tennant in U.K. (1803); Greek *iris,* rainbow; silvery-yellow solid

Iron (Fe)   known since antiquity; Anglo-Saxon *iron;* Latin *ferrum;* silvery solid

Krypton (Kr)   William Ramsay and Morris William Travers in U.K. (1898); Greek *kryptos,* hidden; colorless gas

Lanthanum (La)   Carl Gustav Mosander in Sweden (1839); Greek *lanthano,* to lie unseen; silvery solid

Lawrencium (Lw)   Albert Ghiorso, Torbjørn Sikkeland, Almon Larsh, Robert Latimer in U.S. (1961); Ernest O. Lawrence; only trace amounts obtained

Lead (Pb)   known since antiquity; Anglo-Saxon *lead;* Latin *plumbum;* silvery-gray solid

Lithium (Li)   Johan August Arfvedson in Sweden (1817); Greek *lithos,* stone; silvery solid

Lutetium (Lu)   Georges Urbain in France, Karl Auer von Welsbach in Austria, and Charles James in U.S. (1907); Latin *Lutetia,* Paris; silvery solid

Magnesium (Mg)   Joseph Black in Scotland (1755) and isolated by Humphry Davy in U.K. (1808) and by Antoine Alexandre Bussy in France (1828); Greek *Magnesia* (a district of Greece); silvery solid

Manganese (Mn)   Johann Gottlieb Gahn and Karl Wilhelm Scheele in Sweden (1774); Latin *magnes,* magnet; gray-pink solid

Mendelevium (Md)   Albert Ghiorso, Bernard Harvey, Gregory Choppin, Stanley Thompson, Glenn Seaborg in U.S. (1955); Dmitri Mendeleev; only trace amounts obtained

Mercury (Hg)   known since antiquity; planet Mercury (named for the Roman messenger of the gods); hydrargyrum from Greek *hydro* + *argyros,* water-silver (liquid silver); silvery liquid

Molybdenum (Mo)   Karl Wilhelm Scheele and Peter Jacob Hjelm in Sweden (1782); Greek *molybdos,* soft and lead-like; silvery solid

Neodymium (Nd)   Karl Auer von Welsbach in Austria (1885); Greek *neos* + *didymos,* new twin; silvery-yellow solid

Neon (Ne)   William Ramsay and Morris William Travers in U.K. (1898); Greek *neos,* new; colorless gas

Neptunium (Np)   Edwin McMillan and Philip Abelson in U.S. (1940); planet Neptune (named for Neptune, Roman god of the seas); silvery solid

Nickel (Ni)   Axel Cronstedt in Sweden (1751); German *nickel,* devil; silvery solid

Niobium (Nb)   Charles Hatchett in U.K. (1801); Greek Niobe (goddess of tears and daughter of Tantalus); gray-blue solid

Nitrogen (N)   Daniel Rutherford in Scotland and Karl Wilhelm Scheele in Sweden (1772); Greek *nitron* + *genes,* nitre (potassium nitrate) forming; colorless gas

Nobelium (No)   Albert Ghiorso, Torbjørn Sikkeland, John Walton, Glenn Seaborg in U.S. (1958); Alfred Nobel; only trace amounts obtained

Osmium (Os)   Smithson Tennant in U.K. (1804); Greek *osmé,* odor; gray-blue solid

Oxygen (O)   Joseph Priestley in U.K. (1774) and Karl Wilhelm Scheele in Sweden (1772); Greek *oxys + genes,* acid forming; colorless gas

Palladium (Pd)   William Hyde Wollaston in U.K. (1803); asteroid Pallas (discovered in 1802 and named after Pallas, Greek goddess of wisdom); silvery solid

Phosphorus (P)   Hennig Brand in Germany (1669); Greek *phos + phero,* bringer of light; yellow, red, or violet-black solid (several forms)

Platinum (Pt)   Antonio de Ulloa (1735) and Charles Wood (1741) in South America and isolated by William Hyde Wollaston in U.K. (1803); Spanish *platina,* diminutive form of *plata* (silver); silvery solid

Plutonium (Pu)   Glenn Seaborg, Arthur Wahl, Joseph Kennedy in U.S. (1940); planet Pluto (named after Pluton, Greek god ruling over the lower world); silvery solid

Polonium (Po)   Marie and Pierre Curie in France (1898); Latin *Polonia,* Poland; gray solid

Potassium (K)   Humphry Davy in U.K. (1807); English potash (a mineral); Latin *kalium;* Arabic *qali,* alkali; silvery solid

Praseodymium (Pr)   Karl Auer von Welsbach in Austria (1885); Greek *praseios + didymos,* green twin; silvery-yellow solid

Promethium (Pm)   Jacob Marinsky, Lawrence Glendenin, Charles Coryell in U.S. (1947); Prometheus, Greek demigod who stole fire from the gods for use by humans; gray solid

Protactinium (Pa)   Kasimir Fajans, Oswald Göhring in Germany (1913); Otto Hahn, Lise Meitner in Germany and Frederick Soddy, John Cranston, Alexander Fleck in Scotland (1917); Greek *protos,* first (protoactinium); gray solid

Radium (Ra)   Pierre and Marie Curie in France (1898); Latin *radius,* ray; silvery solid

Radon (Rn)   Friedrich Ernst Dorn in Germany (1900); named after radium (originally called "radium emanation"); colorless gas

Rhenium (Re)   Walter Noddack, Ida Tacke Noddack, Otto Berg in Germany (1925); Latin *Rhenus,* the river Rhine; silvery solid

Rhodium (Rh)   William Hyde Wollaston in U.K. (1804); Greek *rhodon,* rose; silvery solid

Rubidium (Rb)   Robert Bunsen and Gustav Kirchoff in Germany (1861); Latin *rubidus,* dark red; silvery solid

Ruthenium (Ru)   Emile Osann in Russia (1827) and rediscovered by Karl Klaus in Estonia (1844); Latin *Ruthenia,* Russia; silvery solid

Samarium (Sm)   Paul Émile Lecoq de Boisbaudran in France (1879); samarskite (a mineral named after a Russian mine official, V. E. Samarski); silvery solid

Scandium (Sc)   Lars Nilson in Sweden (1876); Latin *Scandia,* Scandinavia; silvery solid

Selenium (Se)   Jöns Jakob Berzelius in Sweden (1817); Greek *selene,* moon; red or gray-black solid (different forms)

Silicon (Si)   Jöns Jakob Berzelius in Sweden (1823); Latin *silex,* flint; gray solid

Silver (Ag)   known since antiquity; Anglo-Saxon *soelfor;* Latin *argentum* from Sanskrit *argunas,* shining; silvery solid

Sodium (Na)   Humphry Davy in U.K. (1807); English soda (a mineral); Latin *natrium;* silvery solid

Strontium (Sr)   Adair Crawford in Scotland (1790); isolated by Humphry Davy in U.K. (1808); *Strontian,* Scotland; silvery solid

Sulfur (S)   known since antiquity; Sanskrit *sulveri,* sulfur; Latin *sulpur;* yellow solid

Tantalum (Ta)   Anders Ekeberg in Sweden (1802); Tantalus, father of Niobe and son of Jupiter in Greek mythology; gray-black solid

Technetium (Tc)   Carlo Perrier and Emilio Segré in Italy (1937); Greek *technetos,* artificial; silvery solid

Tellurium (Te)   Franz Müller von Reichenstein in Romania (1782); Latin *tellus,* earth; silvery solid

Terbium (Tb)   Carl Gustav Mosander in Sweden (1843); Ytterby, Sweden; silvery solid

Thallium (Tl)   William Crookes in U.K. (1861) and isolated by Claude August Lamy in France (1862); Latin *thallus,* green twig; gray-blue solid

Thorium (Th)   Jöns Jakob Berzelius in Sweden (1829); Thor, Scandinavian god of thunder; silvery solid

Thulium (Tm)   Per Teodor Cleve in Sweden (1879); Thule, ancient name for Scandinavia; silvery solid

Tin (Sn)   known since antiquity; Anglo-Saxon *tin;* Latin *stannum;* silvery solid

Titanium (Ti)   William Gregor in U.K. (1791) and Martin Heinrich Klaproth in Germany (1795); Titans, giants of Greek mythology who were the sons of Uranus; silvery solid

Tungsten (W)   Juan José d'Elhuyar and Don Fausto d'Elhuyar in Spain (1783); Swedish *tung + sten,* heavy stone; gray solid

Unnilennium (Une)   Peter Armbruster and coworkers in Germany (1982); proposed name based on atomic number (109); only trace amounts obtained

Unnilhexium (Unh)   Georgy N. Flerov and coworkers in Russia (1974) and Albert Ghiorso and coworkers in U.S. (1974); proposed name based on atomic number (106); only trace amounts obtained

Unniloctium (Uno)   Peter Armbruster and coworkers in Germany (1984); proposed name based on atomic number (108); only trace amounts obtained

Unnilpentium (Unp)   Georgy N. Flerov and coworkers in Russia (1970) and Albert Ghiorso and coworkers in U.S. (1970); proposed name based on atomic number (105); only trace amounts obtained

Unnilquadium (Unq)   Georgy N. Flerov and coworkers in Russia (1964) and Albert Ghiorso and coworkers in U.S. (1969); proposed name based on atomic number (104); only trace amounts obtained

Unnilseptium (Uns)   Georgy N. Flerov and coworkers in Russia (1976) and Peter Armbruster and coworkers in Germany (1981); proposed name based on atomic number (107); only trace amounts obtained

Uranium (U)   Martin Heinrich Klaproth in Germany (1789) and isolated by Eugène-Melchior Peligot in France (1841); planet Uranus (discovered in 1781 and named after Uranus, Greek god of heaven); silvery solid

Vanadium (V)   Andrés Manuel del Rio in Mexico (1801); rediscovered by

Nils Gabriel Sefström in Sweden (1830); Vanadis, Scandinavian goddess of beauty; silvery solid

Xenon (Xe) William Ramsay and Morris William Travers in U.K. (1898); Greek *xenos*, stranger; colorless gas

Ytterbium (Yb) Jean Charles de Marignac in Switzerland (1878); Ytterby, Sweden; silvery solid

Yttrium (Y) Johan Gadolin in Finland (1794); Ytterby, Sweden; silvery solid

Zinc (Zn) known in India and China about 1000; German *zinke*, spike; gray-blue solid

Zirconium (Zr) Martin Heinrich Klaproth in Germany (1789) and isolated by Jöns Jakob Berzelius in Sweden (1824); zircon (a mineral) from Arabic *zargun*, gold color; silvery solid

APPENDIX E
Cell Structure

Cells are the basic unit of life. All organisms are made up of one or more cells, and—in our experience—new cells come only from other cells that already exist.

There are two main types. **Prokaryotic cells** such as bacteria are relatively simple in structure and don't contain a distinct nucleus or other membrane-bound compartments called *organelles*. The word *prokaryotic* means "before the nucleus."

Animal and plant cells are **eukaryotic,** which means that they do have organelles (Figure E1). A eukaryotic cell is surrounded by a *plasma membrane,* which controls the passage of chemicals and chemical signals into and out of the cell (see Section 20.7).

The largest organelle is the *nucleus*. The nucleus contains most (95% or more) of the cell's DNA, bound to proteins in the form of chromosomes. It is the

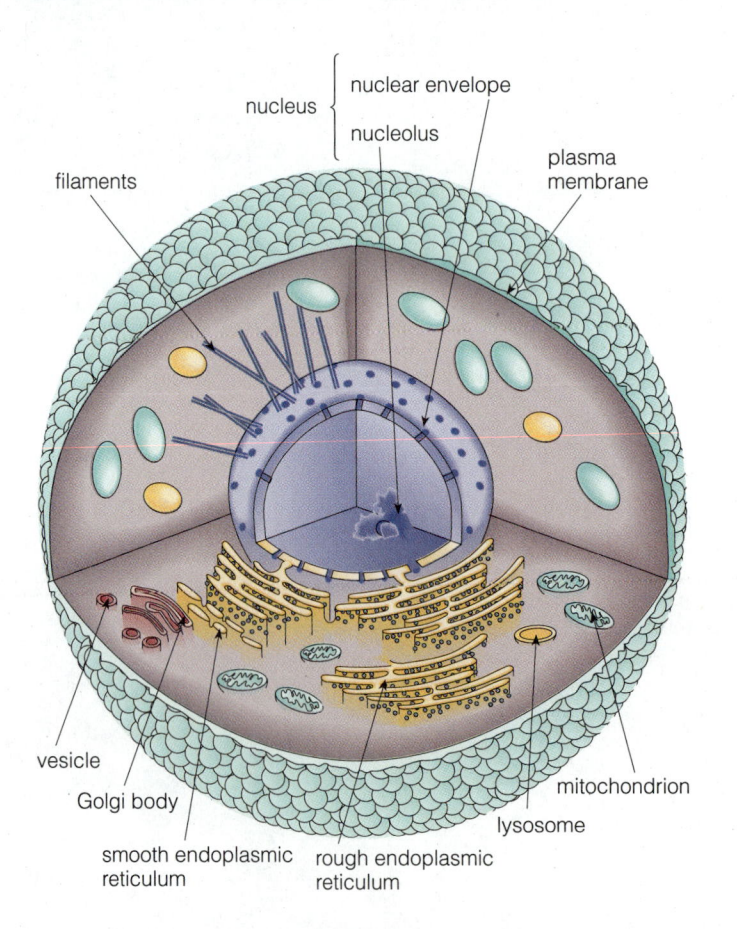

Figure E1 Generalized sketch of an animal cell.

location of both DNA and RNA synthesis (see Sections 23.2 and 23.3). Some RNA leaves the nucleus directly through pores in the nuclear membrane or envelope. Other RNA combines with protein to form *ribosomes,* which then pass out through the pores. The *nucleolus* (plural *nucleoli*) is a dense area in the nucleus where ribosomes are assembled.

Cytoplasm includes everything inside the plasma membrane except the nucleus. The gel-like fluid is often called *cytosol.* Cytoplasm contains other organelles and a network of filaments that provides shape and support inside the cell.

The second largest organelle is the *mitochondrion* (plural *mitochondria*). Eukaryotic cells have dozens to hundreds of mitochondria, each surrounded by a double membrane. The mitochondrion is the site of electron transport and oxidative phosphorylation (Section 24.4). Here oxygen is metabolized and ATP made to provide energy for cell activities. A small amount (5% or less) of a cell's DNA is in mitochondria and provides genetic information to synthesize many mitochondrial proteins.

The *endoplasmic reticulum* (ER) is a network of membranes that forms a continuous space separate from the cytoplasm. Enzymes in ER not associated with ribosomes, or smooth ER, synthesize lipids and detoxify drugs. ER associated with ribosomes is called rough ER and is the site of protein synthesis (Section 23.4). Some proteins are attached to polysaccharide "labels" in rough ER; these are often sent to Golgi bodies for more processing.

A *Golgi body* is a stack of membranes that bulge at the edges. A cell can have one or more Golgi bodies. Proteins and lipids that pass through them are chemically modified or "labeled" for their final destinations. Some substances leave the Golgi body to become components of other organelles within the cell. Other substances, such as modified proteins, pass through the plasma membrane and are secreted from the cell.

Lysosomes are saclike organelles that bud off from Golgi bodies. They consist of digestive enzymes surrounded by a membrane. Lysosomes break down polysaccharides, nucleic acids, proteins, and certain lipids. They also degrade foreign particles, bacteria, and worn-out cells and cell parts.

Microbodies bud from ER membranes. These organelles often contain enzymes that oxidize fatty acids, amino acids, and hydrogen peroxide (H_2O_2). Such microbodies are often called *peroxisomes.*

In addition to these basic components, some eukaryotic cells have other features. Plant cells have *chloroplasts* to carry out photosynthesis. They also have specialized *plastids,* compartments that store starch, pigments, and other materials. Some one-celled organisms are equipped with *flagella* or *cilia,* microtubular projections that propel them in a watery environment.

Answers to Even-Numbered Exercises

Chapter 1

2. a, b, d, g, i **4.** a, d, e, h **6.** hypothesis **8.** Initially it was a hypothesis. As more data were found to fit with Dalton's proposal, it became a theory. **10.** No. A theory explains how or why something occurs. A law is different; it states what happens. **12.** b

Chapter 2

2. (a) 2, **(b)** 4, **(c)** 4, **(d)** 1, **(e)** 4, **(f)** 7 **4. (a)** 1.6×10^2, **(b)** 5.451×10^5, **(c)** 1.010×10^{-3}, **(d)** 1×10^6 or 10^6, **(e)** 4.089×10^{-10}, **(f)** 1.000000 **6. (a)** 2, **(b)** 0, **(c)** 3, **(d)** 3 **8. (a)** 1.8×10^{-3}, **(b)** 1250, **(c)** 8.1×10^2 **10.** All four statements express calculated values (those in parentheses in statements a, b, and c; the value 10,001 in statement d) with figures that aren't actually significant. **12. (a)** 1.7×10^5, **(b)** 2.59×10^{-8}, **(c)** 3.44×10^6 **14.** International System of Units (SI) **16.** Yes. A balance (unlike a scale) will give the same mass regardless of the gravitational attraction. **18. (a)** 0.000000001, 10^{-9}; **(b)** 1000, 10^3; **(c)** 0.01, 10^{-2}; **(d)** 0.000001, 10^{-6}; **(e)** 0.001, 10^{-3} **20.** d **22.** volume **24. (a)** 0.01 (1/100), **(b)** 1000, **(c)** 0.001 (1/1000), **(d)** 0.029, **(e)** 3330, **(f)** 84,000,000 **26.** 4240 mL **28. (a)** 0.344 kg **(b)** 344,000,000 µg **30.** 8 tablets **32.** 64 mg carbocaine **34.** 0.790 g/mL **36.** 2.17 g/cm³ **38.** 1840 g **40.** density = 1.03 g/mL; specific gravity = 1.03; 1.03 is at the upper end of the normal range **42.** Although the runners lost fat during the run, they also lost carbohydrate, protein, and especially water. The water loss caused the percent body fat to increase.
44. potential energy **46.** 1360 kJ **48. (a)** 9200 kJ; **(b)** 2200 Cal **50. (a)** 308 K; **(b)** 94.3°F; Your body temperature is warm enough to boil ether. **52. (a)** 32°C, **(b)** 305 K **54. (a)** −31°C, **(b)** 364 K, **(c)** −28°C, **(d)** −18°F **56.** 98 mg glucose/dL blood

Chapter 3

2. a and d **4.** physical **6.** b and e **8.** You might see mercury in a thermometer; metal objects, such as jewelry, are usually mixtures of metals. **10.** Yes. The ultimate unit of gold is an atom. Subdividing a gold atom destroys its identity as gold. **12.** 22 g of carbon dioxide **14.** Water consists of a fixed ratio of hydrogen and oxygen atoms, each with its own atomic mass. That ratio, plus the fact that O atoms are about 16 times heavier than H atoms, accounts for their proportion by mass in water. **16.** 18 g of water (The remaining oxygen wouldn't react.) **18.** proton **20. (a)** hydrogen, **(b)** nickel, **(c)** krypton, **(d)** barium, **(e)** lead **22. (a)** carbon-13, **(b)** strontium-90, **(c)** iodine-131 **24. (a)** 26 p, 31 n, 26 e; **(b)** 11 p, 12 n, 11 e; **(c)** 54 p, 78 n, 54 e; **(d)** 80 p, 120 n, 80 e; **(e)** 27 p, 33 n, 27 e; **(f)** 92 p, 143 n, 92 e; **(g)** 1 p, 0 n, 1 e **26.** 207.2 **28. (a)** 17, **(b)** 13, **(c)** 14, **(d)** 8, **(e)** 2 **30. (a)** Fe, **(b)** Cl, **(c)** Na, **(d)** K, **(e)** Mg, **(f)** F, **(g)** As **32.** metals **34. (a)** potassium, **(b)** oxygen, **(c)** krypton, **(d)** silver **36.** shiny, conduct electricity, can be shaped **38.** iron (Fe), copper (Cu), and cobalt (Co) **40.** for proper functioning of the thyroid gland **42.** outside the nucleus in specific energy levels, orbiting the nucleus **44. (a)** 18, **(b)** 10, **(c)** 10, **(d)** 2, **(e)** 14, **(f)** 2, **(g)** 2, **(h)** 50 **46.** d **48.** both are spherical; the 4s orbital is a larger sphere **50. (a)** $1s^2 2s^2 2p^2$, **(b)** $1s_2 2s^2 2p^6 3s^2$, **(c)** $1s^2 2s^2 2p^6 3s^2 3p^6 4s^2 3d^2$, **(d)** $1s^2 2s^2 2p^6 3s^2 3p^6 4s^2 3d^{10} 4p^6 5s^2 4d^{10} 5p^2$
52.
(a) 1s [↑↓] 2s [↑↓] 2p$_x$ [↑↓] 2p$_y$ [↑↓] 2p$_z$ [↑↓] 3s [↑↓] 3p$_x$ [↑↓] 3p$_y$ [↑↓] 3p$_z$ [↑↓] 4s [↑↓]
(b) 1s [↑↓] 2s [↑↓] 2p$_x$ [↑] 2p$_y$ [↑] 2p$_z$ [↑]
(c) 1s [↑↓] 2s [↑↓] 2p$_x$ [↑↓] 2p$_y$ [↑↓] 2p$_z$ [↑↓] 3s [↑↓] 3p$_x$ [↑↓] 3p$_y$ [↑↓] 3p$_z$ [↑]

54. (a) ·Ċ·, **(b)** ·Mg·, **(c)** ·Ti·, **(d)** ·Sn· **56. (a)** ·Ca·, **(b)** ·N̈·, **(c)** :C̈l· **58. (a)** IVA, **(b)** IIA, **(c)** IVA **60. (a)** :N̈e:, **(b)** Rb·, **(c)** :S̈e·, **(d)** ·Ra·, **(e)** ·Ṗb·

62. selenium   **64.** manganese, technetium, rhenium, or unnilseptium   **66.** f   **68.** p   **70.** 4   **72. (a)** 1, **(b)** 6, **(c)** 8, **(d)** 1, **(e)** 3   **74.** c

Chapter 4

2. reactants have lower potential energy   **4.** neon   **6.** iron (Fe^{2+} and Fe^{3+})   **8.** 18   **10.** argon   **12.** 8   **14. (a)** P, **(b)** Cl, **(c)** Mg, **(d)** Na, **(e)** S   **16.** nonmetals   **18. (a)** K_2O, **(b)** SrI_2, **(c)** Al_2S_3, **(d)** Mg_3P_2, **(e)** MgS   **20.** AZ_2 **22. (a)** potassium oxide, **(b)** strontium iodide, **(c)** aluminum sulfide, **(d)** magnesium phosphide, **(e)** magnesium sulfide   **24. (a)** NaCl, **(b)** Cu_2S, **(c)** PbSe, **(d)** Cs_3P, **(e)** ZnF_2 **26. (a)** NaCN, **(b)** $SrSO_3$, **(c)** $Mg(HCO_3)_2$, **(d)** lithium nitrite, **(e)** barium phosphate, **(f)** aluminum sulfate   **28. (a)** Na_2S, **(b)** Na_2SO_3, **(c)** Na_2SO_4 **30.** no; no   **32. (a)** 3, **(b)** 1, **(c)** 2, **(d)** 1   **34. (a)** 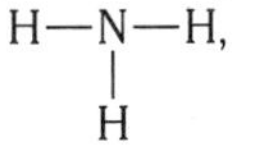 H—N—H with H below, **(b)** N≡N, **(c)** 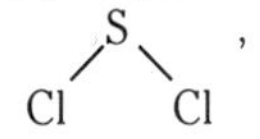 S with two Cl,   **(d)** 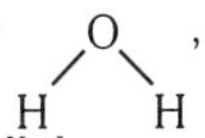 O with two H, **(e)** Cl—Cl   **36.** all the elements except H in all the compounds

38. (a) 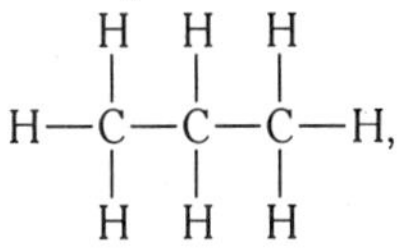 H—C—C—C—H (with H's), **(b)** 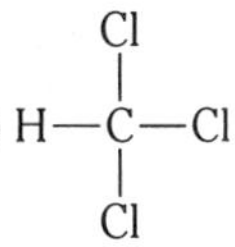 H—C—Cl (with Cl's)

40. 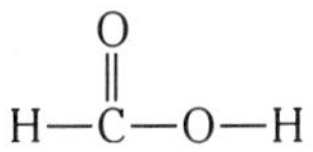 H—C(=O)—O—H

42. (a) 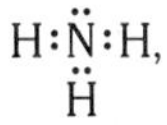 H:N̈:H with H, **(b)**  :N⋮⋮⋮N:, **(c)** 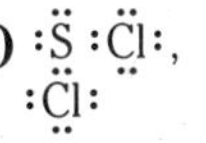 :S̈:C̈l:, :C̈l:,   **(d)** 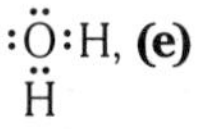 :Ö:H with H, **(e)** 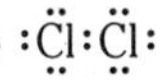 :C̈l:C̈l:

44. (a) 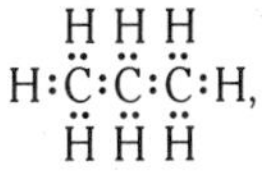 H:C̈:C̈:C̈:H with H's, **(b)** 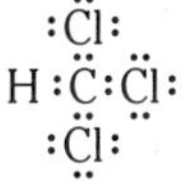 H:C̈:C̈l: with :C̈l:

46. (a) $[:C≡N:]^-$, **(b)** 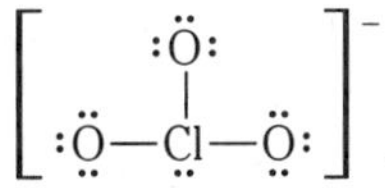 $[:Ö—Cl—Ö:]^-$ with :Ö: above,   **(c)** 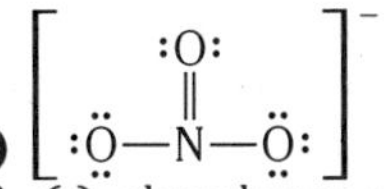 $[:Ö—N—Ö:]^-$ with :O: above (double bond)

48. (a) phosphorus pentachloride, **(b)** dinitrogen trioxide, **(c)** dihydrogen dioxide, **(d)** disulfur decafluoride   **50.** 109°   **52. (a)** pyramidal, **(b)** linear, **(c)** bent, **(d)** bent, **(e)** linear   **54. (a)** ionic, **(b)** covalent, **(c)** covalent, **(d)** covalent, **(e)** ionic   **56. (a)** O, **(b)** Cl, **(c)** Cl, **(d)** F   **58.** c and f   **60.** Carbon and hydrogen are similar enough in electronegativity to form essentially nonpolar covalent bonds with each other.

62. (a) 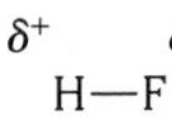 $\overset{\delta^+}{H}—\overset{\delta^-}{F}$, **(b)** 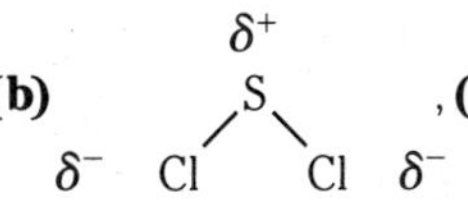 S with two Cl (δ^+ on S, δ^- on Cl), **(c)** 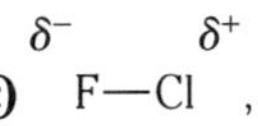 $\overset{\delta^+}{F}—\overset{\delta^-}{Cl}$,   **(d)** 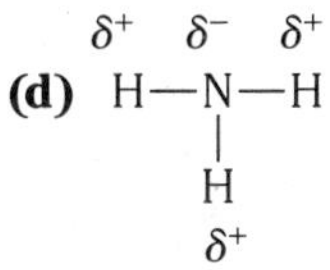 H—N—H with H below (δ^+ on H's, δ^- on N)

64. a, d, and e   **66.** H_2CO is polar; the other compounds are nonpolar.

Chapter 5

2. b and d   **4. (a)** O_2, **(b)** CCl_4, **(c)** $Ca(HCO_3)_2$, **(d)** Ar, **(e)** Cu_2O **6. (a)** $Fe_2O_3 + 2Al \longrightarrow Al_2O_3 + 2Fe$ **(b)** $2NO + O_2 \longrightarrow 2NO_2$ **(c)** $6Li + N_2 \longrightarrow 2Li_3N$ **(d)** $2Al + 6HCl \longrightarrow 3H_2 + 2AlCl_3$ **8. (a)** $2C_4H_{10} + 13O_2 \longrightarrow 8CO_2 + 10H_2O$ **(b)** $H_3PO_4 + 3NaOH \longrightarrow 3H_2O + Na_3PO_4$ **(c)** $P_2O_5 + 5C \longrightarrow 2P + 5CO$ **(d)** $2Cu_2O + Cu_2S \longrightarrow 6Cu + SO_2$ **10. (a)** $2Na + Cl_2 \longrightarrow 2NaCl$ **(b)** $SiO_2 + 4HF \longrightarrow SiF_4 + 2H_2O$ **(c)** $CaCO_3 \longrightarrow CaO + CO_2$ **12.** no   **14.** $2C_8H_{18} + 25O_2 \longrightarrow 16CO_2 + 18H_2O$ **16.** C_8H_{18} is oxidized and is the reducing agent; O_2 is reduced and is the oxidizing agent   **18.** Fe is oxidized and is the reducing agent; O_2 is reduced and is the oxidizing agent   **20. (a)** CO_2 is reduced and H_2O is oxidized; **(b)** Al is oxidized and HCl is reduced; **(c)** C is oxidized and CuO is reduced; **(d)** $C_6H_6O_2$ is oxidized and Ag^+ is reduced   **22. (a)** $Cu(OH)_2$ is the oxidizing agent and CH_2O is the reducing agent; **(b)** Cu^{2+} is the oxidizing agent and Al is the reducing agent; **(c)** O_2 is the oxidizing agent and $C_2H_4O_2$ is the reducing agent; **(d)** $C_{55}H_{94}O_6$ is the oxidizing agent and H_2 is the reducing agent   **24.** oxidized   **26.** O_2 **28. (a)** 48.00 amu; **(b)** 53.50 amu; **(c)** 26.98 amu; **(d)** 176.14 amu; **(e)** 194.22 amu   **30. (a)** 1.4 g; **(b)** 1.6 g; **(c)** 0.78 g; **(d)** 5.1 g; **(e)** 5.6 g   **32. (a)** 1.04 mol; **(b)** 0.935 mol; **(c)** 1.85 mol; **(d)** 0.284 mol; **(e)** 0.258 mol   **34. (a and b)** 8.4×10^{16} particles   **36.** 6.16×10^{-5} mol   **38.** 6.01×10^{21} formula units   **40.** 1.2×10^{22} atoms Hg   **42.** 2.9 g Br_2/mL   **44.** 7.2 mol CO_2 **46.** 6.75×10^3 g H_2O **48.** 4.31 g CO_2 **50.** O_2 is the limiting reactant   **52.** 12% yield   **54.** 0.650 g CO_2 **56.** 1.48×10^{-2} mol CO_2 **58.** $C_{57}H_{104}O_6 + 80O_2 \longrightarrow 57CO_2 + 52H_2O$ respiratory quotient = 0.71   **60.** $N_2 + 3H_2 \longrightarrow 2NH_3$ **62.** H_2 **64.** 4.7 L NH_3 **66. (a)** endothermic, **(b)** exothermic, **(c)** endothermic, **(d)** exothermic   **68.** exothermic, to produce heat to keep the body warm   **70.** 3.81 Cal   **72.** 58% Cal from fat

Chapter 6

2. solid   **4.** b   **6.** Gases have much more space between individual particles.   **8.** solid; strong ionic attractions prevent ions from separating and moving freely   **10. (a)** London forces, **(b)** hydrogen bonds, **(c)** dipole–

dipole forces, **(d)** London forces **12.** H_2 and O_2 are nonpolar and have only London forces between molecules. H_2O molecules are held together by stronger hydrogen bonds. **14. (a)** solid, **(b)** gas, **(c)** solid, **(d)** liquid, **(e)** liquid **16.** radon **18.** mercury (the only metal that is a liquid at room temperature) **20.** Radon (Rn) has the strongest London forces between atoms and requires the most heat (of the noble gases) to cause its atoms to separate. **22.** Heat increases the energy of particles, giving them enough kinetic energy to break free of the forces holding them in a fixed position in a solid. **24. (a)** freezing, **(b)** sublimation, **(c)** condensation, **(d)** vaporization, **(e)** melting **26.** Hexane is nonpolar and has only weak London forces between its molecules, so it vaporizes more readily than does H_2O (which forms hydrogen bonds). **28. (a)** shorter, **(b)** longer **30.** 1.4×10^4 J **32. (a)** 1700 cal; **(b)** 1.02×10^4 cal **34.** polar substance; more heat is needed to break the stronger attractive forces between particles. **36.** hexane; see answer to exercise 26 **38.** 6.2×10^4 cal **40.** H_2 and He **42.** F_2 effuses 1.366 times faster than Cl_2 does **44.** Changing the volume of the accordian chamber inversely changes the pressure of the air inside (Boyle's law). **46.** Increasing temperature causes gas particles to move faster, thus increasing the pressure (Gay-Lussac's law). **48. (a)** 650 torr; **(b)** 1520 torr; **(c)** 743 torr; **(d)** 670 torr **50. (a)** 1.20 atm; **(b)** 912 torr **52.** 248 mL **54.** 3.35×10^{-3} mol H_2 **56.** 1.18×10^{-2} mol He **58.** 44.0 g/mol (molar mass); gas is CO_2 **60.** 4.69 g NH_3 **62.** 587 torr from N_2 **64.** 1.07×10^3 g glucose **66.** 746 L O_2 at STP **68.** 875 L **70.** 1.12×10^{-22} L O_2 **72.** 10.8 mol O_2 **74.** 2.4×10^3 breaths

Chapter 7

2. decreases surface tension **4.** b; Freezing occurs more readily where less salt is present. **6.** Water molecules form a hexagonal arrangement in ice. **8.** No. Water doesn't dissolve nonpolar compounds effectively. **10.** Water is the solvent; ethanol is the solute. **12.** b, d, and e **14.** a and c **16.** b **18.** No, but it increases the rate of dissolving. **20.** more **22.** No. Being ionic, NaCl won't dissolve in hexane, so no charged particles are in solution to conduct electricity. **24.** b, d, and e **26.** No. No ions are mobile to conduct electricity. **28.** because ingredients separate on standing; no **30.** Large particles in milk (a colloid) scatter light. **32.** suspension **34.** d **36.** Add more NaCl to the solution and see if it dissolves. **38.** unsaturated **40. (a)** 16.5 g NaOH; **(b)** 338 g $CuSO_4$; **(c)** 27.7 g $CaCl_2$; **(d)** 0.489 g NaCl **42. (a)** 0.412 mol; **(b)** 2.12 mol; **(c)** 0.25 mol; **(d)** 8.37×10^{-3} mol **44. (a)** 0.824 mol; **(b)** 4.24 mol; **(c)** 0.75 mol; **(d)** 1.67×10^{-2} mol **46. (a)** 1.3 M NaCl; **(b)** 0.0215 M $CoCl_2$; **(c)** 0.308 M $Mg(HCO_3)_2$; **(d)** 1.72×10^{-5} M caffeine **48. (a)** 7.6% (w/v); **(b)** 0.279% (w/v); **(c)** 4.50% (w/v); **(d)** 3.33×10^{-4}% (w/v) **50. (a)** 0.17 g; **(b)** 30.0 g; **(c)** 22 g solute; **(d)** 60.0 g solute **52. (a)** 45.0 mL solute; **(b)** 2.15 mL solute; **(c)** 3.6×10^3 mL solute **54.** 12.9% (w/v) glucose **56.** 95 ppm Ca^{2+} **58.** 4.8×10^{-2} g SO_4^{2-}/L **60.** 0.128 g Cu_2CO_3 **62. (a)** 1150 mL H_2O; **(b)** 10 L H_2O; **(c)** 94 mL H_2O; **(d)** 59.4 mL H_2O; **(e)** 677 mL H_2O **64. (a)** 6.28×10^{-8} M; **(b)** 0.625 M; **(c)** 0.399% (w/v); **(d)** 2.00% (v/v); **(e)** 3.00 meq/L **66. (a)** 1.0 M NaCl, **(b)** 1.0 M glucose, **(c)** 1.0 M NaCl, **(d)** 1.0 M NaCl **68. (a)** 2.0 M NaCl, **(b)** 1.0 M NaCl, **(c)** 1.0 M $CaCl_2$, **(d)** 0.8 M Na_3PO_4 **70.** No. Osmotic pressure depends on the concentration of particles. Because it has a higher formula weight, KCl has a lower concentration of particles in solution (and lower osmotic pressure) than NaCl does when both are present at 0.9 g/100 mL solution. **72.** elevation of boiling point, depression of freezing point, osmotic pressure **74.** increase (due to a higher concentration of dissolved particles) **76.** both increase urine volume

Chapter 8

2. Chemical potential energy is converted into kinetic energy. Energy is not created or destroyed. **4.** When energy is used to do work, some of it changes into a less useful form (usually heat). **6.** Chemical potential energy is changed into mechanical energy and some is lost (in terms of being useful) as heat. **8.** exothermic **10.** measure the disappearance of color at an appropriate wavelength **12.** No. Activation energy affects reaction rate, not the net energy change. **14.** lowers activation energy **16.** 3.0×10^{-4} mol/L per min **18.** radium **20.** fluorine **22.** b; greater surface area causes a faster reaction **24.** The line (representing reaction rate) in the graph would rise faster. **26.** Some collisions lack the energy or proper alignment of reactants for reactions to occur. **28.** reactant is used up (its concentration decreases); accumulation of product might interfere with the reaction; the catalyst (enzyme) might be damaged at higher temperatures **30.** Finely divided Pt provides more surface area to interact with the reactants. **32.** Reactants and products will leave an open container; they also may react with other substances outside the container to form other products. **34.** no (see Table 8.3, for example)

36. (a) $K = \dfrac{[H_2O]^2}{[H_2]^2[O_2]}$ **(c)** $K = \dfrac{[NO_2]^2}{[NO]^2[O_2]}$

(b) $K = \dfrac{[H_3O^+][HSO_4^-]}{[H_2SO_4][H_2O]}$ **(d)** $K = \dfrac{[H_2]^3[AlCl_3]^2}{[Al]^2[HCl]^6}$

38. (a) $K = \dfrac{[H_2]^2[O_2]}{[H_2O]^2}$ **(c)** $K = \dfrac{[NO]^2[O_2]}{[NO_2]^2}$

(b) $K = \dfrac{[H_2SO_4][H_2O]}{[H_3O^+][HSO_4^-]}$ **(d)** $K = \dfrac{[Al]^2[HCl]^6}{[H_2]^3[AlCl_3]^2}$

Answer Section **641**

40. $2KClO_3 \longrightarrow 2KCl + 3O_2$ 42. No. The only necessary relationship is that the ratio of products to reactants, as defined in *K*, equals the equilibrium constant.
44. Yes. Same explanation as for exercise 42.   46. $K = 4.25 \times 10^{-6}$ 48. $[PCl_5] = 1.33 \times 10^{-2}$ *M*   50. left
52. (a) right, (b) right, (c) left, (d) left   54. add NO
56. More product (nylon) forms after it is removed.
58. High pressure shifts the equilibrium to the right, favoring NH_3 formation, because the products have fewer molecules than the reactants.   60. exothermic; left

Chapter 9

2. a substance that provides hydrogen ions (H^+) in water solutions   4. c and e   6. a and f   8. OH^-
10. NO_3^- 12. litmus, other indicators, measure pH, ability to neutralize a known acid or base, taste (dangerous), feel (dangerous)   14. (a) phosphoric acid, (b) carbonic acid, (c) potassium hydroxide, (d) nitric acid   16. (a) HCN, (b) $Mg(OH)_2$, (c) HNO_3, (d) HNO_2
18. see Table 9.2   20. see Table 9.3   22. No. A solution of 0.1 *M* KOH would have a higher pH because it provides more OH^- in solution.

24. (a) $K_a = \dfrac{[H_3O^+][HCO_3^-]}{[H_2CO_3]}$, $K_a = \dfrac{[H_3O^+][CO_3^{2-}]}{[HCO_3^-]}$;

(b) $K_a = \dfrac{[H_3O^+][C_7H_5O_2^-]}{[HC_7H_5O_2]}$

26. carbonic acid   28. $[H_3O^+] = 1.2 \times 10^{-9}$ *M*
30. more H_3O^+ 32. (a) 1.35, (b) 8.11, (c) 7.00, (d) 4.3   34. (a) 3.55×10^{-8} *M*; (b) 4.27×10^{-4} *M*; (c) 1.950×10^{-13} *M*, (d) 1.58×10^{-6} *M*   36. 3.98×10^{-8} *M*   38. (a) neutral, (b) basic, (c) acidic, (d) basic
40. $Mg + 2HCl \longrightarrow H_2 + MgCl_2$
42. (a) $H_3PO_4 + 3KOH \longrightarrow K_3PO_4 + 3H_2O$
(b) $2HC_2H_3O_2 + CaO \longrightarrow H_2O + Ca(C_2H_3O_2)_2$
(c) $H_2CO_3 + 2NaOH \longrightarrow Na_2CO_3 + 2H_2O$
(d) $3HBr + Al(OH)_3 \longrightarrow AlBr_3 + 3H_2O$
44. 7.00
46. (a) $3Na_2CO_3 + 2H_3PO_4 \longrightarrow 3H_2O + 3CO_2 + 2Na_3PO_4$
(b) $CaCO_3 + 2HC_2H_3O_2 \longrightarrow H_2O + CO_2 + Ca(C_2H_3O_2)_2$
(c) $2NaHCO_3 + H_2SO_4 \longrightarrow 2H_2O + 2CO_2 + Na_2SO_4$
(d) $NH_3 + HC_2H_3O_2 \longrightarrow NH_4C_2H_3O_2$
48. increased   50. b and e   52. (a) 31.02 g; (b) 26.00 g, (c) 63.02 g, (d) 40.00 g   54. (a) 9.6×10^{-8} *N*, (b) 0.148 *N*, (c) 0.122 *N*, (d) 3.50 *N*   56. $[Al(OH)_3] = 5.3 \times 10^{-4}$ *M*   58. 35.0 mL   60. (a) 3.8, (b) 10.3, (c) 6.31, (d) 1.32   62. c (Its pK_a value is closest to 7.40.)   64. b (because it is a weak acid)   66. 7.2 (equal amounts of acid and salt occur at pH = pK_a; $K_a = 6.2 \times 10^{-8}$ (Table 9.4), so $pK_a = 7.2$)   68. lower   70. During metabolism, cells produce acids (citric acid, lactic acid, pyruvic acid, etc.), as you will learn in Chapters 24–27.
72. right; the high concentration of O_2 74. left; the high concentration of CO_2 76. causes hyperventilation to expel CO_2 and thus remove H^+ from blood   78. low

Chapter 10

2. nucleus   4. Certain nuclear combinations of protons and neutrons are stable; those isotopes are not radioactive. Isotopes with unstable combinations are radioactive.   6. gamma < beta < alpha   8. gamma
10. gamma   12. a < c < b < e < d
14. (a) $^{12}_{5}B \longrightarrow ^{0}_{-1}\beta + ^{12}_{6}C$
(b) $^{242}_{94}Pu \longrightarrow ^{4}_{2}\alpha + ^{238}_{92}U$
(c) $^{27}_{12}Mg \longrightarrow ^{0}_{-1}e + \gamma + ^{27}_{13}Al$
(d) $^{210}_{84}Po \longrightarrow ^{4}_{2}He + \gamma + ^{206}_{82}Pb$
16. $^{209}_{83}Bi + ^{54}_{24}Cr \longrightarrow ^{262}_{107}Uns + ^{1}_{0}n$ 18. sulfur
20. Chemical reactions don't change half-life.   22. No. The half-life of carbon-14 is too short; almost none would remain after 800,000 years.   24. 19,240 years   26. 10 hrs, 32.0 g; 20 hrs, 16.0 g; 30 hrs, 8.0 g; 40 hrs, 4.0 g; 50 hrs, 2.0 g   28. 5.0 days   30. 7.4×10^4 emissions/sec
32. roentgen   34. gamma, because it is the most penetrating   36. alpha   38. One exposure causes massive damage, which is unlikely to be repaired adequately.
40. DNA   42. During radiation, cancer cells are more likely to be dividing, which makes them more vulnerable to radiation damage.   44. gamma, because they can penetrate the skin more effectively to reach internal tumors   46. short; then the radioactivity ceases after a short time in the body   48. beta and gamma (penetrating enough to be detected by a monitor outside the body)   50. b; being uncharged, neutrons aren't repelled by nuclei   52. c   54. nuclear mass is converted into energy   56. iron   58. uranium-235 and plutonium-239   60. absorb neutrons and thus regulate the rate of fission   62. advantages—abundant fuel, more energy/g fuel, less radioactive waste; disadvantage —technical problems haven't been solved, such as keeping the fuel together long enough at a high enough temperature

Chapter 11

2. No. The source of a molecule doesn't affect its properties.   4. Organic chemistry is the study of carbon-containing compounds; biochemistry is the study of the chemistry of organisms. Most of the substances needed for organisms to live are organic.   6. Ionic compounds have stronger attractions (ionic) than covalent compounds and require more energy (higher temperatures) to break those attractions and cause a solid to melt.   8. H, B, Si, and P   10. Polar. The four atoms bonded to C are not equal in electronegativity (F is much more electronegative than H), so there is a net polarity in the molecule.
12. The sp^3 hybrid orbitals provide a tetrahedral shape with 109.5° bond angles.   14. (a) and (b) tetrahedral; (c) and, (d) triangular   16. (a) and (b) sp^3; (c) and, (d) sp^2

642 Answer Section

18. (a) Cl—C(Cl)(Cl)—Cl, (b) H₂C=O, (c) H—C(H)(H)—N(H)—H,

(d) H₂C=CH₂

20. C₂H₄   **22.** Nothing prevents atoms from rotating about sigma bonds. The side-to-side overlap of p orbitals prevents rotation about pi bonds.   **24.** a, c, and d; one other type of chain, which is branched (shown in b, e, and f)   **26.** Single bond angles are 109.5°, whereas a flat hexagon requires angles of 120°.   **28.** Less stable because the bonds are much farther from 109.5° than in a five-carbon ring, which has angles of 108°.   **30.** No. Graphite is more like sheets of interlocked, six-carbon rings.   **32.** diamond   **34.** alcohols, carboxylic acids, mercaptans (thiols), amines, and amides. Hydrogen bonds increase melting point, boiling point, and solubility in water.   **36.** Functional groups provide the distinctive properties of organic compounds.

38. (a) R—Cl, (b) R—O—R′, (c) R—C(=O)—O—R′, (d) R—C(=O)—O—H

40. (a) organic halide; (b) aldehyde, three alcohol groups; (c) amide group (on both sides); (d) alkene, ester

Chapter 12

2. c and e

4. (a) H—C(H)(H)—Cl, (b) H—C≡C—H, (c) H—C(H)(H)—H,

(d) H₂C=CH₂,

(e) H—(C(H)(H))₁₂—H

6. (a) sp^3, (b) sp, (c) sp^3, (d) sp^2, (e) sp^3

8. C₃H₈, H—C(H)(H)—C(H)(H)—C(H)(H)—H

10. c   **12.** C₄H₁₀

14. H—C(H)(H)—C(H)(H)—C(H)(H)—C(H)(H)—C(H)(H)—C(H)(H)—H
hexane

2-methylpentane   3-methylpentane

2,3-dimethylbutane   2,2-dimethylbutane

16. yes; both have the formula C₆H₁₄

18. (a) ☐, (b) ☐ with Cl, (c) CH₃CH₂CH₂CH₂CH₃ or CH₃(CH₂)₃CH₃, (d) CH₃CHCH₂CH₂CH₃ with CH₃,
(e) CH₃CH₂CHFCH₃ or CH₃CH₂CHCH₃ with F,
(f) CH₃CH₂CH(CH₂CH₃)CH₂CH(CH₂CH₃)CH₃

20. d and f   **22.** CH₃CH₂CH₂CH₂CH₂CH₃ (other isomers, shown in the answer to exercise 14 are also correct written as condensed structural formulas)

24. (a) ☐, (b) ⬡

26. (a) CH₃CHClCH₃ (b) CH₃CH(CH₃)CH₂CHCH₂CH₂CH₃
(c) CH₃CHCH₂I with CH₃ (d) CH₃C(CH₃)₂CH₂CH(CH₂CH₃)CH(CH₃)(CH₂)₄CH₃

(e) cyclohexane with CH₃ and CH₃ substituents

28. (a) cyclobutane, (b) chlorocyclobutane (IUPAC), cyclobutyl chloride (common); (c) pentane (IUPAC), n-

pentane (common); **(d)** 2-methylpentane (IUPAC), isohexane (common); **(e)** 2-fluorobutane (IUPAC), *sec*-butyl fluoride (common); **(f)** 6-ethyl-3-methyloctane or 3-ethyl-6-methyloctane (both IUPAC), no common name **30. (a)** 2-methylpentane, **(b)** 1-chloro-2-methylpropane, **(c)** bromocyclohexane, **(d)** trichloromethane **32. (a)** 6-*sec*-butyl-2,3,5-trimethyldecane; **(b)** 6-*t*-butyl-3-chloro-4-isopropyldecane **34.** None. All are nonpolar. **36.** Gasoline dissolves protective oils from the skin, allowing water to evaporate. **38.** $2C_5H_{10} + 15O_2 \longrightarrow 10CO_2 + 10H_2O$
40. (a) $CH_3CH_2CH_2CH_2Cl$ (1-chlorobutane), $CH_3CHClCH_2CH_3$ (2-chlorobutane)

(b) CH_3CHCH_3 with CH_2Cl (1-chloro-2-methylpropane),

CH_3CCH_3 with CH_3 and Cl (2-chloro-2-methylpropane)

42. No reactions would occur. **44.** Pentane molecules have weaker London forces of attraction than decane molecules do, so pentane more readily exists as a gas and goes higher (cools more) before condensing. **46.** energy **48.** CH_2BrCH_2Br; 1,1-dibromoethane

50. CH_3CCH_2Cl (with two CH_3), $CH_3CHCH_2CH_3$ (with CH_2Cl), $CH_3CCH_2CH_3$ (with CH_3 and Cl),

$CH_3CHCHCH_3$ (with two CH_3 and Cl), $CH_3CHCH_2CH_2Cl$ (with CH_3),

$CH_3CH_2CH_2CH_2CH_2Cl$, $CH_3CH_2CH_2CHCH_3$ (with Cl)

$CH_3CH_2CHCH_2CH_3$ (with Cl)

52. CHI_3; triiodomethane

Chapter 13

2. b and e **4.** b and e **6. (a)** C_4H_8, **(b)** C_4H_8, **(c)** C_8H_{16}, **(d)** C_6H_{12}, **(e)** C_6H_{10}
8. (a) $CH_2=CClCH_2CH_2CH_3$, **(b)** $CH_2=CHCH=CH_2$,

(c) cyclohexene with CH_3 substituent

(d) $CCl_2=CHCl$

10. (a) $CH_2=C(CH_3)-(CH_2)_4CH_3$, **(b)** $CH_3CH=CCl-CH_3$ (with Cl and CH_3)

Neither can exist because each has a carbon atom with five bonds. (In addition, the name for compound b—even if it could exist—is wrong because you would number the carbon chain from the right.) **12. (a)** 1-butyne, **(b)** 3-methyl-1-butene, **(c)** 4-methylcyclopentene, **(d)** 3,4-dimethyl-1-pentene, **(e)** 2-isopropyl-1-pentene, **(f)** 1-bromo-3-methyl-1-butene **14.** b **16.** none **18.** f
20. (a) CH_3CH_2 and $CH_2CH_2CH_3$ on C=C with H's

(b) CH_3 and H on C=C with H and $CHCH_2CH_3$ bearing CH_3

(c) Cl_2CH and H on C=C with H and CH_3

(d) H and H on C=C with CH_3 and $CH-CH_2-CH(CH_2)_3CH_3$ bearing $CH(CH_3)_2$ groups

22. (a) 4 sigma, 0 pi; **(b)** 3 sigma, 1 pi **24. (a)** sp^2, **(b)** sp **26.** c **28.** lower. 2-Methyl-2-butene molecules are smaller than heptane molecules and thus have weaker London forces of attraction. **30.** b, d, a, c, e. With increasing molecular size and weight, the strength of London forces increases so it takes more heat to boil. Phenol forms hydrogen bonds, so it requires the most heat to boil. **32.** Add either $KMnO_4$ solution or Br_2 in CCl_4 solution. If it is an alkane, no color change will occur. If it is an alkene, the color will change from purple to brown with $KMnO_4$ solution; with Br_2, it will change from red-brown to colorless.
34. $2C_6H_{10} + 17O_2 \longrightarrow 12CO_2 + 10H_2O$

36. (a) $CH_3CH_2CH=CH_2 + H_2O \xrightarrow{H^+} CH_3CH_2CHCH_3$ (with OH)

(b) phenyl-$CH=CH_2 + Br_2 \longrightarrow$ phenyl-$CHBrCH_2Br$

(c) no reaction (needs heat and a catalyst such as Pt)

38. a

40. (a) $CH_2=CHCH_2CH_2CH_3 + HCl \longrightarrow CH_3CHClCH_2CH_2CH_3$

(b) $CH_3CH=CHCH_2CH_3 + Cl_2 \longrightarrow CH_3CHClCHClCH_2CH_3$

(c) $CH\equiv CCH_2CH_2CH_3 + 2HCl \longrightarrow CH_3CCl_2CH_2CH_2CH_3$

42. acetylene **44.** The compound must be unsaturated, so it is a chain rather than a cyclic compound (which would have to be saturated to have the formula C_5H_{10}). **46.** Benzene doesn't participate in normal addition reactions of alkenes.

48. (a) phenol (OH on benzene ring), **(d)** $CH_3CH(CH_2)_5CH_3$ with phenyl, **(b)** 2,4,6-trinitrotoluene structure (CH_3 on ring with three NO_2 groups), **(e)** ethylbenzene (CH_2CH_3 on ring), **(c)** 1,4-dichlorobenzene (Cl, Cl para), **(f)** 2-chlorophenol (OH and Cl ortho)

50. (a) aniline, **(b)** *m*-nitrotoluene, **(c)** *p*-xylene, **(d)** iodobenzene, **(e)** 1,3,5-tribromobenzene **52.** b, a, c; They are in order of increasing molecular size and weight, so they have increasingly strong London forces. In addition, nitrobenzene has dipole–dipole attractive forces. **54.** Addition reactions would destabilize the pi electron arrangement in benzene; substitution reactions don't do this.

56. (a) phenyl–CH_2Cl, **(c)** phenyl–NO_2, **(b)** *p*-chlorotoluene (CH_3 and Cl on ring), **(d)** *t*-butylbenzene ($C(CH_3)_3$ on ring)

14. (a) $CH_3\underset{OH}{C}HCH_3$, **(b)** $CH_3\underset{OH}{C}HCH_3$, **(c)** 2-chlorophenol (OH and Cl on benzene)

16. (a) a is a phenol, not an alcohol; **(b)** primary; **(c)** tertiary; **(d)** secondary **18.** secondary **20.** At concentrations above 12–15% *(v/v)* ethanol kills yeasts during fermentation. **22.** Because it has three —OH groups, glycerin forms multiple hydrogen bonds to water molecules.

24. (a) no reaction, **(b)** $CH_2{=}C(CH_2)_3CHCH_2Cl$ with CH_2CH_3 and CH_3 substituents, **(c)** $CH_2{=}C(CH_3)_2$, **(d)** $CH_3CH{=}CH(CH_2)_4CH_3$

26. (a) $CH_2{=}CHCH_3$, **(b)** same as a, **(c)** no reaction

28. $2\ CH_3CH_2CH_2OH \xrightarrow[140°C]{H_2SO_4}$
$CH_3CH_2CH_2{-}O{-}CH_2CH_2CH_3 + H_2O$

30. (a) complex product, **(b)** $HO{-}\overset{O}{\underset{\|}{C}}CH(CH_2)_3CHCH_2Cl$ with CH_2CH_3 and CH_3,
(c) no reaction, **(d)** $CH_3\overset{O}{\underset{\|}{C}}(CH_2)_5CH_3$

32. (a) $CH_3\overset{O}{\underset{\|}{C}}CH_3$, **(b)** same as a, **(c)** complex product
34. any substance that can be oxidized (for example, alkenes, alkynes, or any primary and secondary alcohols)

36. (a) $CH_3\underset{CH_3}{C}HCH_2CH_2SH$, **(b)** $CH_3(CH_2)_3\underset{CH_2CH(CH_3)_2}{C}HCH_2SH$,
(c) $CH_2ClCHClCHCH(CH_2)_2CH_3$ with SH and CH_3 substituents

38. (a) 3,3-dichloro-2-butanethiol, **(b)** 2-isopropyl-1-pentanethiol, **(c)** methanethiol

40. (a) cyclohexyl–S–S–cyclohexyl,
(b) $CH_3(CH_2)_3{-}S{-}S{-}(CH_2)_3CH_3$,
(c) $(CH_3)_2CHCH_2{-}S{-}S{-}CH_2CH(CH_3)_2$

42. (a) phenyl–O–phenyl,
(b) $(CH_3)_2CH{-}O{-}CH(CH_3)_2$,
(c) $CH_3CH_2{-}O{-}CH_3$

44. (a) ethyl phenyl ether, **(b)** methyl propyl ether, **(c)** (di)ethyl ether
46. structural isomers **48.** b (Diethyl ether is quite nonpolar, so it doesn't dissolve well in water but does dissolve nonpolar materials such as natural oils.)

Chapter 14

2. $CH_3CH_2\overset{\delta^-}{-}O\overset{\delta^+}{-}H$ **4.** Ethanol molecules attract each other by hydrogen bonds, which are stronger attractions than those between alkane molecules (London forces only). **6.** b, c, a **8.** higher (Since ethylene glycol molecules have two —OH groups, they have more hydrogen bonds attracting the molecules, making it harder to separate them (more heat is required) when the substance melts or boils.) **10. (a)** *p*-cresol, **(b)** 7-chloro-2-ethyl-6-methyl-1-heptanol, **(c)** 2-methyl-2-propanol, **(d)** 2-octanol **12.** *t*-butyl alcohol

50. (a) $2\ CH_3CH_2SH \xrightarrow{(O)} CH_3CH_2-S-S-CH_2CH_3$

(b) $CH_3CH_2CH_3 + Cl_2 \xrightarrow{uv\ light} CH_3CH_2CH_2Cl$
(and other products; Section 12.5)

(c) ⟨benzene⟩ + $CH_3Cl \xrightarrow{AlCl_3}$ ⟨benzene⟩$-CH_3$

(d) $CH_3CH=CH_2 + H_2O \xrightarrow{H^+} CH_3\overset{OH}{\underset{|}{CH}}CH_3$
(Section 13.4)

Chapter 15

2. 1 sigma, 1 pi **4.** (a) ethanal, (b) 3-iodo-2-methylbutanal, (c) *p*-chlorobenzaldehyde, (d) 3-bromo-2-ethylbutanal **6.** acetaldehyde

8. (a) $CH_3CH_2CH_2CHO$, (b) CH_3CHO, (c) 3,4-dichlorobenzaldehyde,

(d) $CH_3CHBrCH_2\overset{CH_3}{\underset{|}{CH}}CHCH_2CHO$ with $CH(CH_3)_2$ substituent

10. (a) 3-hexanone, (b) 3-methylbutanone, (c) benzophenone, (d) 5-chloro-2-methyl-3-hexanone
12. (b) isopropyl methyl ketone, (c) diphenyl ketone

14. (a) $CH_3-\overset{O}{\underset{\|}{C}}-$⟨phenyl⟩, (c) $CH_3CH_2\overset{O}{\underset{\|}{C}}-$⟨cyclohexyl⟩,

(b) $CH_3\overset{CH_3}{\underset{|}{CH}}\overset{O}{\underset{\|}{C}}CHCH_3$ with CH_3 substituent, (d) cyclohexanone

16. Butanone cannot have any carbon but number 2 as the carbonyl carbon. **18.** b (Acetone has the lowest molecular weight and is the most polar.) **20.** (a) dipole–dipole forces, (b) dipole–dipole forces, (c) London forces, (d) hydrogen bonds **22.** b (Hexane is the most nonpolar solvent, and hexanal is quite nonpolar.) **24.** acetone **26.** 2-butanol

28. $CH_3CH=CH_2 + H_2O \xrightarrow{H^+} CH_3\overset{OH}{\underset{|}{CH}}CH_3$
$CH_3\overset{OH}{\underset{|}{CH}}CH_3 \xrightarrow{(O)} CH_3\overset{O}{\underset{\|}{C}}CH_3$

30. (a) $H-\overset{OH}{\underset{\underset{OH}{|}}{\overset{|}{C}}}-H$, (b) $CH_3CH_2\overset{OH}{\underset{|}{CH}}CH_3$, (c) no reaction

32. a; a **34.** The results are normal. Glucose in blood gives a positive test, and urine normally contains nothing that gives a positive test. **36.** butanal (butyraldehyde) **38.** (a) hemiketal, (b) hydrate, (c) acetal, (d) none

40. (a) $H-\overset{OH}{\underset{\underset{OCH_2CH_3}{|}}{\overset{|}{C}}}-H$, (b) $H-\overset{OCH_2CH_3}{\underset{\underset{OCH_2CH_3}{|}}{\overset{|}{C}}}-H$

42. b (The carbon next to the carbonyl C is not bonded to H.)

Chapter 16

2. the H bonded to O **4.** sp^2 **6.** (a) 2-methylpropanoic acid, (b) *o*-hydroxybenzoic acid, (c) 2-chlorohexadecanoic acid **8.** (a) α-methylpropionic acid, (b) salicylic acid, (c) α-chloropalmitic acid

10. (a) $CH_3CH_2CHClCOOH$,

(b) $CH_3CH_2CH_2\overset{CH_3}{\underset{|}{CH}}CHCH_2COOH$ with CH_3 substituent

(c) 2-chloro-4-phenylbenzoic acid with COOH, (d) phthalic acid (1,2-COOH benzene)

12. 2-chlorobutanoic acid **14.** b

16. $CH_3\overset{}{\underset{\underset{CH_3}{|}}{CH}}\overset{O}{\underset{\|}{C}}-O^-K^+$

18. (a) ⟨phenyl⟩$-\overset{O}{\underset{\|}{C}}-O^-Na^+$,

(b) $(CH_3CH_2CH_2CH_2CH_2\overset{O}{\underset{\|}{C}}-O^-)_2Ca^{2+}$,

(c) $CH_3(CH_2)_2\overset{CH(CH_3)_2}{\underset{\underset{CH(CH_3)_2}{|}}{CH}}(CH_2)_2CCH_2CH_2\overset{O}{\underset{\|}{C}}-O^-K^+$

20. (a) potassium 2-bromopropanoate, (b) sodium methanoate, (c) calcium ethanoate **22.** (a) potassium α-bromopropionate, (b) sodium formate, (c) calcium acetate **24.** c, a, e, b, d **26.** A white solid would come out of solution as sodium benzoate changes into less-soluble benzoic acid.

646 Answer Section

28. CH₃(CH₂)₁₄C(=O)—O⁻Na⁺
30. (a) acetic anhydride, **(b)** valeryl bromide, **(c)** formyl chloride   **32.** e
34. (a) CH₃CH₂CH₂C(=O)—O—CH₂CH₂CH₃,
(b) CH₃C(=O)—O—CH₂CH₂CH₃,
(c) CH₃CH₂C(=O)—O—CH₂CH₂CH₃
36. (a) C₆H₅—C(=O)—O—CH₂CH(CH₃)₂,
(b) CH₃CH₂CH(CH₂CH₃)CH₂C(=O)—O⁻K⁺, **(c)** no reaction
38. (a) propyl butanoate, propyl butyrate (common); **(b)** propyl ethanoate, propyl acetate; **(c)** propyl propanoate, propyl propionate   **40. (a)** octyl methanoate, octyl formate (common); **(b)** isopropyl dodecanoate, isopropyl laurate (common); **(c)** methyl ethanoate, methyl acetate (common)   **42. (a)** 1-octanol (octyl alcohol) and methanoic (formic) acid, **(b)** 2-propanol (isopropyl alcohol) and dodecanoic (lauric) acid, **(c)** methanol (methyl alcohol) and ethanoic (acetic) acid
44. (a) C₆H₅—C(=O)—O—CH₂CH₃,
(b) CH₃(CH₂)₄C(=O)—O—C(CH₃)₃,
(c) (CH₃)₂CH—O—CH
46. c < b < a < d
48. (a) CH₃CH₂—O—C(=O)CH₃ + H₂O $\xrightarrow{H^+}$ CH₃CH₂OH + CH₃COOH
(b) C₆H₅—C(=O)—O—(CH₂)₃CH₃ + H₂O $\xrightarrow{H^+}$ C₆H₅—COOH + CH₃(CH₂)₂CH₂OH,
(c) CH₃(CH₂)₃—C(=O)—O—C₆H₅ + H₂O $\xrightarrow{H^+}$ CH₃(CH₂)₃COOH + C₆H₅—OH

50. (a)
CH₂—O—C(=O)—(CH₂)₁₀CH₃
|
CH—O—C(=O)—(CH₂)₁₄CH₃ + 3NaOH ⟶
|
CH₂—O—C(=O)—(CH₂)₁₆CH₃

CH₂OH   CH₃(CH₂)₁₀COO⁻Na⁺
|
CHOH + CH₃(CH₂)₁₄COO⁻Na⁺
|
CH₂OH   CH₃(CH₂)₁₆COO⁻Na⁺

52. (a) CH₃CH₂CH₂OH + (CH₃)₂CHCH₂COOH;
(b) CH₂ClCH₂C(=O)—O⁻K⁺ + (CH₃)₃COH
54. (a) CH₃OH + HO—P(=O)(OH)—OH ⟶ CH₃—O—P(=O)(OH)—OH + H₂O
(c) CH₃OH + HO—P(=O)(OH)—O—P(=O)(OH)—OH ⟶
CH₃—O—P(=O)(OH)—O—P(=O)(OH)—OH + H₂O

Chapter 17

2. ammonia, NH₃   **4. (a)** tertiary, **(b)** primary, **(c)** primary, **(d)** secondary   **6. (a)** *t*-butylamine, 2-amino-2-methylpropane (IUPAC); **(b)** isopropylamine, 2-aminopropane (IUPAC); **(c)** dimethylamine; **(d)** *N,N*-dimethylaniline
8. (a) (CH₃)₂CHCH₂NH₂,
(b) (CH₃)₂CH—NH—CH(CH₃)₂,
(c) N(CH₂CH₃)₂ (on benzene ring), **(d)** 2-chloroaniline (NH₂ and Cl on benzene ring)

10. 2-propanol (Since O is more electronegative than N, the partial charge in O—H is greater than in N—H; as a result, hydrogen bonding is stronger in alcohols than in amines.)   **12.** tertiary   **14.** cocaine   **16.** b
18. (a) *t*-butylammonium chloride, **(b)** isopropylammonium chloride, **(c)** dimethylammonium chloride, **(d)** dimethylphenylammonium chloride   **20.** The solubility of an amine in water increases when it reacts with HCl solution to form a salt.

22. (a) $(CH_3)_3\overset{+}{N}H\ Br^-$, (b) $(CH_3)_2CHCH_2-\overset{H}{\underset{H}{\overset{|}{N^+}}}-H\ Cl^-$

24. (a) $(CH_3CH_2)_3\overset{+}{N}H$, (b) $CH_3CH_2\overset{O}{\overset{\|}{C}}-N(CH_3)_2$

26. (a) ethanamide, acetamide (common); (b) N-butylethanamide, N-methylacetamide; (c) m-chlorobenzamide (3-chlorobenzamide), (d) N,N-dimethylpropanamide, N,N-dimethylpropionamide

28. (a) $CH_3\overset{O}{\overset{\|}{C}}-NHCH_3$, (b) $CH_3CH_2CH_2\overset{O}{\overset{\|}{C}}-NH_2$,
(c) $CH_3\overset{O}{\overset{\|}{C}}-NHCH(CH_3)_2$, (d) $\text{Ph}-\overset{O}{\overset{\|}{C}}-N(CH_3)_2$

30. a, c, d, b 32. (a) hydrogen bonds (partial ionic), (b) London forces, (c) hydrogen bonds, (d) dipole–dipole forces 34. amines are bases; amides are neutral

36. (a) $\text{Ph}-COOH + CH_3NH_3^+$,
(b) $(CH_3)_2CHCOO^- (+ NH_3)$

38. (a) $CH_3COOH + CH_3NH_3^+Cl^-$,
(b) $CH_3CH_2CH_2COOH + NH_4^+Cl^-$,
(c) $CH_3COOH + (CH_3)_2CHNH_3^+Cl^-$,
(d) $\text{Ph}-COOH + (CH_3)_2NH_2^+Cl^-$

40. (a) $CH_3COO^-Na^+ + NH_3$,
(b) $CH_3COO^-Na^+ + CH_3(CH_2)_3NH_2$,
(c) $\text{(3-Cl-Ph)}-COO^-Na^+ + NH_3$,
(d) $CH_3CH_2COO^-Na^+ + (CH_3)_2NH$

42. e

Chapter 18

2. a, c 4. b 6. b 8. (a) H:H, (b) He:, (c) :Cl:Cl:, (d) :Cl·, d is a free radical 10. reaction of the chain with an inhibitor or another free radical 12. All atoms in monomer units remain when polymerization occurs.
14. (a) $CH_2=CHCN$, (b) $CH_2=CCl_2$
16. $\left(\text{-}CH_2-\underset{CH_3}{\overset{|}{C}Cl}\text{-}\right)_n$
18. PVC is mostly nonpolar hydrocarbon material. Cl isn't electronegative enough for PVC to attract H_2O molecules effectively; H_2O molecules hydrogen bond with each other instead. 20. b and c 22. a, b, c 24. No. Acetic acid has only one carboxylic acid group.
26. b 28. d 30. extensive cross-links 32. Foam rubber is a condensation polymer (polyurethane) and natural rubber is an addition polymer 34. c
36. Polyester molecules, being large, have relatively strong London forces in addition to dipole–dipole forces and thus don't vaporize readily. We don't detect aromas when little or no vapor is present. 38. (a) Si is larger, (b) Si is less electronegative 40. yes (they contain C)

Chapter 19

2. b 4. L 6. Flat five- and six-member rings have bond angles of 108° and 120°, respectively, which aren't too different from the 109° angles around tetrahedral carbon atoms. The difference in six-member rings, however, is great enough that those rings aren't flat. 8. fructose

10. [furanose ring structure with HOCH$_2$, OH, OH, OH]

12. (a) α, (b) β, (c) β

14. [furanose ring with HOCH$_2$, CH$_2$OH, HO, OH, OH]

16. fructose 18. No. They don't have chiral carbons and don't rotate plane-polarized light.
20. $-14.0°$
22. dextrorotatory; cannot tell from the information given

24. $\begin{array}{c} CHO \\ | \\ HO-C-H \\ | \\ H-C-OH \\ | \\ HO-C-H \\ | \\ CH_2OH \end{array}$ L-xylose

26. (a) diastereomers, (b) diastereomers, (c) enantiomers, (d) none of the above, (e) anomers 28. glyceraldehyde 30. no (fructose has one less chiral carbon) 32. b 34. a, c, e 36. β-1,4 38. sucrose (Both anomeric carbons are tied up in a glycosidic bond, so neither ring can open and react.) 40. Yes. A positive reducing sugar test indicates the anomeric carbon can open up to form some of the open-chain aldehyde (or ketone). This quality also is needed for mutarotation.
42. (a) glucose and fructose, (b) glucose 44. Before treatment with saliva—positive iodine test; not a reducing sugar. After treatment with saliva—negative iodine test; is a reducing sugar. 46. (a) yes; (b) yes; (c) amylopectin

Chapter 20

2. b **4.** *trans* (The *cis* arrangement puts a bend in the chain, keeping molecules from packing together; this gives the *cis* isomer a lower melting point because less heat energy is needed to separate the molecules.)

6.
$$CH_2-O-\overset{O}{\underset{\|}{C}}-(CH_2)_7CH=CH(CH_2)_5CH_3$$
$$CH-O-\overset{O}{\underset{\|}{C}}-(CH_2)_7CH=CH(CH_2)_5CH_3$$
$$CH_2-O-\overset{O}{\underset{\|}{C}}-(CH_2)_7CH=CH(CH_2)_5CH_3$$

8. oil **10.** Coconut oil has a high content of short chain (C_{12}) fatty acids. **12.** 145 **14.** Saponification produces salts of fatty acids, which are soaps.

16. $CH_2OH + 3CH_3(CH_2)_5CH=CH(CH_2)_7COOH$
$|$
$CHOH$
$|$
CH_2OH

18. nitrogen **20.** sphingomyelin **22.** d **24.** cerebroside and ganglioside **26.** They lack an enzyme to break down gangliosides. **28.** omit methyl group from junction of two rings on left; attach acetylene group to the five-carbon ring **30.** a **32.** longer **34.** 25%; yes **36.** A, D, E, and K **38.** asymmetric; the layers differ in peripheral proteins and other substances **40.** nonpolar because the interior of the membrane is nonpolar (like dissolves like) **42.** Being ionic, Na^+ won't dissolve in the nonpolar interior of membranes.

Chapter 21

2. β-amino acid (amino group is bonded to the β carbon) **4.** glycine

6. (a) $^+H_3N-CH-COOH$, **(b)** $^+H_3N-CH-COO^-$,
$||$
$CH_2CH(CH_3)_2CH_2CH(CH_3)_2$

(c) $H_2N-CH-COO^-$
$|$
$CH_2CH(CH_3)_2$

8. $^+H_3N-CH-COO^-$
$|$
CH_2COOH

10. (a) anode, **(b)** anode, **(c)** cathode **12.** any pH between 5.07 and 7.59 **14.** methionine

16. $^+H_3N-\underset{\underset{\displaystyle CH_2}{|}}{\overset{\overset{\displaystyle ^+NH_3(CH_2)_4}{|}}{CH}}-\overset{O}{\underset{\|}{C}}-\underset{\underset{\displaystyle H}{|}}{N}-\underset{\underset{\displaystyle CH_2-C_6H_5}{|}}{CH}-COO^-$

18. (a) octapeptide, **(b)** protein, **(c)** polypeptide **20.** carbonyl C—N, carbonyl C=O **22.** any pH below 4.8 **24.** hemoglobin (because its pI is the highest or most alkaline) **26.** antiparallel **28.** parallel **30.** c **32.** dimer **34.** A protein is nutritionally suitable if digestion produces its individual amino acids. Denaturing a protein by cooking doesn't prevent digestion from doing that. **36.** Heavy metals bind to sulfhydryl groups (—SH) and thus alter protein shapes. **38. (a)** aspartic acid; **(b)** Valine is nonpolar, so it tends to be inside proteins. **40.** chromoprotein and metalloprotein **42.** Fibrous proteins have more exposed nonpolar amino acids, which are not soluble in water. **44.** polypeptide **46.** two disulfide bonds **48.** The carboxyl group of glutamic acid joined in a peptide bond is the side-chain carboxyl group, which is bonded to the γ carbon. **50.** c < b < a < d **52.** morphine

Chapter 22

2. decrease energy of activation; increase rate **4.** induced-fit model **6.** 2-Methyl butanoic acid has a chiral carbon (#2) and thus exists as a pair of enantiomers. $KMnO_4$ would produce equal amounts of both enantiomers; an enzyme could selectively produce one member of the pair. **8.** A coenzyme changes back and forth between two forms; a substrate may be converted into many other substances.

10. $CH_3-\overset{O}{\underset{\|}{C}}-COO^- + NADH + H^+ \longrightarrow$
OH
$|$
$CH_3-CH-COO^- + NAD^+$

12. (a) biotin, **(b)** FAD (or NAD^+, $NADP^+$, or FMN), **(c)** pyridoxal phosphate **14. (a)** niacin (B_5), **(b)** riboflavin (B_2) **16.** oxidoreductase **18.** low substrate concentration **20.** a **22.** no; at high temperature enzymes denature, slowing the reaction rate **24.** no **26.** competitive, because mannose-6-P structurally resembles glucose-6-P **28.** percent inhibition will remain constant; noncompetitive inhibition is unaffected by substrate because inhibitor binds to a different site **30.** binding to a different site may alter the conformation at the active site **32.** b and c **34.** a and c **36.** Not necessarily. The inhibitor binds to the regulatory site, not the active site. **38.** yes **40.** CK levels increase sooner and more dramatically than LDH levels, but don't remain elevated for as long (Figure 22.23).

Chapter 23

2. DNA contains A, C, G, and T; RNA contains A, C, G, and U.

4. (a) [structure of adenosine nucleoside] **(b)** [structure of cytidine monophosphate]

6. uracil **8.** inside; perpendicular **10.** no particular pattern, because RNA is a single strand **12.** The phosphate groups in DNA are proton donors. **14.** Neither ^{14}N nor ^{15}N is radioactive. **16.** They prevent DNA replication, and thus block cell division. **18.** 23% A, 30% C, 26% G, 21% U **20.** d **22.** 3'ACC end **24.** reverse transcriptase **26.** A gene is a section of DNA in a chromosome. Human cells normally have 46 chromosomes and perhaps 50,000–100,000 genes. **28.** protein; mRNA **30.** four (one codon functions as a start signal and three are stop signals for protein synthesis) **32.** mRNA; tRNA **34.** No. Animal cells have nuclei, which separate transcription from translation. **36. (a)** transcription occurs to produce proteins, **(b)** transcription occurs to produce repressor, **(c)** repressor not binding effectively, **(d)** RNA polymerase binds and transcribes structural genes **38.** repression couldn't occur; proteins would be synthesized continuously **40.** Mutation of a start or stop codon would change the length of the protein chain. Other mutations would produce either no change (if the new codon coded for the same amino acid) or an amino acid substitution in the protein. **42.** A mutagen causes a mutation; a carcinogen causes cancer. Not all mutagens are carcinogens. **44.** They don't contain a weakened form of the infectious agent. **46.** supply is steady; human insulin is produced instead of beef or pork insulin; less risk of contamination by infectious agents **48.** No. Gene therapy by recombinant DNA provides additional genes. People with Down's syndrome have extra genetic material, not a shortage.

Chapter 24

2. a, c, and e are anabolism; b and d are catabolism **4.** carbohydrates, fats, and proteins **6.** $C_{18}H_{36}O_2 + 26O_2 \longrightarrow 18CO_2 + 18H_2O$ **8.** No. Carbohydrates contain chemical potential energy that, when released during metabolism, can be measured as kJ or Cal. **10. (a)** stimulate, **(b)** inhibit **12.** Compounds in the citric acid cycle may be used to synthesize glucose and amino acids. **14.** glucose → acetyl CoA → fatty acids (The body does change excess carbohydrates into fats and stores the fat.) **16.** b, d, and f **18.** citrate, cis-aconitate, isocitrate **20.** isocitrate, malate **22.** low (ATP is formed from ADP and AMP.) **24.** ATP is an end product of the Krebs cycle (coupled to oxidative phosphorylation) that inhibits allosteric enzymes catalyzing reactions leading to its production. This is the definition of feedback inhibition. **26.** Acids are proton donors. At neutral pH, as occurs in the body, acids are in their neutralized, salt form and occur as negative (carboxylate) ions. **28.** no **30.** No. Some components, such as cytochromes, don't directly carry protons. **32.** An uncoupler allows electron transport to continue but blocks ATP synthesis by preventing formation of a proton gradient across the inner mitochondrial membrane. **34.** Disruption of the inner mitochondrial membrane prevents generation of a H$^+$ gradient across that membrane; this causes uncoupling. **36.** oxidized **38.** +2; +3 **40.** 3; 2 **42.** An enzyme doesn't alter the equilibrium of a reaction. **44.** exothermic **46.** b (High NADH concentrations inhibit the citric acid cycle.)

Chapter 25

2. photosynthesis; oxidation of glucose **4.** Cellulose would pass through the body undigested because humans lack the enzyme to hydrolyze the β-1,4 glycosidic bonds joining glucose units in cellulose. **6.** e **8.** Reactions 9 and 12 produce ATP directly; reaction 8 produces NADH, which is used to produce ATP by electron transport and oxidative phosphorylation. **10.** oxidized; glyceraldehyde-3-P gains O and loses H

12. [structure of 1,3-bisphosphoglycerate]

14. (a) transferase, **(b)** hydrolase, **(c)** isomerase, **(d)** lyase, **(e)** oxidoreductase **16.** less rapidly; high levels of ATP slow glycolysis by feedback inhibition of phosphofructokinase **18.** Lactate is oxidized to pyruvate, which can be oxidized in the citric acid cycle or used for gluconeogenesis. **20. (a)** 2, **(b)** 36 **22. (a)** catabolism, **(b)** anabolism, **(c)** catabolism, **(d)** anabolism, **(e)** catabolism, **(f)** anabolism **24.** liver; into blood **26.** increase glycogen levels because glycogen could be synthesized but not broken down **28. (a)** decrease, **(b)** increase **30.** Caffeine keeps cyclic AMP levels high, which promotes glycogenolysis, which makes stored carbohydrate available for energy during a long run (marathon). **32.** Lack of glucose-6-phosphatase blocks the normal metabolism in liver of glycogen → glucose-6-p → glucose (which goes into blood). As a result, glycogen is not broken down and accumulates. **34.** Pentose phosphate pathway produces NADPH, which is needed for lipid synthesis. **36.** glucagon (increases blood glucose), insulin (decreases blood glucose) **38.** increased **40.** glucagon, to stimulate liver glycogenolysis to maintain blood glucose levels **42.** gluconeogenesis and glycogenolysis produce glucose; amino acids are the main source for gluconeogenesis, and glycogen is the source for glycogenolysis

Chapter 26

2. They are all nonpolar and thus dissolve in nonpolar solvents. Glycerol is polar and is thus not a lipid.

4. CH_2OH $R—COOH$
     $CHOH$ + $R'—COOH$ ⟶
     CH_2OH $R''—COOH$

$$\begin{array}{l} CH_2—O—\overset{O}{\overset{\|}{C}}—R \\ CH—O—\overset{O}{\overset{\|}{C}}—R' \\ CH_2—O—\overset{O}{\overset{\|}{C}}—R'' \end{array}$$

6. By activating adipose tissue lipase, caffeine helps release fatty acids from storage and makes them available as fuel during exercise. **8.** yes; metabolize glycerol to dihydroxyacetone-P, then by glycolysis to pyruvate, and then to acetyl CoA **10.** 146 ATP
12. $CH_3CH_2\overset{O}{\overset{\|}{C}}—SCoA$ **14.** Acetyl CoA in the citric acid cycle cannot increase the amount of oxaloacetate, which is used for glucose synthesis. **16.** During glycogen depletion, the body lacks glucose and uses fats for energy, producing ketone bodies and causing ketoacidosis.

18. (a) $CH_3—\overset{OH}{\overset{|}{CH}}—CH_2—COO^-$,
(b) $CH_3—\overset{O}{\overset{\|}{C}}—CH_2—COO^-$

20. During diabetes, cells cannot metabolize glucose so they oxidize fatty acids for energy, producing acetyl CoA which forms ketone bodies. Ketones then appear in urine (ketonuria). **22.** cytoplasm **24.** c (After eating, the body converts excess carbohydrate into fat.)
26. $CH_3(CH_2)_4CH=CH—\overset{O}{\overset{\|}{C}}—SACP$
28. 7 ATP (7 of the 8 acetyl CoA are made into malonyl CoA) **30.** acetyl CoA carboxylase; citrate and isocitrate **32.** biotin **34.** HMGCoA reductase; cholesterol **36.** People with diabetes produce excess acetyl CoA; some forms ketone bodies, but some forms cholesterol.

Chapter 27

2. (a) arg, **(b)** phe and trp **4.** Proenzymes will not digest the cell synthesizing them. **6.** phenylalanine (tyr is made from phe) **8.** dopamine; phenylalanine (and tyrosine)
10. $CH_3—\overset{O}{\overset{\|}{C}}—COO^-$; pyruvate
12. The keto group involves the α carbon (the carbon adjacent to the carbonyl carbon). **14.** pyridoxal phosphate **16.** Isoleucine can be metabolized both to acetyl CoA (making it ketogenic) and to succinyl CoA (making it glucogenic). **18.** N **20.** They make less tyrosine, which is needed to make melanin. **22.** yes (especially early in life) **24.** yes, because high levels of phenylketones in maternal blood can harm the fetus **26.** anabolism; requires energy (4 ATP) **28.** high concentration of ammonium ion in blood **30.** Caffeine is a purine. People with gout have problems due to accumulation of the purine breakdown product, uric acid. **32.** c **34.** Heme products that normally provide pigments for feces don't reach the intestine.

Chapter 28

2. stomach contents **4.** fibrinogen **6.** increases osmotic pressure **8.** H^+, released when Hb binds O_2, reacts with HCO_3^- and $HbCO_2^-$ to produce CO_2. **10.** HbO_2^- is bright red **12.** No. A (defective) α chain would also be in fetal hemoglobin. **14.** HPO_4^{2-}; on the alkaline side of pK_a, the less protonated form prevails. **16.** A 120-lb (55-kg) person, for example, has 34 L H_2O **18.** B lymphocytes **20.** binds to specific antigens **22.** a, c **24.** conversion of prothrombin into thrombin **26.** its negative charge attracts Ca^{2+} **28. (a)** stimulate, **(b)** inhibit, **(c)** stimulate **30.** insulin **32.** vasopressin inhibits diuresis (urine production) **34.** increase **36.** decrease blood pressure **38.** H_3O^+, NH_4^+, $H_2PO_4^-$ **40.** to prevent loss of glucose from blood **42.** Tyrosine metabolism is blocked in albinism, so high levels of tyrosine accumulate in the body and appear in urine.

Chapter 29

2. c, d, f **4.** Antibiotics attack chemical features in bacterial cells that don't occur in normal body cells. **6.** Sulfa drugs block bacterial synthesis of folic acid. **8.** Sulfa drugs aren't classified as antibiotics because they aren't synthesized by organisms. **10.** e **12.** c

14. reverse transcriptase   **16.** a, c, d, e   **18.** a form of folic acid   **20.** Histamine reduces blood volume, allowing water to leave blood vessels.

22. [structure: CH₂CH₂NH⁺(CH₃)₂ group attached via O-CH(phenyl)(phenyl), with Cl⁻ counterion]

24. (a) dendrite, (b) synapse, (c) axon   **26.** Nerve impulses continue to cross the synapse, causing paralysis.   **28.** 4-aminobutanoate   **30.** At high concentrations, atropine is a poison that blocks acetylcholine action.   **32.** d   **34.** hypothyroidism   **36.** Insulin lowers blood glucose levels (produces hypoglycemia) by enabling glucose in blood to enter cells.   **38.** RU-486 prevents implantation of an embryo in the uterus.   **40.** treat anemia and cancer patients receiving chemotherapy   **42.** Beta blockers give shooters an unfair advantage by slowing the pulse.

Appendix B

1. (a) 0.64; (b) 51,600,000; (c) 0.000000851; (d) 9,000,000,000; (e) 0.001   **2.** (a) 5.32×10^6; (b) 8.8×10^1; (c) 1.16×10^{-12}; (d) 1.0030×10^{-1}; (e) 6.331×10^{23}; (f) 1×10^2 (or 10^2); (g) 1×10^{-6} (or 10^{-6})   **3.** (a) 2.1×10^{-3}; (b) 2.22×10^{20}; (c) 3.6×10^{-6}; (d) 2.15×10^{-1}; (e) 3.5×10^{13} **4.** (a) 1.4×10^{-7}; (b) 3×10^3; (c) 2×10^{19}; (d) 3.9×10^{-8}; (e) 1.56×10^6 **5.** (a) 1.60×10^3; (b) 6.60×10^{-12}; (c) 4.0×10^1; (d) 1.6×10^{22}; (e) 5.46×10^{-6} **6.** (a) 1.7×10^9; (b) 3.00×10^{-7}; (c) 2.5×10^{-3}; (d) 1.4×10^4; (e) 8×10^1

Appendix C

1. (a) 49,000 mg (4.9×10^4 mg); (b) 0.0148 L (1.48×10^{-2} L); (c) 35,800 cm (3.58×10^4 cm); (d) 68,000 μg (6.8×10^4 μg); (e) 23,000 nL (2.3×10^4 nL); (f) 0.065 g (6.5×10^{-2} g); (g) 86.4 cm (8.64×10^1 cm); (h) 12.5 dL (1.25×10^1 dL); (i) 104 cal (1.04×10^2 cal); (j) 1160 kJ (1.16×10^3 kJ)   **2.** (a) 3.94 in; (b) 1.83 m; (c) 60.4 L; (d) 30.8 oz; (e) 172 km; (f) 56.8 kg; (g) 13.0 kg; (h) 943 mL
3. $3100
4. 3.18 L
5. 5,220,000 J; 5220 kJ

Index

Entries and page numbers in **bold face** type refer to definitions of key terms in the text. Page numbers in *italics* refer to molecular structures.

Absolute zero, 23, 24
Accuracy, **14**, 15
Acetal, **340**, 341, 419
Acetaldehyde (ethanal), *328*, *484*, 485, *546*
Acetaminophen, *360*, 379
Acetate, 204
Acetic acid (ethanoic acid), 189, 192, *350*, *484*
Acetic anhydride, *355*
Acetoacetic acid, *562*
Acetone, *328*, *330*, 333, 335, 551, *562*
Acetyl ACP, 562, 563
Acetylcholine, *359*, 608, 609, 610
Acetylcholinesterase, 609, 610
Acetyl CoA carboxylase, 562, 563, 565
Acetyl coenzyme A (acetyl CoA), 525, *527*, *528*, 548, 562, 575
  in amino acid metabolism, 575–576
  in cholesterol synthesis, 565–566
  in fatty acid synthesis, 562–565
  in ketoacidosis, 562
  from pyruvate, 545
Acetylene, 279, *281*, 283, 286
N-Acetylgalactosamine, *585*
Acetylsalicylic acid, *see* Aspirin
Acid, **186**
  amino, **450**
  Arrhenius, 186
  Brønsted-Lowry, 186–187
  carboxylic, 249, **348**, 349–356
  fatty, 349, 358, **430**, 431–435
  naming of, 187–188
  properties of, 187, 188
  reaction with active metals, 197
  reaction with base, 197–200
  strength, 189–192, 350–351
  strong, **189**, 190–192, 350–351
  table of, 189
  weak, **189**, 190–192, 350–351
Acid anhydride, **355**, 356, 376, 377
Acid-base reaction, **187**, 197–200
Acid-base titration, 200–203
Acid deposition, 198–199
Acid dissociation constant (K_a), **190**, 191, 192, 205, 317, 351
Acidic solution, 194
Acid ionization constant (K_a), **190**, 191, 192, 205, 317, 351
Acidosis, **207**, 589
  metabolic, **207**
  respiratory, **207**

Acid phosphatase, 489
Acid rain, 133, 198–199
Aconitase, 528, 533
cis-Aconitate, *527*, *528*
ACP (acyl carrier protein), 562, 563, 564
Acquired Immune Deficiency Syndrome, *see* AIDS
Acrilan, *393*, 394
Acrylic acid, *353*
Acrylonitrile, *393*
Actin, 464
Actinium, 630
Actinomycin D, 605
Activation energy, **167**, 171–172, 475
Active site, of enzyme, **475**
Actual yield, 106–108
Acyclovir, *606*
Acyl carrier protein (ACP), 562, 563, 564
Acyl CoA, *560*, 561, 563
Acyl CoA dehydrogenase, 560
Acyl group, **355**
Acyl halide, **355**, 356, 376, 377, 397
Addition polymer, **392**, 393–396
Addition reaction, **286**, 297, 434–435
  of aldehydes, 337–342
  of alkenes, 286–291, 434–435
  of alkynes, 286–291
  of ketones, 337–341
Adenine, *473*, *495*, *499*, 607
Adenosine, *493*, *495*, 496
Adenosine deaminase, 493, 516
Adenosine diphosphate (ADP), 523–524, 530
Adenosine monophosphate (AMP), 523–524, 530, 560, 561, 579
Adenosine triphosphate (ATP), *496*, 508, 522, 523–524, 534, 612, 637
  in citric acid cycle, 528, 529, 530, 533, 546
  in fatty acid oxidation, 561
  in glycolysis, 541–545, 546–547
  high-energy phosphate bonds in, 524
  in urea cycle, 577, 578
Adenylate cyclase, 549, 550
ADH, *see* Antidiuretic hormone
Adipic acid, *398*
Adipose tissue, 559
ADP, *see* Adenosine diphosphate
Adrenal gland, 548, 571
Adrenalin, 372, 548, 549, 550, *574*, 592, 593 (*see also* Epinephrine)

Agent Orange, 297, 298
AIDS (Acquired Immune Deficiency Syndrome), 506, 590, 606
Alanine, *451*, *574*, 575
β-Alanine, 579
Alanine aminotransferase, 489
Albinism, 494, 516, 576, **577**, 597
Albumin, 464, 559, 586
  in blood, 157, 559, 586
Alcohol, 249, **306**, 307–316
  from aldehydes and ketones, 337–338
  classification of, 308
  hydrogen bonding in, 309–310
  naming of, 306–308
  physical properties of, 309–310, 352
  polyhydroxy, 307–308
  reactions of, 312–316
  toxicity of, 316
  uses of, 310–312
Aldehyde, 249, 313, **326**, 327–329, 331–342
  acetals and hemiacetals from, 339–341
  from alcohols, 313–316
  naming of, 327–329
  physical properties of, 328, 331–332, 352
  reactions of, 333–342
  uses of, 332–333
Aldolase, 489, 542
Aldol condensation, **341**, 342
Aldosterone, 438, 593, 594, *595*
Alkali metal, **45**
  properties of, 46
Alkaline earth metal, **45**
Alkaline phosphatase, 489
Alkaline solution, 194
Alkaloid, **373**
Alkalosis, **207**, 589
  metabolic, **207**
  respiratory, **207**
Alkane, **256**, 257–271
  bonding in, 256, 258
  cyclic, 257, 258
  general formula for, 256
  isomers of, 258–260
  naming of, 257, 262–266
  physical properties of, 266–267, 352
  reactions of, 267–269
  sources of, 269–271
  structural formulas of, 257

653

Index

Alkaptonuria, 576, 577, 597
Alkene, 249, **280**, 281–292
  bonding in, 283
  general formula for, 280
  geometric isomerism in, 283–285
  naming of, 280–283
  physical properties of, 285–286
  production from alcohols, 312–313
  reactions of, 286–292
  structural formulas of, 280–281
Alkyd resin, 397
Alkylation, of aromatic rings, 298, 299
Alkyl group, **263**, 264
Alkyl halide, **268**, 269, 272–273
  naming of, 272
Alkyne, 249, **283**
  naming of, 283
  physical properties of, 285–286
  reactions of, 286–292
Allopurinol, 579, *580*, 607
Allose, *415*
Allosteric enzyme, **482**, 486–487
Alloy, **70**
Alpha particle (α), 39, **215**
Alpha radiation, 215–217, 221–222
Aluminum, 3, 65, 66, 100, 197, 630
  production of, 3
Aluminum hydroxide, 185, 190
Alzheimer, Alois, 610
Alzheimer's disease, 226, 610
Americium, 630
Ames, Bruce, 512
Ames test, 512
Amide, 249, **376**, 377–382, 397, 456, 459
  formation of, 376–377
  hydrolysis of, 380–382
  naming of, 378
  physical properties of, 372, 378–379
Amine, 249, **370**, 371–377, 380–382, 397
  basic properties of, 374–376
  classification of, 370
  heterocyclic, **371**
  naming of, 371
  physical properties of, 372, 375, 376
  reactions of, 374–377
  salt formation of, 374–376
  uses of, 379–380
Amine salt, 374–376
Amino acid, 377, **450**, 451–456
  acid-base properties of, 452–454
  classification of, 450–452
  essential, **454**, 455–456, 573
  glucogenic and ketogenic, **575–576**
  metabolism of, 524–526, 548, 573–577
  stereoisomerism of, 452
Aminoacyl-tRNA, 508
Aminoacyl-tRNA synthetase, 508–510
p-Aminobenzoic acid (PABA), *484*, *603*, 604, 606
γ-Aminobutyrate, *608*, 609, 610
Amino group, **371**
Aminopeptidase, 573
β-Aminopropionitrile, *382*
Ammonia, 73, 89, 370
  as a base, 187, 188, 190, 200, 208
  excretion of, 200, 208
  synthesis of, 89
Ammonium cyanate, 237

Ammonium ion, 68, 74, 187, 200, 577, 578, 597
Ammonium salt, 89
Amniocentesis, 43
AMP, *see* Adenosine monophosphate
Amphetamine, 23, *374*, 610, *611*, 614
Ampicillin, *604*
Amu (atomic mass unit), 43
Amylase, 489, 541
Amylopectin, *421*
Amylose, *421*
Anabolic steroid, 429, 614
Anabolism, 522, **523**, 524–526, 534
Anaerobic metabolism, 545–546
Analgesic, 468, 612
Anemia, 46, 213, 480, 612
  hemolytic, 462–463, 580
  pernicious, 46, 480
  sickle cell, 449, 516, 589
  β-thalassemia, 516, 589
Anesthetic, local, 373, 375, 609
Angina pectoris, 382
Angiotensin I, 595
Angiotensin II, 595
Angiotensinogen, 595
Anhydride:
  acid, **355**, 397
  phosphate, 362, **363**
Aniline, *295*, *370*
Anomer, **416**, 417–418, 419
Anomeric carbon, 416
Anorexia nervosa, **23**
Antabuse, *484*, 485
Antacid, 185
Anthracene, *294*
Antibiotic, **604**, 605–606
Antibody, 464, 467, 468, 585, 586, 590
Anticoagulant, 487–488, 607
Anticodon, 503, 505, **508**, 509, 510
Antidepressant, 609–610
Antidiuretic hormone (ADH), 157, 593, 594, 595
Antigen, 467, 468, 585, 590
Antihistamine, 607–608, 612
Antimetabolite, 473, **606**, 607–608
Antimony, 305, 630
Antioxidant, 315, 316
Antiviral drug, 606
Apocrine sweat, 255
Arachidonic acid, *430*, 440
Arginine, *451*, 455, 575, *578*
Arginosuccinate, *578*
Argon, 630
Aristotle, 553
Aromatic compound, 280, **294**
  naming of, 295–297
  polycyclic, **294**
Aromatic hydrocarbon, 293–299
  physical properties of, 297
  reactions of, 297–299
Arrhenius, Svante, 186
Arrhenius' definition of acids and bases, 186
Arsenic, 305, 630
Arsphenamine, *602*, 603
Arthritis, 607
Ascorbic acid (vitamin C), 47, 189, 465–466, 477, *479*, 480

Asparagine, *451*, 575
Aspartate, 573, *578*
Aspartate aminotransferase, 489
Aspartic acid, *451*, 575
Asphalt, 270
Aspirin, *360*, 440, 553, 602, 612
Astatine, 630
Asthma, 207
Asymmetric carbon, 414
Atherosclerosis, 148, 438, 558
Athletic performance, 612–614
Atmosphere, 115
  composition of, 115
  unit of pressure, 128
Atom, **38**
  Bohr model of, 48–49
  characteristics of, 38–43, 47–55
  Dalton's theory of, 37–38
  electronic structure of, 47–54
  Greek concept of, 38
  quantum mechanical model of, 49, **50**, 51–54
  size of, 40
  subatomic particles of, 39–40
Atomic bomb, 228, 230
Atomic mass, 42, **43**, 44
  reference standard for, 43
Atomic mass unit (amu), 43
Atomic nucleus, discovery of, 39
Atomic number, **40**, 41, 44
  and protons in the nucleus, 40, 41
  and radioactivity, 216–217
Atomic orbital, **51**, 52, 53
Atomic theory, 37, **38**
  Dalton's, 37–38
  quantum mechanical model, 49, **50**, 51–54
Atomic weight, *see* Atomic mass
ATP, *see* Adenosine triphosphate
ATPase, 532–533
Atropine, 609
Aufbau principle, 52
Aureomycin, 604, *605*
Autoclave, 123
Avogadro, Amedeo, 98, 99, 131
Avogadro's law, **131**
Avogadro's number, 98, **99**, 100, 149, 152, 202
Axon, 608
AZT (azidothymidine), *506*, 606

Background radioactivity, 221, 222
von Baeyer, Adolph, 380
Bakelite, 332, 398, *399*
Baking soda, 185, 200
BAL (British Anti-Lewisite), 305
Balance, for measuring mass, 17, 18
Barbiturate, 381, 382
Barium, 630
Barometer, 128
Basal metabolic rate (BMR), **22**
Base, **186**
  acid-base reaction, **187**, 197–200
  Arrhenius, 186
  Brønsted-Lowry, 186–187
  naming of, 187, 188
  properties of, 187, 188
  strength of, 190–192
  strong, **190**, 191, 192

Base *(continued)*
table of, 190
weak, **190**, 192
Base dissociation constant (K_b), **192**
Base ionization constant (K_b), **192**
Base pairing in DNA, 497–499
Basic solution, 194
Basophil, 586, 587
Battery, 96, 98
Bauxite, 3
B cell, of immune system, 587, 590
Becquerel, Henri, 213, 215, 221
Becquerel, unit of radioactivity, 221
Beeswax, *255*, *431*
Benadryl, *607*, 608
Bends, 139
Benedict test, 334, **335**, 336, 418
Bent shape, of molecules, 78, 79
Benzaldehyde, *328*, *333*
Benzene, 288, 293, *294*
bonding in, 293–294
Benzenesulfonic acid, 298
Benzo[a]pyrene, *294*
Benzoic acid, 189, *350*, 353
Benzophenone, *328*
Beriberi, 479, 480
Berkelium, 630
Beryllium, 630
Beta blocker, 613, 614
Beta particle (β), 215, **216**
Beta radiation, 215–217, 221–222, 224
BHA (butylated hydroxyanisole), *315*, 316
BHT (butylated hydroxytoluene), *315*, 316
Bicarbonate ion, 68, 185, 206–208, 586, 588
Bile, 581
Bile salt, *438*, 558–559, 565, 566
Bilirubin, 580, 581
Biliverdin, 580
Biochemistry, **408**
Biological sciences, 4
Biotin, 477, 480
Birth control, 279, 439, 611
Bismuth, 630
1,3-Bisphosphoglycerate, 524, *542*, *543*, *544*
Bleach, 95
Blood, 585–590
carrier of CO_2, 206
carrier of O_2, 206
clotting of, 591–592
composition of, 69, 145, 586–588
dialysis of, 156, 596
functions of, 588–590
osmotic pressure of, 156, 586, 587, 594
pH of, 206–208, 586, 588–589, 595–596
typing of, 585
Blood clotting, 487–488, 557, 591–592
Blood doping, 612
Blood glucose, 336, 422, 547, 549, 550–553
Blood pressure, 595
Blood typing, 585
Body temperature, 13
Bohr, Niels, 48
Bohr model of electronic structure, 48–49
Boiling, **121**, 122–123

Boiling point, 122–123, 331
effect of pressure, 122–123
elevation of, 154–155
normal, 122
Bond, *see* Chemical bond
Boric acid, 189
Boron, 630
Botulin toxin, 609
Boyle's law, of gases, **128**
Branched polymer, 391–392
Branching enzyme, 549
Breathalyzer test, 314–315
British Anti-Lewisite (BAL), *305*
Bromine, 35, 100, 630
addition to multiple bonds, 287–288
reaction with benzene, 288, 293
test for unsaturation, 288
Brønsted, Johannes, 186
Brønsted-Lowry definition of acids and bases, 186–187
Buckminsterfullerene, 248
Buffer, **204**, 205
in blood, 206–208, 588
Bulimia, **23**
Buret, 201
Butabarbital, *380*
1,3-Butadiene, *281*
Butadiene polymer, 396
Butanal, *328*
Butane, *257*
Butanoic acid, *350*
2-Butanol, *414*
Butanone, *328*
1-Butene, *281*
cis-2-Butene, *281*
trans-2-Butene, *281*
Butyl group, *263*
sec-Butyl group, *263*, 264
tert (*t*)-Butyl group, *263*, 264
1-Butyne, *281*
Butyraldehyde, *328*
Butyric acid, *350*, 352, 353, 589
Butyryl ACP, 564

Cadaverine, *373*
Cadmium, 47, 630
Caffeine, *373*, 614
Calcium, 46, 185, 347, 630
in blood clotting, 591
Calcium carbonate, 185, 199
Calcium hydroxide, 190
Californium, 630
Caloric value of food, 109
Calorie, 21, 108–110, 612
Calorimeter, 108
Cancer, 407, 473, 489, 512–513
from **carcinogens**, **512**, 513
from radiation, 213, 222, 274
treatment of, 223–224, 473, 502–503, 612
Caproic acid, *350*, 352
Carbamoyl-P, *578*
Carbaryl (Sevin), *359*
Carbocation, 290
Carbohydrate, 407, **408**, 409–412, 414–423, 612
as blood-group markers, 585
classification of, 408–409
D, L isomers of, 414–416

dietary needs, 408, 422
disaccharide, **408–409**, 419–421
drinks, 612
energy value of, 523
metabolism of, 524–525, 539–553
monosaccharide, **408**, 409–412
polysaccharide, **409**, 421–423, 541
reactions of, 418–420, 423
Carbohydrate loading, 539
Carbolic acid, *see* Phenol
Carbon, 35, 100, 238–250, 631 (*see also* Hydrocarbon)
in biochemistry, 238
compounds of, 248
covalent bonding of, 238
forms of, 246–248
hybrid orbitals and, 240–243
tetrahedral structure of, 239
Carbonate ion, 68, 69, 70, 75, 185
Carbon dioxide, 74, 78, 83, 93, 115, 474, 528, 588
and acidosis, 207
in blood, 206–208, 588
in citric acid cycle, 525, 527–530
from combustion, 93, 267–268
greenhouse effect and, 115
Carbon-14 dating, 219–220
Carbonic acid, 189, 192, 198, 206, 474
Carbon monoxide, 82,
in combustion, 82, 268
toxicity of, 82
Carbon tetrachloride, 272
Carbon tetrafluoride, 83
Carbonyl group, **313**, 326, 327, 329, 348
Carboxylate ion, **351**
naming of, 351
Carboxyl group, **348**
Carboxylic acid, 249, 314, 327, 347, **348**, 349–356, 360–362
acidity of, 350–351
from alcohols and aldehydes, 314, 334
hydrogen bonding in, 352
naming of, 327, 349–350
physical properties of, 352–353
reactions of, 354–356, 360–361, 376–377
salts of, 351–354
uses of, 353–354
Carboxypeptidase, 573
Carcinogen, **512**, 513
Carlyle, Thomas, 382
Carotenoid, 442
Carothers, Wallace, 389
Carroll, Lewis, 71
Casein, 459, 464
CAT scan, 226, 227
Catabolism, **522**, 523, 524–526, 534
Catalyst, **171**, 172, 174, 271, 299, 474–477
Catalytic hydrogenation, **286**, 287, 434
Catalytic reforming, 271
Cataract, 553
Cell membrane, **442**, 443–444
Cell structure, 636–637
Cellulose, 422, *423*
Celsius temperature scale, 23–25
Cerebroside, 436, *437*
Cerebrospinal fluid, 586
Cerium, 631
Cesium, 631

Cesium-137, 224
CFC (chlorofluorocarbon), 274
Chain, Ernst, 601
Chain reaction, 228, 229, 230
Change:
  chemical, 5, **36**
  physical, **36**, 90–91
Charles' law, of gases, **128**
Chelating agent, 353–354, 591
Chemical bond, **62**, 64, 70, 74, 79–81
  coordinate covalent, **74**
  covalent, 62, **72**, 73, 74
  ionic, 62, **64**, 123, 124, 461
  metallic, 62, **70**, 123, 124
  minimum potential energy and, 62
  nonpolar covalent, **79**
  pi (π), **242**, 243, 326, 327
  polar covalent, **79**, 80
  predicting, 79
  sigma (σ), **240**, 242, 243
Chemical change, 5, **36**
Chemical equation, 91, **92**, 93
  balancing, 91–93
  format for writing, 91
  symbols used in, 91
Chemical equilibrium, 121, 172–179, 474
  effect of catalysts on, 174, 474
  effect of changes in concentration, 175–177
  effect of pressure on, 178
  effect of temperature on, 178–179
  equilibrium constant and, 173, **174**, 175
Chemical reaction, **90**, 91–98, 164–179
  (*see also* Oxidation-reduction reactions)
  acid-base, **187**, 197–200
  of acids, 197–198
  addition, **286**, 297
  arithmetic of, 102–110
  combustion, **93**, 267–268
  endothermic, **108**, 168, 179
  energy of, 108–110
  exothermic, **108**, 168, 179
  Le Châtelier's principle and, **175**, 176–179
  mechanism of, 171, 290
  oxidation-reduction, 93, **94**, 95–98, 291–292
  physical changes and, 90–91
  rate of, **166**, 167–172, 474
  signs of, 90–91
  substitution, **269**, 297
Chemical reactivity, 169
Chemiosmotic hypothesis, **531**, 532–533
Chemistry, 4
Chemotherapy, 473, **602**
Chernobyl, 230
Chiral carbon, **414**, 416, 452
Chloral hydrate, *339*
Chloramphenicol, 605
Chloride ion, 61, 63–64, 65, 586, 594, 597
Chlorine, 46, 72, 96, 631
1-Chloro-2,3-butanediol, *414*
Chlorofluorocarbon (CFC), 274
Chloroform, 272, 273
Chlorophyll, 442
Chloroplast, 637

Chlorpromazine, *381*, 609
Cholecalciferol, 441
Cholesterol, 395, 433, 437, *438*, 439–440, 565, *566*
  biosynthesis of, 565–566
  in cell membranes, 443–444
  gallstones and, 438
  heart disease and, 438–440, 558, 565–566
Cholestipol, 440, 565–566
Cholic acid, *438*
Choline, *359*, 435, 436, 610
Choline acetylase, 60
Chromium, 631
Chromoprotein, 464
Chromosome, 507, 514, 636
Chylomicron, 439, **558**
Chymotrypsin, 573
Cilia, 637
Cimetidine, *607*, 608
Cinnamaldehyde, 333
***Cis* isomer**, **283**, 284, 285, *395*, 412, 413, 430–431, 433
Cisplatin, 502
Citral, *333*
Citrate, 353–354, *527*, *528*, *564*, 565, 575
Citrate synthase, 528, 530
Citric acid, *350*, 527
Citric acid (Krebs) cycle, 525, 526, **527**, 528–530, 561, 574, 575
  ATP from, 533, 546
  reactions of, 527–530
  regulation of, 528, 529, 530
Citrulline, *578*
Civetone, *333*
CK (creatine kinase), 488, 489
Clotting, of blood, 487–488, 557, 591–592
Cloxacillin, *604*
CoA, *see* Coenzyme A
Coal, 238, 269–270
Coal gas, 269, 270
Coal tar, 269, 270
Cobalamin (Vitamin B$_{12}$), 477, 480
Cobalt, 100, 631
Cobalt-60, 123, 124, 125
Cocaine, *373*, *375*, 609
Codeine, *373*
Codon, **507**, 508
Coefficient, in balancing equations, 92, 102
Coenzyme, **477**, 478–481
Coenzyme A (CoA), 477, 527, 528, 529, 530, 560, 561
Coenzyme Q, 530, *531*
Cofactor, **477**
Coke, 269, 270
Collagen, 464, 465–466, 481, 489
  structure of, 465
Colligative property, 154, **155**, 156
  boiling point elevation, 154–155
  freezing point depression, 154–155
Collision energy, 166–167
Collision frequency, 166–167
Collision orientation, 166–167
Colloid, 140, **146**
  properties, 146
  size of particles, 146
Colloidal dispersion, 146
Color blindness, 516

Combustion, **93**, 267–268, 291
Compazine, *381*
Competitive inhibition, of enzymes, **483**, 484, 606
Complete protein, 454
Compound, 35, **36**
  covalent, **72**, 73–78, 117–119
  definite composition of, 37, 38
  ionic, 61, 63, **64**, 65–70
Computer-assisted tomography (CAT) scan, 226, 227
Concentrated solution, 149
Concentration, of a solution, **149**
  effect on reaction rate, 170
  equivalents per liter, 152–153, 202
  molarity, **149**, 150
  normality, **202**, 203
  parts per million (ppm), **152**
  percent, **150**
    volume/volume, 152
    weight/volume, 150–151
    weight/weight, 152
Condensation, 117, **121**
Condensation polymer, **392**, 396–401
Condensation reaction, 313, 354
Condensed structural formula, **261**, 262
Conjugate acid, 187
Conjugate base, 187
Conjugated protein, 463, **464**
Conservation of energy, law of, 164, 228
Conservation of mass, law of, **37**, 38, 228
Conservation of mass-energy, law of, 228
Contraception, 279, 439, 611
Control rod, 229, 230
Conversion factor, 25–28, 100–110, 618, 625–629
Converting between units, 625–629
Coordinate covalent bond, **74**
Copolymer, **390**, 396
Copper, 45, 96, 100, 631
Copper sulfate, 149
Cortisol, 438, 593
Cortisone, 333
Coumadin, *487*
Covalent bond, 62, **72**, 73, 74
  coordinate, **74**
  multiple, 74
  number elements form, 73
  nonpolar, **79**
  polar, **79**, 80
Covalent compound, **72**, 73–83, 117–119
  formula for, 72–73
  naming of, 76
Cracking, of hydrocarbons, 271
Creatine kinase (CK), 488, 489
Creatine phosphate, 524
Creatinine, 597
Crenation, 156
Cresol, *307*, 308–309, 312
Crick, Francis, 493, 497
Critical mass, 228, 229
Cross-link, **392**, 396
Cross-linked polymer, 392, 396
Cross-linking agent, of DNA, 502–503
CT (computerized tomography) **scan**, **226**, 227

Index   657

Curare, 609
Curie, Marie, 213, 215, 221
Curie, Pierre, 213, 221
Curie, unit of radioactivity, **221**
Curium, 631
Cyanide ion, 68, 521
Cyanide poisoning, 389, 521
Cyano compound, 382
Cyclic AMP, *549*, 550
Cyclohexane, *258*, *261*
Cyclohexanone, *328*
Cyclohexene, *281*
Cyclopropane, 267, *268*
Cyclotron, 225
Cysteine, *451*, 457, 575
Cystic fibrosis, 516
Cytidine, 496
Cytochrome, 464, 530, 531
Cytoplasm, 541, 550, 562, 564, 637
Cytosine, *495*, *499*, *512*, 579
Cytosol, 637

Dacron, *396*, 397, 398, 401
Dalmane, *381*
Dalton, John, 11, 37, 38, 39, 83, 131
Dalton's atomic theory, 37–38
Dalton's law of partial pressures, **131**
Data, **5**
DDI (dideoxyinosine), *506*, 606
DDT, *296*
Deamination, oxidative, 576, 577
Debranching enzyme, 549
Decane, *257*
1-Decanol, *307*
1-Decene, *281*
Definite proportions, law of, **37**, 38
Dehydration reaction, of alcohols, **312**, 313
7-Dehydrocholesterol, *443*
Democritus, 38
Denaturation, of protein, 461, **462**, 463, 483
Denatured alcohol, 311
Dendrite, 608, 609
Density, **19**, 27
    body fat and, 20
    of lipoproteins, 439
2-Deoxyguanosine, *506*
Deoxyribonucleic acid (DNA), 222, 223, 473, **494**, 495–517, 636, 637
    base pairing in, 497–499
    composition of, 494–495, 498
    genetic code and, 507–508
    hydrogen bonding in, 497–498
    mutations and, 512–513
    protein synthesis and, 507–511
    recombinant, **513**, 515
    replication of, **499**, 500–503, 637
    structure of, 497–499
    transcription of, **503**, 504–507
2-Deoxyribose, *409*, *494*, 495
Derepression, 510
Detergent, 148, 362, 434, 438, 463, 558
Dextrorotatory, **416**
Dextrose, 416
Diabetes, 207, 326, 336, 466, **551**, 558, 561–562, 597, 611
    carbohydrate metabolism and, 551–553

ketoacidosis and, 326, 561–562
lipid metabolism and, 326, 558, 561–562
test for, 336, 552
types of, 551–553, 611
Dialysis, **156**, 596
1,6-Diaminohexane, *398*
Diamond, 247
Diarrhea, 207
Diastereomer, **415**, 416, 417
Diazepam, *381*
Dicarboxylic acid, **350**, 396, 397
Dicoumarol, *487*, 488, 607
Dideoxyinosine (DDI), *506*, 606
Diesel fuel, 270
Diet:
    carbohydrates in, 408, 422
    energy in, 22–23, 523
    lipids in, 109, 432–433, 559, 564
    proteins in, 454–456, 573
Diethyl ether, *318*, 319
Diffusion, **126**, 127
Digestion:
    bile salts and, 148, 438, 558–559, 565, 566
    of carbohydrates, 540–541
    of lipids, 558–559
    of proteins, 572–573
Digitalis, 602
Diglyceride, 432, 558, 559
Dihydroxyacetone phosphate, *542*, *543*, 545, *559*
Dillinger, John, 442
Dilute solution, 149
Dilution, **154**
Dimensional analysis, 25–28, 100–110, 149–153, 626–629
N,N-Dimethyl-*m*-toluamide, *379*
Diol, 307–308, 338, 396
Dipeptidase, 573
Dipeptide, 456
Diphenhydramine, *607*, 608
Dipolar ion, 379, **453**
Dipole, 81
Dipole-dipole force, *118*, 119, 331
Dipole moment, 81, 83
Diprotic acid, 189
Disaccharide, **408–409**, 419–421, 540–541
Disinfectant, 96
Disparlure, *320*
Distillation, fractional, **270**
Disulfide, **317**, 457, 460,461,462,466
Disulfiram, *484*, 485
Diuretic, 23, 595, 614
DNA, *see* Deoxyribonucleic acid
DNA fingerprinting, 516
DNA ligase, 500, 502, 513, 515
DNA polymerase, 500, 501
Domagk, Gerhard, 603
L-Dopa, *571*, *574*, 609
Dopamine, *571*, *574*, 608, 609, 610, *611*
Double helix, of DNA, 497–499
Down's syndrome, 516
Drug, 473, 487–488, **601**, 602–614
    antibiotic, **604**, 605, 606
    antidepressant, 609–610
    antihistamine, 607–608, 612
    antimetabolite, **606**, 607–608

antiviral, 600
athletic performance and, 612–614
nervous system and, 608–610
sulfa, **603**, 604, 606
Dry ice, 78, 130
Dwarfism, 467
Dynamic equilibrium, **121**, 172–179, 474
Dynamite, 369
Dynorphin, 468
Dysprosium, 631

Eddington, Arthur S., 55, 179
Edema, 157
EDTA, *354*, 591
Effusion, **127**
Ehrlich, Paul, 602–603
Eicosapentaenoic acid, *557*
Einstein, Albert, 228
Einsteinium, 631
Elastase, 489
Elastin, 464
Elastomer, **391**, 394–396
Electrolysis, 35, 36
Electrolyte, **145**, 186
    in the body, 145
    classes of compounds, 145
Electrolyte drinks, 612
Electromagnetic radiation, 214
Electron, 39, 217
    arrangement in atoms, 47–54
    energy levels of, 48, 50, 51
    gain or loss of, *see* Oxidation-reduction reaction
    orbitals of, 51–53
    order of filling orbitals, 52
    properties of, 39
    valence, **53**, 54, 55
Electron capture, 217
Electron dot structure, **53**, 54–55
    of atoms, 53–55
    periodic table and, 54–55
    of polyatomic ions, 75–76
Electronegativity, **79**, 80, 169
Electron structure of elements, 47–54
    Bohr model of, 47–48
    quantum mechanical model of, **49**, **50**, 51–54
Electron transport, 521, 525, 530, **531**, 532–533, 561, 637
Electrophoresis, **454**, 458–459
Element, **35**, 36, **40**, 630–635
    atomic mass of, 42–43, 44
    atomic number of, 40–41, 44
    classification of, 44–45
    composition in Earth's crust, 93
    composition in human body, 45–46
    discovery of, 630–635
    metal, **44–45**
    metalloid, **45**
    names of, 44, 630–635
    nonmetal, **45**
    periodic table of, **43**, 44, 45, 54–55
    transition, **45**
Elongation, in protein synthesis, 509–510
Embalming fluid, 325
Embden, Gustav, 541
Embden-Meyerhof pathway, 541
Emollient, 255

658 Index

Emphysema, 129, 207
Emulsion, 148
Enantiomer, 415, 416, 417, 452, 476–477
Endoplasmic reticulum, 636, 637
Endorphin, 468, 469, 612
Endothermic reaction, 108, 168, 179
End point, 201
Energy, 21, 164–168, 170–172
   activation, 167, 171–172, 475
   chemical, 21
   classification of, 21
   collision, 166–167
   commercial use of, 269
   conservation of, 164, 228
   conversion and efficiency, 165
   in the diet, 22–23, 523
   kinetic, 21, 22, 23, 116, 117, 121
   mass relationship, 228
   measurement of, 21
   net, 167, 168, 172
   nuclear, 227–232
   potential, 21, 22, 62, 63, 72
   production and use in the body, 533–534
Energy level, of electrons, 48
   principal, 50, 51–53
   sublevel of, **50**, 51–53
Enflurane, *319*
Engels, Friedrich, 469
Enkephalin, 468, 469, 612
Enolase, 542
Enoyl-CoA hydratase, 560
Entropy, 164, 165
Enzyme, 171, 172, 450, 464, 473, **474**, 475–489
   allosteric, 482, 486–487
   as catalysts, 171, 172
   classes of, 481–482
   inhibition of, 483–485
   medical uses of, 487–489
   naming of, 481–482
   pH effect on, 483
   regulation of, 485–487
   restriction, 513, 515
   specificity of, 475–477
   temperature effect on, 482, 483
Enzyme active site, 475
Enzyme cofactor, 477
Enzyme substrate, 474
Enzyme-substrate complex, 482
Eosinophil, 586, 587
Epilepsy, 610
Epinephrine (adrenalin), 372, 548, 549, 550, *574*, 592, 593, 608, 614
Epitestosterone, 614
Equation, *see* Chemical equation; Nuclear equation
Equilibrium, dynamic, *see* Chemical equilibrium
Equilibrium constant (K), 173, **174**, 175, 190, 192
Equivalent, 152, 153, 202
Equivalent weight, 152, 153, 202
Erbium, 631
Erythrocyte, 586, 587
Erythromycin, 605
Erythropoietin, 612
Essential amino acid, 454, 455–456, 573
Essential fatty acid, 431, 564

Ester, 249, 348, 352, **354**, 355–356, 357–362
   naming of, 357
   phosphate, 362, 363
   physical properties of, 352, 357–358
   reactions of, 354–356, 360–362, 377
   saponification of, 361
   uses of, 358–360
Esterification, of carboxylic acids, 354–356, 358
Estradiol, *439*, 593
Estrogen, 438, 611
Ethanal, *see* Acetaldehyde
Ethane, *257*
Ethanoic acid, *see* Acetic acid
Ethanol (ethyl alcohol), *74*, 306, *307*, 310–311, 325, 483, *484*, *546*, 607
   denaturation of, 311
   metabolism of, 316
   physiological effect of, 306
   uses of, 310–311, 325
Ethanolamine, 436
Ethene (ethylene), *281*, 286, *393*
   polymerization of, 392–393
Ether, 249, 306, 313, **318**, 319–320
   anesthetic, 319
   hydrogen bonding of, 319
   naming of, 318
   physical properties of, 319–320, 352
   production from alcohols, 313
   uses of, 319, 320
Ethyl alcohol, *see* Ethanol
Ethylamine, 372
Ethyl chloride, 272, 273
Ethylene, *see* Ethene
Ethylene glycol, *307*, 311, 316, *396*, 483, *484*, 607
Ethyl ether, 318, 319
Ethyl group, *263*
Ethyne, *see* Acetylene
Eukaryotic cell, 636, 637
Europium, 631
Evaporation, 126
Excited state, of electron, 48, 49, 50
Exon, 505, 507
Exothermic reaction, 108, 168, 179
Exponent, 621–624
Exponential number, 621–624
Extracellular fluid, 586

Fact, 5
Factor-label method, 24, 26–28, 100–110, 149–153, 626–629
FAD and FADH$_2$ (flavin adenine dinucleotide), 522, *525*
   in citric acid cycle, 525, 527, 529, 530, 533
   in electron transport, 525, 530–533
   in fatty acid oxidation, 560, 561
Fahrenheit temperature scale, 23–25
Fat, 432, 433
   composition of, 432
   content in body, measurement of, 20
   content in food, 109
   digestion of, 148, 558–559
   energy content of, 432–433, 559
   metabolism of, 109, 433, 558–566
Fatty acid, 349, 350, 358, **430**, 431–435, 523, 548

   essential, 431, 564
   n-3 types of, 557
   oxidation of, 433, 560, **561**
   physical properties of, 430–431
   synthesis of, **562**, 563–565
Feedback inhibition, 486, 487, 524
Fehling test, 334, **335**, 336, 418
Fermentation, 310, 474, **546**
Fermium, 631
Fetal alcohol syndrome, 306
Fetal hemoglobin, 589
Feynman, Richard, 252
Fiber, 391
Fibrin, 591, 592
Fibrinogen, 586, 591, 592
Fibrous protein, 463, **464**
Fingerprinting, DNA, 516
First law of thermodynamics, 164, 165
Fischer projection formula, 414
Fission, nuclear, **227**, 228–232
Flagella, 637
Flavine adenine dinucleotide, *see* FAD
Flavine mononucleotide (FMN), 530, 531
Fleming, Alexander, 601
Florey, Howard, 601
Fluid mosaic model, of membranes, 444
Fluoride ion, 47, 65
Fluorine, 65, 631
Fluoroacetate, 533, 534
Fluoroacetyl CoA, *533*
Fluorocitrate, *533*
5-Fluorouracil, *473*, 502, 607
Fluosol, *273*
Fluothane, *268*
Flurazepam, *381*
FMN and FMNH$_2$ (flavin mononucleotide), 530, 531
Foaming, 399
Foam rubber, 399
Folic acid, 477, 480, 484, *603*, 604, 606, 607
Formaldehyde (methanal), *78*, 316, 325, *328*, 331, 335, 338–339, *399*, *484*
Formalin, 332, 338–339
Formic acid, *350*, 353
Formula mass, 99
Formula unit, 65
Formula weight, 99, 132
Forward reaction, 173
Fossil fuel, 269, 270–271
Fractional distillation, 270
Francium, 631
Franklin, Rosalind, 497
Free radical, 269
Freezing, 117, **119**, 120
Freezing point, depression of, 154–155
Freon, 274
Fructose, 340, *409*, *411*
   metabolism of, 545
Fructose 1-phosphate, 545
Fructose 1,6-bisphosphatase, 542
Fructose 1,6-bisphosphate, *542*, *543*
Fructose 6-phosphate, *481*, *541*, *542*, *543*
Fuel rod, 228, 230
Fullerene, 248
Fumarase, 529
Fumarate, *527*, *529*, *575*, *576*, *578*
Functional group, 248, 249, 250

Index

Fusion, heat of, 124, 125
Fusion, nuclear, 228, **232**

Gadolinium, 631
Galactose, *409, 411, 415,* 549, *550,* 585, 597
Galactosemia, 516, 549, 597
Galactose-1-P, 549, *550*
β-Galactosidase, 511
Gallium, 631
Gallstone, 438
Gamma-aminobutyrate, *608,* 609, 610
Gamma globulin, 464, 586
Gamma radiation (γ), 214, 215, **216,** 217, 222, 224
Ganglioside, 436, *437*
Gas:
  Avogadro's law, 131
  Boyle's law, 128
  Charles' law, 128
  Dalton's **law of partial pressures, 131**
  Gay-Lussac's law, 128
  Graham's law, 127
  ideal, 131
  noble, 45
  physical properties of, 116
  standard temperature and pressure (STP), 131–132
  stoichiometry and, 133
  universal gas law, 131, **132,** 133
Gas constant, universal, 133
Gasohol, 310
Gasoline, 256, 270–271
  octane number of, 271
Gaucher disease, 516
Gay-Lussac's law, of gases, **128**
GD, *359*
Geiger counter, 221, 222
Gene, 463, 494, **507**
  regulation of, 510–511
  regulator, 510, 511
  structural, 510, 511
Genetic code, 507–508
Genetic disorder, 493, 512, 516 (*see also* names of specific disorders)
Genetic engineering, 466, 467, 493, **513,** 514–517, 591, 612
Genetic therapy, 493, 515–517
Geometric isomer, 283, 284–285, *412, 413*
Germanium, 631
Globular protein, 463, 464
Globulin, 464
  gamma, 464, 586
Glucagon, 548, 549, 550, 551, 592, 593
Glucogenic amino acid, **575,** 576
Glucokinase, 541, 542
Gluconeogenesis, 547, 548, 561, 593
Gluconic acid, 336
Glucose, 98, 336, *340, 409, 411, 415, 481,* 540, *541, 542,* 597
  concentration in blood, 336, 422, 547, 549, 550–553
  conversion into fat, 564–565
  from gluconeogenesis, 547–548
  from glycogen, 422
Glucose 1-phosphate, 524
Glucose oxidase, 336, 488

Glucose 6-phosphatase, 481, 542, 548
Glucose 6-phosphate, *481, 483,* 524, *541, 542*
Glucose 6-phosphate dehydrogenase, 481
Glucose tolerance test, 552
Glutamate, *574, 576, 577,* 578
Glutamic acid, *451,* 575
Glutamine, *451,* 573, 575
Glutaraldehyde, *332*
Glutathione, 466
Glyceraldehyde, *452,* 545
Glyceraldehyde-3-P dehydrogenase, 542
Glyceraldehyde 3-phosphate, *542, 543,* 545
Glycerol (glycerin), 311, 325, 432, *476, 559*
  from hydrolysis of fats, 433, *558,* 559
  in phosphoglycerides, 435–436
  in triglycerides, 432
  uses of, 311, 325
Glycerol-3-phosphate, *476, 559*
Glycine, *74, 451,* 465, 573
Glycogen, *421,* 422, 539, 541
  anabolism of, 548–550
  catabolism of, 548–550
  storage of, 422, 539
Glycogenesis, 548, 549–550, 592
Glycogen loading, 539
Glycogenolysis, 548, 549, 550, 593, 614
Glycogen synthetase, 549, 550
Glycolipid, 437, 444
Glycolysis, 541, 542–547, 559, 592
Glycoprotein, 444, 464, 467, 590
Glycoside, 419
Glycosidic bond, 419, 420, 421, 422, 423, 436, 495
Goiter, 46, 47, 61, 611
Gold, 631
Gold foil experiment, 39
Golgi body, 636, 637
Goodyear, Charles, 396
Gout, 347, *579,* 580, 607
Graham's law, of diffusion, **127**
Graham, Thomas, 127
Grain alcohol, *see* Ethanol
Gram, unit of mass, 16, 17
Graphite, 247
Graves' disease, 611
Gray, unit of radioactivity, 221
Grease, 270
Greenhouse effect, 77, 115
Ground state, of electrons, 48, 49, 50
Group, of elements, **44**
Growth hormone, human, 246, 467, 515
GTP, *see* guanosine triphosphate
Guanine, *495, 499,* 579
Guanosine, 496, *606*
Guanosine monophosphate (GMP), *579*
Guanosine triphosphate (GTP), 506, 527, 529, 533, 534, 547
Gulose, *415*
Gypsy moth, 320

Haber, Fritz, 89
Haber process, for ammonia, 89, 94
Hafnium, 631
Hahn, Otto, 227
Hair curl, 317–318

Half-life, of radioisotope, **218,** 219–220
Halide, organic, 249
Hall, Charles M., 3, 7
Hallucinogen, 373–374, 610
Halogen, 45, 46
Halogenation:
  of alkanes, 268–269
  of alkenes and alkynes, 287–288
  of aromatic rings, 298, 299
Halothane, *268*
Handler, Philip, 489
Haworth structure, 411
Haworth, Walter, 411
HDL (high density lipoprotein), 433, 439–440, 464
Heart disease, 488, 489, 557
Heat, 21
Heat energy, 21
Heating oil, 270
Heat of fusion, 124, 125
Heat of reaction, 108
Heat of vaporization, 124, 125, 126
Helium, 631
α-**Helix,** in protein, **460**
Heme, 531, 573, 579, *580,* 581
Hemiacetal, 339, 340, 341, 410, 419
Hemiketal, 339, 340, 341, 410, 419
Hemoglobin, 449,, 460, 464, 521, 579, 580, 588, 589, 612
  effect of carbon monoxide, 82
  oxygen binding to, 176, 206, 588, 589, 612
  sickle cell, 449, 589
  structure of, 460, 461, 462, 589
  β-thalassemia, 589
Hemolysis, 156
Hemolytic anemia, 462–463, 580
Hemophilia, 516, 591
Henderson-Hasselbalch equation, 205, 453
Hepatitis, 489
Heptane, *257,* 271
Heptose, 409
Heroin, *358*
Herpes, 606
Heterocyclic amine, 371
Heterocyclic ring, 371
Heterogeneous matter, 34, 35
Hexachlorophene, 311, *312*
Hexane, 83, *257,* 288
1,6-Hexanediamine, *398*
Hexanoic acid, *350*
1-Hexanol, *307*
1-Hexene, *281,* 288
Hexokinase, 476, 541, 542
Hexose, 409
4-Hexylresorcinol, *311*
1-Hexyne, *281*
High density lipoprotein (HDL), 433, 439–440, 464, 565
Histamine, *607,* 608
Histidine, *451,* 455, 575, 588
Histone, 464
HIV (human immunodeficiency virus), 506, 606
HMG-CoA, *566*
HMG-CoA reductase, 565, 566
Hoffman, Felix, 360
Holmium, 631

660 Index

Homogeneous matter, 34, 35
Homogentisate, *576*, 577, 597
Homopolymer, 390
Hormone, 592, 593, 611
   steroid, 438–439
Human body:
   composition of, 45–46
   pH of, 205–208
   water balance in, 156–157
Human growth hormone (HGH), 246, 467, 515
Hund, Frederick, 52
Hund's rule, 52
Huntington's disease, 516
Huxley, Aldous, 273
Hybrid orbital, 240–243
Hydrate, 338, 339
Hydration reaction, 291, 338–339
   of aldehydes and ketones, 338–339
   of alkenes and alkynes, 291
Hydride ion, 97, 338
Hydrocarbon, 255, **256**
   alkane, **256**, 257–271
   alkene, **280**, 281–292
   alkyne, **283**
   aromatic, 293–299
   saturated, **256**
   unsaturated, **280**
Hydrochloric acid, 145, 185, 189, 192, 197, 608
Hydrocortisone, 438
Hydrocyanic acid, 189, 192
Hydrogen, 36, 72, 197, 632
   in Haber process, 89
   molecule of, 72
   reaction with oxygen, 166, 168, 170, 172
Hydrogenation:
   of aldehydes and ketones, 337–338
   of alkenes, 286–287
   of alkynes, 286–287
   of vegetable oils, 287, 288, 433, 434
Hydrogen bond, 117, **118**, 119
   in alcohols, 309
   in amides, 379
   in amines, 372
   in carboxylic acids, 352
   in DNA, 497–498
   in nylon, 398
   in proteins, 460–461
   in water, 140–141
Hydrogen bromide:
   reaction with alkenes and alkynes, 289
Hydrogen chloride, 79–80
   reaction with alkenes and alkynes, 289
Hydrogen cyanide, *74*
Hydrogen fluoride, 81
Hydrogen gas, 72
Hydrogen peroxide, 95, 637
Hydrogen phosphate ion, 68, 588
Hydrolase, 481
Hydrolysis, 341, **360**
   of acetals and ketals, 341
   of amides, 380–382
   of carbohydrates, 420, 475
   of esters, 360
   of proteins, 572–573
   of triglycerides, 558, 559
Hydrometer, 21, 596

Hydronium ion, 68, **186**
   concentration in water, 193–196
Hydrostatic (underwater) weighing, 20
Hydroxide ion, 68, 186
   from bases, 186
   concentration in water, 193
β-Hydroxyacyl CoA dehydrogenase, 560
β-Hydroxybutyric acid, *562*
β-Hydroxy-β-methylglutaryl CoA (HMG-CoA), *566*
Hyperammonemia, 577–578
Hyperglycemia, 46, **550**, 551–553, 611
Hyperthyroidism, 223, 611
Hypertonic solution, 156
Hyperventilation, 207
Hypervitaminosis, 441
Hypnotic, 379–380
Hypoglycemia, **550**, 551–553
Hypoglycemic drug, 552–553, 611
Hypophysis, 467
Hypothesis, **6**, 7, 8
Hypothyroidism, 611
Hypotonic solution, 156
Hypoventilation, 207
Hypoxanthine, *579*, 607

Ibuprofen, *360*, 440, 553, 612
Ice, 120, 134, 141
   heat of fusion of, 124
   structure of, 141
Ideal gas, 131
Idose, *415*
Immune system, 467, 493, 515, **590**
Immunoglobulin, 467, 468, 590
Incomplete protein, 454–455
Indicator, acid-base, 187, 188, 196
Indium, 632
Induced-fit model, 475, 476
Inducer, 510, 511
Induction, of protein synthesis, **510**, 511
Infrared (IR) radiation, 214
Inhibition, of enzymes, 483–485, 486–487
Initiation, of protein synthesis, 508–509
Initiator, of polymerization, 392, 393
Inner transition element, 45
Inosine, *493*
Inosine monophosphate (IMP), *579*
Insulin, 466, 5515, 592, 593, 611
   effect on glucose levels, 551–553, 592, 593, 611
   effect on metabolism, 592, 593
   structure of, 466
Integral protein, in membrane, 444
Interferon, 464, 515
Interleukin-2, 515
International System of Units (SI), **15**, 625–626
Interstitial fluid, 586
Intracellular fluid, 586
Intron, 505, 507
Iodide ion, 47,, 611
Iodine, 120, 310–311, 434, 632
   thyroid gland and, 47, 223, 224
Iodine number, 432, **434**, 435
Iodine-131, 223, 225, 611
Iodine-123, 224, 225
Iodine test, for starch, 424
Iodized salt, 61, 611
Ion, 64

   charge, and group number, 65
   composition in blood, 69
   polyatomic, 68, **69**, 70, 74–76
Ionic bond, 62, **64**, 123, 124, 461
Ionic compound, 61, 63, **64**, 65–70
   formulas for, 65–67
   naming of, 67–68
   properties of, 117, 118, 142–143
   uses of, 68
Ionic crystal lattice, 64, 65
Ionization energy, 63, 169
β-Ionone, *333*
Ion-product constant of water (K_w), 193
Iridium, 632
Iron, 46, 47, 91, 632
   rusting of,
Iron (II) chloride, 68
Iron (III) chloride, 68
Iron (III) oxide, 91
Irreversible inhibition, of enzymes, **485**
Isaiah, 444
Isobutane, *258*
Isobutyl group, *263*
Isocarboxazid, 610
Isocitrate, *527*, *528*, 565, 575
Isocitrate dehydrogenase, 529, 530
Isoelectric point, **453**, 458–459
Isoenzyme, **486**, 488
Isoleucine, *451*, 455, 486, *487*, 575, 576
Isomerase, 481
Isomer:
   geometric, **283**, 284–285, 412, 413, 433
   optical, 412, **413**, 414–418, 452
   stereo-, **412**, 413–418,
   structural, **258**, 259–260, 412, 413
Isooctane, *see* 2,2,4-Trimethylpentane
Isopentenyl PP, 565, 566
Isoprene, *394*, 395–396, 442, 565
Isopropyl alcohol, *307*, 311, 316
Isopropyl group, *263*
Isotonic solution, 156
Isotope, **41**, 42
   symbolism for, 41
Isozyme, *see* Isoenzyme
IUPAC, 262
IUPAC name, 262

Jaundice, 580, 581
Jewett, Frank F., 3
Joule, 14, 21, 108

K, *see* Equilibrium constant
K_a, *see* acid ionization constant
K_b, *see* base ionization constant
K_w, *see* ion-product constant of water
Kekulé, August, 293
Kelvin, Lord, 27
Kelvin temperature scale, 23–25
Keratin, 464
Kerosene, 270
Ketal, 340, 341, 419
α-Keto acid, 574, 575, 576
Ketoacidosis, 326, 422, **562**, 611
Ketogenic amino acid, **575–576**
α-Ketoglutarate, *527*, *528*, *529*, *574*, 575, *576*, *577*, 578
α-Ketoglutarate dehydrogenase, 529

Index

Keto group, 329
Ketone, 249, 314, 326, **329**, 330–342
  from alcohols, 314–315
  ketals and hemiketals from, 339–341
  naming of, 329–330
  physical properties of, 328, 331–332, 352
  reactions of, 333–341
  uses of, 333
Ketone body, 562
Ketonemia, 562
Ketonuria, 562
Ketosis, 333, 561, **562**
Kidneys, 593–596
  buffering and, 595–596
  dialysis and, 593, 596
Kidney stones, 347
Kilocalorie (Cal), 21
Kinase, 541
Kinetic energy, 21, 22, 116, 117, 121
Kinetic molecular theory, 116, 117
Knockout drops, 339
Krebs, Hans, 527, 528
Krebs cycle, *see* Citric acid cycle.
Krypton, 632
Kwashiorkor, 455–456

Lac operon, 511
Lactase, 540, 541
Lactate dehydrogenase, 486, 488, 489
Lactate, 486, *545*, 546
Lactic acid, 97
Lactose, *420*, 421, 511, 540, 541
Lactose intolerance, 540–541
Lactose operon, 511
Lanthanum, 632
Laughing gas, 77
Lauric acid, *350*, *430*
Lavoisier, Antoine, 37, 110
Law, 5, 6, 7, 8
  Avogadro's, 131
  Boyle's, 128
  Charles', 128
  of conservation of energy, 164, 228
  of conservation of mass, 37, 38, 228
  of conservation of mass-energy, 228
  of definite proportions, 37, 38
  Gay-Lussac's, 128
  Graham's, 127
  of gravity, 5, 6
  of partial pressures, 131
  of thermodynamics:
    first law, 164, 165
    second law, 164, 165, 179
Lawrence, Ernest, 225
Lawrencium, 632
Laxative, 23
LDL (low density lipoprotein), 433, 439–440, 464
Lead, 632
Lead poisoning, 305, 462, 463
Lead-206, 218, 219
Le Châtelier, Henri, 175
Le Châtelier's principle, 175, 176–179, 206, 340, 355, 418, 453
Lecithin, 435, *436*, 610
Lenin, Vladimir, 325
Lesch-Nyhan syndrome, 579
Leucine, *451*, 455, 575, 576

Leukemia, 213, 607
Leukocyte, 586, 587, 588
Levorotatory, 416
Levulose, 416
Levi, Primo, 71, 208
Lewis, Gilbert N., 53
Lewis structure, 53, 75–76
Lidocaine, *373*
Ligase, 481
Light, plane polarized, 412–413
Lime, 199
Limiting reactant, 106
Limiting reagent, 106
Limonene, *443*
Linear accelerator, 225
Linear polymer, 391–392
Linear shape, of molecules, 78, 79
Line spectrum, 49
Linoleic acid, *430*, 431
Linolenic acid, *430*, 431
Lipase, 558, 559, 592, 593
Lipid, 430, 431–444, 523
  metabolism of, 524–526, 558–566
Lipid storage disease, 437
Lipoic acid, 477, 529
Lipoprotein, 148, 433, **439**, 449, 464, 565
Liquid:
  boiling of, 122–123
  freezing of, 117, 119, 120
  physical properties of, 116
  vapor pressure of, 121–122
Liquid mosaic model, of membranes, 444
Lister, Joseph, 311, 602
Liter, unit of volume, 16, 18, 19
Lithium, 70, 632
Lithium aluminum hydride, 337, 338
Lithium carbonate, 70
Litmus paper, 187, 188
Liver, 546, 547, 548, 549, 551, 565, 580
Local anesthetic, 373, 375, 609
Lock-and-key model, 475, 476
Logarithm, 195
London force, 117, 118, 119, 258, 267, 461
London, Fritz, 117
Lovastatin, 440, 566
Low density lipoprotein (LDL), 433, 439–440, 464, 565
Lowry, Thomas, 186
LSD, *374*, 610
Lubricating oil, 270
Lucite, *393*, 394
Luminal, *380*
Lutetium, 632
Lyase, 481
Lycra, 399
Lye, 434
Lymph, 558, 586
Lymphocyte, 586, 587, 590
Lysine, *451*, 455, 465, 575, 576
Lysosome, 636, 637

Macromolecule, 390
Macrophage, 590
Mad Hatter, 71
Magnesium, 46, 632
Magnesium hydroxide, 185, 190
Magnetic resonance imaging (**MRI**) **scan, 227**

Malate dehydrogenase, 530
Malate, *527*, *529*, *530*, 575
Malonate, *483*
Malonyl ACP, 562, 563
Malonyl CoA, 563
Maltase, 540, 541
Maltose, *420*, 421, 540, 541
Manganese, 305, 632
Manhattan Project, 231
Manic depression, 70, 226, 609
Mannose, *415*
MAO (monoamine oxidase), 609–610
Marasmus, 455–456
Marfan syndrome, 516
Markovnikov's rule, 289, 290, 291, 311, 312
Markovnikov Vladimir, 289
Mass, 17
  conservation of, 37, 228
  energy relationship, 228
  measurement of, 16–17
  and weight, 17
Mass-mass calculations, 105
Mass number, 41
  and radioactivity, 216–217, 228
Mathematical review, 618–629
Matter, 4
  classification of, 34–36
  heterogeneous, 34, 35
  homogeneous, 34, 35
  measurement of, 15–25
  physical states of, 36, 116–126
  ultimate particles of, 40
McCord, David 580
Measurement, 14, 15–25
  significant figures and, 13, 14–15
Mechanism, reaction, 171, 290
Medical drugs, *see* Drug
Melamine, 399
Melanin, 576, 577
Melting, 117, **119**, 120
Melting point, 118, 120, 123
Membrane, 442, 443–444
  fluid mosaic model of, 444
  functions of, 442
Mendeleev, Dmitri, 33, 43, 44
Mendelevium, 632
Mepivacaine, *373*
Mercaptan, 317 (*see also* Thiol)
Mercaptoethanol, *463*
6-Mercaptopurine, *473*, 502, 607
Mercury, 71, 100, 107, 632
  in barometer, 128
  toxicity of, 71, 305, 462, 463
Mescaline, *374*, 610, *611*
Messenger RNA (mRNA), 514, 505–507, 508–510
Mestranol, *438*
Metabolic acidosis, 207
Metabolic alkalosis, 207
Metabolism, 177, 474, **522**
  amino acid, 573–577
  anabolism, 522, 523
  carbohydrate, 524–525, 539–553
  catabolism, 522, 523
  fatty acid, 559–565
  overview of, 524–526
  protein, 524–526, 572–573
  purine and pyrimidine, 578–579

Metabolism (continued)
triglyceride, 559, 561
Metal, **44–45**, 63, 70
  alkali, **45**
  alkaline earth, **45**
  properties of, 44–45, 70, 117, 118
  toxic, 71, 305
Metal crystal lattice, 70
Metallic bonding, 62, **70**, 123, 124
Metalloid, **45**
Metalloprotein, 71, 464
Meter, unit of length, 16, 17, 18
Methanal, see Formaldehyde
Methane, 257
Methanoic acid, see Formic acid
Methanol (methyl alcohol), 307, 311, 316, 483, 484, 607
  synthesis of, 311
  toxicity of, 311, 316, 483–484, 607
  uses of, 311
Methionine, 451, 455, 509, 510, 575
Methotrexate, 607
Methoxyflurane, 319
Methyl alcohol, see Methanol
Methylamine, 372
Methyl bromide, 273
Methylene chloride, 272
Methyl group, 263
Methyl methacrylate, 393
Metric system, **15**
  conversion between units, 625–629
  prefixes used, 16
  units used, 15–25, 625–626
Mevalonic acid, 566
Mevinolin, 566
Mevinphos, 359
Meyer, Lothar, 43
Meyerhof, Otto, 541
Microbody, 637
Microwaves, 214
Miescher, Friedrich, 494
Mifepristone, 611
Milk of magnesia, 185
Milk sugar, see Lactose
Minamata Bay, Japan, 71
Mineral, in diet, 45–46
Mirror image, 414–415
Mitchell, Peter, 531, 532
Mitochondrion, 494, 526, 530–533, 545, 560, 561, 564–565, 636, 637
Mittasch, Alwyn, 299
Molarity, **149**, 150
  titration and, 201–203
Molar mass, **100**, 132
Mole, **100**, 101–110
  and Avogadro's number, 100
  and molar mass, 100
Molecular formula, **256**
Molecule, **72**
  forces between, 117–119, 123, 124
  nonpolar, **80**, 81, 83
  polar, **81**, 82, 83
  shapes of, 77–79, 239–246
Mole-mass calculations, 104–105
Mole-mole calculations, 103–104
Mole ratio, **103**, 105
Molybdenum, 632
Monoamine oxidase (MAO), 609–610
Monocyte, 586, 587, 590

Monoglyceride, 432, 558, 559
Monomer, **390**, 461
Monoprotic acid, 189
Monosaccharide, **408**, 409–412, 540
Morphine, 358, 373, 469, 602, 612
Morton, William, 319
Moss, Norman, 534
Moyers, Bill, 9
MRI (magnetic resonance imaging) **scan**, **227**
mRNA (messenger RNA), **504**, 505–507, 508–510
Muscone, 333
Muscular dystrophy, 516
Mutagen, **512**, 513
Mutarotation, 418, **419**, 423
Mutation, 222, **512**, 513
Mylar, 397
Myocardial infarction, 488
Myoglobin, 461, 464
Myosin, 464
Myrcene, 443
Myristic acid, 430

NAD⁺ and NADH (nicotinamide adenine dinucleotide), 97, 477, 478, 522
  in citric acid cycle, 525, 527–530, 533
  in electron transport, 525, 530–533
  in fatty acid oxidation, 560, 561
  in glycolysis, 542, 545, 546
NADP⁺ and NADPH (nicotinamide adenine dinucleotide phosphate), 477, 478, 522, 534
  in cholesterol synthesis, 565–566
  in fatty acid synthesis, 562–564
  in pentose phosphate pathway, 550, 551, 564
Naming inorganic compounds, 67–68
Nandrolone decanoate, 429
Naphthalene, 294
Narcotic, 612
Nash, Ogden, 363
Natural gas, 266, 269
Nembutal, 380
Neodymium, 632
Neon, 632
Neoprene, 395
Nephron, 594
Neptunium, 632
Nerve cell, see Neuron
Nerve gas, 359, 609
Nerve impulse, 359
Nerve poison, 359, 609
Nervous system, 608–609
Nervous-system drugs, see Drug; names of specific drugs
Neuron, 608, 609, 610
Neurotoxin, 609
Neurotransmitter, 359, 374, **608**, 609, 610
Neutralization, **197**, 198–200, 351, 354, 374
Neutron, **39**, 217
  isotopes and, 41–42
  properties of, 39
Neutrophil, 586, 587
Newton, Isaac, 5, 6
Niacin, 440, 477, 479, 480, 565
Nickel, 305, 632

Nicotinamide Adenine Dinucleotide, see NAD⁺
Nicotinamide Adenine Dinucleotide Phosphate, see NADP⁺
Nicotine, 373
Niemann-Pick disease, 437
Night blindness, 441, 442
Niobium, 632
Nitrate ester, 382
Nitrate ion, 68
Nitration, of aromatic ring, 298, 299
Nitric acid, 189, 192, 198, 199, 298
Nitric oxide, 199
Nitrile, 370, **382**
Nitrite ion, 68, 521
Nitrocellulose, 382
Nitro compound, 382
Nitrogen, 74, 370, 632
  in atmosphere, 130, 370
  in Haber process, 89
  liquid, 130
  supersaturation, 139
Nitrogen balance, 454, 573
Nitrogen dioxide, 127, 128, 179
Nitrogen mustards, 502, 503
Nitroglycerin, 369, 382
Nitro group, 370, **382**
Nitromethane, 382
Nitrous acid, 512
Nitrous oxide, 77
Nobel, Alfred, 369
Nobelium, 633
Noble gas, **45**, 62, 63
Nonane, 257
Noncompetitive inhibition, of enzymes, **484**, 485
Nonmetal, **45**, 63
Nonpolar covalent bond, **79**
Nonpolar molecule, **80**, 81, 83
Norepinephrine, 374, 574, 608, 610
Norethindrone, 279
Norethynodrel, 439
Normality, **202**, 203
Novocaine, 373
Nuclear energy, 228–232
Nuclear equation, 216–218
Nuclear **fission**, **227**, 228–232
Nuclear **fusion**, 228, **232**
Nuclear medicine, 223–227
Nuclear power plant, 228–232
Nucleic acid, **494**, 495–517, 573 (see also Deoxyribonucleic acid; Ribonucleic acid)
Nucleolus, 636, 637
Nucleoprotein, 464
Nucleoside, **495**, 496
Nucleotide, **495**, 496, 573
Nucleus, atomic, **39**
  charge of, 39
  composition of, 39
  discovery of, 39
  mass of, 39
Nucleus, cell, 494, 503, 593, 636, 637
Nupercaine, 609
Nylon, 389, 391, 397, 398

Octane, 257
Octane number, **271**
Octet rule, 62, **63**, 64

Index

Oil, as fuel, 256, 267, 270–271
Oil, and fat, **432**, 433
    composition of, 432
    polyunsaturated, 282
Oleic acid, *430*, *431*
Oncogene, 513
Operator, site on DNA, 510, 511
Operon, 510, 511
Opium, 468, 601
Oppenheimer, J. Robert, 231, 232
Optical activity, **413**, 419
Optical isomer, 412, **413**, 414–418
Oral contraceptive, 299, 439, 611
Orbital, **51**, 52, 53
    hybrid *sp*, 242–243
    hybrid *sp*^{2}, 241–242
    hybrid *sp*^{3}, 240–241
    shape of *p*, 51, 52
    shape of *s*, 51, 52
Orbital hybridization, 240
Organic acid, *see* Carboxylic acid
Organic chemistry, **238**
Organic compound, 238
    comparison with inorganic compounds, 239
    functional groups, **248**
    molecular shapes, 239–246
Organic halide, 249
Orlon, *393*, 394, 398
Ornithine, *578*
Osmium, 633
Osmosis, 155, **156**
Osmotic pressure, 155, **156**, 586, 587, 594
Osteoporosis, 46, 185
Ostwald, Wilhelm, 597
Oxalate, 353–354
Oxalic acid, 95, 316, 347, *350*, *484*
Oxaloacetate, *527*, *530*, *547*, 548, 562, *564*, 565, 575
Oxandrolone, *429*
Oxidation, 93, **94**, 95–98, 109, 291
    of alcohols, 313–316, 334, 354
    of aldehydes, 314, 334, 354
    of alkenes, 291–292
    of amino acids, 574–575
    of carbohydrates, 418
    of fatty acids, 560–562
β**-Oxidation**, of fatty acids, 560, **561**, 562
Oxidation-reduction reaction, 93, **94**, 95–98, 477, 478
Oxidative deamination, 576, 577
Oxidative phosphorylation, 530, **531**, 532–533, 544, 561, 637
Oxidizing agent, **94**, 95–98
Oxidoreductase, 481
Oxygen, 36, 93, 97, 98, 115, 129, 163, 612, 633
    in atmosphere, 115
    combustion reactions and, 93
    electron transport and oxidative phosphorylation, 521, 530–533, 637
    transport by hemoglobin, 176, 588, 589, 612
    treatment with, 163
Oxygen debt, 97
Ozone, 274
Ozone layer, 274

Palladium, 633
Palmitic acid, *350*, *430*, 561, *563*, 564
Palmitoleic acid, *430*
Pancreas, 466, 489
    digestive enzymes of, 558, 573
    hormones of, 548, 551, 592, 593
Pantothenic acid, 477, 480
Para-aminobenzoic acid (PABA), *484*, *603*, 604, 606
Parathion, *609*
Parkinson's disease, 571, 609
Partial charge, 79, 80
Partial pressure, 131
    law of, **131**
Parts per million (ppm), **152**
Pascal, unit of pressure, 128
Pasteur, Louis, 602
Pellagra, 479–480
Penicillin, 601, *604*
Penicillinase, 604
Penicillium mold, 601, 604
Pentane, *257*
Pentanoic acid, *350*
2-Pentanone, *328*
1-Pentene, *281*
Pentobarbital, *380*
Pentose, 409
Pentose phosphate pathway, **550**, 551, 564, 565
PEP carboxykinase, 547, 548
Pepsin, 485, 573
Pepsinogen, 485
Peptide, **456**, 592, 593
Peptide bond, **456**, 459, 509, 510
Peptidyl transferase, 509, 510
Percentage yield, 106, **107**, 108
Percent concentration, **150**
    volume/volume, 152
    weight/volume, 150–151
    weight/weight, 152
Perchlorate ion, 611
Perfluorodecalin, *273*
Period, of elements, **44**
Periodic table of the elements, 33, **43**, 44, 45, 54–55
    development of, 33, 43, 44
    electronic structure and, 54–55
    groups of elements in, **44**
    properties of elements in, 46
Peripheral protein, in membrane, 444
Permanganate ion, 68
Pernicious anemia, 46, 480
Peroxisome, 637
Pesticide, 273, 359, 609
PET (positron emission tomography) **scan**, **226**, 227
Petrochemical, **269**
Petroleum, 269–271
Petroleum ether, 270
pH, **194**, 195–197
    of blood, 194, 206–208, 586, 588–589, 595–596
    and buffers, 204–205
    and enzyme activity, 483
    and salt solutions, 196–197
Phenanthrene, *294*
Phenobarbital, *380*
Phenol, *295*, 306, *307*, **308**, 309, 325, *399*
    acidity of, 316–317

    naming of, 308–309
    physical properties of, 309–310
    uses of, 311–312, 325
Phenolphthalein, 187, 188, 201
Phenylalanine, *451*, 455, 575, *576*, 577
Phenyl group, 296, 297
Phenylketonuria (PKU), 326, 516, **576**, 577, 597
Phenylpyruvate, *577*, 597
Pheromone, 320
pH meter, 196
Phosgene, 191
Phosphate anhydride, 362, **363**, 524
Phosphate buffer, 204
Phosphate ester, **362**, 363, 496
Phosphate ion, 68, 204, 206, 597
Phosphatidyl choline (lecithin), 435, *436*
Phosphatidyl ethanolamine, *436*
Phosphatidyl serine, *436*
3′,5′-Phosphodiester bond, 496, 497
Phosphoenolpyruvate (PEP), 524, *542*, *544*, *547*, 548
Phosphofructokinase, 542, 543
Phosphoglucomutase, 542, 549
6-Phosphogluconate, *551*
6-Phosphogluconolactone, *481*
Phosphoglucose isomerase, 481, 542
2-Phosphoglycerate, *542*, *544*
3-Phosphoglycerate, *542*, *544*
Phosphoglycerate kinase, 542
Phosphoglyceride, **435**, 436, 442–443, 523
Phospholipid, **435**, 436, 439
Phosphoric acid, 189, 192, 362–363
    anhydride of, 362–363
    ester of, 362–363
Phosphorus, 35, 46, 633
Phosphorylase, 549, 550
Phosphorylation:
    of enzymes, 486
    oxidative, **531**, 532–533, 544, 637
    substrate-level, 531, 544
Photosynthesis, 115, **540**
pH paper, 196
pH scale, 194–196
Physical change, **36**, 90–91
Physical sciences, 4
Physical state, of matter, 36, 116–119
    changes in, 119–126
Physostigmine, *610*
pI (**isoelectric point**), **453**, 458–459
Pi (π) bond, **242**, 243, 326, 327
Pi (π) electron, **242**, 293–294, 297
Pituitary gland, 467, 593
pK$_a$, **205**, 351, 453, 483, 588
PKU (phenylketonuria), 326, 516, **576**, 577, 597
Plane-polarized light, 412–413
Plasma, **586**
Plasma membrane, 636, 637
Plasmid, **513**, 514, 515
Plasmin, 488, 489
Plasminogen, 488
Plastic, **391** (*see also* Polymer)
Plastid, 637
Platelet, 587, **588**
Platinum, 171, 172, 633
β-**Pleated sheet**, in protein, **460**
Plexiglas, *393*, 394

Index

Plutonium, 633
Polar covalent bond, 79, 80
Polarimeter, 412, 413
Polarized light (plane), 412–413
Polar molecule, 81, 82, 83
Polonium, 213, 633
Polyacrylonitrile, *393*
Polyamide, 397, 398
Polyatomic ion, 68, **69**, 70, 188
   electron dot structure of, 74–76
Polycyclic aromatic compound, 294
Polydeoxynucleotide, 497
Polyester, 358, **396**, 397
Polyethylene, *393*
Polyisoprene, *394*, *395*
Polymer, 390, 391–401
   addition, **392**, 393–396
   branched, 391–392
   condensation, **392**, 396–401
   cross-linked, 392
   linear, 391–392
   silicone, 400, 401
   thermoplastic, 392
   thermosetting, 392, 399
Polymerization, 392–394, 396, 398–400
   addition, 392–394
   condensation, 396, 398–400
Polynucleotide, 496–497
Polypeptide, 456
Polypropylene, *393*, 394
Polyribosome, 510
Polysaccharide, 409, 421–423, 541
Polysome, 510
Polystyrene, *393*
Polyunsaturated compound, 282
Polyunsaturated fat, 96
Polyurethane, *399*
Polyvinyl chloride (PVC), *393*, 394
Polyvinylpyrrolidene (PVP), *393*, 394
Positron, 217, 232
Positron emission tomography **(PET) scan, 226**, 227
Potassium, 46, 61, 633
Potassium cyanide, 521
Potassium dichromate, 292, 315
Potassium hydroxide, 190
Potassium iodide, 611
Potassium ion, 534, 597, 608, 612
Potassium permanganate, 95, 292
Potassium sorbate, *353*
Potential energy, 21, 22, 62, 63, 72
   minimum, 62
Praseodymium, 633
Precipitate, 90, 91
Precision, 14, 15
Preservative, 325, 339, 353
Pressure, 127–133
   atmospheric, 128
   effect on equilibrium, 178
   effect on solubility, 144
   partial, 131
   standard, 131–132
   units of, 128
Pressure cooker, 123
Priestley, Joseph, 77, 395
Primary alcohol, 308
Primary amine, 370
Primary carbocation, 290
Primary carbon atom, 260

Primary structure, of protein, **456**, 457–459, 463, 510
Principal energy level, 50, 51–53
Procaine, *373*, 609
Proenzyme, 485, 486, 572, 573, 591
Progesterone, *279*, 333, *439*, 593, 611
Progestin, 611
Prokaryotic cell, 636
Proline, *451*, 465, 481, 575
Promazine, *381*
Promethium, 633
Promoter, site in DNA, 503, 510, 511
Prontosil, *603*
Proof, of alcoholic beverages, 152, 310
Propanal (propionaldehyde), *328*
Propane, *257*
1,2,3-Propanetriol, *see* Glycerol
1-Propanol, *307*
2-Propanol, *307*, 311, 316
Propanoic acid, *see* Propionic acid
Propanone, *see* Acetone
Propene (propylene), *281*, 393
Propionic acid, *350*
Propranolol, 614
Propylene, *see* Propene
Propylene glycol, *see* 1,2-Propanediol
Propyl group, *263*
Protactinium, 633
Prostaglandin, 440, 612
Protein, 377, **450**, 451–469, 507–511
   acid-base properties of, 458–459
   amino acids in, 451
   classification of, 463–464
   complete, 454
   denaturation of, 461, **462**, 463, 483
   digestion of, 462, 472–473
   energy from, 454, 523
   functions of, 450
   incomplete, 454–455
   metabolism of, 524–526, 572–573
   structure of:
      primary, **456**, 457–459, 463, 572
      quaternary, **461**, 462, 463
      secondary, **460**, 463
      tertiary, **460**, 461, 463
   synthesis of, 507–511, 573, 605–606
Protein kinase, 550
Protein malnutrition, 455–456
Prothrombin, 591, 592
Proton, 39, 217 (*see also* Hydrogen ion)
   properties of, 39
Proton acceptor, *see* Base
Proton donor, *see* Acid
Proton gradient, 530, 531–533
Proust, Joseph, 37
Psilocybin, *374*
Pure substance, 34, 35, 36
Purine, *371*, 495, 573, 578
   catabolism of, 579
   synthesis of, 573, 578
Puromycin, *605*, 606
Putrescine, *373*
PVC (polyvinyl chloride), *393*, 394
Pyramidal shape, of molecules, 78, 79
Pyridoxal phosphate, 477, 529, 574
Pyridoxamine, *479*
Pyridoxine, 477, 480
Pyrimidine, *371*, 495
   catabolism of, 579

   synthesis of, 578–579
Pyrophosphoric acid, *363*, 508, 523, 524
Pyrrole, *371*
Pyruvate carboxylase, 547, 548
Pyruvate kinase, 542
Pyruvate, *544*, *547*, 548, *574*, 575
   from amino acids, 575
   from glycolysis, 542, 544, 545
   metabolism to acetyl CoA, 545
   metabolism to lactate, 545–546
Pyruvic acid, 97

Quantum, 50
Quantum mechanical model, of atomic structure, 49, **50**, 51–54
Quark, 40
Quaternary carbon atom, 260
Quaternary structure, of protein, **461**, 462, 463

Rad, unit of radioactivity, **221**
Radiation, electromagnetic, 214
Radical, free, *see* Free radical
Radioactive waste, 231–232
Radioactivity, 39, 213, **214**, 215–232
   background, 221, 222
   discovery of, 215
   effects of, 213, 221–223
   half-life of, 218–220
   nuclear equations and, 216–218
   types of, 214–216, 217, 227–232
   units of, 221
   uses of, 218–220, 223–226, 611
Radio and TV waves, 214
Radiocarbon dating, 219–220
Radioisotope, 215
   uses in medicine, 223–226
Radium, 213, 633
Radon, 633
Radon-222, 221, 222
Rain, acid, 133, 198–199
Rate of reaction, *see* Reaction rate
RDA, *see* Recommended Daily Allowances
Reaction, *see* Chemical reaction; Nuclear reaction
Reaction mechanism, 171, 290
Reaction rate, 166, 167–172, 474
   activation energy and, 166–168, 171–172
   chemical reactivity and, 169
   effect of catalyst, 171–172
   effect of concentration, 170
   effect of physical subdivision, 169–170
   effect of temperature, 170–171
Recombinant DNA, 513, 515, 612
Recommended Daily Allowances (RDA), 45–46, 185, 441, 480
Red blood cell (erythrocyte), 449, 462–463, **586**, 587, 612, 613
Reducing agent, 94, 95–98
Reducing sugar, 418, 423
Reduction, 93, **94**, 95–98, 291
   of aldehydes, 337–338
   of alkenes and alkynes, 286–287
   of ketones, 337–338
Regulator gene, 510, 511
Regulatory site, of enzyme, 482, 486
Rem, unit of radioactivity, **221**

Renin, 595
Replication, of DNA, **499**, 500–503
Repression, of protein synthesis, 487, **510**, 511
Repressor protein, 510, 511
Respirator, 129
Respiratory acidosis, 207
Respiratory alkalosis, 207
Respiratory quotient, 104
Restriction enzyme, 513, 515, 516
Retinal, *285*
Retinoic acid, *442*
Retinol, *441*
Retrovirus, 506
Reverse reaction, 173
Reverse transcriptase, 506, 513
R group, 248
Rh antigen, 585
Rhenium, 633
Rheumatoid arthritis, 607
Rhodium, 633
Riboflavin, 477, *479*, 480
Ribonucleic acid (RNA), 494, 495–497, 503–511
   composition of, 494–495
   genetic code and, 507–508
   messenger RNA (mRNA), 504, 505–507
   protein synthesis and, 507–511
   ribosomal RNA (rRNA), 504
   transcription and, **513**, 504–507, 637
   transfer RNA (tRNA), 503, 505
   translation and, **507**, 508–511
Ribose, *409*, 494, 495, 550
Ribose-5-P, *551*
Ribosomal RNA (rRNA), 504
Ribosome, 494, **504**, 505, 508–510, 637
Rickets, 441, 442, 443
Rifamycin, 605
RNA, *see* Ribonucleic acid
RNA polymerase, 503, 504, 510, 511
Roentgen, unit of radiation, **221**
Roentgen, Wilhelm, 221, 224
Rosenberg, Barnett, 502
Rounding off numbers, 619
rRNA (Ribosomal RNA), 504
Rubber, 394–396
Rubbing alcohol, 311, 316
Rubidium, 633
RU-486, *611*
Ruthenium, 633
Rutherford, Ernest, 39

Salicylic acid, *350*, *360*, 602
Saline solution, 150, 151
Salinity, 199
Salt, 61, 64
   pH of solutions, 196–197
Salvarsan, *602*, 603
Samarium, 633
Sandburg, Carl, 134
Saponification, 361, 362, 434
   of esters, 361
   of triglycerides, 361–362, 434
Saran, 391, *393*
Sarin, *609*
Saturated hydrocarbon, 256
Saturated solution, 147, **149**

Scandium, 633
Schizophrenia, 226, 609
Schrödinger, Erwin, 50, 51
Science, 4
   limitations of, 8, 9
   methods of, 5–7
Scientific **data, 5**
Scientific **hypothesis, 6**, 7, 8
Scientific **law, 5**, 6, 7, 8
Scientific method, 6, 7, 8
Scientific notation, 15, 622–624
Scientific **theory, 6**, 7
Scrubber, 199
Scurvy, 480–481
Sebaceous gland, 255
Secobarbital, *380*
Seconal, *380*
Secondary alcohol, 308
Secondary amine, 370
Secondary carbocation, 290
Secondary carbon atom, 260
Secondary structure, of protein, **460**, 463
   α-**helix, 460**
   β-**pleated sheet, 460**
Second law of thermodynamics, 164, 165, 179
Selenium, 633
Serine, *451*, *452*, 486, 575
Serotonin, *374*, *608*, 610
Serturner, Friedrich, 601
Shape, of molecule, 77–79, 239–246
   orbital hybridization, 240–243
   VSEPR theory, 77, 78, 239–242
SI, *see* International System of Units
Sickle cell anemia, 449, 516, 589
Sievert, unit of radioactivity, 221
Sigma (σ) bond, 240, 242, 243
Significant figure, 13, **14**, 15, 618–620
Silicon, 400, 633
Silicone implant, 400
Silicone polymer, **400**, 401
Silver, 70, 634
Silver mirror test. *See* Tollens test
Silver nitrate, 91
Sinsheimer, Robert, 517
Skin, 255
Snake venom, 532, 609
Snowflake, 141
Soap, 361–362, 434
Sobrero, Ascanio, 369
Social sciences, 4
Sodium, 46, 634
Sodium benzoate, *353*
Sodium bicarbonate, 185, 200
Sodium borohydride, 337, 338
Sodium butyrate, *353*
Sodium chloride, 61, 142–143, 145
Sodium dichromate, 292
Sodium hydroxide, 149, 190, 434
Sodium hypochlorite, 95
Sodium ion, 63–64, 65, 534, 586, 593, 594, 597, 608, 612
Sodium nitrate, 89
Sodium nitrite, 521
Sodium propionate, *353*
Sodium thiosulfate, 95, 521
Sodium urate, 347, 579, 580

Solid:
   physical properties of, 116
Solidification, 120
Solubility, 144
   effect of pressure, 139, 144
   effect of temperature, 144
   general solubility rules, 142
   of hydrocarbons, 267
Solute, 141
Solution, 34, 35, **141**
   acidic, 194
   alkaline, 194
   basic, 194
   colligative properties of, 154–156
   concentrated, 149
   concentration, 149
      equivalents per liter, 152–153, 202
      hypertonic, 156
      hypotonic, 156
      isotonic, 156
      molarity, 149, 150
      normality, 202, 201
      parts per million (ppm), 152
      percent v/v, 152
      percent w/v, 150–151
      percent w/w, 152
      saturated, 147, **149**
      standard, 200
      supersaturated, 149
      types of, 142
      unsaturated, 147, **149**
   dilute, 149
   dilution of, 153, **154**
   process of, 142–143
   properties of, 146
Solvent, 141
Sorbitol, *551*, 553
Spandex, 399
sp hybrid orbital, 242–243
sp^2 hybrid orbital, 241–242, 283, 293–294, 326–327, 348, 459
sp^3 hybrid orbital, 240–241, 258
Specific gravity, 19, 21, 596
Specific heat, 124, 125, 126
Specificity, of enzymes, 475–477
Specific rotation, 413
Sphingolipid, 436, 437, 443–444, 585
Sphingomyelin, *436*, 437
Sphingosine, *436*
Sports drink, 612
Stain remover, 95
Standard solution, 200
Standard temperature and pressure (STP), 131–132
Starch, *421*, 424, 483, 541
   iodine test for, 424
Starvation, 207, 561–562
States of matter, physical, 36
Steady state, 177
Stearic acid, *350*, *430*, *431*
Stereoisomer, 412, 413–418
   maximum number of, 414
Sterling silver, 70
Steroid, 429, **437**, 438–439, 593
   anabolic, 429, 614
   hormone, 593
Stevens, Dorothy Mae, 24
Stimulant, 374

Stoichiometry, 102, 103–108, 133
  gases and, 133
STP (standard temperature and pressure), 131–132
Strassmann, Fritz, 227
Stratum corneum, 255
Streptokinase, 488
Strong acid, 189, 190–192
Strong base, 190, 191, 192
Strontium, 634
Structural formula, 258
  condensed, 261, 262
Structural gene, 510, 511
Structural isomer, 258, 259–260, 412, 413
Styrene, *393,* 396
Subatomic particle, 39–40
Sublevel, electron, **50,** 51–53
Sublimation, 120
Substitution reaction, 269, 297
Substrate, 474, 482
Substrate-level phosphorylation, 531, 544
Succinate, *483, 527, 529,* 575
Succinate dehydrogenase, 529
Succinyl CoA, *527, 529,* 575, 579
Succinyl CoA synthetase, 529
Sucrase, 540
Sucrose, 145, *420,* 421, 475, 540
  sweetness of, 411, 412
Sugar, 37
Sulfadiazine, *603*
Sulfa drug, 298, **603,** 614, 606
Sulfanilamide, 298, *484, 603,* 604
Sulfate ion, 68, 597
Sulfhydryl group, 457, 462
Sulfide ion, 65, 66, 521
Sulfisoxazole, *613*
Sulfite ion, 68
Sulfonation, of aromatic ring, 298
Sulfur, 35, 65, 100, 634
Sulfur dioxide, 133, 198, 199
Sulfuric acid, 189, 191, 192, 198, 199, 291, 298
Sulfur oxide pollutants, 198, 199
Supersaturated solution, 149
Supersaturation, nitrogen, 139
Surface tension, 140
Suspension, 146, 147
Sweat:
  apocrine, 255
  eccrine, 255
  pH of, 586
Sweetener, artificial, 407
Sweetness, of sugars, 407, 411, 412
Synapse, 608
Synovial fluid, 586
Szent-Györgyi, Albert, 342, 480

Table salt, *see* Sodium chloride
Tacrine, 610
Tagamet, 607, 608
Talose, *415*
Tantalum, 634
Tar, 270
Tartaric acid, *414*
Tay-Sachs disease, 437, 516
T cell, of immune system, 587, 590
Technetium, 634
Technetium-99, 224, 225

Teflon, 273, 390, *393,* 395
Tellurium, 634
Temperature, 23, 24–25, 127–133
  effect on equilibrium, 178–179
  effect on gases, 127–133
  effect on solubility, 144
  measurement of, 13, 23, 24
  reaction rate and, 170–171, 482, 483
  regulation in the body, 589–590
  standard, 131–132
Temperature scale, 13, 23, 24
  conversion between, 23, 24
Terbium, 634
Terephthalic acid, *396*
Termination, of protein synthesis, 510
Terpene, 442, 443
Terramycin, 604, *605*
Tertiary alcohol, 308
Tertiary amine, 370
Tertiary carbocation, 290
Tertiary carbon atom, 260
Tertiary structure, of protein, **460,** 461, 463
Testosterone, 333, *429,* 593, 614
Tetracaine, 609
Tetracycline, 604–605
Tetrafluoroethylene, *393,* 395
Tetrahedral carbon, 239–240, 258
Tetrahedral shape, of molecules, 78, 79, 239, 240
Tetrahydrofolic acid, 477, 607
Tetramer, 461, 462
Tetrapeptide, 456
Tetrose, 409
β-Thalassemia, 516, 589
Thallium, 634
Theoretical yield, 107–108
Theory, 6, 7
Thermodynamics, 164, 165
  first law of, 164, 165
  second law of, 164, 165, 179
Thermometer, 13, 23–24
Thermoplastic polymer, 392
Thermosetting polymer, 392, 399
Thiamine, 477, *479,* 480
Thiocyanate ion, 521, 611
Thiol, 249, 305, 306, **317,** 318, 457
  naming of, 317
  properties of, 317
  reactions of, 317, 457
Thiolase, 560
Thiosulfate, ion, 521
Thiouracil, *611*
Thiourea, *611*
Thomson, William, 27
Thorazine, *381,* 609
Thorium, 634
Three Mile Island, 230
Threonine, *451,* 455, 486, *487,* 575
Thrombin, 591, 592
Thulium, 634
Thymidine, 496, *506*
Thymine, *473,* 495, *499,* 579
Thyroid gland, 47, 223, 224, 611
  hyperthyroidism, 223, 224, 611
  hypothyroidism, 611
Thyroxin, 47, 372, *574,* 593, 611
Time, unit of, 25
Tin, 634

Tissue plasminogen activator (tPA), 488, 515
Titanium, 634
Titration, 200, 201–203
TNT (2,4,6-Trinitrotoluene), 382
α-Tocopherol (vitamin E), *441,* 442
Tollens test, 334, 335, 418
Tolstoy, Leo, 567
Toluene, *295*
Torricelli, Evangelista, 128
Torr, unit of pressure, 128
tPA, *see* Tissue plasminogen activator
Tranquilizer, 380, 381
Transamination, 574, 575, 576, 577
Transcription, of DNA, **503,** 504–507, 605, 606
Transferase, 481
Transfer RNA (tRNA), 503, 505, 508–510
Trans **isomer, 283,** 284, 285, *395,* 412, 413, 433
Transition element, 45
Translation (protein synthesis), **507,** 508–511, 605, 606
Transuranium element, 225
Tranylcypromine, 610
Triacylglycerol, *358,* 361–362, **432,** 433, 439, 558–562
  catabolism of, 559–562
  composition of, 432
  digestion of, 558–559
  hydrogenation of, 433
  hydrolysis of, 433–434, 558, 559
  saponification of, 361–362, 434
Triangular shape, of molecules, 78, 79
Tricarboxylic acid, 350
Tricarboxylic acid (TCA) cycle, *see* Citric acid cycle
Triethylamine, 376
Triglyceride, *see* Triacylglycerol
2,2,4-Trimethylpentane ("isooctane"), *271*
2,4,6-Trinitrotoluene (TNT), *382*
Triol, 307–308
Triose, 409
Triosephosphate isomerase, 542
Tripeptide, 456, 466
Triphosphoric acid, 363
Triprotic acid, 189
tRNA, *see* Transfer RNA
Tropocollagen, 465
Trypsin, 489, 573
Tryptophan, *451,* 455, 480, 575, 576
Tungsten, 634
Tyndall effect, 146, 147
Tyndall, John, 146
Tyrosine, *451,* 469, 486, *574,* 575, *576,* 597, 611

UDP-galactose, 550
UDP-glucose, 549, 550
UDP-glucose pyrophosphorylase, 549
Ulcer, 185, 608
Ultraviolet (UV) radiation, 214, 269, 274
Uncertainty principle, 49–50
Uncertainty, significant figures and, 14–15
Uncoupler, of oxidative phosphorylation, 532

Unit conversion factor, 25–28, 624–629
Unit conversion method, 25–28, 625–629
Units of measurement, 15–25
   density, 19, 20
   energy, 21
   length, 17
   mass, 16–17
   pressure, 128
   radioactivity, 221
   temperature, 23–25
   time, 25
   volume, 17–18
Universal gas constant, 133
Universal gas law, **132**, 133
Unnilennium, 634
Unnilhexium, 634
Unniloctium, 634
Unnilpentium, 634
Unnilquadium, 634
Unnilseptium, 634
Unsaturated hydrocarbon, **280**
Unsaturated solution, 147, **149**
Updike, John, 401
Uracil, *473*, *494*, *512*, 579, 607
Uranium, 634
Uranium-235, 227, 228, 229
Uranium-238, 218, 219, 220
Urea, *74*, 237, 379, 463, *523*, 577, *578*, 593, 597
Urea cycle, 575, **577**, 578
Uric acid, 347, 578, *579*, 597
Uridine, 496
Urine, 586, 593, 596–597
   composition of, 597
   pH of, 586
   specific gravity of, 21, 596
Urinometer, 21, 596
Urokinase, 488

Vaccine, 515, 590
Valence electron, **53**, 54, 55
Valence-shell electron-pair repulsion (VSEPR), **77**, 78, 239–242
Valeric acid, *350*

Valine, *451*, 455
Valium, *381*
Vanadium, 635
Vanillin, *333*
Vaporization, 117, **121**
   heat of, **124**, 125, 126
Vapor pressure, **121**, 122
Vasopressin, 593, 594, 595
Vegetable **oil**, 282, 287, 288, **432**, 433
Very low density lipoprotein (VLDL), 439, 464
Vinegar, 188, 190, 200
Vinyl chloride, *282*, *393*
Vinyl group, *282*, 394
Vinylidene chloride, *393*
Vinylidene group, 394
Vinyl pyrrolidene, *393*
Virus, 506, 513, **606**
Visible light, 214
Vital force, 237
Vitamin, 373, 440–442, 477, **479**, 480, 481
   classes of, 440, 479
   deficiency diseases and, 441–442, 479–481
   fat-soluble, 440–442
   water-soluble, 477, 479–481
Vitamin A, 280, 285, 395, *441*, 442
Vitamin B$_1$, *see* Thiamine
Vitamin B$_2$, *see* Riboflavin
Vitamin B$_6$, *see* Pyridoxine
Vitamin B$_{12}$, 477, 480
Vitamin C. *See* Ascorbic acid
Vitamin D, *441*, 442, *443*
Vitamin E. *See* α-Tocopherol
Vitamin K, *441*, 442, *487*, 488, 591, 607
VLDL (very low density lipoprotein), 439, 464
Volume, **17**
   gases and, 117–123
   measurement of, 17–18
VSEPR, *see* valence-shell electron-pair repulsion
Vulcanization, 396

Wald, George, 285, 417
Warfarin, *487*, 488, 607
Water, 72–73
   in the body, 156–157, 586, 589–590, 594, 596
   boiling point of, 122
   density of, 19, 20, 141
   hydrogen bonding in, 119, 140–141
   ionization of, 192–193
   K_w value, 193
   polarity of, 81, 83
   properties of, 140–141, 145
   surface tension of, 140
Watson, James, 493, 497
Wavelength, 214
Wax, **431**
Weak acid, **189**, 190–192
Weak base, **190**, 192
Weight, **17**
   reduction in body, 22–23
Weinberg, Steven, 9
White blood cell (leukocyte), **586**, 587, 588
Wilkins, Maurice, 497
Wine, 152, 310, 311, 314
Wöhler, Friedrich, 237, 379
Wood alcohol, *see* Methanol
Wunderlich, Carl, 13

Xanthine, *579*
Xenon, 635
X ray, 214, 215, 216, **224**, 225, 226, 227
Xylene, *295*
Xylitol, *409*
Xylose, *409*

Ytterbium, 635
Yttrium, 635

Zinc, 96, 635
Zirconium, 635
Zwitterion, 450, 453
Zymogen, 485–486, 591

Photo, Art, and Table Credits

This page constitutes an extension of the copyright page. We have made every effort to trace the ownership of all copyrighted material and to secure permission from copyright holders. Should there be any question about the use of any material, we will be pleased to make the necessary corrections in future printings.

Part I, p. 1: French/Stock Imagery. **Chapter 1**: **p. 3** (all), courtesy, ALCOA; **p. 4**, Axel-Jacana/Photo Researchers; **p. 5** (top), Sam Zarember/The Stock Shop; **p. 5** (bottom), Lawrence Migdale/Stock Boston; **p. 6** (top), The Bettmann Archive; **p. 6** (bottom), Stephen Frisch/Photo 20-20; **p. 7**, Gayna Hoffman/Stock Boston; **p. 8**, courtesy, NASA; **p. 9**, Greg Gawlowski/Photo 20-20. **Chapter 2**: **p. 13**, Frank Siteman/Stock Boston; **p. 14**, courtesy, Larry Watson/Central Washington University; **p. 17**, Mark Richards; **p. 18**, Aaron Haupt/Stock Boston; **p. 19**, courtesy, Larry Watson/Central Washington University; **p. 21**, courtesy, Larry Watson/Central Washington University; **p. 22**, Richard Kolar/Earth Scenes; **p. 23**, Michael Newman/PhotoEdit. **Chapter 3**: **p. 33**, The Bettmann Archive; **p. 34** (top), TomMcHugh/Photo Researchers; **p. 34** (bottom left, right), Mark Richards; **p. 35**, E.R. Degginger/Earth Scenes; **p. 36** (top), Stephen Frisch/Photo 20-20; **p. 36** (bottom), Grushow/Grant Heilman; **p. 37** (top), The Burndy Library; **p. 37** (middle), Mark Richards; **p. 37** (bottom), Smithsonian Institution; **p. 38**, courtesy, IBM; **p. 40**, Omikron/Photo Researchers; **p. 45**, Barry L. Runk/Grant Heilman; **p. 47** (top), Rich Treptow/Photo Researchers; **p. 47** (bottom), Bob Daemmrich/Stock Boston; **p. 48**, AIP/Niels Bohr Library; **p. 49** Richard Laird/The Stock Shop; **p. 51**, The Bettmann Archive. **Chapter 4**: **p. 61** (top), Bill Pierce/Rainbow; **p. 61** (bottom), Barbara Von Hoffman/Tom Stack & Associates; **p. 62** (left), Dan McCoy/Rainbow; **p. 62** (right), E.R. Degginger/Earth Scenes; **p. 68**, Matt Meadows/Peter Arnold; **p. 71**, Dan McCoy/Rainbow; **p. 77**, The Bettmann/Archive; **p. 79**, Janice M. Sheldon/Photo 20-20; **p. 81** (both), Stephen Frisch/Photo 20-20. **Chapter 5**: **p. 89** (top), Phyllis Picardi/Stock Boston; **p. 89** (bottom), The Bettmann Archive; **p. 90**, Bob Daemmrich/Stock Boston; **p. 91** (left), Stephen Frisch/Photo 20-20; **p. 91** (middle), Peter Menzel/Stock Boston; **p. 91** (right), Jim McGrath/Rainbow; **p. 95**, courtesy, Christine A. Pullo; **p. 97**, Bob Daemmrich/Stock Boston; **p. 99**, The Bettmann Archive; **p. 100**, Stephen Frisch/Stock Boston; **p. 104**, Jim Olive/Peter Arnold; **p. 107**, Stephen Frisch/Photo 20-20. **Chapter 6**: **p. 115**, Ted Streshinsky/Photo 20-20; **p. 116**, Mark Richards; **p. 120** (top), Fritz Goro/Life Magazine; **p. 120** (bottom), Stephen Frisch/Photo 20-20; **p. 122**, Barry L. Runk/Grant Heilman; **p. 123**, SIU/Peter Arnold; **p. 126**, Michael P. Gadonski/Photo Researchers; **p. 127**, Stephen Frisch/Photo 20-20; **p. 129** Yoav Levy/Phototake; **p. 130**, Jim Olive/Peter Arnold; **p. 132** Stephen Frisch/Photo 20-20. **Chapter 7**: **p. 139**, courtesy, John Colt/Montgomery Watson Co.; **p. 140**, Hermann Eisenbeiss/Photo Researchers; **p. 141**, courtesy, N.O.A.A.; **p. 144**, Mark Richards; **p. 145**, Stephen Frisch/Photo 20-20; **p. 146** (both), Stephen Frisch/Photo 20-20; **p. 147**, Ernest Braun/Photo 20-20; **p. 149** (both), Stephen Frisch/Photo 20-20; **p. 151**, Yoav Levy/Phototake; **p. 152**, Mark Richards; **p. 155** (left), Mark Richards; **p. 155** (right), Stephen Frisch/Photo 20-20; **p. 156**, Stephen Frisch/Photo 20-20. **Chapter 8**: **p. 163**, C.C. Duncan/Medical Images; **p. 164**, Agence Vandystadt/Photo Researchers; **p. 165**, Larry Lefever/Grant Heilman; **p. 166**, UPI/Bettmann; **p. 170** UPI/Bettmann; **p. 171**, Table 8.1, Data from J. Loeb and J. H. Northrop, 1917, *Journal of Biological Chemistry*, 32, 103-121; **p. 175** The Smithsonian Institution; **p. 178** (all), Stephen Frisch/Photo 20-20; **p. 179** (all), Stephen Frisch/Photo 20-20. **Chapter 9**: **p. 185**, Mark Richards; **p. 186**, Chemical Heritage Foundation; **p. 188** (all), Stephen Frisch/Photo 20-20; **p. 190**, Mark Richards; **p. 196** (both), Stephen Frisch/Photo 20-20; **p. 197** (both), Stephen Frisch/Photo 20-20; **p. 200**, Stephen Frisch/Photo 20-20; **p. 201**, Stephen Frisch/Photo 20-20; **p. 204**, Stephen Frisch/Photo 20-20. **Chapter 10**: **p. 213**, The Granger Collection; **p. 214**, Richard Palsey/Stock Boston; **p. 215**, The Granger Collection; **p. 221**, Yoav Levy/Phototake; **p. 222**, Table 10.3, Based on Report 93, 1987, National Council on Radiation Protection; **p. 223** (top), CNRI/Phototake; **p. 223** (bottom), courtesy, Varian Associates; **p. 224** (all), Dr. I. Ross McDougall/Stanford University; **p. 226** (top), Susan Leavines/Science Source/Photo Researchers; **p. 226** (bottom), CNRI/GJLP/Phototake; **p. 227**, NIH/Science Source/Photo Researchers; **p. 230**, courtesy, National Atomic Museum; **p. 231**, Sonnee Gottlieb/UPI/Bettmann. **Part II, p. 237**, Lonnie Duka/Tony Stone Images. **Chapter 11**: **p. 239** (top), Beckman Center for the History of Chemistry. **p. 239** (bottom), John Cancalosi/Stock Boston; **p. 239**, Figure 11.4, Based on *Chemistry: A Basic Introduction*, 3rd ed., by G. Tyler Miller, p. 552. Copyright © 1991 by Wadsworth Publishing Company; **p. 240**, Ray Ellis/Photo Researchers; **p. 243**, Figure 11.9, Based on *Chemistry: A Basic Introduction*, 3rd ed., by G. Tyler Miller, p. 559. Copyright © 1991 by Wadsworth Publishing Company; **p. 248**, courtesy, Abraham M. de Vos/Genentech; **p. 249** (top), courtesy, G.E. Superabrasives; **p. 249** (bottom), courtesy, CarboMedics. **p. 257**, Tony Kent/Sygma; **p. 258**, Jacques Jangoux/Peter Arnold; **p. 269** (top), *The Cleveland Plain Dealer*; **p. 269** (bottom) NOAA /Center for Marine Conservation; **p. 270**, Figure 12.8, Based on a figure in *Chemistry and Petroleum*. The American Petroleum Institute, New York; **p. 273**, courtesy, Christine A. Pullo; **p. 275** (left), Gebauer Co., Cleveland, OH.; **p. 275** (middle), Stacy Pick/Stock Boston; **p. 275** (right), courtesy, Dr. Leland C. Clark/Antioch College. **Chapter 13**: **p. 281**, courtesy, Wyeth-Ayerst Laboratories; **p. 282**, Raymond Reuter/Sygma; **p. 284**, Mark E. Gibson/Photo 20-20; **p. 285**, Tom Tracy/The Stock Shop; **p. 287**, UPI/Bettmann; **p. 288**, Henry Pohs; **p. 290** (all), Stephen Frisch/Photo 20-20; **p. 294** (both), Stephen Frisch/Photo 20-20; **p. 295** The Bettmann Archive; **p. 298**, UPI/Bettmann; **p. 300**, AP/Wide World Photos. **Chapter 14**: **p. 307**, Karl Hartmann/Sachs/Phototake; **p. 308**, Yoav Levy/Phototake; **p. 313** (top), Ted Streshinsky/Photo 20-20; **p. 313** (bottom), The Bettmann Archive; **p. 316**, courtesy, Intoximeters, Inc.; **p. 321**, courtesy, National Library of Medicine; **p. 322**, E.R. Degginger/Photo Researchers. **Chapter 15**: **p. 327**, Sovfoto/Eastfoto; **p. 328**, Marco Polo/Phototake; **p. 337**, Stephen Frisch/Photo 20-20; **p. 338** (top), Stephen Frisch/Photo 20-20; **p. 338** (bottom), Mark Richards; **p. 341**, Runk/Schoenberger/Grant Heilman. **Chapter 16**: **p. 349**, Runk/Schoenberger/Grant Heilman **p. 350**, Marco Polo/Phototake; **p. 355** (both), Stephen Frisch/Photo 20-20. **Chapter 17**: **p. 371** (top), courtesy, Swedish Information Center; **p. 371** (bottom), Jim Zipp/Photo Researchers; **p. 372**, Denise Marcotte/Stock Boston; **p. 377** (top left & right), courtesy, D.E.A; **p. 377** (bottom), courtesy, National Library of Medicine; **p. 378**, Stephen Frisch/Photo 20-20; **p. 381** (both), courtesy, USDA. **Chapter 18**: **p. 389**, courtesy, DuPont; **p. 390**, courtesy, Du Pont; **p. 391** (top), Michael Abbey/Science Source/Photo Researchers; **p. 391** (bottom), courtesy, Dow Chemical Co; **p. 394** (top), courtesy, B.F. Goodrich, SP & C Division; **p. 394** (bottom), courtesy, Delco Remy; **p. 397** (top), courtesy, Du Pont; **p. 397** (bottom), Jonathan Selig/Photo 20-20; **p. 400**, Nubar Alexanian/Stock Boston. **Part III, p. 404** Stuart Wetmorland/Stock Imagery. **Chapter 19**: **p. 407**, Ted Streshinsky/Photo 20-20; **p. 408** (top), Luiz C. Marigo/Peter Arnold; **p. 408** (bottom), Jeanne Heiberg/Peter Arnold; **p. 421**, Figure 19.17, From *Biology*, 6th ed., by Cecie Starr and Ralph Taggart, 1992, p. 39, figure 3.7. Copyright © 1992 by Wadsworth Publishing Company. Redrawn by permission of the publisher; **p. 422**, Biophoto Associates/Science Source/Photo Researchers; **p. 423**, Biophoto Associates/Science Source/Photo Researchers. **Chapter 20**: **p. 423**, Figure 19.19, From *Biology*, 6th ed., by Cecie Starr and Ralph Taggart, 1992, p. 39, figure 3.8. Copyright © 1992 by Wadsworth Publishing Company. Redrawn by permission of the publisher; **p. 429**, Bruce Curtis/Peter Arnold;

Credits

p. 431, Stephen Frisch/Photo 20-20; **p. 434**, Photo 20-20; **p. 437**, Yoav Levy/Phototake; **p. 438** (both), Biophoto Associates/Science Source/Photo Researchers; **p. 442**, courtesy, New England Journal of Medicine; **p. 443**, courtesy, Dr. Michael Holick. **p. 443**, Figure 20.19, From *Biology*, 5th ed., by Cecie Starr and Ralph Taggart, 1992, pp. 85, 86, figure 6.1. Copyright © 1989 by Wadsworth Publishing Company. Redrawn by permission of the publisher; **p. 444**, Figure 20.20, From *Biology*, 5th ed., by Cecie Starr and Ralph Taggart, 1992, p. 87, figure 6.2. Copyright © 1989 by Wadsworth Publishing Company. Redrawn by permission of the publisher; **Chapter 21**: **p. 449** (both), Bill Longcore/Photo Researchers; **p. 450**, Cary Wolinsky/Stock Boston; **p. 455**, courtesy, The World Health Organization; **p. 458**, courtesy, Beckman Instruments; **p. 460**, Figure 21.11, From *Biology*, 6th ed., by Cecie Starr and Ralph Taggart, 1992, p. 44, figure 3.18. Copyright © 1992 by Wadsworth Publishing Company. Redrawn by permission of the publisher; **p. 460**, Figure 21.12, From *Biology*, 6th ed., by Cecie Starr and Ralph Taggart, 1992, p. 45, figure 3.19a. Copyright © 1992 by Wadsworth Publishing Company. Redrawn by permission of the publisher; **p. 462**, Figure 21.14, From *Biology*, 6th ed., by Cecie Starr and Ralph Taggart, 1992, p. 45, figure 3.19b. Copyright © 1992 by Wadsworth Publishing Company. Redrawn by permission of the publisher; **p. 465**, J. Gross/SPL/Photo Researchers; **p. 465**, Figure 21.16, From *Biology*, 5th ed., by Cecie Starr and Ralph Taggart, 1992, p. 59, figure 4.17. Copyright © 1989 by Wadsworth Publishing Company. Redrawn by permission of the publisher; **p. 467**, The Bettmann Archive; **p. 468**, R. Feldman & Dan McCoy/Rainbow. **Chapter 22**: **p. 473**, J. Berndt/Stock Boston; **p. 474**, Matthew Borkoski/Stock Boston; **p. 476**, courtesy, Dr. Thomas A. Steitz/Yale University. **Chapter 23**: **p. 493**, Baylor College of Medicine/Peter Arnold; **p. 494**, Lynn McLaren/Photo Researchers; **p. 500**, Figure 23.11, From *Biology*, 6th ed., by Cecie Starr and Ralph Taggart, 1992, p. 211, figure 13.8. Copyright © 1992 by Wadsworth Publishing Company. Redrawn by permission of the publisher; **p. 501**, Figure 23.12, From *Biology*, 6th ed., by Cecie Starr and Ralph Taggart, 1992, p. 210, figure 13.7. Copyright © 1992 by Wadsworth Publishing Company. Redrawn by permission of the publisher; **p. 502**, courtesy, Doris Beck & Tim Highman/Bowling Green State University; **p. 504**, Figure 23.15, From *Biology*, 6th ed., by Cecie Starr and Ralph Taggart, 1992, p. 220, figure 14.4. Copyright © 1992 by Wadsworth Publishing Company. Redrawn by permission of the publisher; **p. 505**, Figure 23.16 From *Biology*, 5th ed., by Cecie Starr and Ralph Taggart, 1992, pp. 214, 215, Figures 6.9 and 16.10. Copyright © 1989 by Wadsworth Publishing Company. Redrawn by permission of the publisher; **p. 505**, Figure 23.17, From *Biology*, 6th ed., by Cecie Starr and Ralph Taggart, 1992, p. 223, figure 14.9. Copyright © 1992 by Wadsworth Publishing Company. Redrawn by permission of the publisher; **p. 505**, Figure 23.18, From *Biology*, 6th ed., by Cecie Starr and Ralph Taggart, 1992, p. 221, figure 14.5. Copyright © 1992 by Wadsworth Publishing Company. Redrawn by permission of the publisher; **p. 507**, courtesy, City of Hope National Medical Center, Cytogenetics Lab; **p. 509**, Figure 23.22, From *Biology*, 5th ed., by Cecie Starr and Ralph Taggart, 1992, pp. 216-217, figure 16.12. Copyright © 1989 by Wadsworth Publishing Company. Adapted by permission of the publisher; **p. 511**, Figure 23.24, From *Biology*, 6th ed., by Cecie Starr and Ralph Taggart, 1992, p. 235, figure 15.2. Copyright © 1992 by Wadsworth Publishing Company. Redrawn by permission of the publisher; **p. 510**, Omikron/Science Source/Photo Researchers; **p. 512**, courtesy, Dr. Bruce N. Ames; **p. 514** (top), Dr. Donald R. Helinski/UCSD; **p. 514** (bottom), courtesy, David Dressler & Huntington Potter/Harvard Medical School; **p. 515**, courtesy, Dr. R.L. Brinster; **p. 516**, courtesy, Cellmark Diagnostics. **Chapter 24**: **p. 521**, Diane Rawson/Photo Researchers; **p. 522**, Gunter Ziesler/Peter Arnold; **p. 526**, Dennis Kunkel/CNRI/Phototake; **p. 526**, Figure 24.7, From *Biology*, 5th ed., by Cecie Starr and Ralph Taggart, 1992, p. 76. Copyright © 1989 by Wadsworth Publishing Company. Adapted by permission of the publisher; **p. 528**, The Bettmann Archive; **p. 532** (top), Arthur Sirdofsky/The Stock Shop; **p. 532** (bottom), UPI/Bettmann; **p. 534**, Ken Highfill/Photo Researchers. **Chapter 25**: **p. 539**, Miro Vinton/Stock Boston; **p. 540**, Doug Wechsler/Earth Scenes. **Chapter 26**: **p. 557**, Mark Kelley/Stock Boston; **p. 558** (left), Dan McCoy/Rainbow; **p. 558** (right), Friend/The Stock Shop. **Chapter 27**: **p. 571**, Yoav Levy/Phototake; **p. 572**, Barry L. Runk/Grant Heilman; **p. 580**, American College of Rheumatology. **Chapter 28**: **p. 586**, Fred Bavendam/Peter Arnold; **p. 587**, Ed Reschke/Peter Arnold; **p. 587**, Figure 28.3, From *Biology*, 6th ed., by Cecie Starr and Ralph Taggart, 1992, p. 669, figure 38.14. Copyright © 1992 by Wadsworth Publishing Company. Redrawn by permission of the publisher; **p. 590**, Manfred Kage/Peter Arnold; **p. 591**, Prof. P. Motta/Dept. of Anatomy, University "La Sapienza", Rome/Photo Researchers; **p. 594**, Figure 28.11, From *Biology*, 6th ed., by Cecie Starr and Ralph Taggart, 1992, p. 721, figure 41.3. Copyright © 1992 by Wadsworth Publishing Company. Redrawn by permission of the publisher; **p. 596**, David York/Medichrome. **Chapter 29**: **p. 601**, courtesy, Bristol-Myers Squibb Archives; **p. 602**, courtesy, D.E.A.; **p. 604**, courtesy, Pfizer; **p. 605** (left), Dr. Tony Brain/Science Source/Photo Researchers; **p. 605** (right), courtesy, Bristol-Myers Squibb Archives; **p. 606**, Figure 29.12, From *Biology*, 6th ed., by Cecie Starr and Ralph Taggart, 1992, p. 693. Copyright © 1992 by Wadsworth Publishing Company. Redrawn by permission of the publisher; **p. 608**, Sacchi-Lecaque/Russel-UCLA/CNRI/SPL/Photo Researchers; **p. 613** (top), Renee Lynn/Photo Researchers; **p. 613** (bottom), Peter Miller/Photo Researchers.